CAMBRIDGE MONOGRAPHS ON PHYSICS

GENERAL EDITORS

M. M. WOOLFSON, D.SC.
Professor of Theoretical Physics, University of York

J. M. ZIMAN, D.PHIL., F.R.S.
Melville Wills Professor of Physics, University of Bristol

NEGATIVE IONS

NEGATIVE IONS

SIR HARRIE MASSEY
Emeritus Professor of Physics, University College, London

THIRD EDITION

CAMBRIDGE UNIVERSITY PRESS
CAMBRIDGE
LONDON · NEW YORK · MELBOURNE

CAMBRIDGE UNIVERSITY PRESS
Cambridge, New York, Melbourne, Madrid, Cape Town,
Singapore, São Paulo, Delhi, Tokyo, Mexico City

Cambridge University Press
The Edinburgh Building, Cambridge CB2 8RU, UK

Published in the United States of America by Cambridge University Press, New York

www.cambridge.org
Information on this title: www.cambridge.org/9780521283175

© Cambridge University Press 1976

This publication is in copyright. Subject to statutory exception
and to the provisions of relevant collective licensing agreements,
no reproduction of any part may take place without the written
permission of Cambridge University Press.

First published 1938
Second edition 1950
Third edition 1976
First paperback edition 2011

A catalogue record for this publication is available from the British Library

Library of Congress Catalogue Card Number: 74–31792

ISBN 978-0-521-20775-1 Hardback
ISBN 978-0-521-28317-5 Paperback

Cambridge University Press has no responsibility for the persistence or
accuracy of URLs for external or third-party internet websites referred to in
this publication, and does not guarantee that any content on such websites is,
or will remain, accurate or appropriate.

CONTENTS

Preface — page xv

CHAPTER 1
Introduction – the negative ion of hydrogen (H⁻)

- 1.1 General theoretical considerations — 1
- 1.2 Calculation of negative-ion structure by quantum methods — 5
- 1.3 The hydrogen negative ion — 6
  - Calculation of the electron affinity — 6
  - The charge distribution in H⁻ — 11

CHAPTER 2
Ground states of complex atomic negative ions – theoretical considerations

- 2.1 The states of complex atoms — 16
  - Electron configurations — 16
  - States arising from a given configuration — 17
- 2.2 Calculation of the structure and energy of complex atomic negative ions — 19
  - The Hartree–Fock method — 19
  - The Hartree–Fock–Roothan (HFR) procedure — 23
  - Inclusion of correlation — 25
  - The fluorine negative ion — 27
  - Electron affinities of alkali metal atoms — 28

CHAPTER 3
The electron affinities of the elements

- 3.1 Experimental methods for measuring electron affinities — 31
  - From photodetachment — 32
  - From the threshold frequency for polar photodissociation — 32

	From threshold energies for dissociative attachment processes	page 33
	From threshold energies for polar dissociation of molecules by electron impact	33
	From the variation with frequency of the affinity continuum	33
	From measurements of ion currents arising from reactions at a hot filament	34
	From the energy relations involved in cyclic processes	46
3.2	Empirical methods	48
	Isoelectronic extrapolation	49
	Horizontal extrapolation	49
	Results for atoms of the first two short periods	60
	Electron affinities of heavier elements	61

CHAPTER 4
Atomic negative ions – excited states – autodetachment; general account

4.1	Bound excited states of negative ions	66
4.2	Continuum states of negative ions	67
	Relation to elastic scattering by a centre of force	68
	Convergence of the phase shifts	71
	Relation of phase shifts to time delay in scattering	72
	Wave functions describing electron scattering by atoms	73
4.3	Negative-ion formation by excited atoms – autodetachment	77
	Lifetime of an autodetaching state	79
	The observation of non-metastable autodetaching states	81
	Autodetaching states and elastic scattering of electrons by atoms	82
	Autodetaching states and inelastic scattering of electrons by atoms	90
	Autodetaching states and photodetachment	92
	Calculation of the energy of doubly-excited states of negative ions	92
	Summary	95
4.4	One-body or shape resonances	97
4.5	Siegert states	104

CONTENTS

4.6	Type I and type II resonance states	page 107
4.7	Experimental study of autodetaching and shape resonance states from resonance effects in electron–atom collisions	107
	Transmission experiments	109
	Measurements of energy and angular distributions	112
	From energy distributions of detached electrons	114
	Measurement of optical excitation functions	114
4.8	Experimental study of autodetaching states through photodetachment	115
4.9	Doubly excited states which decay through emission of radiation – change of parity	115

CHAPTER 5
Autodetaching states of specific atomic negative ions

5.1	H^-	116
	The $(2p)^2\, {}^3P_e$ state	120
5.2	He^-	122
	He^- states below the excitation threshold (19.8 eV)	122
	The $1s2s^2\, {}^2S$ state	122
	The $1s2s2p\, {}^4P$ state	126
	States between the excitation and ionization thresholds – 2^2P and 2^2D	135
	Triply-excited states of He^-	138
5.3	Negative ions of heavier rare gases	142
5.4	O^-	143
5.5	N^-	144
5.6	Cl^-	146
5.7	Negative ions of the alkali metal atoms	148
5.8	Doubly-charged negative ions	150
	H^{2-}	151
	Other doubly-charged negative ions	151

CHAPTER 6
Molecular negative ions – ground states

6.1	The quantum states of diatomic molecules – potential-energy curves	156

6.2	Enumeration and properties of the electronic states of diatomic molecules	page 160
6.3	The vibrational and rotational energy	162
6.4	The Franck–Condon principle	165
	Electron affinity and vertical detachment energy	166
6.5	Theoretical calculation of the properties of the ground states of diatomic molecules	166
6.6	The observation of autodetaching levels associated with the ground states of diatomic negative ions	168
6.7	Electron affinities of diatomic molecules and structure of the ground states of the negative ions	172
	The determination of electron affinities of diatomic molecules	172
	Determination of structural properties of diatomic molecular ions	174
	H_2^-	174
	OH^- and SH^-	179
	Other diatomic hydrides	183
	O_2^-	183
	SO^- and S_2^-	190
	NO^-	191
	N_2^-	194
	CO^-	204
	C_2^-	207
	CN^-	210
	F_2^-, Cl_2^-, Br_2^-, I_2^-	210
	Other diatomic molecules	213
6.8	Polyatomic negative ions	214
	O_3^-	215
	SO_2^-	215
	NO_2^-	216
	N_2O^- and CO_2^-	218
	NH_2^-, PH_2^-, AsH_2^-	218
	NO_3^-	219
	O_4^-	220
	CO_3^- and CO_4^-	220
	SF_6^- and TF_6^-	221

CHAPTER 7
Excited electronic states of molecular negative ions

7.1	Introduction	page 222
7.2	H_2^-	222
7.3	N_2^-	228
7.4	CO^-	235
7.5	O_2^-	237
7.6	NO^-	237

CHAPTER 8
Modes of formation of negative ions – formation by radiative processes – radiative attachment and polar photodissociation

8.1	Radiative capture of electrons – electron affinity spectrum	242
8.2	Direct observation of affinity spectra	248
	Experimental methods	248
	Observed results	252
8.3	Dielectronic attachment	254
8.4	Polar photodissociation	255

CHAPTER 9
Modes of formation of negative ions – formation by three-body collisions and by collisions of electrons with molecules – dissociative attachment and polar dissociation

9.1	Capture of electrons in three-body collisions	264
9.2	Formation of negative ions on collisions of electrons with molecules – theoretical introduction	266
	Processes involving electron capture	266
	Non-capture collisions – dissociation into ions	272
9.3	Formation of negative ions on collision of electrons with molecules – experimental methods of study using homogeneous electron beams	274
	Introductory remarks – early techniques	274
	Measurement of total attachment cross-sections	278
	Measurement of the velocity distribution and composition of the ions	282
	Measurements employing mass analysis but not energy analysis	284

CONTENTS

	Measurement of the angular distribution of the ionic momenta	page 288
9.4	Formation of negative ions in collision of electrons with molecules – attachment experiments with electron swarms	290
	Introductory remarks	290
	General description and theory of swarm experiments	291
	Experimental methods	292
9.5	Discussion of results of experiments on attachment of electrons to diatomic molecules	309
	H_2, HD and D_2	309
	O_2	316
	CO	333
	NO	338
	Halogen molecules	341
9.6	Attachment to polyatomic molecules	345
	H_2O	347
	NH_3	351
	CO_2	352
	N_2O	358
	NO_2	365
	O_3	366
	SF_6	368
	Other halogen-containing molecules	376

CHAPTER 10

Formation of negative ions by capture of bound electrons

10.1	Formation of negative ions in collisions between neutral systems	383
10.2	Low-energy experiments on the measurement of cross-sections and threshold energies	389
	Some typical experimental arrangements	391
	Results obtained in low-energy experiments	394
10.3	High-energy experiments – experimental methods	399
	Electron capture by H atoms	399
	Production of He^-	405
10.4	Negative-ion formation by double capture of electrons	414

CHAPTER 11
Detachment of electrons from negative ions – photodetachment, field detachment and detachment by electron impact

11.1	Introduction	page 416
11.2	Photodetachment – introductory theoretical considerations	417
11.3	Calculated photodetachment cross-sections	421
	H^-	421
	Other atomic negative ions	424
11.4	Photodetachment – experimental methods	430
	Introduction	430
	Crossed-beam methods	432
	Measurement of the energy and angular distribution of the detached electrons	438
	Absorption by shock-heated alkali halides	442
	Photodetachment studies using an ion cyclotron resonance spectrometer	445
	Photodetachment from ions aged in a drift tube	448
11.5	Application to different atomic negative ions	449
	H^-	449
	He^-	450
	O^-	450
	S^- and Se^-	454
	C^-	458
	Halogen negative ions	459
	Negative ions of the noble metals	461
	Negative ions of the alkali metal atoms	466
	Other atomic ions	472
11.6	Application to different molecular negative ions	473
	OH^-, OD^- and SH^-	473
	CH^-	477
	NH^-	478
	O_2^-	478
	NO^-	482
	C_2^-	484
	S_2^-	484
	SO^-	484
	NH_2^-, PH_2^-, AsH_2^-	486

	SO_2^-	page 487
	NO_2^-	488
	Other negative ions of atmospheric interest	489
11.7	Multiphoton detachment	489
11.8	Detachment from negative ions in electrostatic fields	495
	Theoretical considerations	495
	The measurement of the rate of field detachment from H^-	497
	Field detachment from excited negative ions	500
11.9	Detachment by electron impact	500
	Theoretical considerations	500
	Measurement of detachment cross-sections	503
	Results obtained for single detachment	509
	Double detachment	512

CHAPTER 12

Detachment, charge transfer and other reactions between negative ions and neutral systems at low and intermediate energies

12.1	Classification of types of ionic reaction	513
12.2	Orbiting collisions	514
12.3	The theory of detachment reactions	516
12.4	Ionic mobilities	519
12.5	Experimental methods for measuring mobilities and/or reaction rates for negative ions in gases	521
	The combined drift-tube and mass-spectrometer technique	522
	The flowing afterglow method	526
	Experiments in static afterglows	528
	Extension of the pulse method	530
	Experiments with low-energy ion beams	534
12.6	The determination of threshold energies for endothermic charge transfer reactions	537
12.7	Discussion of observed results	540
	Mobilities and reactions of oxygen ions in oxygen	540
	Reactions of oxygen negative ions with atoms and molecules of other species	549
	Reactions of negative ions other than those of oxygen	560

CHAPTER 13
Detachment, charge transfer and other reactions involving negative ions – collisions at high impact energies

13.1 Symmetrical charge transfer – theoretical considerations *page* 579
    Effect of detachment on charge transfer cross-sections 584
13.2 Unsymmetrical charge transfer – theoretical considerations 586
13.3 Calculation of detachment cross-sections 587
13.4 The measurement of charge transfer and detachment cross-sections 590
    The condenser plate method 591
    Crossed-beam method 597
13.5 Observed results 599
    H^-–H collisions 599
    Charge transfer reactions involving atomic negative ions and neutral atoms 602
    Charge transfer reactions involving molecules 603
    Detachment reactions involving H^- ions and other neutral species 605
    Detachment from He^- 610
    Detachment reactions involving other negative ions 612
    Double detachment 614

CHAPTER 14
Recombination of negative and positive ions – mutual neutralization

14.1 Theoretical considerations 618
    Mutual neutralization 618
    Recombination in three-body collisions 621
14.2 The measurement of recombination and mutual-neutralization rates 625
    The medium pressure range 625
    Measurements at high pressures 628
    Measurements at low pressures – rate of mutual neutralization 628

xiv   CONTENTS

  Measurements by merging and inclined-beam techniques   *page* 632
14.3 Results obtained for specific mutual-neutralization reactions   635
  H^+–H^- 635
  N^+–O^-, N_2^+–O_2^-, O_2^+–O_2^- 638
  He^+–H^- 639

CHAPTER 15
Negative ions in electric discharges, planetary and stellar atmospheres, trace analysis and tandem accelerators

15.1 Negative ions in glow discharges   640
  General effects of negative ions in discharges   640
  Analysis of a discharge in oxygen   642
15.2 The effect of negative ions on current build-up and electrical breakdown in gases   653
  Current build-up and breakdown in air and oxygen   657
  Current build-up and breakdown in halogen-containing substances   660
15.3 Negative ions in the terrestrial ionosphere   663
  Introduction   663
  The lower ionosphere   666
15.4 Negative ions in the atmospheres of the sun and stars   673
  H^- and the continuous emission spectrum of the sun   673
  The emission spectrum in terms of atmospheric absorption coefficients   673
  The contribution from free–free absorption   680
  Negative ions in stellar atmospheres   682
15.5 The application of electron attachment to the qualitative and quantitative analysis of trace samples – the electron capture detector   682
15.6 The use of negative ions in particle accelerators   690
  Tandem accelerators   690
  H^- ions in cyclotrons and synchrocyclotrons   691

References   693
Author index   717
Subject index   727

Preface to Third Edition

Since the first two editions were prepared there has been an explosive increase in knowledge, from both experimental and theoretical sources, of the properties and behaviour of negative ions in gases. Thus in 1949 when the second edition was in preparation no experimental measurements had been made of radiative processes involving negative ions. Since that time the experimental study of photodetachment has become a most important source of information about negative ions and provides values of electron affinities of reliability and precision unmatched by any other technique. The availability of laser sources of light of appropriate frequency has extended the scope of the work so, at the time of writing this preface, it is one of the liveliest branches of negative-ion study.

Again only a paragraph or two appears in the earlier editions about autodetaching states of negative ions. Not only has the identification of such states and determination of their properties become a very fruitful and active branch of both experimental and theoretical atomic physics but, thanks to the pioneer work of Herzenberg and Mandl the importance of autodetachment in determining the rates of dissociative attachment reactions of electrons with gas molecules has been realized and exploited. The experimental study of these reactions has also developed so that earlier difficulties of interpretation have been recognized and obviated by the introduction of improved techniques.

Collision processes involving negative-ion production and destruction have been investigated using charged and neutral beams of several keV energy while the determination of threshold energies for charge transfer reactions has yielded valuable information about electron affinities. Measurements of the rates of mutual neutralization and other ionic reactions and of detachment by electron impact and by electrostatic fields, have also been made. On the theoretical side, the considerable development of atomic and molecular structure calculations made possible through the availability of high-capacity electronic computers, has naturally included negative ions both atomic and molecular.

Because of the expansion of knowledge about negative ions their influence on the nature of physical phenomena in which they are concerned has been understood in much greater detail. This applies to absorption in stellar, including solar, atmospheres, to electric discharge and breakdown phenomena in electronegative gases and to the terrestrial ionosphere. Negative ions are employed with advantage in tandem and circular particle accelerators while the most sensitive detectors of halogen-containing substances, such as many pesticides, important in environmental studies, depend on negative-ion formation.

Inclusion of this new material, both important and abundant, has meant that, in order to prevent the new edition growing to an excessive bulk, it has not been possible to aim at a comprehensive account of all the subjects as could be done without difficulty in earlier editions. Nevertheless it is hoped that no significant subject has been omitted while emphasis has naturally been placed on the most significant developments. Also, because so many branches of physics are involved it has no longer been possible to provide the physical background for all the different sources of information about negative ions. In particular, for the discussion of autodetaching states, this would have meant inclusion of a considerable part of atomic collision theory and practice. Nevertheless, in cases such as this, enough background has been included to make comprehensible the subsequent specific discussion of applications to negative ions.

Because information comes from so many sources, problems of notation sometimes become acute. In one field a particular symbol acquires a well-recognized significance which may be quite different in another, both of which contribute to knowledge of negative ions. As far as possible a more or less uniform notation has been adopted throughout the book so that in a number of cases symbols unfamiliar to the specialist in a particular field may be used in the interest of uniformity. However, a degree of flexibility has been retained in these matters.

The rapid expansion of the subject has necessarily led to a

somewhat higher degree of sophistication in the treatment but, relative to the whole field of atomic and molecular physics, the level has been kept much the same. Elaborate mathematics has been avoided and theory introduced in terms of simple physically illustrative examples instead of elaborate general formalism. No attempt has been made to follow the historical order of developments but in many cases the history has been summarized particularly when it is of special interest as in that of the determination of the electron affinity of atomic oxygen. In the same spirit, in discussing techniques the most recent versions alone are described if they incorporate real advantages over earlier methods. However, the reasons why they do present such advantages are usually given, and in some cases when an obsolete method has played a major role at an early stage, such as the Lozier method for studying dissociative attachment, at least a brief description is included. No new results which have appeared later than April 1974 have been included.

My thanks are due to Drs Lovelock, Moores, Norcross, Pagel and Parkes for providing me with information on recent developments in different aspects of the subject and to Dr G. Herzberg, F.R.S. for supplying me, for reproduction, with a print of a C_2^- band spectrum obtained in his experiments with Dr Lagerqvist. I have also benefited very much from discussions with Professor J. D. Craggs and with Professor M. J. Seaton, F.R.S. The contribution made by my secretary Mrs M. Harding in assisting in the preparation of the material for press, including the conversion of unreadable manuscript into typescript, has been of key importance, and I am extremely grateful for her willing cooperation. The Press, as usual, have done a splendid job.

London H.S.W.M.

CHAPTER 1

Introduction – the Negative Ion of Hydrogen (H⁻)

1.1 General theoretical considerations

Applying the general ideas of quantum theory we may regard a negative ion as a system with a number of quantum states. In general the ion will be observed in its normal state and, if it is to be stable, the energy of this normal state must be less than that of the normal state of the corresponding neutral atom. The energy difference $E_0 - E_-$ between the normal states of the atom and ion is called the electron affinity of the atom. This energy will also be equal to that necessary to detach an electron from the ion, so may also be called the detachment energy of the ion. A positive electron affinity indicates a stable negative ion.

To discuss the stability of an ion we must endeavour to determine the conditions under which $E_0 - E_-$ is positive. As a first approximation this quantity may be regarded as the binding energy of the added electron; hence the question arises as to the nature of the force which holds this electron bound to the atom. The fact that the ion has a negative charge does not mean that the net force acting on the electron is that of a negative charge. The electron is not influenced by its own field but only by that of the remaining atomic electrons and the nucleus. That this field is attractive may be seen as follows.

Consider for simplicity a hydrogen atom. The nucleus can be regarded as a point charge $+e$, while the charge of the atomic electron is spread out in a spherically symmetrical cloud around the nucleus as centre. Let us determine the potential energy of a charge $-e$ placed in this distribution at a distance r_1 from the centre. Suppose α is the fraction of the total charge of the atomic electron contained within a distance r_1 of the nucleus. The charge $e(1-\alpha)$ within the sphere of radius r_1 acts as if concentrated at the nucleus and so gives rise to a potential $e(1-\alpha)/r_1$. If the remainder, $-e(1-\alpha)$, of the atomic electronic charge were concentrated at r_1 it would give rise to a potential $-e(1-\alpha)/r_1$, but it is actually distributed throughout space from r_1 to ∞ and so can only give rise to a potential $\beta e(1-\alpha)/r_1$, where β is less than unity. The total potential energy of the charge $-e$ is therefore

$$-(1-\beta)(1-\alpha)e^2/r_1,$$

and so is negative.

This shows that the additional electron will be acted on by a net attractive force. Further, since $(1-\alpha)(1-\beta) \to 0$ as $r_1 \to \infty$, the effective field of force acting on the electron falls off much more rapidly with distance than a Coulomb field. This difference between the asymptotic behaviour of the effective field acting on an electron in a negative ion and in a neutral atom (which behaves in the same way as a Coulomb field) has important consequences for the theory of negative ions, as we shall see below.

The calculation of the potential energy of an electron in the field of a hydrogen atom may be carried out to a first approximation as follows. If we neglect polarization effects (which are actually decisive in determining the stability of H⁻) the probability of finding the atomic electron at a distance between r and $r+dr$ from the nucleus is $4\pi r^2 \psi^2 \, dr$, where $\psi = (\pi a_0^3)^{-1/2} \exp(-r/a_0)$ is the wave function of the ground state of hydrogen. The potential energy due to the atomic electron at a point distant r_1 from the nucleus is

$$\frac{4\pi e^2}{r_1}\int_0^{r_1} \psi^2 r^2 \, dr + 4\pi e^2 \int_{r_1}^{\infty} \frac{\psi^2 r^2}{r} \, dr,$$

the first term arising from the charge within r_1, the second from that without. Carrying out the elementary integrations involved gives

$$e^2 \left\{ \frac{1}{r_1} - \exp(-2r_1/a_0)\left(\frac{1}{r_1} + \frac{1}{a_0}\right) \right\}.$$

Adding the potential energy $-e^2/r_1$ due to the nucleus we find for the total potential energy

$$-e^2 \exp(-2r_1/a_0)\left(\frac{1}{r_1} + \frac{1}{a_0}\right). \tag{1.1}$$

INTRODUCTION – THE NEGATIVE ION OF HYDROGEN

This gives a field of force falling off exponentially with distance, but polarization effects probably modify this to an inverse fifth-power law. This is still qualitatively distinct from a Coulomb field.

To a first approximation we see then that the allowed energy values of a negative ion may be considered in terms of the stationary states of an electron in an attractive field of force falling off rapidly with distance. In such a field the number of stationary states is finite and not infinite as for the Coulomb field. We may illustrate this by taking a simple case. Represent the potential energy V by taking $V = -V_0$, a constant, for distances r less than r_0 and zero elsewhere. The wave equation for the states with zero angular momentum in this field is

$$\frac{d^2\psi}{dr^2} + \frac{2}{r}\frac{d\psi}{dr} + \frac{8\pi^2 m}{h^2}(E+V_0)\psi = 0 \quad (r < r_0), \tag{1.2a}$$

$$\frac{d^2\psi}{dr^2} + \frac{2}{r}\frac{d\psi}{dr} + \frac{8\pi^2 m}{h^2} E\psi = 0 \quad (r > r_0). \tag{1.2b}$$

The substitution $\psi = ru$ reduces the equation to the form

$$\frac{d^2 u}{dr^2} + \lambda^2 u = 0, \tag{1.3}$$

where

$$\lambda^2 = \frac{8\pi^2 m}{h^2}(E+V_0) = \nu^2 \quad (r < r_0), \tag{1.4a}$$

$$\lambda^2 = \frac{8\pi^2 m}{h^2} E = -\kappa^2 \quad (r > r_0). \tag{1.4b}$$

The boundary conditions to be satisfied are that ψ be finite, continuous and single-valued everywhere. This requires that $u = 0$ at $r = 0$.

Equation (1.3) is the familiar equation for simple harmonic motion, and the solution which satisfies the boundary conditions for $E < 0$ is

$$u = A \sin \nu r \quad (r < r_0) \tag{1.5a}$$

$$u = B \exp(-\kappa r) \quad (r > r_0). \tag{1.5b}$$

Continuity of the charge and current density requires that u and du/dr be continuous at $r = r_0$, so

$$A \sin \nu r_0 = B \exp(-\kappa r_0), \tag{1.6a}$$

$$A\nu \cos \nu r_0 = -\kappa B \exp(-\kappa r_0). \tag{1.6b}$$

Eliminating A/B gives

$$\frac{\tan \nu r_0}{\nu} = -\frac{1}{\kappa}. \quad (1.7)$$

The roots of this equation give the proper energy values for $E < 0$. We see at once that, if $\nu r_0 < \frac{1}{2}\pi$, there are no roots and no negative energy levels, while, if $\frac{1}{2}\pi < \nu r_0 < \frac{3}{2}\pi$, there is one root and one energy level, and so on.

Similar considerations apply to the states with non-vanishing angular momenta. For example, it is easy to see that, if there is no energy level with zero angular momentum, there will be none with angular momentum $\{l(l+1)\}^{1/2}\hbar$ (l a positive integer), for an angular momentum of this magnitude introduces a centrifugal force corresponding to a potential energy $l(l+1)\hbar^2/2mr^2$. This additional energy, being positive, cannot give rise to a stronger binding than that of the deepest level with $n = 0$.

We do not suggest that this crude model has any quantitative significance as applied to negative ions, but it illustrates the qualitative features of short-range fields.

When attempting to apply this to a survey of the possibilities of negative-ion formation by the various elements we must take into account the key principle of atomic structure – the Pauli principle. This makes it impossible for more than two electrons (with opposite spins) to occupy the same orbital quantum state and, in conjunction with the limited number of possible stationary states for the attached electron, severely restricts the number of elements which can form stable negative ions.

It is now possible to understand the behaviour of different atoms towards electron attachment. It is only necessary to consider which of the limited number of stationary states for the attached electron are excluded by the Pauli principle. For the hydrogen atom, no states are excluded and a stable negative ion will result if the effective attractive field is strong enough. An electron can only be attached to a helium atom in a two-quantum state, for the one-quantum level is fully occupied. Although the effective field acting on the third electron will be greater than in hydrogen, it is unlikely to be strong enough to give two stationary states. The non-existence of stable He^- is thus understood.

These considerations can be extended to the other atoms of the periodic table. The general rule is that those atoms with completely filled electron shells will be unlikely to form negative ions. For, in such cases, the attached electron must be bound in a state with total quantum number greater by one than that of the outer atomic electrons. The effective field is usually too weak to give such a stationary state. Of the atoms with incomplete outer shells it is to be expected that those (F, Cl, Br, I) in which the shells are most nearly complete will form the most stable negative ions. For, in these, the outer atomic electrons are least effective in screening the nuclear charge from the extra electron. Of the four halogen atoms, fluorine might be expected to have the greatest electron affinity as the attachment takes place to a shell with the lowest total quantum number but, in fact (see Table 3.3), it is a little smaller than that of chlorine.

All of these qualitative considerations are in agreement with observations, but they are not accurate enough to lead to any quantitative predictions of the structure of negative ions. For this, more detailed calculations must be carried out.

1.2 Calculation of negative-ion structure by quantum methods

The approximation we have been making in the preceding section of regarding the determination of the energy of a negative ion as a one-body problem is not adequate for our present purposes. In fact the mean static field such as (1.1) for the H atom is not strong enough to produce a bound state. For heavier atoms there may be bound states in the static field but these are excluded by the Pauli principle, as described earlier. The formation of a stable negative ion depends on correlation effects between the attached and atomic electrons which reduce the mean interelectronic repulsion. Any theoretical calculation of electron affinities must take these effects fully into account. However, because the electron affinity is the very small difference between the total energy of the negative ion and the corresponding neutral atom, it is not possible as yet to calculate it to any desired precision for any atom other than hydrogen. For a few other atoms, results of quite high accuracy may be obtained as will be described in the following chapter. Nevertheless, use may be made of calculations of negative-ion structure by self-consistent field and related methods to determine effective radii and other properties of negative ions and also to provide a basis for extrapolation of ionization and excitation energies to obtain reliable information about electron affinities. The development of effective empirical methods, based on some form of extrapolation, is the only procedure available at the time of writing (but see Chapter 2, p. 28) for reasonably accurate estimation of electron affinities for most elements.

We now describe the results of calculations of negative-ion structure, concentrating particularly on H$^-$, for which very extensive and precise calculations have been carried out.

1.3 The hydrogen negative ion

Calculation of the electron affinity

The calculation of the electron affinity of the simplest negative ion, H$^-$, involves the solution of a two-body problem analogous to that of the helium atom. Exact solution is not possible but variational methods may be used to obtain very accurate values for the electron affinity.

These methods, which were soon exploited in detail by Hylleraas (1930), make use of the minimal properties of the ground state wave function. The true wave function, which cannot be found exactly, is represented by an analytical approximation ψ_t containing a number of parameters. These are adjusted so the energy E given by

$$E = \int \psi_t^* H \psi_t \, d\tau / \int \psi_t^* \psi_t \, d\tau, \qquad (1.8)$$

where H is the Hamiltonian for the system, is a minimum. This minimum represents the best approximation to the true proper energy attainable with a wave function of the assumed form. By arranging the trial wave function to embody the physical features of the problem, it is possible to obtain a reasonably accurate value of the energy which is always an upper bound.

Even when the approximation to the energy value is very good, the accuracy of the approximate wave functions cannot be guaranteed at all points in the configuration space. The energy is determined mainly by the value of the function in a limited region of space. Inaccuracy in the function at other parts of the space may

INTRODUCTION – THE NEGATIVE ION OF HYDROGEN 7

not influence the calculated value appreciably, but may be important in other applications of the function. Nevertheless such elaborate wave functions have been used that they are likely to be adequate for most purposes such as the calculation of the absorption coefficient of H⁻ ions for optical radiation (see Chapter 11, p. 421).

The first calculations carried out by Hylleraas (1929) showed that the undisturbed atomic field of the hydrogen atom (see (1.1)) is not strong enough to produce a stable state. However, the strong interaction between the electrons in the same 1s level greatly modifies the charge distribution. When account is taken of this, it is found that each electron moves in an effective attractive field strong enough to give a total binding energy greater than that of the neutral atom.

The most elaborate calculations are those of Pekeris (1958, 1962), which are likely to be so accurate as to provide an effectively exact result for the electron affinity which can be used as a basis for comparison with results obtained using simpler functions which are more practicable for other applications.

If r_1, r_2 and r_{12} are the coordinates of the two electrons with respect to the nucleus and to each other respectively, the Hamiltonian of the system is

$$H = -\frac{\hbar^2}{2m}(\nabla_1^2 + \nabla_2^2) - e^2\left(\frac{1}{r_1} + \frac{1}{r_2} - \frac{1}{r_{12}}\right). \quad (1.9)$$

The ground state function will depend only on r_1, r_2 and r_{12}. Since the ground state is a singlet, the wave function must be symmetrical with respect to interchange of the two electrons. Most authors, following Hylleraas (1929), have worked with trial functions expressed in terms of $s, = r_1 + r_2, t, = r_1 - r_2$, and r_{12}. Thus Hart and Herzberg (1957) used a 20-parameter function

$$\psi_t = Ne^{-1/2\kappa s}\left(1 + \sum_0^{j+k+2l=4} a_{jkl} u^j s^k t^{2l}\right), \quad (1.10)$$

which has proved to be a useful compromise between accuracy and practicability.

Pekeris (1958) used instead of the coordinates s, t and r_{12}, which are not independent because of the triangular condition, the perimetric coordinates u, v, w given by

$$u = r_2 + r_{12} - r_1, \quad v = r_1 + r_{12} - r_2, \quad w = 2(r_1 + r_2 - r_{12}), \quad (1.11)$$

which are independent and range from 0 to ∞. In terms of these coordinates the trial wave function was taken to be of the form, in atomic units,

$$\psi_t = \exp\{-(u + v + w)\} F(u, v, w) \quad (1.12)$$

with

$$F(u, v, w) = \sum_{l,m,n=0} A(l, m, n) L_l(u) L_m(v) L_n(w), \quad (1.13)$$

the functions L_n being the normalized Laguerre polynomials of order n

$$L_n(w) = \sum_{k=0}^{n} \binom{n}{k} \frac{(-w)^k}{k!}. \quad (1.14)$$

Numerical calculations were carried out for values of $\omega = l + m + n$ up to 16 inclusive, involving 444 parameters $A(l, m, n)$. Values found for the total energy $-\epsilon$ of H⁻ for $\omega =$ 10, 12, 14 and 16 are given in Table 1.1 together with an extrapolated value.

TABLE 1.1 *Total energy $-\epsilon$ of H⁻ in atomic units calculated by Pekeris (1962)*

ω	10	12	14	16	
Number of parameters	125	203	308	444	Extrapolated
ϵ	0.527 750 610	0.527 750 936	0.527 750 991	0.527 751 006	0.527 751 014

The accuracy worked to here is so great that account must be taken of the finite mass of the proton and of relativistic effects which are neglected in the Hamiltonian (1.9). The electron affinity taking into account these effects can be written, in atomic units

$$E_a = (\epsilon - \tfrac{1}{2})\left(1 - \frac{m}{M}\right) + \epsilon_M + \epsilon_R, \quad (1.15)$$

INTRODUCTION – THE NEGATIVE ION OF HYDROGEN

where ϵ_M is a further correction due to the finite value of M and ϵ_R is a relativistic correction. The final values obtained by Pekeris are given in Table 1.2.

TABLE 1.2 *Electron affinity E_a of H^-, in atomic units, calculated by Pekeris*

ω	10	12	14	16	
Number of parameters	125	203	308	444	Extrapolated
$10(\epsilon - \tfrac{1}{2})\left(1 - \tfrac{m}{M}\right)$	0.277 355 0	0.277 355 3	0.277 355 9	0.277 356 1	0.277 356 8
$10\,\epsilon_M$	−0.000 179 0	−0.000 178 9	−0.000 178 9	−0.000 179 0	−0.000 179 0
$10\,\epsilon_R$	−0.000 013 8	−0.000 013 8	−0.000 013 8	−0.000 013 8	−0.000 013 8
$10\,E_a$	0.277 159 2	0.277 162 6	0.277 163 1	0.277 163 3	0.277 164 0

To examine the importance and nature of the correlation effects in leading to a stable negative ion, we now compare the results obtained for the total energy, using trial wave functions including no correlation, with those including correlation in varying degrees (Banyard, 1968). For this purpose we select the following trial functions.

(a) The simple product function

$$\psi_t = N \exp\{-Z(r_1 + r_2)\}, \tag{1.16}$$

with $Z = 0.6875$ in atomic units. This includes no correlation.

(b) The Hartree–Fock (HF) function

$$\psi_t = N\phi(r_1)\phi(r_2), \tag{1.17}$$

where $\phi(r)$ is the self-consistent field wave function (see p. 19). This also includes no correlation.

(c) The so-called (1s 1s') function of Shull and Löwdin (1956),

$$\psi_t = N\{\exp(-ar_1 - br_2) + \exp(-ar_2 - br_1)\}, \tag{1.18}$$

with $a = 1.0392$, $b = 0.2832$, in atomic units. This includes radial but not angular correlation.

(d) The function of Green, Lewis, Mulder, Wyeth and Woll (1954)

$$\psi_t = N\{1 + \alpha r_{12} + \beta(r_1 - r_2)^2\}$$
$$\times \{\exp(-ar_1) + \gamma \exp(-cr_2)\}$$
$$\times \{\exp(-ar_2) + \gamma \exp(-cr_1)\} \tag{1.19}$$

with $\alpha = 0.3746$, $\beta = 0.1043$, $\gamma = 0.0775$, $a = 0.8660$ and $c = 0.4330$, in atomic units. This includes some allowance for both radial and angular correlation.

In Table 1.3 we compare the energy values calculated using these relatively simple functions with those obtained with the more elaborate trial functions of Hart and Herzberg (1957) and of Pekeris (1962). For stable H⁻ the magnitude of the total energy must exceed 0.5000. It will be seen from the table that both the functions (a) and (b), which include no allowance for correlation, fail to predict stability. On the other hand, even the simple function (c), which includes some radial but no angular correlation, does yield a stable negative ion. Taking the HF function (b) as the best obtainable when correlation is ignored, the contribution from correlation obtained using any other function can be taken approximately as the excess binding energy given above that obtained with the HF function. Assuming that the Pekeris function gives exact results within the accuracy of the values given in Table 1.1, we can assign to each trial function a percentage contribution to the exact correlation. On this basis it will be seen that, while the function (c), which includes allowance for radial correlation only, gives 63.6% of the actual correlation energy, the function (d), which

TABLE 1.3 *Energy of the ground state of* H⁻ *calculated using different trial functions*

Trial function	−Energy (a.u.)	−ΔE	% correlation
(a)	0.4727	—	—
(b)	0.4880	0	—
(c)	0.5133	0.0253	63.6
(d)	0.5273	0.0393	98.7
Hart and Herzberg	0.5276	0.0396	99.5
Pekeris	0.5278	0.0398	100

INTRODUCTION – THE NEGATIVE ION OF HYDROGEN

includes allowance for angular correlation, in addition, actually yields 98.7% of the full affect, while the much more complex 20-term function of Hart and Herzberg gives 99.5%.

The charge distribution in H⁻ (See Banyard, 1968.)

We turn now to consider the electron charge distributions as given by the various functions. The two-particle radial charge density is given by

$$\rho(r_1, r_2) = \iint \psi_t^*(r_1, r_2) \psi_t(r_1, r_2) \, d\Omega_1 \, d\Omega_2, \quad (1.20)$$

where $d\Omega_i = \sin\theta_i \, d\theta_i \, d\phi_i$, θ_i and ϕ_i being the usual spherical polar angles. Integrating over r_2 we obtain the one-particle radial density distribution

$$D(r_1) = 2 \int_0^\infty \rho(r_1, r_2) r_1^2 r_2^2 \, dr_2. \quad (1.21)$$

In Fig. 1.1 we show $D(r)$ as a function of r calculated using the different trial functions and compared with the corresponding density function for neutral atomic hydrogen. It will be seen that all trial functions for H⁻ give a more diffuse charge distribution than

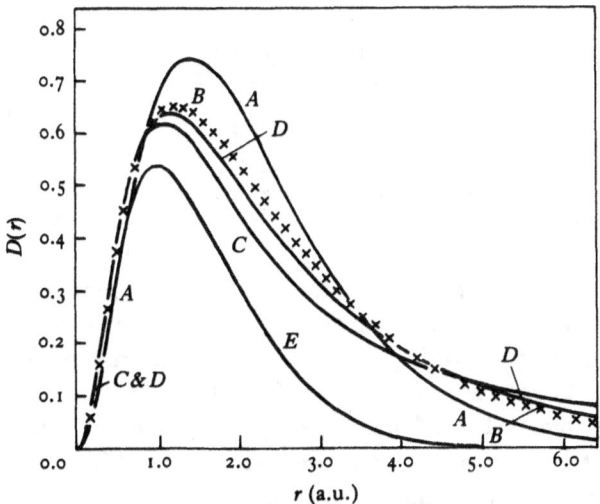

Fig. 1.1 The single-particle radial density formation $D(r)$ for H⁻, calculated with various approximate wave functions. Curves A, B, C, D refer to the functions (a), (b), (c) and (d) respectively of the text, pp. 9–10, E to neutral H.

for the neutral atom. The most diffuse distribution is given by the function (c) which allows for radial but not angular correlation. It will be noticed that the HF function gives a distribution remarkably close to the function (d) which includes angular correlations and gives a very good result for the electron affinity.

In general we would expect radial correlation to increase the spread of the charge distribution as it tends to keep the electrons relatively further apart. This may be seen by examining the

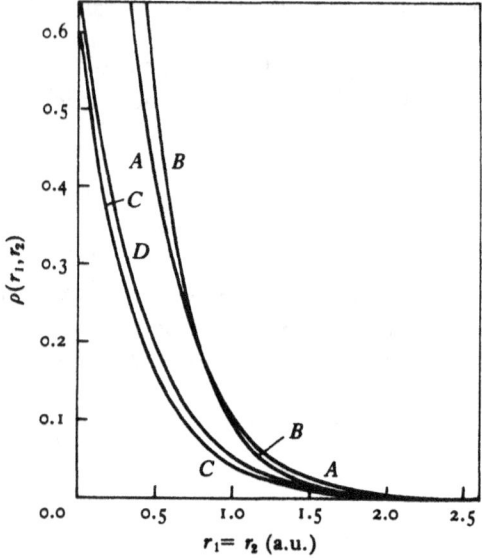

Fig. 1.2. The two-particle radial density function $\rho(r_1,r_2)$, when $r_1 = r_2$, calculated for H^- with various approximate wave functions. Curves A, B, C, D refer to the functions (a), (b), (c) and (d) respectively of the text, pp. 9–10.

two-particle radial densities given by the different functions. In particular we would expect that $\rho(r_1,r_2)$, for $r_1 = r_2$, should be reduced by an increasing extent as more and more radial correlation is introduced. This may be seen from Fig. 1.2. The distributions for the uncorrelated functions (a) and (b) are much the same as compared with those given by (c) and (d), which are also much the same and much smaller at all values of $r_1 = r_2$. Further features are seen in Fig. 1.3 which shows $\rho(r_1,0)$ as a function of r_1. In this case the function (d) gives results which tend to fall between those

INTRODUCTION – THE NEGATIVE ION OF HYDROGEN

for (c) and the uncorrelated functions for $r_1 < 2.0a_0$. This raises the possibility that (c) overestimates the radial correlation, a result which would also be consistent with the evidence from the one-particle distributions shown in Fig. 1.2.

Still further evidence can be obtained by comparing the mean values $\langle r^n \rangle$, where

$$\langle r^n \rangle = \langle r_1^n + r_2^n \rangle,$$

$$= \iint \psi_t^* (r_1^n + r_2^n) \psi_t \, d\tau_1 \, d\tau_2.$$

In Table 1.3 such a comparison is made for $n = -2, -1, 1, 2, 3, 4$ and 8. For $n = 1$ and 2, values obtained with the elaborate Pekeris function are included as well as those obtained using the Hart–Herzberg function (1.10), which also include $n = 3, 4$ and 8. These values, which are presumably quite close to the correct ones, provide a useful basis for comparison.

It will be seen that, in general, the uncorrelated functions which give less diffuse charge distributions also give smaller mean values for $n > 0$. This becomes more marked as n increases. It is also noticeable that the function (d) gives quite good results for $n > 0$ although it does overestimate the mean values by an increasing extent as n increases. The function (c) on the other hand gives mean values for $n > 0$ which are too large, showing that, as indicated from the analysis of data on the one- and two-particle charge distributions, it overestimates the radial correlation and so gives too diffuse distributions. The HF function, while uncorrelated, gives quite good results for $n = 1$.

TABLE 1.4 *Calculated mean values* $\langle r^n \rangle = \langle r_1^n + r_2^n \rangle$, *in atomic units for* H^-

Wave functions	−2	−1	1	2	3	4	8
(a)	1.891	1.375	4.364	12.69	46.16	201.4	2.840×10^5
(b)	2.161	1.371	5.008	18.82	97.38	652.0	81.23×10^5
(c)	2.187	1.322	6.225	33.99	270.2	2749	1313×10^5
(d)	2.241	1.368	5.385	23.41	147.6	1232	320.6×10^5
Hart & Herzberg (1957)	—	—	5.356	22.74	135.1	1023	139.0×10^5
Pekeris (1962)	—	—	5.420	23.82	—	—	—

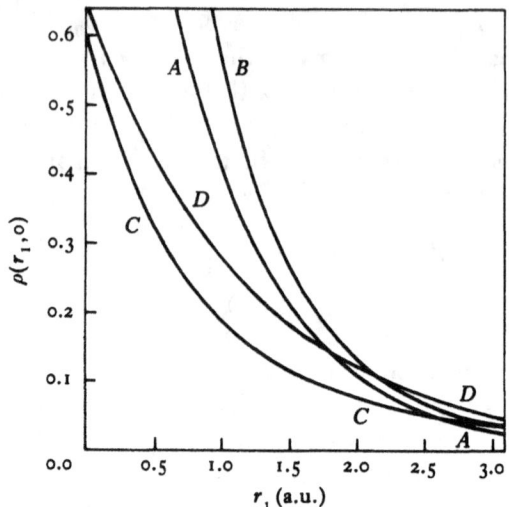

Fig. 1.3. The two-particle radial density function $\rho(r_1, r_2)$, when $r_2 = 0$, calculated for H⁻ with various approximate wave functions. Curves A, B, C, D refer to the functions (a), (b), (c) and (d) respectively of the text, pp. 9–10.

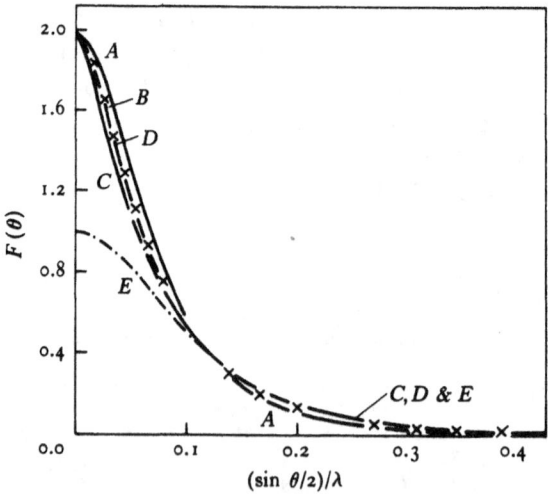

Fig. 1.4. The X-ray scattering factor $F(\theta)$ for H⁻, calculated with various approximate wave functions. Curves A, B, C, D refer to the functions (a), (b), (c) and (d) respectively of the text, pp. 9–10. Curve E is the scattering factor for H.

INTRODUCTION – THE NEGATIVE ION OF HYDROGEN

The diamagnetic susceptibility of H⁻ may be calculated from $\langle r^2 \rangle$ in the usual way to give 18.873×10^{-6} e.m.u. per gram-atom as obtained using the Pekeris function.

Finally, we show in Fig. 1.4 the X-ray scattering factor $F(\theta)$ given by

$$F(\theta) = \int_0^\infty D(r) \frac{\sin(4\pi r \sin\tfrac{1}{2}\theta/\lambda)}{4\pi r \sin\tfrac{1}{2}\theta/\lambda} \, dr, \qquad (1.22)$$

where λ is the wavelength. It will be seen that the results do not depend greatly on the choice of wave function. For $\sin\tfrac{1}{2}\theta/\lambda > 0.15/a_0$ the form factor is indistinguishable on the scale of the diagram from that for the neutral atom.

To summarize, it seems that the relatively simple function (d), which includes both angular and radial correlation, gives remarkably good results not only for the electron affinity but also for the charge distribution out to quite large radial distances. For most purposes it should give results of sufficient accuracy. If in order to make a particular calculation possible it is necessary to use uncorrelated wave functions, the full HF form (b) is certainly to be preferred over the simple exponential form (a).

CHAPTER 2

Ground States of Complex Atomic Negative Ions – Theoretical Considerations

The detailed theory of the ground states of H⁻ shows how important correlation effects are in determining the electron affinity. This can be expected to be a general result which certainly complicates the theoretical calculation of the electron affinities of complex atoms. In this chapter we describe the work which has been carried out in investigating theoretically the properties of the ground states of complex atomic negative ions. To introduce this we begin with a summary of the designation of states of complex atomic systems in general.

2.1 The states of complex atoms

Electron configurations

As a first approximation, a state of a many-electron system can be regarded as made up of an aggregate of single electron states or orbitals. These are essentially hydrogen-like, corresponding to motion in an effective central field due to the nucleus and the mean interaction with the other electrons. They are defined by total and azimuthal quantum numbers n, l but as the central field is no longer exactly Coulomb, the energy will depend on l as well as n. The distribution of electrons among the different nl values is called the electron configuration. It is represented by giving the number of electrons with a particular pair of nl values. Thus, for example

$$(1s)^2(2s)^2(2p)^3$$

denotes a configuration of a seven-electron system in which two electrons occupy a 1s orbital, two a 2s orbital and three a 2p orbital. The subset of electrons occupying a particular nl orbital is said to constitute a shell.

The assignment of electrons to configurations is restricted by the Pauli principle which requires that two electrons with the same

GROUND STATES OF COMPLEX ATOMIC NEGATIVE IONS 17

z-component of spin should not occupy the same orbital. For some purposes it is desirable to include the spin wave function together with the space-dependent part in referring to individual orbitals. It is usual to denote the spin functions, referring to equal and opposite z-components of spin as α, β respectively. The total number of electrons which can occupy an nl orbital is $2l + 1$ in which case the (nl) shell is said to be closed.

The normal or ground configuration of an atom or ion will be the one in which the electrons occupy the lowest accessible orbitals.

States arising from a given configuration

The interaction between the orbital and spin angular momenta of the different electrons leads to a dependence of the energy, for a given configuration, on certain combinations of these angular momenta. In this connection it is worth recalling that the resultant angular momentum arising from the combination of two quantized angular momenta, with magnitude specified by the total quantum numbers L_1 and L_2, must have a magnitude given by one of the quantum numbers

$$L_1 + L_2, L_1 + L_2 - 1, \cdots, |L_1 - L_2|. \tag{2.1}$$

The total quantized angular momentum of a many-electron atom, including both spin and orbital components is accurately conserved. We shall designate the magnitude of this angular momentum by the quantum number J.

In not too heavy atoms the interaction between spin and orbital motion is weak so the total orbital angular momentum is approximately constant and may be designated by a quantum number L, which must satisfy the combination rule above, applied to all electrons. Furthermore, although there is only weak direct coupling between the spins, a relatively strong effective coupling arises because of the Pauli principle. As a result the total spin angular momentum, expressed in terms of the quantum number S, is also approximately conserved.

This approximation, known as Russell–Saunders or LS coupling, is most satisfactory for the low-lying states of light atoms. For heavy atoms spin–orbit interaction becomes so important that neither

L nor S is approximately constant. Instead the total angular momenta for each electron (j,j coupling) are specified. For most cases with which we shall be concerned, LS coupling provides a sufficient approximation and we shall assume it to be valid henceforward.

We now distinguish between the levels of the same configuration with different L, S and J values. In general the order of importance of the angular momenta quantum numbers in determining the energy of an atomic state arising from a given configuration is, in descending order, L, S, J.

The aggregate of levels with the same L and S but different J is referred to as a *term*. Terms are designated by a notation such as ^{3}P where the capital letter specifies the L value according to the same convention as for the l values of individual electrons and the left-hand index gives the multiplicity, $2S+1$. Thus a ^{3}P term has $L = 1$ and $S = 1$.

For a closed shell of electrons L and S must both be zero, so a configuration in which all electrons are in closed shells, as for the normal states of the rare gas atoms or of the negative ions of the halogen atoms, can give rise to a ^{1}S term only. For an incomplete shell, in which no two electrons occupy orbitals with the same n and l value, the terms which can arise are simply obtained from the rules for combination of angular momenta. Thus for a pd configuration the possible L values are 1, 2, 3 and the possible S values 0 and 1, giving rise to ^{1}S, ^{1}P, ^{1}D, ^{3}S, ^{3}P and ^{3}D terms. When the shell contains equivalent electrons, i.e. electrons occupying orbitals with the same n and l, then a number of otherwise possible terms are excluded by the Pauli principle. Simple procedures are available for selecting these terms (Slater, 1929; Hartree, 1957).

The dependence of the energy on the quantum number J is so small that it leads to the fine structure of the term system. For a given L and S there will be $2S+1$ substates with different J if $L > S$, and $2L+1$ if $S > L$. For this reason $2S+1$ is called the *multiplicity* of a term with given S.

The J value is indicated by adding it as a right-hand suffix to the term designation. Thus ^{3}P$_2$ denotes a level with $L = 1$, $S = 1$ and $J = 2$.

For terms arising from a given configuration with equivalent electrons the following rules apply.

GROUND STATES OF COMPLEX ATOMIC NEGATIVE IONS 19

(a) Levels with largest S lie deepest and of these the lowest is the one with the greatest L.

(b) For levels arising from a configuration in which all shells are complete except the outer, which is less than half-filled, the energy for given L and S increases with J and the multiplet is said to be *normal*. This situation is reversed when the shell is more than half-filled, the energy decreasing with increase of J, and the multiplet is said to be *inverted*.

Thus for a p^2 configuration $^3P_1$, ^1S and ^1D terms can arise. Of these the ^3P is the lowest and within the multiplet, in order of increasing energy we have $^3P_0$, $^3P_1$ and $^3P_2$. For a p^4 configuration ^3P is again the lowest term but within the multiplet the order is $^3P_2$, $^3P_1$, $^3P_0$.

In addition, if the nucleus possesses spin, account must be taken of its coupling with the electron spin. This is always weak and leads to hyperfine structure in the energy levels. We shall rarely need to take this into account when dealing with the properties of negative ions so we shall not discuss it further here.

2.2 Calculation of the structure and energy of complex atomic negative ions

The Hartree–Fock (HF) method

The most effective practicable approach to the theoretical determination of the structure of complex atoms and of atomic ions has been that of the self-consistent field, first applied by Hartree in 1928 and then refined by Fock (1930) who took account of electron exchange effects. The HF approximation, which is now available for all atomic systems, still provides the basis on which improved approximations are built. It gives good results when used to calculate mean atomic properties which do not depend on small differences between large quantities. For example, as it does not allow for correlation effects it cannot be used to calculate electron affinities though it will give the *total* energy of a stable negative ion with quite high percentage accuracy. In recent years the problem of calculating the effects of correlations between electronic motion has been tackled energetically and promising results have been obtained. It has not, at the time of writing, yet been developed to such an extent as to yield reliable results directly for electron affinities except in a very few cases (see p. 27). Nevertheless by semi-empirical methods it is possible to use the results to estimate affinities with some accuracy.

Although it would be out of place here to discuss in any detail the theory of the HF method and of the procedures adopted to calculate correlation energies, we shall give a brief outline so that the physical assumptions involved and the techniques employed can be understood.

In the original Hartree method, which essentially determines the mean energy of the terms of a configuration, each atomic electron is considered to move under the combined action of the nuclear attraction and a mean repulsion due to the remaining electrons. This means that, the overall wave function for the ground state of an atom of nuclear charge Ze containing N electrons can be written in the form of the product

$$\Psi = \prod_k \psi(k|nl), \qquad (2.2)$$

where $\psi(k|nl)$ is the wave function for the kth electron which occupies an orbital defined by the quantum numbers n, l. Furthermore $\psi(k|nl)$ will be a solution of the single-particle Schrödinger equation

$$\nabla^2 \psi(k|nl) + \frac{2m}{\hbar^2} \left\{ E(k|nl) + Ze^2 \, r^{-1} - e^2 \sum_{i \neq k} \int r_{ik}^{-1} |\psi(i|st)|^2 \, d\tau_i \right\}$$
$$\psi(k|nl) = 0, \qquad (2.3)$$

in which the term

$$e^2 \int r_{ik}^{-1} |\psi(i|st)|^2 \, d\tau_i \qquad (2.4)$$

is the mean repulsion acting on the kth electron due to the ith electron, r_{ik} being their separation.

In this way the problem becomes one of solving the coupled set of N equations (2.3). This was done by an iterative process of successive approximation, assisted by guesses from experience, so that the wave functions $\psi(k|nl)$ for each electron are self-consistent. In carrying out these calculations the further approximation must be made of averaging the charge distributions of the remaining electrons in each equation (2.3) so as to give a resultant spherically symmetrical field acting on the particular electron concerned.

Although this self-consistent procedure was first based on physical arguments it may be shown that, if it is assumed that the ground state wave function can be written to a good approximation in the form (2.2), then application of the variational principle (1.8) shows that the best form which the single-electron wave functions can take is that given from solution of the self-consistent equations (2.3).

This original Hartree method did not allow for the indistinguishability of electrons although account was taken of the Pauli principle in the proper assignment of the quantum numbers n, l to the different electrons.

As a result, the Hartree approximation to the energy depends on the electron configuration but not on the term value. To determine the separate energies of the terms arising from a configuration, allowance must be made for exchange as in the HF method.

When explicit account is taken of the spin as well as the orbital dependence of the single-electron wave functions, the overall wave function must be antisymmetric with respect to interchange of any pair of electrons. Thus, corresponding to a particular orbital $\psi(k|nl)$ there will be two spin functions α_k, β_k. For atoms or ions with closed shells, the appropriate overall wave function is the determinant

$$\Phi_0 = \begin{vmatrix} \psi(1|nl)\alpha_1, \cdots, \psi(1|st)\alpha_1, \cdots \\ \psi(1|nl)\beta_1, \\ \psi(2|nl)\alpha_2, \\ \psi(2|nl)\beta_2, \\ \vdots \\ \psi(N|nl)\beta_N, \cdots, \psi(N|st)\beta_N, \cdots \end{vmatrix} \quad (2.5)$$

If this wave function is used as a trial function in a variational method, it is found that the functions ψ now satisfy a set of coupled integro-differential equations, the integral terms of which represent the exchange interaction.

If the configuration is not one involving closed shells only, the wave function can no longer be written in general as a single determinant. This is because the value of the z-component of orbital angular momentum m_l for the individual occupied orbitals is not specified. However, in most cases it is still possible to work in terms of single determinants. This is because we may set up individual wave functions for particular terms which refer to specified values of the quantum numbers M_L, M_S which specify the z-components of total orbital and spin angular momentum respectively. Thus for the lowest term which has the largest S and largest L for that value of S, the wave function, with $M_S = S$, $M_L = L$, is a single determinant. Since the energy is independent of M_L and M_S for given L and S, this wave function may be used to calculate it. In other cases sum rules and other devices may be used to extend calculations carried out with wave functions in the form of single determinants (Slater, 1929; Hartree, 1957).

When applied to negative ions in which the additional electron is attached in a shell outside that occupied by the other electrons, as would be the case for He⁻, for example, the energy of the system in the HF approximation will be normally minimized when the outer electron is at infinity with zero binding energy – the self-consistent field acting upon it will usually be inadequate to bind it because even if a stable negative ion of this type actually exists, it will only be through correlation effects which are neglected in the approximation. However, if we are considering a negative ion in which the electron is attached in an orbital for which the quantum numbers n, l are the same as for one or more of the other electrons, then the limitation imposed by the HF method, that the wave functions, for a given term, depend only on n and l, ensures that a self-consistent solution will be obtained with the additional electron attached at a finite mean distance. The existence of this solution does not require that the self-consistent approximation to the total energy of the ion will be less than that for the corresponding neutral atom. Even if it is less, this does not prove that the electron affinity is positive. This will only be implied if the calculated total energy of the ion is less than the actual (observed) total energy of the atom. Such high accuracy is usually beyond the range of the approximation. Nevertheless the HF solution for a negative ion, when this is known from other evidence to be stable, does represent a good approximation to the ionic charge distribution (cf. for H⁻, p. 11).

The Hartree–Fock–Roothan (HFR) procedure

Despite the added complexity of the numerical problem, HF equations were solved by iterative procedures for a number of atoms and ions before the introduction of high speed computers. These included Cl⁻ (Hartree and Hartree, 1936), O⁻ (Hartree, Hartree and Swirles, 1939) and N⁻ (Hartree and Hartree, 1948). An important step forward was taken by Roothan (1951), who suggested an alternative approach to the numerical determination of HF fields based on the use of trial functions with the variational principle (1.8). The individual orbitals are expanded in terms of trial basis functions of the form

$$\chi_t(i,j;l,m) = \exp(-\xi_{i,j} r) r^{j-1} P_l^{|m|}(\cos\theta) \exp(im\phi), \quad (2.6)$$

where $j = 1, 2, ...$, and $P_l^{|m|}(\cos\theta)$ is a spherical harmonic. Thus $\psi(k|nl)$ would be expanded as

$$\sum_i \sum_j c_{ij,l} \exp(-\zeta_{ij,nl} r) r^{j-1} P_l^{|m|}(\cos\theta) \exp(im\phi). \quad (2.7)$$

The exponential parameters $\zeta_{ij,nl}$ are taken to be the same for all orbitals with a particular l value and the coefficients $c_{ij,nl}$ must be such that all of these orbitals are orthogonal to each other. Subject to this condition the $\zeta_{ij,nl}$ and $c_{ij,nl}$ are determined by the variational principle (1.8). Practical application depends on the use of a finite number of terms in the different expansions. The determination of the $\zeta_{ij,nl}$ requires by far the most computer time, and experience has led to reduction in the number of different parameters required. In some cases, for example, $\zeta_{ij,nl}$ can be assumed to be independent of nl (see Table 2.1).

An advantage of this approach is that it provides a more suitable basis for improved approximations. Moreover it is applicable to molecules as well as atoms. The term, HF approximation, now means the best obtainable approximation in which the wave function for the system is represented by an antisymmetrized linear combination of one-electron orbitals. It is no longer thought of primarily in

TABLE 2.1 Parameters defining the HF orbitals for F$^-$ and Cl$^-$

F$^-$

			$n=1, l=0$	$n=2, l=0$		$n=2, l=1$	
i	j	$\zeta_{ij,nl}$	$c_{ij,nl}$	$c_{ij,nl}$	i	j	$\zeta_{ij,nl}$	$c_{ij,nl}$
1	1	8.9165	0.894 85	−0.202 09	1	2	2.0519	0.495 34
2	1	14.7007	0.036 07	−0.008 06	2	2	3.9288	0.309 59
1	2	3.2762	0.003 98	0.629 40	3	2	1.4496	0.052 59
2	2	8.0477	0.085 84	−0.097 02	4	2	0.9763	0.266 64
3	2	1.8485	−0.000 35	0.482 31	5	2	8.2943	0.017 09

Cl$^-$

i	j	$\zeta_{ij,nl}$	$n=1, l=0$ $c_{ij,nl}$	$n=2, l=0$ $c_{ij,nl}$	$n=3, l=0$ $c_{ij,nl}$	i	j	$\zeta_{ij,nl}$	$n=2, l=1$ $c_{ij,nl}$	$n=3, l=1$ $c_{ij,nl}$
1	1	17.2875	0.917 93	−0.256 52	0.075 23	1	2	7.6419	0.656 25	−0.163 76
2	1	28.4472	0.014 95	−0.004 60	0.001 52	2	2	13.9763	0.036 96	−0.009 08
1	2	6.8172	0.003 55	+0.890 68	−0.287 24	1	3	2.8985	0.012 49	+0.328 80
2	2	15.3682	0.081 73	−0.139 56	+0.044 85	2	3	1.0374	0.000 51	+0.323 45
1	3	2.8669	0.000 12	−0.006 30	+0.695 90	3	3	5.9321	0.370 39	−0.110 84
2	3	1.6723	−0.000 06	−0.000 38	+0.452 12	4	3	1.8615	−0.002 71	+0.501 95
3	3	5.9503	−0.000 97	+0.241 36	−0.152 24					

GROUND STATES OF COMPLEX ATOMIC NEGATIVE IONS 25

terms of functions satisfying a self-consistent set of integro-differential equations.

Using the Roothan procedure (referred to henceforward as HFR), Clementi and his collaborators have worked out the HF fields for Li$^-$, B$^-$, C$^-$, N$^-$, O$^-$ and F$^-$ (Clementi and McLean, 1964), Na$^-$, Al$^-$, Si$^-$, P$^-$ and Cl$^-$ (Clementi, McLean, Raimondi and Yoshimine, 1964) and K$^-$, Sc$^-$, Ti$^-$, V$^-$, Cr$^-$, Mn$^-$, Fe$^-$, Co$^-$, Ni$^-$ and Cu$^-$ (Clementi, 1964). To illustrate the orbitals which are obtained in these calculations, we give in Table 2.1 the parameters $\zeta_{ij,nl}$ and $c_{ij,nl}$ determined for F$^-$ and for Cl$^-$.

In these two cases the total energy of the negative ion is less than that of the corresponding neutral atom by 0.0501 and 0.0947 a.u. for F and Cl respectively. These are to be compared with the measured electron affinities, 0.127 and 0.131 a.u. respectively. Similar results were found for C and for S, the HF energy of the ion being less than that for the neutral atom by 0.033 and 0.043 a.u. respectively as compared with measured electron affinities of 0.046 and 0.078 a.u. On the other hand, for O the HF energy of the ion exceeds that of the atom by −0.020 a.u. whereas O is known to have a positive electron affinity of 0.054 a.u. (1.465 eV, see p. 41).

Inclusion of correlation

It is quite clear that, from HF field calculations alone, it is not possible to predict even whether a particular negative ion is stable. Correlation effects must be taken into account (cf. the corresponding situation for H$^-$, p. 10). Clementi (1964) was able to estimate the contribution from these effects by an empirical procedure. We have for the electron affinity

$$E_\mathrm{a} = (E_\mathrm{HF}^0 - E_\mathrm{HF}^-) + (E_\mathrm{C}^0 - E_\mathrm{C}^-), \tag{2.8}$$

where $E_\mathrm{HF}^0, E_\mathrm{HF}^-$ refer to the HF energies for the neutral atom and negative ion respectively, $E_\mathrm{C}^0, E_\mathrm{C}^-$ to the corresponding correlation energies. Clementi estimated $E_\mathrm{C}^0 - E_\mathrm{C}^-$ for atoms of the first short period by extrapolation from values for $E_\mathrm{C}^+ - E_\mathrm{C}^0$ and $E_\mathrm{C}^{2+} - E_\mathrm{C}^+$ which, given the HF energies for the atoms and positive ions, could be obtained from the measured first and second ionization potentials. For atoms in the second short period, insufficiently accurate data were available for this procedure to be effective, so Clementi assumed that $E_C^0 - E_C^-$ is the same as $E_C^+ - E_C^0$ for a system with atomic number greater by unity. For example $E_C^0(\text{Cl}) - E_C^-(\text{Cl}^-)$ is taken to be the same as $E_C^+(\text{Ar}^+) - E_C^0(\text{Ar})$. When these procedures were applied, the electron affinities of F, Cl, S, C and O were estimated to be 0.124, 0.131, 0.078, 0.046 and 0.045 a.u. which are quite close to the respective observed values 0.127, 0.131, 0.076, 0.054 and 0.057 a.u. Results obtained in this way for other atoms are given in Tables 3.5–3.7, in which they are compared with those obtained by other empirical methods.

In the last few years a great deal of attention has been devoted to developing methods for accurate calculation of correlation energies. Considerable success has been achieved (Nesbet, 1969) by application of a method suggested by Brueckner (1954) and developed by Bethe and Goldstone (1957), for calculating the binding energies of atomic nuclei. The basic assumption is that the total correlation energy can be regarded to a close approximation as given by the sum of contributions from different orbital pairs,

$$E_C = \sum_{i,j} \epsilon_{ij}, \qquad (2.9a)$$

terms of higher order involving three or more orbitals being neglected. In addition, the further assumption is made that the ϵ_{ij} can be calculated separately by applying the variational principle to a trial function

$$\Psi_{t,ij} = \Phi_0 + \sum_a c_a \Phi_{i,a} + \sum_b c_b \Phi_{j,b} + \sum\sum c_{ab} \Phi_{ij,cd} \qquad (2.9b)$$

in which the different terms have the following significance. Φ_0 is the HF determinant already calculated (see (2.5)), $\Phi_{i,a}$ differs from Φ_0 in that the orbital ψ_i is replaced by an orbital a of a suitably chosen basis set, while in $\phi_{ij,cd}$ both orbitals i and j are replaced by c and d respectively, which are also members of the chosen set. In this formulation the correlation effect is expressed in terms of the virtual excitation of one or two orbitals. Success in accurate prediction of the correlation energy depends on the choice of the basic set of orbitals $a, b, c, d \ldots$. These are again expressed as linear combinations of functions of the form (2.6) and it is essential that they be orthogonal to each other and to the functions used in deriving the HF approximation. In general it is found that the optimized excited orbitals extend over much the same region of space as the HF orbitals but possess more nodes in order to satisfy the orthogonality condition. This is to be expected because the biggest contribution will come from virtual excitation of orbitals which overlap the initial orbitals as much as possible.

Two alternative ways of proceeding depend on whether the pair correlations refer to individual spin–orbit functions or to pairs with particular coupled angular momenta – so-called symmetry-adapted pairs. The former have been used extensively by Nesbet (1969) and his collaborators while Weiss (1971) has carried out detailed calculations using symmetry-adapted pairs.

The fluorine negative ion

We shall illustrate the results obtained by these two procedures by considering the case of F and F^- to which Moser and Nesbet (1971) have applied the individual spin-orbital excitation procedure and Weiss (1971) that involving symmetry-adapted pairs. Their results are summarized in Table 2.2.

TABLE 2.2 *Energies of F and F^- calculated by Weiss (a) and by Moser and Nesbet (b)*

| | ϵ_{ij} (a.u.) | | | |
| | F | | F^- | |
Pair	(a)	(b)	(a)	(b)
$2p^2$	0.1465	0.1569	0.2315	0.2616
$2s2p$	0.0590	0.0918	0.0793	0.00958
$2s^2$	0.0106	0.0125	0.0110	0.0120
$1s^2$	0.0394	0.0398	0.0397	0.0398
$1s2p$	0.0152	0.0169	0.0154	0.0135
$1s2s$	0.0049	0.0056	0.0047	0.0055
Single correlations	0.0280	0.0022	–	–
Total correlation energy (a.u.)	0.3037	0.3259	0.3817	0.4282
HF energy (a.u.)	−99.4093		−99.4594	
Total energy (a.u.)	−99.7131	−99.7352	−99.8411	−99.8876

The calculated electron affinity of F is thus 0.128 and 0.1452 a.u. according to the two procedures, as compared with the observed value of 0.127 a.u. The use of symmetry-adapted pairs gives remarkably good results while the individual spin-orbital method overestimates the electron affinity. It is difficult to say how general these results are, particularly as both methods are based on physical intuition rather than mathematical derivation. However, Moser and Nesbet (1971) find that their method also overestimates the electron affinity for O and C atoms. They proceeded to estimate, with somewhat lower accuracy, the contributions from third-order terms arising from configurations represented by determinants $\Phi_{ijk,cde}$ and found them to be significant and in the correct sense to reduce the calculated electron affinities to nearly the observed value for C, O and F. That for N becomes negative and that for B, 0.22 eV (see also Schaefer and Harris, 1968). It may be that with the choice of symmetry-adapted pairs the contribution from third-order terms is reduced – the definition of terms of successive order is somewhat different in the two procedures.

Electron affinities of alkali metal atoms

An interesting application of the pair treatment of correlation has been made by Weiss (1968), who estimated the electron affinities of the alkali metal atoms by assuming them to be given by

$$E_a \simeq E_{HF}^0 - E_{HF} + \epsilon(ns^2), \qquad (2.10)$$

where $\epsilon(ns^2)$ is the correlation energy between the outer two orbitals in the negative ion. Table 2.3 shows the convergence of the calculated value of $\epsilon(ns^2)$ for Na$^-$ with the number of excited configurations taken into account.

TABLE 2.3 *Correlation energy $\epsilon(3s^2)$ for* Na$^-$

Configurations	$\epsilon(3s^2)$ (a.u.)
$3s^2 + 3p^2$	0.0201
$3s^2 + 3p^2 + 3d^2$	0.0203
3 config + $3s4s + 4s^2 + 4p^2 + 4d^2 + 4f^2$	0.0236
8 config + $3s5s + 5s^2$	0.0239

Values obtained for the electron affinities of Li, Na and K are given in Table 2.4. A useful check on the reliability of the method is obtained by comparing the predictions made for the ionization

GROUND STATES OF COMPLEX ATOMIC NEGATIVE IONS 29

potentials of isoelectronic systems by following the same procedure with observed values. It is seen from Table 2.4 that the errors are relatively quite small and in fact the calculated electron affinities are remarkably close to the accurate observed values (see Chapter 3, p. 42 and Chapter 11, p. 469) for Li and Na. Even for K the agreement is still quite good.

TABLE 2.4 *Calculated total energies, electron affinities and ionization energies (in atomic units) for* Li, Na *and* K *atoms and some of their isoelectronic ions*

	Total energy		Ionization energy or electron affinity	
	$(ns)^2$ ^1S	ns ^2S	Calculated	Observed
		$n=2$		
Li⁻	−7.4553	−7.4327	0.0226	0.0229
Be	−14.6189	−14.2774	0.3415	0.3426
B⁺	−24.2985	−23.3760	0.9225	0.9241
C²⁺	−36.4832	−34.7261	1.7571	1.7598
		$n=3$		
Na⁻	−161.8787	−161.8589	0.0198	0.0201
Mg	−199.6480	−199.3715	0.2765	0.2810
Al⁺	−241.7143	−241.0304	0.6839	0.6919
Si²⁺	−288.0409	−286.8211	1.2198	1.2309
		$n=4$		
K⁻	−599.1818	−599.1646	0.0182	0.0185
Ca	−676.7863	−676.5698	0.2165	0.2247
Sc⁺	−759.4962	−758.9805	0.5157	0.5333
Ti²⁺	−847.2694	−846.3802	0.8892	0.9088

It is possible that the main cause of the discrepancy for K is neglect of core polarization. This has been taken into account by Norcross (1974) in a calculation essentially similar to that used in the calculation of collision cross sections (see the close-coupling method in Chapter 4, p. 94). The problem is treated as a two-electron one with the effect of the core included as a source of potential energy for each electron given by

$$V_c = V_s(r) + V_p(r), \qquad (2.11)$$

where V_s is a short-range interaction and V_p that due to polarization. The former was represented by a scaled Fermi–Thomas form and the latter by

$$V_p(r) = \alpha_2 r^{-4} W_2(r_c, r) + \alpha_4 r^{-6} W_4(r_c, r), \qquad (2.12)$$

where α_2 and α_4 are the dipole and quadrupole polarizabilities of the core and the W's cut-off functions given by

$$W_m(r_c, r) = 1 - \exp\{-(r/r_c)^m\}. \qquad (2.13)$$

For α_2 the best value derived from observation was used, but α_4, r_c and also the scaling factor λ were adjusted to give the best agreement with the observed electron affinity. Account was also taken of polarization correlation (Chisholm and Öpik, 1964).

With this semi-empirical procedure very good agreement was obtained with observed electron affinities not only for Li, Na and K but also for Rb and Cs (see Chapter 3, p. 64). The application of this method to investigate the p^2 ^{3}P excited state of the negative ions is discussed in Chapter 4, p. 150.

By using semi-empirical methods to determine the energy arising from correlation in more complicated cases, it is possible to obtain quite good estimates of electron affinities, particularly when precise information is available about the ionization and excitation energies of neighbouring atoms and positive ions. We shall discuss these methods in the next chapter, together with other empirical procedures for estimating electron affinities.

CHAPTER 3

The Electron Affinities of the Elements

We have already discussed the meagre information about electron affinities which is obtained by direct application of quantum mechanical methods. The electron affinity of H^- is very accurately known and good results are available for the higher alkali metal atoms. For other atoms we must resort to either experimental or semi-empirical methods which may depend on theory to a considerable extent, and we now proceed to discuss these methods and the information available about the electron affinities of the atoms as a whole.

3.1 Experimental methods for measuring electron affinities

The most important and effective method which is of wide applicability is the determination of the threshold frequency for photodetachment from the ion concerned, a technique which has been greatly reinforced by the availability of lasers which provide highly monochromatic light sources. As a detailed account of the technique and of the results obtained with its use is given in Chapter 11, we shall merely summarize the principles involved here and list the electron affinities obtained in the appropriate comparative tables. The fact that the number of accurately known electron affinities has been substantially increased from laser photodetachment observations has, in turn, improved the effectiveness of semi-empirical procedures through the provision of reliable 'anchor' values which help to fix the adjustable parameters. While we are still far from knowing the electron affinities of the elements with the same accuracy as the ionization energies except in a few cases, we have a much more extensive and reliable knowledge of them than was available at the time of publication of the second edition of this book.

A number of other experimental methods, depending on dynamical phenomena, are only briefly referred to in this chapter although the results obtained by their use will also be listed. Some methods, however, depending on measurements made of equilibrium conditions either in reactive systems or in energy relations involved in cyclic processes will be described in this chapter.

Many of the methods are applicable to molecular negative ions but results for such ions are discussed in Chapter 6.

From photodetachment

Total cross-sections for the photodetachment process

$$A^- + h\nu \to A + e$$

may be measured as functions of photon frequency. This may be carried out with high energy resolution using a tuned dye-laser as light source. The threshold frequency is given by

$$\nu_t = E_a/h,$$

where E_a is the electron affinity of the atom A.

The kinetic energy of the electrons ejected by radiation of frequency is given by

$$E = h\nu - E_a,$$

so that, if it can be measured, E_a may be obtained. This is possible using a fixed frequency laser as light source.

The theory of photodetachment and of the methods for measuring photodetachment cross-sections and photoelectron energy distributions is discussed in Chapter 11, p. 417, as well as the results obtained. Electron affinities measured in this way for H, O, S, Se, P, As, Sb, C, Ge, Sn, Li, Na, K, Rb, Cs, Cu, Ag, Au, Pt, are given in Table 3.7.

Photoabsorption by F^-, Cl^-, Br^- and I^- may be observed in the vapour from shock-excited alkali halide and the electron affinities of the corresponding atoms derived from observation of the threshold frequency. This technique is described in Chapter 11, p. 442. Electron affinities obtained for the halogen atoms in this way are included in Table 3.3.

From the threshold frequency for polar photodissociation

In certain cases absorption of light with frequency above a certain threshold value produces dissociation of a molecule AB into ions

A^+ and B^- (see Chapter 8, p. 255). From observations of the threshold frequency the electron affinity of B may be determined, provided it may be assumed that, at the threshold the ions are formed with zero kinetic energy. Results obtained in this way for Br^- and I^-, derived from photodissociation of Br_2 and I_2 respectively, are given in Table 3.3.

From threshold energies for dissociative attachment processes

Electron impact with molecules AB in certain circumstances lead to dissociative attachment reactions of the form

$$AB + e \longrightarrow A + B^-.$$

From a study of the threshold electron energy for production of ions of different kinetic energy it is possible in many cases to obtain the electron affinity of B. These reactions and their analysis are discussed in Chapter 9, and results obtained in this way for O^-, C^- and Br^- are given in Table 3.3.

From threshold energies for polar dissociation of molecules by electron impact

Polar dissociation by electron impact is discussed in Chapter 9, p. 272. As for dissociative attachment, measurement of the threshold electron energy for production of ions of different kinetic energies yields the electron affinity. Results available for O^- and C^- are given in Table 3.3.

From the variation with frequency of the affinity continuum

If an electron of kinetic energy E_e is captured by a neutral atom A into the ground state of the negative ion

$$e + A \longrightarrow A^- + h\nu,$$

the frequency of the emitted radiation, which is often referred to as the affinity continuum, is given by

$$h\nu = E_e + E_a,$$

where E_a is the electron affinity of the atom A. The threshold frequency for the continuum is therefore E_a/h so its measurement determines E_a. The observation of the affinity continuum is discussed in Chapter 8.

From measurements of ion currents arising from reactions at a hot filament

A number of measurements of electron affinities have been made from observations of the ionized products, both positively and negatively charged, resulting from reactions at the surface of a hot filament.

The reactions concerned either involve dissociation of a molecule MX of which X is an electron acceptor so that X^- ions are reaction products, or the surface ionization of atoms Y say to form Y^+ and Y^- ions. A further possibility is to observe the dissociation of molecules XY so that both X^- and Y^- ions are formed as reaction products.

It is important to be sure that the results do not depend on surface conditions. For this reason the simplest case to consider is that in which the dissociation energy of the molecule MX is small so that it is almost completely dissociated by the hot filament. Under these circumstances the equilibrium constant of the reaction

$$X + e \rightleftharpoons X^-$$

taking place in the neighbourhood of the filament may be measured and the electron affinity derived.

Molecular dissociation

If the concentrations of atoms X, electrons and ions X^-, are n_0, n_e, n_- per cm^3 respectively, when equilibrium is attained, the equilibrium constant K, which is a function of the temperature T_s of the filament only, is given by

$$K = n_0 n_e / n_-. \qquad (3.1)$$

Now it may be shown by the method of statistical mechanics (Fowler, 1929, p. 281) that

$$K = (g_0 g_e/g_-)(2\pi mkT_s/h^2)\exp(-E_a/kT_s), \qquad (3.2)$$

where g_0, g_e, g_- are the statistical weights of the normal states of the atoms, free electrons and negative ions respectively, m the mass of the electron, k Boltzmann's gas constant, h Planck's constant and E_a the energy liberated in the reaction, i.e. the electron affinity of the atom X. Hence if K can be measured, E_a may be calculated. This may be done by observing the currents of electrons and of negative ions issuing from the surface. For, from kinetic theory considerations, when equilibrium is attained, the concentration n may be related to the number of particles Z of the particular species leaving the filament in cm^{-2} s^{-2}. Thus

$$Z = n(kT_s/2\pi m)^{1/2}, \quad (3.3)$$

giving

$$K = (Z_0 Z_e/Z_-)(kT_s/2\pi m)^{1/2}. \quad (3.4)$$

Z_e/Z_- can be directly measured as the ratio of electron current i_e to negative-ion current i_- proceeding from the filament. Z_0 can only be determined from the pressure of the gas X_2 which is dissociated into atoms X on striking the filament. If T_G is the gas temperature and p_{X_2} the pressure of the gas X_2 we have, on the assumption of complete dissociation at the filament,

$$Z_0 = \tfrac{1}{2} p_{X_2} (2\pi M k T_a)^{1/2}, \quad (3.5)$$

where M is the mass of a gas atom.

The determination of K then involves only the directly measured quantities i_-, i_e, p_{X_2}, T_s and T_G. It is important to remember, however, that the method assumes complete dissociation of the gas X_2 by the hot filament, and even for I_2, the dissociation energy of which is only 1.5 eV, conflicting evidence has been obtained as to the validity of this assumption. A great advantage of the method, however, is that the equilibrium constant K does not have to be measured very accurately to give a reasonably accurate value for E_a. Even if only half of the iodine molecules were dissociated per impact with the filament, the error made in actual determinations of E_a would be only about 0.1 eV.

Experimental methods which have been used differ in the means employed to measure the ratio i_e/i_-. Glockler and Calvin (1935) employed space-charge effects while Sutton and Mayer (1935) and several later investigators used the principle of the magnetron. Some time later Bailey (1955) introduced a further technique involving a mass spectrometer.

The space-charge method is based on the space-charge equation

$$\log i = \log C(e/M')^{1/2} + \tfrac{3}{2}\log V, \qquad (3.6)$$

giving the space-charge limited current i between a pair of electrodes in terms of their potential difference V, a geometrical constant C and the specific charge e/M' of the current carriers. To apply this, the current i passing to the plate of a cylindrical diode *in vacuo* is measured, and, since in this case the current carriers are electrons only, e/m is known and C determined. In the presence of a gas X_2 the current is transported partly by ions X^- and partly by electrons. The effective mass M' of the carriers which must be used in equation (3.6) in this case may be related to the ratio i_-/i_e by the formula

$$\frac{i_-}{i_e} = \frac{M'^{1/2} - M^{1/2}}{M^{1/2} - M'^{1/2}}, \qquad (3.7)$$

where m is the electronic mass, M that of an atom X. Knowing C, i_- and V, M' is determined and hence i_-/i_e.

In the magnetron method the apparatus consists of a cylindrical triode with axial tungsten filament, enclosed in a coaxial solenoid, through which a current can be passed. In performing an experiment the plate and grid are maintained at positive potential with respect to the filament, the vapour of molecules MX admitted and the plate current measured. This is effectively equal to i_e since $i_e \gg i_-$. To measure i_- a current is passed through the solenoid to produce an axial magnetic field sufficiently large to prevent the electrons from reaching the plate. Under these conditions the plate current is equal to i_-.

Bailey's method, using a mass spectrograph, is convenient when the molecule MX is such that M has a low ionization potential E_i. In that case if Z_M, Z_{M^+} are the number of atoms M leaving the filament per square centimetre per second respectively as neutral or as singly-charged positive ions, consideration of the equilibrium (see p. 34)

$$M \rightleftharpoons M^+ + e^-$$

shows that

$$Z_{M^+}/Z_M = r = \tfrac{1}{2}\exp\{(\phi - E_i)/kT_s\}, \qquad (3.8)$$

where ϕ is the effective work function of the filament surface. Also

$$Z_{M^+} + Z_M = Z_{X^-} + Z_X \qquad (3.9)$$

so that

$$Z_X/Z_{M^+} = (r+1)/r. \qquad (3.10)$$

We now have

$$K = \frac{r+1}{r} \frac{Z_{M^+} Z_e}{Z_-} \left(\frac{kT_s}{2\pi m}\right)^{1/2}$$

$$= (r+1)(i_{M^+} i_e/i_-)(kT_s/2\pi m)^{1/2}. \qquad (3.11)$$

To determine r, ϕ is required. This may be obtained from the Richardson equation for electron emission according to which the electron current density emitted from the surface is given by

$$j_e = 120.1 T_s^2 \exp(-\phi/kT_s) \text{ A cm}^{-2}. \qquad (3.12)$$

The current ratio i_{M^+}/i was measured with the mass spectrometer while $i_e \ (\gg i_-)$ was directly measured as the total negative current. Care must be taken to ensure that the collecting efficiency of the spectrometer for negative ions is the same as for positive ions.

The most serious difficulty however is that the surface of the filament is likely to be uneven. In this case the positively-charged particles will tend to be emitted from the areas of high work function, the negative from areas of low work function. The value of ϕ obtained from the Richardson equation (3.12) will therefore be a different mean value over the surface from that which applies to ϕ in (3.8).

An alternative procedure is to observe the ion currents of X^- and Y^- ions arising from dissociation at the filament surface of a molecule $X_l Y_n$. In this case we consider the equilibrium constant for the reaction

$$X + Y^- \rightleftharpoons X^- + Y. \qquad (3.13)$$

We have

$$K = (Z_{X^-} Z_Y / Z_{Y^-} Z_X) \qquad (3.14)$$

$$= \frac{g_Y g_{X^-}}{g_{Y^-} g_X} \exp\{(E_{a,X} - E_{a,Y})/kT_s\}, \qquad (3.15)$$

where g_Y, g_X, g_{Y^-} and g_{X^-} are the statistical weights of the normal states of the atoms Y, X and negative ions Y⁻, X⁻ respectively. If there is complete dissociation at the filament, $Z_Y/Z_X = n/l$ so that $E_{a,X} - E_{a,Y}$ may be determined from measurement of $Z_{X^-}/Z_{Y^-} = i_{X^-}/i_{Y^-}$. This may be done with a mass spectrometer and is free from any difficulties arising from non-uniformity of the filament surface.

In the above discussion it is assumed that the dissociation of the molecular species is so complete that no account need be taken of the nature of the surface. Attention was directed to this by the attempts made to determine the electron affinity of atomic oxygen which led to contradictory results. Page (1961) has attempted to overcome these difficulties with a more thermodynamical approach to the problem but when doubt arises it is best to regard a knowledge of the electron affinity as required to interpret the data rather than to attempt to obtain reliable information about the electron affinity.

Surface ionization

Finally we have the methods depending on surface ionization at the hot filament of atoms X to produce positive and negative ions X⁺ and X⁻ respectively. The respective currents i_+, i_- of these ions are given according to the Saha–Langmuir equation (Langmuir and Kingdon, 1925) by

$$i_+ = (g_+/g_0) i_0 \exp\{(\bar{\phi}_+ - E_i)/kT_s\}, \qquad (3.16)$$

$$i_- = (g_-/g_0) i_0 \exp\{(E_a - \bar{\phi}_-)/kT_s\}, \qquad (3.17)$$

provided $|\bar{\phi}_+ - E_i|$ and $|\bar{\phi}_- - E_a| \gg kT_s$. i_0 is the number of neutral atoms incident per second on the filament while E_i, E_a are respectively the ionization energy and electron affinity of X. The introduction of two work functions, a mean $\bar{\phi}_+$ applying to the positive ions and one $\bar{\phi}_-$ to the negative, is necessary because it must be assumed that the filament surface will be patchy (see p. 37). It follows that to obtain E_a from the measured ratio of currents i_+/i_-, observations must be possible over such a range of temperature as to permit separate determination of $\bar{\phi}_+$ and $\bar{\phi}_-$.

One way of reducing errors due to patchiness of the surface is to work at temperatures T_s close to the melting point of the filament as was done by Scheer (1970).

An alternative procedure is to use the method for determination of the difference in the electron affinity of two atoms X, Y. The ratio of the current i_{X^-}, i_{Y^-} of the respective negative ions will be

$$i_{X^-}/i_{Y^-} = (i_X/i_Y) \frac{g_Y g_{X^-}}{g_X g_{Y^-}} \exp\{(E_{a,X} - E_{a,Y})/kT_s\}. \quad (3.18)$$

This difference method may also be applied if one of the negative ions results from attachment to an atom produced by dissociation in the neighbourhood of the filament of a molecule of low bond energy.

A variant of this procedure is to observe both the positive- and negative-ion currents arising from simultaneous ionization of two atomic species. The ratio

$$(i_{X^-}/i_{X^+})/(i_{Y^+}/i_{Y^-}) = C \exp\{(E_{a,X} - E_{a,Y} + E_{i,X} - E_{i,Y})/kT_s\}, \quad (3.19)$$

where C is independent of T_s, regardless of the nature of the filament surface. This may be used when complex ions are produced as well as the simple atomic ions. If the ion currents are mass-analysed then the relation applies with the currents referring to the atomic ions only.

This technique has been applied by Zandberg and Paleev (1970) to measure the electron affinities of Sb and Bi relative to that of Ag despite the formation of Sb_2^-, Sb_3^-, Bi_2^- and Bi_3^- ions on the hot filament (see Table 3.2).

Application to halogen atoms

These methods are most readily applied to atoms with high electron affinities such as the halogen atoms. The first measurements were made for iodine by the equilibrium method using I_2 as the parent molecule. Sutton and Mayer (1935) using the magnetron method found values between 3.06 and 3.25 eV from seven sets of observations, the mean being 3.14 eV, while Glockler and Calvin (1935) obtained a mean of 3.24 eV from 13 sets in which the values ranged from 2.89 to 3.49 eV. Table 3.1 summarizes the results obtained in these and other experiments for iodine as well as for

TABLE 3.1 *Electron affinities of halogen atoms determined from measurements of ion currents arising from reactions at a hot filament*

Atom	Method	Parent molecule	Electron affinity (eV)	Author
F	Equilibrium (magnetron)	F_2	3.56 ± 0.17	Bernstein and Metlay (1951)
	Equilibrium (mass spectrometer)	RbF	3.60 ± 0.07	Bailey (1955)
		KF	3.45 ± 0.10	Bailey (1955)
Cl	Equilibrium (magnetron)	Cl_2	3.72 ± 0.04	McCullum and Mayer (1943)
		HCl	3.84	Page (1961)
	Equilibrium (mass spectrometer)	KCl	3.75s ± 0.09	Bailey (1955)
Br	Equilibrium (space charge)	Br_2	3.81	Glockler and Calvin (1935)
	Equilibrium (magnetron)	Br_2	3.49	Doty and Mayer (1944)
		HBr	3.60	Page (1960)
	Equilibrium (mass spectrometer)	KBr	3.51 ± 0.07	Bailey (1955)
I	Equilibrium (space charge)	I_2	3.23 ± 0.13	Glockler and Calvin (1935)
	Equilibrium (magnetron)	I_2	3.05 ± 0.07	Sutton and Mayer (1934, 1935)
		I_2	3.225	Page (1961)
			Electron affinity difference (eV)	
F–Cl	Equilibrium (mass spectrometer)	ClF_3	−0.24	Bailey (1955)
F–Br	Surface ionization		0.02	Bakulina and Ionov (1959)
Cl–Br	Surface ionization		0.25	Bakulina and Ionov (1959)
I–Br	Equilibrium (mass spectrometer)	IBr	−0.33	Bailey (1955)
	Surface ionization		−0.27	Bakulina and Ionov (1959)

other halogen atoms. Values obtained directly for differences in the electron affinities of different atoms are included, as well as absolute measurements. On the whole the results, including the difference measurements, are reasonably consistent, apart from the rather high values obtained for Br using the space-charge technique. It would seem that, if this measurement is ignored, the mean values should be correct to ±0.1 eV. Comparison with results obtained by other methods is shown in Table 3.3 and discussed on p. 47.

Application to oxygen

It is considerably more difficult to apply these methods successfully to atomic oxygen, partly because of its lower electron affinity and partly because there are few convenient molecules in which oxygen is so weakly bound that the bond will be dissociated in the neighbourhood of a hot filament.

The first measurements were made by Vier and Mayer (1944) using the magnetron method with O_2 as parent molecule. While it was appreciated that O_2 has a dissociation energy as high as 5.09 eV and so is likely to undergo only a very small fractional dissociation at the filament, it was not clear how to allow properly for this. A high value of 3.1 eV was deduced for the electron affinity, 0.9 eV higher than the current value derived from electron impact studies. Four years later Metlay and Kimball (1948) seemed to have cleared up the difficulty by using N_2O as parent molecule which should certainly have been largely dissociated into N_2 and O near the filament. They obtained 2.33 ± 0.03 eV which seemed to be in satisfactory agreement with the current value. However, the development of techniques for quantitative study of photodetachment showed that the apparent consistency between the values derived by the two quite different methods was illusory. From photodetachment the electron affinity was found to be 1.465 ± 0.005 eV (Branscomb, Burch, Smith and Geltman, 1958) (see Chapter 11, p. 451). Page (1960), in the course of a re-examination of the whole question of the analysis which should be applied when only incomplete dissociation occurs near the filament, repeated the N_2O measurements and obtained good agreement with this lower value. The reason for the higher value obtained by Metlay and Kimball is not clear but it may possibly have been due to the presence of impurities.

There is now no doubt that the electron affinity given by the photodetachment studies is correct. The reason why the electron impact studies gave larger values is now understood. It is quite subtle and is discussed in Chapter 9, p. 276.

TABLE 3.2 *Difference $E_a(X) - E_a(Ag)$, in eV, between the electron affinity of an atom X and that of Ag, determined from surface ionization*

X	Sb(a)	Bi(a)	Si(b)	Ge(b)	Pb(c)	Sn(c)	Cu(c)
$E_a(X) - E_a(Ag)$	−0.40 ±0.07	−0.14 ±0.06	−0.06 ±0.02	−0.16 ±0.03	−0.25 ±0.08	−0.14 ±0.06	−0.12 ±0.06
$E_a(X)$ taking $E_a(Ag) = 1.303$ eV	0.90	1.16	1.24	1.14	1.05	1.16	1.18
$E_a(X)$ from photodetachment (See Chapter 11.)	1.05(d)	0.9–1.2(d)	—	1.20(d)	—	1.25(d)	1.226(e)

References

(a) Zandberg and Paleev (1970), (b) Zandberg, Kamenev and Paleev (1972), (c) Zandberg, Kamenev and Paleev (1971), (d) Feldmann, Rackwitz, Heinicke and Kaiser (1973), (e) Hotop, Bennett and Lineberger (1973).

Applications to other atoms

Comparatively few observations have been carried out for other atoms, many of which have electron affinities so low as to render the method unreliable.

Khovstenko and Dukelskii (1960) have obtained a value 0.8 ± 0.1 eV for H using the surface ionization technique, while Bakulina and Ionov (1957) have obtained 1.23 ± 0.05 eV as the excess of the electron affinity of Br over that of S.

Among the remaining cases studied are the interesting ones of Cu, Ag and Au studied by Bakulina and Ionov (1964) with the surface ionization difference technique. Molecular beams of the substances under investigation were directed from two independent evaporators on to the tungsten filament heated to temperatures T_s between 1800 and 2300 °K. In one set of experiments metallic copper, silver or gold were put into one evaporator and potassium iodide into the other. Measurements were made of the current of metallic and iodine negative ions i_{M^-} and i_{I^-} respectively. Using (3.18) the difference between the electron affinities of the metal and of iodine could be determined from the slope of a plot of $\log(i_{M^-}/i_{I^-})$ against $1/T_s$. Values for $E_a(I) - E_a(M)$, where M is the appropriate metal atom, were found of 1.6 ± 0.5, 1.07 ± 0.2 and 0.27 ± 0.1 eV for Cu, Ag and Au respectively. A further less reliable set of observations was carried out using the formula (3.17) which requires a knowledge of the mean positive- and negative-ion work functions $\bar{\phi}_+$ and $\bar{\phi}_-$ respectively. These were determined from observations of the currents of Cu^+ and Ag^+ and of I^- ions as functions of the temperature. Values for $E_a(M)$ of 2.1–2.2 eV were found in this way for Cu and of 2.0–2.3 eV for Ag. These are not inconsistent with the more accurate difference values. Thus if $E_a(I)$ is taken as, say 3.2 eV, $E_a(I) - E_a(M)$ becomes between 1.1 and 1.0 eV for Cu and between 1.2 and 0.9 eV for Ag. Nevertheless the precise measurements made by laser photodetachment techniques shows that the values found for all three metals are too high (see Table 3.3).

The surface ionization method in the form (3.18) has been applied to determine the difference $E_a(X) - E_a(Ag)$ between the electron affinity of an element X and that of Ag. Results are given in Table 3.2. The values obtained for $E_a(X)$, assuming for $E_a(Ag)$

TABLE 3.3 *Well-determined electron affinities (eV) of the elements*

Element	Calc.	Photodetachment threshold	Photoelectron spectrum	Shock wave	Derived from lattice energies	Polar photo-dissociation	Hot wire equilibrium	Surface ionization
H	0.754	—	—	—	—	—	—	—
F	—	—	—	3.448 ±0.005	3.45	—	3.56 ± 0.17	E_a(Cl) − 0.29
Cl	—	—	—	3.613 ±0.003	3.71	—	3.60 ± 0.07	E_a(Br) + 0.02
							3.72 ± 0.04	E_a(Br) − 0.25
Br	—	—	—	3.363 ±0.003	3.49	3.54 ±0.12	3.75₅ ± 0.09	—
							3.49 ± 0.02	
I	—	3.059 ±0.002	—	3.063 ±0.003	3.19	3.073 ±0.014	3.23 ± 0.13	E_a(Br) − 0.35
O	—	1.465 ±0.005	—	—	—	—	1.46	E_a(Br) − 0.27
S	—	2.0772 ±0.005	—	—	—	—	—	—
Se	—	2.0206 ±0.0003	—	—	—	—	—	—
Te	—	1.9 ±0.15	—	—	—	—	—	—
P	—	0.77 ±0.05	—	—	—	—	—	—
As	—	0.80 ±0.05	—	—	—	—	—	—

Sb	—	1.05 ±0.05	—	—	—	—	—	—	$E_a(Ag) - 0.40$
C	—	—	—	—	—	—	1.270 ±0.010	—	—
Ge	—	1.20 ±0.1	—	—	—	—	—	—	$E_a(Ag) - 0.16$
Sn	—	1.25	—	—	—	—	—	—	$E_a(Ag) - 0.14$
Li	0.612	0.61 ±0.05	—	—	—	—	0.620 ±0.007	—	—
Na	0.540	0.543 ±0.010	—	—	—	—	0.548 ±0.004	—	—
K	0.492	0.5012 ±0.005	—	—	—	—	—	—	—
Rb	—	0.4859 ±0.0015	—	—	—	—	—	—	—
Cs	—	0.472 ±0.003	—	—	—	—	—	—	—
Cu	—	—	—	—	—	—	1.226 ±0.010	—	—
Ag	—	—	—	—	—	—	1.303 +0.007 −0.011	—	—
Au	—	2.3086 ±0.0007	—	—	—	—	—	—	—
Pt	—	2.128 ±0.002	—	—	—	—	—	—	—
Cr	—	0.66 ±0.05	—	—	—	—	—	—	—

For references see text.

the precise value obtained from photodetachment measurements (Chapter 11, p. 463), are also included and compared with values obtained directly from such measurements. There is reasonably good agreement but certainly not better than to 0.1 eV.

Scheer and Fine (1969) have attempted to determine the electron affinity of Li from the surface ionization of Li on a heated molybdenum ribbon. While there was no difficulty in determining $\bar{\phi}_+$, the Li⁻ current was so small that $\bar{\phi}_-$ could not be determined from measurements of the variation of this current with temperature. As a result the experiment was only able to show that 0.65 eV $< E_a(\text{Li}) < 1.05$ eV.

Scheer (1970) has measured the electron affinities of Mo, Ta, W and Re by surface ionization measurements near the melting point of the filament (see p. 39). In each case the ions were evaporated from a filament of the same species. Results are given in Table 3.7.

From the energy relations involved in cyclic processes

The simplest illustration of this method is provided by Mayer's (1930) determination of the electron affinity of iodine. Consider a polar molecule XY. We may regard this molecule as capable of formation in two alternative ways:

$$X + Y \longrightarrow XY + D_{XY}; \qquad (3.20)$$

$$X + Y \longrightarrow X^+ + Y^- + E_{a,Y} - E_{i,X}, \qquad (3.21)$$

$$X^+ + Y^- \longrightarrow XY + D_{X^+Y^-}. \qquad (3.22)$$

Here $E_{a,Y}$, is the electron affinity of atom Y, $E_{i,X}$ the ionization energy of atom X, D_{XY} the energy of dissociation of the molecule XY into atoms X, Y, and $D_{X^-Y^+}$ that of dissociation into ions X⁺, Y⁻.

The energies evolved in the two processes must be equal, so

$$D_{XY} = D_{X^+Y^-} + E_{a,Y} - E_{a,X}. \qquad (3.23)$$

In certain cases all these quantities may be directly measured except $E_{a,Y}$, which can therefore be determined. The difficulty usually lies in measuring $D_{X^+Y^-}$, but for the alkali halides this quantity may be found in much the same way as that described on

p. 442 for determining electron affinities. The equilibrium constant of the reaction

$$XY \rightleftharpoons X^+ + Y^- \qquad (3.24)$$

is measured in the gas phase by evaporation of the halide and electrical measurement of the percentage of ions present. From this constant $D_{X^+ Y^-}$ is obtained by using the formula (3.2).

The values of A_Y for the halogens found in this way are given in Table 3.3, column 6 and are in general agreement with values found by the other methods.

This method can be used only for a few atoms. An alternative method, which is potentially of greater applicability, consists in determining $D_{X^+ Y^-}$ by use of a combination of theory and experiment involving consideration of the energy relations concerned in the formation of a crystalline solid from ions X^+, Y^-. This may be carried out in two ways:

$$X^+ + Y^- \longrightarrow XY \text{ crystal} + U; \qquad (3.25)$$

$$X^+ + Y^- \longrightarrow XY \text{ gas} + D_{X^+ Y^-}, \qquad (3.26a)$$

$$XY \text{ gas} \longrightarrow XY \text{ crystal} + S, \qquad (3.26b)$$

where U is the lattice energy and S the heat of sublimation per molecule XY of the crystal. The equivalence of the energy evolution in processes (3.25) and (3.26) gives

$$D_{X^+ Y^-} = U - S. \qquad (3.27)$$

S may be determined experimentally, but U can be determined only by calculation from the properties of the ions X^+, Y^-. Consistency is the principal test of the validity of the results obtained in this way; the electron affinities deduced from crystals containing the same anion but different cations should be the same.

This procedure was first applied by Mayer and Helmholz (1932) and a thorough revision has been carried out, using more recent measurements of the physical properties involved, for all the alkali halides by Cubicciotti (1959, 1960, 1961). The values derived for the electron affinities of the different halogen atoms from the various halides are given in Table 3.3. It will be seen that, in all cases except CsCl, there is very good consistency between values derived for the same halogen from the properties of crystals with different anions. The mean values for each halogen also agree well with electron affinities derived by other methods as may be seen from Table 3.3

An interesting application of the method was made by Mayer and Maltbie (1932) to estimate the second electron affinities of O and S from the alkaline earth oxides and sulphides. They obtained the values

$$O + 2e \longrightarrow O^{2-} - (6.5 \pm 2.2) \text{ eV},$$

$$S + 2e \longrightarrow S^{2-} - (3.9 \pm 2.2) \text{ eV}.$$

3.2 Empirical methods

These methods depend on comparison of spectra of neighbouring atoms and ions with the aim of determining simple empirical relationships which can hopefully be extrapolated so as to obtain information about the spectra of other atoms and ions, including negative ions.

Defining an ion in terms of the atomic number Z and the number of electrons N so that the degree of ionization $q = Z - N$, the most obvious comparisons are:

(*a*) Vertical comparison – Z fixed.
(*b*) Horizontal comparison – q fixed.
(*c*) Isoelectronic comparison – N fixed.

The third of these has theoretical advantages for comparisons involving a neutral atom and the isoelectronic positive ions but this advantage is absent when an extrapolation to the isoelectronic negative ion is attempted. This is because for $q \geqslant 0$ all electrons move in a field which is asymptotically of Coulomb form but for negative ions the field is of much shorter range. Nevertheless the first empirical methods for estimating electron affinities were of this kind and although they cannot be regarded as reliable they gave quite good results for atoms with well-determined electron affinities.

Isoelectronic extrapolation

As early as 1934 Glockler used a quadratic extrapolation of the form

$$I = aZ^2 - bZ + c \tag{3.28}$$

to obtain the ionization energy I for an ion of atomic number Z in an isoelectronic series. His results were revised by Bates (1947) using more accurate determinations of ionization potentials. In fact, in terms of perturbation theory, the ionization energy is expressed in the form (Johnson and Rohrlich, 1959)

$$I = \alpha(Z - \sigma)^2 + \gamma + \sum_{k=1}^{\infty} \frac{a_k}{(Z - \sigma)^k}. \tag{3.29}$$

Edlèn (1960) therefore replaced the simpler quadratic extrapolation formula (3.28) by

$$I = \frac{R}{n^2} \{\zeta^2 + 2a\zeta - b + c(\zeta + a)^{-1}\}, \tag{3.30}$$

where $\zeta = Z - (N - 1) = q + 1$, q being the degree of ionization. R is the Rydberg constant and n the principal quantum number of the electron which is removed in the ionization process. This formula is correct in the limit of large Z and q and includes the first term in the series within the summation sign of (3.29). Edlèn obtained remarkably good results for the atoms whose electron affinities were well known at the time, as may be seen from Table 3.5. However, the reliability of this method was called into question by Edie and Rohrlich (1962), who added a further term $d(\zeta + a)^{-2}$ within the bracket in (3.30). They found considerably poorer results for O (1.82 against 1.47 eV) and noted that, when the additional term was included, the value of c was changed drastically. This throws grave doubts on the convergence of the perturbation expansion and removes most of the theoretical justification of the procedure.

Because of this, alternative methods have been introduced which take advantage of well-known values of the electron affinity to provide the basis.

Horizontal extrapolation

The method of Edie and Rohrlich

The procedure which has been exploited most widely was first introduced by Edie and Rohrlich (1962), who applied it to the atoms of the first two short periods, and extended with some small revision by Zollweg (1969) to most elements of the periodic table. In principle, horizontal extrapolation ((b) of p. 48) is used although recourse has to be made to isoelectronic extrapolation of certain

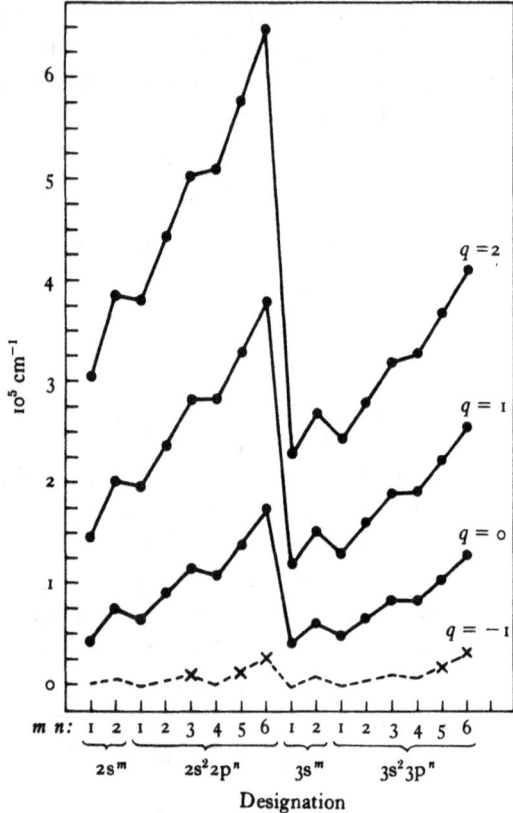

Fig. 3.1. The observed ionization energies for ions of the two short periods, with $q = 0$, 1 and 2, plotted as functions of the atomic number. Measured values indicated by a cross are also given for the electron affinities for ions with $q = -1$. The broken curve shows the expected general trend for such ions.

parameters which are, however, much less sensitive than the electron affinity itself.

Basically the method depends on the regularities which appear when ionization potentials are plotted for fixed values of the degree of ionization but different numbers of electrons in each shell. These may be seen by reference to Fig. 3.1 which shows the results for the first two short periods. The known electron affinities, shown as crosses in the broken curve for $q = -1$, are at least consistent with the preservation of the same shape of the $q = -1$ diagram as for the well-determined $q = 0$, 1 and 2 diagrams. In fact it would be possible to estimate the electron affinities of phosphorus and nitrogen but with only very low percentage accuracy because they are certainly very small in absolute value. The minimum which appears in the p shell at $k = 4$ for all values of q may be understood as follows.

The energy of an atom or ion can be regarded as made up of a part α, due to the energy of the core electrons and of the interaction between the outer electrons and the core, and a part β, due to the interaction between the outer electrons. Of these α will vary smoothly with atomic number but β will not, for the following reasons. The electrostatic repulsion between p electrons would increase smoothly with the number of electrons but for the operation of the Pauli principle. The effect of this is to prevent electrons with the same spin from approaching too closely. It therefore reduces the electrostatic repulsion by an amount roughly proportional to the number of pairs of electrons with the same spin and it is this which does not vary smoothly with atomic number. In the ground state the number of pairs of electrons with the same spin will be a maximum, but there can never be more than three electrons with the same spin as there are only three p orbitals. The number of pairs of electrons with the same spin is therefore 0, 1, 3, 3, 4, 6 for systems with 1, 2, 3, 4, 5, 6 p electrons in their ground states respectively. Thus the apparent attraction due to the Pauli principle does not increase smoothly, leading to a relatively larger total energy for a system with four p electrons than would otherwise be the case. This is reflected in the comparative ease of detaching the fourth p electron from such a system. There is no doubt that both nitrogen and phosphorus will have small, even if not negative, electron affinities but more refined extrapolation procedures must be developed.

One possibility is to attempt to extrapolate for a fixed number k of p electrons from known q to $q = -1$. However, a less severe nearly-linear extrapolation is possible if, instead of dealing with ionization energies which are differences between the total energies of ions with q differing by 1, we consider the differences in the energies of the centres of the ground configurations of the respective ions. Thus the ground configuration will give rise to a number of terms the lowest of which is the ground state. If E_s is the energy of a term reckoned from the ground term as zero and g_s is the

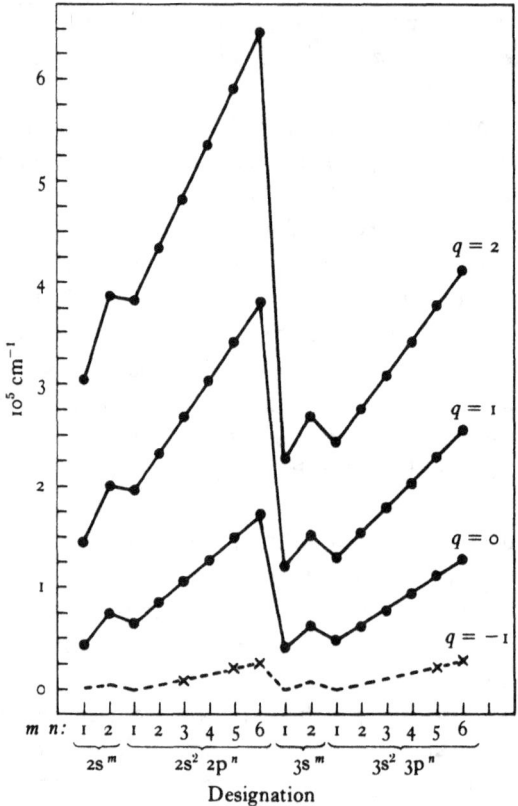

Fig. 3.2. Corresponding plots to those in Fig. 3.1 for the differences of the energies of the centres of the ground configurations of an ion of charge q and the corresponding ion of charge $q-1$.

statistical weight, the energy C of the centre of the configuration relative to that E of the ground state is $\sum_s g_s E_s/g_s$ and the total energy $C+E$. Fig. 3.2 shows the corresponding plot to Fig. 3.1 and it will be seen that now the variation is nearly linear across an entire p shell. This is because, by dealing with all the terms in the configuration we are effectively averaging out the exchange effects which, through the Pauli principle, produce the minima at $k = 4$ in Fig. 3.1. It seems clear that extrapolation to the case $q = -1$ is possible. We would then obtain

$$\Delta C(-1, p^k) = C(0, p^{k-1}) - C(-1, p^k) + E(0, p^{k-1}) - E(-1, p^k), \tag{3.31}$$

where $C(q, p^k)$ is the energy of the configuration centre for the ion of charge q with k p electrons. The electron affinity $E_a(p^{k-1})$ of the atom with $k - 1$ electrons is now given (see Fig. 3.2) by

$$E_a(p^{k-1}) + C(0, p^{k-1}) - \Delta C(-1, p^k) = C(-1, p^k). \tag{3.32}$$

Hence if ΔC is obtained from extrapolation and $C(0, p^{k-1})$ is available from the spectrum of the neutral atom, E_a can be derived if $C(-1, p^k)$ is known. The weakness of the method is that $C(-1, p^k)$ can only itself be derived by isoelectronic extrapolation. However, this is likely to be much more reliable than for the electron affinity directly as there will not be a very marked change in going from $q = 0$ to $q = -1$.

For the horizontal extrapolation to determine $\Delta C(-1, p^k)$ we note that, since the points on Fig. 3.2 for fixed q and variable k lie very closely on straight lines, it is only necessary to extrapolate the slopes and intercepts of these lines. The former are very nearly linear functions of q and there is no difficulty, but the intercepts vary nearly as q^2 and a considerable error is likely to be made in extrapolation. This may be avoided by using the accurately known electron affinities of O and of Cl to fix the intercepts for $q = 0$. Correction may also be made in this way for the small deviations, independent of q, of the ionization potentials from linear variation with k.

Zollweg (1969) found that the slopes for the 2p, 3p, 4p and 5p shells for $q = 2, 1, 0$ extrapolate linearly to nearly the same value for $q = -1$, a result which applies also to the 4s, 5s and 6s + 6p shells (see Fig. 3.3). He therefore chose a mean value (0.84 eV/electron) for all these shells. The intercepts were chosen so as to provide agreement of the extrapolated with the measured values for the electron affinities of O, Cl, Br and I. For the 6p shell the appropriate atom is astatine for which the electron affinity was

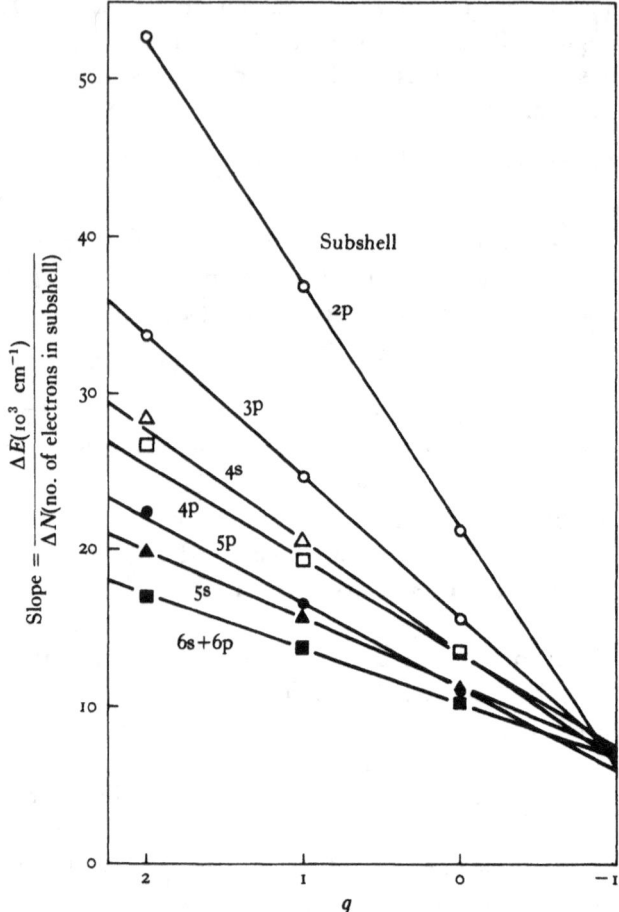

Fig. 3.3 Slopes of plots such as in Fig. 3.1 for different subshells, as indicated, plotted as a function of q, extrapolated to $q = -1$.

estimated as 2.80 ± 0.2 eV by vertical extrapolation from the values for the other halogens.

To complete the estimation of electron affinities it is necessary to know the energy of the configuration centres for the different negative ions. This was done by extending the quadratic isoelectronic extrapolation first carried out by Bates and Moiseiwitsch (1955). As remarked earlier, this should be reliable as there is only a quite gradual variation in passing from neutral atoms to negative

ELECTRON AFFINITIES OF THE ELEMENTS

TABLE 3.4 *Excitation energies of the terms of the ground configurations of different negative ions, determined by quadratic isoelectronic extrapolation*

Ion	Ground configuration	Term	$\mathcal{J}$	Energy (eV)	Energy of configuration centre (eV)
Be$^-$	$(2s)^2 2p^*$	^2P	$\frac{1}{2}$	0	0.003
		^2P	$\frac{3}{2}$	0.004	
B$^-$	$(2s)^2(2p)^2$	^3P	0	0	0.288
		^3P	1	0.002	
		^3P	2	0.005	
		^1D	2	0.608	
		^1S	0	1.250	
C$^-$	$(2s)^2(2p)^3$	$^4S^0$	0	0	1.323
		$^2D^0$		1.404	
		$^2P^0$		1.968	
N$^-$	$(2s)^2(2p)^4$	^3P	2	0	0.636
		^3P	1	0.012	
		$^3P_0$	0	0.017	
		^1D	2	1.340	
		$^1S_0$	0	2.776	
O$^-$	$(2s)^2(2p)^5$	^2P	$\frac{3}{2}$	0	0.009
		^2P	$\frac{1}{2}$	0.028	
Mg$^-$	$(3s)^2(3p)^*$	^2P	$\frac{1}{2}$	0	0.003
		^2P	$\frac{3}{2}$	0.004	
Al$^-$	$(3s)^2(3p)^2$	^3P	0	0	0.219
		^3P	1	0.004	
		^3P	2	0.012	
		^1D	2	0.442	
		^1S	0	1.068	
Si$^-$	$(3s)^2(3p)^3$	$^4S^0$		0	0.914
		$^2D^0$		0.941	
		$^2P^0$		1.479	
P$^-$	$(3s)^2(3p)^4$	^3P	2	0	0.421
		^3P	1	0.027	
		^3P	0	0.038	
		^1D	2	0.840	
		^1S	0	2.00	
S$^-$	$(3s)^2(3p)^5$	^2P	$\frac{3}{2}$	0	0.021
		^2P	$\frac{1}{2}$	0.063	
Cu$^-$	$(3d)^{10}(4s)^2$	^1S	0	0	
Zn$^-$	$(4s)^2 4p^*$	^2P	$\frac{1}{2}$	0	0.010
			$\frac{3}{2}$	0.0145	

TABLE 3.4 – *continued*

Ion	Ground configuration	Term	j	Energy (eV)	Energy of configuration centre (eV)
Ga⁻	$(4s)^2(4p)^2$	^3P	0	0	
			1	0.028	
			2	0.068	0.280
		^1D	2	0.512	
		^1S	0	1.212	
Ge⁻	$(4s)^2(4p)^3$	$^4S^o$	$\tfrac{3}{2}$	0	
		$^2D^o$	$\tfrac{3}{2}$	0.907	
			$\tfrac{5}{2}$	0.954	0.939
		$^2P^o$	$\tfrac{1}{2}$	1.533	
			$\tfrac{3}{2}$	1.593	
As⁻	$(4s)^2(4p)^4$	^3P	2	0	
			1	0.136₅	
			0	0.173	0.583
		^1D	2	1.133	
		^1S	0	2.501	
Se⁻	$(4s)^2(4p)^5$	$^2P^o$	$\tfrac{3}{2}$	0	0.096
			$\tfrac{1}{2}$	0.288	
Br⁻	$(4s)^2(4p)^6$	^1S	0	0	0
Ag⁻	$(4d)^{10}(5s)^2$	^1S	0	0	0
Cd⁻	$(5s)^2 5p^*$	$^2P^o$	$\tfrac{1}{2}$	0	0.038
			$\tfrac{3}{2}$	0.057	
In⁻	$(5s)^2(5p)^2$	^3P	0	0	
			1	0.082	
			2	0.182	0.358
		^1D	2	0.598	
		^1S	0	1.224	
Sn⁻	$(5s)^2(5p)^3$	$^4S^o$	$\tfrac{3}{2}$	0	
		$^2D^o$	$\tfrac{3}{2}$	0.846	
			$\tfrac{5}{2}$	0.848	0.851
		$^2P^o$	$\tfrac{1}{2}$	1.436	
			$\tfrac{3}{2}$	1.415	
Sb⁻	$(5s)^2(5p)^4$	^3P	2	0	
			1	0.343	
			0	0.359	0.481₅
		^1D	2	0.940	
		^1S	0	1.1130	
Te⁻	$(5s)^2(5p)^5$	^2P_o	$\tfrac{3}{2}$	0	0.210
			$\tfrac{1}{2}$	0.630	
I⁻	$(5s)^2(5p)^6$	^1S	0	0	0
Au⁻	$(5d)^{10}(6s)^2$	^1S	0	0	0
Hg⁻	$(6s)^2 6p$	$^2P^o$	$\tfrac{1}{2}$	0	0.159
			$\tfrac{3}{2}$	0.238	

TABLE 3.4 – *continued*

Ion	Ground configuration	Term	J	Energy (eV)	Energy of configuration centre (eV)
Tl⁻	$(6s)^2(6p)^2$	^3P	0	0	0.96 (est)
			1		
			2		
		^1D	2		
		^1S	0		
Pb⁻	$(6s)^2(6p)^3$	$^4S^o$	$\tfrac{3}{2}$		1.96 (est)
		^2D	$\tfrac{3}{2}$		
			$\tfrac{5}{2}$		
		$^2P^o$	$\tfrac{1}{2}$		
			$\tfrac{3}{2}$		
Bi⁻	$(6s)^2(6p)^4$	^2P	2		1.2 (est)
			1		
			0		
		^1D	2		
		^1S	0		
Po⁻	$(6s)^2(6p)^5$	$^2P^o$	$\tfrac{3}{2}$		0.41 (est)
			$\tfrac{1}{2}$		
At⁻	$(6s)^2(6p)^6$	^1S	0	0	0

* See however p. 60 and Table 3.5.

ions in contrast to the electron affinity itself. Also the basic experimental data are more accurate being based on differences between directly measured energy values. Table 3.4 lists the extrapolated terms (Edie and Rohrlich, 1962; Zollweg, 1969) with their designation and energy as well as that of the configuration centre. In some cases the fine structure separations within particular terms are significant, in which case these are included separately.

From these data electron affinities for the two short periods given in Table 3.5 are obtained. In the first period the measured value for O has been used as a basis but the values obtained for C and F are quite close to the measured values of these atoms. Again, for the second period, based on Cl, there is good agreement with the measured value for S.

Before discussing results obtained by Zollweg (1968) for heavier elements, we describe briefly two other empirical procedures which essentially involve horizontal extrapolation.

TABLE 3.5 *Electron affinities of atoms of the first two short periods determined by empirical methods*

Atom	Negative-ion ground state	Electron affinity (eV)					
		Measured	Zollweg	Ginsberg and Miller	Crossley	Clementi	Edlèn
Li	$2s^2\,{}^1S_0$	0.620	–	–	0.59	0.58	0.82
Be	$2p\,{}^2P^0_{1/2}$	–	–0.65	–0.32	–0.68	–	–0.19
	$3s\,{}^2S_{1/2}$	–	+0.38	+0.70	–	–	–
B	$2p^2\,{}^2P_0$	<0.5	0.18	0.36	0.16	0.30	0.33
C	$2p^3\,{}^4S^0_{3/2}$	1.270	1.29	1.39	1.33	1.17	1.24
N	$2p^4\,{}^3P_2$	–	–0.21	–0.30	–0.32	–0.27	0.05
O	$2p^5\,{}^2P_{3/2}$	1.46	(1.46)	1.40	1.39	1.22	1.47
F	$2p^6\,{}^1S_0$	3.45	3.50	(3.42)	(3.45)	3.37	3.50
Na	$3s^2\,{}^1S_0$	0.548	–	–	0.22	0.78	–
Mg	$3p\,{}^2P^0_{1/2}$	–	–0.52	–0.76	–0.69	–	–0.32
	$4s\,{}^2S_{\frac{1}{2}}$	–	–0.22	–0.44	–	–	–
Al	$3p^2\,{}^3P_0$	–	+0.20	+0.23	+0.27	0.49	+0.52
Si	$3p^3\,{}^2P^0_{1/2}$	[1.24]	1.36	1.43	1.40	1.39	1.46
P	$3p^4\,{}^3P_2$	0.77	0.71	0.55	0.62	0.78	0.77
S	$3p^5\,{}^2P^0$	2.077	2.04	2.04	2.03	2.12	2.15
Cl	$3p^6\,{}^1S_0$	3.62	(3.62)	(3.62)	(3.61)	3.56	3.70

Measured values are those obtained from photodetachment, including shock wave absorption (see Chapter 11, pp. 449–73), except for Si which was measured by surface ionization relative to Ag (see Table 3.2).
Bracketed values ( ) are measured values assumed in the extrapolation procedure.

The method of Ginsberg and Miller

The first such method to be proposed was that of Ginsberg and Miller (1958). For an atom or ion of atomic number Z containing N electrons they defined certain quantities $P_{\epsilon,\eta}(Z,N)$ as follows. Each electron in the atom or ion is considered to be assigned values of the 4 quantum numbers n, l, m_e and m_s which we define as the arrangement of the N electrons. $E_\eta(Z,N)$ is defined as the energy required to remove the electron in state η from the (Z,N) system without any other change occurring in the electron arrangement. A quantity $\Delta E_\eta(Z,N)$ is now introduced such that

$$\Delta E_\eta(Z,N) = E_\eta(Z+1,N) - E_\eta(Z,N), \qquad (3.33)$$

the (Z,N) and $(Z+1,N)$ systems being taken to have the same electron arrangement.

Consider now a second electron state ϵ, occupied in the (Z, N) and $(Z + 1, N)$ systems, which differs from η in one or more of n, l and m_l. Systems $(Z, N - \epsilon)$ and $(Z + 1, N - \epsilon)$ with the same electronic arrangement may now be defined, which differ from the (Z, N) and $(Z + 1, N)$ system only in that they lack an electron in state ϵ. Then

$$\Delta E_\eta(Z, N - \epsilon) = E_\eta(Z + 1, N - \epsilon) - E_\eta(Z, N - \epsilon) \quad (3.34)$$

and $P_{\epsilon, \eta}(Z, N)$ is now given by

$$\begin{aligned} P_{\epsilon, \eta}(Z, N) &= \Delta E_\eta(Z, N - \epsilon) - \Delta E_\eta(Z, N) \\ &= E_\eta(Z + 1, N - \epsilon) - E_\eta(Z, N - \epsilon) - E_\eta(Z + 1, N) \\ &\quad + E_\eta(Z, N). \quad (3.35) \end{aligned}$$

Thus if the Z, N system is a negative ion, $P_{\epsilon, \eta}(Z, N)$ involves removal energies from the positive ion $(Z + 1, N - \epsilon)$, the neutral atoms $(Z, N - \epsilon)$ and $(Z + 1, N)$ and the negative ion (Z, N). With proper choice of η the latter removal energy is equal, apart from an energy of order of the fine structure separation at most, which in any case can be estimated, to the electron affinity. Ginsberg and Miller found that, when the (Z, N) systems are neutral atoms, so that all the removal energies in (3.35) can be obtained from observed data, $P_{\epsilon, \eta}(Z, N)$ is practically independent of (Z, N). If it is assumed that the same result holds for negative ions, then the known electron affinity of the halogen atoms can be used to estimate the electron affinities of atoms in the same row of the periodic table. Thus for F$^-$, $P_{\epsilon, \eta}(Z, N)$ may be obtained for $\epsilon = 2s0\tfrac{1}{2}$ and $\eta = 2p1-\tfrac{1}{2}$, for example, both of which states are occupied in the ground configuration of O$^-$. From this $E_\eta(O^-)$ is obtained as 1.47 eV and $E_a(O^-)$ differs from this by 0.01 eV. Very good agreement is thereby obtained with the observed value for O$^-$ but the agreement is less satisfactory for the alternative choice of $\epsilon = 2p0+\tfrac{1}{2}$ and $\eta = 2p1\tfrac{1}{2}$. For this reason results for other atoms in the first row were obtained using the former choice of ϵ, η.

Results obtained by this method are included in Table 3.5 for the elements of the two short periods. The agreement with the measured values for O and S is as satisfactory as that obtained by Zollweg but is not so good for C.

The method of Crossley

Crossley (1964) based his horizontal extrapolation on the following. Referring to the original quadratic formula (3.28) of Glockler, he wrote the ionization energy for an ion of atomic number Z, in a given isoelectronic series, as

$$I = \frac{M}{m + M}(\alpha Z^2 + \beta Z) + \Delta, \quad (3.36)$$

where M, m are the masses of the nucleus and of an electron respectively. Δ is a slowly varying function of Z which he writes as

$$\Delta = \Delta_0 + \nabla, \qquad (3.37)$$

where Δ_0 is the value of Δ for the neutral atom in the series. α, β and Δ_0 are calculated for the series which involves the neutral atoms of the first two short periods. Using the known electron affinities of H, F and Cl, ∇ was determined for systems of three different atomic numbers with charge number $q = -1$. Interpolation between these values to obtain ∇, and hence $I(=E_a)$, for systems with intermediate Z was carried out assuming a linear variation with Z as was found for singly-ionized atoms ($q = 1$).

The results obtained are shown in Table 3.5. On the whole, as judged by the results for O, C and S, Crossley obtains slightly better agreement than do Ginsberg and Miller.

Finally, we refer to the semi-empirical method used by Clementi which was described in Chapter 2, p. 25.

Results for atoms of the first two short periods

The results obtained for the atoms of the first two short periods are given in Table 3.5. Considering the results as a whole for the atoms from Li to C, it seems that we have a fairly clear picture of the variation with Z in each period. Apart from Edlèn's isoelectronic interpolation, all procedures give a negative affinity for N and it is very likely that this is correct. For P, on the other hand, there is little doubt that the affinity is positive and close to 0.77 eV.

An important question, which we have not yet reviewed, is whether it is correct to assume that the additional electron in the ion will occupy the orbital of lowest energy as judged for a neutral atom. Thus, as pointed out by Bates (1947), since the attractive field acting on the additional electron has a much shorter range for a negative ion than for a neutral atom, the extent to which the outer orbital penetrates the core has a much greater effect on its energy. This tends to increase the binding for orbitals of low azimuthal quantum number. In fact Bates found, by carrying out the quadratic extrapolation procedure which was used to obtain the energies of the different terms of the ground configurations of negative ions (see Table 3.4), that for Be⁻ the 3s orbital lies deeper than the 2p and for Mg⁻ the 4s deeper than 3p.

Accordingly, in Table 3.5, two values for the electron affinities of Be and of Mg are given, one based on attachment in the np and one in the $(n+1)$s orbital. It will be seen that, for Be⁻, the ground state is $3s\,^2S_{1/2}$ and is almost certainly stable whereas the $2p\,^2P^0_{1/2}$ is unstable. For Mg⁻, while the $4s\,^2S_{1/2}$ state lies lower, both states are unstable.

In all other cases of the atoms in the first two short periods, the lowest orbital for the negative ion is found to be the same as for the neutral atom.

Electron affinities of heavier elements

We now consider electron affinities for heavier elements, confining ourselves in the first instance to ions with outer p shells. The extrapolation methods of Zollweg (1969) and of Ginsberg and Miller (1958) can be applied out to the 5p shell inclusive using as basis the measured values for Br and I. In addition, Zollweg has estimated the affinity of astatine, by extrapolation down the column of halogen atoms, as 2.80 eV, and used it as a basis for extending the data to the 6p shell. Values obtained are given in Table 3.6. Results obtained by a further empirical procedure due to Politzer, which is particularly appropriate for interpolation down a column, are also given.

Politzer has given reasons why a relation of the form

$$E_i - E_a = k(1/\langle r \rangle)^n \qquad (3.38)$$

should exist between the ionization energy E_i and electron affinity E_a of an atom, $\langle r \rangle$ being the average value of the radial distance r from the nucleus of the electrons of highest energy in the atom. k and n are parameters which must be determined. Because they depend largely on the relative positions of the valence electrons with respect to each other, they should be nearly constant within each vertical family of the periodic table. This was verified for the singly-charged positive ions of groups 2A, 4A, 5A, 6A and 7A, using values of $\langle r \rangle$ tabulated by Froese (1966) from HF functions, E_i being now the second and E_a the first ionization energy of the atom. It was also verified for the halogen atoms using the measured electron affinities (Table 3.3). If then for a particular vertical family two electron affinities are known, k and n may be obtained from the tables of $\langle r \rangle$ for the other atoms in that family. There is no difficulty for the 6A group as the affinities of O and S are known. For the other groups values obtained by Clementi which are given in Table 3.6 were used to fix the parameters. In this way the results given in Table 3.6 were obtained.

TABLE 3.6 *Electron affinities of the A subgroups in the periodic table, determined by empirical methods*

Atom	Negative-ion ground state	Electron affinity (eV)			
		Measured	Zollweg	Ginsberg and Miller	Politzer
Zn	$4p\,^2P^0_{1/2}$	–	–0.67	–1.08	
	$5s\,^2S_{1/2}$	–	+0.09	–0.32	
Ga	$4p^2\,^3P_0$	<0.5	0.37	0.00	0.50
Ge	$4p^3\,^4S^0$	1.20	1.44	1.02	1.37
As	$4p^4\,^3P_2$	–	1.07	0.47	0.74
Se	$4p^5\,^2P^0$	2.077	2.12	2.02	2.11
Br	$4p^6\,^1S_0$	3.36	(3.36)	(3.36)	(3.36)
Cd	$5p\,^2P^0_{1/2}$	–	–0.78	–0.78	–
	$6s\,^2S_{1/2}$	–	–0.27	–0.27	–
In	$5p^2\,^3P_0$	<0.5	0.20	0.02	0.72
Sn	$5p^3\,^4S^0$	1.25	1.03	–	1.47
Sb	$5p^4\,^3P_2$	1.05	0.94	–	0.61
Te	$5p^5\,^2P^0$	1.9	1.96	2.12	2.21
I	$5p^6\,^1S_0$	3.06	(3.06)	(3.06)	(3.06)
Hg	$6p\,^2P^0_{1/2}$	–	–0.67	–	–
	$7s\,^2S_{1/2}$	–	–0.19	–	–
Tl	$6p^2\,^3P_0$	<0.5	0.32	–	1.21
Pb	$6p^3\,^4S^0$	[1.05]	1.03	–	1.79
Bi	$6p^4\,^3P_2$	0.9–1.2	0.95	–	–0.34
Po	$6p^5\,^2P^0$	–	1.32	–	1.97
At	$6p^6\,^1S_0$	((2.80))	(2.80)	–	–

Measured values are those obtained from photodetachment, including shock-wave absorption (see Chapter 11, pp. 454–73). [ ] was measured by surface ionization, relative to Ag (see Table 3.2).
Bracketed values ( ) are measured values assumed in the extrapolation procedure. The double bracketed value for astatine has been obtained by extrapolation down the halogen column.

It is to be noted that, as for Be⁻ and Mg⁻, for Zn⁻ and Cd⁻ the $(n+1)s$ orbital lies lower than the np. Whether or not Zn has a positive affinity is very uncertain though Cd probably has not. For the elements of group 6A, the empirical values are probably quite good as they are anchored to the known value for the neighbouring halogen atom or to the accurately known values for O and S. The

ELECTRON AFFINITIES OF THE ELEMENTS 63

reliability of the results for atoms in the other groups is certainly much lower.

Zollweg has extended his procedure to atoms of the long periods $d^k s^2$. According to the regularities between atomic configurations shown by Catalán, Rohrlich and Shenstone (1954) the ground configurations of the negative ions arise from $d^k s^2$ configurations with $k = 0, ..., 10$. This does not mean that the ground configuration of the corresponding neutral atom is necessarily $d^k s$, though it is so for the first ($k = 0$) and last ($k = 10$) member of the period. In these cases, owing to the large number of terms arising from a given configuration and incomplete knowledge of their energies even for neutral and positively-ionized atoms, isoelectronic extrapolation is not possible. However, it was verified that the energy differences between the ground terms of the $d^k s^2$ and $d^k s$ configurations for the neutral and singly-ionized atoms and for the singly- and doubly-ionized atoms of the long periods, all varied linearly with k, to a fair approximation. However, the accuracy of the data did not permit extrapolation to the difference between the $d^k s^2$ and $d^k s$ configurations for negatively ionized and neutral atoms. Instead, the slopes and intercepts for the plot in these cases were obtained by fixing the points $k = 0$ from the measured affinities of K, Rb and Cs (see Chapter 11, p. 466) and $k = 10$ from those measured for Cu, Ag and Au (see Chapter 11, p. 461). Having obtained the energy difference between the ground state of the negative ion and that of the $d^k s$ configuration of the neutral atom, it was then only necessary to allow for the excess energy of the latter over the ground configuration of the atom (available from spectroscopic data) to derive the electron affinity. This is necessarily less reliable than for the elements of the short periods but gives the results shown in Table 3.7. At the time this work was first carried out by Zollweg, the only directly measured values available were those for Cu, Ag and Au obtained with the hot filament technique (see p. 38) while those for K, Rb and Cs were derived from charge-exchange experiments.

Table 3.7 was prepared using the accurate values for all six atoms provided from laser photodetachment measurements (see Chapter 11).

It will be seen that the extrapolated value for Pt agrees very well with the accurately measured value. No other comparison of

TABLE 3.7 *Electron affinities for atoms of the long periods, determined by empirical methods*

Negative-ion ground states	Atom	Electron affinity (eV)				Atom	Electron affinity (eV)				Atom	Electron affinity (eV)		
		Measured	Zollweg (a)	(b)	Clementi		Measured	Zollweg (a)	(b)			Measured	Zollweg (a)	(b)
$ds^2\,{}^1S$	K	0.50	(0.50)	—	+0.90	Rb	(0.49)	(0.49)	—		Cs	(0.47)	(0.47)	—
$d^9s^2\,{}^2D$	Ca	—	−1.62	—	—	Sr	—	−1.74	—		Ba	—	−0.54	—
$d^2s^2\,{}^3F$	Sc	—	−0.80	—	−0.14	Y	—	−0.76	—		La	—	+0.44	—
$d^3s^2\,{}^4F$	Ti	—	−0.11	—	+0.39	Zr	—	+0.08	—		Hf	—	−0.78	—
$d^4s^2\,{}^5D$	V	—	+0.51	—	0.94	Nb	—	+0.77	—		Ta	[0.8]	−0.05	—
$d^5s^2\,{}^6S$	Cr	0.66	+0.85	—	0.98	Mo	[1.0]	+0.86	—		W	[0.5]	0.98	—
$d^6s^2\,{}^5D$	Mn	—	−1.19	—	−1.07	Tc	—	0.63	—		Re	[0.15]	0.09	—
$d^7s^2\,{}^4F$	Fe	—	−0.14	—	0.58	Ru	—	1.04	—		Os	—	1.10	—
$d^8s^2\,{}^3F$	Co	—	+0.65	—	0.94	Rh	—	1.12	—		Ir	—	1.58	—
$d^9s^2\,{}^2D$	Ni	—	+1.13	0.24	1.28	Pd	—	0.40	1.35		Pt	2.28	2.12	2.7
$d^{10}s^2\,{}^1S$	Cu	(1.23)	—	—	—	Ag	(1.30)	—	—		Au	(2.31)	—	—

( ) bracketed values assumed as basis for extrapolation. Measured values are those obtained from photodetachment (Chapter 11, pp. 461–73) except for those in square brackets which were measured from surface ionization observations (see p. 38).
(a) Obtained by horizontal extrapolation along the row.
(b) Obtained from noble metal atom affinity by extrapolation of energy difference between $d^{10}s$ and d^9s^2 configurations.

this kind is yet possible though the results of Scheer's hot-filament measurements for Ta, W and Re are included. For the atoms from K to Cu there is general but not detailed agreement with the semi-empirical results obtained by Clementi.

An alternative way of estimating the affinities of Ni, Pd and Pt follows from the difference plots for addition of s and p electrons shown in Fig. 3.3. According to these, the energy released in attaching an s electron to the d^{10} configuration to form $d^{10}s$ should be 0.84 eV less than for attachment to the $d^{10}s$ configuration of the respective noble metals (Cu, Ag and Au). The ground states of the negative ions of the transition metals are in the d^9s^2 configuration and the energy difference between this and the $d^{10}s$ is then obtained by quadratic extrapolation along the isoelectronic series. The resulting affinities given in Table 3.7 agree quite well with those derived by horizontal extrapolation for Pt, less so for Pd and very poorly for Ni.

Zollweg has also made some very rough estimates of electron affinities for rare earth atoms and considers that they are usually less than 0.5 eV, even when positive.

CHAPTER 4

Atomic Negative Ions – Excited States – Autodetachment; General Account

4.1. Bound excited states of negative ions

In §1.1 we noted the general result that the number of bound quantum states of a negative ion is not infinite. Indeed the small binding energy of the attached electron in its ground state makes it most unlikely that, apart from fine and hyperfine structure levels associated with the ground state, any bound excited states will exist with appreciable binding energy.

The separations of the terms associated with the ground configurations of different negative ions, obtained by quadratic extrapolation, are given in Table 3.4. Comparison of the energy required to excite the lowest excited term with the electron affinities given in Tables 3.5–3.7 shows that the only ions for which the first excited term might be a bound state are those with three outer p electrons. There is some evidence of the existence of a metastable state of C^- (see Chapter 10, p. 415) with detachment energy of about 0.03 eV which could well be the $^2D_0$ state. According to Table 3.4 this is estimated to lie about 1.40 eV above the ground $^4S^0$ state. This is 0.13 eV greater than the observed electron affinity but the extrapolation could well be incorrect by 0.2 eV. For Si^-, Ge^- and Sn^-, the electron affinities are 1.25, 1.20 and 1.25 eV respectively and the corresponding excitation energies of the first excited ^2D term are 0.94, 0.90_5 and 0.85 eV. It is unlikely that the estimated values are sufficiently in error for the ^2D terms in these three cases to be unbound. Indeed for Si^- and Ge^- the higher ^2P terms are estimated to lie at 1.48 and 1.53 eV which are not far above the corresponding electron affinities (see Chapter 10, p. 415).

The only other cases listed in Table 3.4 where any possibility of a bound excited term exists are those of ions with four p electrons, excluding N^-. For P^-, As^- and Sb^- the estimated electron affinities are 0.77, 1.07 (estimated) and 1.05 eV as compared with the excitation energies of the ^1D terms of 0.84, 1.13 and 0.94 eV. It is not possible because of the inaccuracy of the estimates to decide whether

[66]

or not the ^{1}D terms lie above or below the continuum. There remains the question as to whether, for ions such as those with p^5 or p^6 shells for which there is only one term associated with the ground configuration, any terms of an excited configuration are bound. We have already referred on p. 60 to the results obtained by Bates and Moiseiwitsch (1955) from quadratic extrapolation of the energies between the ground terms of different configurations which led to the conclusion that the outer $(n+1)$s orbitals of the negative ions such as Be$^-$, Mg$^-$, etc., lie lower than the outer np. The same extrapolations applied to other ions indicate that in all of these cases the p^k configuration is lowest and that the energy required to excite the lowest term of any higher configuration is greater than the electron affinity. Thus for O$^-$ the (2p)^{4}3s configuration is estimated to lie about 2.0 eV higher than the (2p)^5, greater than the electron affinity by 0.5 eV. A similar excess is estimated for the energy required to excite the first upper configuration in F$^-$ and Cl$^-$.

Among heavier atoms we note that Zollweg (1969) found empirically that the d^{10}s ^{2}S$_{1/2}$ states of Ni$^-$, Pd$^-$ and Pt$^-$ lie respectively 1.11, 0.8 and 1.59 eV above the ground d^{9}s^2 ^{2}D$_{5/2}$ term. In all cases these are smaller than the estimated electron affinities for Ni and Pd and the observed value for Pt so that again we would expect the ^{2}S$_{1/2}$ states to be bound. There are probably many other cases in which stable excited states of this kind exists for heavy negative ions.

In Table 3.4 extrapolated values of fine structure separations for the ground terms of stable negative ions are given.

4.2 Continuum states of negative ions

The unbound states of negative ions form a continuum just as do the corresponding states of neutral atoms. However, whereas the unbound electron in the latter cases moves in a field which falls off asymptotically at large distances from the nucleus as r^{-1}, for negative ions the interaction falls off much faster. The importance of this in limiting the number of bound states has already been pointed out on p. 3. For the unbound states it has an important effect on the asymptotic form of the wave function which usually plays a vital role in the discussion of the rates of processes in which transitions to or from unbound states occur.

Relation to elastic scattering by a centre of force

Because the interaction falls off at least as fast as r^{-2}, at large r the asymptotic form may be expressed as an incident plane wave plus an outgoing (or in some cases incoming) spherical wave. If the electron energy is E so the wave number $k = (2mE/\hbar^2)^{1/2}$, the wave function for a plane wave of unit amplitude propagating in the z-direction has the form

$$\psi = \exp(ikz) = \exp(ikr\cos\theta). \tag{4.1}$$

Since, according to quantum mechanics, the current flux corresponding to a wave function ψ is given by

$$\mathbf{j} = (ie\hbar/2m)(\psi\,\mathrm{grad}\,\psi^* - \psi^*\,\mathrm{grad}\,\psi), \tag{4.2}$$

the wave function (4.1) represents a beam of electrons of unit density moving with velocity $v = k\hbar/m$ in the z-direction. Similarly the function

$$\psi = f(\theta)\,r^{-1}\exp(ikr) \tag{4.3}$$

represents an outgoing flux in the radial direction equal to $|f(\theta)|^2 v/r^2$.

Thus if the unbound wave function has the asymptotic form

$$\psi = \exp(ikz) + r^{-1}\exp(ikr)f(\theta), \tag{4.4}$$

the ratio of the number of outgoing electrons falling normally per second on a small surface of area δS at a distance r from the nucleus, to the number incident on the area per second is

$$|f(\theta)|^2 v\delta S/vr^2 = |f(\theta)|^2\,\mathrm{d}\omega, \tag{4.5}$$

where $\mathrm{d}\omega$ is an element of solid angle about the direction θ. This ratio is directly measurable. Thus the chance $p(\theta)$ that, in travelling a small distance δz through a gas containing N atoms per unit volume, an electron is scattered through an angle θ into the solid angle $\delta\omega$ about θ will be given by

$$p(\theta) = NI(\theta)\,\delta\omega\,\delta z, \tag{4.6}$$

where

$$I(\theta) = |f(\theta)|^2. \tag{4.7}$$

The chance P that the electron will be scattered through any angle in traversing the distance δz will then be

$$P = \int p(\theta)\,\mathrm{d}\omega$$
$$= NQ\delta z, \tag{4.8}$$

where

$$Q = 2\pi \int_0^\pi I(\theta) \sin\theta \, d\theta. \qquad (4.9)$$

The quantities Q and $I(\theta)$ have the dimensions of area. Q is usually called the *total effective cross-section* for collision between an electron of energy E and the atom, and is such that the scattering probability P is the same as it would be if the atom were a rigid spherical obstacle of cross-section Q. $I(\theta) d\omega$ is the *differential* scattering cross-section which determines the angular distribution of the scattered electrons.

Apart from its usefulness in describing collision phenomena, the asymptotic form (4.4) is convenient if the unbound wave function is to be normalized to unity within a large volume $\mathscr{V}$. If it has the asymptotic form (4.4) then the normalizing factor is simply $\mathscr{V}^{-1/2}$.

We now turn to describe briefly how solutions with the asymptotic form (4.4) are obtained. For this it is convenient to deal first with the simple approximation in which the atom, which is the 'core' of the negative ion, is regarded as a structureless centre of force exerting an interaction $V(r)$ on the electron.

As for the bound states, the angular momentum about the centre is quantized as $\{l(l+1)\}^{1/2} \hbar$. For states with $l = 0$, the wave function $\psi(r)$ for the electron will be a function of r only and must satisfy the Schrödinger equation

$$\frac{1}{r^2} \frac{d}{dr}\left(r^2 \frac{d\psi}{dr}\right) + (k^2 - U)\psi = 0, \qquad (4.10)$$

where $k^2 = 2mE/\hbar^2$, $U = 2mV/\hbar^2$. Writing $u = r\psi$ then (4.10) becomes

$$\frac{d^2 u}{dr^2} + (k^2 - U) u = 0. \qquad (4.11)$$

It may be shown that if $V(r)$ falls off at least as fast as r^{-2} as $r \to \infty$, then for large r, u will satisfy

$$\frac{d^2 u}{dr^2} + k^2 u = 0, \qquad (4.12)$$

i.e.
$$u \sim \alpha \sin kr + \beta \cos kr$$
$$= C \sin(kr + \eta), \quad (4.13)$$

where $\eta = \arctan(\beta/\alpha)$ and is a phase shift produced by the central potential. Thus if $V = 0$, then $u = A \sin kr$ for all r in order that ψ should be finite at $r = 0$.

Taking as illustration the same interaction as on p. 3, namely $V = -V_0$, a constant, for $r < r_0$ and zero elsewhere, we have

$$\frac{d^2 u}{dr^2} + k^2 u = 0, \quad r > r_0, \quad (4.14)$$

$$\frac{d^2 u}{dr^2} + \lambda^2 u = 0, \quad r < r_0, \quad (4.15)$$

with $\lambda^2 = 2m(E + V_0)/\hbar^2$. Since $u = 0$, $r = 0$, we have

$$u = C \sin(kr + \eta), \quad r > r_0. \quad (4.16)$$
$$= A \sin \lambda r, \quad r < r_0. \quad (4.17)$$

Continuity of charge and current density at $r = r_0$ requires

$$C \sin(kr_0 + \eta) = A \sin \lambda r_0. \quad (4.18)$$
$$Ck \cos(kr_0 + \eta) = A\lambda \cos \lambda r_0, \quad (4.19)$$

from which, by eliminating the ratio A/C, we have

$$\eta = \arctan\left(\frac{k}{\lambda} \tan \lambda r_0\right) - kr_0. \quad (4.20)$$

If the angular momentum quantum number is l, then the wave function is of the form

$$R_l(r) P_l(\cos \theta), \quad (4.21)$$

where $P_l(\cos \theta)$ is the zonal harmonic of order l. Writing $R_l = r^{-1} u_l$ we now have

$$\frac{d^2 u_l}{dr^2} + \left\{ k^2 - U - \frac{l(l+1)}{r^2} \right\} u_l = 0, \quad (4.22)$$

the extra term arising from the effective potential energy $\hbar^2 l(l+1)/2mr^2$ of the centrifugal force.

The asymptotic form of the solution of (4.22) can again be written

$$u_l \sim C_l \sin(kr - \tfrac{1}{2}l\pi + \eta_l), \quad (4.23)$$

ATOMIC NEGATIVE IONS – EXCITED STATES

the additional term $-\tfrac{1}{2}l\pi$ being introduced because in the limit when $V \to 0$ the asymptotic form of u_l is

$$u_l \sim C_l \sin(kr - \tfrac{1}{2}l\pi). \qquad (4.24)$$

We may therefore write the asymptotic form of an unbound wave function in the form

$$\Psi = \sum_l C_l r^{-1} u_l(r) P_l(\cos\theta), \qquad (4.25)$$

where

$$u_l \sim \sin(kr - \tfrac{1}{2}l\pi + \eta_l). \qquad (4.26)$$

Since the plane wave function e^{ikz} can be expanded in the form

$$e^{ikz} = (kr)^{-1} \sum_l i^l (2l+1) S_l(r) P_l(\cos\theta), \qquad (4.27)$$

where

$$S_l(r) \sim \sin(kr - \tfrac{1}{2}l\pi), \qquad (4.28)$$

we see that (4.25) will have the asymptotic form (4.4) if

$$C_l \sin(kr - \tfrac{1}{2}l\pi + \eta_l) = i^l k^{-1}(2l+1)\sin(kr - \tfrac{1}{2}l\pi) + A_l e^{ikr} \qquad (4.29)$$

giving

$$C_l = k^{-1} \exp(i\eta_l) i^l (2l+1), \qquad (4.30)$$

$$A_l = (2ik)^{-1}\{\exp(2i\eta_l) - 1\}(2l+1). \qquad (4.31)$$

Thus

$$f(\theta) = (2ik)^{-1} \sum_l (2l+1)\{\exp(2i\eta_l) - 1\} P_l(\cos\theta), \qquad (4.32)$$

and since

$$\int_0^\pi P_l(\cos\theta) P_{l'}(\cos\theta) \sin\theta\, d\theta = \frac{2}{2l+1}\delta_{ll'},$$

$$Q = 2\pi \int_0^\pi |f(\theta)|^2 \sin\theta\, d\theta$$

$$= (4\pi/k^2) \sum_l (2l+1) \sin^2\eta_l. \qquad (4.33)$$

Convergence of the phase shifts

Thus the observable scattering cross-sections are expressed in terms of the phase shifts η_l. An important question is the convergence of the series in (4.32). According to classical ideas, a particle of angular momentum J will pass the centre of force at a distance J/mv if the deviating effect of the scatterer is ignored. Thus if the range of the interaction is a, particles with angular momentum $> mva$ will not be influenced by the interaction. In quantum mechanical terms this would mean that the phase shift will be small if

$$\{l(l+1)\}^{1/2}\hbar \gg ka, \qquad (4.34)$$

where a, the effective range of interaction, is such that $V(a) = E$. This may be confirmed by detailed consideration of the form of the wave functions u_l when $V \to 0$. These are found to be small for $r < r_m$, where $kr_m \simeq \{l(l+1)\}^{1/2}$, so that the overlap with the interaction which determines the phase shift will be small if $r_m \gg a$. Thus, if $V(r) = \alpha r^{-s}$ we have

$$a = (\alpha/E)^{1/s},$$

and η_l will be small if

$$l\hbar/k \gg (\alpha/E)^{1/s}$$

or

$$l \gg E^{(1/2 - 1/s)} \alpha^{1/s} (2m/\hbar^2)^{1/2} \qquad (4.35)$$

This shows that, if $s > 2$, all phases except possibly η_0 will tend to zero as E tends to zero. Also, the greater the value of E the larger the number of phase shifts which must be taken into account. For small E the cross-section can be written

$$Q = \frac{4\pi}{k^2} \sin^2 \eta_0. \qquad (4.36)$$

Relation of phase shifts to time delay in scattering

The behaviour of the phase shifts as functions of the wave number k or kinetic energy E has a special significance in relation to the time delay suffered by the incident electron in traversing the region of interaction with the scatterer.

Consider the scattering of a wave packet of zero angular momentum and velocity v. This packet may be made up of two beams of

ATOMIC NEGATIVE IONS – EXCITED STATES

nearly equal energies $E \pm \delta E$ and wave numbers $k \pm \delta k$. The incident packet then has the asymptotic form

$$\psi_{\text{inc}} \sim r^{-1} \langle \exp[-i\{(k+\delta k)r + \hbar^{-1}(E+\delta E)t\}] + \\ + \exp[-i\{(k-\delta k)r + \hbar^{-1}(E-\delta E)t\}]\rangle. \quad (4.37)$$

The centre of the packet at time t is located by the conditions that the two waves are in phase there, i.e.

$$r\delta k - \hbar^{-1} t \delta E = 0. \quad (4.38)$$

Through the scattering, phase shifts $\eta \pm \delta\eta$ will be introduced in the waves of wave number $k \pm \delta k$ respectively, so the wave function for the outgoing packet will have the form

$$\psi_{\text{out}} \sim r^{-1} \langle \exp[-i\{-(k+\delta k)r + \hbar^{-1}(E+\delta E)t - 2(\eta+\delta\eta)\}] + \\ + \exp[-i\{-(k-\delta k)r + \hbar^{-1}(E-\delta E)t - 2(\eta-\delta\eta)\}]\rangle, \quad (4.39)$$

and its centre will be located where

$$r\delta k - \hbar^{-1} t \delta E + 2\delta\eta = 0. \quad (4.40)$$

Hence

$$r = \hbar^{-1} t \frac{dE}{dk} - 2 \frac{\partial \eta}{\partial k}$$

$$= \hbar^{-1} \frac{dE}{dk}\left(t - 2\hbar \frac{\partial \eta}{\partial E}\right). \quad (4.41)$$

This means that the packet has been retarded, through scattering, by a time

$$2\hbar \frac{\partial \eta}{\partial E} = \frac{2}{v} \frac{\partial \eta}{\partial k}. \quad (4.42)$$

Wave functions describing electron scattering by atoms

We must now consider how this analysis must be modified to deal with the motion of an unbound electron in the field of a neutral atom. This is no longer a one-body problem as the motion of the atomic electrons will be affected by interaction with the incident electron. The full wave function Ψ will now depend on the coordinates $\mathbf{r}_a$ of the atomic electrons as well as those $\mathbf{r}$ of the incident. It can always be expanded in a series of unperturbed atomic wave functions $\psi_i(\mathbf{r}_a)$ in the form

$$\Psi(\mathbf{r}, \mathbf{r}_a) = \sum \psi_i(\mathbf{r}_a) F_i(\mathbf{r}), \qquad (4.43)$$

where it is important to remember that the unbound atomic states must also be included so that the sum really includes an integral over this continuum of states.

Considering in particular the unbound states of H⁻, we now have that Ψ must satisfy

$$(H - E - \epsilon_0)\Psi = 0, \qquad (4.44)$$

where

$$H = -\frac{\hbar^2}{2m}(\nabla_1^2 + \nabla_2^2) + e^2\left(-\frac{1}{r_1} - \frac{1}{r_2} + \frac{1}{r_{12}}\right) \qquad (4.45)$$

and ϵ_0 is the energy of the ground state of hydrogen. We have changed the notation so r_1, r_2 refer to the unbound and bound electrons, respectively. The wave functions $\psi_i(\mathbf{r}_2)$ satisfy

$$\left(\frac{\hbar^2}{2m}\nabla_2^2 + \frac{e^2}{r_2}\right)\psi_i = -\epsilon_i \psi_i, \qquad (4.46)$$

where ϵ_i is the energy of the ith state of hydrogen. Hence we have on substitution of (4.43) for Ψ

$$\sum_i \psi_i(\mathbf{r}_2)\left(\frac{\hbar^2}{2m}\nabla_1^2 + \frac{e^2}{r_1} + E_i\right) F_i(\mathbf{r}_1) +$$
$$+ \sum \psi_i \frac{e^2}{r_{12}} F_i(\mathbf{r}_1) = 0, \qquad (4.47)$$

where

$$E_i = E + \epsilon_0 - \epsilon_i. \qquad (4.48)$$

Multiplication on the right by ψ_j^* and integration over all $\mathbf{r}_2$ gives, because of the orthogonal properties of the ψ_i,

$$\left(\frac{\hbar^2}{2m}\nabla_1^2 + E_i + \frac{e^2}{r_1}\right) F_i(\mathbf{r}_1) = \sum_j V_{ij}(\mathbf{r}_1) F_j(\mathbf{r}_1), \qquad (4.49)$$

where

$$V_{ij}(\mathbf{r}_1) = e^2 \int r_{12}^{-1} \psi_i(\mathbf{r}_2) \psi_j^*(\mathbf{r}_2)\, d\mathbf{r}_2. \qquad (4.50)$$

ATOMIC NEGATIVE IONS – EXCITED STATES

This infinite set of coupled equations must be solved to give solutions for F_i which satisfy the appropriate boundary conditions.

If $\epsilon_1 - \epsilon_0 > E$ so the incident electron has insufficient energy to excite the atom, then $k_i^2 = 2mE_i/\hbar^2 < 0, = -\lambda_i^2$ for $i \neq 0$ and the conditions are

$$F_0 \sim \exp(ik_0 r \cos\theta) + f_0(\theta) \exp(ik_0 r) r^{-1}, \qquad (4.51)$$

$$F_i \sim \exp(-|\lambda_i|r) r^{-1} g_i(\theta), \quad i \neq 0. \qquad (4.52)$$

Under these circumstances all electrons incident with a given energy and angular momentum are scattered with the same energy and angular momentum so that incoming and outgoing radial fluxes are equal. It may then be shown that $f_0(\theta)$ in (4.51) may be expressed in the form (4.32) with the η_l all real, so the differential and total cross-sections I and Q are given as before by (4.7) and (4.33) respectively.

If $\epsilon_1 - \epsilon_0 < E$, then

$$F_0 \sim \exp(ik_0 r \cos\theta) + f_0(\theta)\exp(ik_0 r) r^{-1},$$

$$F_i \sim f_i(\theta) \exp(ik_i r) r^{-1}, \qquad k_i^2 > 0,$$

$$\sim g_i(\theta) \exp(-|\lambda_i| r) r^{-1}, \; k_i^2 < 0 = -\lambda_i^2. \qquad (4.53)$$

In such cases because inelastic collisions are energetically possible (the differential cross-section for excitation of the ith state is $(k_i/k_0)|f_i(\theta)|^2 d\omega$), the outgoing flux of electrons of initial energy and angular momentum is less than the incoming. This situation may be represented formally by taking

$$\eta_l = \xi_l + i\zeta_l, \qquad (4.54)$$

where ξ_l and ζ_l are real and $\zeta_l > 0$.

It is quite clear that we cannot hope to solve the infinite set of equations (4.49) and that to proceed further it is necessary to make quite drastic simplifications. One of the simplest from our point of view is to ignore all terms in (4.43) except the first so that we take simply

$$\Psi \simeq \psi_0(r_2) F_0(\mathbf{r}_1), \qquad (4.55)$$

where, referring to (4.49), we have
$$(\nabla_1^2 + k_0^2 - U_{00}) F_0 = 0 \qquad (4.56)$$
with
$$U_{00} = 2mV_{00}/\hbar^2.$$

The problem is then reduced to one of the scattering of electrons by a centre of force exerting an interaction equal to the mean field of the unperturbed atom in its ground state.

In practice, for unbound states in which the kinetic energy of the incident electron is less than the energy of excitation of the first excited state, the approximation (4.55) is a poor one. It is important to take account of the Pauli principle so the function Ψ, including spin coordinates, is rendered antisymmetrical in the coordinates of every pair of electrons. This allows for exchange of electrons between the atom and the incident beam during the collision. Again, the distortion of the atom by the field of the slowly moving electron is important. The higher terms in (4.43) include this effect but to take it properly into account in practice involves approximate procedures based on the physics of the phenomenon. They include allowance for long-range near-adiabatic polarization of the atom and also for the electron correlation effects which are so important in determining the electron affinity (see p. 25).

In general, discussion of the methods used for obtaining reasonably accurate solutions for the functions Ψ for the continuum states of negative ions would take us too far from the main theme of this book, far into atomic scattering theory. (For a detailed account see Mott and Massey, 1965, Chapters XII–XVIII, and Massey, Burhop and Gilbody, 1969, Chapters 6–9.) However, we shall have occasion to refer to different approximations when discussing the continuous absorption coefficient of the H$^-$ ion for radiation in Chapter 15, and we must also consider a little further what is involved if the approximation (4.55) is modified by adding a further term from the full series (4.43). For simplicity we suppose that the incident electron has insufficient energy to excite the atom, and write

$$\Psi = \psi_0(r_2) F_0(\mathbf{r}_1) + \psi_1(r_2) F_1(\mathbf{r}_1). \qquad (4.57)$$

Referring to (4.49) we see that, in place of the single equation (4.56), we now have two coupled equations

$$(\nabla^2 + k_0^2 - U_{00})F_0 = U_{01}F_1, \qquad (4.58)$$

$$(\nabla^2 - \lambda_1^2 - U_{11})F_1 = U_{10}F_0 \qquad (4.59)$$

with

$$F_0 \sim \exp(ik_0 r \cos\theta) + f_0(\theta) r^{-1} \exp(ik_0 r), \qquad (4.60)$$

$$F_1 \sim r^{-1} \exp(-|\lambda_1|r) g_1(\theta) \qquad (4.61)$$

and $U_{ij} = 2mV_{ij}/\hbar^2$.

We shall see that, from consideration of coupled equations of this form, we can examine the effect of the interaction of doubly-excited states of negative ions with continuum states. In particular we shall see how it is possible to obtain information about these doubly-excited states from observations of scattering cross-sections as functions of energy and angle.

Before doing this we must introduce the subject of doubly-excited states and the phenomenon of their break-up by auto-detachment. Meanwhile, it is as well to point out that there are circumstances in which good solutions may be obtained for Ψ by including further terms in (4.43) while still making it possible to solve the resulting equations by means of a high-speed computer.

4.3 Negative-ion formation by excited atoms – autodetachment

As our arguments have been applied only to singly-excited states of the ions, we must consider the possibility of an excited atom forming a negative ion, with overall reduction of energy, by electron attachment. Such a system would correspond to a doubly-excited state of the ion.

Attention was called to this question by the calculations of Ta-You Wu (1936). However, in ions of this kind the electron affinity of the excited states will normally be less than the excitation energy so that the doubly-excited ion will be unstable towards dissociation into a normal atom and a free electron

$$X^{-\prime} \longrightarrow X + e. \qquad (4.62)$$

This is usually referred to as *autodetachment* because of its essential similarity to autoionization which occurs with doubly-excited atoms or positive ions. If E_{ex} is the excitation energy and E_a' the electron affinity of the excited state, the electron in (4.62) will be freed with a kinetic energy at infinite separation equal to $E_{ex} - E_a'$.

A degeneracy therefore exists between the doubly-excited configuration of the ion and the unbound state in which an electron moves in the field of the neutral atom so as to escape to infinity with finite kinetic energy, i.e. an unbound continuum state of the negative ion. The properties of such states are normally studied through the observation of scattering processes (of electrons or of heavy particles) by the atom, or of radiation processes, in which electron ejection occurs by photon absorption by the ion or electron capture occurs with photon emission by the atom. Continuum states in which the free electron energy is close to that which would result from autodetachment from some doubly- or multiply-excited states are likely to be modified by the existence of these states. It is from the nature of these modifications, their observation and theoretical interpretation that we now have a considerable amount of information available about autodetaching states.

Thus the cross-sections for collisions of electrons with atoms will be modified by the possibility of occurrence of the two processes

$$e + A \longrightarrow A^{-**}, \qquad (4.63)$$

$$A^{-**} \longrightarrow e + A, \qquad (4.64)$$

where A represents the normal atom and A^{-**} a doubly-excited ion. Such processes will only be important when the electron energy is very close to that resulting from break-up of the doubly-excited ion. However, if $\hbar/\Gamma$ is the mean lifetime, the uncertainty principle requires that the energy be uncertain by an amount Γ. We therefore expect that any effects on collision cross-sections will be confined to an energy range of order Γ about the definite energies E_r, say, which are the energies with which electrons are ejected in the autodetachment of each doubly-excited state which arises.

In some cases a doubly-excited state may break up in more than one way. Thus, if it is based on an atom in the nth excited state, it may break up into neutral atoms which are excited in the $(n-1)$th or lower states as well as the ground state. In such cases the inelastic scattering leading to excitation of the states up to the $(n-1)$th inclusive will also be modified in the same qualitative way as the elastic scattering.

The observability of these effects depends on the order of magnitude of the mean lifetime. This in turn determines the range of energy over which the effect will be significant and the energy resolution of the observing equipment must be high enough so that the effect is not averaged out. An energy range of 0.1 eV corresponds to a lifetime 10^{-14} s. We therefore consider next the rate at which autodetachment occurs. It will be found that, in many cases, it is of the order of magnitude which would lead to observable effects in the scattering measurements. The nature of the effects expected will then be discussed.

Lifetime of an autodetaching state

According to time-dependent perturbation theory, the rate at which autodetachment from a doubly-excited to a singly-excited continuum state occurs is given by

$$\frac{4\pi^2}{h} |H_{ac}|^2. \qquad (4.65)$$

H_{ac} is the transition matrix element

$$H_{ac} = \int \chi_a V \psi_c \, d\tau, \qquad (4.66)$$

where χ_a is the wave function of the doubly-excited state, ψ_c of the singly-excited continuum state and V is the interelectronic interaction which causes the transition.

For H$^-$, denoting the coordinates of the two electrons by $\mathbf{r}_1, \mathbf{r}_2$ respectively, we have approximately

$$\chi_a = \psi_i(\mathbf{r}_1) \psi_j(\mathbf{r}_2), \qquad (4.67)$$

where ψ_i and ψ_j are normalized hydrogen atomic wave functions for the ith and jth bound excited state. ψ_c, according to the simple approximation (4.55), is given by

$$\psi_c = N\psi_0(r_2) F_0(r_1), \quad (4.68)$$

where F_0 is as given in (4.56) and (4.51), and N is a normalizing factor which is best determined by regarding the doubly-excited ion as confined within a very large but finite volume $\mathscr{V}$. In that case, with F_0 having the asymptotic form (4.60), N will be $(\rho/\mathscr{V})^{1/2}$, where ρ is the number of final states per unit energy range about the energy of the final state. This is given from phase-space considerations by

$$\frac{4\pi p^2}{h^3} \frac{\mathrm{d}p}{\mathrm{d}E} \mathscr{V} = 2km\mathscr{V}, \quad (4.69)$$

where p is the momentum of the ejected electron. We thus have for the rate of autodetachment

$$(8\pi^2 me^4 k/h^3)|\iint \psi_i(\mathbf{r}_1)\psi_j(\mathbf{r}_2) r_{12}^{-1} \psi_0(r_2) F_0(r_1) \,\mathrm{d}\mathbf{r}_1 \,\mathrm{d}\mathbf{r}_2|^2. \quad (4.70)$$

This is a very crude formula which ignores, among other things, the Pauli principle and also the very approximate character of the functions χ_a and ψ_c which we have assumed.

In particular, in the H$^-$ case we need to consider whether the doubly-excited state is a member of the singlet or triplet series. With the interaction we have assumed, which is independent of the spin coordinates, autodetachment can only occur to states of the continuum of the same multiplicity. This is always possible for H$^-$ (but see § 4.9). Rough estimates of autodetachment rates give lifetimes of order 10^{-14}s or more. This is to be compared with the time taken (1.6×10^{-15}s) for an electron of 1 eV energy to travel 10^{-7} cm. It is seen that the lifetimes are significantly longer than the normal travel times of electrons across the interacting region. Furthermore, the energy resolution of experimental equipment is high enough to observe effects arising from the decay of states with lifetimes of this order.

As only doublet states arise from the interaction of an electron with the ground, singlet state of He, autodetachment from the 1s2s2p ^{4}P state of He$^-$ in this case can only arise through spin-dependent interactions. Another case we shall be referring to later is that of the $2p^4\,^1D$ state of N^- produced by attachment of an electron to the metastable $2p^3\,^2D$ state of N. As the ground state of N is a $2p^3\,^4S$ state, the only states which can arise through its interaction with an unbound electron are quintet and triplet states. Transitions from a singlet doubly-excited state to either of these states can only occur through spin-dependent interactions.

Because of the weakness of these interactions the lifetimes of doubly-excited states which can only suffer autodetachment through them are relatively very long, of the order 10^{-6} s or even longer provided they cannot make allowed radiation transitions (see p. 115). This is long enough for these ions to pass, without break-up, through experimental equipment of normal dimensions provided their velocities are of order 10^7 cm s^{-1}, and it is possible to measure their lifetimes directly. For many purposes they behave as stable negative ions (Chapter 5, pp. 126–35 and Chapter 15, p. 690).

The observation of non-metastable autodetaching states

Normally, autodetachment will occur far too rapidly for the doubly-excited ions to be observed directly. One method which has been used successfully is to pass a beam of normal negative ions through a gas so that excitation of autodetaching states occurs. These break up before traversing a distance much greater than 10^{-6} cm. Each break-up releases an electron with an energy determined by that of the autodetaching doubly-excited state concerned. Hence if the energy distribution of electrons produced in collisions of the ion beam with the gas atoms is measured, peaks should occur at these energies (see p. 120). An account of experiments of this kind with H^- is given in Chapter 5, p. 119.

The most effective method of studying the autodetaching states is from their effects on either the variation with angle or energy of collision cross-sections. Thus we have noted that the lifetime of an autodetaching state is long compared with the travel time of an electron across the interaction region. Also we showed that the variation with energy of the phase shift determining the elastic scattering of electrons of given angular momentum depends on the delay introduced by the scattering. If an electron has the appropriate energy, it is possible that it be captured into a doubly-excited state on collision, in which case it will suffer an extra time delay equal to the lifetime of the state towards autodetachment. This should produce a marked effect on the phase shift at energies close to the 'resonance' energy for capture. We shall now show how this arises in terms of the theory we sketched out, and derive explicit formulae for the effect on the cross-sections.†

Autodetaching states and elastic scattering of electrons by atoms

We consider for simplicity the two-electron case and return to the two-state approximation (4.57),

$$\Psi = \psi_0(r_2) F_0(\mathbf{r}_1) + \psi_1(r_2) F_1(\mathbf{r}_1)$$

for the wave function describing the impact of a beam of electrons of energy E below that, $\epsilon_1 - \epsilon_0$, necessary to excite the first excited state of the atom. As shown in §4.2, F_0 and F_1 satisfy the coupled equations

$$(\nabla^2 + k^2 - U_{00}) F_0 = U_{01} F_1, \qquad (4.71)$$

$$(\nabla^2 - \lambda^2 - U_{11}) F_1 = U_{10} F_0. \qquad (4.72)$$

U_{00}, U_{11} are the mean interactions of the electron with the undisturbed atom in its ground and first excited state respectively, while

$$\lambda^2 = -k^2 + \frac{2m}{\hbar^2}(\epsilon_1 - \epsilon_0) = -k^2 + k_0^2 > 0. \qquad (4.73)$$

U_{01} is equal to $(2m/\hbar^2) V_{01}$, where V_{01} is given by (4.50). F_0 and F_1 have the respective asymptotic forms (4.51), (4.52). For brevity and clarity we have dropped the suffices 0 and 1 from k and λ respectively.

The case of zero angular momentum

For simplicity we suppose that the excited state as well as the ground state is an s state and that the incident electron has zero angular momentum. In that case F_0 and F_1 are functions of r only and if we write

$$F_0 = r u_0, \quad F_1 = r u_1, \qquad (4.74)$$

† For an alternative treatment see Fano (1961) and Fano and Cooper (1964).

then
$$\frac{d^2 u_0}{dr^2} + (k^2 - U_{00}) u_0 = U_{01} u_1, \qquad (4.75)$$

$$\frac{d^2 u_1}{dr^2} + (-\lambda^2 - U_{11}) u_1 = U_{10} u_0. \qquad (4.76)$$

If the interaction V_{01} is switched off so the equations are uncoupled, then u_1 is a function representing bounded motion in the field of the excited atom. In general there will exist one or more bound states in this field of energy, $-\hbar^2 \lambda_1^2/2m, -\hbar^2 \lambda_2^2/2m, \ldots$, with corresponding wave functions $\chi_1, \chi_2, \ldots$. There will also be a continuum of unbound states of energy ϵ_c and wave functions χ_c. With the interaction switched off, when

$$-k^2 + \frac{2m}{\hbar^2}(\epsilon_1 - \epsilon_0) = \lambda_1^2 \text{ or } \lambda_2^2 \text{ or } \ldots, \qquad (4.77)$$

we would have two wave functions corresponding to the same total energy describing the state of the system. One, $r_1^{-1} u_0(r_1)\psi_0(r_2)$, corresponds to a singly-excited continuum state of the negative ion while the other corresponds to a doubly-excited state (with an electron attached to the excited atom).

From this point of view the coupling U_{01} is responsible for transitions between these degenerate states, i.e. to autodetachment. To pursue this further (Feshbach, 1958, 1962) we expand the function u_1 in a series of the eigenfunctions χ so that

$$u_1 = \sum_j a_j \chi_j + \int a(\epsilon) \chi_c(\epsilon) \, d\epsilon. \qquad (4.78)$$

On substitution in (4.76) we have, since

$$\frac{d^2 \chi_j}{dr^2} + (-\lambda_j^2 - U_{11}) \chi_j = 0, \qquad (4.79)$$

$$\sum_j (\lambda_j^2 - \lambda^2) a_j \chi_j - \int (k_c^2 + \lambda^2) a(\epsilon) \chi_c(\epsilon) \, d\epsilon = U_{10} u_0. \qquad (4.80)$$

Multiplying both sides by χ_j^* and integrating over all r gives, because the χ form an orthogonal set,

$$a_j = \int \chi_j^* U_{10} u_0 \, dr / (\lambda_j^2 - \lambda^2). \qquad (4.81)$$

Similarly
$$a(\epsilon) = -\int \chi_c^*(\epsilon) U_{10} u_0 \, dr / (k_c^2 + \lambda^2). \qquad (4.82)$$

Suppose now that $\lambda^2 \simeq \lambda_j^2$, a_j will then be much larger than all the other coefficients in the expansion (4.78), and we may take

$$u_1 \simeq \chi_j \int \chi_j^* U_{10} u_0 \, dr/(\lambda_j^2 - \lambda^2). \tag{4.83}$$

Substituting in (4.75) now gives

$$\frac{d^2 u_0}{dr^2} + (k^2 - U_{00}) u_0 = A U_{01} \chi_j, \tag{4.84}$$

where

$$A = \int \chi_j^* U_{10} u_0 \, dr/(\lambda_j^2 - \lambda^2). \tag{4.85}$$

We are now faced with the problem of obtaining a solution of this equation which has the correct asymptotic form (4.51), in the course of which we can determine A. To do this we use the method of the Green's function, according to which the solution of the equation

$$\frac{d^2 u}{dr^2} + (k^2 - U) u = h(r) \tag{4.86}$$

is expressed in terms of those of the homogeneous equation

$$\frac{d^2 v}{dr^2} + (k^2 - U) v = 0. \tag{4.87}$$

Let $v_{s,c}$ be the solutions of (4.87) with asymptotic form

$$v_s \sim \sin(kr + \eta), \tag{4.88}$$

$$v_c \sim \cos(kr + \eta). \tag{4.89}$$

Then the appropriate solution of (4.86) with the asymptotic form appropriate to the scattering problem is (Mott and Massey, 1965, Chapter 4)

$$u(r) = e^{i\eta} v_s(r) + \int G(r, r') h(r') \, dr', \tag{4.90}$$

with

$$G(r, r') = -k^{-1} \{v_s(r) v_c(r') + i v_s(r) v_s(r')\}, \quad r' > r. \tag{4.91}$$

Applying this to (4.84) we find,

$$u_0(r) = e^{i\eta} v_s(r) + A \int G(r, r') U_{01}(r') \chi_j(r') \, dr'. \tag{4.92}$$

Also, from (4.85), we have

$$(\lambda_j^2 - \lambda^2) A = \int \chi_j^* U_{10} u_0(r) \, dr. \tag{4.93}$$

ATOMIC NEGATIVE IONS – EXCITED STATES

Hence by multiplying both sides of (4.92) by $\chi_j^* U_{10}$ and integrating we find

$$(\lambda_j^2 - \lambda^2) A = e^{i\eta} \int v_s(r) \chi_j^* U_{10} \, dr$$
$$+ A \iint \chi_j^*(r) U_{10}(r) G(r,r') U_{01}(r') \chi_j(r') \, dr \, dr'. \quad (4.94)$$

Referring to the form (4.91) for the Green's function, $G(r,r')$ we see that we may write

$$A = e^{i\eta} \gamma_j / (\lambda_j^2 - \lambda^2 - \Delta\lambda_j^2 + i|\gamma_j^2|/k), \quad (4.95)$$

where

$$\gamma_j = \int \chi_j^* U_{10} v_s \, dr,$$
$$\Delta\lambda_j^2 = \iint \chi_j^*(r) U_{10}(r) \operatorname{re} G(r,r') U_{01}(r') \chi_j(r') \, dr \, dr', \quad (4.96)$$

it being supposed that $U_{01} = U_{10}$, as will be the case when both are real.

The asymptotic form of the solution $u_0(r)$ can now be found since

$$G(r,r') \sim -k^{-1} \exp\{i(kr + \eta)\} v_s(r'). \quad (4.97)$$

We find

$$u_0 \sim e^{i\eta} \sin(kr + \eta) - \frac{k^{-1} e^{ikr} e^{2i\eta} |\gamma_j|^2}{\{\lambda_j^2 - \lambda^2 - \Delta\lambda_j^2 + i|\gamma_j|^2/k\}}$$
$$= \sin kr + e^{ikr} \left\{ \frac{e^{2i\eta} - 1}{2i} - e^{2i\eta} \frac{\tfrac{1}{2}\Gamma}{E - E_r + \tfrac{1}{2}i\Gamma} \right\}, \quad (4.98)$$

where

$$\Gamma = (4m/k\hbar^2) |\int \chi_j^* V_{01} v_c \, dr|^2, \quad (4.99)$$

$$E = k^2 \hbar^2 / 2m = -\lambda^2 + \frac{\hbar^2}{2m}(\epsilon_1 - \epsilon_0), \quad (4.100)$$

$$E_r = (k_0^2 - \lambda_j^2 + \Delta\lambda_j^2) \hbar^2 / 2m. \quad (4.101)$$

The partial elastic cross-section, contributed by incident electrons of zero angular momentum, is then

$$Q^{(0)} = \frac{4\pi}{k^2} \left| \frac{e^{2i\eta} - 1}{2i} - \frac{\tfrac{1}{2} i \Gamma e^{2i\eta}}{E - E_r + \tfrac{1}{2} i\Gamma} \right|^2. \quad (4.102)$$

If $\Gamma = 0$, this reduces to the expression (4.33) for $l = 0$, with η determined by the mean field of the undisturbed atom in its ground state. Writing Γ in full in the form

$$\Gamma = (4mk/\hbar^2) |\int \psi_1^*(r_2) \phi_j(r_1) e^2 r_{12}^{-1} \psi_0(r_2) \mathscr{F}(r_1) \, \mathrm{d}r_1 \, \mathrm{d}r_2|^2, \qquad (4.103)$$

where $\phi_j = \chi_j(r_1) r_1^{-1}$, $\mathscr{F}(r_1) = v_c(r_1)/kr_1$, we see by comparison with (4.70) that, apart from the difference between ϕ_j and ψ_j, $\Gamma/\hbar$ is equal to the rate of break-up of the doubly-excited state. Indeed with the form (4.103), $\Gamma/\hbar$ gives a somewhat better approximation than (4.70).

Returning to (4.102), if we write

$$\tfrac{1}{2}\Gamma/(E_r - E) = \tan \sigma, \qquad (4.104)$$

we have

$$-i\Gamma \, e^{2i\eta}/(E - E_r + \tfrac{1}{2}i\Gamma) = -e^{2i\eta}(1 - e^{2i\sigma}). \qquad (4.105)$$

so

$$Q^{(0)} = \frac{4\pi}{k^2} \left| \frac{e^{2i(\eta+\sigma)} - 1}{2i} \right|^2 \qquad (4.106)$$

$$= \frac{4\pi}{k^2} \sin^2(\eta + \sigma). \qquad (4.107)$$

The possibility of formation of the doubly-excited state therefore introduces an additional phase shift

$$\sigma = \arctan\{\tfrac{1}{2}\Gamma/(E - E_r)\}.$$

$$\frac{\mathrm{d}\sigma}{\mathrm{d}E} = \frac{\tfrac{1}{2}\Gamma}{(E - E_r)^2 + \tfrac{1}{4}\Gamma^2}, \qquad (4.108)$$

so the time delay introduced is, according to (4.42),

$$\hbar^{-1} \frac{\mathrm{d}\sigma}{\mathrm{d}E} = \frac{\tfrac{1}{2}\Gamma/\hbar}{(E - E_r)^2 + \tfrac{1}{4}\Gamma^2/\hbar^2}. \qquad (4.109)$$

When $E = E_r$ this is $2\hbar/\Gamma$, corresponding to a time $\hbar/\Gamma$ required to form the complex, and a similar time for it to break up.

In the neighbourhood of the energy E_r, the contribution from the interaction with the doubly-excited state will be dominant and

ATOMIC NEGATIVE IONS – EXCITED STATES

we have

$$Q^{(0)} \simeq \frac{\pi}{k^2} \frac{\Gamma^2}{(E - E_r)^2 + \tfrac{1}{4}\Gamma^2}. \qquad (4.110)$$

This is of the form of a damped resonance curve of width Γ, exactly as expected from the uncertainty principle (see p. 78).

The energy E_r differs by a small amount $\hbar^2 \Delta\lambda_j^2/2m$ from the energy

$$E_j = \epsilon_1 - \epsilon_0 - \hbar^2 \lambda_j^2/2m$$

which the free electron would possess to be captured without energy change to form the doubly-excited state. Thus the formula (4.110) is very similar to spectrum line broadening by collisions in which the broadening is also accompanied by a relatively small energy shift.

In order that a resonance effect can be observed by observation, say, of the differential cross-section at a fixed angle of scattering, or of the total cross-section Q as a function of the kinetic energy of the incident electron, it is necessary that the energy-resolving power of the experimental apparatus should be high enough to observe variations within an energy range smaller than the line width Γ. In terms of electron-volts, a lifetime of 10^{-14} s corresponds to a line width of 0.15 eV. The best resolution attainable at the time of writing is around 0.05 eV which would make it possible to observe resonance peaks due to autodetaching states with lifetimes less than 3×10^{-13} s. This falls within the range of lifetimes of many doubly- or multiply-excited states of negative ions as we shall see.

The case of unrestricted angular momentum

Although the formula (4.102) has been derived for the case in which the total angular momentum is zero, there is no difficulty in extending it to other quantized angular momenta. Thus if an autodetaching resonance state with total angular momentum quantum number L occurs with an energy $E_r + \epsilon_0 - \epsilon_1$ say, then the angular distribution of electrons, with energy close to E, scattered by the atom, will be given by

$$2ikf(\theta) = \sum_{l \neq L} (2l+1)\{\exp(2i\eta_l) - 1\}P_l(\cos\theta) + \\ + (2L+1)[\exp\{2i(\eta_L + \sigma_L)\} - 1]P_L(\cos\theta), \qquad (4.111)$$

with σ_L given by (4.104) applied to the particular resonance state. The resonance peak, which will, in general, dominate the scattering when $E \simeq E_r$, will vary with angle as $\{P_L(\cos\theta)\}^2$, so that observations of the angular distribution can be used to determine the angular momentum of a resonance state.

The total cross-section Q can be written

$$Q = Q_r + Q_b, \qquad (4.112)$$

where

$$Q_b = \frac{4\pi}{k^2} \sum_{l \neq L} (2l+1) \sin^2 \eta_l, \qquad (4.113)$$

$$Q_r = \frac{4\pi}{k^2} (2L+1) \sin^2(\eta_L + \sigma_L)$$
$$= Q_a \sin^2(\eta_L + \sigma_L)/\sin^2 \eta_L, \qquad (4.114)$$

with Q_a the partial cross-section for angular momentum quantum number L in the absence of the resonance state. Hence

$$Q = Q_a \frac{(q_L + \epsilon)^2}{1 + \epsilon^2} + Q_b \qquad (4.115)$$

with

$$q_L = -\cot \eta_L, \quad \epsilon = (E - E_r)/\tfrac{1}{2}\Gamma. \qquad (4.116)$$

Finally, (4.115) may be written in the form

$$Q = Q_a + Q_b + Q_a(q_L^2 + 2q_L\epsilon - 1)/(1 + \epsilon^2),$$
$$= Q_B + Q_a(q_L^2 + 2q_L\epsilon - 1)/(1 + \epsilon^2), \qquad (4.117)$$

where Q_B denotes the cross-section which would be obtained if the resonance state did not exist. Q_B can be expected to vary smoothly and gradually with electron energy E. The variation of the function

$$(q_L^2 + 2q_L\epsilon - 1)/(1 + \epsilon^2), \qquad (4.118)$$

with ϵ for different values of q_L, is shown in Fig. 4.1.

If the energy-resolving power of the observing equipment is high enough, q_L may be determined from the observed shape of the resonance peak. In many cases however, the resolving power will be insufficient and the measurements are integrals over an energy

ATOMIC NEGATIVE IONS – EXCITED STATES

band large compared with the line width. We have

$$\int Q \, dE = \int (Q_a + Q_b) \, dE + \tfrac{1}{2}\Gamma \int Q_a \frac{q_L^2 + 2q_L \epsilon - 1}{(1 + \epsilon^2)} \, d\epsilon, \tag{4.119}$$

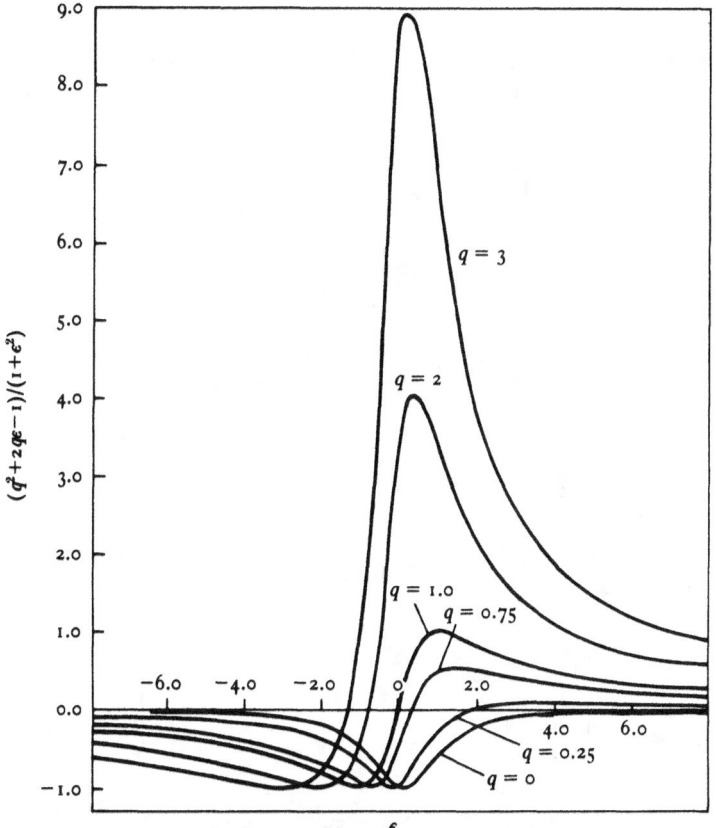

Fig. 4.1. Form of function $(q^2 + 2q\epsilon - 1)/(1 + \epsilon^2)$.

where the integrals are taken over a width $> \Gamma$, embracing the resonance. As far as the second integral is concerned, the limits may be taken as between $\pm \infty$ without serious error, giving

$$\int_{\text{res}} Q \, dE = \int_{\text{res}} Q_B \, dE + (2\pi/k^2) \, \Gamma(2L+1) \, \pi(q_L^2 - 1) \sin^2 \eta_L$$

$$= \int_{\text{res}} Q_B \, dE + (2\pi/k^2) \, \Gamma(2L+1) \cos 2\eta_L. \tag{4.120}$$

The resonance will then produce either a peak or depression of the observed cross-section at the resonance energy, according as $\cos 2\eta_L$ is $>$ or $<$ 0 respectively.

Autodetaching states and inelastic scattering of electrons by atoms

So far we have considered only the interaction of a doubly-excited state with one singly-excited continuum. There will be many

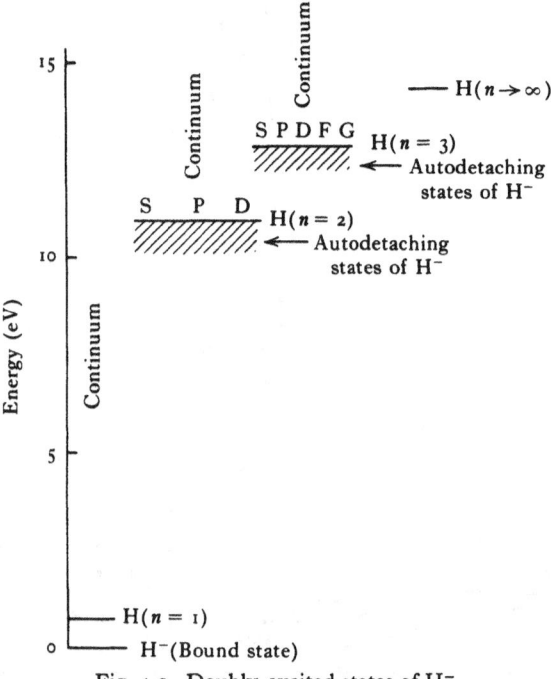

Fig. 4.2. Doubly-excited states of H⁻.

occasions in which interaction can occur with more than one such continuum. Thus we can expect doubly-excited states to arise from attachment of an electron to a three-quantum excited state of atomic hydrogen to form say a (3s)² state. This state will interact not only with the continuum associated with the 1s state of H but also that associated with the 2s and 2p states as may be seen from Fig. 4.2. The existence of the doubly-excited state in this case will affect not only the cross-section for elastic scattering by normal H atoms but also that for excitation of the two-quantum states. Thus, once the doubly-excited state is formed it may break up, making a transition to the 1s continuum, in which case it produces elastic scattering, or by making a transition to the 2s or 2p continua, in which case it produces inelastic scattering.

Ignoring the contribution from background scattering which takes place without formation of the resonance state, the elastic cross-section is

$$Q^{el} = \frac{\pi}{k^2} \Gamma_0^2 / \{(E - E_r)^2 + \tfrac{1}{4}\Gamma^2\}, \quad (4.121)$$

and the inelastic, for excitation of each state j

$$Q_j^{in} = \frac{\pi}{k^2} \Gamma_0 \Gamma_j / \{(E - E_r)^2 + \tfrac{1}{4}\Gamma^2\}. \quad (4.122)$$

In these formulae, $\Gamma = \Gamma_0 + \Sigma_j \Gamma_j$ and Γ_j/Γ_0 may be interpreted as the chance that, once formed, the complex will break up through interaction with the continuum associated with the excited state j. Γ_j is essentially similar to Γ of (4.103) except that $\psi_0(r_1)\mathscr{F}(r_1)$ is replaced by the appropriate function $\psi_j(r_2)\mathscr{F}_j(r_1)$.

From this point of view we can regard the cross-sections (4.121) and (4.122) as arising in the following way. They may be written

$$Q^{el} = (\Gamma_0/\Gamma) Q_c, \quad Q_j^{in} = (\Gamma_j/\Gamma) Q_c, \quad (4.123)$$

where

$$Q_c = \frac{\pi}{k^2} \frac{\Gamma_0 \Gamma}{(E - E_r)^2 + \tfrac{1}{4}\Gamma^2}, \quad (4.124)$$

so we may take Q_c as the cross-section for formation of the doubly-excited state by capture of the incident electron.

Examples of resonance effects in inelastic scattering due to formation of autodetaching states are described in Chapter 5, pp. 135-43. In general we find that, in the neighbourhood of a single resonance, the cross-section for the process concerned behaves as in (4.115) though the significance of the parameter q is more complex than for the case of elastic scattering at electron energies below the excitation threshold.

Autodetaching states and photodetachment

Autoionizing states of neutral atoms have been studied extensively through their effects on optical absorption spectra. In a similar way it is possible to observe such effects in cross-sections for photodetachment as a function of frequency. Thus, at a frequency close to that E_r/h required to excite an autodetaching state, the photodetachment cross-section can be written

$$Q_d = Q_d^a + Q_d^b \frac{(q+\epsilon)^2}{1+\epsilon^2}, \qquad (4.125)$$

as in (4.115). Here $\epsilon = (E - E_r)/\tfrac{1}{2}\Gamma$ where E_r is the photon energy and Γ the level width of the autodetaching state, while the line profile index q is a real number which may have any positive or negative value depending on the nature of the autodetaching state.

Since at a great frequency separation from the state ($\epsilon \gg 1$)

$$Q_d = Q_d^a + Q_d^b,$$

the effective contribution of the autodetaching line is

$$\int_0^\infty (Q_d - Q_d^a - Q_d^b)\, dE = \tfrac{1}{2}\pi\Gamma(q^2 - 1)\, Q_d^a. \qquad (4.126)$$

When $|q| > 1$ the level will lead to increased absorption over the line, but when $|q| < 1$ the absorption will be reduced over this region. In the latter case we have a so-called 'window' resonance, many examples of which have been observed in the absorption of ultra-violet light by rare gases. The introduction of laser techniques has made it possible to search for quite narrow resonance effects in photodetachment, and a number of examples of window resonances have already been observed. These are discussed in Chapter 5, p. 150, in relation to the levels concerned, and in Chapter 11, p. 467, in connexion with the photodetachment measurements and their interpretation.

Calculation of the energy of doubly-excited states of negative ions

In the simple example discussed on p. 82, in which not only exchange effects are ignored but also all terms in the expansion (4.43) of the

ATOMIC NEGATIVE IONS – EXCITED STATES

complete wave function other than those corresponding to the ground and first excited state of the atom, the energies of the doubly-excited states arising from attachment of an electron atom in its first excited state are first obtained by calculating the allowed bound state energies $-(\hbar^2/2m)\lambda_1^2, -(\hbar^2/2m)\lambda_2^2, \ldots$ for the potential energy V_{11} of an electron in the mean field of the excited atom. The total energy of a doubly-excited ion is then one of $\epsilon_1 - (\hbar^2/2m)\lambda_1^2, \ldots$, where ϵ_1 is the energy of the excited atom.

The projection method

This will in fact be a gross oversimplification for a case such as hydrogen in which the 2s and 2p states are of equal energy and at least the 2p as well as the 2s states must be included in the expansion (4.43). This gives rise to more than two coupled equations and the doubly-excited states arise from the allowed bound states in the coupled motion of an electron in the mean fields of the atom in the 2s and 2p states. However, it is possible, effectively, to include not only such obviously important coupling as this but also to a large extent that with higher states, by using a modified variational method. It is only necessary to use trial functions which exclude any contribution from the state in which the electron is moving in an unbound orbital in the field of the normal atom. Thus, ignoring the indistinguishability of the electrons, suppose we begin with a trial function $\Psi_t(\mathbf{r}_1, \mathbf{r}_2)$. This may be expanded in the form

$$\Psi_t(\mathbf{r}_1, \mathbf{r}_2) = \sum \psi_s(\mathbf{r}_2) G_s(\mathbf{r}_1), \qquad (4.127)$$

where

$$G_s(\mathbf{r}_1) = \int \Psi_t(\mathbf{r}_1, \mathbf{r}_2) \psi_s^*(\mathbf{r}_2) \, d\mathbf{r}_2. \qquad (4.128)$$

We now exclude the term $\psi_0(\mathbf{r}_2) G_0(\mathbf{r}_1)$ and take as our first trial function

$$\Psi_t'(\mathbf{r}_1, \mathbf{r}_2) = \Psi_t - \psi_0(r_2) G_0(\mathbf{r}_1). \qquad (4.129)$$

Having done this, the variational method gives upper bounds to the energies of the doubly-excited states just as for normal atoms (see p. 6). There is no difficulty in extending this procedure to allow for the Pauli principle. For relatively simple systems such as H$^-$, quite elaborate trial functions including correlation can be used but, as for the calculation of electron affinities of neutral atoms, the problem grows very rapidly in complexity when systems of more than two electrons are considered. Elaborate calculations have, however, been carried out for the metastable doubly-excited He⁻ ion. In this case, because of the symmetry conditions which must be satisfied by the wave function, and hence of any trial functions, there is no difficulty in applying the variational method directly as in the calculation of energies of bound states. Accounts are given of the results for H⁻ and He⁻ on pp. 116 and 126 respectively.

The use of close-coupling expansions

The energy obtained by these methods will be the energy unmodified by the level shift $(\hbar^2/2m)\Delta\lambda_j^2$ of (4.96). This only arises because of the coupling with the ground or lower excited states. Having obtained Ψ_t by a variational procedure, it is possible to estimate the level shift by a perturbation calculation. An alternative procedure which gives the resonance energies, including the level shift, is to work directly with the coupled equations. For the reasons outlined above it is usually necessary to work with more than two equations to obtain satisfactory results. In addition, it is necessary to solve the equations for a large number of neighbouring energies in the neighbourhood of a resonance. Burke and Taylor (1966) have combined a variational method with the eigenfunction expansion approach so that they use as a trial function

$$\Psi_t = \sum_{i=0}^{s} \psi_i(\mathbf{r}_2) F_i(\mathbf{r}_1) + \chi(r_1, r_2, r_{12}), \qquad (4.130)$$

where χ is an exponentially decaying function of r_1 and r_2 including correlation terms in r_{12}. The variational method here is one appropriate to dealing with continuum states (Mott and Massey, 1965, Chapter VI) and essentially determines the phase shifts $\eta_L + \sigma_L$ of (4.111). From these, the resonance energies E_r, including the level shifts, are then obtained. It is also possible, in principle, to determine the lifetimes $\hbar/\Gamma$.

The stabilization method

A powerful method which gives the true resonance energy, including the level shift, and also an estimate of the level width, has been developed by Taylor and his associates (Eliezer, Taylor and Williams, 1967). It is known as the stabilization method and is very similar to the variational methods used to determine the energy of the bound excited states.

Near a resonance energy the scattering wave function has a relatively large amplitude within the region of interaction with the target, and the procedure is directed to determining this part of the function with good accuracy at energies near the resonance energy. The success of the method depends to a great extent on the choice of suitable trial functions. The first step is to choose a basic configuration wave function Φ made up of a combination of bound functions with the proper symmetry, spin functions, etc. In addition, a suitable set of bound functions χ_i is chosen. Eigenenergies are then determined by a variational procedure corresponding to functions

$$\Psi_t(n) = \Phi + \sum_{i=1}^{n} c_i \chi_i. \qquad (4.131)$$

The behaviour of each energy as a function of n is then examined. If it is found that for $n > N$, where N is not too large, any particular eigenenergy is nearly independent of n, as also are the corresponding trial wave functions, then this energy is a close approximation to the unshifted resonance energy. Again if E_r is the energy concerned, $\partial E_r/\partial n$ is proportional to the level width and should also be independent of n. In practice it is possible to estimate the level width quite well from $\partial E_r/\partial n$, or it may be calculated from $\Psi_t(n)$, which is practically independent of n. Thus, referring to (4.70), in the two-electron case this amounts to replacing $\psi_i(r_1)\psi_j(r_2)$ by $\Psi_t(n)$. The generalization to a many-body system is obvious.

The justification for this procedure has been discussed very fully by Taylor (1970) and has been shown to give very good results for certain model cases (see p. 102). As emphasized earlier however, the method will not automatically yield good results no matter how inappropriate the choice of the functions Φ and χ_i may be.

Summary

We may summarize the results of this analysis with particular reference to the possibilities they offer for the laboratory study of autodetaching states of negative ions as follows.

Electrons may attach to neutral atoms in excited states with energy release of the order of 1 eV or so comparable with the magnitudes of electron affinities of normal atoms. Because this energy release is smaller than the excitation energy of the atom in almost all cases, these doubly-excited states are unstable towards break up into a normal atom and an unbound electron – a process of autodetachment. The lifetime before break up in this way is usually around 10^{-14} s which, while short compared with radiative lifetimes for optically allowed transitions ($\sim 10^{-8}$ s), is nevertheless considerably longer than the mean time for an electron, moving with the energy of the ejected electron in autodetachment, to traverse a distance over which the interaction with the atom is appreciable. The cross-sections for collision of electrons with the atoms concerned can therefore be expected to be modified by the possibility of occurrence of the two processes

$$e + A \longrightarrow A^{-**},$$
$$A^{-**} \longrightarrow e + A,$$

where A represents the normal atom and A^{-**} a doubly-excited ion. Such processes will only be important when the electron energy is very close to that resulting from break up of the doubly-excited ion. However, if $\hbar/\Gamma$ is the mean lifetime, the uncertainty principle shows that the energy range over which the scattering will be influenced will be around Γ. Thus from the location of the peak effect in energy the energy of the doubly-excited ionic state may be determined, and from the energy range over which the effect is significant it is possible to determine the lifetime of the state.

Below the threshold for inelastic scattering, the elastic scattering differential and total cross-sections are given by

$$I(\theta)\,d\omega = |f(\theta)|^2\,d\omega$$
$$= \left| f_B(\theta) - \frac{e^{2i\eta}}{4ik} \frac{\Gamma}{E - E_r + \tfrac{1}{2}i\Gamma} (2L+1) P_L(\cos\theta) \right|^2, \tag{4.132}$$

$$Q = Q_B + Q_a(q_L^2 + 2q_L\epsilon - 1)/(1 + \epsilon^2), \tag{4.133}$$

where

$$q_L = -\cot\eta_L, \quad \epsilon = (E - E_r)/\tfrac{1}{2}\Gamma. \tag{4.134}$$

Here we have assumed that the orbital angular momentum quantum number for the doubly-excited state is L, that for the normal atom being taken as zero. f_B and Q_B are background values of the scattered amplitude and of the total scattering cross-section. E_r differs from the energy of the electron ejected by break up of the complex by a small level shift ΔE which is usually smaller than Γ. The η_l are the phase shifts which would arise in the absence of the doubly-excited state.

If the energy resolution of the apparatus is poor, then the average value of the total cross-section as a function of electron energy passing through the resonance is given by

$$\overline{Q}_{\text{res}} = \overline{Q}_B + \frac{2\pi}{k^2} \Gamma(2L+1) \cos 2\eta_L. \qquad (4.135)$$

From these relations it is possible to derive E_r and L for a doubly-excited state producing a resonance effect. Examples are discussed in Chapter 5.

Doubly-excited states of negative ions based on the nth excited states of neutral atoms will, in addition, produce qualitatively similar resonance effects in the cross-sections for excitation of all lower atomic states (see pp. 135-42).

4.4 One-body or shape resonances

The resonance states we have been discussing which decay by autodetachment depend essentially on correlated effects involving two or more electrons. It is possible, however, for resonance states to occur for the case of an electron moving in a central field of force although in general the lifetimes of these states are much shorter.

Consider the case of an electron moving under the influence of a potential of the form shown in Fig. 4.3 consisting of an inner attractive potential well surrounded by a potential barrier which vanishes as the distance r from the centre tends to infinity. For the s waves moving in such a field the wave equation (4.11) applies. This is formally the same as for the one-dimensional motion of a particle in the x-direction under the action of the potential $V(x)$ but there is the special condition that u must vanish at $r = 0$. We now follow a discussion due to Herzenberg (1973).

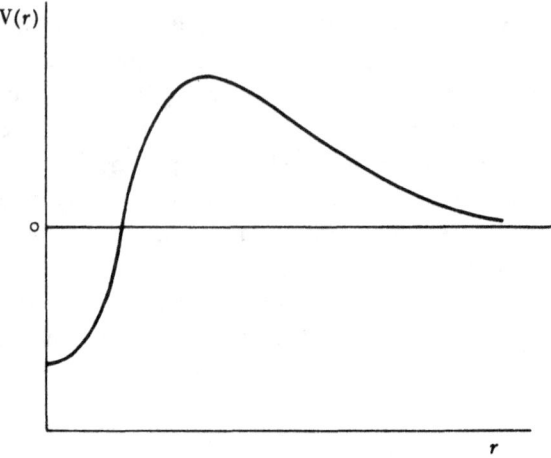

Fig. 4.3.

Consider a wave $\exp(-ikr)$ of wave number k impinging on the barrier from outside. The wave penetrating to the inner well will be $-T\exp(-ikr)$, where T is the amplitude transmission coefficient through the barrier which will in general be complex to allow for phase changes. Because $u = 0$ at $r = 0$ this wave must be totally reflected there so that the outgoing wave $T\exp(ikr)$ exactly cancels the incoming at $r = 0$. As this outgoing wave reaches the barrier, partial reflection takes place with a reflection coefficient R such that $R^2 + T^2 = 1$. This gives rise to an ingoing wave

$$R\,e^{2ika}(-T\,e^{ikr}), \tag{4.136}$$

the factor $\exp(2ika)$ taking account of phase change involved in traversing the well. This wave will in turn be totally reflected at $r = 0$, then partially at the barrier and so on. The total wave within the barrier will therefore be given by

$$(e^{-ikr} - e^{ikr})T\{1 + R\,e^{2ika} + (R\,e^{2ika})^2 + \ldots\}$$
$$= -2i\sin kr\,T/(1 - R\,e^{2ika}). \tag{4.137}$$

If the barrier is strong, $|R| \simeq 1$, $|T| \ll 1$, and the amplitude within the barrier is generally of order $|T|$ and so will be much smaller than that outside. However, if k is such that

$$R\,e^{2ika} = |R|, \tag{4.138}$$

ATOMIC NEGATIVE IONS – EXCITED STATES

the amplitude inside the well is given by

$$\frac{T}{1-|R|} \simeq \frac{T}{\frac{1}{2}(1-|R|^2)}. \qquad (4.139)$$

But $|R|^2 + |T|^2 = 1$, so the amplitude inside the well becomes approximately $2/T^*$, the magnitude of which is $\gg 1$. This is because, when the condition (4.138) is satisfied, the multiply-reflected waves are in phase with each other and build up a large resultant amplitude. The conditions in fact are just those which we expect to associate with a resonant state.

To pursue this a little further in terms of observable scattering we can regard the scattered wave as arising from two sources, (a) the initial reflection on impact on the barrier and (b) the penetration through the barrier of the outgoing wave within the well. The first of these gives rise to a scattered amplitude of the same order as that of the incident wave since the magnitude of the reflection coefficient is near unity. The second contributes an amplitude of order

$$T^*T/\{1 - R\exp(2ika)\}, \qquad (4.140)$$

where T^* is the amplitude transmission coefficient from left to right. This will be of order $|T|^2$ and $\ll 1$ unless the build-up resonance condition (4.138) applies, in which case it becomes of order $|T|^2/|T|^2 = 1$, comparable with the directly-scattered component.

Moreover, we may expand the denominator about a resonance energy E_r in the form

$$1 - R\exp(2ika) \simeq 1 - R\exp(2ik_r a) + 2iaR\exp(2ik_r a)\left(\frac{\partial k}{\partial E}\right)_{k_r} \times$$
$$\times (E - E_r) + ..., \qquad (4.141)$$

where $k_r^2 = 2mE_r/\hbar^2$. Since (4.138) is satisfied at $E = E_r$ the right-hand side of (4.141) may be written

$$1 - |R| - 2i|R|a\left(\frac{\partial k}{\partial E}\right)_{k_r}(E - E_r) + ... \qquad (4.142)$$

Further, as

$$1 - |R| = (1 - |R|^2)/(1 + |R|)$$
$$\simeq \tfrac{1}{2}|T|^2$$

and $|R| \simeq 1$, the amplitude (4.140) can be written to a good approximation as

$$|T|^2 / \left\{ \tfrac{1}{2}|T|^2 - 2\mathrm{i}a \left(\frac{\partial k}{\partial E}\right)_{k_\mathrm{r}} (E - E_\mathrm{r}) \right\}, \qquad (4.143)$$

or, writing

$$\Gamma/\hbar = |T|^2 / 2a \left(\frac{\partial k}{\partial E}\right)_{k_\mathrm{r}}, \qquad (4.144)$$

it becomes

$$\mathrm{i}\Gamma / \{E - E_\mathrm{r} + \tfrac{1}{2}\mathrm{i}\Gamma\}. \qquad (4.145)$$

This is of the same form as for resonance scattering by an autodetaching level. Thus the contribution to the scattering cross-section will be

$$\frac{\pi}{k^2} \frac{\Gamma^2}{(E - E_\mathrm{r})^2 + \tfrac{1}{4}\Gamma^2}, \qquad (4.146)$$

which is of the same form as (4.110).

$\hbar/\Gamma$ can thus be interpreted as the lifetime of the resonance state produced through trapping by the potential barrier. Since

$$\frac{\partial k}{\partial E} = \frac{\hbar}{v},$$

where v is the velocity of the incident particle, we have for the lifetime

$$\frac{\hbar}{\Gamma} = \frac{2a}{v} \frac{1}{|T|^2}. \qquad (4.147)$$

The delay time due to the trapping is therefore increased by a factor $1/|T|^2$ above the time $2a/v$ required to make a double traverse of the potential well.

Resonances of this kind are known as *shape resonances*. While an interaction of the form shown in Fig. 4.3 does not arise in electron–atom collisions, for collisions in which the electron has a finite angular momentum $\{l(l+1)\}^{1/2}\hbar$ about the centre a barrier is present due to centrifugal force which is equivalent to a repulsive potential $\{\hbar^2/2m\}\{l(l+1)/r^2\}$. Shape resonances may arise due to this barrier but these will not be observed for s electrons.

In most cases of atomic interactions, either with atoms in their ground or excited states, the barrier due to centrifugal force will not exceed a few eV so that any shape resonance will lie within a few eV above the atomic state concerned. Furthermore, the width Γ will usually be of the same order unless the resonance lies very close to this atomic state so that the barrier is thick and $|T|^2 \ll 1$.

While there are no cases known in which a narrow shape resonance occurs just above the ground state of a neutral atom, such resonant states play an important part in low-energy collision phenomena associated with molecules, such as N_2 and CO (see Chapter 6, p. 194). Cases of shape resonances associated with excited atomic states are known.

Because of their relatively large width it is often difficult to distinguish effects due to shape resonances from other causes of broad extrema in observed cross-sections. Thus we may write for the partial cross-section given in (4.36)

$$Q^{(0)} = 4\pi/|ik - M|^2, \qquad (4.148)$$

where

$$M = k \cot \eta_0. \qquad (4.149)$$

When η_0 passes through an odd multiple of $\pi/2$, at say $E = E_0$, $M = 0$ so that, for E near E_0,

$$M \simeq (E - E_0) \left(\frac{\partial M}{\partial E}\right)_{E_0}.$$

Hence, writing

$$\left(\frac{\partial M}{\partial E}\right)_{E_0} = \frac{2k_0}{\Gamma}, \qquad (4.150)$$

we have, at energies near E_0,

$$Q^{(0)} = \frac{\pi}{k^2} \frac{\Gamma^2}{(E - E_0)^2 + \frac{1}{4}\Gamma^2},$$

which is again of the resonance form but, as may be shown from the detailed analysis of the behaviour of the phase shift in the complex E plane, does not mean that a resonance state exists at

$E = E_0$. On the other hand, since

$$\left(\frac{\partial M}{\partial E}\right)_{E_0} = \left[\left(k\cosec^2\eta_0 \frac{\partial \eta_0}{\partial k} + \cot\eta_0\right)\frac{\partial k}{\partial E}\right]_{\eta_0 = \frac{1}{2}\pi}$$

$$= k\left(\frac{\partial \eta_0}{\partial k}\right)_{k_0} \frac{\partial k}{\partial E}$$

$$= (m/\hbar^2)\left(\frac{\partial \eta_0}{\partial k}\right)_{k_0},$$

we have that

$$\frac{\hbar}{\Gamma} = \frac{m}{2k_0\hbar}\frac{\partial \eta_0}{\partial k}.$$

Referring to (4.42) we see that the right-hand side is the time delay τ associated with the scattering at the energy at which the partial cross-section is a maximum. Under these conditions we have, in general, a pseudo-resonance of width $\hbar/\tau$ at this energy, which will lead to enhancement of certain observable cross-sections.

The stabilization method described on p. 95 for the determination of the energies of autodetaching resonance states may be applied to determine the energies of shape resonances although, because of the diffuse nature of the resonance in these cases, much more care must be taken in choosing the basic trial function Φ and the series of functions χ_i. As an illustration and check of the method, Hazi and Taylor (1970) have considered the one-dimensional motion of a particle in a field of potential $V(x)$ given by

$$V(x) = \tfrac{1}{2}x^2, \quad x \leqslant 0, \tag{4.151}$$

$$= \tfrac{1}{2}x^2 \exp(-\lambda x^2), \quad x \geqslant 0. \tag{4.152}$$

which has the shape shown in Fig. 4.4.

The functions χ_i were taken to be the eigenfunctions of the harmonic oscillator which are exact solutions when $\lambda = 0$. Thus

$$\chi_i(x) = (2^i i! \pi^{1/2})^{-1/2} H_i(x)\exp(-\tfrac{1}{2}x^2), \quad i = 0, 1, 2, \ldots,$$

where H_i is the ith Hermite polynomial. Φ was taken to be equal to χ_0.

Following the procedure outlined on p. 95, the eigenvalues corresponding to the trial functions $\sum_0^n c_i\chi_i(x)$ were calculated

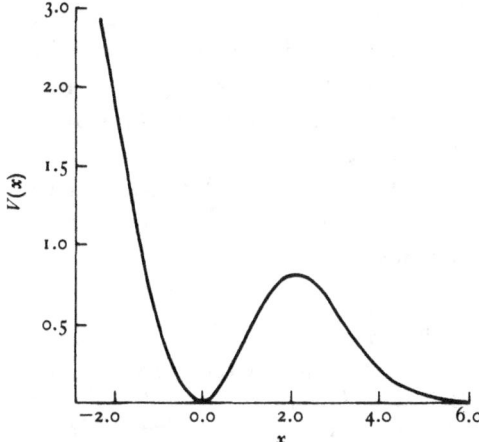

Fig. 4.4. The model potential used by Hazi and Taylor in their model calculations on the effectiveness of the stabilization method for determining the energies of shape resonances.

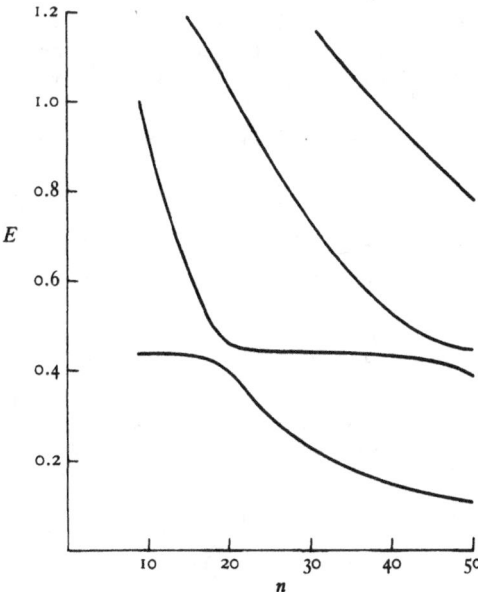

Fig. 4.5. The four lowest eigenvalues in the model calculations of Hazi and Taylor, considered as functions of the number n of basis functions included.

as functions of n. Fig. 4.5 shows the results for the four lowest eigenvalues for $\lambda = 0.225$. The stable region extending from $n = 10$ to an eigenvalue close to $k^2 = 0.8826$ is apparent. The curve crossing discontinuity at $n = 20$, when the stable region jumps from the lowest to the second lowest root, is not important. Comparison with accurate numerical solutions of the wave equation shows complete agreement, to the accuracy given, for the energy of the shape resonance.

4.5 Siegert states

A further useful way of regarding resonances states is in terms of a definition due to Siegert (1939). We shall describe this in terms of a one-particle model but it can readily be generalized to apply to the many-particle case.

The radial wave equation for motion of a particle of mass m about a centre of force of potential $V(r)$ with zero angular momentum is

$$\frac{d^2 f_E}{dr^2} + \frac{2m}{\hbar^2}(E - V)f_E = 0, \qquad (4.153)$$

where $f_E = r\psi, = 0$ for $r = 0$.

If the potential $V(r)$ vanishes for $r > r_0$, then the solution representing the zero angular momentum component of the incident plane wave of unit amplitude, and the corresponding scattered wave, will be, for $r > r_0$,

$$f_E = k^{-1}\sin kr + S\exp(ikr), \qquad (4.154)$$

where $k = (2mE/\hbar^2)^{1/2}$. The partial scattering cross-section is then $4\pi|S|^2$.

Siegert now considers S as a function of the energy E, including complex values of E. Because of the continuity of f and df/dr $(=f')$ at $r = r_0$ we have

$$S = \{f_E(r_0)\cos kr_0 - k^{-1}f'_E(r_0)\sin kr_0\}\exp(-ikr_0)\{f'_E(r_0) - ikf_E(r_0)\}^{-1}. \qquad (4.155)$$

The energies at which the denominator vanishes are determined by the eigenvalues of the equation

$$\frac{d^2 f_n}{dr^2} + \frac{2m}{\hbar^2}(W_n - V)f_n = 0, \qquad (4.156)$$

with $f_n = 0$, $r = 0$ and $f'(r_0) - ik_n f(r_0) = 0$ at $r = r_0$, k_n being given by

$$k_n^2 = 2mW_n/\hbar^2.$$

The value of S in the neighbourhood of a singular energy W_n can be obtained as follows. Multiplying (4.153) by f_n and (4.156) by f_E and subtracting gives

$$\left(\frac{d^2 f_n}{dr^2} f_E - \frac{d^2 f_E}{dr^2} f_n\right) + (k_n^2 - k^2) f_n f_E = 0. \quad (4.157)$$

Integrating between $r = 0$ and r_0 and using the boundary conditions gives

$$f_n(r_0)\{ik_n f_E(r_0) - f_E'(r_0)\} + (k_n^2 - k^2) \int_0^{r_0} f_n f_E \, dr = 0. \quad (4.158)$$

The denominator of (4.155) then becomes

$$f_E'(r_0) - ikf_E(r_0) = \{(k_n^2 - k^2)/f_n(r_0)\}\left\{\int_0^{r_0} f_n f_E \, dr + i\frac{f_n(r_0) f_E(r_0)}{k_n + k}\right\}$$

$$\rightarrow \{(k_n^2 - k^2)/f_n(r_0)\}\left\{\int_0^{r_0} f_n^2 \, dr + if_n^2(r_0)/2k_n\right\}. \quad (4.159)$$

as $E \rightarrow W_n$.

Similarly, for the numerator of (4.155)

$$f_E(r_0) \cos kr_0 - k^{-1} f_E'(r_0) \sin kr_0 \longrightarrow f_n(r_0) \exp(-ik_n r_0). \quad (4.160)$$

We now have

$$S \longrightarrow (k_n^2 - k^2)^{-1} \frac{f_n^2(r_0) \exp(-2ik_n r_0)}{\int_0^{r_0} f_n^2 \, dr + i\frac{f_n^2(r_0)}{2k_n}} + g(E), \quad (4.161)$$

where $g(E)$ is a regular function in the neighbourhood of W_n. This expression may be recast in the form (4.98) provided E is close to W_n. Thus, writing

$$f_n = |f_n(r)| \exp\{i\delta(r)\} \quad (4.162)$$

and supposing that δ varies very little from a mean value $\bar{\delta}$ over the range of r which contributes appreciably to the integral in the denominator, we have

$$S = (k_n^2 - k^2)^{-1} \frac{|f_n(r_0)|^2 \exp(2i\eta)}{\int_0^{r_0} |f_n(r)|^2 \, dr + \frac{i}{2k_n}|f_n(r_0)|^2 \exp\{2i(\delta(r_0) - \bar{\delta})\}} + g(E),$$

$$(4.163)$$

where

$$\eta = \delta(r_0) - k_n(r_0) - \bar{\delta}.$$

If we now write

$$W_n = E_n - \tfrac{1}{2}i\Gamma_n,$$

where E_n and Γ_n are real, then it follows from (4.156) and its complex conjugate that

$$\frac{\hbar^2}{2m}|f_n(r_0)|^2 = \frac{\Gamma_n}{k_n + k_n^*}\int_0^{r_0}|f_n(r)|^2\,dr, \qquad (4.164)$$

so we obtain

$$S = \frac{\exp(2i\eta)\,\Gamma_n}{E_n - \tfrac{1}{2}i\Gamma_n - E}\left[k_n + k_n^* + \frac{i\Gamma_n m}{k_n \hbar^2}\exp\{i(\delta(r_0)-\bar\delta)\}\right]^{-1} + g(E). \qquad (4.165)$$

Since $|\Gamma_n|/E_n$ is taken to be $\ll 1$ we finally have

$$S = \frac{\exp(2i\eta)\,\Gamma_n}{E_n - E - \tfrac{1}{2}i\Gamma_n}\frac{1}{2k_n} + g(E), \qquad (4.166)$$

which is of exactly the same form as (4.98) with $g(E)$ representing the background scattering.

It may readily be shown that the result does not depend on the choice of r_0 provided $V(r) = 0$ for $r = r_0$. In practice, of course, $V(r)$ is usually finite for all finite r and the problem is to choose a value r_0 for which it can be assumed that $V(r)$ is sufficiently close to 0 for all greater values of r.

Herzenberg and Mandl (1963) used this definition, generalized to the many-particle case, in their pioneer study of the potential-energy curves for autodetaching molecular negative-ion states. For practical purposes the complex energy eigenvalues of the equation (4.156) may be determined approximately by a variational procedure (Herzenberg and Mandl, 1963). We define

$$D[f_n] = \int_0^a f_n(r)\,Hf_n(r)\,dr + \frac{\hbar^2}{2m}f_n(a)\left\{\frac{df_n}{dr} - ikf_n(r)\right\}_{r=a}, \qquad (4.167)$$

$$N[f_n] = \int_0^a f_n^2(r)\,dr, \qquad (4.168)$$

where $H = -(\hbar^2/2m)(d^2/dr^2) + V(r)$ and a is any value of r such that $V(r) = 0$, $r > a$. The variational principle depends on the condition

$$\delta(D_t/N_t) = 0$$

if D_t and N_t refer to (4.167) and (4.168) with a trial function f_n^t, containing adjustable parameters, substituted for f_n, and k_n is treated as fixed during

ATOMIC NEGATIVE IONS – EXCITED STATES 107

the variation. In contrast to the variation principle for bound states no minimum principle exists.

This variational method has been used by Bardsley, Herzenberg and Mandl (1966) to calculate approximately the real and imaginary parts of the energies of the two lowest states of H_2^- as functions of nuclear separation. Their results are described in Chapter 6, p. 176 (see Figs. 6.4 and 6.6).

4.6 Type I and type II resonance states

It will be convenient to distinguish henceforward between autodetaching doubly- or multiply-excited resonance states and one-body shape resonances. This distinction is particularly important in dealing with the states of molecular negative ions. The autodetaching states will be referred to as type I states and the shape resonances as type II.

In general, the former will be found quite close below excited states of the neutral atom or molecules and are usually of comparatively long lifetime (level widths > 0.1 eV), while the latter lie close above a ground or excited state of the neutral system and are of much shorter lifetime (level widths of order 1 eV) unless they possess very little energy in excess of the neutral state concerned.

4.7 Experimental study of autodetaching and shape resonance states from resonance effects in electron–atom collisions

Although it would be out of place here to give a comprehensive account of the experimental techniques involved in the study of autodetaching states through their effects in electron scattering by neutral atoms and molecules, we shall describe some of the typical experimental arrangements which have yielded a wealth of information about these states. For further details see Massey, Burhop and Gilbody (1969), Chapters 1–5.

The measurements which are made can be classified under the following headings:

(*a*) The transmission of a monochromatic beam of electrons through a gas as a function of electron energy.

(b) The total cross-sections for collisions of electrons in a gas, also as functions of electron energy.

(c) The energy distribution of electrons scattered at a fixed angle by a gas (or molecular beam) as a function of electron energy.

(d) The angular distribution of electrons scattered by a gas (or molecular beam), either elastically or with fixed energy loss, as a function of angle of scattering.

(e) The optical excitation functions of spectrum lines excited by electron impact. These are equivalent to measurement of cross-sections for excitation of specific excited states as functions of electron energy.

(f) Special methods are available in some cases for measurement of cross-section for excitation of metastable states as functions of electron energy.

(g) The ionization cross-sections of atoms or molecules as functions of electron energy.

Of these, most attention has been concentrated on (a), (c) and (d), and more recently on (e) and (g) though measurements of type (b) are important for absolute calibration of cross-sections. Usually measurements as in (c) and (d) can be carried out with the same equipment. For the detailed analysis of a resonance effect and hence for the determination of the lifetime and the quantum numbers of the autodetaching state as well as its energy, experiments which measure (c) and (d) are essential but, for a general survey of the location of resonances, transmission experiments (a) and (b) are very useful.

By measuring the intensity of light emitted in a particular transition (i → f) say due to impact of electrons, as a function of their kinetic energy, the variation of the cross-section for excitation of the ith state may be determined, at least in principle, as a function of energy. This may involve correction for cascade and other effects (Massey et al., 1969, Chapter 4) but if carried out successfully with high enough energy resolution, resonance effects due to autodetachment

$$A + e \longrightarrow A^{-*},$$
$$A^{-*} \longrightarrow A'' + e,$$

can be distinguished. This method is especially valuable for the study of higher autodetaching states.

Cross-sections for excitation of metastable atoms should, as functions of electron energy, also exhibit fine structure due to autodetaching states. While these cross-sections cannot be measured by the optical methods referred to under (*e*), it is possible in some cases to determine them by measuring the flux of metastable atoms produced by collision in the gas as a function of electron energy, these atoms being detected by the electron current produced on impact with a suitable surface (Massey *et al.*, 1969, Chapter 4, §2.1.2).

The interpretation of data on ionization cross-sections is complicated by the fact that there are two distinct sources of structure in the cross-section–energy curve. One of these is through autodetachment in which decay takes place with the emission of two electrons

$$A + e \longrightarrow A^{-*},$$
$$A^{-*} \longrightarrow A^{+} + 2e,$$

while the other involves excitation of an autoionizing state of the neutral atom

$$A + e \longrightarrow A' + e,$$
$$A' \longrightarrow A^{+} + e.$$

In all experiments it is essential that high energy resolution is available both for the incident and, for experiments under (*c*) and (*d*), the scattered electrons. As mentioned earlier, the full width at half-height of the energy distribution can be reduced to between 25 and 50 meV, for which purpose three different types of energy analyser have been mainly used – concentric hemispherical electrostatic, 127° cylindrical electrostatic, and trochoidal. In all cases it is important that the transmission of the monochromator, used as a detector, should be independent of electron energy.

Transmission experiments

Fig. 4.6 shows the experimental arrangement developed by Simpson (1964) which has been applied extensively. It employs identical hemispherical analysis for the incident and transmitted beams.

Electrons from the hot filament F are accelerated towards the

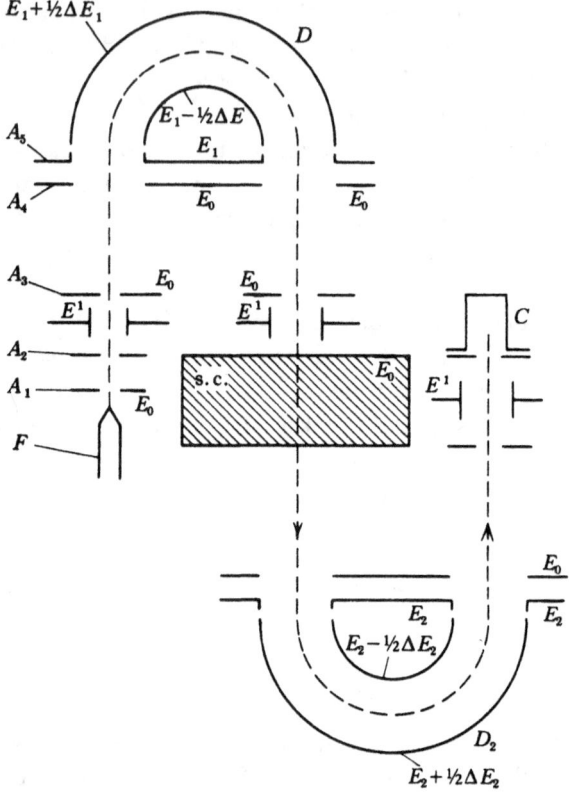

Fig. 4.6. Arrangement of apparatus used by Simpson to measure the transmission of electrons of homogeneous energy through gases.

anode A_1 to an energy E_0 and then pass through a series of electrodes A_3, A_4 and A_5 which focus the beam on to the entrance of the hemispherical analyser D after deceleration through A_5 which is maintained at a potential $E_1 < E_0$. By means of the voltage ΔE_1 between the hemispherical electrodes the beam is focused after deflexion through 180° and reaccelerated to the energy E_0 to pass through the scattering chamber. The transmitted electrons are analysed by passage through a similar analysing system to enter the collector C. The electrodes E^1 make it possible to centre the beam electrostatically.

With the background pressure of 10^{-9} torr the full energy spread at half-height is about 0.04 eV and the collected current 3×10^{-10} A.

Electrons which have been scattered through an angle greater than 0.02 rad or suffered energy loss greater than 0.05 eV are not collected. No attempt is made in this apparatus to measure the pressure in the scattering chamber but it was monitored by measuring the pressure in the vacuum envelope.

Fig. 4.7 illustrates the arrangement used by Schulz and his collaborators (Stamatovic and Schulz, 1970) in which the electrons are rendered homogeneous in energy by means of a trochoidal monochromator. Unlike earlier methods, it measures directly not the transmitted current but the derivative of this current with respect to electron energy.

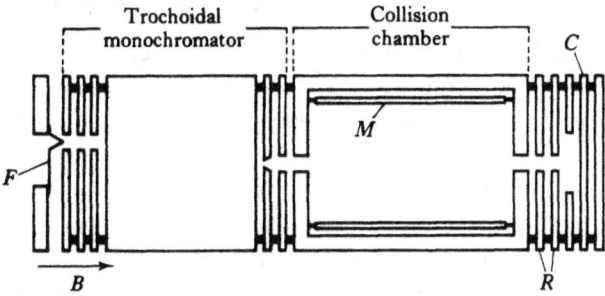

Fig. 4.7. Arrangement of apparatus used by Schulz and his collaborators to measure the rate of change of the transmission of electrons through gases, with electron energy.

Electrons from a thoria-coated iridium filament F, after alignment by an axial magnetic field of 130 G and passage through the trochoidal monochromator, emerge as a beam of about 3×10^{-9} A with an energy spread, at half-height, between 25 and 40 meV. The electrons are then accelerated to the desired energy to enter the collision chamber in which the gas under study is maintained at a pressure of about 0.01 torr. Electrons reaching the exit aperture of the collision chamber are decelerated to nearly zero energy by the retarding electrodes R before reaching the collector C. This is to prevent collection of scattered electrons which do not possess a sufficient component of velocity in the axial direction to penetrate the steady field.

To obtain directly the energy derivative of the transmitted intensity, a sinusoidal voltage of variable amplitude (5–50 meV) is applied between the collision chamber and an insulated coaxial cylinder M enclosed within. The resulting modulation of the transmitted current signal is amplified and measured in phase with the modulating signal by a phase-sensitive detector. During each sweep in electron energy the modulating signal is kept constant so that the output signal is proportional to the energy derivative of the transmitted current.

This arrangement has the advantage of reducing the signal-to-noise ratio which is more than ten times as favourable as in the experiments carried out with the apparatus shown in Fig. 4.6.

Measurements of energy and angular distributions

Fig. 4.8 illustrates the arrangement developed by Ehrhardt and Willmann (1967) which has yielded excellent results. It employs 127° electrostatic analysers for energy selection and analysis. The target gas takes the form of a molecular beam in which the pressure is around 10^{-3} torr.

The electrons are emitted from a hot cathode of thoriated tungsten of hairpin shape so that it is effectively a point source across which the potential drop is small. The reflector and electrode 1 focus the electrons on electrode 2 so that a beam of about 10^{-6} A of 1 eV energy is injected with small angular divergence into the selector. Electrodes $1a$ and $1b$ are deflector plates for adjustment.

After passage through the selector the electrons are accelerated, or decelerated, to the desired energy and focused on the molecular beam by the lens electrodes 5 and 6. Electrons scattered through a chosen angle are accelerated, or decelerated, to 1 eV energy by the corresponding lens electrodes 8 and 9 and focused on to the entrance aperture of the analyser. The current of transmitted electrons is detected by a multiplier, the output pulses of which are amplified and, after removal of background, are shaped and recorded electronically. Count rates range from 20 to 5×10^4 s^{-1}.

The selector electrodes of the electrostatic analysers are made from woven mesh with 90% transmission. To reduce the concentration of stray electrons, and hence the space charge, between the electrodes, two concentric solid plate electrodes are included outside the mesh. These are covered with porous gold black and held at a positive potential of 10 V with respect to the neighbouring electrodes. To correct for fringing fields the actual deflection angle in the analysers is 119° instead of 127°.

It is important to ensure that the transmission is independent of electron energy. Arrangements were made to vary the potentials V_4, V_5 on the electrodes 4 and 5 and V_8, V_9 with electron energy V_e so that

$$V_k = V_{0k} + C_k V_e, \quad k = 4, 5, 8, 9,$$

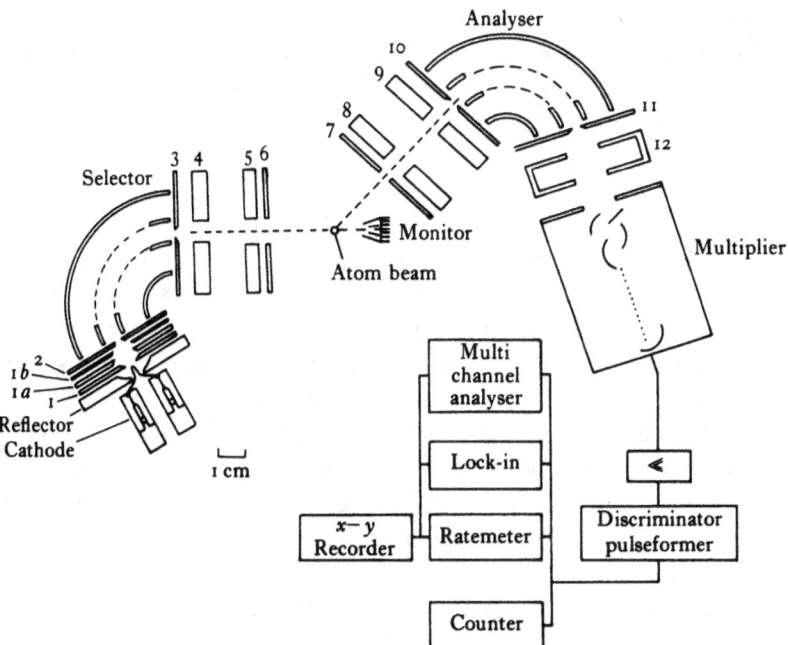

Fig. 4.8. Arrangement of apparatus used by Ehrhardt and his collaborators for measurements of elastic and inelastic differential cross-sections for scattering of electrons of homogeneous energy in gases.

where V_{0k} and C_k are constants determined experimentally so that the beam is focused on the scattering centre. As a check on the constancy of the transmission the elastic differential cross-section was measured as a function of electron energy and compared with previous measurements.

The magnetic field near the scattering centre is compensated to about 2 mG. All electrodes are gold-plated and before operation are baked to 400 °C.

The full width at half-height of the energy distribution of the primary beam is about 50 meV, the angular resolution about 2°. By rotating the analyser system, measurements can be made at angles of scattering ranging from 0 to 110°.

From energy distributions of detached electrons

This technique is applicable for negative ions A^- with stable ground states. If a beam of such ions with a few keV energy is fired through a gas such as helium, electrons will be detached both directly and through excitation of an intermediate autodetaching excited state

$$A^- + B \longrightarrow A^{-*} + B,$$
$$A^{-*} \longrightarrow A + e.$$

The kinetic energy of the electron produced in such a two-stage process relative to the A atoms gives directly the energy of the resonant state of A^- concerned.

In applying this technique, measurements are made of the energy distribution of electrons ejected in a chosen direction relative to the incident beam. Correction must be made for the kinetic energy of the ions. Some details of the method are given in Chapter 4, p. 119, including a description of the procedures adopted to calibrate the energy scale in the experiments.

Measurement of optical excitation functions

In experiments specially designed to observe optical excitation functions with high energy resolution, Heddle, Keesing and Kurepa (1973) used a hemispherical analyser for the electron beam, of a type very similar to that illustrated schematically in Fig. 4.3. The energy resolution of the emerging beam was between 20 and 33 mV. It was fired through helium at a pressure between 1 and 3×10^{-3} torr and the light emitted analysed by a monochromator and detected by a photomultiplier operated as a photon counter. The signal rates were from 1 to 10 s^{-1} as compared with a dark rate of 1–2 s^{-1} and data were accumulated over many hours. The overall energy resolution was 35–45 mV which is to be compared with the energy spread, 30 mV, due to thermal energy of the helium atoms.

Results obtained in these experiments are illustrated in Figs. 5.13 and 5.14.

4.8 Experimental study of autodetaching states through photodetachment

Observations of the cross-section for photodetachment from a negative ion as a function of frequency made with high frequency resolution will reveal the presence of autodetaching states as discussed on p. 92. Experimental techniques of this kind are described in Chapter 11.

4.9 Doubly-excited states which decay through emission of radiation – change of parity

It was pointed out on p. 80 that, within the accuracy of LS coupling, autodetachment involving change of multiplicity is relatively improbable. Autodetachment is also forbidden if a change of parity† is involved but in such cases an allowed radiative transition, which requires a parity change, will usually be possible, so that the doubly, or multiply, excited state concerned will only have a lifetime of 10^{-7}–10^{-8} s.

An example is the $(2p)^{2\,3}P_e$ state of H^-. Decay by autodetachment to the ground 1s state of H requires that the released electron be in a p state. Since the former is of even and the latter odd parity, the overall parity will be odd so the transition is forbidden. Nevertheless, allowed radiative transitions can occur from the state as described in Chapter 5, p. 120. A further example is the $1s2p^{2\,2}P_e$ state of He^-.

---

† The parity is even or odd according as the wave function for the state remains unchanged or changes in sign on reflection about the origin.

CHAPTER 5

Autodetaching States of Specific Atomic Negative Ions

5.1 H⁻

The degeneracy of the angular momentum states with the same total quantum number n for a hydrogen atom introduces some special features into the system of doubly-excited states of H⁻ as compared with, say, that for the isoelectronic He atom. Because of the coupling between the degenerate states, the effective interaction acting on the attached electron falls as slowly as r^{-2} at large r (Gailitis and Damburg, 1963). The energy-level system would then consist of an infinite set of doubly-excited states associated with each value of n, the total quantum number of the excited atomic state, but only for certain values of the total angular momentum quantum number. Thus we have such infinite series of states for

$$n = 2, \quad L = 0, 1, 2,$$
$$n = 3, \quad L = 0, 1, 2, 3, 4,$$
$$n = 4, \quad L = 0, 1, 2, 3, 4, 5, 6,$$

and so on.

The energies of the different levels below the ionization limit to which they tend form a geometrical progression so that if $E_{nL}^{(1)}$ is the energy of the lowest state of a series associated with particular values of nL, the energies of the higher levels will be $E_{nL}^{(1)}(\alpha, \alpha^2, \alpha^3, ...)$, where $\alpha < 1$.

This system is modified in practice by the Lamb shift which removes much of the degeneracy between levels of the same total quantum number. Because of this, a particular series of terms nL will terminate beyond a level $E_{nL}^{(s)}$ for which

$$E_{nL}^{(s)} \geqslant \tfrac{1}{2}(L+1)\Delta\epsilon,$$

where $\Delta\epsilon$ is the appropriate energy separation produced by the Lamb shift (about 4×10^{-6} eV for $n = 2$).

In Table 5.1, calculated values of the energies of a number of doubly-excited levels based on the excited states of H with $n = 2$

are given. These include results obtained by the projection method (O'Malley and Geltman, 1964) which do not allow for the level shift ΔE_r so that they give $E_r - \Delta E$, as well as results obtained by use of the 1s-2s-2p eigenfunction expansion with inclusion of correlation (Burke and Taylor, 1966) and by a variation method appropriate to continuum states (Seiler, Oberoi and Callaway, 1970) which gives E_r and also the level width Γ directly. There is quite good agreement between the theoretical values even for Γ, and this agreement extends also to the observed results which are given in Table 5.1.

The application of electron scattering techniques to atomic hydrogen encounters difficulties due to the instability of the atomic species. The first evidence of fine structure, near 9.7 eV, was observed by Schulz (1963) in the electron current transmitted through partially dissociated hydrogen. This was followed by the observations of Kleinpoppen and Raible (1965), who used a modulated crossed-beam technique to monitor the variation with electron energy of the intensity of scattering through 94°, and also detected structure near 9.7 eV.

A much more detailed experiment was carried out a little later by McGowan, Clarke and Curley (1965), who observed the intensity of scattering of electrons normal to the plane of intersection of an electron beam with a highly-dissociated hydrogen beam. They used a 127° electrostatic selector so that their energy resolution was quite good.

Finally, Sanche and Burrow (1972), using the transmission technique described on p. 111, measured the rate of change of transmitted current with energy with high resolution, obtaining the results shown in Fig. 5.1(a). To analyse these data, to obtain resonance energies and level widths, the shape of the curve was taken to be of the theoretical form but with the energies and level widths as adjustable parameters. Phase shifts associated with the non-resonant background scattering (see p. 87) were taken to have the theoretical values obtained by Burke and Taylor (1966) with the 1s-2s-2p close-coupling approximation plus correlation. Allowance was made for the electron energy distribution which was assumed to be of Gaussian form with a width at half-height of 70 meV. Doppler broadening due to thermal motion of the atoms

TABLE 5.1 Energies and level widths in eV of autodetaching states of H^-

State	$E_r - \Delta E$ (calc.) (a)	E_r (calc.) (b)	(c)	Γ (calc.) (b)	(c)	E_r (obs.) (d)	(e)	(f)	Γ (obs.) (d)
^1S	9.559	9.560	9.574	0.0475	0.0544	9.56	9.558	9.61	0.043
	10.178	10.178	10.178	2.19×10^{-3}	2.31×10^{-3}				
	—	—	10.203	—	1.3×10^{-4}				
^3S	10.149	10.150	10.151	2.06×10^{-5}	1.90×10^{-5}				
	10.202	—	10.201	—	1.36×10^{-6}				
	10.204	—	—	—	8.16×10^{-8}				
^1P	10.178	10.177	10.185	4.0×10^{-5}	2.42×10^{-5}				
	10.204	—	10.204	—	2.06×10^{-7}			(e)	(e)
^3P	9.727	9.740	9.768	5.94×10^{-3}	7.98×10^{-3} (d) 9.71 (g) 9.73	9.738	9.78	5.6×10^{-3}
	10.198	—	10.202	—	4.28×10^{-5}	(i)	(e)	(f)	(e)
^1D	—	10.125	10.160	8.8×10^{-3}	7.74×10^{-3}	10.13	10.128	10.20	7.3×10^{-3}

References

(a) O'Malley and Geltman (1964), (b) Burke and Taylor (1966), (c) Seiler, Oberoi and Callaway (1970), (d) McGowan (1967), (e) Sanche and Burrow (1972), (f) Edwards, Risley and Geballe (1971), (g) Kleinpoppen and Raible (1965), (h) Schulz (1963), (i) Ormonde et al. (1969).

was also allowed for. The points shown in Fig. 5.1(a) represent the best fit obtained in this way using the values of Γ and E_r given in Table 5.1.

No special difficulty is associated with H⁻ in experiments which observe the energy distribution of electrons detached from a beam of ions of some keV energy in passing through a gas.

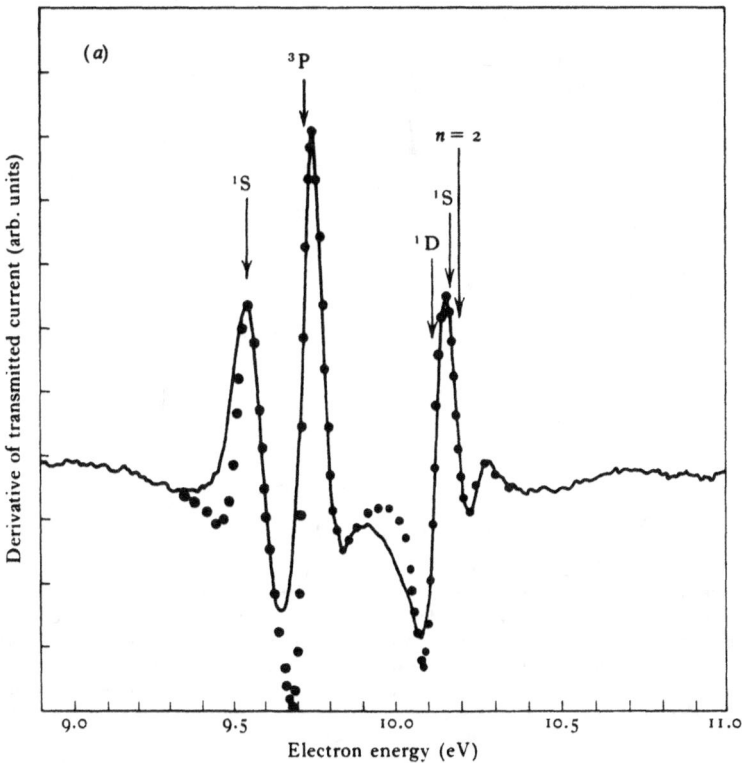

Fig. 5.1(a). For legend see p. 120.

Results obtained by Edwards *et al.* (1971) for electrons detached at an angle of 9° to the incident beam with energies in the range 28–31 eV by collision of H⁻ ions with kinetic energy 10 keV in an H_2–N_2 mixture are shown in Fig. 5.1(b).

It was also found that electrons of sharply defined energy arise from the charge transfer reaction

$$H^- + Ar \longrightarrow H + Ar^-,$$

which leads initially to the production of an autodetaching excited state of Ar⁻. As for the excited H⁻, this decays rapidly to produce the observed electrons. The energy of the most prominent excited state of Ar⁻ is 11.08 ± 0.05 eV (see p. 142). Using this as a calibration, and allowing for energy shift due to the kinetic energy of the primary ion, the lowest energy peak involves an excitation

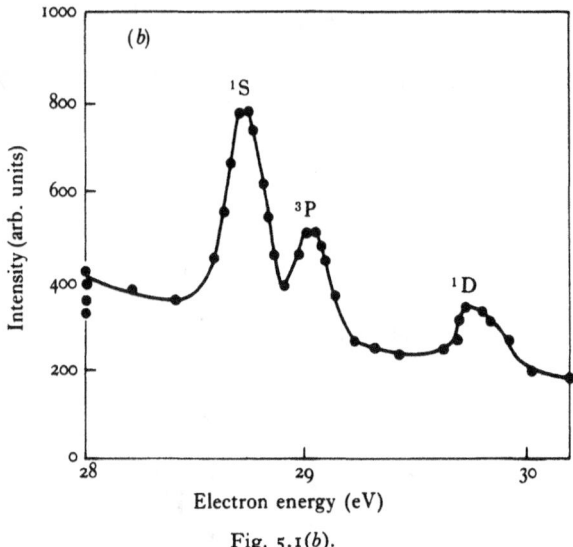

Fig. 5.1(b).

Fig. 5.1.(a) Energy derivative of the transmitted electron current as a function of electron energy in atomic hydrogen, observed by Sanche and Burrow (1972). —— observed, ● calculated, as described in text. The arrows indicate the location of the resonant peaks according to the calculations of Burke and Taylor (1966). (b) Energy distribution of electrons ejected at 9° to the direction of incidence of an H⁻ beam of 10 keV energy, passing through a mixture of H₂ and N₂. The correspondence between these peaks and those in (a) is indicated by the term designations of the autodetaching states involved.

energy of 9.611 ± 0.046 eV which agrees well with that for the H⁻(2s ²¹S) state. The higher energy peaks lie 0.1710 ± 0.0005 and 0.5869 ± 0.02 eV above the ¹S and correspond to the 2s2p ³P and 2p² ¹D states.

The $(2p)^{2\,3}P_e$ state

As explained in Chapter 4, p. 115, the $(2p)^{2\,3}P_e$ state of H⁻ is metastable towards autodetachment but it can decay by making allowed radiative transitions to the 2s2p ^{3}P^0 autodetaching state or to the 1skp ^{3}P^0 continuum. Elaborate calculations of the energy of the state have been made by Drake (1970) and by Bhatia (1970) using trial functions of the same general form as that used by Hart and Herzberg (1957) (see Chapter 1, (1.10)) but with the appropriate symmetry and angular dependence. With a 50-parameter function of this kind the energy of the state is found to be 0.0095 eV below the 2p level of H.

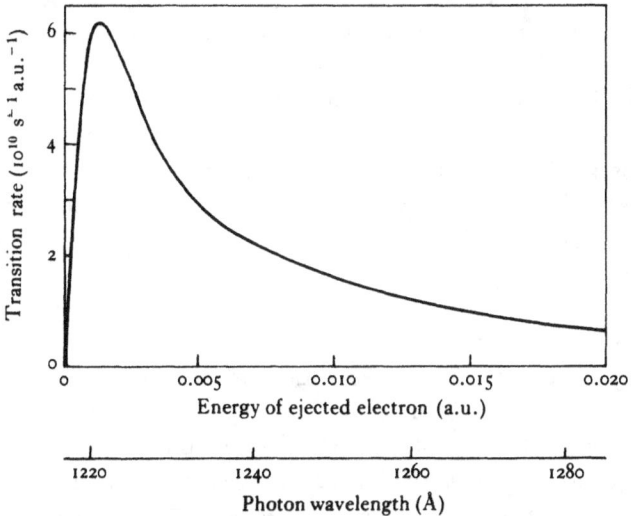

Fig. 5.2. Differential rate per unit energy range for transitions between the 2p^2 ^{3}P$_e$ state of H$^-$ and the 1skp ^{3}P^0 continuum, calculated by Drake (1973).

Drake (1973) has calculated the rate of radiative decay both to the autodetaching ^{3}P^0 state and to the continuum. For the former, a comparatively slow rate of 2.5×10^6 s^{-1} was obtained. Fig. 5.2 shows the transition rate per unit energy range of the ejected electron for transitions to the continuum, the energy being expressed in atomic units. This gives a total decay rate of 5.77×10^8 s^{-1}. It will be seen that the peak intensity of the emitted radiation lies close to 1222 Å. It will be difficult to detect it in, for example, solar radiation, because of the interference by the strong Ly α line of H at 1215.7 Å.

5.2 He⁻

He⁻ states below the excitation threshold (19.8 eV)

Attachment of an electron in a 2-quantum orbital to helium atoms in $1s2s\,^3S$ excited states will lead to He⁻ ions in $1s2s^2\,^2S$ and $1s2s2p\,^2P$ and ^4P states. Of these the first will be unstable towards autodetachment through Coulomb interactions as described on pp. 77–80. This will also be true for the ^2P state but, as pointed out on p. 80, the ^4P state is metastable towards autodetachment as this can only occur through spin–orbit and spin–spin interaction. He⁻ ions in this state will survive long enough to be observed directly in many experiments and indeed can be made use of in tandem accelerators (see Chapter 15, p. 690).

The ^2S state shows up strongly in resonance effects in electron scattering experiments and indeed this was the first such effect to be observed. We shall first describe the information available about this state both from experiment and theory and then discuss the metastable state. There is no definite evidence that the ^2P state exists.

The $1s2s\,^{2\,2}S$ state

Fig. 5.3 shows the first observation of the resonance due to this state. It takes the form of a sharp decrease in the intensity of electrons scattered elastically at 72° in helium by about 14% at an energy close to 19.4 eV. These first results, obtained by Schulz (1963),* were followed by a number of other observations which confirmed the existence of the resonance and provided precise data about its properties. Thus Simpson and Fano (1964) observed it in a transmission experiment (see p. 107) using the apparatus described on p. 109. The same apparatus was then employed by Kuyatt, Simpson and Mielczarek (1965) in a systematic study of other rare gases as well as helium. Golden and Bandel (1965) observed the resonance in measurements of the total cross-section for electrons in helium. Fig. 5.4 shows the appearance of the resonance in the later experiments of Sanche and Schulz (1972) in which the rate of change of the transmitted electron current with energy is measured

---

* Evidence of a resonance effect was also obtained independently by Fleming and Higginson (1963) in the course of a study of excitation cross-sections using the method of Maier-Leibnitz (1935).

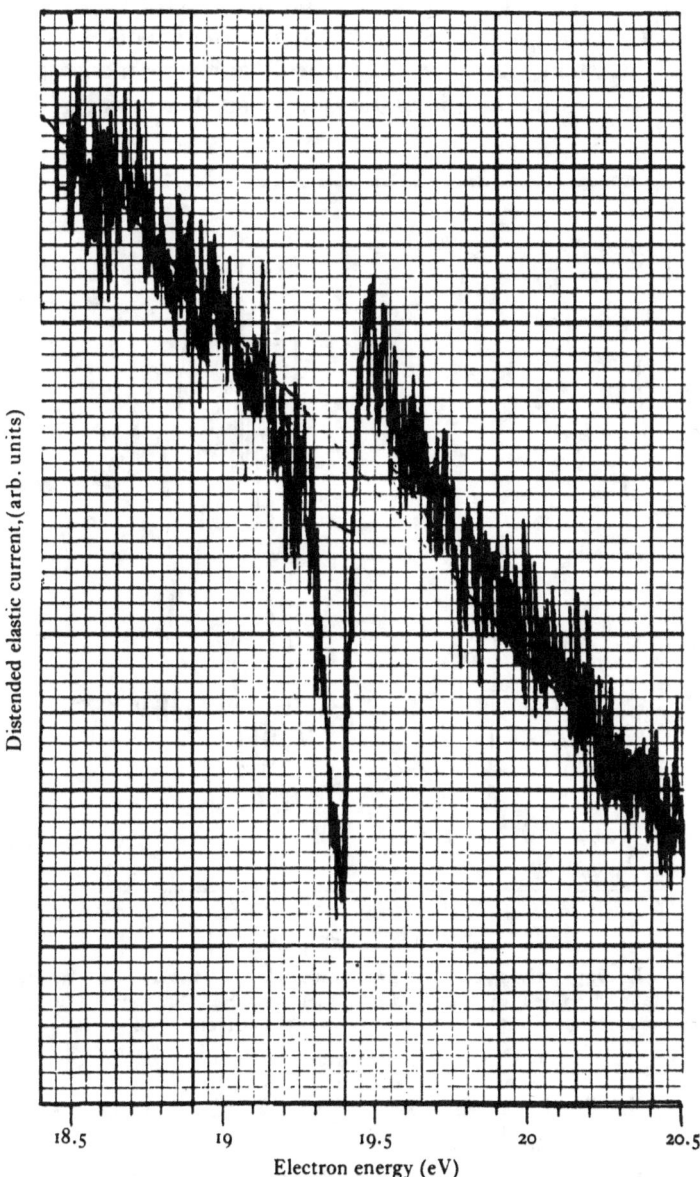

Fig. 5.3. The first evidence of the $1s2s^2\,^2S$ resonance in helium, observed by Schulz (1963) in the transmission of electrons.

directly (see p. 111 and Fig. 4.7). Table 5.2 lists the values for the resonance energy observed in different experiments which are almost consistent with 19.35 eV. The width of the resonance is more difficult to measure as it is comparable or smaller than that of the

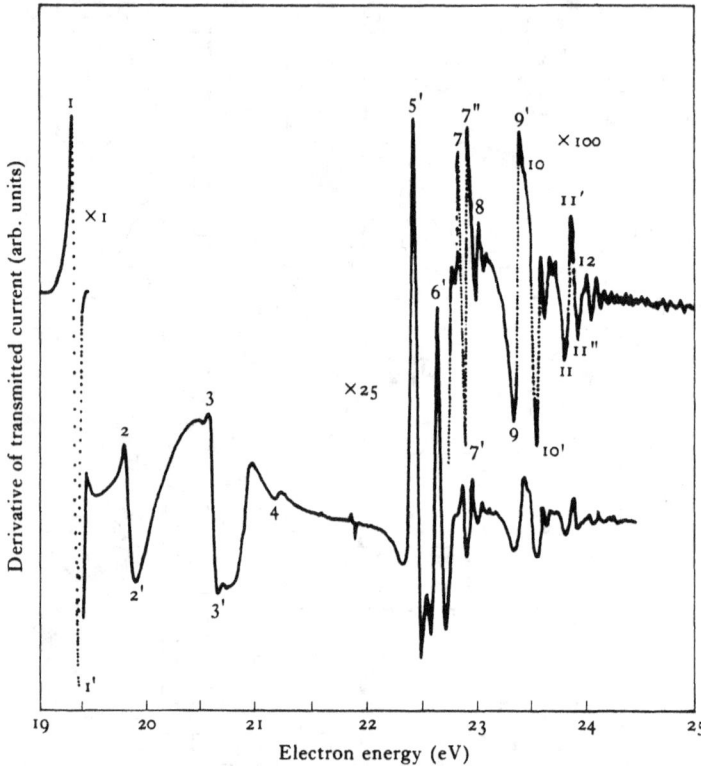

Fig. 5.4. Energy derivative of the transmitted electron current as a function of electron energy in helium, observed by Sanche and Schulz (1972). The 19.3 eV resonance corresponds to a change in transmitted current of about 10%.

energy distribution of the electrons used in the experiments. Some values are given in Table 5.2.

It is well established from the experiments that the resonant state is an S state. Thus Fig. 5.5. shows observed results obtained by Andrick and Ehrhardt (1966) at different angles of scattering using the apparatus described on p. 112. It will be seen that the resonance structure is clearly visible at 90°, at which angle it would vanish if it arose from capture of a p-electron. Detailed analysis in terms of the theory outlined on p. 87 shows that the resonance state may be interpreted as an S state with a width close to 8 meV. This analysis includes background phase shifts η_0, η_1 and η_2 which

TABLE 5.2 *Properties of the* $1s2s\,^2S$ *state of helium*

Method	Energy (eV)	Width (eV)
Observed results		
Transmission	19.31 ± 0.03(a)	–
	19.30 ± 0.01(b)	0.008
	19.34 ± 0.02(c)	–
Total cross-section	19.285 ± 0.025(d)	
Scattering at 72°	19.30 ± 0.05(e)	–
Scattering at 90°	19.3 ± 0.1(f)	
	19.35 ± 0.02(g)	
Differential cross-section	–	0.008(h)
	19.355 ± 0.08(i)	
Calculated values		
Variational	19.5(j)	–
	19.67(k)	
	19.4(l)	0.015(l)
Close coupling	19.33(m)	0.039(m)
Stabilization	19.3(n)	–
Quasi-projection	19.363(o)	0.014(o)
Bound-state	19.34	–

References
(a) Kuyatt et al. (1965), (b) Golden and Zecca (1970, 1971), (c) Sanche and Schulz (1972), (d) Golden and Bandel (1965), (e) Schulz (1963), (f) McGowan et al. (1965), (g) Mazeau, Grestau, Joyez, Reinhardt and Hall (1972), (h) Gibson and Dolder (1969), (i) Cvejanovic, Comer and Read (1973), (j) Kwok and Mandl (1965), (k) Young (1968), (l) Sinfailam and Nesbet (1972), (m) Burke and Taylor (1966), (n) Eliezer and Pan (1970), (o) Temkin, Bhatia and Bardsley (1972).

are those derived from analysis of observed scattering under conditions in which resonance effects are unappreciable (Gibson and Dolder, 1969).

Table 5.2 also lists theoretical values for the energy of the ^2S state calculated by the methods indicated. The agreement between these values and the observed results is very satisfactory and leaves little doubt of the validity of the identification of the state.

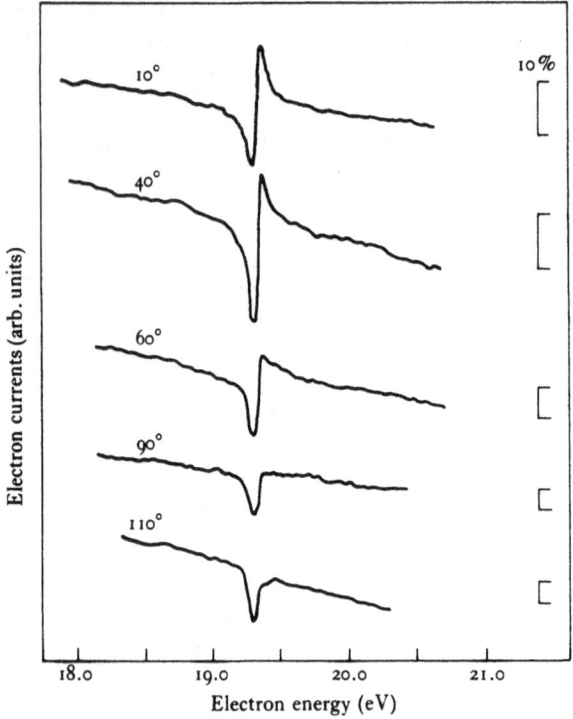

Fig. 5.5. Variation of the differential cross-section for elastic scattering of electrons in helium with electron energy at different scattering angles, observed by Andrick and Ehrhardt (1966).

The 1s2s2p ^{4}P state

The existence of an He$^-$ ion with a lifetime at least comparable with the time of passage through a mass spectrometer was first observed by Hiby (1939) although it was realized at the time that the ground $1s^2 2s\,^2S_{1/2}$ state of the ion was almost certainly not stable. A little later Ta-You Wu (1936) suggested that the observed ions were in the (1s2s2p)^{4}P state but it was some time before variational calculations were of sufficient accuracy (Holøien and Midtal, 1955) to establish that the energy of the state was less than that of the $2\,^3$S state of He. This being so ions in the ^{4}P state would be metastable towards autodetachment. Experimental evidence of a finite lifetime for the He$^-$ ions was first obtained by Riviere and Sweetman (1960). The first measurements of the lifetime (Nicholas, Trowbridge and Allen, 1968) did not give quantitative results because it was not realized that the different fine structure levels of the ^{4}P state would decay at different rates. In fact while the ^{4}P$_{1/2}$ and ^{4}P$_{3/2}$ levels can decay through spin–orbit and spin–spin interaction, the ^{4}P$_{5/2}$ can only do so through the latter, considerably weaker, interaction.

Later experiments have taken this into account and have measured the separate lifetimes and also the fine structure separations.

The energy of this metastable state has been determined directly from observations of the energy distribution of electrons produced by photodetachment from a beam of He$^-$ ions, the details of the experiment being described in Chapter 11, p. 450. In this way the state is found to lie 19.68 eV above the ground ^{1}S state of He, more than 0.3 eV above the 1s^2 2s^2 ^{2}S state. This is at first sight surprising because, of the terms arising from a given configuration, the one of highest multiplicity usually is the lowest. In seeking an explanation, account should be taken of the fact that attachment of a 2s electron to an He atom in the 1s 2s 2 ^{1}S excited state will also lead to an He atom in a 1s 2s^2 2 ^{2}S state, the energy of which will be close to that arising from attachment to He(2^{3}S). Interaction between these two states will be strong, tending to depress one and raise the other. The state at 19.35 eV therefore lies deeper than would otherwise be the case. No such effect arises for the 2^{4}P state. Some evidence of the excitation of a second 2^{2}S state, just above the threshold for excitation of He(2^{1}S) has been noted (see Fig. 5.4.).

The energy of the 2^{4}P state has been calculated by Weiss (1968), using a wave function of the type discussed in Chapter 2, p. 26. It involved the superposition of a number of configurations of the type (2.9b), covering 28 different choices of the quantum numbers n and l for the orbitals. He found a binding energy of 0.067 eV relative to the 2^{3}S state as compared with 0.080 eV observed.

The fine structure and the decay rate

Manson (1971) has discussed the fine structure theoretically. Departures from LS coupling will result in coupling between the ^{4}P$_{1/2}$ and ^{4}P$_{3/2}$ levels and the ^{2}P$_{1/2}$. As the latter decays rapidly through Coulomb interaction this coupling reduces the lifetime of the former levels. However, the ^{4}P$_{5/2}$ level remains uncoupled and as in any case it can only decay through the weak spin–spin interaction, its lifetime must be much longer than for the levels with $\mathcal{J} = \frac{1}{2}$ and $\frac{3}{2}$. Furthermore, because it is not affected by coupling with the doublet states, the calculation of the decay rate is relatively simple.

The most accurate calculations are those of Estberg and La Bahn (1970). They used the multiconfiguration wave function of Weiss (1968) for the ⁴P state.

The final state is a $^2F_{1/2}$ state in which an electron in a continuum f orbital moves in the field of a ground state helium atom. For this, a wave function calculated by collision theory which takes account of electron exchange and atom-distortion effects (Callaway, La Bahn, Pu and Duxler, 1968) was used. The resulting lifetime came out to be 4.55×10^{-4} s. Even a comparatively crude final state wave function ignoring atom distortion gave a value 5.44×10^{-4} s which did not differ greatly.

Calculation of the decay rates for the levels with $\mathcal{J} = \frac{1}{2}$ and $\frac{3}{2}$ is much more difficult because of the need to include coupling with the ²P state. Manson (1971), using effectively an HF wave function less accurate than those of Estberg and La Bahn (1970), obtained 10^{-3} s for $^4P_{5/2}$ and 0.33×10^{-4} s for $^4P_{3/2}$ but, owing to serious cancellation in the matrix elements concerned, was unable to make useful predictions for $^4P_{1/2}$.

Manson (1971) predicted that, for the fine structure levels, the energy would increase with decreasing $\mathcal{J}$ and that the energy excesses Δ_{53}, Δ_{51} of the $\frac{5}{2}$ level above the $\frac{3}{2}$ and $\frac{1}{2}$ respectively would be -0.068 and -0.314 cm^{-1} respectively.

In the presence of a uniform magnetic field, substates with the same value of M_J, the quantum number specifying the component of the total electronic angular momentum about the field direction, are mixed. This is mainly important between the $\mathcal{J} = \frac{5}{2}$ and $\frac{3}{2}$ substates because they are much closer in energy than either is to the $\frac{1}{2}$ substate. Thus, in the presence of a field H the wavefunction $|\frac{5}{2}, M\rangle_0$ for a sublevel $\frac{5}{2}$, M of the substate $\mathcal{J} = \frac{5}{2}$ changes to the linear combination

$$|\tfrac{5}{2}, M\rangle_H = \cos\theta_M |\tfrac{5}{2}, M\rangle_0 + \sin\theta_M |\tfrac{3}{2}, M\rangle_0, \qquad (5.1)$$

where θ_M is a function of $\mu_0 H/\Delta_{53}$, μ_0 being the Bohr magneton and Δ_{53} the energy separation of the $\mathcal{J} = \frac{5}{2}$ and $\frac{3}{2}$ substates. For small fields $\sin^2\theta_M$ is proportional to H^2. The $(\tfrac{5}{2}, \pm\tfrac{5}{2})$ sublevels are not affected.

Because of the mixture, the decay rate of the $(\tfrac{5}{2}, M)$ sublevel becomes

$$\Gamma_M = \cos^2\theta_M \gamma_{5/2} + \sin^2\theta_M \gamma_{3/2}, \qquad (5.2)$$

where $\gamma_{5/2}$ and $\gamma_{3/2}$ are the decay rates for the $\mathcal{J} = \tfrac{5}{2}$ and $\tfrac{3}{2}$ substates in the absence of the field. Since $\gamma_{3/2} \gg \gamma_{5/2}$ this means that for all

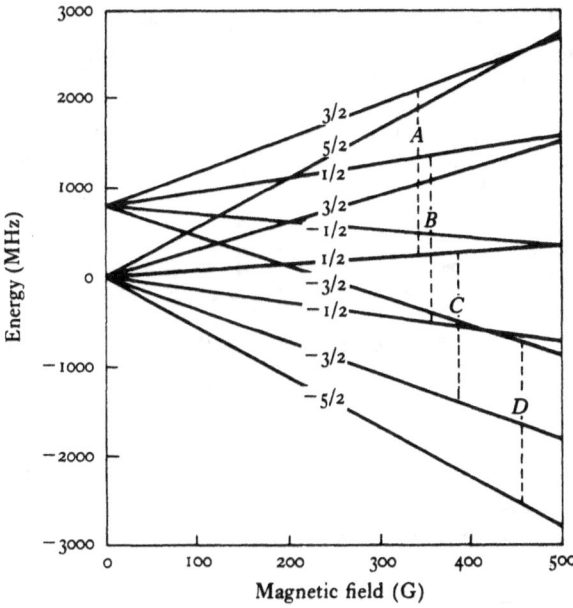

Fig. 5.6. Variation with magnetic field of the energy of the sublevels of the $\mathcal{J} = \tfrac{5}{2}$ and $\mathcal{J} = \tfrac{3}{2}$ levels of the ^4P state of He^-.

sublevels except the $(\tfrac{5}{2}, \pm\tfrac{5}{2})$ the lifetime of the $\tfrac{5}{2}$ substates will fall off rapidly with the strength of the applied magnetic field.

Fig. 5.6 shows the variation of the energies of the sublevels for $\mathcal{J} = \tfrac{5}{2}$ and $\tfrac{3}{2}$ with the magnetic field, the energy values being normalized to agree with the experimental results described below.

Although differing in detail these theoretical conclusions are in general agreement with the experimental measurements which we now describe.

Measurement of the lifetimes of the separate fine structure states

Fig. 5.7 illustrates the arrangement of the experiments carried out by Blau, Novick and Weinflash (1970) and later extended by Novick and Weinflash (1970).

He⁻ ions were produced through double charge transfer (see Chapter 10, p. 414). A beam of He⁺ ions of 3000 eV energy extracted from a r.f. discharge was passed through potassium vapour. A fraction of these ions captured electrons into the $2\,^3S$ state (see p. 405), and of these a further fraction captured a second electron to produce He⁻($2\,^4P$) ions. The charge components, He⁺, He⁰

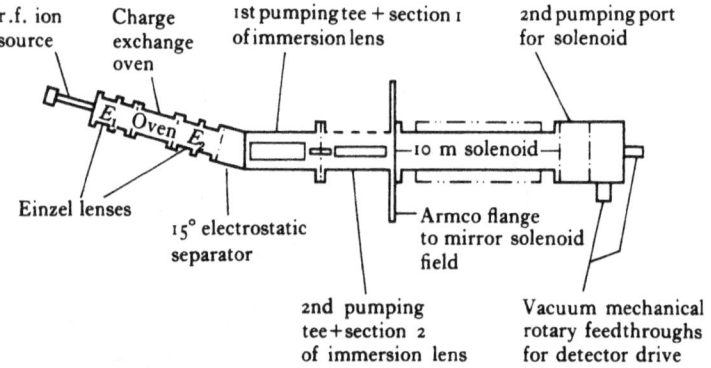

Fig. 5.7. Arrangement of apparatus used by Blau et al. (1970) for measurement of the lifetimes of the ⁴P fine structure levels of He⁻.

and He⁻ from the mixed beam emerging from the potassium vapour were separated electrostatically. In this way an He⁻ beam of about 10^{-7} A was obtained. This was passed through an electrostatic lens system and decelerated to any chosen energy above 50 eV. It then passed into a drift tube 10 m long throughout which there was present an axial magnetic field which could be as large as 1500 G. The intensity of the beam after passage through the drift tube was measured by a Faraday cup, movable under vacuum conditions along nearly the full length of the drift tube. To prevent electrons produced by autodetachment from the He⁻ beam from entering the cup, a grid of high transparency, maintained at a retarding potential of 20 V, was placed across the entrance. It was necessary to maintain a low pressure ($< 10^{-7}$ torr) in the drift tube in order to minimize loss of ions by collision.

The lifetimes of the different states were determined from measurements of the variation of the beam intensity at the detector with the distance from the source, the magnetic field, the beam velocity and the residual gas pressure, but most of the data were obtained at a beam energy of 100 eV and a field of 400 G.

The chief difficulty in analysing the data is to distinguish the three exponential decay rates which are superposed, particularly because the two faster rates are not very different. In fact the observed variation of the He$^-$ current with flight distance could be fitted in all cases observed to 1% accuracy as a weighted mean of two exponentials with very different decay rates. Of these the fast rate was effectively independent of magnetic field, beam velocity and residual gas pressure but the slow rate decreased with all of these variables so it was necessary to extrapolate to zero gas pressure.

This behaviour is consistent with the interpretation of the fast rate as due to decay of the $^4P_{3/2}$ and $^4P_{1/2}$ substates and the slow to $^4P_{5/2}$. Furthermore, when extrapolated back to zero drift time the relative intensities of the two components were found to be equal, just as expected from the ratio of the statistical weight of the $J = \frac{5}{2}$ substate to the sum $4 + 2$ of the weights of the respective $\frac{3}{2}$ and $\frac{1}{2}$ substates.

Having made this identification the zero field lifetime of the $^4P_{5/2}$ substate may be obtained from the observed dependence of the current on magnetic field. Thus, according to (5.2), at sufficiently long flight times (> 30 μs in practice) only $J = \frac{5}{2}$ substates remain in the beam so that, at such times, the beam current will be given by

$$I(t) = \sum_{M_J} \exp(-\Gamma_{M_J} t). \tag{5.3}$$

At high magnetic fields only levels with $M_J = \pm\frac{5}{2}$ will remain so that the current will be $\frac{1}{3}$ of that obtained in zero field. In less extreme cases, at any time t the slow decay rate will be given by

$$\gamma_L = -\frac{1}{I}\frac{dI}{dt} = \sum \exp(-\Gamma_{M_J} t)\, \Gamma_{M_J} / \sum \exp(-\Gamma_{M_J} t). \tag{5.4}$$

Using (5.2), which may be rewritten as

$$\Gamma_M = \gamma_{5/2} + (\gamma_{3/2} - \gamma_{5/2}) \sin^2 \theta_M$$

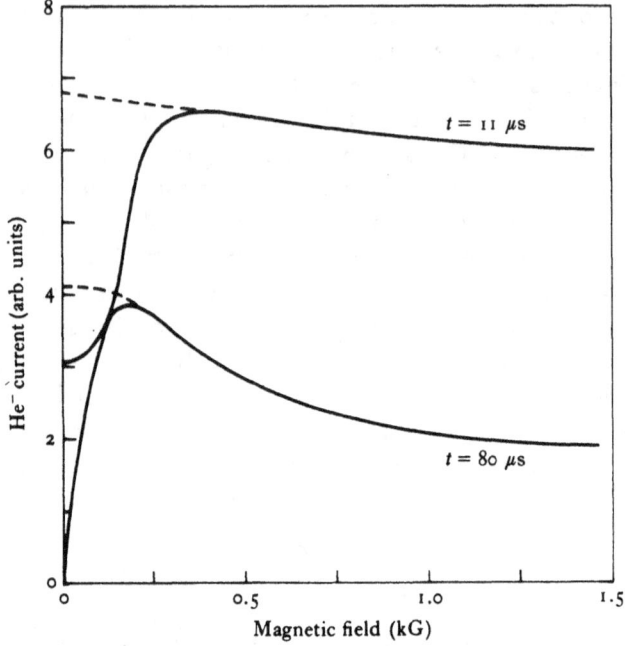

Fig. 5.8. Typical observed variation of He⁻ current with magnetic field for two different flight times t, as indicated (from Novick and Weinflash, 1970). The beam energy was 115 eV.

it follows that

$$\gamma_L = \gamma_{5/2} + f(\gamma_{3/2} - \gamma_{5/2}; \Delta_{53}/H; t)$$

where f is a known function. From the observed variation of γ_L for fixed t as a function of H, $\gamma_{3/2} - \gamma_{5/2}$ and Δ_{53} may be determined. $\gamma_{5/2}$, and hence $\gamma_{3/2}$, may then be obtained from the absolute value of γ_L. Finally, given $\gamma_{3/2}$, the component decaying as $\gamma_{1/2}$ may be obtained by subtraction from the observed fast rate decay curve.

Typical observed examples of the variation of He⁻ current with magnetic field for two different flight times are shown in Fig. 5.8, while Fig. 5.9 shows the analysis of the variation of the He⁻ current with flight time, at a fixed magnetic field, into a long-lived and two short-lived components.

AUTODETACHING STATES OF SPECIFIC ATOMIC IONS 133

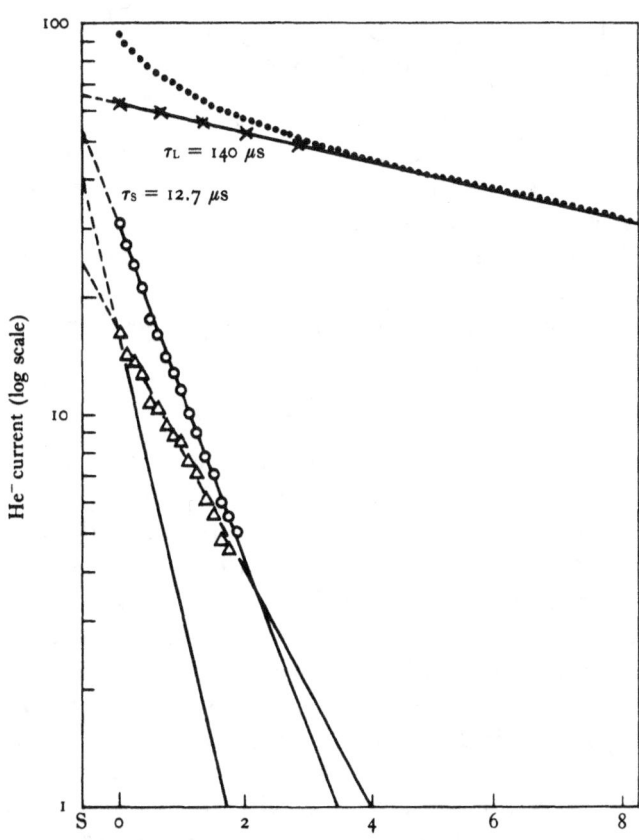

Fig. 5.9. Illustrating the analysis of typical observed data on the variation of He⁻ current with time-of-flight, in the experiments of Novick and Weinflash (1970). The magnetic field was 400 G, the beam energy 135 eV and the drift-tube pressure 2.3×10^{-7} torr. ● observed points. Analysis into two lifetimes –×– long (140 μs) and –○– short (12.7 μs). Further analysis into two short lifetimes –△– (16 μs) and ——— (10 μs).

In this way the following values were obtained

$$\tau_{5/2} = 1/\gamma_{5/2} = 500 \pm 200 \,\mu\text{s}, \; \tau_{3/2}$$
$$= 10 \pm 2 \,\mu\text{s}, \; \tau_{1/2} = 16 \pm 4 \,\mu\text{s},$$
$$\Delta_{53} = \pm 36 \pm 9 \times 10^{-3} \text{ cm}^{-1}. \tag{5.5}$$

Referring back to p. 128, it is seen that the value for $\tau_{5/2}$ agrees quite well with the best calculated value. For $\tau_{3/2}$ and $\tau_{1/2}$ no theoretical results of comparable accuracy are available but the observed values are of the order expected.

Precision measurement of fine structure separations

Mader and Novick (1972) have used magnetic resonance methods to make precision measurements of the fine structure separations $\varDelta_{53}$ as well as, with less accuracy, $\varDelta_{51}$.

The principle of the method depends on the fact that magnetic dipole transitions between the ^4P_J substates in a uniform magnetic field can occur under the action of an r.f. field parallel to the steady field when the selection rules

$$\varDelta J = \pm 1, \quad \varDelta M = 0,$$

are obeyed.

Examples of such transitions are indicated as A, B, C and D in Fig. 5.6. If in some way a transition from a $J_{5/2}$ to a $J_{3/2}$ level is induced, there will be a fall in collected ion current due to enhancement of the component of the beam that decays at a fast rate. To induce such a transition an r.f. field of suitably chosen frequency ν is applied in the axial direction to the beam in the drift tube. If the uniform magnetic field is varied, then when $h\nu$ is equal to the energy difference associated with one of the transitions A, B, C or D, this transition will be excited with consequent reduction in the detected current.

Fig. 5.10 illustrates typical resonance effects of this kind observed at a frequency of 1835 MHz. The resonance D, associated with a $(\frac{5}{2}, -\frac{5}{2})$ to $(\frac{3}{2}, -\frac{3}{2})$ transition, is unaffected by any coupling with $^4P_{1/2}$ and so from it $|\varDelta_{53}|$ is directly determined. Both resonances B and C include appreciable contributions from coupling with $^4P_{1/2}$ and so yield values for $\varDelta_{51}$. The remaining resonance A is determined mainly by $|\varDelta_{53}|$ and only to a slight extent by $|\varDelta_{51}|$. The measurements do not determine the signs of $\varDelta_{53}$ and $\varDelta_{51}$, but show that the energies of the substates either decrease or increase as J increases from $\frac{1}{2}$ to $\frac{5}{2}$. In this way it was found that

$$|\varDelta_{53}| = 825.23 \pm 0.82 \text{ MHz},$$

$$|\varDelta_{51}| = 8663 \pm 56 \text{ MHz},$$

compared with the values 2030 and 9410 MHz calculated by Manson (p. 128) and with 1080 ± 270 MHz observed for $|\varDelta_{53}|$ by Zeeman quenching (see p. 133).

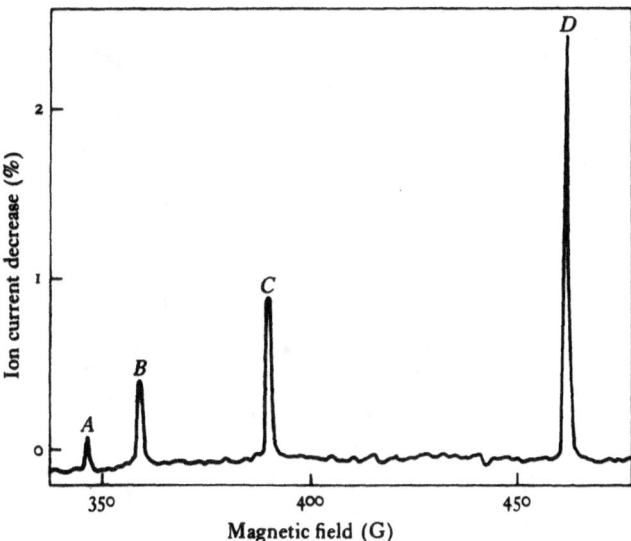

Fig. 5.10. Decrease in He⁻ current as a function of magnetic field when an r.f. field of 1835 MHz is applied, as observed by Mader and Novick (1972).

States between the excitation and ionization thresholds – 2^2P and 2^2D

These states appear only as very broad resonance effects which show up in the energy dependence of the differential cross-sections for excitation of the 2^3S and 2^1S states (Chamberlain and Heideman, 1965; Ehrhardt, Langhans and Linder, 1968). Thus in the results shown in Fig. 5.11, obtained by Ehrhardt et al. (1968) using the apparatus described on p. 112, a broad peak is seen in the scattering at 70° for 2^3S excitation at an energy below the 2^1S threshold. At 90° this peak has disappeared. This suggests that the peak arises from capture of a p electron, a result confirmed when the full angular distribution of the scattered electron is measured as seen from Fig. 5.12. It is natural then to identify the peak as due to capture into the 2^2P state.

Again a second broad peak is clearly visible at an energy just above the 2^1S threshold that remains strongly present at 90°. The full angular distribution observed is shown in Fig. 5.12 and is of the form expected when capture of a d electron is involved. We therefore identify the broad peak as due to capture into the 2^2D state.

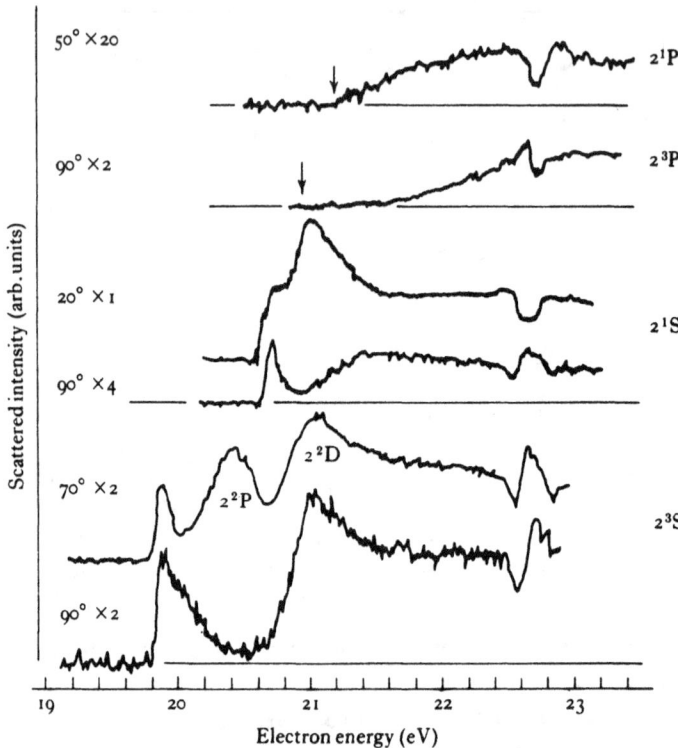

Fig. 5.11. Variation with electron energy of differential cross-sections for excitation of the 2^3S, 2^1S, 2^3P and 2^1P states of helium at different scattering angles as indicated. Resonance peaks due to the 2^2P and 2^2D autodetaching states are indicated. From Ehrhardt et al. (1968).

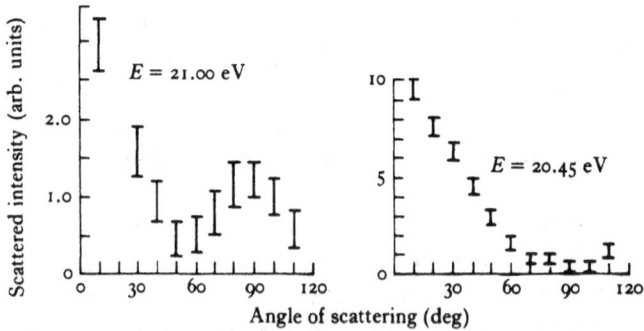

Fig. 5.12. Angular distributions of electrons scattered after excitation of the 2^3S state of helium at electron energies 20.45 and 21.00 eV respectively, corresponding to the 2^2P and 2^2D peaks shown in Fig. 5.11.

Three-quantum states

Referring again to Fig. 5.11 we note the existence of structure in the energy range 22.4–22.6 eV in the differential cross-sections for excitation of all the two-quantum states of He.

Structure in this energy range also appears in the transmission data of Sanche and Schulz (1972) shown in Fig. 5.4 as well as the earlier data of Kuyatt et al. (1965) and in the observations of the total cross-section for metastable He production as a function of energy made by Pichanick and Simpson (1968).

Detailed consideration of the structure leads to the tentative identification of three states near 22.42, 22.66 and 22.8 eV respectively. The first two may probably be identified as the $(1s3s^2)3\,^2S$ and $(1s3s3p)3\,^2P$ states.

Higher states

Resonance states at energies near 23.4 and 23.8 eV have been observed in the transmission experiments of Sanche and Schulz (1972) and of Kuyatt et al. (1965). The former has also been observed by Pichanick and Simpson (1968) in their observations of the total cross-section for metastable He production. Of the other two states the first may be taken to be one in which the quantum numbers of each of the two-outer occupied orbitals is 4 and the second 5.

Much more evidence concerning higher states is obtained from measurements of optical excitation functions. Thus Fig. 5.13 shows the optical excitation function for the 5047 and 4713 Å lines of He arising from impact excitation of the 4^1S and 4^3S states of He, observed by Heddle, Keesing and Kurepa (1973). Clear evidence is seen of peaks due to excitation of resonance states. Structure of this kind is also clearly seen in the results for the 5876 and 4472 Å lines arising from excitation of the 3^3D and 4^3D states (Fig. 5.14). In each figure the locations of the various singly-excited states of He are shown. The peak in the 4^1S and 4^3S cases near 23.94 eV is very probably due to capture into a five-quantum level and associated with the 23.8 eV feature observed in transmission. Effects due to four-quantum states of He$^-$ are strongly present in the curves of Fig. 5.14.

A great deal of effort will clearly be required before all of the rich variety of states observed in optical excitation experiments can be interpreted. Meanwhile we give in Fig. 5.15 an energy-level diagram which summarizes the information at present available about the location of the states of He⁻ which lie below the ionization threshold of He.

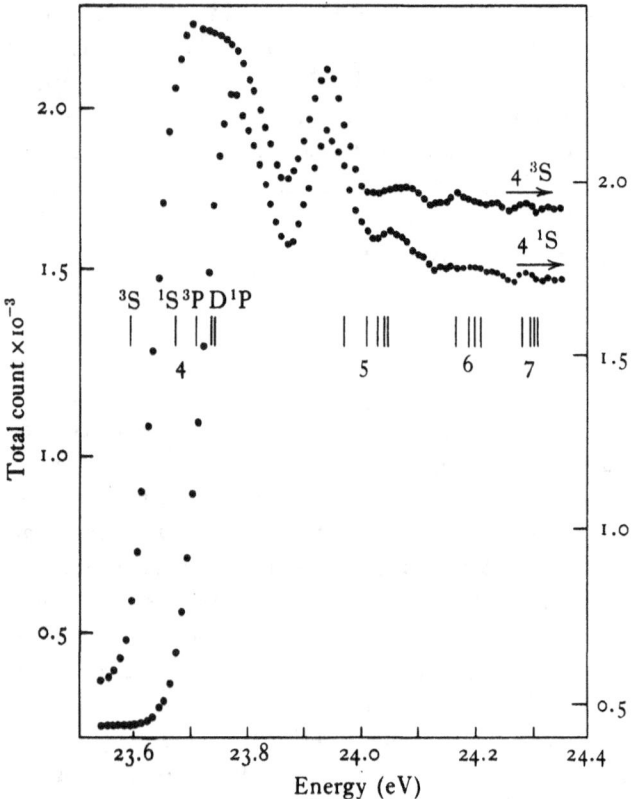

Fig. 5.13. Optical excitation functions for the 4^1S and 4^3S states of He, observed from the intensities of the 5047 and 4713 Å lines respectively, by Heddle et al. (1973).

Triply-excited states of He⁻

Evidence for the existence of doubly-excited autoionizing states of He was first obtained by Priestley and Whiddington (1934). They were observing the energy loss spectra of electrons of 200 eV energy in helium and found a peak in the spectra at an energy loss near 64 eV. This was interpreted as due to excitation of a doubly-excited state or states of helium. Confirmation of this was provided from the observations of Silverman and Lassettre (1964) which were analysed in terms of the resonance theory outlined on pp. 90–1, by Fano (1961). Since then a great deal of data on the doubly-excited states has become available both from theory and experiment (Massey, Burhop and Gilbody, 1969).

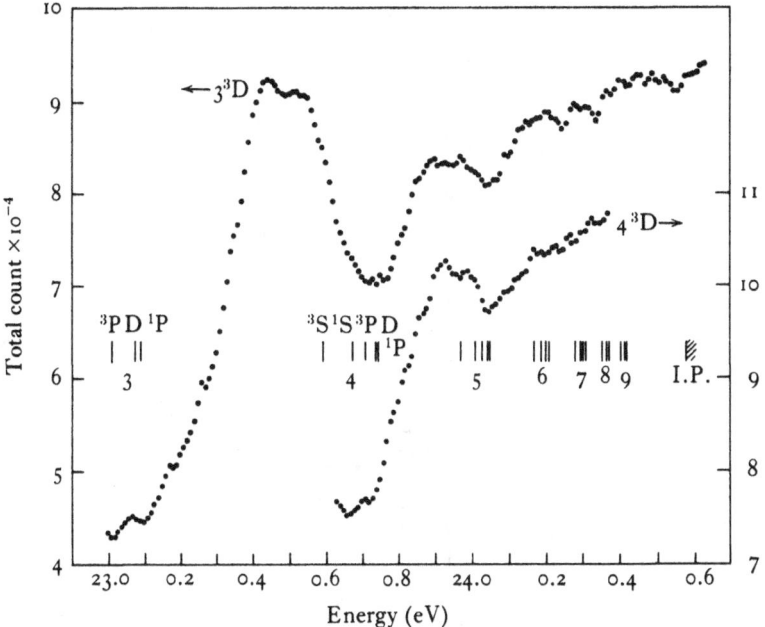

Fig. 5.14. Optical excitation functions for the 3^3D and 4^3D states of He observed from the intensities of the 5876 and 4472 Å lines respectively, by Heddle et al. (1974).

The lowest of these states are the $(2s^2)\,^1S$ at 57.82 eV and $(2s^2\,2p)\,^3P$ at 58.34 eV. Kuyatt et al. (1965) found resonance effects in their transmission experiments in helium at electron energies close to these values and their results were confirmed by Golden and Zecca (1970) and by Sanche and Schulz (1972). Fig. 5.16 illustrates results obtained by the latter authors. The states concerned were identified by Fano and Cooper (1965) as $(2s^2\,2p)\,^2P$ near 57.16 eV and $(2s\,2p^2)\,^2D$ near 58.25 eV. This assignment is in agreement with theoretical calculations by the stabilization method (Eliezer and Pan, 1970) and by a variational method including correlation (Nicolaides, 1972). It places the first state 0.66 eV below $(2s^2)\,^1S$ and the second 0.09 eV below $(2s\,2p)\,^3P$.

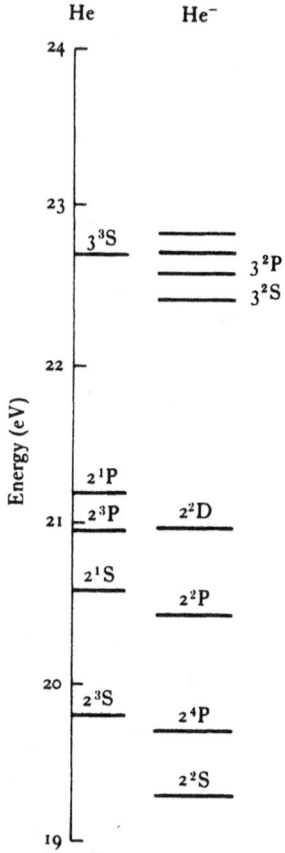

Fig. 5.15. Energy-level diagram of well-established states of He⁻ lying below the ionization threshold of He.

Excitation of these triply-excited states is also apparent in observed differential cross-sections for excitation of the 2^3S, 2^1S and 2^1P states (Simpson, Menendez and Mielczarek, 1966) as well as in the ionization cross-sections observed by Grissom, Compton

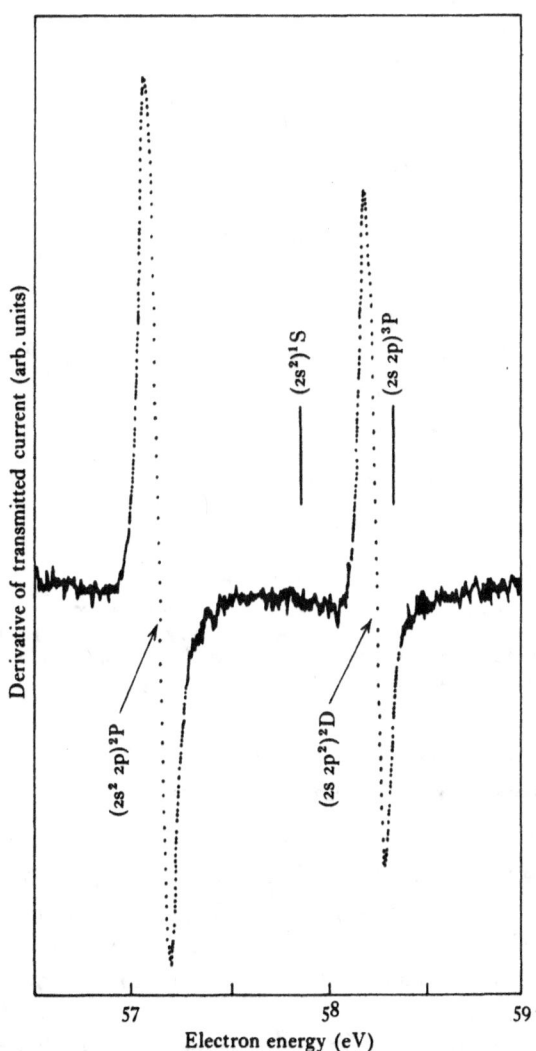

Fig. 5.16. Energy derivative of the transmitted electron current as a function of energy in helium in the region of the doubly-excited states whose locations are indicated. From Sanche and Schulz (1972).

and Garrett (1969) and by Quéméner, Paquet and Marmet (1971) (see Fig. 5.17). Analysis of the results obtained by the latter investigators gives for the respective level widths of the ^{2}P and ^{2}D states, 0.045 and 0.025 eV.

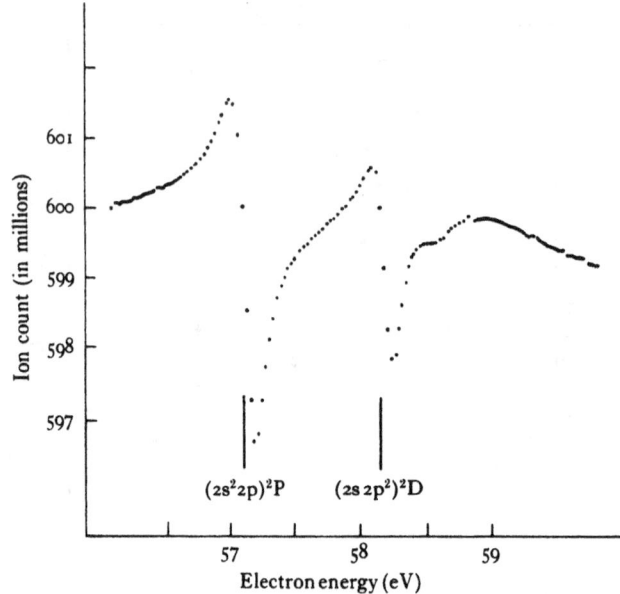

Fig. 5.17. Resonance effects, due to triply-excited states of He⁻, observed by Quéméner et al. (1971) in the cross-section for ionization of He by electron impact. The background ionization has been subtracted out. The location of the He⁻ states concerned is indicated.

5.3 Negative ions of heavier rare gases

The lowest resonance states of Ne⁻, Ar⁻, Kr⁻ and Xe⁻ are ^{2}P states with configuration $2n\mathrm{p}^5(n+1)\mathrm{s}^2$, n being 2, 3, 4, and 5 for the respective cases. Thus they consist of two s electrons attached to the $n\mathrm{p}^5$ positive-ion core. We would therefore expect the doublet splitting in the resonance states to be close to that for the ground state of the ion core.

The doublet splittings of the resonance states have been observed by Kuyatt et al. (1965) and by Sanche and Schulz (1972) using the transmission technique (see p. 111). Their results agree very well, giving 0.095, 0.172, 0.65 and 1.25 eV for Ne⁻, Ar⁻, Kr⁻ and Xe⁻, which are to be compared with 0.097, 0.177, 0.666 and 1.306 for Ne⁺, Ar⁺, Kr⁺ and Xe⁺ respectively.

The $^2P_{3/2}$ state is found to lie 0.51, 0.44, 0.40, and 0.415 eV below the lowest $^3P_2$ excited state for the respective cases of Ne⁻

to Xe⁻. For Ne⁻ the width of the state has been derived from experimental data as 8.95 meV (Haselton, 1973).

A number of other resonance states have been observed below the ionization energy for each negative ion (Sanche and Schulz, 1972; Kuyatt *et al.*, 1965; Pichanick and Simpson, 1968).

A further set of resonance states have been detected at energies above the ionization threshold (Sanche and Schulz, 1972; Grissom, *et al.*, 1969; Boldue, Quéméner and Marmet, 1972). These states occur in the energy ranges 42–50 eV in Ne, 24–32 eV in Ar, 22–27 eV in Kr and 18–20 eV in Xe.

5.4 O⁻

The excitation of autodetaching excited states of O⁻ has been observed by Edwards *et al.* (1971) using the technique described on p. 119 and in Chapter 4, in which the energy distribution of electrons detached in a particular direction from a beam of O⁻ ions in passing through helium is observed.

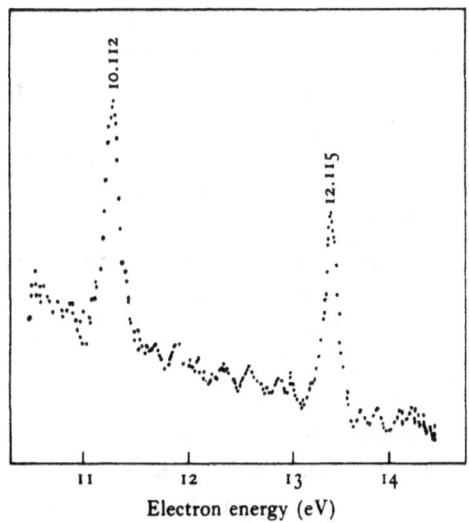

Fig. 5.18. Energy spectrum of electrons produced by impact with He of a beam of O⁻ ions of 2 keV kinetic energy. The energies indicated are in the centre of mass system. The electrons were observed in a direction making an angle 9° with the incident beam. From Edwards *et al.* (1971).

Fig. 5.18 illustrates the results obtained for O⁻ of 2 keV energy, the electrons being observed at an angle of 9° to the direction of the ion beam.

Two peaks are clearly seen at energies 10.112 and 12.115 eV in the centre of mass system. The calibration of the energy scale was carried out by observing the energy distributions of the electrons from the charge transfer

$$O^- + Ar \longrightarrow O + Ar^{-*} \longrightarrow O + Ar + e,$$

where Ar^{-*} refers to an autodetaching state. This state could be identified from the doublet splitting as the ^2P state referred to above. As its energy is well known from electron scattering measurements, this provided the required calibration.

The two O⁻ states which they observed were tentatively identified by Edwards *et al.* (1971) as $2p^3(^4S)3p^2(^2P)$ and $2p^3(^2D)3s^2\,^2D$ lying 0.628 and 0.424 eV below the parent $2p^3(^4S)3p$ and $2p^3(^2D)3s$ states of O. Theoretical estimates (Ormonde *et al.*, 1973; Matese, Rountree and Henry, 1973*a*) are not inconsistent with this.

5.5 N⁻

We have already referred on p. 81, to the existence of a state of N⁻ metastable towards autodetachment. This is the ^1D term associated with the ground $1s^2\,2s^2\,2p^4$ configuration.

Evidence for the existence of this state was first provided by the experiments of Fogel *et al.* (1959) who studied the ions resulting from charge transfer on passage of N⁺ ions of 34 keV energy through krypton and observed N⁻ ions formed by double charge transfer (see Chapter 10, p. 414) the cross-section being only about 3×10^{-22} cm². Since the lowest ^3P term of N⁻ has an energy above that of the ^4P ground term of N, the observed ions could only have been in a metastable state, presumably the ^1D.

Further evidence has been forthcoming from experiments on the affinity spectrum of radiation emitted from arc discharges in nitrogen (see Chapter 8, p. 254) which could only arise from capture of electrons by N atoms. These results are compatible with the assumption that the electrons are captured by excited $N(^2D)$ atoms to form $N^-(^1D)$ with an energy release of 1 eV.

Schaefer and Harris (1968) have calculated this energy difference by a method in which correlation is taken account of by the procedure outlined on p. 26, only configurations in which at most one electron is assigned to an orbital outside the valence shell being included. Correction to the calculated values were made by comparison with corresponding results for the isoelectronic positive ion and with the known electron affinities of C, O and F (see pp. 44–5). It was finally estimated that the $N^-(^1D)$ state lies 0.844 eV below $N(^2D)$ whereas $N^-(^4S)$ lies 0.213 eV above $N(^4P)$.

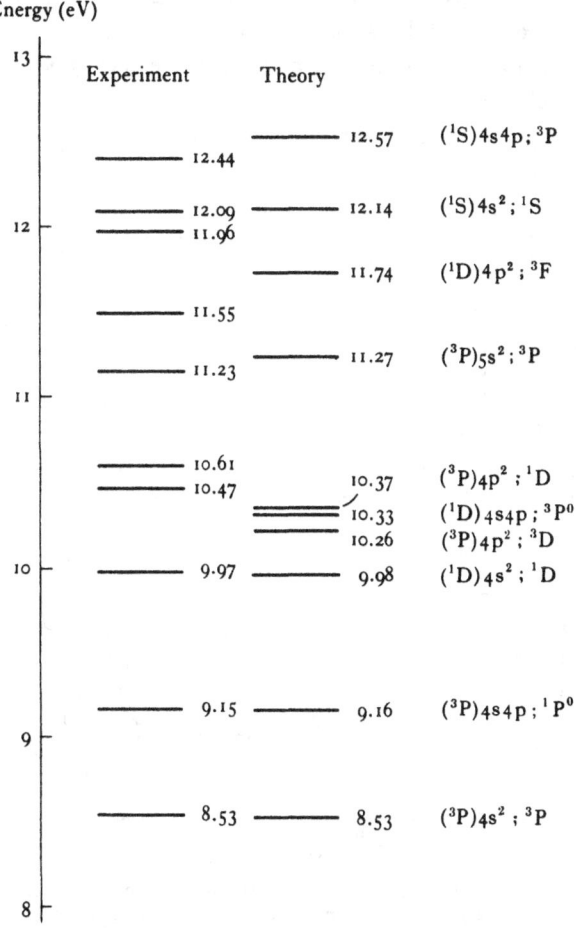

Fig. 5.19. Energy-level diagram of Cl^- showing observed and calculated resonance states. From Matese et al. (1973b).

5.6 Cl⁻

The energy distribution of electrons detached from Cl⁻ ions in collision with gaseous targets of He and H_2 has been measured by Cunningham and Edwards (1973) using the same technique as for H⁻ and O⁻. The ground term of Cl⁻ is $3s^2 3p^6\, ^1S$ and the lowest doubly-excited resonance states will arise from excitation of two of the 3p electrons to four- or five-quantum states. They then populate excited orbitals in the field of a $3s^2 3p^4$ core of Cl⁺. Matese, Rountree and Henry (1973b) have calculated the energies of the resonance states which lie less than 12.6 eV above the ground state of Cl assuming that this core remains unchanged throughout but including a number of two-electron configurations for the two excited electrons.

Fig. 5.19 shows the calculated energies and term designations of the resonant states which they found.

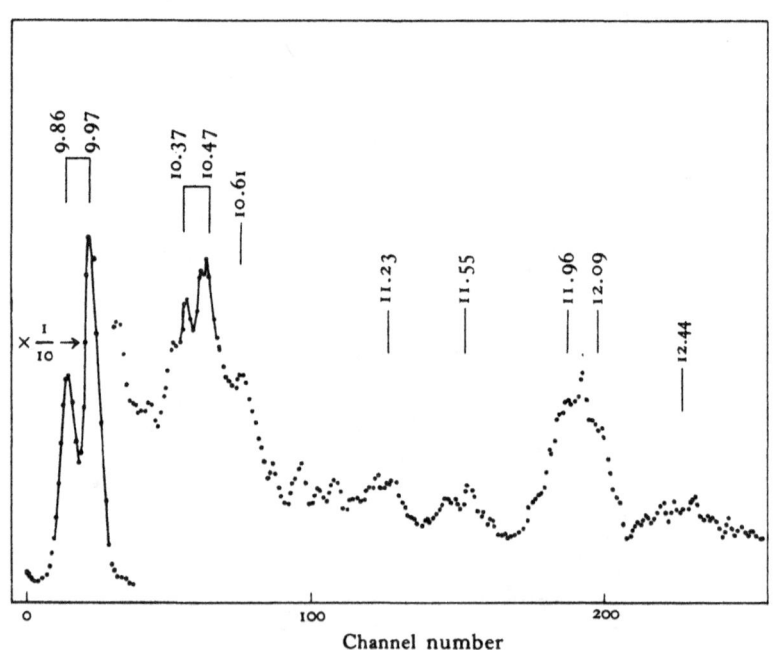

Fig. 5.20. Energy spectrum of electrons produced by impact with He of a beam of Cl⁻ ions of 1.5 keV kinetic energy. The energies indicated are in the centre of mass system. The electrons were observed in a direction making an angle of 15° with the incident beam. From Cunningham and Edwards (1973).

Since the ground states of Cl⁻ and of He are singlets, excitation of triplet states of Cl⁻ can only occur if the He atom is also excited to a triplet state. This requires an additional 20 eV energy and at the impact energies concerned (2 keV or less) this will be very improbable (see p. 389). We would therefore expect only singlet states to be excited in He. For impact with H_2, triplet excitation requires considerably less energy (~ 11 eV) and so will be more probable.

Fig. 5.20 shows a typical energy spectrum in He, observed with Cl⁻ ions of 1.5 keV and a scattering angle of 15°. The dominant features are the two peaks at 9.86 and 9.97 eV the separation of which, 0.106 ± 0.006 eV, is close to the fine structure separation 0.109 eV between the ground $^2P_{1/2}$ and $^2P_{3/2}$ substates of neutral Cl. Similar doublet separations are apparent in other features of the energy spectrum. It seems clear that the peaks correspond to transitions from autodetaching states to the two ground substates of Cl.

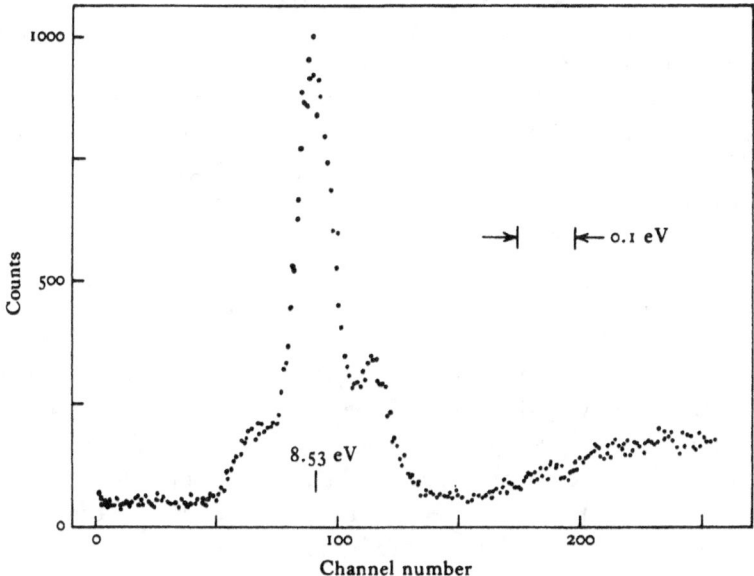

Fig. 5.21. Energy spectrum of electrons produced by impact with H_2 of a beam of Cl⁻ ions of 1 keV kinetic energy, the electrons being observed in a direction making an angle of 15° with the incident beam. The peak at 8.53 eV in the centre of mass system can be identified with the lowest ^3P state of Cl⁻. From Cunningham and Edwards (1973).

Comparison of the observed with the calculated energies is shown in Fig. 5.19. Identification of the experimental resonance states at 9.15, 9.97 and 12.09 eV with calculated singlet states seems to be established, but, as expected, none of the calculated triplet states appear from the observed data in He. However, the lowest such state was observed in the energy spectrum in H_2 as shown in Fig. 5.21. Further triplet levels were not identified because of the presence of broad peaks due to H^- autodetaching states arising from break-up of H_2.

5.7 Negative ions of the alkali metal atoms

The most conspicuous effects due to resonance states of the negative ions of the alkali metal atoms occur immediately below the first excited ^2P state of the neutral atom.

For sodium the effects are seen clearly in measurements of the differential cross-section for elastic scattering made by Andrick, Eyb and Hofmann (1972) using the method described in Chapter 4, p. 112. Fig. 5.22 illustrates results which they obtained showing the structure observed when the electron energy is close to the ^2P excitation threshold of 2.1 eV. This structure is actually compounded of contributions from the ^1P and ^1D autodetaching states of Na^- and from the 'cusp' effect arising from the opening of a new channel which appears in the s-wave scattering (Wigner, 1948; Mott and Massey, 1965, p. 379).

Moores and Norcross (1972) have calculated differential cross-sections for scattering by Li, Na and K using a close-coupling expansion involving the lowest s, p and d states and in some cases the lowest excited s state. The problem was treated as a two-electron one with the effect of the core represented by a scaled Fermi–Thomas potential modified by a term which allows for core polarization (see Chapter 2, p. 29). This is found to give quite good agreement with the observed results for Na, including the structural features near the ^2P threshold. The same close-coupling expansion has also been used to calculate cross-sections for photodetachment and again structural features are present near the ^2P excitation threshold which are found experimentally (see Chapter 11, pp. 425 and 466).

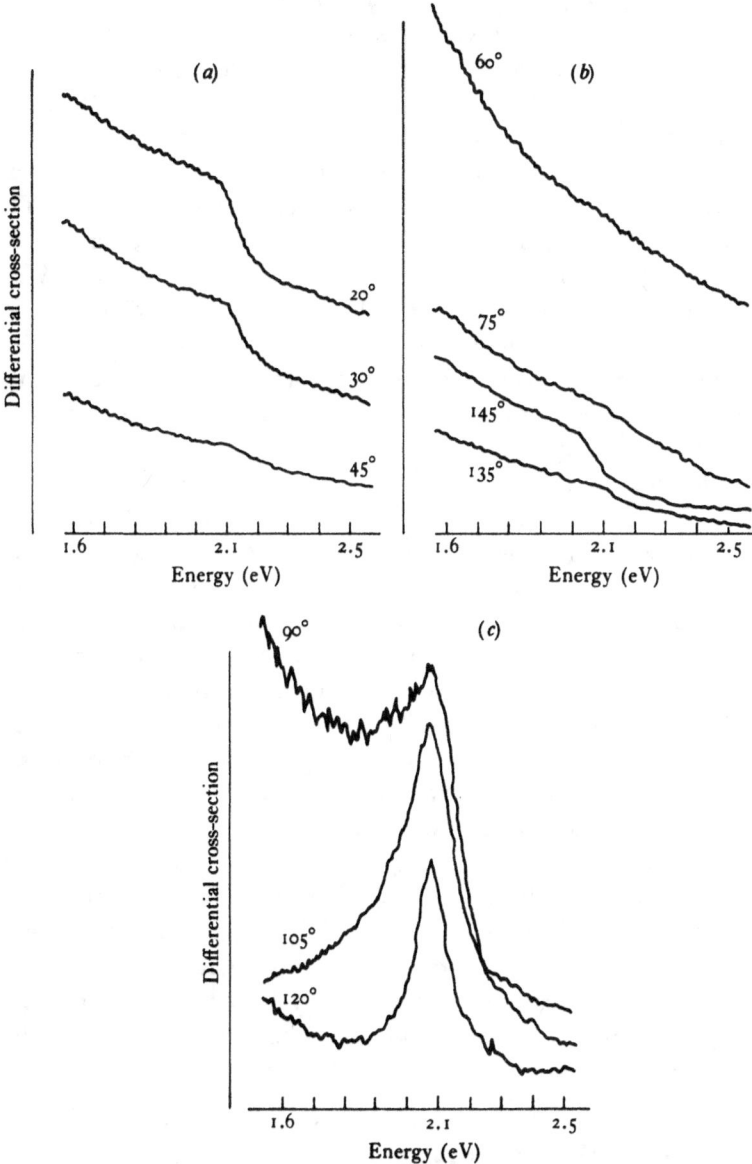

Fig. 5.22. Variation with electron energy of differential cross-sections for elastic scattering of electrons by sodium atoms at different angles of scattering, as observed by Andrick et al. (1972). The scales of (b) and (c) are respectively 3.7 and 25 times larger than that of (a).

For the heavier alkali metal ions, Rb⁻ and Cs⁻, the thresholds for excitation of the $^2P_{1/2}$ and $^2P_{3/2}$ states may be distinguished separately in photodetachment experiments with tunable dye-laser light sources (Chapter 11, pp. 469–71). For both ions a 'window' resonance is observed very close to the excitation threshold for $^2P_{1/2}$ which is very narrow (0.15 meV for Rb⁻) and has, as far as can be observed, a zero minimum (see Figs. 11.28 and 11.29). The nature of the resonance state concerned is not yet clear but for caesium may well have a predominantly 6s 7p configuration. Window resonances are also observed very close to the threshold for the $^2P_{3/2}$ substate. These are wider than for $^2P_{1/2}$ and have non-zero minima (see Fig. 11.28). Details of the methods used to measure the photodetachment cross-sections and of the analysis of the data are given in Chapter 11.

Norcross (1974) has applied the mathematical technique described in Chapter 3, p. 29, to search for $n_0^* p^{2\,3}P_e$ resonance states analogous to the $2p^{2\,3}P_e$ state of H⁻. As discussed on p. 115, this latter state can only autoionize through departures from LS coupling and so its fastest decay mode is an allowed radiative transition to either the $2s\,2p\,^3P^0$ autodetaching state or to the $1s\,kp\,^3P^0$ continuum. The same considerations should apply to the alkali ion states if they exist. Norcross found they do indeed exist for Na⁻, K⁻, Rb⁻ and Cs⁻ with energies 0.062, 0.119, 0.144 and 0.166 eV below the $n_0\,^2P$ state of the respective neutral atom. For Li, however, no such state was found.

Fung and Matese (1972) have searched for resonance states of Li⁻ derived from the 3^2S, ^2P and ^2D states of the neutral atom as parent states. They used a multiconfiguration wave function in conjunction with the projection method (see Chapter 4, p. 93) and obtained three resonance states closely below the 3^2S threshold (3.36 eV) and four lying between this threshold and that of 3^2P (3.84 eV).

5.8 Doubly-charged negative ions

Evidence of the existence of resonant states of doubly-charged negative ions has been obtained in recent years.

* n_0 is the quantum number of the outermost occupied orbital in the neutral atom.

AUTODETACHING STATES OF SPECIFIC ATOMIC IONS

H^{2-}

As described in Chapter 11, p. 511, Walton, Peart and Dolder (1970), in the course of their measurements of the cross-section for detachment of electrons from H^- ions by electron impact as a function of electron energy, observed structure centred at an energy of 14.2 eV (see Fig. 11.50) which could be interpreted as a resonance effect.

Taylor and Thomas (1972) applied the stabilization technique (Chapter 4, p. 95) to investigate the matter theoretically. Clearly if an H^{2-} complex is involved, all three electrons must occupy excited orbitals. The $(2s)^2 2p$ and $(2p)^3\, ^2P_0$ configurations were chosen as the most probable, and elaborate multiconfiguration calculations were carried out. A resonance state with energy 14.8 eV above the ground state of H^- was found. This state was largely a mixture of 66% $(2s)^2 2p$ and 29% $2p^3$. The level width was also estimated from the range of energies about the resonant energy for which a stabilized wave function could be found. It is around 1 eV which is compatible with the observations.

It appeared also from the calculations that another resonance state, probably largely $2p^3$ in character, exists at a higher energy and, as described in Chapter 11, p. 512, Peart and Dolder (1973), using the inclined-beam technique, found evidence of a higher resonance near 17.2 eV.

Other doubly-charged negative ions

In the course of mass analysis of negative ions extracted from an ion source, peaks have been observed which correspond to ions of mass one-half that of a known negative ion X^-. Such peaks could arise either from doubly-charged ions X^{2-} or from ions X^- produced by dissociation of X_2^- ions before entering the analysing magnet. Thus, with a 60° sector magnet, as in Fig. 5.23, providing a magnetic field H, the deflection θ produced in a beam of ions of charge ne, mass M and kinetic energy E is given by

$$\sin \theta = nelH/(2ME)^{1/2}, \tag{5.6}$$

where l is the distance traversed by the ion in the magnetic field. If V is the extraction voltage at the ion source then for X^{2-}, $E = 2eV$ and

$$n(ME)^{1/2} = 2^{1/2}(MV)^{1/2}. \tag{5.7}$$

For X^- arising from dissociation of X_2^-, $E = \tfrac{1}{2}eV$ and $n = 1$ so that $n/(ME)^{1/2}$ has the same value (5.7).

To distinguish between these possibilities Baumann, Heinicke, Kaiser and Bethge (1971) introduced a further electrostatic analysis as in the experimental arrangement shown in Fig. 5.23. With the sector magnet the mass resolution attained was about $\tfrac{1}{150}$. The electrical analysis was carried out by passage between two parallel plates between which a suitable voltage could be applied. With no

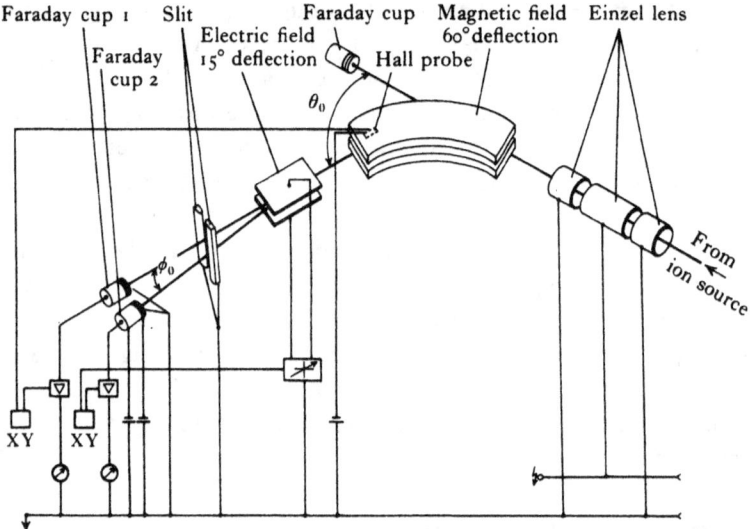

Fig. 5.23. Arrangement of apparatus used by Baumann *et al.* (1971) for investigating the production of doubly-charged negative ions.

voltage on, the beam issuing from the analyser magnet was collected in the Faraday cup 1 but with an appropriate voltage difference it was deflected through an angle ϕ into a second cup as shown. ϕ is given by

$$\tan\phi = neUs/2dE,$$

where U is the voltage difference, s the distance traversed between the plates and d the plate separation.

In this case if U_2 is the voltage difference between the plates which deflects X^{2-} ions into the second cup for which $\phi = \phi_0$, we have

$$\tan\phi_0 = \tfrac{1}{2}U_2 s/dV.$$

AUTODETACHING STATES OF SPECIFIC ATOMIC IONS

To produce the same deflection of X^- ions arising from dissociation of X_2^- requires a voltage U_1, where

$$U_1 = dV \tan \phi_0 / s$$
$$= \tfrac{1}{2} U_2.$$

Fig. 5.24 shows a typical magnetic analysis of the mass spectrum of negative ions from a Penning source in iodine. The peak at an apparent mass of 63.5 is produced by the I^- ions arising from the dissociation of I_2^- before analysis as well as any I^{2-} ions. Fig. 5.25 shows the electrical analysis of the ions producing the peak. It

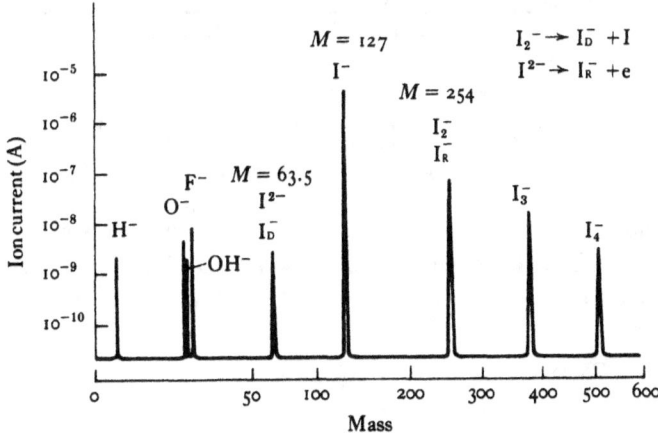

Fig. 5.24. Mass spectrum of negative ions from iodine, observed by Baumann et al. (1971).

will be seen that both I^{2-} and I^- are clearly present. Also as the pressure in the chamber increases the former component decreases and the latter increases as would be expected. There is a further small flat maximum in Fig. 5.25 at a voltage difference twice that for collection of I^{2-}. This is probably due to I^{2-} ions which have decayed to I^- after magnetic analysis but before reaching the plates. For such ions the voltage for collection will be twice that for the I^{2-}.

Furthermore the peak in the magnetic spectrum, at an apparent mass of 254, can be ascribed both to I_2^- and to I^- which arises from decay of I^{2-} in flight before analysis. Fig. 5.26 shows the electrical

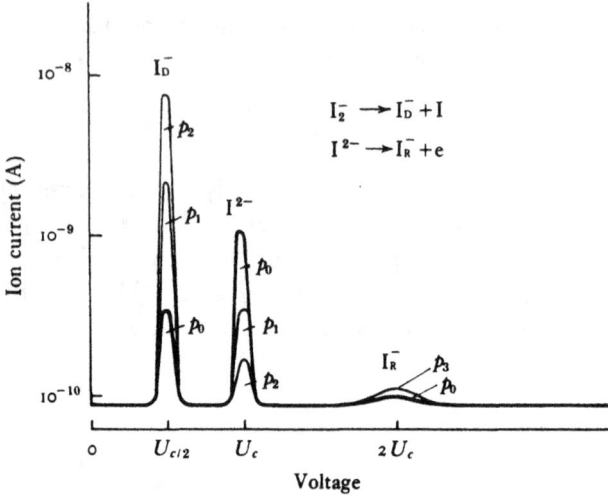

Fig. 5.25. Electrical analysis of the peak in the mass spectrum at a mass of 63.5 shown in Fig. 5.24. The observations by Baumann *et al.* (1971) refer to three different pressures p_0, p_1 and p_2 in the beam-handling system with $p_0 < p_1 < p_2$.

analysis of this peak. In this case if the collection field for I_2^- is F_3, that for I^- from I^{2-} decay will be $2F_3$ while I^- ions resulting from dissociation of I_2^- after magnetic analysis but before reaching the

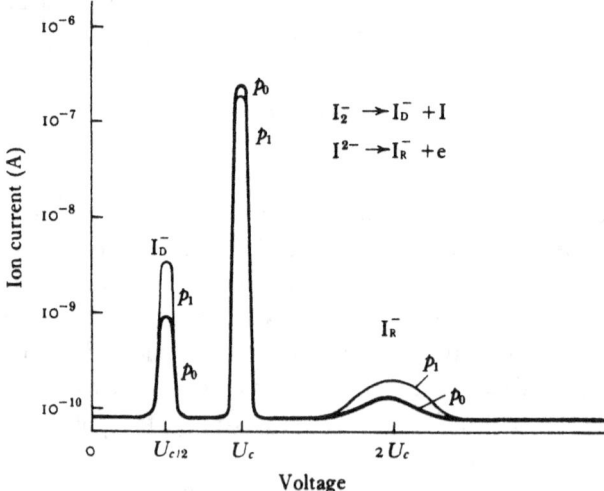

Fig. 5.26. Electrical analysis of the peak in the mass spectrum at a mass of 127 shown in Fig. 5.24. The observations by Baumann *et al.* (1971) refer to two different pressures p_0 and p_1 in the beam-handling system with $p_0 < p_1$.

AUTODETACHING STATES OF SPECIFIC ATOMIC IONS

plates will be $\frac{1}{2}F_3$. Peaks corresponding to all these are clearly seen in Fig. 5.26 and these vary in the expected sense with change of pressure.

In this way doubly-charged ions of O, F, Cl, Br, I, Te and Bi were observed. The attainable current strengths, at a pressure of 4×10^{-6} torr in the analysing chamber, were around 0.2 nA for O^{2-} and F^{2-}, 0.6 nA for Cl^{2-} and Bi^{2-} and 2 nA for Br^{2-}, I^{2-} and Te^{2-}.

The nature of the electronic states of these ions, which must have lifetimes of order 10^{-6}–10^{-7} s to be observed in this way, is not established. It is possible that they owe their long lifetime to high multiplicity as suggested by Fano (1973). Thus the Cl^{2-} state might be a sextet with configuration $3p^4(^3P)4s4p^2$ and O^{2-} a septet with configuration $2p^3(^4S)3s3p^2$.

CHAPTER 6

Molecular Negative Ions – Ground States

We now consider the structure and properties of molecular negative ions. This introduces a great number of new possibilities and we begin by dealing with diatomic ions. After a short introductory account of the structure of diatomic molecules in general and of the theoretical methods for investigating it we shall discuss the special features which are associated with the corresponding negative ions. In later sections the discussion will be extended, in less detail, to polyatomic negative ions.

6.1 The quantum states of diatomic molecules – potential-energy curves

When discussing the structure of molecules, charged or uncharged, the motion of the nuclei must be taken into account. Since the electronic motion is much faster, we can, as a first approximation, regard the nuclei as at rest and calculate the energy of a particular electronic state for various relative positions of the nuclei. This energy can then be regarded as nuclear potential energy for the calculation of the levels of the nuclear motion. Corresponding to a given electronic state there will then be a series of vibrational and rotational nuclear states. In higher approximations the effect of the nuclear motion on the electron states must be taken into account, but for most descriptive purposes these can be neglected. The usual graphical method of representing molecular states is illustrated in Fig. 6.1. In (a) the electron motion gives rise to a potential-energy curve with a minimum. The limit for infinitely separated nuclei is the sum of the energies of the atoms into which the molecule dissociates on adiabatic separation of the nuclei. The nuclear states form a discrete set tending to a limit at this energy, above which they form a continuum. They are indicated by the level lines such as *ab*, *cd* in Fig. 6.1(*a*). In Fig. 6.1(*b*) the potential-energy curve has no minimum and the nuclear levels form only the continuum shown.

MOLECULAR NEGATIVE IONS – GROUND STATES 157

To the accuracy of our approximation the wave function for a given molecular state can be expressed as the product

$$\psi(r, R)\chi(R), \qquad (6.1)$$

where $\psi(r, R)$ is the electronic wave function for fixed values R of the nuclear coordinates and $\chi(R)$ is the nuclear wave function representing motion with the effective potential energy arising from the electron motion. The wave functions $\chi(R)$ are large only in the regions of allowed classical motion. Thus, for the vibrational

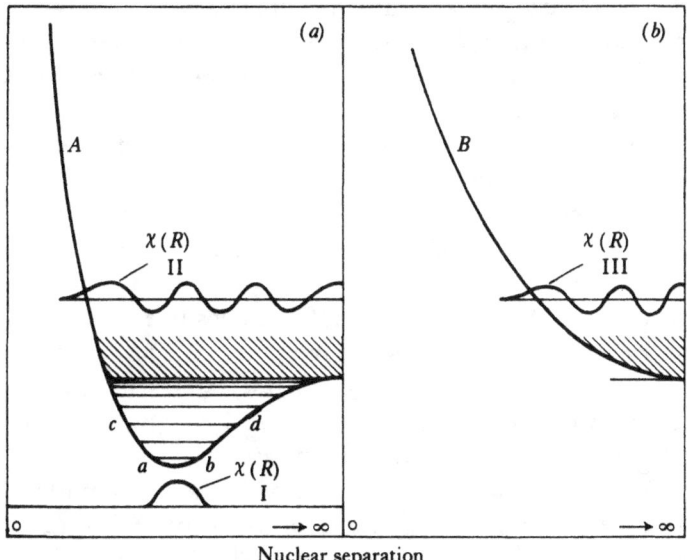

Fig. 6.1. Graphical representation of molecular states. A and B are effective potential-energy curves, $\chi(R)$ 'vibrational' wave functions: I corresponding to level ab; II, III corresponding to states of the continuum.

level represented by ab in Fig. 6.1(a), $\chi(R)$ is concentrated between a and b and is very small outside these limits. For a state in the energy continuum, $\chi(R)$ is negligible inside the classical closest distance of approach and oscillates rapidly outside this distance. These forms for $\chi(R)$ are illustrated diagrammatically in Fig. 6.1.

If a molecule is excited to a state in which the nuclear energy level lies in the continuum, the relative motion of the nuclei is no longer confined within definite limits and the molecule dissociates.

In applying this description to molecular negative ions we must take account of the fact that, for one or more ranges of values of the nuclear separation, the strength of the field acting on the electrons may not be great enough to give rise to a bound state for the additional electron. This is a feature which is not encountered for neutral or positively-ionized molecules. At each value of the nuclear separation we can in principle investigate the possible states of the negative ions with 'frozen' nuclei. These may be similar to those for an atomic ion such as H^-, consisting of a single bound state and a number of autodetaching doubly-excited states with lifetimes $\hbar/\Gamma_i$. Under other circumstances the situation may be as in He^- with no bound state but again with a number of autodetaching states.

In this chapter we are concerned with the ground electronic states of molecular negative ions. Figs. 6.2(a) and (b) illustrates possible forms of the electronic potential-energy curves for these states of an ion AB^- in relation to those for the ground state of the neutral molecule.

In both cases (a) and (b) the atom B has a positive electron affinity so at large nuclear separations a bound state exists for the additional electron. However, whereas in (a) such a state exists at all separations and we have a well-defined potential-energy curve for AB^- which is always below that for AB, in case (b) at separations R less than $R < R_c$ no such state exists. Under these circumstances a well-defined potential-energy curve for AB^- exists only for $R > R_c$. At smaller R the lifetime of a transitory bound system will depend on the total energy of the system. If it has a maximum at a particular energy, as will be the case if a 'shape' or type II resonance in the sense discussed in Chapter 4, p. 107 occurs, and the lifetime $\tau(R)$ is long compared with the vibrational period of the nuclei, it is still sensible to think of the potential energy for the ion for $R < R_c$ as defined by the shape resonance energies at each R but blurred by an amount of order $\Gamma(R) = \hbar/\tau(R)$ due to the finite lifetime. This is the situation shown in Fig. 6.2(b).

If no resonant energy maximum exists, or if it does but the lifetime is short compared with the vibrational period, this picture is no longer useful but it is still possible to derive formulae for transition rates by working through collision theory in which dis-

MOLECULAR NEGATIVE IONS – GROUND STATES 159

tortion of free electron motion by the molecular field is taken into account.

C_2^- is an example of case (a) (see Fig. 6.28), H_2^- of case (b) (see Fig. 6.3). O_2^- (Fig. 6.12) and NO^- (Fig. 6.16) are intermediate in that the potential minimum for the ion lies below that for the neutral molecule so the intersection occurs for R_c less than the equilibrium separation for the molecule.

Other cases might arise in which neither of the atoms A or

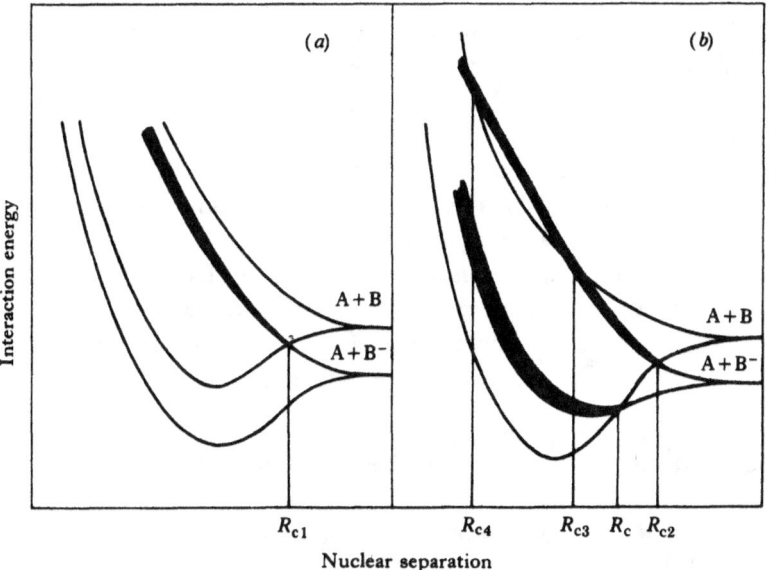

Fig. 6.2. Illustrating potential-energy curves for electronic states of molecular negative ions.

B have positive electron affinities but bound states for an additional electron exist over a finite range of nuclear separation. Such cases are less likely to occur in practice but examples of importance, such as N_2^-, are known in which over a wide range of nuclear separation a narrow shape (type II) resonance exists over a considerable range of R, defining a significant potential-energy curve with adequate precision.

We also illustrate in Figs. 6.2(a) and (b) possibilities arising in relation to repulsive states which tend to normal atoms A and ions B⁻. In Fig. 6.2(a) the repulsive curve for AB⁻ lies below that for AB at all nuclear separations but intersects the ground state curve for AB at $R = R_{c1}$. Autodetachment to the ground state can therefore occur for $R < R_{c1}$ so the ion curve is blurred for such values of R. For (b) the repulsive curve for AB⁻ intersects that for AB at $R = R_{c3}$ and $R = R_{c4}$, while it intersects the ground state curve for AB at $R = R_{c2}$. This means that the negative-ion state is stable towards autodetachment for $R > R_{c2}$. For $R_{c3} < R < R_{c2}$ it may undergo autodetachment but only to the ground state of AB. Finally for $R < R_{c4}$ it is again restricted to autodetaching transitions to the ground state of AB. In terms of the notation of Chapter 4, p. 107, the repulsive negative-ion state will be of type II for $R_{c4} < R < R_{c3}$ and of type I for $R < R_{c4}$, the repulsive state of AB being the parent.

Similar, as well as additional, possibilities arise in connexion with potential-energy curves for excited electronic states but we shall defer consideration of them until the next chapter.

Before proceeding to consider what progress has been made in calculating the properties of the ground states of diatomic negative ions we summarize the nomenclature for electronic states of diatomic molecules and their classification as stable or unstable.

6.2 Enumeration and properties of the electronic states of diatomic molecules

We adopt the molecular orbital method for this purpose as it is most convenient to use. The electrons in a molecule are each assigned to separate quantum states or orbitals. These orbitals are distinguished by three quantum numbers and at most two electrons, with opposing spins, can be assigned to each.

One of the quantum numbers λ refers to the angular momentum of the electron round the nuclear axis. Orbitals in which this angular momentum is 0, ± 1, ± 2, ... quantum units are designated $\sigma, \pi, \delta, \ldots$ states. The remaining two quantum numbers cannot be simply specified in the same way as for atoms. Instead it is usual to give the quantum states of the separated atoms from which the orbitals are derived, or those of the 'united' atom to which they tend when the nuclear separation tends to zero. The notation usually adopted is to write the designation of the atomic states of the separated atoms after the $\sigma, \pi, \delta, \ldots$, and that of the united atom before. Thus the orbital $2p\sigma$ of a molecule XY is one with zero angular momentum about the nuclear axis which, in the limit of

MOLECULAR NEGATIVE IONS – GROUND STATES   161

vanishing nuclear separation, tends to the 2p state of the united atom. The designation $\sigma 2p^x$, on the other hand, refers to a 2p orbital which is derived from the 2p state of the atom X. For homonuclear molecules, each state of the separated atom gives rise to two molecular states, one symmetrical with respect to the nuclei, the other antisymmetrical. These are distinguished by the suffixes g, u.

The complete electronic state of the molecule is then specified by giving the number of electrons occupying each orbital, the total angular momentum of the system about the nuclear axis, and, for a homonuclear molecule, the nuclear symmetry of the whole. Capital Greek letters $\Sigma, \Pi, \Delta, \ldots$ are used to specify the total angular momentum about the nuclear axis. A Σ state of a homonuclear molecule may be either symmetrical or antisymmetrical with respect to the mid-plane through the nuclei. These are distinguished as Σ^+, Σ^- respectively.

The total electron spin associated with the state is indicated as for atoms by a pre-superscript denoting the multiplicity. Thus $^3\Sigma$ denotes a triplet Σ state.

We note that a σ orbital can accommodate at most two electrons with opposite spin, but a π orbital can take four (two for angular momentum $+1$ unit, two for angular momentum -1 unit). As an example of the nomenclature we have, for the ground state of O_2,

$$(\sigma_g 2s)^2(\sigma_u 2s)^2(\sigma_g 2p)^2(\pi_u 2p)^4(\pi_g 2p)^2 \, ^3\Sigma_g^+, \tag{6.2}$$

the K electrons of each atom being supposed to play no part in the structure of the molecule as such. The indices give the number of electrons in each orbital.

The advantage of this method of describing molecular states lies chiefly in the possibility of classifying given orbitals as bonding, antibonding or inactive. Bonding orbitals tend to assist molecule formation and usually arise from a state of the separated atoms having the same total quantum number as that of the united atom (non-promoted orbitals). Antibonding orbitals tend to oppose combination of the atoms and usually arise from a state of the separated atoms with total quantum number less, by one or more, than that of the united atom (promoted orbitals). Inactive orbitals have little effect either way.

For homonuclear molecules formed by atoms in the first row of the periodic table the order of the states is

$$(\sigma_g 2s)^2, \quad (\sigma_u 2s)^2, \quad (\sigma_g 2p)^2 \quad \text{or} \quad (\pi_u 2p)^4, \quad (\pi_g 2p)^4, \quad (\sigma_u 2p)^2. \tag{6.3}$$

The s shells of the separated atoms play little part in the binding of such molecules as N_2, O_2, etc. and we will not concern ourselves any further about orbitals arising from them. Of the remaining states $\sigma_g 2p$ and $\pi_u 2p$ are bonding, $\pi_g 2p$ and $\sigma_u 2p$ strongly antibonding, the latter particularly so. From the number of bonding and antibonding orbitals which make up a given molecular state, a good idea of the nature of the corresponding nuclear potential-energy curve may be obtained. An excess of bonding orbitals will usually give a curve with a minimum (stable state), while an excess of antibonding orbitals will have the reverse effect.

In terms of the united atom specification it may be shown that

$$\sigma_g 2s \longrightarrow 2s\,\sigma, \quad \sigma_u 2s \longrightarrow 3p\,\sigma, \quad \sigma_g 2p \longrightarrow 3s\,\sigma,$$
$$\pi_u 2p \longrightarrow 2p\,\pi, \quad \pi_g 2p \longrightarrow 3d\,\pi, \quad \sigma_u 2p \longrightarrow 4p\,\sigma. \quad (6.4)$$

It will be noted that, while for bonding orbitals the total quantum number is the same in the separated and united atom limits, for antibonding orbitals promotion occurs.

As a rough guide the binding energy of a homonuclear molecule composed of atoms of the first long row of the Periodic Table is 2.5 eV per valence bond, defined as the difference between the number of pairs of bonding and antibonding electrons. This rule is due to Mulliken (1932).

For heteronuclear molecules, in which the atoms are not too dissimilar, the g, u symmetry is no longer present and the order of the states is

$$(\sigma\,2s)^2, \quad (\sigma^*\,2s)^2, \quad (\sigma 2p)^2 \text{ or } (\pi\,2p)^4, \quad (\pi^*\,2p)^4(\sigma^*\,2p)^2. \quad (6.5)$$

In terms of the united atom specification

$$\sigma 2s \longrightarrow 2p\,\sigma, \quad \sigma^*\,2s \longrightarrow 3s\,\sigma, \quad \sigma\,2p \longrightarrow 3p\,\sigma,$$
$$\pi\,2p \longrightarrow 2p\,\pi, \quad \pi^*\,2p \longrightarrow 3p\,\pi, \quad \sigma^*\,2p \longrightarrow 3d\,\sigma. \quad (6.6)$$

Of these states $\sigma 2s$ and $\pi 2p$ are bonding, $\sigma^*\,2s$, $\sigma^*\,2p$ and $\sigma^*\,2p$ antibonding while $\sigma 2p$ tends to be antibonding at large and bonding at small nuclear separations.

If the atoms are very dissimilar the order of the states may differ very much from that given in (6.5). For the diatomic hydrides it is best to specify the states exclusively in terms of the united atom limit.

6.3 The vibrational and rotational energy

We consider first the vibrational and rotational energy levels associated with $^1\Sigma$ molecular states which possess no net electronic orbital or spin angular momentum.

Let $V(R)$ be the nuclear potential energy, R being the nuclear separation. The equilibrium separation R_e is defined by $V'(R_e) = 0$. To the approximation we first make, of complete independence, the rotation will take place about an axis through the centre of mass perpendicular to the nuclear axis, the moment of inertia C_0 about this axis being MR_0^2, where M is the reduced mass of the two atoms. The allowed energy levels for this simple rotator are given by

$$E^r = \frac{h^2}{8\pi^2 C_0} J(J+1), \quad J = 0, 1, 2, \ldots. \quad (6.7)$$

MOLECULAR NEGATIVE IONS – GROUND STATES   163

This formula may now be corrected to allow for the lack of rigidity of the molecule. The centrifugal force due to the rotation will tend to increase the nuclear separation. This effect will increase with the angular momentum and hence with J. It can be taken into account by introducing a correction term proportional to $J^2(J+1)^2$, giving

$$E^r = E_0^r J(J+1) - E_1^r J^2(J+1)^2, \qquad (6.8)$$

where $E_0^r = \dfrac{h^2}{8\pi^2 C_0}$, $E_1^r > 0$.

A further correction can now be introduced to allow for the effect of vibration on the rotational energy. Vibration will affect the mean moment of inertia to an extent which will depend on the amplitude of vibration. Both coefficients E_0^r and E_1^r must then be regarded as functions of v, the quantum number specifying the vibration. We shall be able to consider this in more detail after consideration of the vibrational energy.

Provided the amplitude of vibration is not too large we may write the nuclear potential energy in the form

$$V(R) = V(R_0) + \tfrac{1}{2}(R - R_0)^2 \left(\frac{d^2 V}{dR^2}\right)_{R=R_e}. \qquad (6.9)$$

Classically this gives simple harmonic vibrations of frequency

$$\nu_0 = \frac{1}{2\pi}\left\{\frac{1}{M}\left(\frac{d^2 V}{dR^2}\right)_{R=R_e}\right\}^{1/2}$$

and in quantum theory a correspondingly simple formula for the energy

$$E^v = (v + \tfrac{1}{2})h\nu_0, \quad v = 0, 1, 2, \ldots \qquad (6.10)$$

This formula is satisfactory for small values of v but for larger values the neglect of higher terms in the expansion of $V(R)$ becomes serious – classically, the vibrations become anharmonic. One method of dealing with this effect is to expand E^v in powers of $v + \tfrac{1}{2}$ so that

$$E^v = (v + \tfrac{1}{2})h\nu_0\{1 - x_e(v + \tfrac{1}{2}) + \cdots\}. \qquad (6.11)$$

An alternative procedure is to attempt to represent the shape of the nuclear potential-energy curve by an analytical function of sufficient simplicity for the vibrational energy to be obtainable as an analytical expression. The most useful function of this kind was proposed by Morse (1929) and has the form

$$V(R) = D[1 - \exp\{-a(R - R_0)\}]^2. \qquad (6.12)$$

D is chosen to agree with the energy of the separated atoms to which the molecular state tends at large separations and a is adjusted to give the correct frequency of oscillation with small amplitude about the equilibrium separation R_0. The vibrational energy is then given by

$$E^v = (v + \tfrac{1}{2}) h\nu_0 \left\{ 1 - \frac{ah}{(8\mu D)^{1/2}} (v + \tfrac{1}{2}) \right\},$$

where μ is the reduced mass of the system. It is noted that the number of vibrational energy levels for this potential is finite. This will be so whenever the attractive potential energy falls off more rapidly than R^{-2} at large distances.

Fig. 6.1(a) illustrates the way in which the separations of successive vibrational levels for the Morse potential become smaller as the degree of excitation increases.

The Morse potential is remarkably successful in providing a very good initial approximation to the shape of a nuclear potential-energy curve. It is particularly useful in working back to the curve from a knowledge of the vibrational energy levels.

The presence of rotational motion will, in the first instance, affect the vibration by adding an additional repulsive potential energy due to centrifugal force. Thus, if the rotational quantum number is J, this additional energy is given by $h^2 J(J+1)/8\pi^2 C_0 R^2$. Except for very large values of J this is usually quite small for molecules of present interest though there are some effects, such as dissociation due to high rotation, of which it is the prime cause.

In gathering together the formulae for the energy of combined nuclear vibration and rotation we shall use the standard notation of molecular spectroscopy and work in terms of term values rather than energy. The term value is given by E/hc. We write then

$$E^r/hc = B_v J(J+1) - D_v J^2 (J+1)^2, \qquad (6.13)$$

where B_v and D_v depend on the vibrational quantum number v. This dependence may be expressed in terms of a power series

$$B_v = B_e + B_1 v + \ldots, \qquad (6.14a)$$

$$D_v = D_e + D_1 v + \ldots, \qquad (6.14b)$$

where

$$B_e = \frac{h}{8\pi^2 c C_0}. \qquad (6.15)$$

Similarly we have

$$E^v/hc = (v + \tfrac{1}{2})\omega_e - \omega_e x_e (v + \tfrac{1}{2})^2 + \omega_e y_e (v + \tfrac{1}{2})^3 \qquad (6.16)$$

where $\omega_e = \nu_0/c$ and x_e and y_e depend on the shape of the potential-energy curve.

For molecules which are not in $^1\Sigma$ states account must be taken of the coupling of the nuclear with the electronic angular momentum. It would be out of place here to discuss the various possibilities which arise as there are few cases in which they affect the behaviour of molecular negative ions that we shall be discussing.

6.4 The Franck–Condon principle

Transitions between electronic states of molecules will normally take place so rapidly that there will be insufficient time for the nuclei to have moved from their initial positions. Thus in diatomic molecules those transitions will be most probable which leave the nuclear separation unaltered. This may be expressed in terms of the wave functions for the molecular energy states as follows.

If we denote the coordinates of the electrons relative to the nuclei by $\mathbf{r}$, those of the nuclei by $\mathbf{R}$, the wave function for the electronic motion will be of the form $\psi(\mathbf{r},\mathbf{R})$ and those for the nuclear motion $\chi(\mathbf{R})$. The probability of any transition between two molecular states in which the electronic state changes from n to m and the nuclear from s to t will be determined mainly by the square of the perturbation energy $V(\mathbf{r},\mathbf{R})$ causing the transition, averaged over the wave functions of the initial and final states, viz.

$$\left| \iint \psi_m^*(\mathbf{r},\mathbf{R}) \chi_t^*(\mathbf{R}) V(\mathbf{r},\mathbf{R}) \psi_n(\mathbf{r},\mathbf{R}) \chi_s(\mathbf{R}) \, d\mathbf{R} \, d\mathbf{r} \right|^2.$$

The probability will be greatest for given n, m, s, when the overlap between χ_t and χ_s is a maximum. Now these wave functions are greatest in the region of allowed classical motion and χ_s is therefore only appreciable for nuclear separations $R_1 < R < R_2$ confined within the amplitude of classical oscillations. The probability will be a maximum when χ_t is also large in this region so that the range of nuclear separations involved in the transition is effectively between R_1 and R_2. In general χ_s will be a maximum at the classical turning points so, for given m, n and s the transition probability considered as a function of t will have two maxima. If, however, χ_s is the ground state vibrational function corresponding to the electronic state m it will have a single maximum (see Fig. 6.1). In such a case the transition probability will have a single maximum only as t varies (see Chapter 11, p. 480).

The Franck–Condon principle will lose its strict validity if χ_s

or χ_t are appreciable outside the regions of classically allowed motion, but even for the lightest molecule it presents a sufficiently good approximation for qualitative argument (James, Coolidge and Present, 1936).

Electron affinity and vertical detachment energy

Just as for atoms and atomic ions the electron affinity of a molecule or radical is defined as the energy excess of the ground state of the neutral system over that of the negative ion. However, whereas the electron affinity of an atom is also equal to the minimum energy necessary to detach an electron from the ion this is not necessarily so, and in fact will usually not be correct, for diatomic and polyatomic systems. Thus detachment of an electron will usually take place without change of the nuclear coordinates, the Franck–Condon principle being applicable. It is therefore convenient to define the *vertical detachment energy* as the minimum energy required to detach an electron from the negative ion without change in the nuclear coordinates. This is analogous to the vertical ionization energy of a normal molecule, a concept introduced by Mulliken (1934). The vertical detachment energy may be greater than, equal to or less than the electron affinity.

6.5 Theoretical calculation of the properties of the ground states of diatomic molecules

Most detailed calculations of the properties of diatomic molecules have been carried out by an extension of Roothan's method for determining the HF self-consistent fields for atoms. Thus for molecules with closed shells the trial wave function is a determinant of the form (2.5) but with the functions $\psi(k|nl)$ now replaced by molecular orbitals $\psi(k|i\lambda)$. In the spirit of the Roothan method for atoms the molecular orbitals are built up of a combination of functions centred on the respective nuclei which have the form (2.6) and the appropriate molecular symmetry. Thus $\psi(k|i\lambda)$ is written in the form

$$\psi(k|i\lambda) = \phi_{i\lambda}^A(\mathbf{r}_A) + \phi_{i\lambda}^B(\mathbf{r}_B), \tag{6.17}$$

where $\phi_{i\lambda}^A$, $\phi_{i\lambda}^B$ are functions centred on the respective nuclei A and B. These functions are then expanded in the form (cf. (2.7))

$$\phi_{i\lambda}^{A;B} = \sum_p \sum_q \sum_l C_{pq,\,i l \lambda}^{A,\,B} \exp\{-\zeta_{pq,\,l\lambda}^{A,\,B} r\} r^{q-1} P_l^{|\lambda|}(\cos\theta) \exp(i\lambda\phi),$$

(6.18)

which has the appropriate symmetry about the molecular axis if this is taken as the axis of polar coordinates. The exponents ζ and coefficients C are then optimized by use of the variational principle consistent with the requirement of orthogonality of the different molecular orbitals.

Although following essentially the same principle as for atoms, it is clear that the need to include functions centred around two separated nuclei greatly complicates the calculation so that it is difficult to include enough terms of the full expansion (6.18) to yield a close approximation to the true self-consistent field. This is usually referred to as the truncation error.

In addition there is the difficulty that, in general, functions of the form (6.17) do not tend to the correct limit as the nuclear separation tends to infinity. Determinants representing other configurations must be added to the trial wave function to eliminate this. In fact the correction to the self-consistent field results arising from correlation between electrons will depend on the nuclear separation. Because of this, even an accurate solution for the self-consistent field will not yield a good value for the dissociation energy. However, it will usually give a good approximation to the equilibrium nuclear separation and to the charge distribution. It may also give quite good results for the curvature of the potential-energy curve near the minimum.

Improvement of the HFR basic function requires the introduction of further configurations but, as for atoms, the convergence will be very slow unless these are properly chosen. There is usually no very great difficulty in adding the minimum number of configurations necessary for the wave function to tend to the correct asymptotic limit but the problem of calculating to useful accuracy the variation of the correlation energy with nuclear separation is more serious. Two methods, one referred to as that of pseudo-natural orbitals (PNO), the other as that of optimized valence configurations (OVC) have been used with considerable success for a number of neutral diatomic molecules. The PNO method is basically an extension to molecules of Nesbet's adaption of the Bethe–Goldstone procedure to atoms (see Chapter 2, p. 26). In the OVC method the basic trial function Φ_0 is taken to be that given by the HFR approximation plus the minimum number of configurations added to give the correct asymptotic limit at large nuclear separations. To this is added a number of configurations in which pairs of electrons outside the atomic cores are excited to other orbitals.

These methods are immediately applicable to molecular negative ions with stable ground states and the same limitations apply. Moreover, the methods will, in general, yield potential-energy curves at all nuclear separations even for those ions which are unstable towards autodetachment over a certain range of separations. Over these ranges the methods yield approximations to the real part of the potential-energy curve in much the same way as does the stabilization procedure. The latter method has been used effectively for autodetaching states of H_2^- (see pp. 174–9) as has also been the variation procedure based on the concept of a Siegert resonance (see Chapter 4, p. 104). The alternative approach via scattering theory may also be used and has been applied particularly to N_2^-.

6.6 The observation of autodetaching levels associated with the ground states of diatomic negative ions

The possibilities which arise in the relations between the electronic energy of the ground state of a diatomic negative ion and of the corresponding neutral molecule have been discussed in §6.1 and illustrated in Fig. 6.2.

In the case illustrated in Fig. 6.2(b) the ground state of the ion is unstable towards autodetachment for a considerable range of nuclear separation. The effects which arise because of this depend very much on whether the lifetime towards autodetachment is much larger than, much smaller than, or comparable with the vibrational period of the ion; or in terms of the level width Γ and the vibrational frequency ν, whether $\Gamma \ll$, $\simeq$ or $\gg h\nu$.

If $\Gamma \ll h\nu$ many vibrational oscillations can occur during the lifetime of the ion and it will be a good approximation to determine the vibrational energy spectrum as if the ion were completely stable. Consider now the collision of electrons with the neutral molecule. Resonance effects will occur when the electron energy E is such that

$$E + E_0 = E_{v,r},$$

where E_0 is the energy of the ground state of the neutral molecule and $E_{v,r}$ is that of the negative ion in the vibrational state with quantum number v. Thus the transmission of electrons through the molecular gas should show resonance peaks at these energies corresponding to the different vibrational levels of the ion. The spectrum of resonance peaks will be the same, as far as energy separation is concerned, as that of the vibrational energy levels of the molecular ion.

This resonance energy spectrum will also appear in the total and differential cross-sections for excitation of different vibrational states of the neutral molecule by electron impact. Furthermore, if $\Gamma \ll h\nu$, they will occur at the same incident electron energies for all final states, including the case of elastic scattering.

When these conditions prevail, the shape of the potential-energy curve of the negative ion may be determined in the usual way once the resonance spectrum, and hence the vibrational energies of the ion, have been observed. This applies to O_2^- and NO^- as described on pp. 189 and 193 respectively.

In the other extreme case, in which $\Gamma \gg h\nu$, there is no time for vibration of the ion to develop and no resonance structure associated with it will be observable. This is the case for H_2^- (see p. 178).

The intermediate case, in which Γ and $h\nu$ are comparable, is one in which resonance effects associated with vibration occur but the peaks are located at different energies for different collision processes. This situation applies to N_2^- and CO^- (see pp. 198–207).

We have not, in this discussion, taken account of rotation of the molecule. In many cases the rotational period will be long compared with the lifetime of the ion but, for O_2^-, this is not so and account must be taken of the rotation in determining, for example, the angular distribution of the electrons scattered after exciting a particular vibrational state of O_2.

In other cases, in which rotational effects are unimportant, the nature of the electronic state of the ion which is responsible for the resonance effect can be determined from observations of the angular distribution of the scattered electron at energies close to resonance, just as for atoms (see Chapter 4, p. 87). Allowance must of course be made for the non-central character of the interaction.

To see, in descriptive terms, what is to be expected when the period of rotation is long compared with the lifetime of the autodetaching state, let Θ, Φ be the polar angles of the direction of the molecular axis with respect to the direction of incidence and ϑ, ϕ those of the direction of scattering relative to the molecular axis. We then have, if θ is the angle of scattering

$$\cos\theta = \cos\Theta\cos\vartheta + \sin\Theta\sin\vartheta\cos(\Phi - \phi)$$

and

$$P_l(\cos\theta) = P_l(\cos\Theta) P_l(\cos\vartheta)$$
$$+ 2\sum \frac{(l-|m|)!}{(l+|m|)!} P_l^m(\cos\Theta) P_l^m(\cos\vartheta) \cos\{m(\Phi - \phi)\}.$$

To the approximation in which the scattering potential can be regarded as central, the scattered amplitude for electrons of wave number k will take the form (4.32) which may be written in terms of the coordinates $(\Theta, \Phi), (\vartheta, \phi)$ as

$$\frac{1}{2ik}\sum_l (2l+1)\{\exp(2i\eta_l) - 1\} \sum_{m=-l}^{l} \frac{(l-|m|)!}{(l+|m|)!}$$
$$\times P_l^m(\cos\Theta) P_l^m(\cos\vartheta) \cos m(\Phi - \phi), \qquad (6.19)$$

or, in terms of normalized spherical harmonics,

$$Y_l^m(\theta, \phi) = (4\pi)^{-1/2}\left\{(2l+1)\frac{(l-|m|)!}{(l+|m|)!}\right\}^{1/2} P_l^m(\cos\theta) \exp(im\phi), \quad (6.20)$$

as

$$(2\pi/ik)\sum_l \{\exp(2i\eta_l) - 1\} \sum_{m=-l}^{l} Y_l^m(\Theta, \Phi) Y_l^{-m}(\vartheta, \phi). \qquad (6.21)$$

As a next approximation, to allow for the axial symmetry we allow for a dependence of the phase shift η on $|m|$ as well as on l so that, in place of (4.32), we have

$$(2\pi/ik)\sum_{l,m} \{\exp(2i\eta_{l|m|}) - 1\} Y_l^m(\Theta, \Phi) Y_l^{-m}(\vartheta, \phi). \qquad (6.22)$$

For scattering due to a resonance state which, in the united atom limit is a state with quantum numbers $l, |m|$, the angular distribution will be given to this approximation, for a fixed direction of the nuclear axis, by

$$|Y_l^m(\Theta, \Phi) Y_l^{-m}(\vartheta, \phi) + Y_l^{-m}(\Theta, \Phi) Y_l^m(\vartheta, \phi)|^2. \qquad (6.23)$$

To relate to observation this must be averaged over all orientations of the molecular axis. Thus, to take a simple example, for a pσ resonance with $l = 1$, $m = 0$ we need to average $\cos^2\Theta \cos^2\vartheta$ in this way. Now
$$\cos\vartheta = \cos\Theta \cos\theta + \sin\Theta \sin\theta \cos(\Phi - \alpha),$$
where α is the azimuthal angle of the direction of scattering relative to that of incidence. It may easily be shown that

$$\frac{1}{4\pi} \int_0^{2\pi} \int_0^{\pi} \cos^2\Theta \{\cos\Theta \cos\theta + \sin\Theta \sin\theta \cos(\Phi - \alpha)\}^2 \sin\Theta \, d\Theta \, d\Phi$$

$$= \tfrac{2}{15}(2\cos^2\theta + 1). \tag{6.24}$$

In other states the calculation is more complicated but may be carried out using the transformation formulae for spherical harmonics referred to different axes. In particular we find for pπ states the characteristic angular distribution is given by
$$1 + 7\cos^2\theta, \tag{6.25}$$
and for dπ by
$$1 - 3\cos^2\theta + (14/3)\cos^4\theta. \tag{6.26}$$
For both the p states, the distribution generally resembles that for p states in the central case with a minimum at 90°, while for dπ it has a maximum at 90° with minima near 50° and 130°, closely similar to that for d states in the central case.

These formulae are still only approximate because in the non-central case the angular functions will not remain unmodified as we have assumed. However, for the applications we are considering the electron will normally be attached in an antibonding orbital localized at a considerable distance from the centre of the molecule so that on the average it feels a nearly central potential.

It must be remembered also that the angular distributions we have been discussing are those which will be found if the resonance state dominates the scattering process concerned. Otherwise there will be contributions from direct scattering which will modify the distribution. Such modifications will usually be present to a significant extent in the elastic scattering but less so for inelastic collisions involving excitation of vibration.

Application of these considerations to N_2^- and CO^- is discussed on pp. 198 and 206 respectively.

The experimental methods for observing resonant effects are essentially the same as for atomic ions, the energy resolution in modern experiments being sufficient to resolve clearly vibrational energy losses. However, except for H_2 and D_2, rotational energy losses are not resolved.

6.7 Electron affinities of diatomic molecules and structure of the ground states of the negative ions

The determination of electron affinities of diatomic molecules

The problem of determining electron affinities of diatomic molecules is considerably more difficult than for atoms.

Empirical procedures are harder to apply for several reasons. Thus it is not possible to compare isoelectronic sequences of similar molecules in different states of ionization – few doubly-ionized molecules are stable and this applies *a fortiori* to more highly-ionized systems. Apart from the relatively greater difficulty of obtaining a good approximation to the HF self-consistent field, the estimation of correlation errors is more complicated. In certain cases, as for the diatomic hydrides for example, it appears that a good estimate may be made by comparison with the corresponding correction for the united atom and its negative ion. Another method is to compare with isoelectronic atomic or molecular systems for which estimates are available.

The determination of electron affinities from lattice energies and the Born–Haber cycle is possible in principle for diatomic radicals such as OH though is less reliable because of the anisotropic character of the negative ion. It is rarely useful for the electron affinities of stable molecules though there are exceptions such as O_2 which forms super-oxides with the alkali metals.

Of experimental methods, measurements of the threshold energy for photodetachment, which is the most accurate method for atomic ions, is complicated for molecules by the possibility that the neutral molecule may be left in an excited vibrational state. In certain cases, such as ON and SH, this is unimportant and a direct measurement can be made of the threshold energy for the $0 \to 0$ vibrational transition from ion to atom. For other cases, such as O_2, it is very difficult to disentangle the vibrational transitions. However, these difficulties may be overcome if it is possible to measure the energy and angular distributions of the photoelectrons from detachment by laser light of an accurately known wavelength. This method, which has been very successfully employed for O_2 and NO and for a number of other molecules is undoubtedly the most accurate and definite one available and has great potentialities. It also provides information, not only about the electron affinity but also about the equilibrium separation R_e and curvature of the potential-energy curve of the ion at that separation, i.e. about R_e and ω_e. Details of photodetachment techniques and applications are discussed in Chapter 11.

Although a number of experiments have been carried out to determine electron affinities of molecules by hot-wire equilibrium measurements, the results are often uncertain due to difficulties of interpretation. The variety of possible processes which can occur on the surface is usually much greater than for atoms.

Apart from rather special cases, such as that of O_2, for which a quite accurate value of the electron affinity has been obtained from observations of the equilibrium constant for the reaction

$$O_2 + e \rightleftharpoons O_2^-$$

in pulse experiments on attachment and detachment in O_2 (see Chapter 12, p. 530), further information about molecular electron affinities has come from the observation of thresholds for suitably chosen reactions arising from electron or heavy particle impact.

On p. 41 we have already drawn attention to some of the difficulties associated with the interpretation of appearance potentials for atomic negative ions produced by electron impact. These are enlarged upon in Chapter 9, p. 276. While the same difficulties naturally arise also when the ions are molecular, there is the additional problem of allowing for the fact that these ions may be formed in different vibrational states.

Considerable progress has been made in obtaining reliable values of electron affinities by determining the threshold for a suitable exothermic reaction. Examples are charge transfer reactions such as

$$Cl_2 + I^- \longrightarrow Cl_2^- + I,$$

which are discussed in Chapter 12, p. 566, and such as

$$K + Br_2 \longrightarrow K^+ + Br_2^-,$$

discussed in Chapter 10, p. 394. In measurements of this type the chief difficulty is that of determining the true threshold by proper allowance for temperature and possibly other effects.

It is also possible in some cases to provide useful upper and lower limits to the electron affinity by determining whether certain reactions are exothermic or not. This may be done by measuring the mean cross-section for the reaction as a function of temperature. Although this seems to be a relatively simple procedure, inconsistent results have sometimes been obtained in practice (see for example, the case of O_2 on p. 184).

Determination of structural properties for diatomic molecular ions

We have already pointed out that laser photodetachment experiments make it possible to determine the equilibrium separation and fundamental vibrational frequency of the ion. These quantities may also be determined in certain cases such as for O_2 and NO in which the upper vibrational states of the ground electronic state of the ion are unstable towards autodetachment and hence may be located from resonance peaks in collision experiments (see Chapter 4, p. 90).

Also, again as pointed out above, calculations of HFR type probably give more reliable information about these properties than about electron affinities.

We now go on to consider specific molecules.

H_2^-

The ground state is the

$$(1s\,\sigma_g)^2 (2p\,\sigma_u)\,^2\Sigma_u^+$$

state involving two bonding orbitals and one antibonding orbital. We would therefore expect only a weakly-bound molecule as compared with H_2 for which the ground state is a $(1s\,\sigma_g)^2\,^1\Sigma_g$ state with no antibonding orbital. This suggests that the potential-energy curve for H_2^- is of the form shown in Fig. 6.2(b) so that for

MOLECULAR NEGATIVE IONS – GROUND STATES 175

nuclear separations $R < R_c$ the molecular ion is unstable towards autodetachment.

Early attempts to calculate the potential-energy curve of the ground state were carried out by the variation principle without taking into account the possibility of autodetachment. As the complexity of the trial wave functions increased, the energy, at a nuclear separation equal to the equilibrium separation in H_2, decreased until Taylor and Harris (1963) showed that the lowest energy was simply that of the ground state of H_2 plus an unbound electron of zero energy. They found, however, that for $R > 3a_0$ the energy

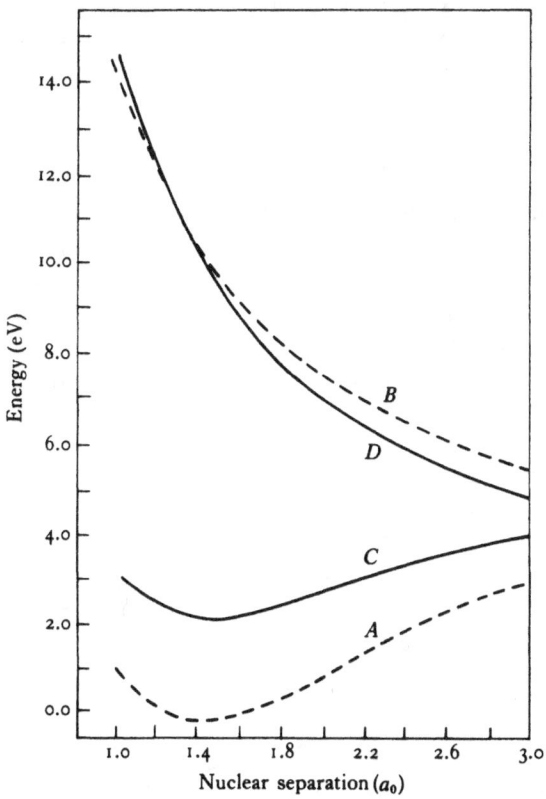

Fig. 6.3. Real parts of potential-energy curves for H_2^- which tend in the limit of large nuclear separations to ground state H and H$^-$, calculated by Eliezer et al. (1967). The corresponding curves for H_2 are included for comparison. For H_2 (broken lines) A, $X^1\Sigma_g^+$ ground state; B, $^3\Sigma_u^+$. For H_2^- (full lines) C, $^2\Sigma_u^+$ ground state; D, $^2\Sigma_g^+$.

of H_2^- is lower than that for H_2 as it must of course be in the limit $R \to \infty$.

Eliezer, Taylor and Williams (1967) later applied the stabilization method (Chapter 4, p. 95) to seek for the resonance energy as a function of R for $R < R_c$. Starting from a ground state wave function for H_2 built up from the three configurations $\sigma_g 1s \sigma_g 1s'$, $\sigma_u 1s \sigma_u 1s'$ and $\pi_u^c 2p_{+1} \pi_u^c 2p'_{-1}$, in which the orbitals labelled ^{c} have the same average radial extent as the $\sigma_g 1s$ orbital, a third σ_u electron was added to each so as to give H_2^- configurations of $^2\Sigma_u^+$ symmetry. Fig. 6.3 illustrates the real part of the potential-energy curve for H_2^- obtained in this way.

Bardsley, Herzenberg and Mandl (1966) made detailed calculations for the same range of nuclear separations but took into account explicitly that the energies of the H^- states are complex. Defining the complex energy according to the Siegert definition (see Chapter 4, p. 104) they determined it by using the variational formalism described in Chapter 4, p. 106. Their results depend somewhat on the range assumed for the interaction (see (4.167)) particularly for small nuclear separations. However, the real part of the energy is quite close to that obtained by Eliezer *et al.* near the equilibrium separation.

Fig. 6.4 shows the imaginary part obtained with a truncation range of $15a_0$ for the interaction. In the neighbourhood of the

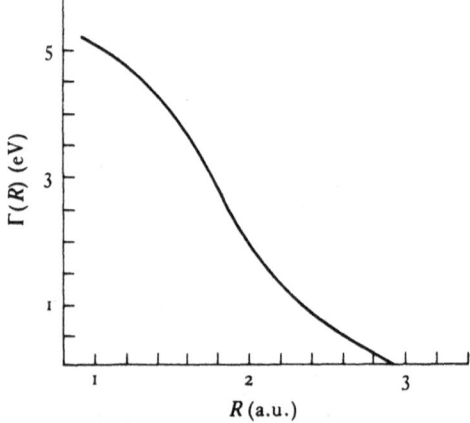

Fig. 6.4. Width Γ of the $^2\Sigma_u^+$ state of H_2^- as a function of nuclear separation, calculated by Bardsley *et al.* (1966).

equilibrium separation the level width is nearly 2.5 eV which corresponds to a lifetime of only 2×10^{-16} s. This is much smaller than the vibrational period as may be seen from the fact that the vibrational quantum is only about 0.5 eV. In fact the resonance states are so broad as to be barely recognizable as such.

The existence of a $2p\sigma_u$ resonance at an energy nearly 2.3 eV

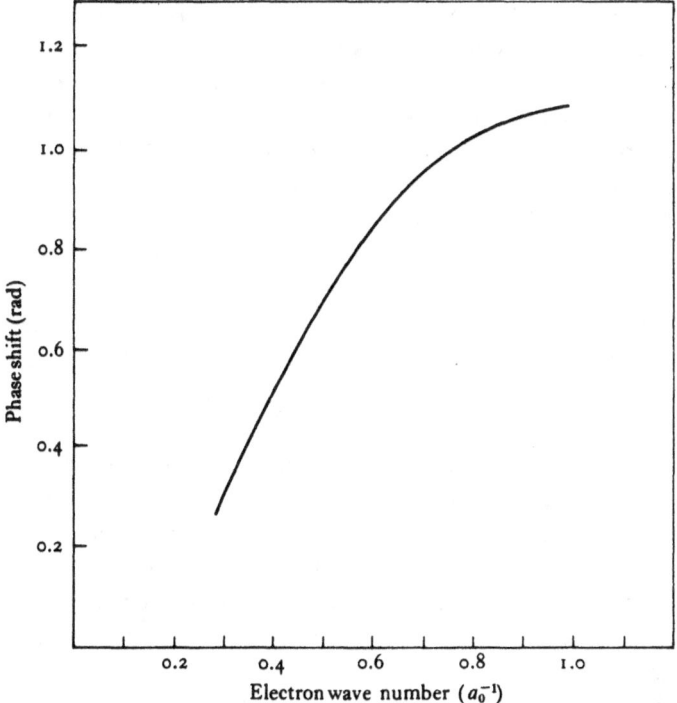

Fig. 6.5. Variation with electron energy of the $2p\sigma$ phase shift for scattering of electrons by H_2, calculated by Hara (1969).

above the ground state of H_2 should be apparent in the elastic scattering of electrons by H_2. In fact the corresponding phase shift which is close to η_{10} in the notation of p. 170 behaves as a function of electron energy as shown in Fig. 6.5, which reproduces the results of calculations using spheroidal coordinates. The time delay due to scattering imposed on an electron in a $p\sigma$ state, which according to (4.42) of Chapter 4 is given by $2\hbar\partial\eta_{10}/\partial E$, has a maximum of about 10^{-16} s at an energy near 2.5 eV. This corresponds to the Siegert resonance state discussed above. The differential and total elastic cross-sections calculated with the same assumptions agree fairly well with the observed results and this goes some way to confirm the existence of the Siegert resonance. This is a type II resonance in which the additional electron is temporarily trapped by an effective p-wave potential barrier (see Chapter 4, p. 97).

Vibrational excitation of H_2 may be regarded as arising through capture into the resonance state as an intermediate step. However, the lifetime of the state is so short that the process may equally well be treated by procedures which include the relatively gradual variation of electron hold-up time in the neighbourhood of the molecule but do not consider it in terms of a resonance state. We shall see, on p. 186, the opposite extreme case of O_2 in which the intermediate complex has a lifetime long compared with the vibrational period and the resonance aspect is dominant.

One further configuration of H_2^- tends in the limit of large nuclear separation R to ground state H and H^-. This is the

$$1s\,\sigma_g (2p\,\sigma_u)^2\ ^2\Sigma_g^+$$

state obtained by attachment of a $2p\,\sigma_u$ electron to the $1s\,\sigma_g\,2p\,\sigma_u\ ^3\Sigma_u^+$ excited configuration of H_2 which, in the limit of large R, tends to two ground state H atoms. Fig. 6.3 shows the calculated energy curve of the resonance state obtained by Eliezer et al. (1967) using the stabilization method. At some large separation R_1, around $6a_0$–$7a_0$ and beyond, the energy of the $^2\Sigma_g^+$ state falls below that of the ground state of H_2.

It will be seen that the curve lies below that for the $^3\Sigma_u^+$ state of H_2 for $R > 1.4a_0$ and above for $R < 1.4a_0$. Thus the state would be of type I for $R > 1.4a_0 < R_1$ and type II for $R < 1.4a_0$. As this result is sensitive to the trial wave functions employed it cannot be regarded as definitely established. Even if a cross-over occurs it may not be at the calculated separation. In fact Bardsley et al. (1966), determining the location in the complex plane of the appropriate Siegert resonance, find the resonance to be of type II at all separations $< R_1$. Fig. 6.6 shows the level widths which they obtain using the same truncated range of interaction as for the $^2\Sigma_u^+$ state. Again they are large.

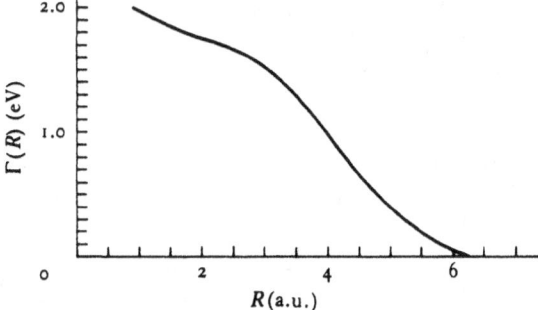

Fig. 6.6. Width Γ of the $^2\Sigma_u^+$ state of H_2^- as a function of nuclear separation, calculated by Bardsley et al. (1966).

The properties of the $^2\Sigma_u^+$ state may be investigated from experimental observation of dissociative attachment of electrons in H_2,

$$H_2 + e \longrightarrow H + H^-.$$

This is discussed in Chapter 9, p. 312, where it is shown that the general description of the real and imaginary parts of the energy of the state as functions of nuclear separation, given above, is in agreement with the observed results.

OH⁻ and SH⁻

The electron affinities of OH and SH are well determined from the observed variation with frequency of the cross-sections for photodetachment as discussed in detail in Chapter 11, p. 474. There is clear evidence from the observations, which also included a study of photodetachment from OD⁻, that no appreciable excitation of transitions, other than between the ground vibrational states of the ion and neutral molecule, occurs.

For OH, the difference in equilibrium nuclear separation R_e between an ion and neutral molecule could be placed at less than 0.001 Å. From analysis of the threshold behaviour, allowing for the doublet character of the ground state of OH and for rotational transitions, the electron affinity of OH came out to be 1.825 ± 0.002 eV, differing from that of OD by 0.002 eV. From this latter result it may be deduced that the zero point energy of OH⁻ differs from that of OH by less than 0.008 eV (see p. 476).

The theoretical description of OH and OH⁻ is also relatively complete and satisfactory and agrees well with the conclusions from the photodetachment experiments.

The ground $X\,^1\Sigma^+$ state of OH⁻ has the configuration

$$1\sigma^2\,2\sigma^2\,3\sigma^2\,1\pi^4,$$

obtained by adding an antibonding π electron to that of the ground $X\,^2\Pi$ state of OH. The most elaborate calculations of the properties

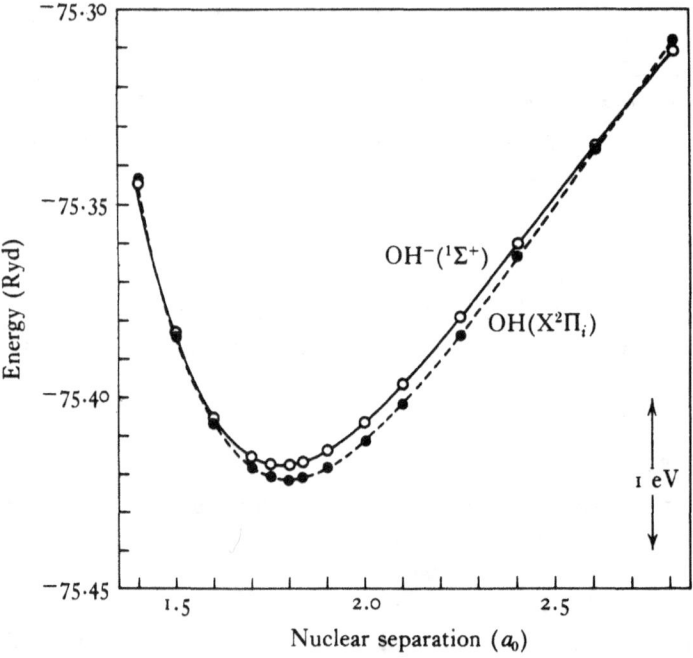

Fig. 6.7. Potential-energy curves for the ground states of OH and OH⁻ calculated using the HFR method by Cade (1967a).

of the $X\,^1\Sigma^+$ state are those of Cade (1967a) using the HFR method with the same basis set as used in the corresponding case of the neutral radical. The exponents ζ as well as the linear parameters were reoptimized but, whereas the latter varied with nuclear separation, the ζ were taken to be the same as at the equilibrium separation R_e.

Fig. 6.7 shows the calculated potential-energy curves for OH and OH⁻.

It will be seen that the calculated potential-energy curves are indeed very similar. For OH and OH⁻ the calculated values of R_e are 1.795 and 1.781a_0 respectively and those of ω_e 4062.0 and 4087.9 cm⁻¹ respectively. However, the calculated electron affinity is negative (−0.10 eV) which is not surprising as no account is taken of the correlation energy. This was estimated by Cade in two different ways.

The first, probably more reliable, procedure takes account of the fact that, for neutral hydrides, the correlation energy decreases from the united atom to the molecule at the equilibrium separation by an amount between 0.3 and 0.5 eV which does not depend to an important extent on the nuclear charge. Assuming the same behaviour for the negative ions it is possible to use the estimated correlation energies for atomic negative ions (see Chapter 2, p. 25) to estimate those for the negative hydride ions. In this way the correlation energy for OH⁻ is estimated to be less than that of OH by 2.09 eV at the equilibrium separation. Allowance for this gives an estimated electron affinity of 1.91 eV, not very different from that 1.83 eV determined experimentally.

The second procedure compared correlation energies for different isoelectronic systems, irrespective of nuclear charge. Thus, if the change (1.23 eV) in correlation energy for

$$HF(X\,^1\Sigma^+) \longrightarrow HF^+(X\,^2\Pi) + e$$

is taken to be the same as that for

$$OH^-(X\,^1\Sigma^+) \longrightarrow OH(X\,^2\Pi) + e,$$

the estimated electron affinity for OH is 1.13 eV, rather too small.

By comparison with observed spectroscopic data for OH and also between observed and calculated data for other hydrides, it was possible to correct other calculated results for OH⁻.

Fig. 6.8 shows how the calculated dipole and quadrupole moments as a function of nuclear separation differ between the negative ion and neutral molecule.

Before leaving OH attention should be drawn to the electron affinity (2.1 eV) derived by Goubeau and Klemm (1937) from analysis of the lattice energies of the alkali hydroxides and (1.8 eV) by Kay and Page (1966) from a study of the negative ions resulting from dissociation of H_2O_2 on a hot filament.

Results of comparable accuracy are available for SH⁻. Measurements of the photodetachment cross-section as a function of frequency show again no appreciable contribution from transitions other than that between the ground vibrational states. For both SH⁻ and SD⁻, R_e differs by less than 0.02 Å from the values for the respective neutral molecule while ω_e for SH⁻ differs from that for SH by less than 0.035 eV (300 cm⁻¹). The electron affinity is determined as 2.319 ± 0.010 eV.

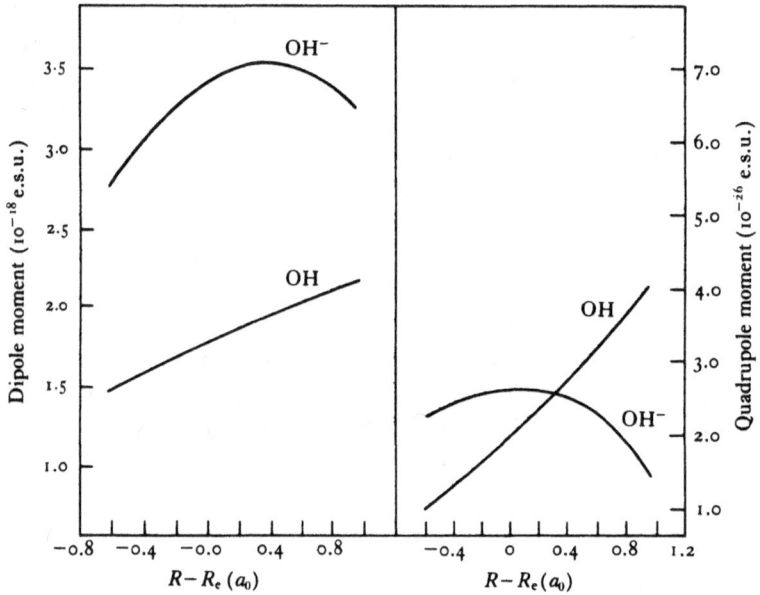

Fig. 6.8. Dipole and quadrupole moments of OH and OH⁻ calculated as a function of nuclear separation by Cade (1967a). R_e is the equilibrium separation.

Cade's calculations, carried out as for OH, give potential-energy curves for SH and SH⁻ which exhibit the same close similarity. In contrast with OH, the electron affinity given by the HFR approximation is already positive (1.21 eV). Empirical corrections for correlation energy add 1.04 and 0.90 eV according to the first and second of the two methods applied to estimate this energy difference for OH and OH⁻. The final estimated values of the electron affinity are thus 2.25 and 2.11 eV which are quite close to the observed values and also closer to each other than for OH.

Other diatomic hydrides

In view of the success of the empirical methods of correcting the HFR calculations for correlation energy Cade (1967b) also made estimates of the electron affinities of CH, NH, SiH and PH. Table 6.1 summarizes the results, together with those discussed above for OH and SH.

TABLE 6.1 *Semi-empirical electron affinities (in eV) for diatomic hydrides*

	E_a(HFR)	SE(1)	SE(2)	E_a(corr, 1)	E_a(corr, 2)	Observed
CH	0.28	1.33	0.28	1.61	0.56	0.74
NH	−1.55	2.01	1.92	0.46	0.37	0.38 ± 0.03
OH	−0.10	2.09	1.23	1.99	1.13	1.825 ± 0.002
SiH	0.64	0.82	0.85	1.46	1.49	–
PH	−0.23	1.16	1.09	0.93	0.86	–
SH	1.21	1.04	0.90	2.25	2.11	2.319 ± 0.010

E_a(HFR) is the electron affinity calculated in the HFR approximation, SE(1), SE(2) are the correlation energy corrections estimated by the first and second methods described above, E_a(corr, 1) and E_a(corr, 2) the corresponding final values of the electron affinity.

Since the semi-empirical estimates were made, further measurements have become available. Celotta, Bennett and Hall (1974), from the energy spectra of the electrons detached from the ions by argon ion laser light (Chapter 11, p. 478), find E_a(NH) = 0.38 ± 0.003 eV while Feldmann (1970), from observation of the photo-detachment threshold (Chapter 11, p. 477), finds E_a(CH) = 0.74 eV. In both cases these results agree better with the values given by Cade's second method in contrast to OH and SH which agree better with the first.

O_2^-

A great deal of experimental study of this negative ion has been carried out, partly because of its importance in atmospheric physics but also because there are a number of special features associated with the processes which lead to its formation.

The ground state of O_2^- has the electron configuration

$$(\sigma_g\,2s)^2(\sigma_u\,2s)^2(\sigma_g\,2p)^2(\pi_u\,2p)^4(\pi_g\,2p)^3\ {}^2\Pi_g.$$

This differs from that of O_2 in possessing an additional electron in the antibonding ($\pi_g 2p$) orbital so a weaker binding would be expected. According to empirical rules formulated by Mulliken (1932), the binding energy of a homonuclear diatomic molecule, comprised of atoms in the first long row of the periodic table, is 2.5 eV per valence bond, this being defined as the difference between

TABLE 6.2 *Observed value of the electron affinity $E_a(O_2)$ of O_2*

Authors	Technique	$E(O_2)$ (eV)
Burch, Smith and Branscomb (1958, 1959)	Threshold behaviour of total photodetachment cross-section	0.15
Curran (1961)	Appearance potential for O_2^- in electron impact dissociation of O_3	$\geqslant 0.58$
Pack and Phelps (1961, 1966)	Equilibrium constant for $O_2 + e \rightleftharpoons O_2^-$	0.43 ± 0.02
d'Orazio and Wood (1965)	Lattice energy of alkali superoxides	0.65 ± 0.22
Fischer, Neuert, Peuckert-Kraus and Vogt (1966); Dunkin, Fehsenfeld and Ferguson (1970)	Exothermicity of $H^- + O_2 \rightarrow H + O_2^-$	> 0.75; < 0.75
Fehsenfeld, Albritton, Burt and Schiff (1969)	Exothermicity of $O_2(a^1 \Delta_g) + O_2^- \rightarrow 2O_2 + e$	< 0.98
Lacmann and Herschbach (1970)	Threshold for $K + O_2 \rightarrow K^+ + O_2^-$	0.5 ± 0.02
Johnsen, Brown and Biondi (1971)	Rate of $O^- + O_2^- \rightarrow O_4 + O_2^-$ in a drift tube	$\simeq 0.47$
Nalley and Compton (1971)	Threshold for $Cs + O_2 \rightarrow Cs^+ + O_2^-$	0.46 ± 0.05
Chantry (1971)	Threshold for $O^- + O_2 \rightarrow O + O_2^-$	0.5 ± 0.01
Berkowitz, Chupka and Gutman (1971)	Threshold for $I^- + O_2 \rightarrow I + O_2^-$	0.48 ± 0.01
Celotta, Bennett, Hall, Siegel and Levine (1972)	Energy and angular distribution of photoelectrons from O_2^- photodetachment	0.44 ± 0.008

the number of pairs of bonding and antibonding electrons. As O_2^- possesses one-half a valence bond less than O_2 it would be expected to have about 1.2 eV less binding energy. Now

$$E_a(O_2) = D(O_2^-) - D(O_2) + E_a(O), \qquad (6.27)$$

where $E_a(X)$ is the electron affinity of X and $D(XY)$ the dissociation energy of XY. Since $E_a(O) = 1.46_5$ eV (see Chapter 3, p. 41), equation (6.27) suggests that $E_a(O_2)$ will be small, at any rate considerably less than 1 eV.

In fact very good experimental information is now available not only about $E_a(O_2)$ but also about the equilibrium separation R_e, zero point 'energy' ω_e and anharmonicity x_e of the ground state of O_2^-.

Much of this information has come from laser photodetachment measurements which are described in detail in Chapter 11, p. 478. These are of such precision that a thorough analysis of the separate contributions from different vibrational transitions can be made. From the energy of the photoelectrons produced in the (0,0) transition, the electron affinity is found to be 0.440 ± 0.008 eV.

Previous to these experiments a number of other determinations of $E_a(O_2)$ had been made. These are listed in Table 6.2.

It will be seen that most of the later measurements (1970 or later), by whatever technique, agree quite well with the results obtained from the photoelectron spectra. Of earlier measurements the outstanding one is that of Phelps and Pack (1961) using the technique described in Chapter 12, p. 530. As refined in their later work their result is of high accuracy. A number of determinations based on the observed lattice energies of alkali superoxides were made prior to the one listed, but the latter used the most recent values of the various quantities required. Perhaps surprisingly, it is consistent with the accurately measured affinity.

Of the other measurements which do not agree well, that based on the threshold behaviour of the photodetachment cross-section failed because the frequency resolution was inadequate to resolve the different vibrational transitions. The electron impact appearance potential is perhaps not so far from correct as it was not primarily designed to determine $E_a(O_2)$ and no effort was made to improve the analysis to obtain it. On the other hand, it is surprising that one group of experimenters considered a particular charge transfer reaction to be exothermic, whereas it is in fact endothermic. This shows how much care must be taken in using evidence of this kind to bracket electron affinities (see also Chapter 12, p. 571). It is gratifying that the heavy particle reaction thresholds all give good results, as this should be a technique of wide generality.

As far as other properties of the O_2^- ground state potential-energy curve are concerned, additional information is available both from the observation of resonant effects in the elastic and

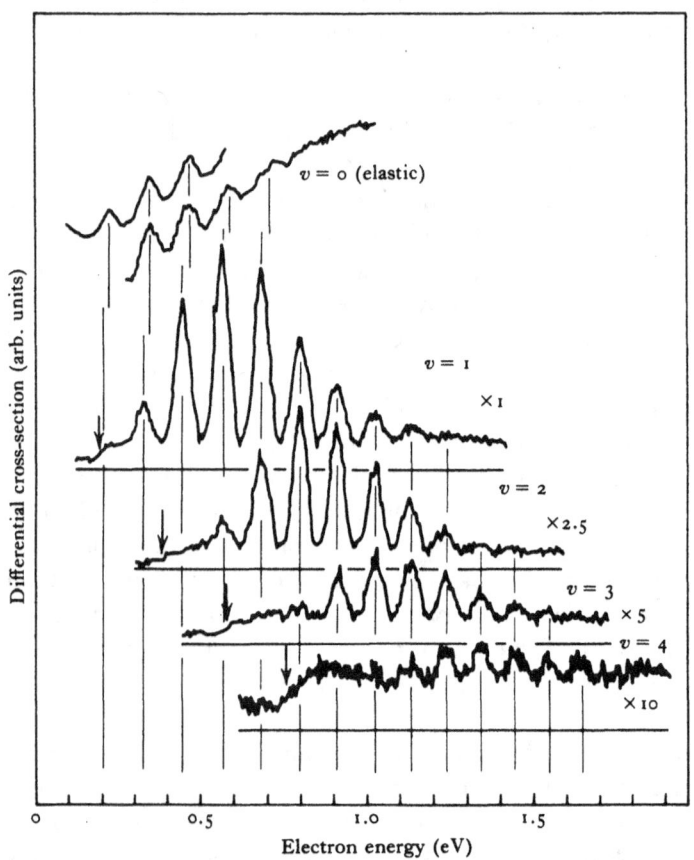

Fig. 6.9. Differential cross-sections as a function of electron energy for excitation of vibrational states of O_2, with quantum number v as indicated, observed by Linder and Schmidt (1971) at a scattering angle of 60°.

inelastic scattering of slow electrons from O_2, and from photoelectron spectra of O_2^-.

The resonance effects arise because the higher vibrational states ($v_r > 4$) of O_2^- are unstable towards autodetachment. Fig. 6.9 shows the differential cross-sections as a function of electron energy, for excitation of vibrational levels from $v = 0$ to 4 of O_2, as observed by Linder and Schmidt (1971) using the apparatus illustrated on p. 113. It will be seen that the resonance peaks occur at the same electron energies independently of the final vibrational state. The small departure for $v = 0$, the elastic scattering, can be ascribed to interaction with the background elastic scattering (see

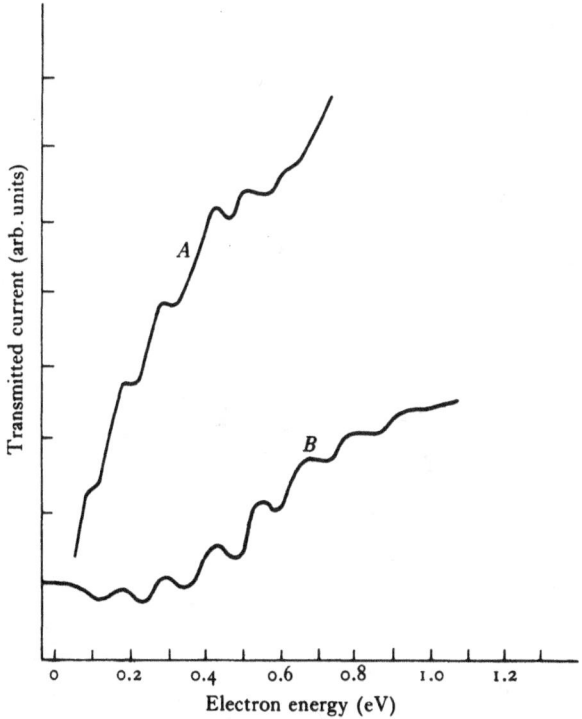

Fig. 6.10. Transmission of electrons through O_2 as a function of electron energy in the range 0–1.4 eV. A, total transmission (Boness and Hasted, 1966); B, transmission without energy change (Hasted and Awan, 1969). The voltage scale has been shifted to lower energies by 0.4 eV for A and 0.25 eV for B, so the peak energies agree with those derived from observed differential cross-sections.

p. 87). A similar situation prevails with the data shown in Fig. 6.10 which refer to transmission and to total elastic scattering.

Since the ground electronic state of O_2^- is obtained from that of O_2 by addition of a $3d\pi_g(\pi_g 2p)$ electron the angular distribution of electrons scattered at a resonance peak (see p. 87), if rotation can be neglected, should be given by (6.26) which is of a form possessing minima at 50° and 130° and a maximum at 90°. The observed results of Linder and Schmidt (Fig. 6.11) show that this is not a good approximation. However, if it is assumed that the

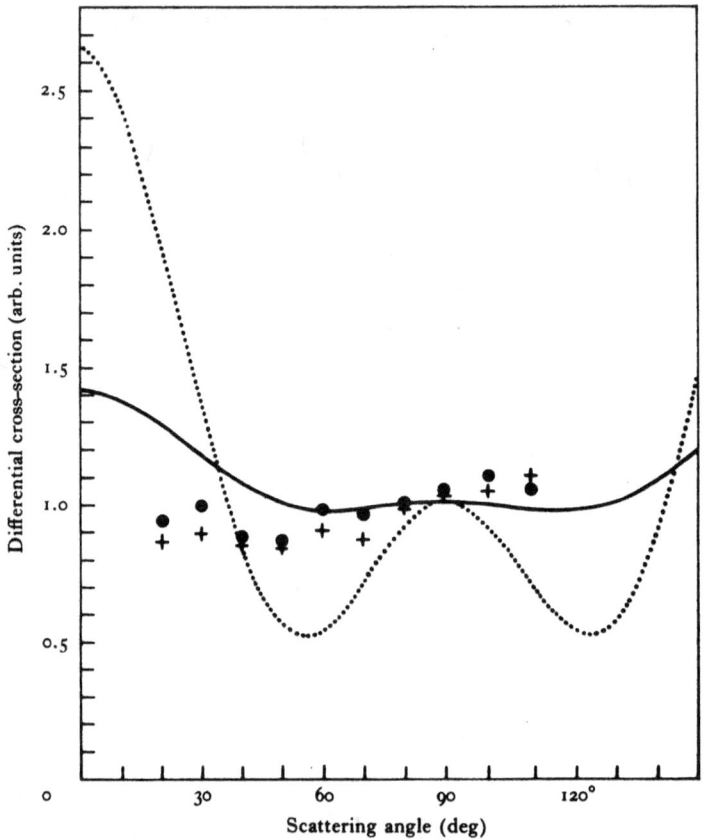

Fig. 6.11. Angular distribution of electrons of initial energy 0.680 eV after excitation of the $v = 1$ and $v = 2$ vibrational levels of the ground electronic state of O_2. + observed $v = 1$, ● observed $v = 2$, ····· calculated ignoring the possibility of rotation, ——— calculated allowing for rotation (from Linder and Schmidt, 1971).

lifetime of the autodetaching states is long compared with the rotation period, the angular distribution is smoothed out to resemble closely the observed form. As the rotation period in O_2^- is about 2×10^{-11} s, the lifetimes of the autodetaching levels are likely to be of order 10^{-10} s, very long compared with the vibrational period which is close to 2×10^{-15} s. For such long-lived states the width of the resonance is only of the order of 10^{-5} eV, which is much less than the instrumental width.

If during the lifetime of the vibrationally excited O_2^- formed by electron capture a collision occurs, which reduces the energy of the system below that of the ground state of O_2, stable O_2^- is produced. The rate of production of stable O_2^- in three-body collisions (see Chapter 9)

$$e + O_2 + X \longrightarrow O_2^- + X \qquad (6.28)$$

will therefore also show the resonance effects as a function of electron energy. These have been observed (see Fig. 9.36).

From these data the vibrational spectrum with $v > 3$ can be built up. Taken in conjunction with the known electron affinity and with the data available from the photoelectron energy spectra (Chapter 11, p. 481) it is found that

$$R_e = 1.341 \pm 0.010 \text{ Å}, \quad \omega_e = 1089 \text{ cm}^{-1}, \quad \omega_e x_e = 12.1 \text{ cm}^{-1}.$$

There is a near-coincidence between the energies of the vibrational levels $v = 8$ of O_2^- and $v = 3$ of O_2.

Fig. 6.12 shows the potential-energy curves of the ground states of O_2 and O_2^- in relation to each other with some of the vibrational levels indicated.

Zemke, Das and Wahl (1972) have carried out a detailed calculation, using the OVC multiconfiguration method, of the dissociation energy of O_2^- which they obtain as 4.14 eV. This gives $E_a(O_2) = 0.42$ eV, remarkably close to the best observed value. In a later paper Krauss, Neumann, Wahl, Das and Zemke (1973) have extended the calculations to the excited states which lead to normal O and O⁻ in the limit of infinite nuclear separation. Their results are shown in Fig. 6.12. It will be seen that a number of the states are attractive with minima occurring at rather large nuclear separations. Included also in the figure is the repulsive part of the $^2\Pi_u$ curve derived by O'Malley (1967) (see Chapter 9, p. 323) from observed data on dissociative attachment in O_2.

Of these excited states, the only one about which there is any experimental information is the $^2\Pi_u$. As described in Chapter 9, p. 320, there is evidence that the dissociative attachment reaction

$$O_2 + e \longrightarrow O + O^-$$

proceeds via capture to this state. In that case, the location of the $^2\Pi_u$ curve above that of the ground state of O_2 at nuclear separations close to the equilibrium value for O_2 ($1.26a_0$) is much as would be expected from Fig. 6.12. Some evidence about the slope of the $^2\Pi_u$ curve is available from the observed variation of the dissociative attachment cross-section with temperature (Chapter 9, p. 321).

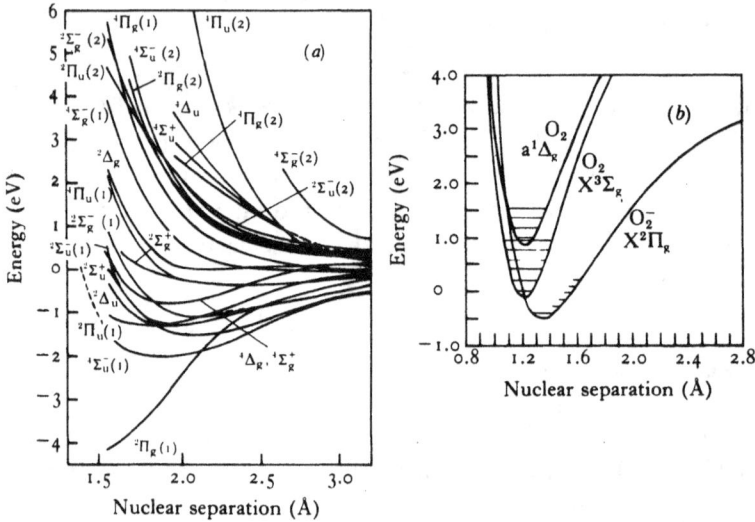

Fig. 6.12. (a) Potential-energy curves of low-lying states of O_2^-, calculated by Krauss et al. (1973) (full-line curves). Broken curve is the $^2\Pi_u$ curve derived by O'Malley from observed data in O_2 (see Figs. 9.29 and 9.30). (b) Comparison of potential-energy curves for the ground ($X^3\Sigma_g^-$) and a $^1\Delta_g$ states of O_2 with that for the ground ($X^2\Pi_g$) state of O_2^-, vibrational levels being indicated.

SO⁻ and S_2^-

Electron affinities of both these molecules have been measured from observations of the energy spectra of electrons detached from the ions by argon ion laser light (see Chapter 9, p. 438). For S_2^- (Chapter 9, p. 484) Celotta et al. (1974) found

$$E_a(S_2) = 1.663 \text{ eV},$$

without allowing for rotational and fine structure effects, while for SO⁻, Bennett (1972) (Chapter 9, p. 484) obtained $E_a(SO) = 1.14$ eV. Measurements of $E_a(SO)$ have also been made by Feldmann (1970) from observations of the photodetachment threshold. He finds $E_a(SO) = 1.09$ eV.

NO^-

In some respects NO^- resembles O_2^-, particularly in the fact that low energy resonance effects occur in the scattering of electrons by NO and that NO^- may be formed by three-body reactions similar to (6.28).

According to (6.27), since $D(NO) = 6.49$ eV,

$$E_a(NO) = D(NO^-) - 5.03 \text{ eV}.$$

We would not expect $D(NO^-)$ to be greater than that of the isoelectronic homonuclear molecule O_2 (5.119 eV), so that it would be surprising if $E_a(NO)$ exceeded a few tenths of an eV. In fact NO was the first molecule to be investigated by laser photodetachment electron spectroscopy and it was found that $E_a(NO) = 24 ^{+10}_{-50}$ meV. The details of the experiment and their analysis are discussed in Chapter 11, p. 482.

Other determinations which agree with this are those of Lacmann and Herschbach (1970) ($\sim$ 0 eV, see Chapter 10, p. 398) from the threshold for the charge transfer reaction

$$K + NO \longrightarrow K^+ + NO^-,$$

and of Berkowitz et al. (1971) (0.09 ± 0.1 eV, see Chapter 12, p. 568) from that for

$$I^- + NO \longrightarrow I + NO^-.$$

Also Gilmore (1965) estimated $E_a(NO)$ as 0.1 eV by extrapolating across isoelectronic series. On the other hand Farragher, Page and Wheeler (1964) obtained 0.9 ± 0.1 eV from observations of the ion equilibrium arising from reactions of N_2O in the neighbourhood of a hot filament. This shows how difficult it is to be sure that the correct interpretation of the observed data is made when polyatomic systems interact on a hot surface.

In contrast to O_2^-, only the ground vibrational level of NO^- lies below the ground level of NO. As for O_2^-, information about the structure of NO^- is available from experiments on the elastic and inelastic scattering of electrons from NO and on the photoelectron spectrum of NO^-.

Fig. 6.13 illustrates the resonance peaks observed in different experiments on the elastic and total scattering, while Fig. 6.14 shows the corresponding peaks in the cross-sections of excitation of

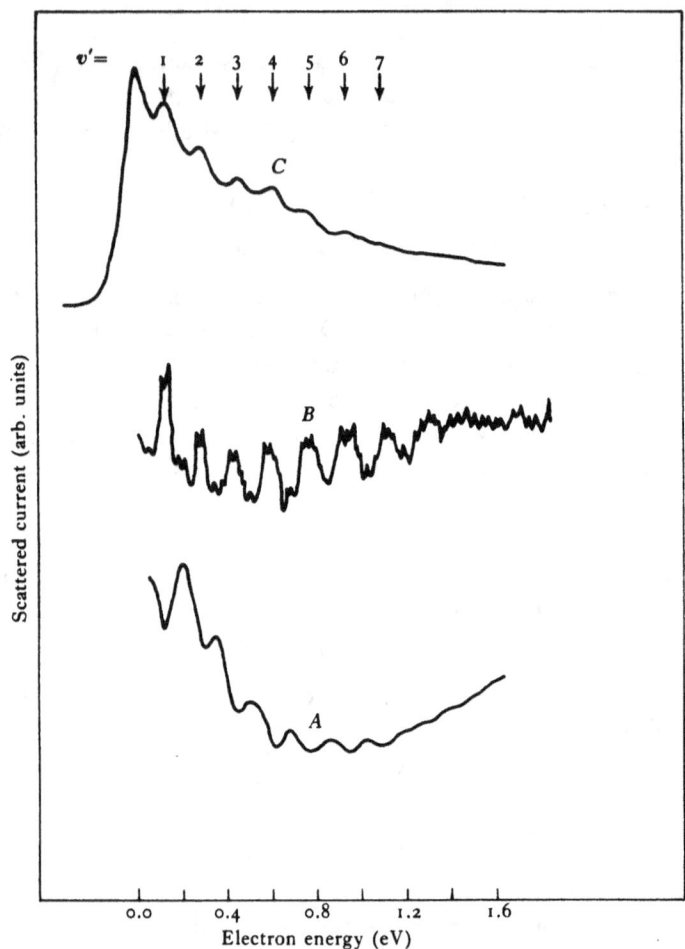

Fig. 6.13. Observed data on total and elastic scattering of low-energy electrons by NO, showing the resonance structure with associated vibrational quantum numbers v' for NO$^-$. Curves A, transmission of electrons without energy loss (Boness et al., 1968); B, differential elastic cross-sections at 20° (Ehrhardt and Willmann, 1967); C, total scattering at wide angles (Spence and Schulz, 1971).

different vibrational states. As for O$_2$, the peaks occur at the same incident electron energies for all elastic and inelastic processes, showing that the NO$^-$ ion has a lifetime much greater than the vibrational frequency. However, in contrast to O$_2^-$ the lifetime is short enough for the resonance widths to be comparable with the instrumental, so it is possible to disentangle the true widths from the observed by a deconvolution procedure. When this is done, the natural widths for elastic scattering are found to decrease with the vibrational quantum number of the ion as shown in Fig. 6.15.

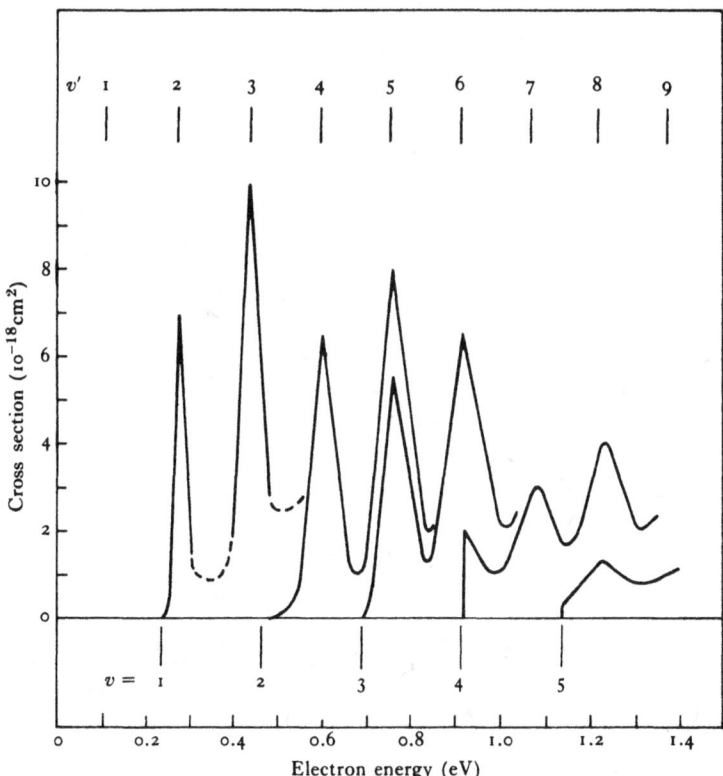

Fig. 6.14. Observed cross-section for excitation of different vibrational levels of the ground state of NO by electron impact. The quantum numbers v of these levels are indicated, while the locations of the vibrational levels of NO⁻ are shown above, for v' ranging from 1 to 9 (Spence and Schulz, 1971).

For the lowest vibrational state the lifetime is 3×10^{-14} s as compared with a vibrational period of 1.7×10^{-15} s.

From the vibrational spectrum for NO⁻ obtained from the scattering experiments, ω_e and x_e were determined as

$$\omega_e = 1371 \text{ cm}^{-1} \text{ (0.170 eV)}, \quad \omega_e x_e = 0.01 \text{ eV}.$$

The photoelectron spectrum measurements described in Chapter

11, p. 483, give $\omega_e = 1470 \pm 200$ cm^{-1} (0.182 ± 0.023 eV) and also the equilibrium separation R_e as 1.258 ± 0.010 Å.

Fig. 6.16 illustrates the potential-energy curves for NO and NO$^-$ with the vibrational states indicated. It will be noted that the $v = 4$ state of NO is very nearly coincident in energy with the $v = 6$ state of NO$^-$.

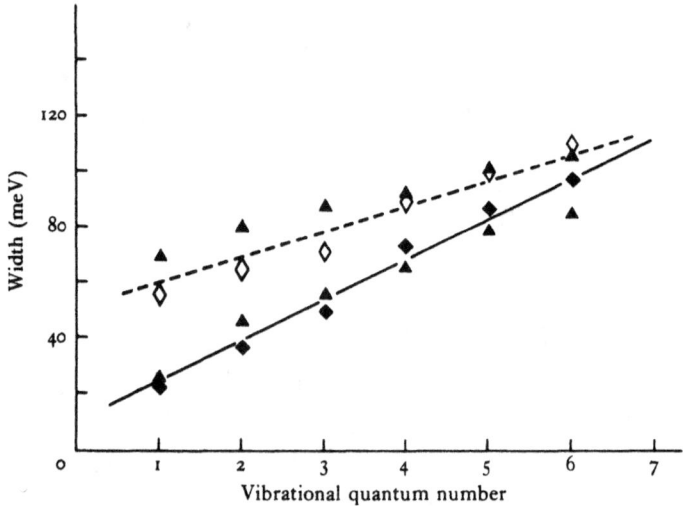

Fig. 6.15. Observed width of resonances in elastic scattering of electrons by NO as functions of the vibrational quantum number v of the NO$^-$ level concerned. ---- without allowance for instrumental width, ——— with allowance for instrumental width, observed, ▲ by Spence and Schulz (1971), ◇, ♦ by Ehrhardt and Willmann (1967).

N$_2^-$

The ground state of N$_2$ has the closed shell electron configuration.
$$(\sigma_g\, 2s)^2 (\sigma_u\, 2s)^2 (\sigma_g\, 2p)^2 (\pi_u\, 2p)^4.$$

The corresponding configuration for N$_2^-$ is obtained then by adding an additional electron in an antibonding $\pi_g 2p$ orbital. According to Mulliken's rule (see p. 162) this reduces the binding energy by about 1.25 eV but, in contrast to O$_2^-$, in the separated atom limit the ion N$_2^-$ is not stable so that we must expect that the lowest level of N$_2^-$ will be at least 1.25 eV above that for N$_2$, as well as occurring at a larger nuclear separation. Thus all vibrational

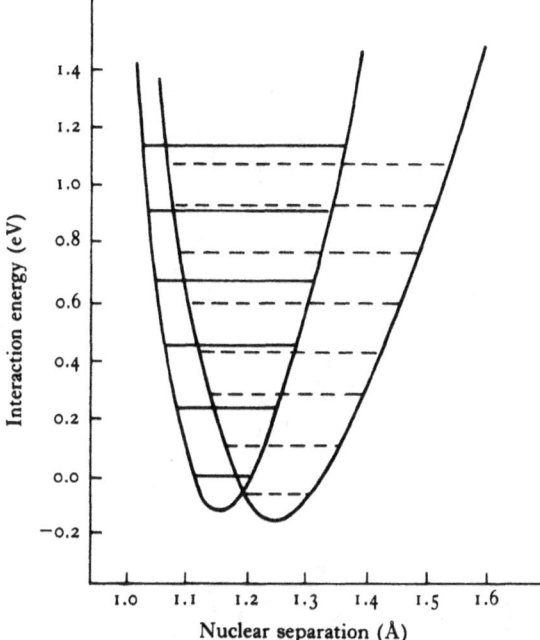

Fig. 6.16. Potential-energy curves and vibrational levels for the ground states of NO and NO⁻, derived from observation.

levels of the N_2^- ground state must be unstable towards autodetachment. N_2^- thus represents a further step in going from O_2^- to NO⁻. In O_2^- four vibrational levels are stable, in NO⁻ only one. Again, the lifetime of the autodetaching levels in O_2^- is much higher than in NO⁻. We would expect the levels for N_2^- to be even more ephemeral, possibly with lifetimes comparable with the vibrational periods so that vibrational quantization is not even well developed.

In the united atom limit the $\pi_g 2p$ orbital tends to $3d\pi_g$ (see p. 162). As it is antibonding, it is localized at a relatively large distance from the valence electrons so that the form of the wave function resembles that of a d state with 1 unit of angular momentum about the nuclear axis. The effective interaction which such an electron has with the molecule will have the form, at large distance,

$$V(r) \sim 6^{1/2} \hbar^2 / 2mr^2 \qquad (6.29)$$

due to the centrifugal force. This gives rise to the possibility of a shape or type II resonance as discussed in Chapter 4, p. 97.

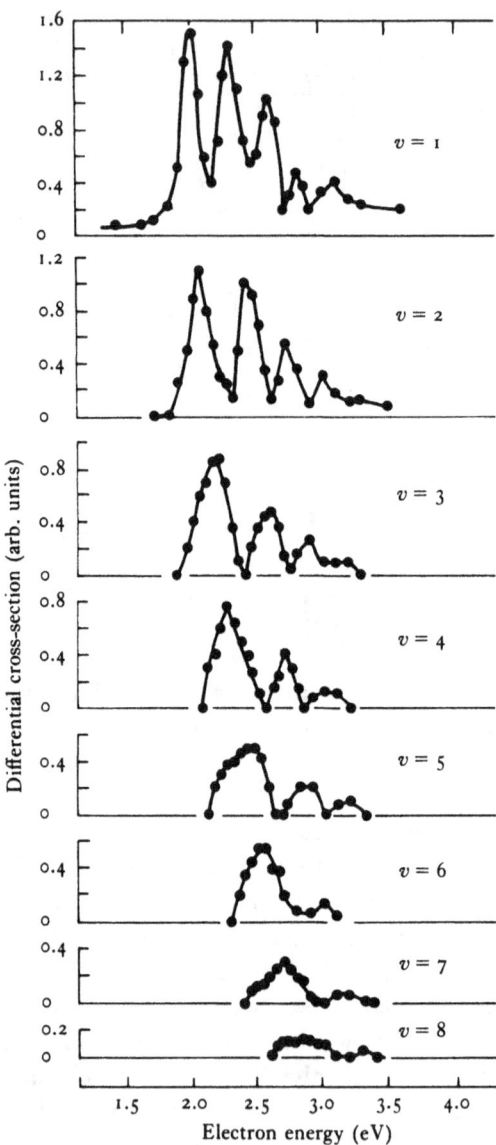

Fig. 6.17. Variation with electron energy of differential cross-sections for excitation of vibrational levels of the ground state of N_2 by electron impact, observed at 72° angle of scattering by Schulz (1964).

The first suggestion that an intermediate N_2^- ion plays a role in processes involving the collision of slow electrons in N_2 was made by Haas (1957). As long ago as 1927, Harries and Hertz obtained evidence from electron diffusion experiments in N_2 that electrons of about 5 eV energy have a chance of about 1 in 79 of producing vibrational excitation. Using more-modern equipment Haas (1957) repeated their experiments and extended them to energies of 2 eV. He found that the chance of exciting vibration per collision increases rapidly at lower energies, reaching 0.15 at 2 eV. Even the much lower probability observed in the earlier experiments was much higher than expected so Haas suggested that the process takes place through the formation of an intermediate N_2^- ion. Further evidence that some such possibility must be involved was provided by Schulz (1959, 1962) who found evidence that there is a considerable probability of exciting more than one vibrational quantum in a collision.

The first direct evidence that an intermediate N_2^- ion is formed was obtained by Schulz (1964) who used the apparatus referred to on p. 122 of Chapter 5 to observe, with high resolution, the energy loss spectrum of electrons scattered at 72° in N_2. Energy losses corresponding to excitation of 2, 3, ..., 7 vibrational quanta were observed as shown in Fig. 6.17. In all cases resonance peaks were observed but, in contrast to O_2 and NO, the peaks did not appear at the same incident electron energies for the different inelastic processes, a result of considerable significance which will be discussed further below.

Experimental study of the N_2^- ground (autodetaching) state
Fig. 6.18 gives the variation with electron energy of the transmission of electrons through N_2 observed by Boness and Hasted (1966) showing a well-defined series of resonances. The resonances are also clearly seen in the differential elastic cross-sections observed as a function of electron energy at a number of scattering angles by Ehrhardt and Willmann (1967) (see Fig. 6.21) using the apparatus described in Chapter 4, p. 112.

Fig. 6.19 shows the variation of the energy of the successive peaks with vibrational quantum number, obtained from the results of Schulz at 72° (see Fig. 6.17) and of Ehrhardt and Willmann at 20°.

Although there is disagreement in detail, both sets of data agree in showing clearly that the peak locations depend on the final state.

Fig. 6.20 shows the angular distribution of electrons scattered after exciting specific vibrational states, for a number of different incident energies covering the resonance region. The most sig-

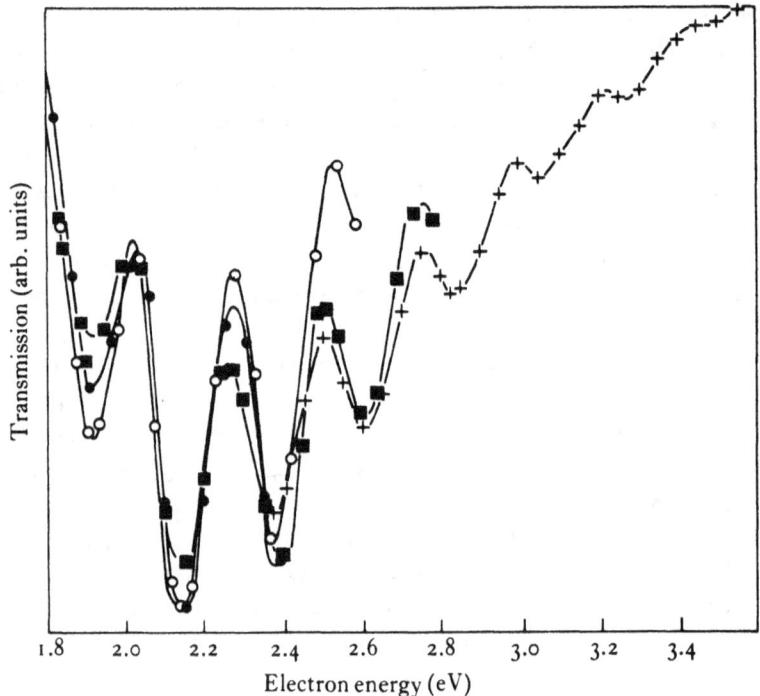

Fig. 6.18. Transmission of electrons in the energy range 1.8–3.6 eV through N_2, observed by Boness and Hasted (1966). The separate curves represent observations taken on different occasions.

nificant feature of these results is the occurrence in all of a maximum at 90° and a minimum near 50° as for a $d\pi$ resonance state (cf. p. 171).

Theoretical interpretation

Krauss and Mies (1970) calculated the wave function for the $^2\Pi_g$ state of N_2^- in terms of an expansion in Gaussian-type orbitals centred at both nuclei and also at the centre of mass of the molecule. With full freedom in the choice of these orbitals the variational

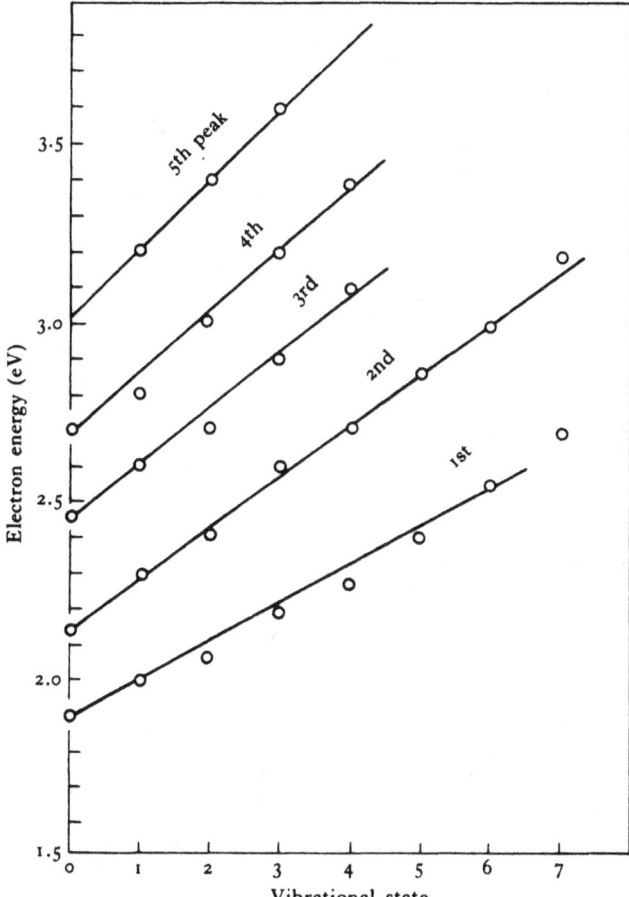

Fig. 6.19. Observed positions of the peaks (as indicated) in the cross-section for excitation of different vibrational levels of the ground state of N_2 by impact of slow electrons (from Schulz, 1964).

method will yield a function which represents a ground state N_2 molecule and an electron in a continuum state of zero energy (cf. Chapter 1, p. 22). However, if the half-widths of the Gaussians are chosen to be less than $6a_0$, the additional electron is found to occupy a bound orbital not very different from the $3d\pi_g$ orbital for an excited π_g state of N_2 consistent with the angular distribution data. Assuming this to give a model for the resonance state, Krauss and Mies obtained the potential-energy curve for this state in the

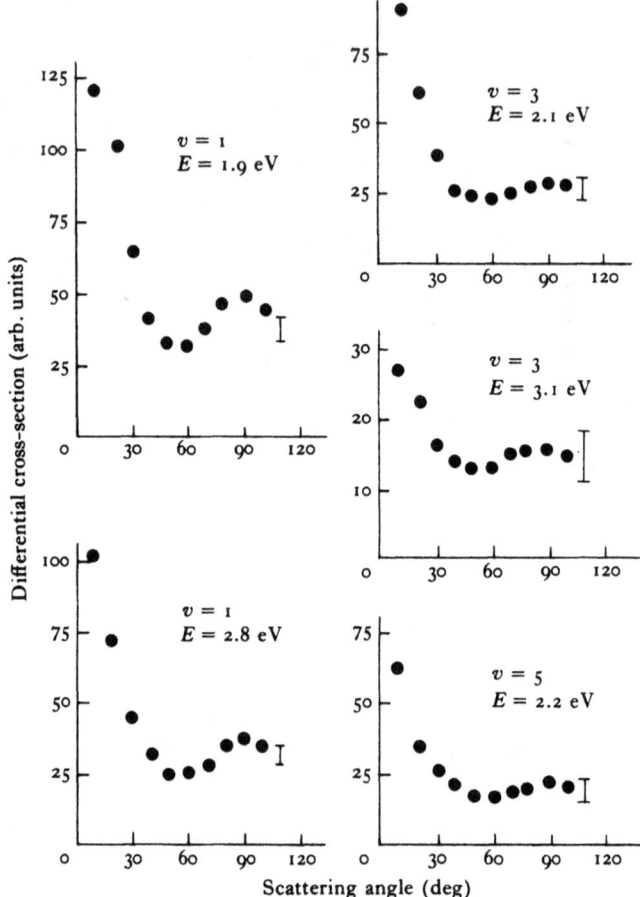

Fig. 6.20. Angular distributions of electrons of various incident energies scattered after excitation of different vibrational levels of the ground state of N_2. ● observed by Ehrhardt and Willmann (1967).

neighbourhood of the equilibrium separation which they found to be $2.27a_0$ as compared with $2.09a_0$ for N_2. The calculated energy of the lowest vibrational state was 2.53 eV above that of the neutral molecule. This value is 0.67 eV greater than the observed location of the resonance. Much of this discrepancy can be ascribed to the neglect of correlation – the difference in the correlation energies of O_2 and O_2^- is about 0.8 eV.

These calculations ignore the fact that the N_2^- state is unstable towards autodetachment so they refer only to the real part of the energy. Krauss and Mies also estimated the level width. They first determined an effective central potential V_{eff} acting on the attached electron which would yield the calculated resonance $3d\pi_g$ orbital. This was then joined smoothly to the centrifugal potential (6.29) and the wave phase shift η_2 for scattering of an electron in this total potential was calculated as a function of electron energy E. The level width was then determined from the condition that at the resonance energy

$$\eta_2 = \frac{\pi}{2}, \qquad \frac{\partial E}{\partial \eta_2} = \Gamma,$$

i.e. the time delay in traversing the molecule, as given by (4.42) of Chapter 4, is $1/\Gamma$ when $\eta_2 = \pi/2$.

The level width calculated in this way was found to increase from 0.13 to 0.8 eV as the nuclear separation decreased from $2.3a_0$ to $2.0a_0$. These level widths are comparable with the vibrational

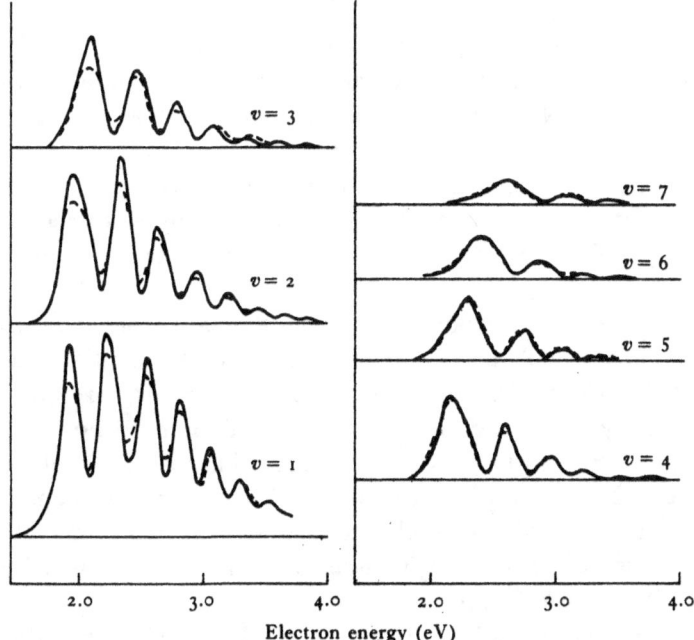

Fig. 6.21. Comparison of observed and calculated differential cross-sections for vibrational excitation of N_2 by electrons scattered through 20°. - - - - calculated by Birtwistle and Herzenberg (1971), ——— observed by Ehrhardt and Willmann (1967).

quanta, which are approximately 0.24 eV, showing that the lifetimes are barely long enough for a single complete vibration to occur. This is the essential reason why the location of peaks in the observed cross-sections for vibrational excitation depends on the final vibrational state concerned (see Fig. 6.19).

Herzenberg (1968) and later Birtwistle and Herzenberg (1971) developed the theory of vibrational excitation through an intermediate state so as to take not only this effect into account but also to allow for the strong dependence of level width on nuclear separation. They applied this theory to the excitation of the vibrational states with $v = 1, 3, 5$ and 7, treating the real and imaginary parts of the energy of N_2^- as adjustable. The good agreement with observation shown in Fig. 6.21 was obtained with the following values of the parameters characterizing the N_2^- state.

$$R_e^- - R_e = 0.18a_0, \quad E_e^- - E_e = 1.925 \text{ eV},$$
$$h\nu^- = 0.244 \text{ eV}, \quad \Gamma(R_e) = 0.57 \text{ eV}.$$

R_e^-, R_e are the equilibrium separations, E_e^-, E_e the lowest energies of N_2^- and N_2 respectively and $h\nu^-$ is the ground vibrational quantum for N_2^-. $\Gamma(R)$ was taken to vary with R as shown in Fig. 6.22.

It will be seen that the values derived by Birtwistle and Herzenberg (1971) are quite close to those calculated by Krauss and Mies (1970) and leave little doubt that the interpretation is generally correct.

An alternative method for studying the nature of the resonance effect is to calculate the partial cross-section for elastic scattering of electrons by a nitrogen molecule. Since a shape resonance is essentially a one-body effect, the calculations need not take into account the possibility of virtual excitation of higher electronic states of N_2. Apart from the non-central character of the interaction, such a calculation is similar to that for the scattering of slow electrons by atoms in which the collision wave function is written in the simple form (4.55).

Burke and Sinfailam (1970) carried out such calculations in terms of an expansion of the collision wave function in spherical polar coordinates about the centre of the molecule. The mean interaction with the undisturbed field of the molecule was obtained

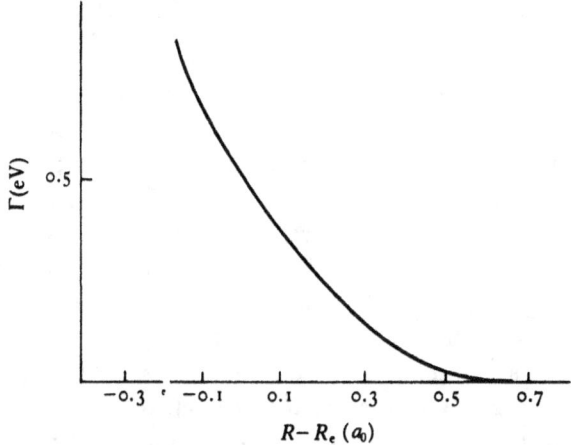

Fig. 6.22. Variation of level width Γ with nuclear separation for the ground state of N_2^-, assumed by Birtwistle and Herzenberg (1971). R_e is the equilibrium separation.

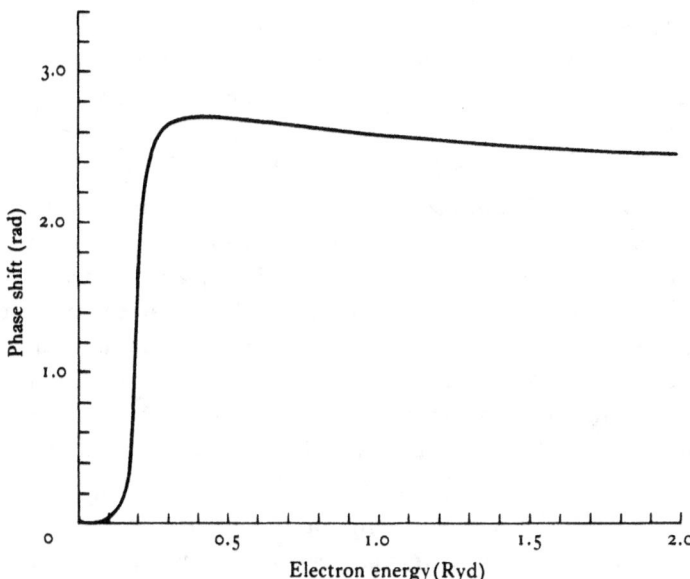

Fig. 6.23. Calculated behaviour of the phase shift for elastic scattering of electrons by N_2 molecules, which is associated with the shape resonance. From Burke and Sinfailam (1970).

from the HFR wave function for N_2 derived by Nesbet (1964), the first two terms in the expansion of which, in spherical harmonics round the molecular centre, were retained. Allowance was made for electron exchange which proved to be quite important. Fig. 6.23 shows the calculated behaviour of the phase shifts corresponding closely to η_{20}. It exhibits the expected behaviour. Thus it increases from 0.2 to 2.7 rad as the electron energy increases from 1.1 to 3.8 eV. In terms of (4.55) of Chapter 4 this corresponds to a time delay of nearly 10^{-15} s and an effective level width of 0.76 eV, quite close to that determined by Birtwistle and Herzenberg for the equilibrium separation. However, while these results confirm the general interpretation, they cannot be relied upon quantitatively as the effect of molecular distortion by the incident electron is considerable (Burke and Chandra, 1972) and the maximum value of the calculated total elastic cross-section is about twice as large as that observed. These discrepancies are not surprising when account is taken of the complexity of the problem.

CO^-

CO^- is isoelectronic with N_2^- so would be expected to behave in a very similar way. One important difference however, is the absence of nuclear symmetry so that the electron configuration of the ground state of CO^- will be

$$(\sigma\,2s)^2(\sigma^*\,2s)^2(\sigma\,2p)^2(\sigma^*\,2p)^2(\pi\,2p)^4(\pi^*\,2p).$$

This arises by addition of an antibonding π^*2p orbital, in the notation of p. 162. In the united atom limit this tends to $3p\pi$ instead of $3d\pi$ as for N_2^- (see p. 195). If a shape resonance exists as for N_2^- the angular distribution of electrons scattered at the resonance energy will be characteristic of a $p\pi$ rather than a $d\pi$ state.

Referring to p. 171, (6.25) we see that, when averaged over all orientations of the molecular axis, this gives a distribution

$$1 + 7\cos^2\theta,$$

where θ is the angle of scattering, predominantly of p wave form.

Also, because the centrifugal barrier holding in the captured electron in the $p\pi$ resonance state will be smaller than for the $d\pi$ state of N_2^-, we would expect the lifetime to be shorter in CO^- so that the resonance effects should be broader.

Finally, dissociative attachment of electrons may occur in CO so providing a means of obtaining information about some of the excited states of CO^- which is not available for N_2^-.

Fig. 6.24 shows the variation with electron energy of the transmission of electrons through CO as observed by Boness and Hasted

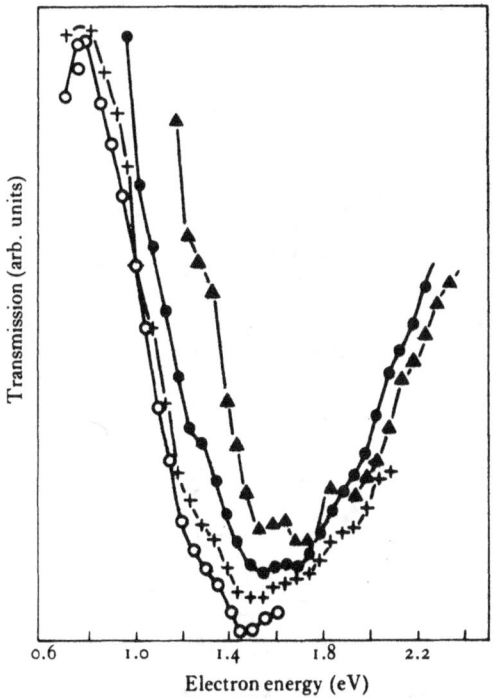

Fig. 6.24. Transmission of electrons in the energy range 0.6–2.4 eV through CO, observed by Boness and Hasted (1966). The separate curves represent observations taken on different occasions.

(1966). Comparison with the corresponding results for N_2 shows that the resonance effects are much less pronounced. Fig. 6.25 gives the absolute total cross-sections for excitation of different vibrational states as functions of electron energy measured by Ehrhardt, Langhans, Linder and Taylor (1968). These are to be compared with the results in Fig. 6.21 for N_2 and show clearly much broader resonance peaks.

Finally, in Fig. 6.26, the angular distributions for electrons scattered elastically and after excitation of different vibrational states, observed by Ehrhardt et al. (1968) are shown. For the inelastic collisions they are close to the form (6.25) over the energy range from 1.55 to 3.0 eV covering the resonance region. The elastic

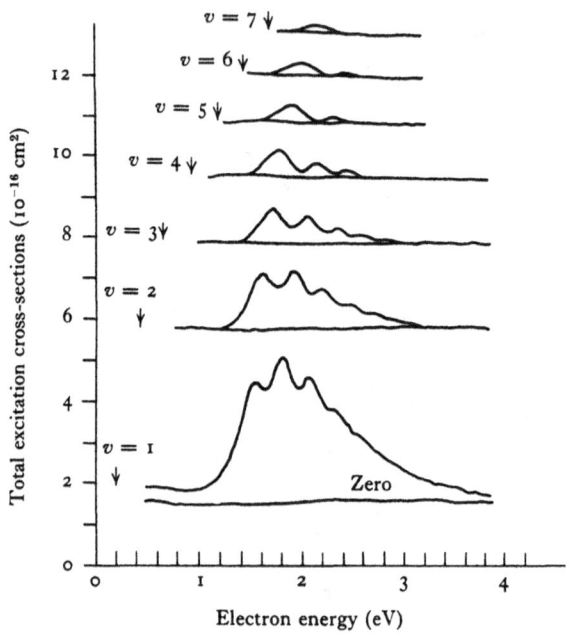

Fig. 6.25. Total cross-sections for excitation of different vibrational levels of the ground state of CO by slow electrons, observed by Ehrhardt et al. (1968).

scattering is complicated by the contribution from potential scattering (see Chapter 4, p. 87) and it is not surprising that it varies considerably over this energy range.

There seems little doubt that the ground state of CO^- has an even shorter lifetime than that of N_2^-. No detailed theoretical calculations of its properties have been carried out at the time of writing.

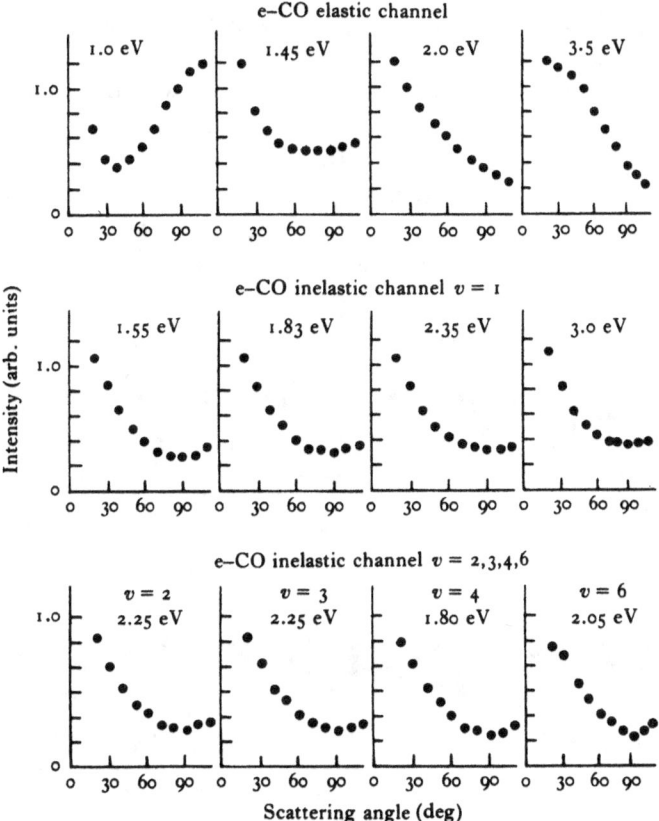

Fig. 6.26. Angular distributions of electrons of various incident energies scattered after excitation of different vibrational levels of the ground state of CO, observed by Ehrhardt *et al.* (1968).

C_2^-

The ground state of C_2 has the configuration

$$(\sigma_g\, 2s)^2(\sigma_u\, 2s)^2(\pi_u\, 2p)^4\ ^1\Sigma_g^+,$$

so that we would expect that for C_2^- to be

$$(\sigma_g\, 2s)^2(\sigma_u\, 2s)^2(\pi_u\, 2p)^4\sigma_g\, 2p\ ^2\Sigma_g^+,$$

the same as for the isoelectronic homonuclear molecule N_2^+. This molecule has a binding energy of 8.7 eV so we would expect that for C_2^- to be comparable. Since the binding energy of C_2 is 6.2 eV (Brewer, Hicks and Krikorian, 1962) and the electron affinity of C is 1.27 eV, the ground state of C_2^- is likely to lie some 3.5 eV or so below that of C_2.

The photodetachment experiments of Feldmann (1970) (see Chapter 11, p. 484) do indeed yield an electron affinity for C_2 of 3.5 eV, provided the equilibrium separations in C_2 and C_2^- are not very different. With such a high value it would not be surprising if there existed excited states of C_2^- lying below the ground state of C_2 and so being stable towards autodetachment.

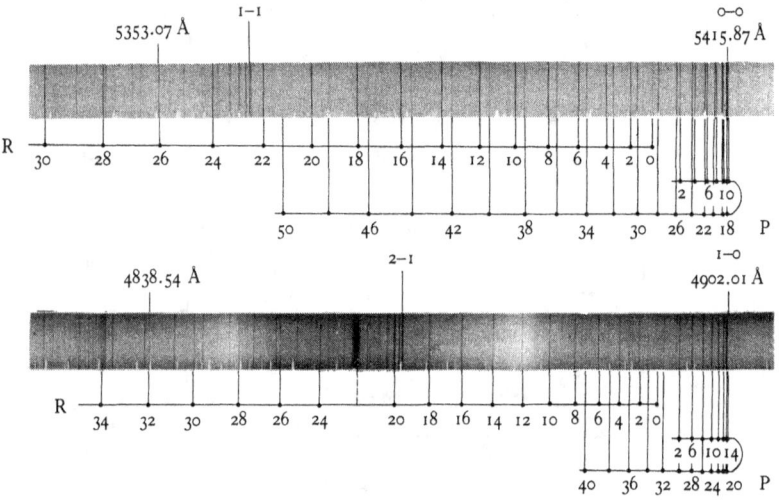

Fig. 6.27. Spectrogram showing the 0–0 and 1–0 bands of the system observed by Herzberg and Lagerquist (1968), emitted from a flash discharge in $^{12}CH_4$.

Direct evidence of the existence of such a state lying 2.0 eV above the ground state of C_2^- has been obtained. Herzberg and Lagerquist (1968) observed a band spectrum in the range 4800–6000 Å, emitted from a high intensity flash discharge in methane, which was positively identified as arising from a diatomic molecule containing two C atoms. This was proved by comparing the intensity variation observed in the bands when the source was $^{12}CH_4$ with that when ^{13}C was substituted. Fig. 6.27 shows observed 0–0 and 1–0 bands emitted from a discharge in $^{12}CH_4$. The odd lines which are strongest in $^{13}CH_4$ are missing. It proved difficult to interpret the spectrum in terms of either C_2 or C_2^+ as

MOLECULAR NEGATIVE IONS – GROUND STATES   209

the emitting molecules but C_2^- seemed to be a real possibility. The presence of a relatively strong concentration of C_2^- in the discharge was verified by mass analysis. As the transition involved was clearly of Σ–Σ type the upper state of C_2^- would be the

$$(\sigma_g\,2s)^2(\sigma_u\,2s)^2(\pi_u\,2p)^3(\sigma_g\,2p)^2\ ^2\Sigma_u^+$$

state. This being so, this state would be about 2 eV above the ground state, well below that of C_2. For N_2^+ the corresponding state lies about 3.2 eV above the ground state (see Fig. 7.5).

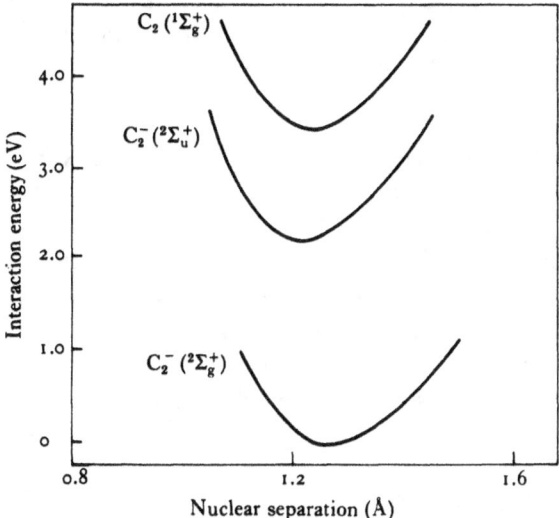

Fig. 6.28. Potential-energy curves for the $^2\Sigma_g^+$ and $^2\Sigma_u^+$ states of C_2^- in relation to the ground $^1\Sigma_g^+$ state of C_2.

The same band system has also been observed in other conditions (Milligan and Jacox, 1969; Frosch, 1971) but the most direct evidence that it really is due to C_2^- comes from the two-photon photodetachment experiments of Lineberger and Patterson (1972), described in Chapter 11, p. 493. In these experiments evidence was found of a further band system at longer wavelengths which could be ascribed to transitions from a $^2\Pi_u$ state of C_2^-. This corresponds to the $A^2\Pi_u$ state of N_2^+ (see Fig. 7.5) which lies about 2 eV above the ground state.

Fig. 6.28 illustrates the potential-energy curves for the ground $^1\Sigma_g^+$ state of C_2 and for the $^2\Sigma_g^+$ and $^2\Sigma_u^+$ states of C_2^- based on the data from the C_2^- bands and the electron affinity of C_2 measured by Feldmann (1970).

Barsuhn (1974) has calculated the potential-energy curves for the low-lying states of C_2^- using the HFR method plus configuration interaction, with Gaussian radial functions. Only valence-excited configurations were included. The calculated constants R_e and ω_e for the upper and lower states agree reasonably well with those derived from the observed bands but the calculated electron affinity is too small by more than 1 eV.

The calculations also predict the existence of two further bound states, a $^2\Pi_u$ state lying about 0.7 eV above the ground state and $^4\Sigma_u^+$ lying about 0.6 eV higher than the $^2\Sigma_u^+$.

CN⁻

This negative ion is isoelectronic with CO and with NO⁺, both of which have the very high dissociation energies of 11.11 and 11.7 eV respectively. It would therefore be very surprising if that of CN⁻ were not comparable. Since the dissociation energy of CN is 7.68 eV and the electron affinity of C is 1.27 eV, we would expect the electron affinity of CN to be at least as large as 3.5 eV.

Berkowitz, Chupka and Walter (1969) have measured $E_a(CN)$ from the observed threshold for the polar photodissociation

$$h\nu + HCN \longrightarrow H^+ + CN^-,$$

as described in Chapter 8, p. 261, and find it to be 3.82 ± 0.02 eV, the largest for any monatomic or diatomic species. Bakulina and Ionov (1959) have compared $E_a(CN)$ with $E_a(S)$ by the hot-wire surface ionization method (see Chapter 3, p. 38) from which $E_a(CN) = 3.4 \pm 0.2$ eV, while Inoue (1966) obtains 3.6 ± 0.3 eV from the threshold for

$$HCN + e \longrightarrow CN^- + H.$$

Using the value of $E_a(CN)$ obtained by Berkowitz et al., the dissociation energy of CN⁻ comes out to be 10.23 eV, falling between that of N_2 and of CO.

F_2^-, Cl_2^-, Br_2^-, I_2^-

The electronic structure of the ground state of a typical member of the series, F_2^-, is

$$(\pi_u\, 2p)^4 (\pi_g\, 2p)^4 (\sigma_u\, 2p)\, \Sigma_u.$$

TABLE 6.3 *Electron affinities of homonuclear halogen molecules*

Authors	Technique	F_2	Cl_2	Br_2	I_2	Text reference
Chupka, Berkowitz and Gutman (1971)	Threshold for $X^- + Y_2 \to X + Y_2^-$, X and Y halogens	3.08 ±0.1	2.38 ±0.10	2.51 ±0.10	2.58 ±0.10	Chapter 12, p. 569
Hughes, Lifshitz and Tiernan (1973)		—	2.32 ±0.1	2.62 ±0.2	2.42 ±0.2	
de Corpo and Franklin (1971)	Dissociative attachment appearance potential	2.9 ±0.22	2.52 ±0.17	2.87 ±0.14	2.6 ±0.1	Chapter 9, p. 381
Baede, Auerbach and Los (1973)	Threshold for $A + X_2 \to A^+ + X_2^-$ (A = Li, K, Na)	—	—	2.55 ±0.1	2.52 ±0.1	Chapter 10, p. 396
Helbing and Rothe (1969)	Threshold for $Cs + Br_2 \to Cs^+ + Br_2^-$	—	—	2.23 ±0.10	—	Chapter 10, p. 397
Mulliken (1932)	Semi-empirical	2.07	2.04	2.03	2.04	
Gilbert and Wahl (1971)	HFR	2.45	2.19	—	—	

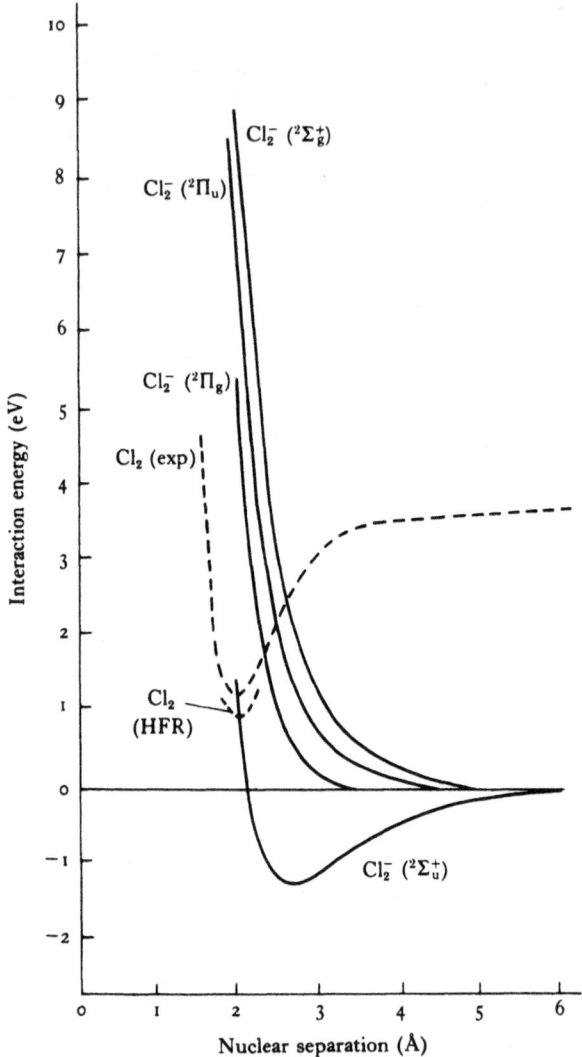

Fig. 6.29. Potential-energy curves (———) for states of Cl_2^- which tend to normal atoms and ions in the limit of large nuclear separations, calculated to the HFR approximation by Gilbert and Wahl (1971). The ground state curves for Cl_2 derived from experiments and from the HFR approximation are also shown (---).

Owing to the strong antibonding character of the $(\sigma_u 2p)$ orbital these molecules will be much less stable than O_2^- or S_2^-. Nevertheless it is clear from (6.27) that all four halogen molecules have positive electron affinities because $E_a(X) > D(X_2)$ for X = F, Cl, Br or I. Thus for I_2, $E_a = 3.22$ eV, $D(I_2) = 1.5$ eV.

The best measured values of the electron affinities have been obtained from the observation of the thresholds for certain charge transfer reactions. These are listed in Table 6.3. Also included are estimates based on an empirical formula of Mulliken (1950).

HFR calculations have been carried out by Gilbert and Wahl (1971) for F_2^- and Cl_2^-. Fig. 6.29 shows the calculated ground state Cl_2^- potential-energy curve compared with that for Cl_2. The calculated electron affinities, without any correction for correlation energy, are 2.45 and 2.19 eV for F_2 and Cl_2 respectively. These values are much closer to the measured value, particularly for Cl_2, suggesting that differential correlation effects are small for these molecules. The calculated values of R_e and ω_e are

$$R_e = 3.6a_0, \quad \omega_e = 510 \text{ cm}^{-1} \text{ for } F_2^-,$$
$$R_e = 3a_0, \quad \omega_e = 260 \text{ cm}^{-1} \text{ for } Cl_2^-.$$

The vertical detachment energy is very sensitive to details of the approximations and it is not possible to determine it from the HFR calculations.

In Fig. 6.29, calculated potential curves for those excited states which dissociate to ground state atoms and ions are also shown. All of these are repulsive.

Other diatomic molecules

O'Hare and Wahl (1970, 1971) have carried out HFR calculations of quite high accuracy for CF^-, OF^- and SiF^- as part of an extensive programme of investigation of the properties of the diatomic fluorides. They estimated the electron affinities in a similar way to that used by Cade (1967b) for the diatomic hydrides, and obtained the values 1.06 ± 0.2, 1.20 and 1.0 ± 0.2 eV for CF, OF and SiF respectively.

O'Hare (1971) has also applied the HFR method to NS and NS^-. Assuming the same correlation correction as for NO, for which the electron affinity is accurately known (see p. 191), he obtains an electron affinity 1.3 ± 0.3 eV for NS.

6.8 Polyatomic negative ions

A new feature, which has to be taken into account in dealing with polyatomic, as distinct from diatomic, negative ions and their energy relationships with the corresponding neutral species, is the geometry of the system. It by no means follows that the geometrical configuration of the ground state of a polyatomic ion will be the same as that for the neutral molecule. For triatomic systems, not involving H, rules set up by Mulliken (1942, 1958) and by Walsh (1953) may be used to predict the equilibrium configuration. According to these rules, molecules with 16 or less valence electrons are linear in their ground state. With an additional electron the equilibrium configuration is a bent one with a bond angle near 135°. This is the case with NO_2 for which the angle is 134.1°. Presumably this will apply to both N_2O^- and CO_2^-. A further valence electron leads to a more strongly bent geometry with bond angle near 120°, as for example O_3 (116.8°), SO_2 (119.5°), S_2O (118°), NF_2 (104.2°) and ClO_2 (117.6°). This can be expected to apply to NO_2^-. Indeed an HFR calculation for this ion does show this feature as may be seen from Fig. 6.30.

Different geometrical configurations for the ground state of the ion and neutral molecules will in general mean that the electron affinity will differ markedly from the vertical detachment energy. In fact the latter may be negative while the former is positive. Under these circumstances it may be difficult to produce the negative ion in its ground state. Thus, whereas N_2O^- and CO_2^- are bent, the neutral molecules N_2O and CO_2 are linear. To produce stable negative ions from the neutral molecules it is necessary to supply sufficient energy to distort the configuration of the latter to the bent form. As explained further on p. 218 and also in Chapter 9, p. 358, this has interesting experimental consequences.

Particularly because of long-range polarization forces, ions, both positive and negative, tend to attach certain other molecules to produce cluster ions. In particular, water molecules, possessing a large permanent dipole moment, may be attached to form quite large clusters. However, other molecules may also attach. Cluster formation is important in many practical situations as, for example, in the D region of the earth's ionosphere (see Chapter 15, p. 671).

The determination of the electron affinity of polyatomic negative ions is even more difficult than for diatomic species. Threshold measurements are complicated by the need to allow for the existence of more modes of internal motion than in diatomic cases. Nevertheless, useful results have been obtained from threshold measurements for charge transfer and other reactions as well as from photodetachment thresholds. The most promising technique, for not too complex molecules, is that of laser photoelectron spectroscopy which opens the possibility, as in O_2^- and NO^-, of distinguishing separate vibrational transitions in photodetachment. In many cases the only estimates of electron affinities of polyatomic radicals have been made from lattice energies and the Born–Haber cycle, but this evidence is rather unreliable.

We now consider some specific polyatomic ions which are of special interest or about which significant information is available.

O_3^-

This ion is of considerable importance in the lower ionosphere. By determining the threshold for the reaction

$$I^- + O_3 \longrightarrow O_3^- + I,$$

Berkowitz et al. (1971) obtained a close lower limit of 1.96 eV to the electron affinity. This is consistent with the fact that Ferguson (1969) found that the charge transfer reaction

$$O^- + O_3 \longrightarrow O_3^- + O$$

occurred rapidly at ordinary temperatures, showing it to be exothermic, and hence that $E_a(O_3) \geqslant E_a(O) = 1.465$ eV.

Both Byerly and Beaty (1971) and Burt (1972a), using somewhat different techniques (see Chapter 11, p. 489), obtained a value 2.1 eV for the photodetachment threshold which is also consistent.

Finally, Wood and d'Orazio (1965), revising an earlier analysis of the lattice energy of potassium ozonide (KO_3), obtained $E_a(O_3) = 1.9 \pm 0.4$ eV.

SO_2^-

This molecule is the first polyatomic negative ion to be studied by laser photodetachment spectroscopy (see Chapter 11, p. 487). SO_2 is triangular with a bond angle of 119.5° and it is to be expected that for SO_2^- the structure will not be very different. For such molecules the three fundamental vibrational modes may be distinguished as symmetrical stretch(ν_1) bending (ν_2) and asymmetrical stretch (ν_3) according to which of these respective modes they tend to for a linear molecule (such as CO_2) in the limit when the bond angle tends to 180°.

It was found possible to identify the peaks in the electron energy spectrum arising from transitions between different vibrational levels. The electron affinity of SO_2 was determined as 1.097 ± 0.036 eV and the bending frequency ν_2 as 989.4 cm^{-1}.

NO_2^-

This ion is also of importance in the lower ionosphere and is often found to appear in laboratory experiments. Fig. 6.30 shows the variation of the energy of the ion with bond angle, for the ground

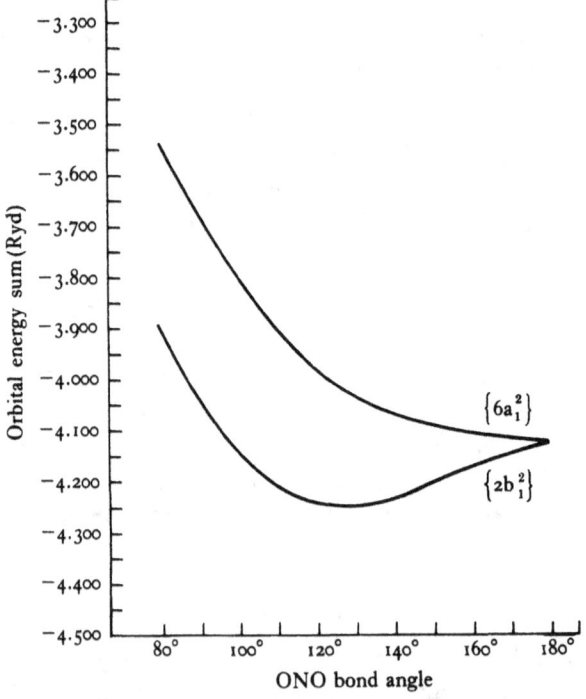

Fig. 6.30. Variation with the bond angle of the total energies of the valence electrons for the ground ($2b_1^2$) and excited ($6a_1^2$) states of NO_2^-, calculated by Pfeiffer and Allen (1969).

and first excited state, as calculated by the HFR method by Pfeiffer and Allen (1969). It will be seen that the equilibrium configuration according to this calculation has a bond angle of 118.9°. This is to be compared with 134.1° for the neutral molecule.

Difficulties are encountered in determining $E_a(NO_2)$ from the threshold energies for the charge transfer reactions with the halogen negative ions. This is because of the considerable extension, over 0.5–0.6 eV, of the 'tail' of the production cross-section, reminiscent of that encountered in analysing the photodetachment of O_2^- and possibly arising from the difference between the equilibrium geometries of NO_2^- and NO_2. However, Hughes et al. (1973) obtained 2.29 ± 0.1 eV for $E_a(NO_2)$ from the threshold for charge transfer reactions of NO_2 with I^-, Br^- and F^- (Chapter 12, p. 570). They also found that the reaction with SH^- was exothermic and with Cl_2^- endothermic so that $2.32 < E(NO_2) < 2.4$, which is consistent.

Measurements were made from the threshold for reactions
$$NO_2 + X \longrightarrow NO_2^- + X^+,$$
where X is an alkali metal atom, and are discussed in Chapter 10, p. 389. They give values close to 2.5 eV for $E_a(NO_2)$, somewhat larger than obtained from the charge transfer experiments.

From his photodetachment measurements Warneck (1969) estimates $E_a(NO_2) = 3.10 \pm 0.05$ eV, quite incompatible with these results. This applies also to the early lattice energy calculations of Yatsimirskii (1947) which gave 1.6 eV and the hot-wire equilibrium which gave 3.9–4 eV.

Resonance structure has been observed in NO_2 in the electron energy range 0–2 eV in transmission (Boness, Hasted and Larkin, 1968; Sanche and Schulz, 1973) and in the total electron-scattering cross-section (Larkin and Hasted, 1972). It seems probable that this structure arises in the same way as in O_2. Thus the ground state of NO_2^- lies 2.5 eV or so below that of NO_2 but when sufficient vibrational excitation is present the energy of the ion will be above that of the ground state of the neutral molecule and autodetachment can occur. Sanche and Schulz have analysed the observed structure in terms of symmetrical stretch and bending vibrations of NO_2^-. By extrapolation they find 162 meV as the spacing between the first two vibrational levels.

N_2O^- and CO_2^-

As mentioned above, these negative ions have a bent equilibrium configuration with bond angle near 134° whereas the corresponding neutral molecules are linear. This has the consequence of rendering very unlikely the production of the negative ions from the neutral species unless an activation energy of the order of a few eV is available.

N_2O^- was first identified in mass spectra by Paulson (1966) and he also was the first to observe CO_2^- a few years later. The N_2O^- ions were produced at a rate proportional to the cube of the pressure by a beam of electrons in N_2O (Chantry, 1971), while the CO_2^- ions were formed in the reactions of O^-, NO^- and O_2^- ions of a few eV kinetic energy with CO_2.

The geometry of N_2O^- has important consequences for the rate of the dissociative attachment reaction

$$N_2O + e \longrightarrow N_2 + O^-,$$

as discussed in Chapter 9, p. 358, and also for the corresponding inverse reaction (see Chapter 12, p. 550).

The true electron affinities of N_2O and CO_2 are probably around 1 eV but little is known about them at the time of writing (see however Chapter 9, p. 363).

NH_2^-, PH_2^-, AsH_2^-

Photodetachment thresholds for these negative ions have been observed using the ion cyclotron resonance technique described in Chapter 11, p. 445. Values obtained which are identified with the electron affinities are 0.744 ± 0.022, 1.26 ± 0.03 and 1.27 ± 0.03 eV for NH_2, PH_2 and AsH_2 respectively. Earlier measurements for NH_2 have given the following results.

Feldmann (1970) using the crossed-beam photodetachment technique found $E_a(NH_2) = 0.76 \pm 0.04$ eV, consistent with the ion-cyclotron measurements, while Celotta et al. (1974), from measurement of the energy spectrum of the electrons detached by an argon ion laser beam, found 0.779 ± 0.037 eV. All of these measurements made through photodetachment are consistent and can be relied upon within the stated accuracy. Earlier measurements gave values greater than 1 eV, as for example 1.12 eV by the hot-wire equilibrium technique (Page and Goode, 1969). However, Young, Lee-Ruff and Bohme (1971) found the reaction

$$H_2 + NH_2^- \longrightarrow H^- + NH_3$$

to be exothermic showing that $E_a(NH_2) \leqslant 1.03$ eV.

Robb and Csizmadia (1971) have carried out calculations of the structure of NH_2^- assuming that the bond angle is close to the tetrahedral value (109°28'). They introduced a contribution from electron-pair correlation in addition to the HFR approximation. Unfortunately, comparable calculations for NH_2 are not available so no estimate may be made of the electron affinity.

NO_3^-

This ion is of especial interest as the most abundant negative species, either free or clustered with H_2O, in the lower ionosphere.

As NO_3 is not stable, the electron affinity cannot be obtained by determining the threshold energy for charge transfer reactions with halogen ions. However, Ferguson, Dunbar and Fehsenfeld (1972) obtained useful results by studying reactions of NO_3^- ions with HCl and HBr at room temperature (see Chapter 12, p. 575). The reaction

$$NO_3^- + HCl \longrightarrow Cl^- + HNO_3$$

was not observed so, assuming it to be endothermic, $E_a(NO_3) > 3.5 \pm 0.2$ eV. On the other hand, with HBr, evidence was obtained that the reaction is nearly thermoneutral (see Chapter 12, p. 576) in which case $E_a(NO_3) = 3.9 \pm 0.2$ eV.

These results are consistent with those obtained by Berkowitz et al. (1971) from measurement of the threshold energy for the reaction

$$I^- + HNO_3 \longrightarrow NO_3^- + HI.$$

They find 2.68 eV $< E_a(NO_3) <$ 4.34 eV.

Fehsenfeld, Ferguson and Bohme (1969) had obtained evidence earlier that the reaction

$$NO_2^- + NO_2 \longrightarrow NO_3^- + NO$$

is exothermic, in which case

$$E_a(NO_3) - E_a(NO_2) > 0.9 \text{ eV},$$

which again is consistent with the results given above.

The high value of the electron affinity is essentially the reason why NO_3^- is dominant in the lower ionosphere (see Chapter 15, p. 670).

O_4^-

Direct evidence of the existence of this ion, which can be regarded as a simple clustered ion $O_2^- \cdot O_2$, was obtained by Conway and Nesbitt (1968). They noted that the known dissociation energy of O_4^+ into O_2^+ and O_2 (0.44 eV) exceeds that (0.24 eV) of $N_2O_2^+$ into N_2 and O_2^+ by a factor of nearly two. If the binding arose in both cases from interaction between the ion and the induced dipole moment of the molecule, there should be little difference as the polarizabilities of N_2 and O_2 are nearly the same. It seems reasonable to ascribe the difference to resonance energy due to the identity of the $O_2^+ \cdot O_2$ and $O_2 \cdot O_2^+$ systems. If this is so it should also be operative in $O_2^- \cdot O_2$.

To search for O_4^- they irradiated O_2 with a tritium source of 2Ci strength and analysed the mass spectra of the ions produced.

As described in Chapter 12, p. 541, they not only observed O_4^- but were able to determine the equilibrium constant for the reactions

$$O_2 + O_2^- \rightleftharpoons O_4^-$$

over a temperature range from 273–353 °K. From this they found, by applying the formula of statistical mechanics, that the enthalpy change associated with the reaction is -13.55 ± 0.16 kcal mol^{-1} (0.588 ± 0.005 eV) in agreement with their prediction.

CO_3^- and CO_4^-

CO_3^- ions, both free and hydrated ($CO_3^- \cdot H_2O$), are important negative ionic constituents of the lower ionosphere. Burt (1972b) (see Chapter 11) has observed the photodetachment threshold energies for these ions as 1.8 and 2.1 eV respectively.

CO_4^- ions regarded as $O_2^- \cdot CO_2$ have been observed in many conditions and in particular in the detachment experiments of Pack and Phelps (1966) and of Moruzzi and Phelps (1966). In the former, evidence was obtained (Chapter 12, p. 549) of the formation of CO_4^- through the reaction

$$e + O_2 + CO_2 \rightleftharpoons CO_4^-.$$

Under certain experimental conditions this process reached equilibrium and the equilibrium constant could be determined. From this the energy of dissociation of CO_4^- into O_2^- and CO_2 is found to be 0.8 ± 0.1 eV.

SF_6^- and TeF_6^-

These ions are of considerable importance because of the high probability of their formation by attachment of electrons of nearly zero energy to SF_6 (see Chapter 9, p. 368). Lifshitz, Hughes and Tiernan (1970) attempted to estimate the electron affinity of SF_6 from the threshold energy for the charge exchange reaction with Cl^-. Difficulty was encountered because of the presence of an even more pronounced tail than for NO_2 (see p. 217).

However, in later measurements, Lifshitz, Tiernan and Hughes (1973), from the thresholds for charge transfer reactions of S^- and O^- with SF_6 (Chapter 12, p. 572), found a mean value of 0.63 eV for $E_a(SF_6)$. This is compatible with measurements of the threshold for the reaction

$$SF_6 + Cs \longrightarrow Cs^+ + SF_6^-, \qquad (6.30)$$

made by Nalley, Compton, Schweinler and Anderson (1973) (see Chapter 10, p. 399), which give $E_a(SF_6) = 0.54^{+0.1}_{-0.17}$ eV. Lifshitz et al. also found evidence that the change transfer reaction between O_2^- and SF_6 is exothermic, in which case $E_a(SF_6) > 0.44$ eV.

Nalley et al. (1973) also obtained $E_a(TeF_6) = 3.34^{+0.1}_{-0.17}$ eV from the observed threshold for the reaction corresponding to (6.30).

Lifshitz et al. (1973) (Chapter 12, p. 573) and Nalley et al. (1973) (Chapter 10, p. 399) both found that $E_a(SF_5) \geqslant 2.8 \pm 0.2$ eV.

CHAPTER 7

Excited Electronic States of Molecular Negative Ions

7.1 Introduction

It seems likely that almost all excited electronic states of molecular negative ions will be unstable towards autodetachment and can be regarded as either type I or type II resonance states. In fact the nature of the state may change with nuclear separation.

If the excited state occurs by temporary attachment of an electron to an excited valence state of a molecule it will usually be of type II, as in the case of the ground states of N_2^-, CO^- and H_2^-. In these cases the electron is temporarily trapped by a centrifugal barrier. On the other hand if the electron is attached to an excited Rydberg state the resonance state will often be of type I. The states will then be very similar to the corresponding doubly-excited resonance states of atomic ions and are often referred to as *core-excited* resonances. In the molecular case the outer two electrons in the temporary negative ion will be in Rydberg orbitals of large radius which will have little effect on the relative motion of the nuclei. We would then expect that the vibrational level systems of these resonance states will be very closely the same as for the positive-ion core which is often referred to as the 'grandparent'. Moreover, the Franck–Condon intensity distribution for transitions from the resonance state to the ground state of the neutral molecule should be very much the same as for transitions from the grandparent state. These expectations are justified by observed results for H_2, N_2, CO, O_2 and NO, as described below, and assist in the identification of resonance states. The experimental techniques employed are essentially the same as for the study of the autodetaching states of atomic negative ions (see Chapter 4, p. 107).

We now proceed to discuss the results for specific molecular ions, confining ourselves to diatomic species only.

7.2 H_2^-

The $^2\Sigma_g^+$ state which dissociates into normal H and H^- has already been described in the preceding chapter (p. 178). A number of core-excited resonance states have been observed in an energy range extending from 11.15 eV above the ground state of H_2. These were discovered by Kuyatt, Mielczarek and Simpson (1964) using their transmission technique (Chapter 4, p. 110). Since then a number of experiments have been carried out observing the resonance effects in inelastic scattering, involving electronic and/or vibrational excitation, as well as further transmission measurements, particularly by Sanche and Schulz (1972).

The resonance states have been analysed into at least five series of vibrational levels. It is possible also that two other series exist but their separate identity has not yet been fully established. The seven series are distinguished by the letters a to g.

Fig. 7.1 shows the analysis of the derivative of the transmitted current as a function of electron energy observed by Sanche and Schulz (1972) using the equipment described in Chapter 4, p. 111. Four of the band systems, a, c, f and g are clearly distinguished. The b series is only observed in measurements of cross-sections for excitation of high vibrational levels of the ground state of H_2. Fig. 7.2 shows such measurements made by Comer and Read (1971a) at a scattering angle of 85°. For final vibrational quantum numbers v ranging from 1 to 8 the resonance energies are all associated with band a only. At higher values of v the b series appears at the low energy end.

Table 7.1 summarizes the data available for the energies of the levels of the different series. These include data obtained from measurements of differential cross-sections for excitation of the $b^3\Sigma_u^+$ continuum and of the $B^1\Sigma_u^+$ and $C^1\Pi_u$ states of H_2 made by Weingartshofer, Ehrhardt, Hermann and Linder (1970) using the apparatus described in Chapter 4, p. 112.

The separate existence of the bands designated d and e is in doubt. d is indistinguishable in energy from a but, as will be seen below, it is expected that the a series is associated with a $^2\Sigma_u^+$ state so the angular distribution of the scattered electrons at the resonance energies should be isotropic. This was not confirmed from the angular distribution measurements of Weingartshofer et al. (1970). On the basis of these measurements Comer and Read (1971a) considered that a $^2\Pi_u$ state must also be involved and they distinguished the series, presumably arising from this state, as the d series.

The separate existence of the series e, as distinct from c, depends

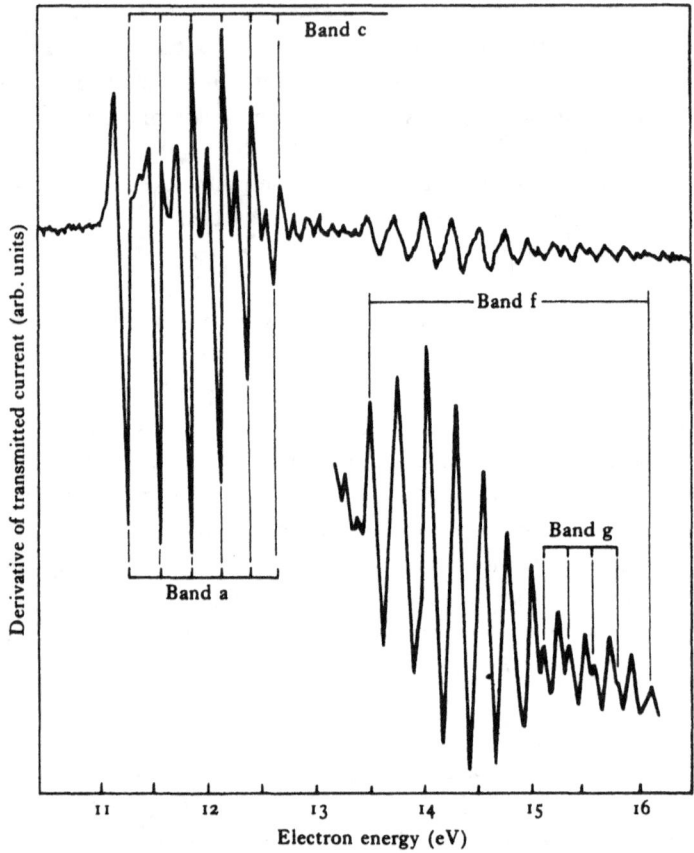

Fig. 7.1. Energy derivative of the transmitted electron current as a function of electron energy in H_2, observed by Sanche and Schulz (1972). For the lower portion of the figure, showing bands f and g, the gain has been increased by a factor of 7.

on the fact that Weingartshofer *et al.* (1970) obtained a series of resonance energies in their electronic excitation measurements which differed appreciably from the c series determined from other types of experiment.

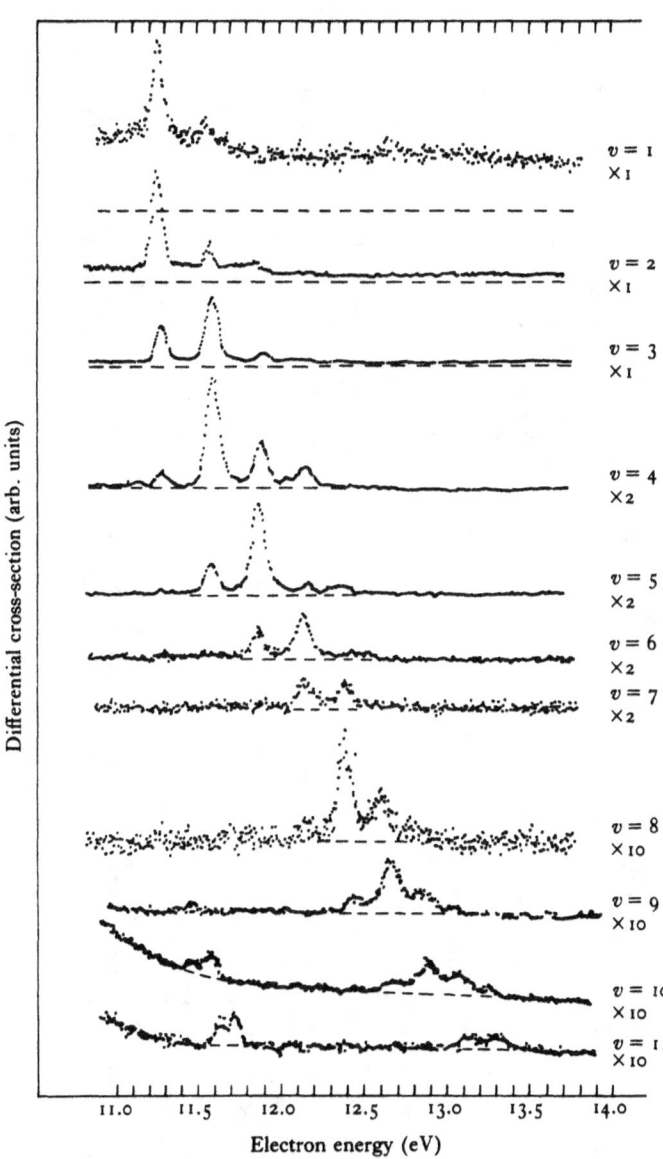

Fig. 7.2. Variation with electron energy of the differential cross-sections for excitation of different vibrational levels of normal H_2, at a scattering angle of 85°, observed by Comer and Read (1971a).

TABLE 7.1 *Vibrational series associated with autodetaching states of H_2^- and D_2^-*

Series designation	a and d				b	c and e					f		g
Probable electronic state	$(\sigma_g 1s)(\pi_u 2p)(\pi_u 2p')^2\Sigma_g^{+*}$				?	$(\sigma_g 1s)(\pi_u 2p)(\pi_u 2p')^2\Sigma_g^+$† or $(\sigma_g 1s)(\pi_u 2p)(\sigma_g 2s)^2{}^2\Pi_u$‡					?		?
Highest vibrational quantum number observed $(=v_m)$	8				7	6					11		4

Energy (eV) above the ground states of level with

		(1)	(2)	(3)	(4)	(5)	(3)	(1)	(2)	(3)	(4)	(5)	(1)	(6)	(4)	(1)
$v = 0$	H_2^-	11.32,	11.28,	11.39,	11.30,	11.07	—	11.43,	11.46,	11.37	11.50,	11.46	13.66,	13.62,	13.63	15.09
	D_2^-	11.34,	11.28,	11.32,	—	—	—	—	—	—	—	—	13.66,	—	—	15.05
$v = 1$	H_2^-	11.62,	11.56,	11.62,	11.62,	11.37	—	11.74,	11.72,	11.71,	11.79,	11.75	13.94,	13.91,	13.93	15.32
	D_2^-	11.56,	11.48,	11.54,	—	—	—	11.67,	—	11.65,	—	—	13.86,	—	—	15.29
$v = v_m - 1$	H_2^-	—	—	13.10,	—	—	11.85	12.83,	12.77,	—	—	12.84	—	16.07	—	15.57
	D_2^-	12.71,	12.64,	12.61,	—	—	—	—	—	—	—	—	—	—	—	15.55
$v = v_m$	H_2^-	—	—	13.28,	—	—	11.96	13.06,	12.97	—	—	—	—	16.26	—	15.77
	D_2^-	12.88,	12.85,	12.75,	—	—	—	—	—	—	—	—	—	—	—	15.71

* Series a only, † best theoretical fit, ‡ suggested from angular distribution measurements (Comer and Read, 1971a).
(1) Sanche and Schulz (1972), transmission, (2) Kuyatt, Simpson and Mielczarek (1966), transmission, (3) Comer and Read (1971a), vibrational excitation, (4) Weingartshofer et al. (1970), electronic excitation, (5) Eliezer et al. (1967), theoretical, (6) Golden (1971).

Table 7.1 also includes observed results for D_2^- for all but the series b.

As far as the lifetimes of the different autodetaching states associated with the different vibrational series are concerned, the only measurements which refer to natural level widths are those of Joyez, Comer and Read (1973). For series b they found a width of 30 meV and in a high resolution experiment in H_2 found that, for series c, it is not greater than 16 meV. The corresponding lifetimes, around 10^{-12} s, are long compared with the vibrational periods, consistent with the clearly-developed vibrational series.

Eliezer, Taylor and Williams (1967) have investigated the autodetaching states to be expected in the energy range concerned, using the stabilization technique. For the basis orbitals they used essentially the same forms, in spheroidal coordinates, as in their calculations on the ground and first excited states of H_2^- described on p. 176. They found two type I autodetaching states both of $(\sigma_g 1s)(\pi_u 2p)(\pi_u 2p')\,^2\Sigma_g^+$ configurations. These are mainly derived through attachment of a π_u electron to an H_2 molecule in the $c^3\Pi_u$ and $C^1\Pi_u$ states, both of which have the configuration $\sigma_g 1s\pi_u 2p$. The real parts of the energies of these states as functions of nuclear separation are shown in Fig. 7.3 in relation to the parent states. The electron affinities of the $c^3\Pi_u$ and $C^1\Pi_u$ states are estimated to be 0.8488 and 0.8435 eV respectively.

The calculated energy values for the upper state agree very well with those observed for series c (see Table 7.1) but, as already indicated, there are problems remaining in the interpretation of these bands. Thus Comer and Read (1971a) do not find the isotropic angular distribution expected for a Σ resonance state and would instead identify it as a $^2\Pi_u$ state. Furthermore, they found evidence of a lower member of the c and e series at 11.19 eV which would be 0.27 eV below that calculated for the $^2\Sigma_g^+$ state.

Eliezer et al. (1967) found no other type I states arising from the $c^3\Pi_u$, $a^3\Sigma_g^+$, $C^1\Pi_u$ and $E^1\Sigma_g^+$ excited states of H_2. The identification of the f and g series, which are at considerably higher energies, remains uncertain at the time of writing.

Both the $^2\Sigma_g^+$ resonance states arise by addition of two electrons in Rydberg states to the $\sigma_g 1s\,^1\Sigma_g^+$ state of H_2^+, which is therefore the 'grandparent' in these cases. For the lowest resonance state the binding energy of the two electrons to the grandparent is 4.1 eV (see p. 240).

7.3 N_2^-

In Chapter 6, p. 194, we discussed the type II resonance states associated with temporary attachment of electrons to the ground $X^1\Sigma_g^+$ state of N_2. Much of the structure observed in transmission experiments at electron energies between 7 and 11 eV (Fig. 7.4) can be ascribed to type II resonance states in which an electron is attached to low-lying excited states of N_2, particularly $B\,^3\Pi_g$ (see

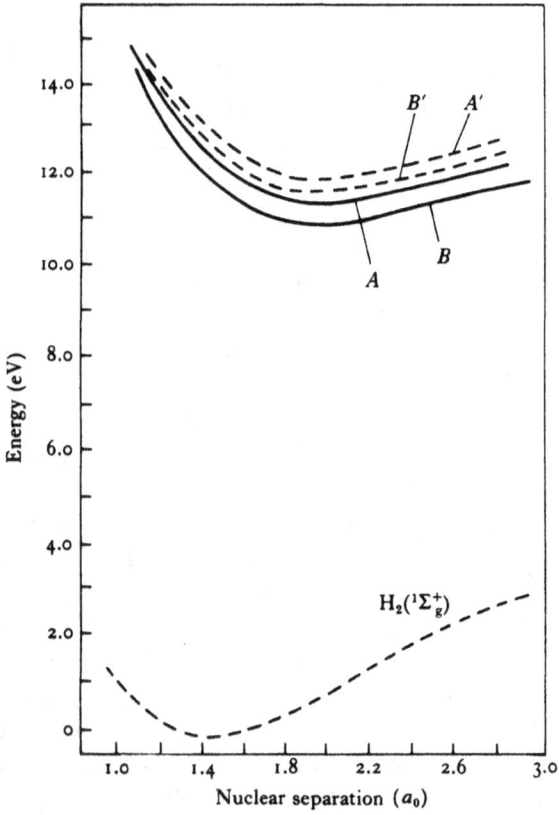

Fig. 7.3. Real parts of potential-energy curves for excited states of H_2^- in relation to the curves for parent states of H_2, calculated by Eliezer et al. (1967). Full-line curves relate to H_2^-, broken curves to H_2, states. A, $^2\Sigma_g^+$ state of H_2^- derived from $C^1\Pi_u$ of H_2 (curve A'); B, $^2\Sigma_g^+$ state of H_2^- derived from $c^3\Pi_u$ of H_2 (curve B').

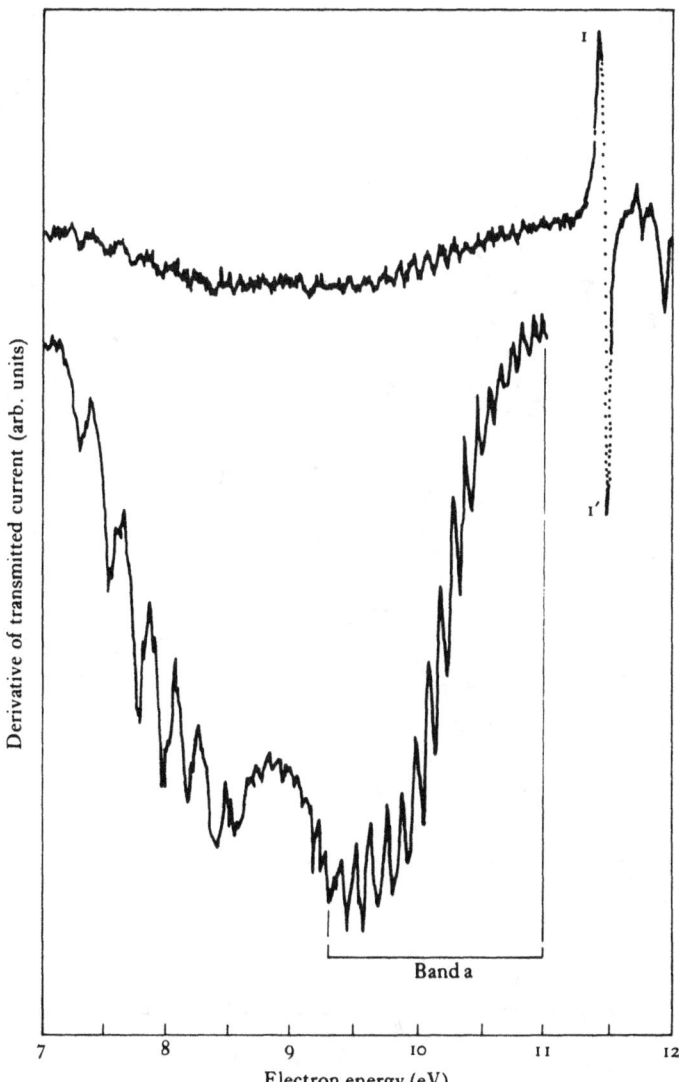

Fig. 7.4. Energy derivative of the transmitted electron current in N_2 in the range 7–11 eV, observed by Sanche and Schulz (1972). The lower curve is taken with higher intensity. Band a arises from type II resonances associated with the $B^3\Pi_g$ state. The structure at lower energies is due to excitation of vibrational levels of $B^3\Pi_g$.

Fig. 7.5). In interpreting the data it is important to remember that the lifetimes of type II states are relatively short and may be

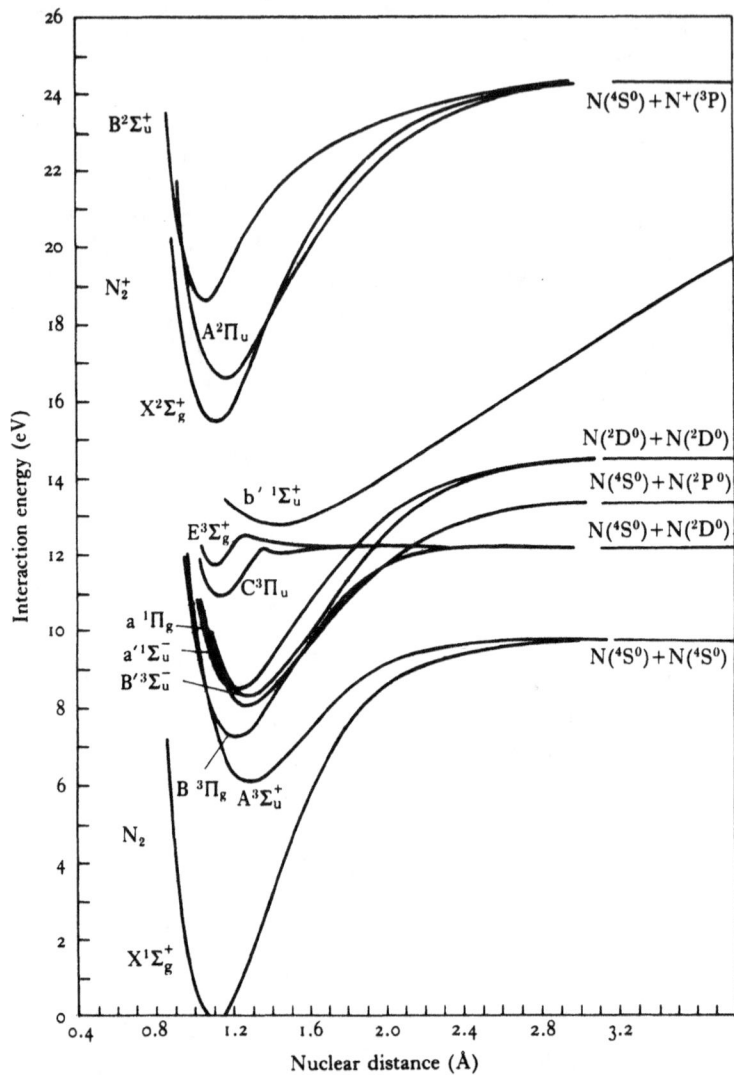

Fig. 7.5. Potential-energy curves of low-lying states of N_2 and of N_2^+.

comparable with the vibrational period as for the resonances associated with the ground state of N_2. If this is so the resonance peaks will occur at different energies for different inelastic processes.

Fig. 7.6 shows differential cross-sections at an angle of scattering of 90° for excitation of different vibrational levels of the $B\,^3\Pi_g$ and $A\,^3\Sigma_u^+$ states as functions of electron energy observed by

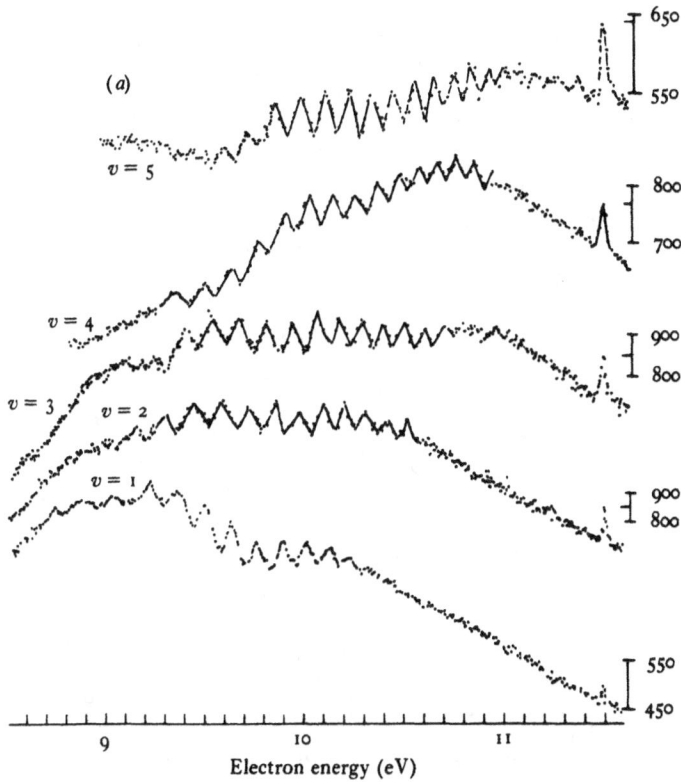

Fig. 7.6(a). For legend see p. 232.

Mazeau, Gresteau, Hall, Joyez, Reinhardt and Hall (1972). Figs. 7.6(a) and (b) show the resonance effects arising from the type II states associated with the $B^3\Pi_g$ and $A^3\Sigma_u^+$ states of N_2 respectively. When analysed in detail the variation of the peak energies with final vibrational quantum number v is apparent. Thus the energy of the fourth peak for $v = 1$ excited through the $B^3\Pi_g$ sequence is 9.500 eV. For $v = 2$, 3 and 4 it is found at 9.445, 9.400 and 9.365 eV respectively.

The energy of the lowest vibrational level of the resonance state lies about 1.8–2 eV above the parent state, which is comparable with the corresponding value 2.3 eV for the type II state associated with the ground state of N_2.

The most prominent resonant feature in the energy region between 11 and 15 eV is due to a type I state arising from attachment

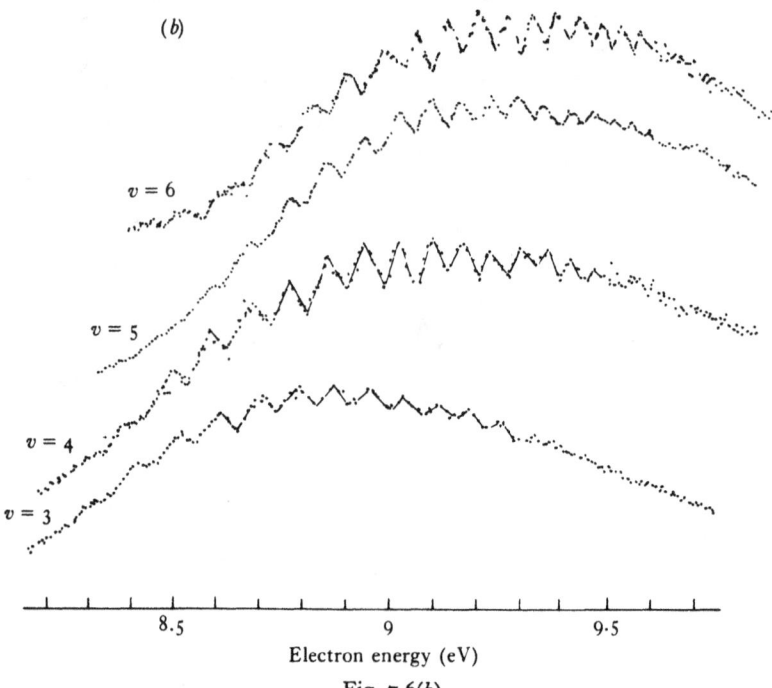

Fig. 7.6(b).

Fig. 7.6. Variation with electron energy of the cross-sections for excitation of different vibrational levels of (a) the $B^3\Pi_g$ and (b) the $A^3\Sigma_u^+$ state of N_2, observed at a scattering angle of 90° by Mazeau et al. (1973).

of an electron to the $E^3\Sigma_g^+$ state of N_2 (see Fig. 7.5) with a binding energy with respect to this state of about 0.4 eV. This was first observed by Heideman, Kuyatt and Chamberlain (1966) using their transmission technique (see Chapter 4, p. 110). The sharp peak at 11.48 eV above the ground state of N_2 which they found is illustrated in Fig. 7.7. It also shows up very clearly in the transmission measurements of Sanche and Schulz (1972) (see Fig. 7.8). Comer and Read (1971b) observed additional members of a vibrational series associated with the resonance in their measurements of differential cross-sections for vibrational excitation of the ground state of N_2. The observed angular distribution of electrons which excited the $v = 1$ level of N_2 was found to be nearly isotropic showing that the resonance is of Σ_g^+ type. Comer and Read not only measured the resonance energy very accurately as 11.48 eV above the ground state of N_2 but also determined the resonance

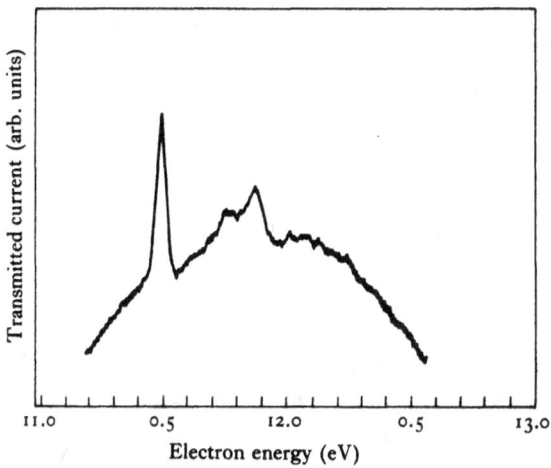

Fig. 7.7. Transmission of electrons, in N_2, as a function of energy in the range 11–13 eV, observed by Heideman et al. (1966). The sharp resonance at 11.48 eV is due to the $^2\Sigma_g^+$ state of N_2^- of which the parent is the $E\,^3\Sigma_g^+$ state of N_2.

width as 6×10^{-4} eV. Kisker (1972) has also observed the resonance in measurements of the optical excitation function of the $C\,^3\Pi_u$ state.

From analysis of the vibrational series the equilibrium nuclear separation in the resonance state was determined as 1.115 ± 0.01 Å, and the energy of a substate with the vibrational quantum number v by

$$E_0 + a(v+\tfrac{1}{2}) - b(v+\tfrac{1}{2})^2,$$

where $E_0 = 11.345$ eV, $a = 0.270 \pm 0.02$ eV and $b = 0.002 \pm 0.002$ eV.

The resonance state is likely to arise by attachment of two $3s\,\sigma_g$ electrons to the $X\,^2\Sigma_g^+$ state of N_2^+ (see Fig. 7.5) as grandparent. This is supported by the fact that the separation of the vibrational levels in the N_2^+ state is 0.271 eV, very nearly the same as for the resonance state. Furthermore, the relative intensity of the peaks arising from the $v = 0$ and $v = 1$ sublevels of the resonance state is about 10:1 which is very close to that (9.96:1) for Franck–Condon transitions from the ground state of N_2 to the $v = 0$ and $v = 1$ sublevels of the $X\,^2\Sigma_g^+$ state of N_2^+.

The binding energy of the lowest member of the series relative to the grandparent state is 4.1 eV, very nearly the same as for other resonance states in which two 3sσ electrons are bound in Rydberg states to an inner core (see p. 240).

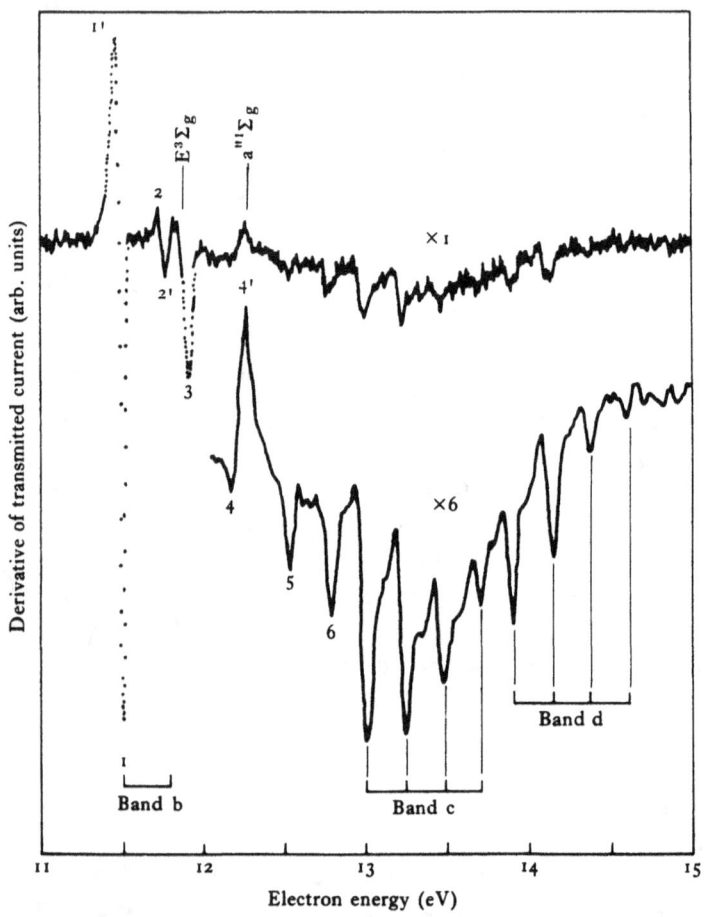

Fig. 7.8. Energy derivative of the transmitted electron current as a function of electron energy in N_2, in the range 11–15 eV, observed by Sanche and Schulz (1972). The structure 1' is due to the 11.48 eV resonance state. The resonance structures marked 5 and 6 as well as bands c and d arise from the $A\,^2\Pi_u$ state of N_2^+ as grandparent.

A similar core-excited resonance state might be expected to arise through binding of two $3s\sigma_g$ electrons to the excited $A\,^2\Pi_u$ valence core of N_2^+ (see Fig. 7.5) which lies 1.14 eV above the ground state. Assuming the same binding energy for the two Rydberg electrons to this core, the lowest member of the corresponding resonance series should lie at $11.48 + 1.14 = 12.62$ eV. Referring to Fig. 7.8, we see that a resonance is observed at 12.64 eV (marked 5 in the figure) which may probably be identified with this state. The separation of the $v = 0$ and $v = 1$ vibrational levels of the grandparent state in this case is 0.23 eV and it is noteworthy that a second feature (marked 6) is apparent in Fig. 7.8 at 12.87 eV, suggesting that this is the second member of the resonance series.

Two further short series of resonance states (bands c and d) are apparent in Fig. 7.8 starting at 13.00 and 13.88 eV respectively. The first series has also been observed (Mazeau et al., 1973) in differential cross-sections for excitation of the $E\,^3\Sigma_g^+$, $a''\,^1\Sigma_g^+$ and $C\,^3\Pi_u$ states. Although it is difficult to disentangle the overlapping resonances, the average spacings in the two series are 0.230 and 0.225 eV, quite close to that for the $A\,^2\Pi_u$ state of N_2^+ which may well be the grandparent for these states also.

Evidence of overlapping resonance states of N_2^- lying above the ionization limit of N_2 has been obtained by Pavlovic, Boness, Herzenberg and Schulz (1972) in observed differential cross-sections for excitation of the $v = 1$, 2 and 3 vibrational levels of the ground state of N_2 by electrons with energy in the range 20–24 eV.

7.4 CO⁻

As pointed out in Chapter 6, the structure of the electronic state of CO⁻ should be similar to that for the isoelectronic molecule N_2^- except for the absence of the nuclear symmetry and the additional possibility of states dissociating into $C + O^-$ or $C^- + O$. In that section the resonance series associated with the type II resonance representing the ground state of CO⁻ was described.

Structures presumably arising from type II resonances in which an electron is temporarily attached to an excited valence state of CO have been observed (as, for example, 3 and 4–4′ in Fig. 7.9) but not yet clearly identified.

Corresponding to the strong $^2\Sigma_g^+$ resonance state in N_2^- there is a strong resonance at 10.04 eV in CO. This was first observed in transmission by Sanche and Schulz (1971) (see Fig. 7.9) and angular distribution measurements (Mazeau et al., 1972) confirm its Σ character. The width, about 45 meV, is much greater than for the corresponding state of N_2^-, presumably because of the possible decay into $O^-(^2P) + C(^3P)$, evidence for which comes from experiments on dissociative attachment of electrons in CO (see Chapter 9, p. 333).

Once more the resonance state arises from the attachment of two

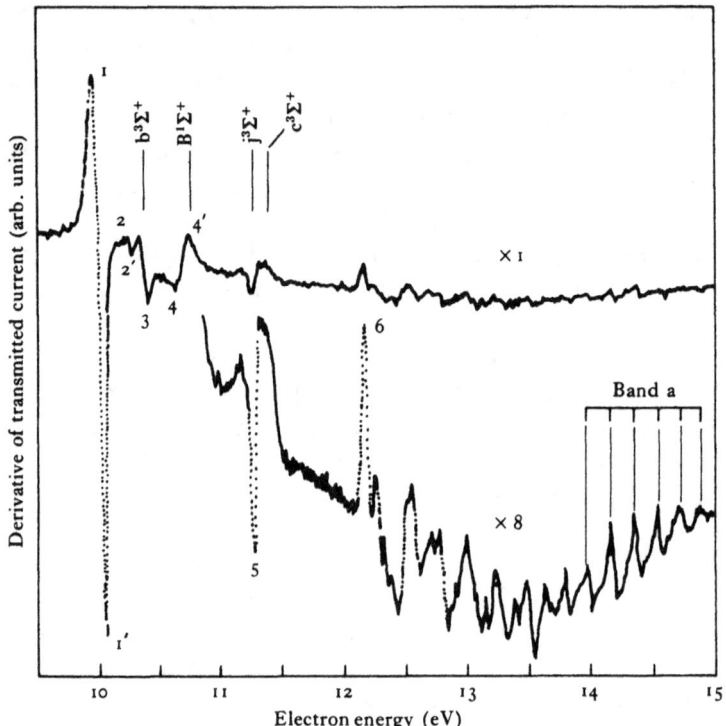

Fig. 7.9. Energy derivative of the transmitted electron current as a function of electron energy in CO, observed by Sanche and Schulz (1971).

$3s\,\sigma$ electrons to the grandparent $X\,^2\Sigma^+$ state of CO^+ with a binding energy of 4.1 eV.

Some evidence exists of type II states arising from the state of CO^+ as grandparent. Thus in Fig. 7.9 the series labelled band a starting at 13.95 ± 0.05 eV may well be associated with such a state.

Evidence of overlapping resonances above the ionization thres-

MOLECULAR NEGATIVE IONS – EXCITED STATES

hold of CO has also been obtained (Chutjian, Truhlar, Williams and Trajmar, 1972; Truhlar, Trajmar and Williams, 1972).

7.5 O_2^-

The resonance series associated with the upper vibrational levels of the ground state of O_2^- has already been discussed in Chapter 6, p. 183. Evidence concerning the $^2\Pi_u$ excited state, provided from measurements of dissociative attachment, is discussed in Chapter 9, pp. 319–25.

There is relatively little data available about higher core-excited resonance states. Fig. 7.10 shows the results obtained by Sanche and Schulz (1972) from transmission measurements in O_2 in the range 8–13 eV.

The vibrational spacing in the $O_2^+(X\,^2\Pi_g)$ ground state, which is a potential grandparent, is 0.232 eV, not far from the spacing between the features labelled 1–1' and 3–3' in Fig. 7.10. Also the energy at which the 1–1' feature occurs is 4.1 eV above that of $O_2^+(X\,^2\Pi_g)$, close to that typical of type I states in which two $3s\sigma$ electrons are attached to a positive-ion core.

There is a clearly visible progression of as many as 10 resonances, referred to as band a in Fig. 7.10, which starts at 11.81 ± 0.005 eV. It was found by Sanche and Schulz (1972) that the vibrational spacings and the Franck–Condon intensity distributions for transitions from the ground state of O_2 are very closely similar to the corresponding quantities for the $a\,^4\Pi_u$ state but no other states of O_2^+. Considered again as a type II state in which two $3s\sigma$ electrons are attached to the excited core, the binding energy is 4.4 eV, again close to the usual value.

A further series of four levels has been observed above the ionization threshold of O_2. The spacings are close to those of the $b\,^2\Sigma^-$ state of O_2^+ suggesting that this state is the grandparent.

7.6 NO⁻

The resonance effects associated with all but the lowest vibrational level of the ground electronic state of NO⁻ have been discussed in Chapter 6, p. 191. Although resonance levels associated with higher electronic states have only been investigated by transmission methods, it was from the application of their technique, described in Chapter 4, p. 111, that Sanche and Schulz (1972) obtained the first substantial evidence on the interpretation of core-excited type I resonances in terms of grandparent states.

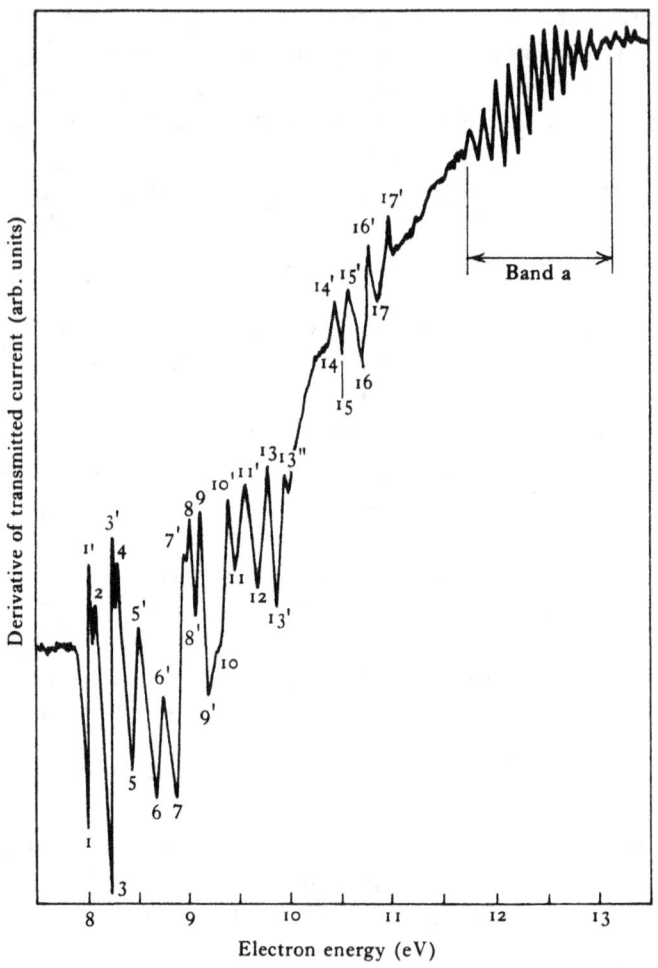

Fig. 7.10. Energy derivative of the transmitted electron current as a function of electron energy in O_2, observed by Sanche and Schulz (1972).

Fig. 7.11 shows the variation with energy of the energy derivative of the transmitted current for electrons with energies in the range 5–7.5 eV in NO. Four vibrational series, referred to as a, b, c and d

Fig. 7.11, can be distinguished. The first members of the series are located at 5.04, 5.41, 5.46 and 6.45 eV respectively and their vibrational spacings are very close to those for the $X^1\Sigma^+$ state of NO^+ as may be seen from Table 7.2. The Franck–Condon intensity distributions for transitions from the ground state of NO

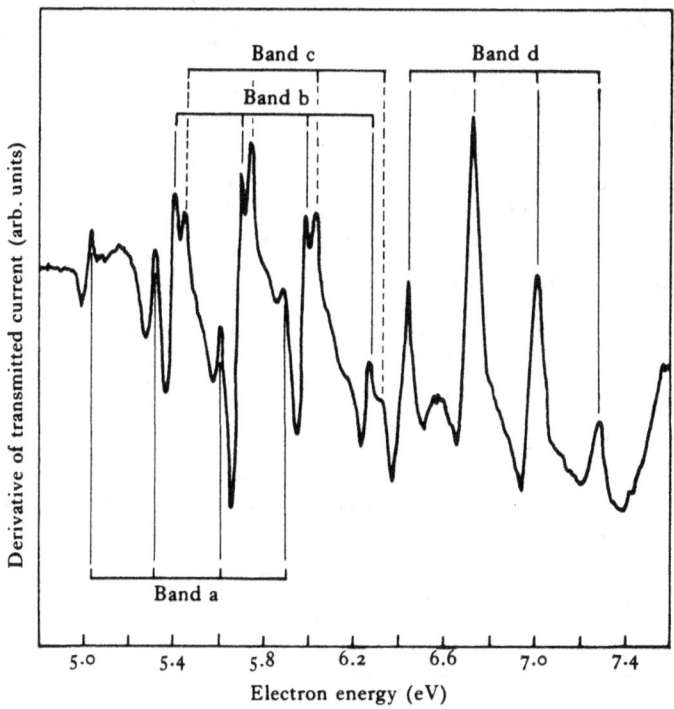

Fig. 7.11. Energy derivative of the transmitted electron current as a function of electron energy in the range 5–7.5 eV in NO, observed by Sanche and Schulz (1972).

also are quite similar. The lowest, a, series suggests that the electronic state is one in which two $3s\sigma$ electrons are attached to the $X^1\Sigma^+$ core. The parent state is probably the $A^2\Sigma^+$ Rydberg state of NO which lies 5.48 eV above the ground state and arises from attachment of an electron in a $3s\sigma$ orbital to the $X^1\Sigma^+$ state of NO^+.

The series b, c and d probably arise from attachment of an electron to the $C^2\Pi$ and $D^2\Sigma^+$ states of NO, which lie at 6.49 and 6.60 eV above the ground state, but for the former two series the

TABLE 7.2 *Comparison of vibrational spacings and Franck–Condon intensity distribution for four resonance series of NO⁻ with the corresponding values for the $X^1\Sigma^+$ ground state of NO⁺*

Vibrational quantum numbers	Vibrational spacings (eV) NO⁻ series				
	a	b	c	d	$NO^+(X^1\Sigma_g^+)$
0–1	0.286	0.290	0.292	0.282	0.290
1–2	0.286	0.290	0.288	0.284	0.287
2–3	0.282	0.284	0.286	0.275	0.283
3–4				0.275	0.278
4–5				0.275	0.273

Franck–Condon intensity distribution for transitions from the ground level of NO

					Observed	Theoretical
0	0.84	0.57	–	0.66	0.7	0.58
1	1	1	–	1	1	1
2	0.62	0.64	–	0.7	0.7	0.92
3	0.16	0.24	–	0.19	0.5	0.49

TABLE 7.3 *Relation of resonance states of NO⁻ to grandparent NO⁺ states*

	Grandparent state	Resonance state		
Specification	Energy above ground state of NO (eV)	Energy above ground state of NO (eV)	Vibrational quantum number	Binding energy (eV)
$X^1\Sigma^+$	9.27	5.04	0	4.23
$b^3\Pi$	16.56	12.36	0	4.20
		12.57	1	
$A^1\Pi$	18.32	14.19	0	4.13
$B^1\Pi$	21.72	17.51	0	4.21

electronic state could be one in which a $3p\sigma$ or $3p\pi$ electron is attached to the $A^2\Sigma^+$ state.

At higher energies, as shown in Fig. 7.12, a number of resonance states arise, most of which are of type I arising from excited states of NO⁺ as grandparents. In Table 7.3 the probable identifications of such states are given. It will be noted that, in all of these cases, the binding energy of the two attached electrons relative to the

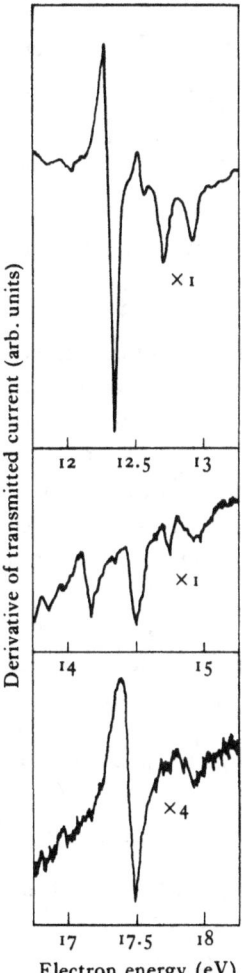

Fig. 7.12. Energy derivative of the transmitted electron current as a function of electron energy in the range 12–18 eV in NO, observed by Sanche and Schulz (1972).

grandparent state is close to 4.2 eV as it should be if the electrons concerned are in $3s\sigma$ orbitals (see pp. 234, 236 and 237). All of the likely grandparent states with equilibrium separations close to the ground state of NO^+ do participate except for $C\,^3\Pi$.

CHAPTER 8

Modes of Formation of Negative Ions – Formation by Radiative Processes – Radiative Attachment and Polar Photodissociation

In this and the following two chapters we consider the processes which lead to negative-ion formation, discussing the nature of the processes involved, the methods, experimental and theoretical, for investigating them and the results obtained.

The simplest manner in which negative ions can be formed is by the direct capture of a free electron by a neutral atom. If the electron has a kinetic energy E before the encounter and the electron affinity of the atom is E_a, an amount of energy $E + E_a$ is released by the capture and must be dissipated in some way. This may occur by radiation or by transmission of the surplus energy to a third body.

In this chapter we discuss the first of these possibilities, deferring consideration of the second to Chapter 9.

A further radiative process, involving absorption instead of emission, which leads to negative-ion production, is one in which a molecule XY is dissociated by absorption of a photon into charged fragments

$$XY + h\nu \longrightarrow X^+ + Y^-.$$

We refer to this process as *polar photodissociation* and discuss it in the second part of this chapter.

8.1 Radiative capture of electrons – electron affinity spectrum

We have given reasons for believing that there is in most cases only one bound negative energy electronic state of a negative ion, so the capture of electrons by a neutral atom will give rise to a continuous emission spectrum extending indefinitely from a long wave limit at

$$\lambda = hc/E_a.$$

NEGATIVE ION FORMATION – RADIATIVE PROCESSES 243

The order of magnitude of the probability of emission of a light quantum on collision may be estimated in the following way. The time taken for a 10 eV electron to traverse the atomic field is about 10^{-15} s. During this period it will have about the same chance of radiating as a bound electron in an excited (non-metastable) state of an atom, i.e. about 10^8 s^{-1}. The total chance of radiation is therefore about 10^{-7} per collision and is very small. This result is confirmed by detailed calculations which we now describe.

The intensity of emission by a radiating electric dipole is given in quanta per second by

$$\frac{64\pi^4 e^2 \nu^3}{3hc^3} |\mathbf{M}|^2, \tag{8.1}$$

where $\mathbf{M}$ is the amplitude of the oscillating electric moment, ν the frequency of the oscillations and e, h, c have their usual significance. For an atomic system where the radiation arises from a transition of an electron from state m to state n, M will be the value of the electric displacement $\mathbf{r}$ measured from the nucleus of the atom and averaged over the wave functions ψ_m, ψ_n of the initial and final states, i.e.

$$\mathbf{M} = \int \psi_m^* \mathbf{r} \psi_n \, d\tau. \tag{8.2}$$

To apply this formula to our particular problem in which the initial state is not a discrete stationary state but one of the continuous energy spectrum, we must be careful to normalize the wave function ψ_m correctly. We wish to obtain finally a formula which will give the intensity of radiation $I_\nu d\nu$ with frequency between ν and $\nu + d\nu$ emitted per unit volume of a gas containing n_0 atoms per unit volume due to capture of electrons from a swarm with a given velocity distribution. If we normalize ψ_m to represent a stream of electrons of such intensity that one electron crosses unit area normally per second, and ψ_n to represent unit density as usual, the resulting value of $\mathbf{M}$ will give, on substitution in (8.1), the number of transitions produced per second per atom by such a stream of electrons. But, if capture were to take place when an electron hits a target of cross-section Q_a^r the unit stream would just produce Q_a^r transitions per atom per second. We therefore call this quantity the effective cross-section $Q_a^r(v)$ for radiative capture of electrons of velocity v. Then

$$I_\nu \, d\nu = n_0 h\nu Q_a^r(v) \, vN(v) \, dv, \tag{8.3}$$

where $N(v)\,dv$ is the number of electrons per unit volume with velocities between v and $v+dv$ and $d\nu/dv = mv/h$.

The calculation of $Q_a^r(v)$ for a given atom involves a knowledge of the wave functions for the motion of electrons with positive energy $\tfrac{1}{2}mv^2$ in the atomic field and also of the wave function for the stationary state of the electron when attached in the ion. The azimuthal quantum number of this stationary state has a marked influence on the form of the cross-section Q_a^r for low electron velocities. If the state is a p state, Q_a^r behaves normally like v^{-1} for small v, but if it is an s state it behaves normally like v. The reason for this difference may be understood in qualitative terms as follows.

Applying the usual selection rules for radiation we see that, for an electron to be captured into a p state, it must be initially in either an s or d state, but for capture into an s state it must be initially in a p state. To avoid confusion we will henceforth distinguish the initial state by a suffix i, so an s_i electron refers to an electron initially in an s state. Now s_i and p_i electrons differ in that the former possess zero angular momentum about the atomic nucleus while the latter possess an amount $2^{1/2}\hbar$. As a consequence s_i electrons make head-on collisions at all velocities, whereas p_i electrons, when possessing a velocity v, pass the nucleus on the average at a distance $r_0 = 2^{1/2}\hbar/mv$. At low velocities the p_i electrons will therefore hardly penetrate the atomic field at all but s_i electrons will do so at all velocities. The overlap of the function ψ_m with ψ_n, which is only large in the neighbourhood of the atom, will therefore be small for low-energy p_i electrons but will not change much with energy for s_i electrons. We may represent this effect by introducing a factor into M, $f(r_0/a)$, where a is a length of atomic magnitude, and f is a function which tends to zero when r_0 tends to infinity, i.e. when v tends to zero. Finally we must introduce a factor v^{-1} for all electrons, because the cross-section will be directly proportional to the time spent in the atomic field. This gives

$$\left.\begin{aligned}Q_a^r(v) &= Av^{-1} \text{ for } s_i \text{ electrons, i.e. capture to a p state,}\\ Q_a^r(v) &= v^{-1}|f(r_0/a)|^2 \\ &= v^{-1}|g(v)|^2 \end{aligned}\right\} \text{for } p_i \text{ electrons, i.e. capture to an s state,} \quad (8.4)$$

where A is practically independent of v, and g, an increasing function of v which vanishes when v tends to zero, approximately as v.

Exceptional cases arise under so-called resonance conditions. If the atomic field gives rise to a stationary state of zero binding energy, the amplitude of an s_i electron in the neighbourhood of the atom increases as the velocity decreases, with the result that the overlap of ψ_m with ψ_n increases approximately as $v^{-1/2}$. Similarly, if a stationary p state of zero binding energy exists, the amplitude of the p_i electron in the neighbourhood of the atom does not fall with velocity as otherwise, so that the overlap of ψ_m with ψ_n remains approximately constant as v decreases. We have then, in place of (8.4):

$$\left.\begin{array}{l} Q_a^r \simeq Av^{-2} \text{ for } s_i \text{ electrons, i.e. capture to a p state} \\ Q_a^r \simeq Bv^{-1} \text{ for } p_i \text{ electrons, i.e. capture to an s state} \end{array}\right\} \begin{array}{l}\text{Exact}\\ \text{resonance}\end{array} \quad (8.5)$$

In practice, it is extremely unlikely that exact resonance would occur in any particular case but, for approximate resonance, the cross-section would tend to vary as in (8.5) down to quite low velocities before finally behaving as in (8.4). Thus, for capture to an s state, the velocity variation of Q_a^r for small v might be written $B/(a + bv)$, where b will be negligible except near resonance and a will vanish for exact resonance.

A more quantitative basis for the above considerations may be obtained by reference to the simplified version of the atomic field used already in Chapter 1, p. 3 and Chapter 4, p. 70. We take for the potential acting on the electron,

$$V = -U \quad (r < a),$$
$$V = 0 \quad (r > a),$$

and consider the capture of s_i electrons.

The wave function ψ_m will then satisfy the equation

$$\left.\begin{array}{l} \dfrac{d^2}{dr^2}(r\psi_m) + k'^2(r\psi_m) = 0 \quad (r < a), \\[2mm] \dfrac{d^2}{dr^2}(r\psi_m) + k^2(r\psi_m) = 0 \quad (r > a), \end{array}\right\} \quad (8.6)$$

where

$$k = mv/\hbar, \qquad k'^2 = k^2 + \mu^2, \qquad \mu^2 = 2mU/\hbar^2.$$

This gives

$$r\psi_m = A \sin k'r \quad (r < a),$$
$$= C \sin(kr + \eta), \quad (8.7)$$

the internal solution being regular at $r = 0$. The external solution must be normalized to represent the component of a unit incident flux of electrons, in the direction of incidence, which have zero angular momentum about the centre of the atom. This makes

$$C = v^{-1/2} k^{-1} e^{i\eta}.$$

A may now be determined by the condition that ψ_m and $d\psi_m/dr$ should remain continuous at $r = a$. We find

$$|A| = v^{-1/2} \{k^2 \sin^2 k'a + k'^2 \cos^2 k'a\}^{-1/2}.$$

Since the bound atomic wave function will be small for $r > a$, the matrix element M will be approximately proportional to $|A|$ for small v. As v and hence $k \to 0$, $k'a \to \mu a$ so

$$|A| \longrightarrow v^{-1/2}/\mu \cos \mu a,$$

provided $\cos \mu a \neq 0$. This gives the normal result (8.4). On the other hand, when

$$\mu a = (2s+1)\pi/2 \quad (s = 0, 1, 2, \ldots),$$

$$|A| \longrightarrow \frac{v^{-3/2} \hbar}{m \sin \mu a},$$

and this is the resonance case (8.5). Referring back to p. 4, it will be confirmed that $\mu a = (2s+1)\pi/2$ is the condition that an s level of zero binding energy exists.

The capture of p_i electrons may be discussed in a similar way. The equations for ψ_m differ from (8.6) only in the inclusion on the left-hand side of each, of the centrifugal term $-2(r\psi_n)/r^2$. The external solution is now

$$3v^{-1/2} k^{-1} \exp(i\eta_1)\{\cos(kr+\eta_1) - (kr)^{-1} \sin(kr+\eta_1)\}$$

and the internal

$$A\{\cos k'r - (k'r)^{-1} \sin k'r\}.$$

A may be determined from continuity conditions as before and, after a little calculation, the conclusions of (8.4) and (8.5) may be verified for this case also. (The condition for the existence of a p level of zero binding energy is that $\mu a = s\pi$ $(s = 1, 2, \ldots)$.)

In practice there is considerably more interest in the inverse process of absorption of light by negative ions which is discussed in Chapter 11, p. 417. The absorption or photodetachment cross-section Q_d^r is related to that for emission, or radiative attachment, Q_a^r through the principle of detailed balancing so that

$$Q_d^r = \frac{m^2 v^2 c^2}{h^2 \nu^2} \frac{g_0}{g_-} Q_a^r, \qquad (8.8)$$

where ν is the frequency of the quantum emitted in the capture and g_0, g_- the statistical weights of the atomic and negative-ion states concerned. Q_d^r is accurately known from experiment over a considerable frequency range for a number of atoms and hence, through (8.8), Q_a^r. Comparison may be made with direct but less precise measurements of Q_a^r from observations of affinity continua.

In all cases except that of atomic hydrogen, observed cross-sections Q_d^r are more accurate than theory can provide. However,

TABLE 8.1 *Calculated cross-sections Q_a^r for capture of electrons by atomic hydrogen*

Electron energy (Rydbergs)	0.01	0.03	0.05	0.1	0.2	0.4	0.6	0.8
Electron energy (eV)	0.135	0.406	0.677	1.354	2.708	5.416	8.124	10.832
Cross-section (10^{-22} cm^2)	0.456	0.570	0.583	0.547	0.455	0.315	0.300	0.270

because of the great astrophysical importance of absorption by H$^-$, a great deal of effort has been devoted with considerable success to obtaining accurate theoretical cross-sections for this comparatively simple case. We shall defer detailed discussion of these calculations to Chapter 11 where they will be considered primarily in relation to absorption by H$^-$. Table 8.1 gives the emission cross-section Q_a^r derived from these results (see Fig. 11.1).

It will be seen that the cross-section is indeed of order 10^{-7} of atomic dimensions. Cross-sections of similar magnitude are obtained for other cases. Thus Fig. 8.1 shows Q_a^r and the attachment rate coefficient $v Q_a^r$ for atomic oxygen derived using (8.8) from observed photodetachment cross-sections (see Fig. 11.15). Whereas for H, capture occurs to an s state, for O it is to a p state so that, as given by (8.5), Q_a^r tends to zero with the electron velocity for H but for O behaves as A/v for small v.

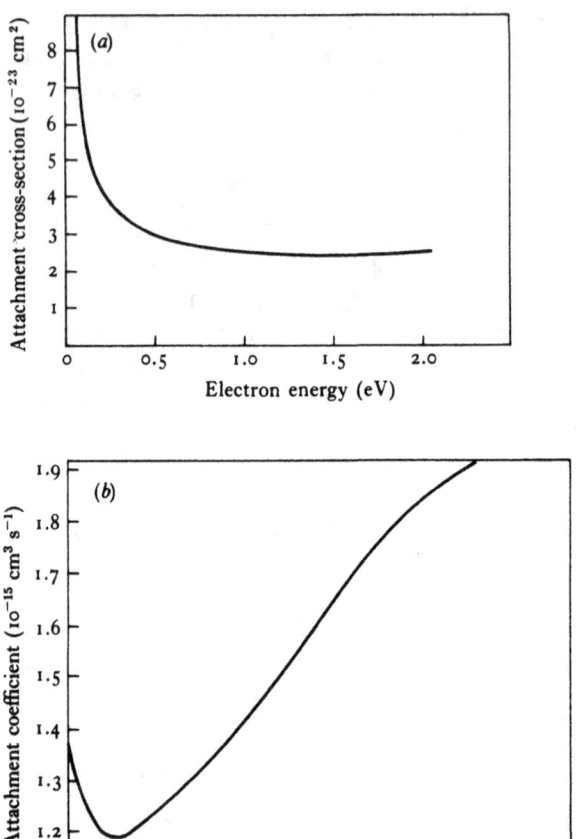

Fig. 8.1. Cross-sections and rate coefficients for radiative attachment of electrons to oxygen atoms. (a) Attachment cross-section. (b) Attachment rate coefficient.

8.2 Direct observation of affinity spectra

Experimental methods

Experimental detection of affinity spectra proved to be very difficult. To obtain a sufficiently high concentration of free electrons, the gas under study must be heated to a high temperature but this also increases the intensity of the radiation arising from recombination or bremsstrahlung transitions involving the free electrons

NEGATIVE ION FORMATION – RADIATIVE PROCESSES 249

and positive ions. It is necessary to secure conditions in which these radiations do not swamp that from capture by neutral atoms.

High-pressure arcs and shock waves have been used to secure high temperatures and it has been possible in some recent experiments to obtain spectra relatively free from background due to recombination and bremsstrahlung.

The first successful observations were made, of the affinity spectrum of H, in Lochte-Holtgreven's laboratory at Kiel in 1951. A number of different sources were used including a spark discharge in hydrogen in a Geissler tube at pressures ranging from 20 to 60 atm (Fuchs, 1951), a spin-stabilized arc discharge through hydrogen at a pressure between 0.4 and 1.0 atm (Lochte-Holtgreven and Nissen, 1952), a high-pressure hydrogen arc (10–140 atm) (Nissen, 1954) and an arc in water vapour at a pressure between 25 and 150 atm (Peters, 1953).

In later experiments in oxygen and nitrogen a wall-stabilized arc was used but, with an arc temperature of 11 000 °K in oxygen the relative intensities of the affinity, recombination and bremsstrahlung emission were as 71:28:1 so that the latter two sources were of considerable intensity. To disentangle them the following procedure was adopted.

It can be assumed that in an arc of this kind the ionized gas is in thermodynamical equilibrium at some temperature T. This may be determined by measuring the intensity I_{nm} of a suitably chosen multiplet, emitted per unit solid angle from a length l of the arc. If A_{nm} is the transition probability for the multiplet, supposed known from other experiments, then

$$I_{nm} = (4\pi)^{-1}(hc/\lambda) A_{nm} n_m(T) l. \qquad (8.9)$$

$n_m(T)$ is the concentration of atoms in the upper, mth state, which is given by

$$n_m = \{n_0/Z_0(T)\} g_m \exp(-E_m/kT), \qquad (8.10)$$

where n_0 is the concentration of normal atoms, Z_0 the partition function for the neutral atom and g_m, E_m the respective statistical weight and excitation energy of the mth state. It follows that, from measurement of I_{nm}, T may be obtained.

Having obtained T, the concentrations n_e, n_+ and n_- of electrons, positive ions and negative ions respectively are related to n_0 by Saha's equations

$$\frac{n_+ n_e}{n_0} = S_0(T), \qquad \frac{n_0 n_e}{n_-} = S_-(T), \qquad (8.11)$$

where

$$S_0 = 2(2\pi mkT/h^2)^{3/2}\{Z_+(T)/Z_0(T)\}\exp(-\chi_0/kT), \quad (8.12)$$

$$S_- = 2(2\pi mkT/h^2)^{3/2}\{Z_0(T)/Z_-(T)\}\exp(-\chi_-/kT). \quad (8.13)$$

Z_+ and Z_- are the partition functions for the positive and negative ions respectively, χ_0, χ_- the ionization energy and electron affinity of the neutral atom.

From the known values of n_+, n_e and T, the intensities of the recombination and bremsstrahlung emission may be calculated and subtracted from the observed intensity to give the affinity continuum. Even when this is done it is necessary to allow for the fact that this continuum will also include a contribution from free–free transitions of electrons in the fields of neutral atoms (see p. 680). Allowance can be made for this by taking advantage of the fact that the affinity continuum has a long wavelength threshold at a wavelength E_a/hc, where E_a is the electron affinity, whereas the neutral free–free continuum will vary gradually with frequency from very long wavelengths.

By using as a source a cylindrically symmetrical spark discharge in the gas at atmospheric pressure and a current below 10 A so that the temperature was relatively low (between 6000 and 8000 °K), Mück and Popp (1968) were able to reduce the contribution from recombination and bremsstrahlung in chlorine to about 3% of the affinity continuum. Fig. 8.2 shows schematically the arrangement of the spark source in these experiments. The light emitted was viewed end-on along the axis of the spark. Because Cl_2 has a strong ultraviolet absorption continuum it was necessary to remove any undissociated Cl_2 from the cool regions of the discharge. This was done by introducing a flow of argon near each electrode. With the chlorine admitted midway between the argon inlets and with outflow openings between the argon and chlorine inlets, it was possible to separate the central chlorine spark from the argon sparks near each electrode and hence to avoid any appreciable concentration of Cl_2 in the line of sight. Results obtained are discussed below and illustrated in Fig. 8.4.

The first observations of an affinity continuum from shock-heated gas were made by Weber (1958) using a mixture of 80% Kr and 20% H_2. With his shock tube this mixture could be heated to temperatures of 6000 °K or more. In this work, conditions could be achieved in which the affinity continuum between 4400 and

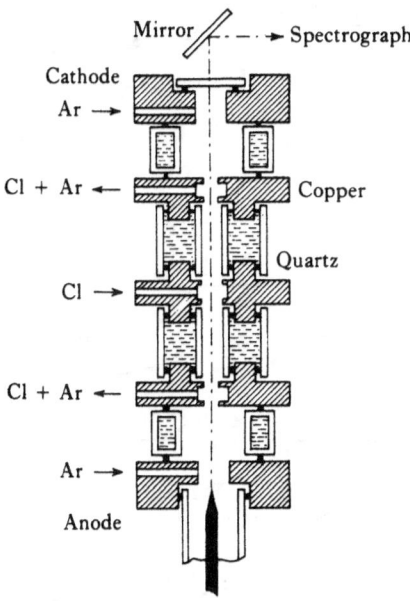

Fig. 8.2. Arrangement of the spark source in the experiments of Mück and Popp (1968) on the affinity spectrum of Cl.

5500 Å was about 30 times as intense as any other source. A little later Berry and David (1964) observed the affinity continuum for Cl, Br and I emitted from shock-heated vapour of alkali metal halides.

The results obtained from these experiments, apart from their intrinsic interest, have been useful in providing a further method for determining electron affinities and also in providing evidence of the existence of metastable N^- ions. We now describe the results which have been obtained for specific atoms.

Observed results

Atomic hydrogen

Fig. 8.3 shows some typical results obtained by Weber with a shock-heated mixture of krypton and hydrogen as referred to above. While the measured absolute intensity is somewhat higher than that calculated it seems clear that at wavelengths longer than the Balmer limit the intensity observed is due to the affinity continuum.

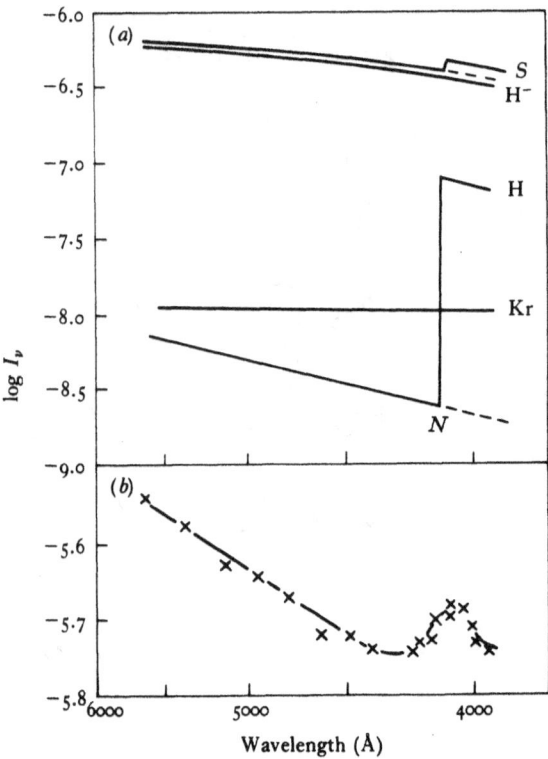

Fig. 8.3. Emission spectra from a shock-heated mixture of hydrogen and krypton. (a) Calculated intensity in photons cm^{-3} s^{-1} per unit frequency range. The contributions from H, Kr and H$^-$ continua are indicated, as well as the total S. N denotes the Balmer limit. (b) Observed intensity. From Weber (1958).

Halogen atoms

Fig. 8.4 shows the continuum observed by Mück and Popp (1968) for chlorine using the spark source described above. The continuum rises quite sharply from the background. By linear extrapolation to zero of the tangent to the intensity curve at the point of inflexion the threshold is determined at 3428 Å, independent of the spark temperature. This gives the electron affinity of Cl as 3.616 eV, in good agreement with the value obtained by Berry and David (1964), from their observations with shock-heated gas and with determinations by other methods (see Table 3.3).

A second threshold is apparent at a somewhat shorter wavelength which, following the same procedure as for the first, is found to occur at 3326.5 Å. This corresponds to an energy difference of 887 cm^{-1} which agrees well with the known fine structure separation of the doublet ground state of Cl. Thus the two thresholds

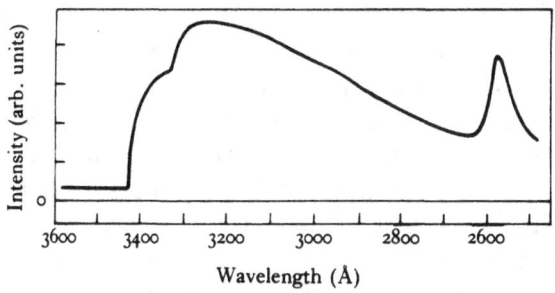

Fig. 8.4. Affinity spectrum of Cl$^-$, observed by Mück and Popp (1968).

arise from capture into the respective fine structure levels of the neutral atom.

Similar results were obtained for F by Popp (1967) and for Br and I by Berry and David (1964). The derived electron affinities are given in Table 3.3.

Atomic oxygen

Boldt (1959a) determined the intensity of the affinity continuum of atomic oxygen using a wall-stabilized arc source. It was necessary to disentangle a considerable contribution from the recombination and bremsstrahlung continuum. In Fig. 11.15 the absorption cross-section for O$^-$ derived from these results is compared with more accurate directly measured values.

Atomic nitrogen

Similar experiments carried out by Boldt (1959b) in nitrogen revealed the presence of a continuum, in addition to that due to recombination and bremsstrahlung, which was of comparable intensity to that observed in oxygen. It could be understood in terms of capture to form a negative ion with an electron affinity of around 1 eV. In view of the arguments discussed in Chapter 2, p. 28, it seems very unlikely that a stable N^- ion is involved. On the other hand the process could well be one of capture by N atoms in the $2p^3\,^2D$ metastable state to form the metastable $2p^4\,^1D$ N^- ion with release of about 1 eV energy. Further evidence concerning this ion has been discussed in Chapter 5, p. 144.

8.3 Dielectronic attachment

One further attachment process to a neutral atom involving emission of radiation is possible. In Chapter 4 we discussed the doubly-excited states of negative ions which are unstable towards autodetachment with lifetimes τ, short compared with radiative lifetimes τ_R. Attachment may therefore take place in the following way.

If the incident electron has an energy such that the total energy of atom plus electron is within the level width of a doubly-excited state of the negative ion, and if certain selection rules are satisfied, it may be captured, without emission of radiation, into this state, the process being one of inverse autodetachment. Once captured, there is a chance $\tau/(\tau + \tau_R)$ that the doubly-excited state will revert to the ground state by emission of radiation.

Bates and Massey (1943) have discussed the probability of this process for capture by atomic oxygen. They find that it is negligible compared with direct radiative capture of electrons, with average energies corresponding to temperatures up to 2000 °K, unless the energy of the doubly-excited state responsible is much less than 0.25 eV above that of the normal state of the neutral atom. The only doubly-excited state which could be effective is that with configuration $(1s)^2(2s)(2p)^6$, no other low-lying state combining optically with the ground configuration. Empirical studies place even this state as much as 10 eV above the ground state of atomic oxygen. It is therefore most unlikely that dielectronic attachment is at all important in this case.

8.4 Polar photodissociation

As early as 1932 Terenin and Popow showed, from a mass analysis of the ions produced by absorption of radiation near 2000 Å in thallium halide vapour, that they were produced by polar photodissociation into Tl$^+$ and halogen negative ions.

The possibility of the process

$$I_2 + h\nu \longrightarrow I^+ + I^- \qquad (8.14)$$

was first suggested by Mulliken (1940) in connexion with the analysis of the absorption spectrum of I_2. The threshold photon energy is given by

$$E_t = D(I_2) + E_i(I) - E_a(I),$$

where the symbols have their usual meaning. Taking $D(I_2) = 1.54$ eV (Verma, 1960), $E_a(I) = 3.08$ eV (see Chapter 3, p. 44), $E_i(I) = 10.45$ eV, E_t comes out to be 8.91 eV, whereas $E_i(I_2) = 9.400$ eV (Venkateswarlu, 1969). The threshold for (8.14) therefore lies 0.49 eV below that for photoionization

$$I_2 + h\nu \longrightarrow I_2^+ + e. \qquad (8.15)$$

The existence of an absorbing process below the ionization limit, $E_i(I_2)$, was observed by Watanabe (1957) but the first mass analysis showing that I$^+$ ions were produced in this process was made by Morrison, Hurzeler, Inghram and Stanton (1960) who also investigated absorption by Br_2. These were the first experiments in which both a monochromator and mass analyser were used.

The light source was a capillary discharge in hydrogen, isolated from the monochromator by a lithium fluoride window 1 mm thick which limited the range of wavelengths observed to greater than 1050 Å. The entrance slit to the ion chamber was of 0.02 in. width and that of the exit slit 0.04 in., which resulted in an energy spread in the photon beam of about 0.05 eV at a wavelength of 1200 Å. The mass analyser was a 60° sector magnet with a mass resolution of 1/300.

Fig. 8.5 shows the results obtained for the rates of production of the atomic and molecular ions as functions of electron energy in Br_2 and I_2. Determination of the electron affinity from these data is complicated by the fact that the kinetic energy of the ions pro-

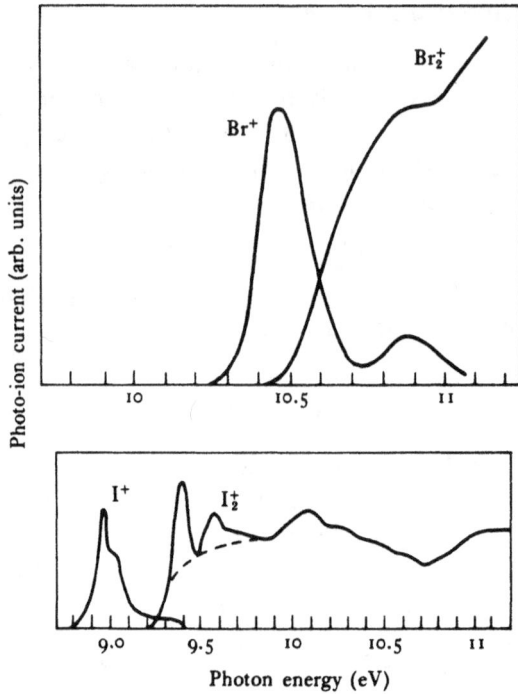

Fig. 8.5. Observed variation with photon energy of the rate of production of atomic and molecular ions of bromine and iodine by photoionization, as observed by Morrison et al. (1960).

duced is unknown. If it is assumed that at the threshold it is zero, then it is found that

$$E_a(\text{Br}) = 3.53 \text{ eV}, \quad E_a(\text{I}) = 3.13 \pm 0.12 \text{ eV},$$

in reasonably good agreement with determinations by other methods (see Chapter 3, p. 44). The second peak in the yield curve for Br$^+$ occurs at a photon energy about 0.4 eV above the main peak. This is very close to the energy separation (0.39 eV) between the $^3P_2$ and $^3P_1$ levels of Br$^+$.

Later measurement of the absorption by I$_2$, due to Myer and Samson (1970), gave the much more precise value

$$E_a(\text{I}) = 3.073 \pm 0.014 \text{ eV}.$$

Special interest attaches to measurements in F$_2$ and HF because of the light they throw on the rather controversial value for the dissociation energy of F_2. The original determination by Stampfer and Barrow (1958), from a study of the equilibrium in partly dissociated gaseous fluorine, gave $D(F_2) = 1.592 \pm 0.006$ eV but doubt was soon thrown on these results from analysis of the vacuum ultraviolet spectrum of F_2 which gave conflicting values 1.63 ± 0.1 eV (Iczkowski and Margrave, 1959) and 1.44 ± 0.07 eV (Stricker and Krauss, 1968).

Evidence in favour of the lower value came from the experiments of Dibeler, Walker and McCulloh (1969a) who determined the appearance potential for the process

$$F_2 + e \longrightarrow F^+ + F + e, \qquad (8.16)$$

and thence obtained $D(F_2) = 1.34 \pm 0.03$ eV. They produced further evidence in support of this value by determining the appearance potentials for the corresponding process in HF (Dibeler et al., 1969b). This gave $D(HF) = 5.74 \pm 0.03$ eV, somewhat lower than the previously determined spectroscopic value (Johns and Barrow, 1959) of 5.86 ± 0.01 eV. $D(HF)$ is related to $D(F_2)$ by

$$D(HF) = \tfrac{1}{2}D(F_2) + \tfrac{1}{2}D(H_2) - \Delta H_{fo}(HF), \qquad (8.17)$$

where $\Delta H_{fo}(HF)$, the heat of formation of HF from its elements, is -2.81 eV. Substituting their value for $D(HF)$, Dibeler et al. obtained $D(F_2) = 1.38 \pm 0.03$ eV, consistent with the value derived directly from their observations in F_2.

To investigate this matter further Berkowitz, Chupka, Guyon, Holloway and Spohr (1971) carried out measurements of photodissociation in both F_2 and HF in which they paid special attention to the determination of the threshold energies, allowing for finite slit widths, temperature effects and complications due to the fact that a contribution to the F^+ yield near the threshold for the process (8.16) comes also from polar photodissociation.

Berkowitz et al. (1971) used the Hopfield continuum of helium generated by a d.c. condensed discharge as the light source, together with a 1 m vacuum u.v. monochromator, and obtained a resolution of 0.04 Å. The ions produced were mass-analysed with a 60° sector magnet. Arrangements were made to cool the ionization chamber with liquid nitrogen so that measurements could be made both at 300 and 75 °K. A further important special feature was that both negative- and positive-ion currents could be measured, although it was necessary to allow for the fact that the sensitivity of the mass analyser was much greater for the positive ions.

Fig. 8.6 shows the observed F^- production as a function of wavelength in two wavelength regions, measured at 78 °K. The structure in the region 770–800 Å is also observed in the F^+ current which is almost certainly due to excitation of states of F_2 which autoionize through vibrational coupling to the $^2\Pi_g$ state of F_2^+.

Dibeler et al. pointed out that polar photodissociation

$$F_2 + h\nu \longrightarrow F^+(^3P) + F^-(^1S) \tag{8.18}$$

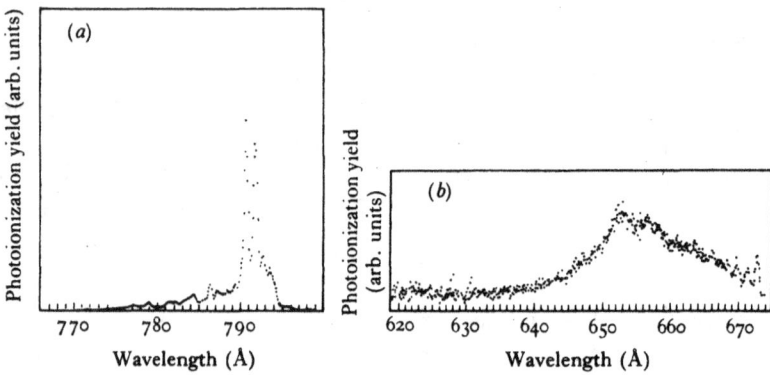

Fig. 8.6. Variation with photon wavelength of the F^- current from polar photodissociation of F_2, observed by Berkowitz et al. (1971).

violates spin conservation and should therefore be very improbable. Berkowitz et al. have given reasons why the process could proceed through crossing of one of the autoionizing states of F_2 with a curve for a state which dissociates into $F^+(^3P)$ and $F^-(^1S)$. Under certain circumstances the probability of a transition near the crossing point could be quite high, even though the interaction between the curves is small (see (10.9) of Chapter 10).

By comparison of the F^+ and F^- currents observed for photon energies for which (8.18) was the only process of ion production, the relative sensitivity of the mass analyser towards the F^+ and F^- ions was obtained. It was then possible, from measurement of the F^- current at photon energies beyond the threshold for (8.18), to obtain and subtract the contribution to the F^+ currents from (8.18).

A typical example is shown in Fig. 8.7. It will be seen that this subtraction removes most of the 'tail' in the F⁺ yield curve, leaving a residual with a much sharper threshold which was determined as 19.001 eV. Allowing for thermal rotational energy of 0.007 eV, this

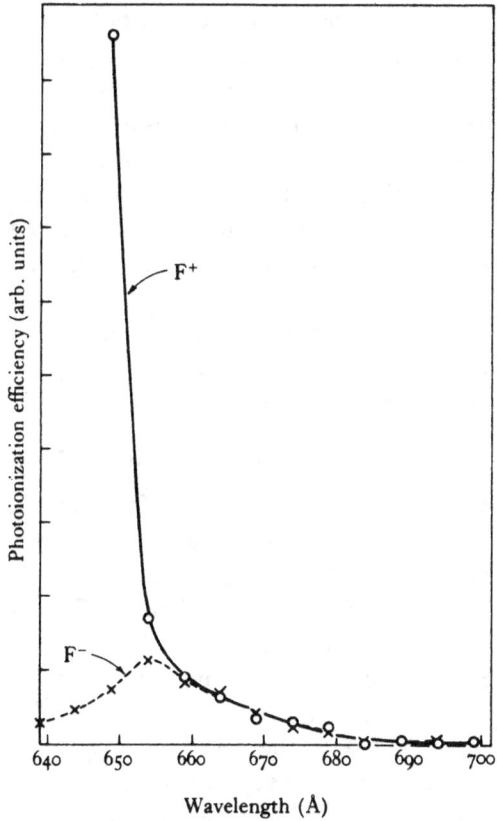

Fig. 8.7. Magnified view of the F⁺ current in the region of the threshold for the process $F_2 + h\nu \to F^+ + F + e$, showing the contribution from polar photodissociation.

becomes 19.01 eV, 0.25 eV greater than determined by Dibeler *et al.*, who made no allowance for polar photodissociation. With this value for the threshold energy Berkowitz *et al.* obtained $D(F_2) = 1.59$ eV, agreeing with the original result of Stampfer and Barrow (1958).

Berkowitz *et al.* also re-examined the photodissociation of HF and DF, paying particular attention to allowance for finite energy resolution and thermal effects. They found for HF a threshold 0.105 eV higher than did Dibeler *et al.* and this leads to a value for $D(F_2)$ consistent with 1.59 eV.

As a further check Chupka and Berkowitz (1971) measured the kinetic energies of the ions produced in polar photodissociation of both F_2 and HF. In these experiments they used a double ionization chamber. Ions were formed under nearly field-free conditions in the first chamber and then accelerated or decelerated by a pair of grids before passing through a second field-free chamber and thence, after further acceleration, to the mass analyser. By varying the potential on the grids a retarding potential analysis of ion energies could be made.

If W^+ is the most probable energy of the positive ions say, resulting from polar photodissociation by photons of frequency ν, then the threshold for the process is

$$E_t = h\nu - \{(M_1 + M_2)/M_2\} W^+,$$

where M_1, M_2 are the respective masses of the positive and negative ions. A small correction must be made to E_t to allow for the thermal rotational energy of the target molecule.

When applied to HF, it was found that $D(HF) = 5.91 \pm 0.002$ or 5.87 ± 0.02 eV according as $E_a(F)$ was taken as 3.448 ± 0.005 eV (Berry and Reimann, 1963, see Chapter 11, p. 461) or 3.400 ± 0.002 eV (Popp, 1967, see p. 253). The smaller value for $D(HF)$ is in very good agreement with the spectroscopic value of Johns and Barrow (1959) and this lends support to Popp's value for $E_a(F)$.

It was also confirmed that the kinetic-energy distributions of F^+ ions found near the threshold for polar photodissociation, 794.7 Å, on the assumption of $D(F_2) = 1.58$ eV and $E_a(F) = 3.400$ eV, differ little from that of thermal F_2^+ ions used for calibration (see Fig. 8.8). If the lower value 1.34 eV of $D(F_2)$ were correct, the ions could have a most probable energy of 0.18 eV which would certainly have been observed. Analysis of the data at higher photon energies is complicated by the need to allow for the fine structure of the F^+ ion produced but it was found to be consistent with the assumed values for $D(F_2)$ and $E_a(F)$.

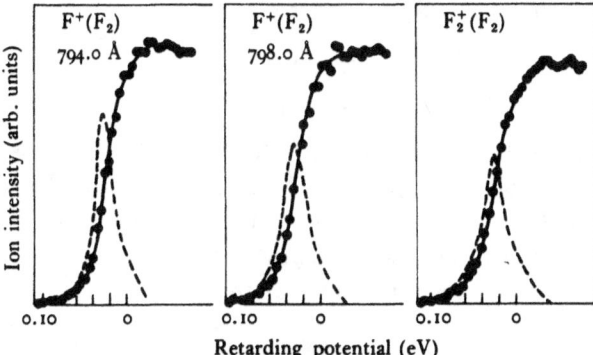

Fig. 8.8. Energy distributions of F^+ ions resulting from polar photodissociation of F_2 at 798.0 and 794.0 Å, compared with that for F_2^+ ions from F_2, observed by Chupka and Berkowitz (1971). —— retarding potential curves, ---- derived energy distribution.

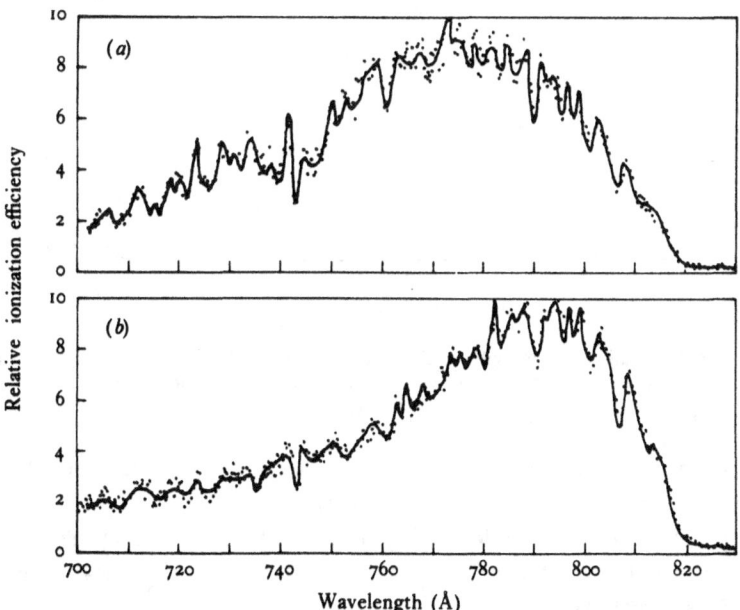

Fig. 8.9. Yields of H^+ and CN^- ions from polar photodissociation of HCN, as a function of photon wavelength, observed by Berkowitz et al. (1969). (a) CN^-, (b) H^+.

A further interesting application has been made by Berkowitz, Chupka and Walter (1969) to the determination of the electron affinity of the CN radical. This they did by measuring the threshold energies of the two reactions

$$\text{HCN} + h\nu \longrightarrow \text{H}^+ + \text{CN}^-, \qquad (8.19)$$

$$\text{HCN} + h\nu \longrightarrow \text{H}^+ + \text{CN} + e. \qquad (8.20)$$

Fig. 8.9 shows the yield curves for H$^+$ and CN$^-$ ions over the photon wavelength range from 700–830 Å. The H$^+$ yield decreases relatively at higher photon energies but this is an instrumental effect due to the increasing kinetic energy carried away by the ionized products. Most of this is taken up by the H$^+$ because of its smaller

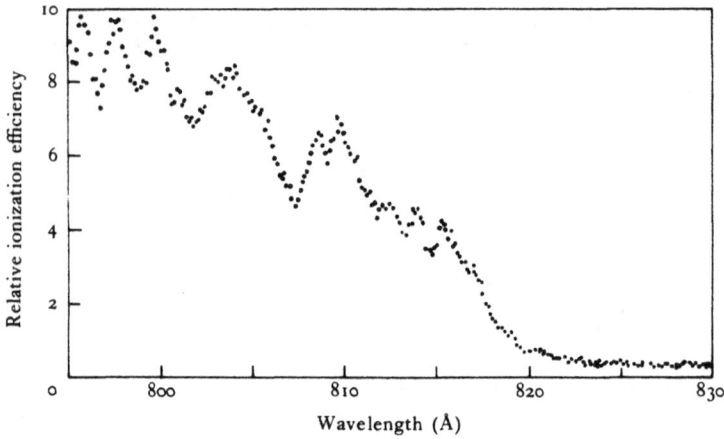

Fig. 8.10. Magnified view of the CN$^-$ yield curve of Fig. 8.9(a) near the threshold for polar photodissociation. From Berkowitz et al. (1969).

mass and, as the sensitivity of the mass analyser decreases with ion energy, this leads to a more marked decrease for H$^+$ than for CN$^-$. Near the threshold the intensities of the two ion currents are approximately equal, as judged from estimates of relative mass-analyser sensitivity.

Fig. 8.10 shows the current of CN$^-$ near the threshold on an expanded wavelength scale. The threshold wavelength was taken as 817 ± 1 Å (15.18 ± 0.020 eV), the observed tailing beyond this value being ascribed to rotational energy and to the presence of some target HCN in the doubly-degenerate ν_2 vibrational state.

For the process (8.20) the threshold was determined from the

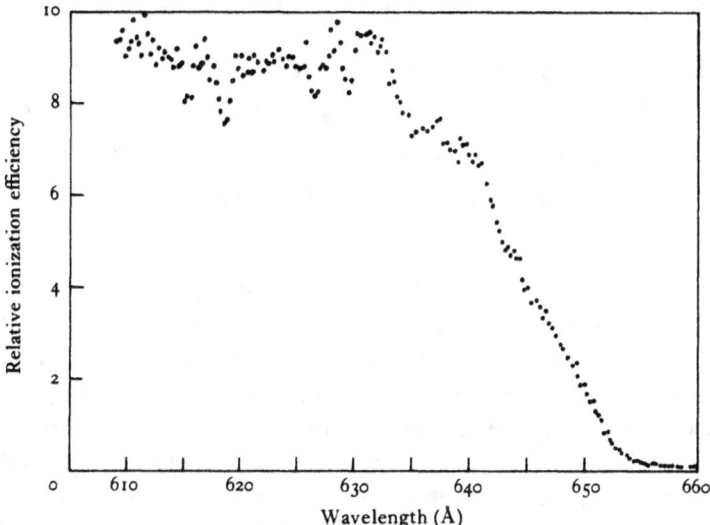

Fig. 8.11. Yield curve for H⁺ ions from the process $h\nu + HCN \rightarrow H^+ + CN + e$ near the threshold wavelength, as observed by Berkowitz et al. (1969).

H⁺ yield curve shown in Fig. 8.11. Extrapolation of the near linear portion to zero gives 653.5 ± 0.2 Å (18.97 eV). Allowance for internal energy of the target raises this to 19.00 eV.

The excess 3.82 eV of this above the threshold for (8.19) gives the electron affinity of CN (see Chapter 6, p. 210).

CHAPTER 9

Modes of Formation of Negative Ions – Formation by Three-Body Collisions and by Collisions of Electrons with Molecules – Dissociative Attachment and Polar Dissociation

In this chapter we consider, in particular, processes leading to negative-ion formation in which the energy given up on capture of a free electron is taken up, not by a photon, but by a third body which may be another electron, an atom or a molecule. After discussing briefly the reactions of this kind in which the third body is free both before and after, a major part of the chapter will be devoted to the discussion of dissociative attachment in which the third body is initially bound in a molecule to which a colliding electron becomes attached. The surplus energy is then taken up by dissociating the molecule. Collisions of this kind between electrons and molecules are of major importance and have been very extensively studied. Opportunity will be taken at the same time to discuss impact of electrons with molecules in which the electron is not captured but negative ions are produced through polar dissociation of the molecule. Thus we have, symbolically

$$XY + e \longrightarrow X + Y^-, \quad \text{dissociative attachment,}$$
$$\longrightarrow X^+ + Y^- + e, \quad \text{polar dissociation.}$$

9.1 Capture of electrons in three-body collisions

We can describe the probability of this process in terms of an effective cross-section Q_a^t per unit concentration (in numbers cm^{-3}) of the third body. The order of magnitude of Q_a^t may be determined by arguments based on information concerning two-body collisions.

The third body (electron, atom or molecule) will be effective in absorbing the energy provided it makes an effective collision with the capturing atom when the latter is at a distance $R_0 \simeq 10^{-8}$ cm (the atomic radius) from the electron. If λ is the mean free path for effective collisions between the atom and third body when the

NEGATIVE ION FORMATION – DISSOCIATIVE PROCESSES 265

concentration of the latter is unity, then this chance is very nearly R_0/λ, giving

$$Q_a^t \simeq \pi R_0^2 R_0/\lambda. \qquad (9.1)$$

It remains to estimate λ. As this will vary considerably with the nature of the third body, it is necessary to consider two cases separately.

(a) Electrons as 'third bodies'. The effective cross-section for energy transfer between an electron and an atom is of the order 10^{-16} cm² (Mott and Massey, 1965, pp. 545–50), so

$$\lambda = 10^{16} \text{ cm}, \quad Q_a^t \simeq 10^{-40} \text{ cm}^5.$$

As the electron concentration n_0 cm⁻³ necessary to produce the same number of non-radiative attachments per second as radiative is given by Q_a^r/Q_a^t, we find that n_0 must be as great as 10^{18} cm⁻³. This is a very large concentration for practical realization.

(b) Atoms and molecules as third bodies. The efficiency in this case depends on whether or not the energy transferred is absorbed wholly as potential energy of the third body. In the first case the energy transfer is a resonance one and the cross-section for such transfer between atomic systems is known to be quite large. Values up to 10^{-14} cm² (Mott and Massey, 1965, p. 645) or even greater may occur, giving

$$\lambda = 10^{14} \text{ cm}, \quad Q_a^t = 10^{-38} \text{ cm}^5.$$

It is therefore possible that concentrations as low as 10^{16} cm⁻³ of third bodies, capable of resonant absorption, may be as effective in producing electron attachment as radiative processes.

If resonant transfer is not possible the excess energy must be partly taken up as kinetic energy of the relative motion of the atomic systems and the corresponding effective area is known to be quite small (Mott and Massey, 1965, p. 650). λ may be 10^{18} cm or more and a concentration of 10^{20} cm⁻³ or more may be necessary to give attachment comparable with that accompanied by radiation.

From these considerations it is clear that molecules, possessing several internal degrees of freedom, will be most effective as third bodies, although it is not always easy to transfer energy to vibrational motion within the molecule (Mott and Massey, 1965, p. 685).

Detailed confirmation of certain of the above arguments has been provided by calculations carried out by Smith (1936) for attachment in which the third body is an electron. He finds values of the order 10^{-40}–10^{-41} cm^5 for Q_a^i, in reasonable agreement with our estimate of 10^{-39}–10^{-40} cm^5. No detailed calculations have yet been carried out for three-body collisions involving atoms or molecules.

9.2 Formation of negative ions on collisions of electrons with molecules – theoretical introduction

Processes involving electron capture

General theoretical considerations

Another possibility, of great importance in practice, which we must now consider, arises when the system capturing the electron is a molecule containing two or more atoms. This provides yet another means of absorbing the energy released on capture of the electron, i.e. by changing the kinetic energy of relative motion of the atomic nuclei. It may be thought of as a three-body attachment process in which the third body is bound to the capturing atom and not freely moving in the surrounding gas.

To discuss the various effects which may be produced by this type of capture we make use of the Franck–Condon principle (see Chapter 6, p. 165) which states that those transitions which leave the nuclear separation unaltered will be most probable.

We now consider the possibilities in terms of the potential-energy curves of the molecular states concerned.

In Figs. 9.1(*a*) and (*b*) potential-energy curves for different electronic states of both neutral and negatively-charged molecules are illustrated. In both cases curve I refers to the normal state of the neutral molecule XY, the remaining curves to states of the molecular ion XY$^-$. These latter curves all tend to a limit for infinite nuclear separation which lies below the curve I by an amount equal to the electron affinity E_a of the atom Y. Over a considerable range of nuclear separation in each case the negative-ion state is unstable towards autodetachment. If $\hbar^{-1}\Gamma(R)$ is the rate at which autodetachment occurs at a separation R, the energy of the state at R will be uncertain by an amount of order Γ so that, over the autodetachment region, the potential energy curve for the negative-ion state should be shown as blurred by this amount. To avoid complexity in the diagram we do not show this blurring, but instead call attention to the instability towards autodetachment by showing the potential-energy curve as broken over this region of R.

The amplitude of oscillation of the neutral molecule in the normal state is from a to b, the range of nuclear separation from R_1 to R_2. The Franck–Condon principle can now be used to show that, in

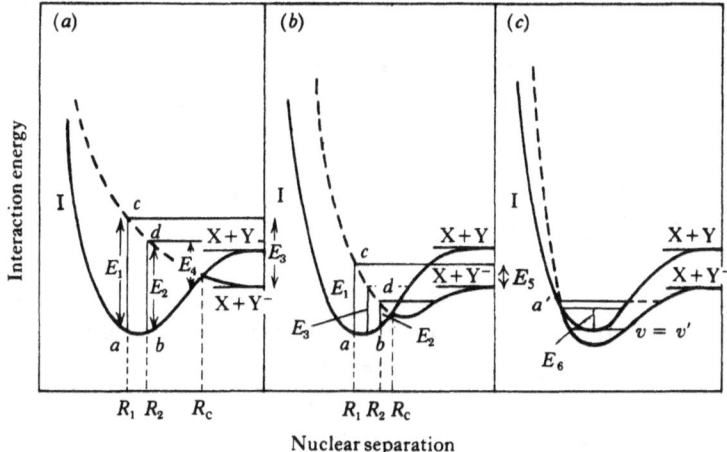

Fig. 9.1. Potential-energy curves illustrating three possible ways in which negative ions may be formed from a molecule XY by electron capture.

the transition consequent on electron capture, the nuclear separation, provided vibrational excitation of the normal molecule does not occur, will remain between R_1 and R_2. Referring to Fig. 9.1(a) we see that this will lead to a final state represented by some point lying between c and d on the potential-energy curve for the state of the molecular ion.

Suppose at first that autodetachment did not occur. Then in case (a) all transitions would result in dissociation into an atom X and ion Y$^-$ with relative kinetic energy between E_3 and E_4. To take account of autodetachment we note that, if the ions are formed at a nuclear separation R_0 and move apart with a velocity of separation

$v(R)$, the fraction remaining when at a separation R will be

$$\exp\left\{-\int_{R_0}^{R}\frac{\Gamma(R')}{\hbar v(R')}\,dR'\right\}. \qquad (9.2)$$

Referring to Fig. 9.1(a) we see that once the nuclear separation passes R_c autodetachment no longer occurs. We may therefore write for the cross-section Q_{da} for production of the ion Y^-, a process known as *dissociative attachment*,

$$Q_{da} = Q_c \exp\left\{-\int_{R_0}^{R_c}\frac{\Gamma(R')}{\hbar v(R')}\,dR'\right\} = Q_c \exp(-\rho),\,\text{say}, \qquad (9.3)$$

where Q_c is the cross-section for the initial capture.

Bardsley, Herzenberg and Mandl (1964) have derived the expression

$$Q_c = \frac{4\pi}{k^2}\frac{2S_c+1}{2(2S_0+1)}\frac{\Gamma_0(R_0)}{\Delta E}\exp[-\{(E-E_c(\bar{R}_0))^2 - \tfrac{1}{4}\Gamma^2(\bar{R}_0)\}/\Delta E^2], \qquad (9.4)$$

in which k is the wave number of the incident electron, S_c and S_0 the total spin quantum numbers of the initial and final molecular states respectively and R_0 is the nuclear separation at which the electron in its particular state is captured. $\hbar^{-1}\Gamma_0(R)$ is the rate at which the negative-ion state, at separation R, decays back to the ground state of XY through autodetachment. $\hbar^{-1}\Gamma(R)$ is the rate of decay to *all* possible lower states of XY. $\bar{R}_0$ is a mean value of R_0 averaged over the Franck–Condon region.

If experiments can be carried out with molecules formed of different isotopes, the order of magnitude at least of the width $\Gamma(R)$ can be obtained. Thus we may write

$$Q_{da} = Q_c \exp(-\bar{\Gamma}\tau/\hbar), \qquad (9.5)$$

where τ is the time taken for X and Y^- to separate from R_0 to R_c. Since the velocity of separation at separation R is given by

$$\tfrac{1}{2}\frac{M_1 M_2}{M_1+M_2}v^2(R) = W(R),$$

where $W(R)$ is the height of the ionic curve at R above the limit of infinite separation, τ is proportional to $\{M_1 M_2/(M_1+M_2)\}^{1/2}$ and

NEGATIVE ION FORMATION – DISSOCIATIVE PROCESSES

will be different for different isotopes for which Q_c and $\bar{\Gamma}$ will be the same.

Having determined $\bar{\Gamma}$, Γ_0 may be estimated from (9.4) and the observed variation of Q_{da} with electron energy. The application of this analysis to dissociative attachment in H_2 and O_2 is discussed on pp. 312 and 323 respectively.

Variation of the attachment cross-section with gas temperature can arise because, as the temperature increases, more and more neutral molecules will be vibrationally excited. The range of nuclear separations over which capture can take place will thereby be increased. If the upper potential-energy curve is very steep over this range, the change in the mean separation lifetime τ can be appreciable, particularly as its effect on the cross-section is magnified by the exponential loss factor. Effects of this kind have been observed in oxygen and are discussed on p. 323 (see Figs. 9.29 and 9.30).

In case (b) electrons captured with energy between E_3 and E_1 can produce dissociative attachment, the Y^- ions and X atoms having relative energy between 0 and E_5. Electrons with energy between E_2 and E_3 can also be captured but this will lead to a vibrationally-excited molecule XY^-. If left to itself such a molecule will dissociate spontaneously by the reverse process to that which led to its formation, but it may be stabilized if the excess vibrational energy can be got rid of by collision.

Let τ be the average time taken for the vibrationally-excited molecule XY^- to transfer its energy to a gas molecule, and θ the time before autodetachment. Then τ is inversely proportional to the gas pressure but θ is independent of it. If now, at time $t = 0$, we have an excited ion, the probability that it will still not have dissociated in time t will be $\exp(-t/\theta)$. The probability of transferring its excess energy at a time between t and $t + dt$ is $\exp(-t/\tau)\,dt/\tau$, so the total probability of the ion making a transfer collision before it dissociated will be

$$\begin{aligned}\rho &= \int_0^\infty \tau^{-1} \exp\{-(t/\theta) - (t/\tau)\}\, dt \\ &= \theta/(\theta + \tau) \\ &= p/(p + p'),\end{aligned} \qquad (9.6)$$

where p is the gas pressure and p' a critical pressure for which $\theta = \tau$. The attachment probability, when the reaction is of type (b) without dissociation into X and Y$^-$, will be independent of pressure for pressures much greater than a certain critical value but for low pressures will be proportional to the pressure.

If the lifetime of the vibrationally-excited molecule is long compared with the period of vibration, the vibrational quantization will show up in the form of resonance peaks in the variation of attachment probability with electron energy. This is because the cross-section Q_c for formation of the complex will exhibit peaks.

In Fig. 9.1(b) the minimum energy of the molecular-ion state is shown as above that for the neutral molecule. However, very similar arguments apply if the situation is as for O_2 or NO, the molecule having a small positive electron affinity and the disposition of potential-energy curves being as in Fig. 9.1(c). In such a case vibrational states of XY$^-$ with $v < v'$ are stable, but those with $v > v'$ are unstable towards autodetachment. Nevertheless they may be stabilized by collision as in case (b) above. Also, as in that case, the variation of attachment probability with electron energy will exhibit resonance maxima associated with the vibrational levels with $v > v'$, provided the lifetime of the complex is long compared with the vibrational period. This situation prevails in O_2 and NO (see Chapter 6, p. 186) and leads to production of O_2^- and NO$^-$ in three-body collisions (see p. 327).

Energy relations in dissociative attachment

Where dissociation occurs on capture we have the following relation between the kinetic energy T of relative motion of the dissociated products, the kinetic energy E of the electron before capture and the difference between the potential energy U_1 of the neutral molecules before impact and that U_2 of its dissociation products.

$$T = E - (U_2 - U_1). \tag{9.7}$$

For a diatomic molecule

$$U_2 - U_1 = D_{XY} - E_a, \tag{9.8}$$

where D_{XY} is the energy of dissociation of the normal state of XY and E_a the electron affinity of Y (see Fig. 9.1). If D_{XY} is known and T and E can be measured, then E_a may be determined.

If the molecule is polyatomic the situation is more complicated because of the possibility of internal excitation of the dissociation products.

Identification of the upper state in dissociative attachment
Information can be obtained about the nature of the upper state in a diatomic dissociative attachment process from the angular distribution of the motion of the ionic product relative to the direction of the incident electron. It is not difficult to see that this distribution is in general anisotropic.

The plane electron wave incident along the direction of the unit vector $\mathbf{n}_0$ is symmetric for rotations about $\mathbf{n}_0$. For a given direction of the nuclear axis of the target molecule the initial molecular state may possess certain symmetries with respect to these operations. Transitions can then only occur to final states possessing the same symmetries.

Arguing on these lines Dunn (1962) showed that, for homonuclear diatomic molecules, within sets of states Σ_g^+, Σ_g^-, Σ_u^+, Σ_u^-, Π_g, Π_u, Δ_u, Δ_g, transitions can only occur between the following states if the nuclear axis is parallel to $\mathbf{n}_0$.

$$\text{Any pair of } \Sigma^+, \quad \Sigma^-, \quad \Pi \text{ or } \Delta \text{ states.} \quad (9.9a)$$

On the other hand if the nuclear axis is perpendicular to $\mathbf{n}_0$ the allowed transitions are as follows

$$\left.\begin{array}{ll} \Sigma_g^+ \text{ to } \Sigma_g^+, \Pi_u, \Delta_g, & \Sigma_g^- \text{ to } \Sigma_g^-, \Pi_u, \Delta_g, \\ \Sigma_u^+ \text{ to } \Sigma_u^+, \Pi_g, \Delta_u, & \Sigma_u^- \text{ to } \Sigma_u^-, \Pi_g, \Delta_u, \\ \Pi_g \text{ to } \Pi_g, \Sigma_u, & \Pi_u \text{ to } \Pi_u, \Sigma_g, \\ \Delta_g \text{ to } \Delta_g. & \Delta_u \text{ to } \Delta_u. \end{array}\right\} \quad (9.9b)$$

Many transitions are allowed for one orientation of the axis and not for the other. The angular distributions for these cases are clearly not isotropic.

For heteronuclear molecules the corresponding allowed transitions are as follows

$$\Sigma^+ - \Sigma^+, \quad \Sigma^- - \Sigma^-, \quad \Pi - \Pi,$$
$$\Delta - \Delta, \text{ axis parallel or perpendicular}, \quad (9.10)$$
$$\Sigma^+ - \Pi, \quad \Sigma^+ - \Delta, \quad \Sigma^- - \Pi, \quad \Sigma^- - \Delta,$$
$$\Pi - \Delta, \text{ axis perpendicular only.} \quad (9.11)$$

O'Malley and Taylor (1968) have carried the analysis further to obtain more definite information about the angular distribution. The dominant term at not too high electron wave numbers is $\{P_L^{|\mu|}(\cos\theta)\}^2$, where θ is the angle the direction of motion of the ion makes with that of the incident electron beam and $P_L^{|\mu|}$ is the usual tesseral harmonic. μ is given by $\varLambda_g - |\varLambda_i|$, where $\varLambda_i$, $\varLambda_g$ are the quantum numbers specifying the axial components of angular momentum of the initial and final molecular electronic states. L is the lowest integer, not less than $|\mu|$, which is even or odd according as the transition does or does not involve a change of parity of the molecular state.

For example, for capture by O_2 the initial state is a $^3\varSigma_g^-$ state with $\varLambda_i = 0$. Of the possible final states, $\varSigma_u^+$ is excluded as a $\varSigma^-$ to $\varSigma^+$ transition is forbidden. This leaves $\varPi_u$, $\varPi_g$, $\varDelta_u$, and $\varSigma_u^-$ for which the dominant terms are $(P_1^1)^2$, $(P_2^1)^2$, $(P_3^2)^2$ and $(P_1)^2$ respectively. This corresponds to angular distributions varying respectively as $\sin^2\theta$, $\sin^2 2\theta$, $\sin^2 2\theta \sin^2\theta$ and $\cos^2\theta$.

In practice, the electron wave number may not be small enough for higher terms in the expansion of the angular distribution function to be ignored and this may limit the usefulness of the method for determining the nature of the upper state.

The experimental evidence for the main dissociative attachment process in O_2 is described and discussed on p. 319.

Non-capture collisions – dissociation into ions

There is one other process of negative-ion formation which may occur on collision of an electron with a molecule. The electron may excite the molecule to an unstable state which dissociates spontaneously into a positive and a negative ion, viz.

$$XY + e \longrightarrow X^+ + Y^- + e.$$

This differs from the type of reaction described on p. 266, as the electron is not captured but merely acts as a source of the energy necessary to produce the electronic transition in the molecule. The potential-energy curves for a case of this type are illustrated in Fig. 9.2. The lower curve is, as usual, that for the ground state of the molecule, ab being the amplitude of nuclear vibration in the ground state of nuclear motion. The other curve represents the potential energy for the upper state. Since this state dissociates spontaneously into ions the curve tends to a limit for infinite nuclear separation which is at a height $E_i - E_a$ above the limit for the normal state, E_i being the ionization energy of X and E_a the

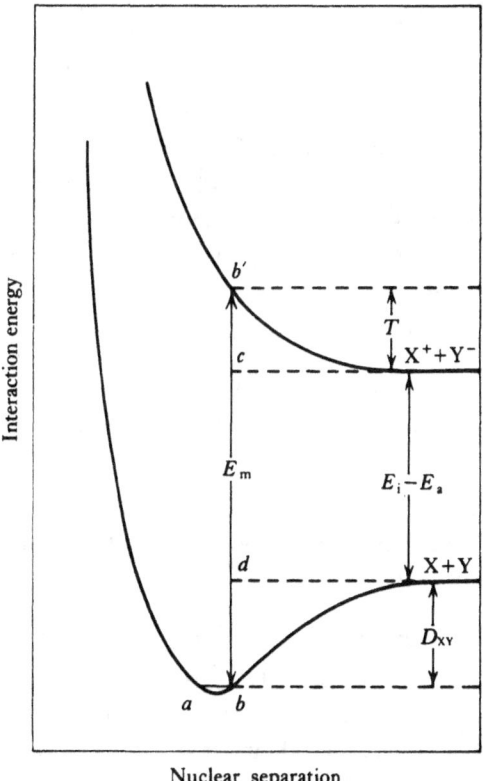

Fig. 9.2. Potential-energy curves illustrating polar dissociation of a molecule XY by electron impact.

electron affinity of Y. Using the Franck–Condon principle, we see that the minimum energy which must be communicated to the molecule to bring about the transition is E_m (bb' in Fig. 9.2). The relative kinetic energy of the ions X^+, Y^- after the collision will be T ($b'c$ in Fig. 9.2) and we have the relation

$$T = E_m + E_a - E_i - D_{XY}, \qquad (9.12)$$

where D_{XY} is the dissociation energy of the molecule XY (bd in Fig. 9.2). Measurement of T and E_m, combined with a knowledge of E_i and D_{XY}, provides a further method of determining the electron affinity E_a of the atom Y.

Since the electron is not captured in this type of collision, the variation of the effective cross-section with velocity is very different from that characteristic of capture reactions. As the process is essentially an electronic excitation of the molecule by electron impact, the velocity variation will be similar to that for inelastic electron collisions with atoms. This is well known and consists in a sharp rise from a zero value at the appearance energy to a maximum at from two to four times this energy, after which it declines steadily as the inverse of the energy, or more slowly if the transition is to a level which combines optically with the normal state (Mott and Massey, 1965, p. 497).

Summarizing, we see that ion pair formation by impact dissociation (which we call *polar dissociation*) will set in at some definite electron energy and persist to quite high energies.

9.3 Formation of negative ions on collision of electrons with molecules – experimental methods of study using homogeneous electron beams

Introductory remarks – early techniques

The beginning of the use of electron beams of approximately homogeneous energy can be dated back to the experiments of Tate and Lozier (1932) and their subsequent exploitation by Lozier (1934). Although it is now realized that many of the data obtained are only of semi-quantitative value, the Tate–Lozier method has been extensively used until quite recently, and it does make it possible to survey extensively the reactions which occur. We shall begin by describing typical apparatus used and then examine the difficulties which arise in the quantitative interpretation of the results obtained.

An electron beam of definite energy is confined by a longitudinal magnetic field to the axis of a cylindrical space containing the gas at low pressure. Ions, formed by the electrons, which pass out in radial directions are collected on the outer metal surface of the cylindrical

NEGATIVE ION FORMATION – DISSOCIATIVE PROCESSES 275

space. By auxiliary electrodes, ions of one sign only may be collected and their velocity distribution examined by retarding potential analysis. Fig. 9.3 is a schematic representation of an actual early apparatus. An electron beam from the filament B issued from the hole E after acceleration of 12 V from B to D and retardation by 10 V from D to E. Between E and F this beam was accelerated to the required energy and it then passed along the axis of the cylindrical space between H and K. The beam which issued at S was collected on the electrode M by a potential of 100 V between L and M. A magnetic field of about 150 G, maintained by a coaxial solenoid, confined the beam to the axis of the cylindrical space. O and P represent concentric sets of thin discs in electrical contact with H and Q respectively. R is a metal cylinder and N a coaxial guard cylinder insulated from R.

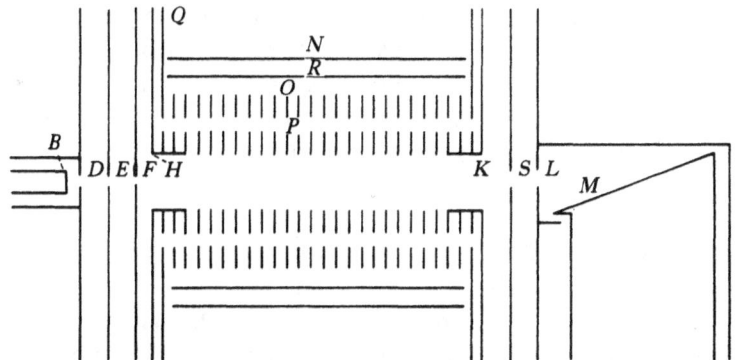

Fig. 9.3. Illustrating Tate and Lozier's apparatus for investigating electron impact with molecules.

The filament assembly was maintained at high vacuum by a separate pumping system, while the gas under investigation was contained in the cylindrical space at pressures varying from 10^{-5}–10^{-4} torr.

Those ions formed by the electron beam which travelled in radial directions, passed between the sets of discs O and P and reached the collector R, where they gave rise to a current which could be measured. Positive and negative ions were separated and the velocity distributions of each examined by applying appropriate potentials between the discs O and P and the collector R.

The entire apparatus was constructed of tantalum and could be baked out at 450 °C, while the purity of the gas used was tested by examination with a mass spectrograph.

The variation of the attachment cross-section with electron energy is perhaps the most reliable of the data obtained with the Tate–Lozier apparatus. Even this may be subject to serious error if the ions are mainly produced with velocity parallel rather than perpendicular to the beam.

Nevertheless a great amount of useful data on the energy variation of cross-sections has been obtained.

The method is not suitable for measurement of absolute attachment cross-sections because no special arrangements are made to collect all the ions produced. The collection efficiency will depend for example on the angular distribution of ion momenta. Furthermore as shown by Tozer (1958), it depends on the energy of the ions, a defect which could only be overcome by choosing the collecting potential to be proportional to the ion kinetic energy.

One of the major applications made of the data obtained in experiments of the Tate–Lozier type was to the determination of dissociative energies and electron affinities via the relations (9.7) and (9.8). The usual procedure was to determine, from a retarding potential analysis, the minimum retarding potential V_r^m necessary to stop collection of ions produced in the gas by electrons of a definite energy E. It was then assumed that

$$eV_r^m = W_i,$$

where $W_i = (1 - \beta)\{E_e - (D - E_a)\}$ is the kinetic energy of the ions, β being the ratio of the mass of the ion to that of the neutral molecule. $eV_r^m/(1 - \beta)$ was then plotted against E_e. Usually the plot was linear over most of the range except close to the threshold $W_i = 0$. It was then customary to extrapolate the linear portion to $V_r^m = 0$, giving a value of E_e which was taken to be equal to $D - E_a$. It was not until 1964 that Chantry and Schulz pointed out that in many cases this procedure can lead to errors of as much as 0.5 eV due to temperature motion of the target molecules.

Because of this motion, ions which for targets at rest would be produced with a definite energy W_0 in the centre of mass system, will be produced with an energy distribution $f(W)\,dW$, where, if $W_0 \gg kT$,

$$kTf(W) \simeq (4\pi w_0)^{-1/2} \exp\{-(w^{1/2} - w_0^{1/2})^2\}. \qquad (9.13)$$

In this formula we have written $w_0 = W_0/\beta kT$, $w = W/\beta kT$, T being the gas temperature. This distribution has a maximum at $W = W_0$ and a full width at half maximum given by

$$W_{1/2} = (11.1\,\beta kTW_0)^{1/2}. \qquad (9.14)$$

For O^- ions of 1 eV energy from O_2 with $\beta = \frac{1}{2}$, at room temperature $W_{1/2} \simeq 0.4$ eV, whereas for H^- of the same energy from H_2O with $\beta = \frac{1}{18}$, $W_{1/2} \simeq 0.13$ eV.

It is quite clear that, if the energy distribution of the ions is broadened to this extent, the procedure adopted for determining $D - E_a$ would lead to errors of comparable magnitude. The best conditions are those in which ions are formed with zero or very small kinetic energy and in which the mass of the ion is small compared with the mass of the parent molecule.

Many problems arise when attempts are made to correct for the temperature motion as this depends very much on the acceptance angle of the retarding potential system and the angular distribution of ionic momenta. All these difficulties are avoided if experiments are carried out which can measure the ion energy distribution resulting from attachment of electrons of well-defined energy. The peak of this distribution gives W_0 which is directly related to $D - E_a$ through (9.13). We describe below experiments which have been carried out on these lines (see p. 316).

The temperature correction was realized after a confusing series of observations by different techniques of the electron affinity of atomic oxygen.

Measurements made by the linear extrapolation Tate–Lozier method applied to dissociative attachment in O_2 gave an electron affinity a little greater than 2.0 eV. Mass analysis of the ions formed confirmed that they were exclusively O^-. Meanwhile measurements made by the hot-wire equilibrium technique (see Chapter 3) applied to O_2 gave the higher value of 3.4 eV. Confidence in the lower value was restored when Metlay and Kimball (1948) repeated the hot-wire experiments, but with N_2O as working gas, and obtained 2.3 eV. Evidence from dissociative attachment in CO and NO was confused by uncertainty in the values of the dissociation energy.

The apparently well-established situation was drastically changed when photodetachment experiments were first carried out, Smith and Branscomb in 1955 obtaining an electron affinity of 1.465 eV from experiments of this type. A little later definite values for the dissociation energies of CO and NO were available from spectroscopic and thermochemical sources. Insertion of these values in the results obtained for $D - E_a$ in attachment experiments in these gases lead to values of E_a compatible with the photodetachment result. In 1961 Page repeated the hot-wire equilibrium experiments (see Chapter 3, p. 41) and obtained a compatible result.

At this stage dissociative attachment in O_2 alone remained anomalous and the most careful experiments failed to give affinities less than 2.0 eV. This outstanding problem was cleared up when the effect of gas temperature was recognized by Chantry and Schulz (1964) who showed that, when taken into account, the attachment data were compatible with an affinity close to 1.5 eV. In CO, O^- ions are formed with zero kinetic energy and temperature effects are unimportant. NO is intermediate between CO and O_2 in this respect.

We now discuss the techniques which have been used for the accurate measurement of absolute attachment cross-sections, of the energy distribution of the ions, their composition and the angular distribution of their momenta.

Measurement of total attachment cross-sections

In principle, the measurement of total cross-sections for negative-ion production is essentially similar to that for measurement of total ionization cross-sections. Attention must be paid however, to certain specific problems which we shall discuss below.

As an example of the experimental arrangements which have been used, we choose the apparatus used effectively by Rapp and Briglia (1965) which is based on that used by Rapp, Englander-Golden and Briglia (1965) to measure ionization cross-sections. Fig. 9.4 shows the general arrangement.

The electron beam from the cathode K was accelerated to the desired energy and collimated by passage between holes in the electrodes 1, 2 and 3. It then passed through the collision chamber C and an electron collection shield ECS before collection on a plate ECP. A magnetic field of around 500 G was applied along the direction of the beam.

Negative ions formed by electron impact with the gas molecules in the chamber C were collected on the plate IC^-, flanked by two guard rings G^-. These electrodes were maintained at a positive potential $V_\parallel$ relative to a parallel system of plates IC^+ and guard rings G^+ on which any positive ions formed were collected. Scat-

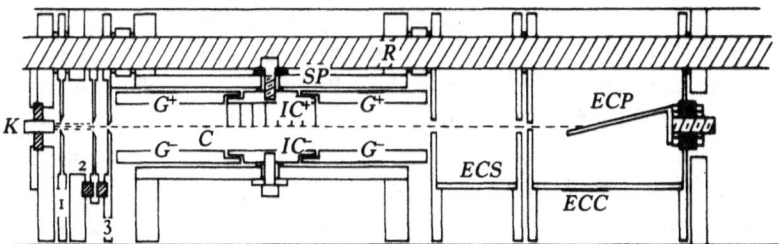

Fig. 9.4. Arrangement of the apparatus used by Rapp and Briglia (1965) to measure total cross-sections for dissociative attachment. 1, 2, 3, apertures; C, oxide-coated cathode; G^+ guard plates; SP, support plate; IC^+, ion collectors; ECS, electron collector shield; ECC, electron collector cylinder; ECP, electron collector plate.

tered or secondary electrons were prevented from reaching the electrodes by the axial magnetic field.

At electron energies below the ionization threshold, no complications due to positive ions arise. It is however essential that the extraction field necessary to saturate the negative-ion collection is not large enough to cause scattered or secondary electrons formed from or by the main beam to reach the collector. This problem is more serious when the electron energy is above the threshold for dissociative ionization which produces energetic positive ions. High extraction fields are required to prevent any of these ions from reaching the negative-ion collector. Rapp and Briglia made a careful study of the shapes of the collector current–voltage characteristics at different electron energies. Fig. 9.5 shows such characteristics obtained for the negative-ion collector current in CO at different electron beam energies.

It will be seen that for an energy of 9.9 eV below the ionization threshold, saturation is reached with a collector voltage of only 2 V. As the voltage increases beyond 8 V the current begins to fall again. This is due to the energy spread across the electron beam due to the transverse field, an effect enhanced by the fact that the dissociative attachment cross-section for CO is a maximum at an electron energy of 9.9 eV.

At an electron energy of 30 eV dissociative ionization occurs and the collector current remains positive until 2 V is applied, while saturation is not reached until the collector voltage is as high as 17 V. This voltage is high enough to produce saturation even when the energy is as high as 50 eV, but at this energy the current remains positive until 8 V is applied. At electron energies of 100 eV saturation cannot be achieved because scattered and secondary electrons are being collected before the energetic positive ions are completely repelled.

On the basis of such measurements Rapp and Briglia (1965) used extraction fields around 3–5 V cm^{-1} for electron energies below the

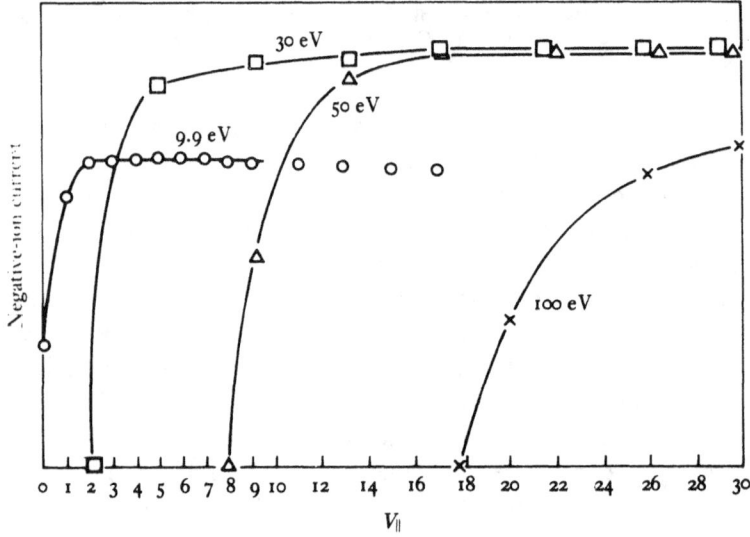

Fig. 9.5. Negative-ion collector current as a function of collector voltage $V_{\parallel}$, observed in CO with electron beams at different energies by Rapp and Briglia (1965).

ionization threshold (usually up to 12 or so eV) and around 25 V cm^{-1} at higher energies, up to about 60 eV.

Total cross-section measurements have also been made by Schulz (1960), by Asundi, Craggs and Kurepa (1963) and by Buchel'nikova (1958) using apparatus operating on basically similar principles. The results obtained by the different experimenters for O_2, CO and CO_2 agree well as regards the maximum cross-section, the electron energy at the peak and the peak width at half-maximum.

Spence and Schulz (1969) have used an arrangement which makes it possible to measure cross-sections in the temperature range from 300–1000 °C. Fig. 9.6 is a schematic diagram of their apparatus. The electron source was a thoria-coated iridium filament used with the retarding potential difference method so that the electron energy spread was small. A magnetic field of between 400 and 1200 G was applied along the direction of the electron beam.

The collision chamber was a cylindrical iridium oven, 10 mm in diameter, 0.05 mm thick, which could be electrically heated – 60 A current yielded a temperature of 950 °K. To minimize heat loss

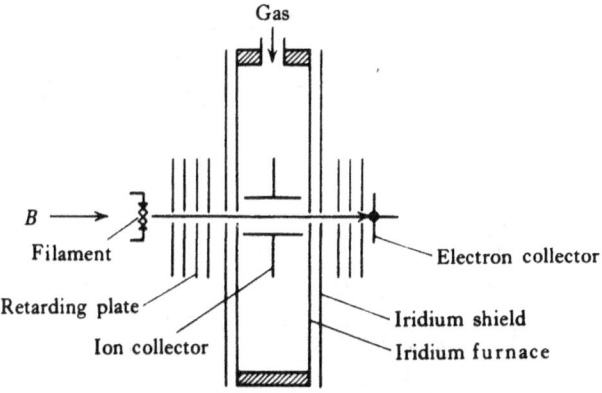

Fig. 9.6. Arrangement of the apparatus used by Spence and Schulz (1969) for measuring dissociative attachment cross-sections in gases at temperatures from 300–1000 °C.

by radiation the collision chamber was surrounded by a cylindrical iridium heat shield. Ions formed in the chamber were collected by application of suitable potentials to a pair of iridium parallel plate collectors, both of which were provided with guard plates. The electron collector was coated with platinum black to minimize electron reflection. Apart from the iridium components, all other electrodes were coated with gold to minimize contact potential difference.

The electrical insulation for the electron gun consisted of sapphire balls and, for the ion collectors, of fused alumina and quartz.

The temperature of the furnace was measured from the positive-ion current produced in helium, the mass flow rate of which was maintained constant. Under these conditions the helium density and hence the positive-ion current varies as $T^{-1/2}$, where T is the absolute temperature. This technique was checked by comparison with observations made with an iridium/iridium–rhodium thermocouple.

The use of this apparatus for measurements up to high temperatures in O_2 and CO_2 is described on pp. 321 and 353 respectively, but it has been used also for room temperature measurements in other gases.

Measurement of the velocity distribution and composition of the ions

Fig. 9.7 illustrates the arrangement used by Chantry and Schulz (1967) which not only measures the variation of the attachment

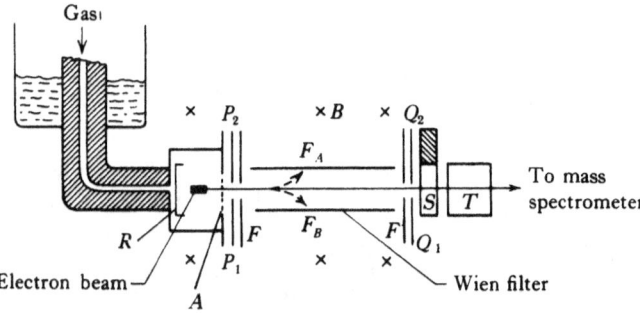

Fig. 9.7. Arrangement of the apparatus used by Chantry and Schulz (1967) for measuring the energy distribution of the negative ions formed in dissociative attachment.

cross-section with electron energy but also observes the energy distribution of the ions and their composition.

The gas under study enters the collision chamber through a copper tube of 1.2 cm external and 0.3 cm internal diameter. To make possible observations at reduced temperatures the outer part of the tube is surrounded by a Dewar into which liquid N_2 may be introduced as required.

The electron beam passes in a direction normal to the plane of the paper. It is produced by a thoria-coated iridium filament and collimated by an electrode system which permits the use of the

NEGATIVE ION FORMATION – DISSOCIATIVE PROCESSES 283

retarding potential difference method for increasing the energy homogeneity of the beam. A uniform magnetic field H of about 600 G normal to the plane of the paper covers the volume indicated by the crosses in Fig. 9.7. It not only aligns the beam but also forms part of the ion energy analyser system.

The collision chamber includes two plane electrodes which are insulated from the chamber walls. One is the repeller electrode R and the other A, mounted opposite to it, is the 'attractor' electrode containing a mesh-covered slit. The electric field between these electrodes attracts the ions produced to the electrode A from which a representative sample passes through the slit for measurement and analysis. By maintaining the chamber walls at a suitable potential between that of A and R the extraction field does not give rise to potential variations along the electron beam. The energy spread transverse to the beam produced by the extraction field, which is usually less than 0.4 V cm^{-1}, is less than 20 mV.

The ions leaving the collision chamber are next energy-analysed by using a Wien filter. This depends upon the fact that charged particles entering a region, in which there are crossed electric and magnetic fields **E** and **H** respectively, along the direction perpendicular to **E** and **H**, will only pass undeviated through the region if their velocity is given by $v = cE/H$.

To ensure that the ions are following the correct trajectory parallel to the magnetic field when they enter the filter F, they first pass through a split plate P between which a small voltage (usually less than 2 V) is applied. The filter is 2.5 cm long and the electric field is applied between the condenser plates F_A and F_B. The split electrode Q, like P, serves to keep the ions passing through the filter on a straight path into the electrode S which acts as a baffle to intercept ions which have passed through the filter at large angles (>8°) to the axis in a plane normal to that of the paper. Such ions are not cut off by the filter which is only sensitive to the velocity of ions moving along the axis.

The electrode T accelerates the ions to about 100 V so that they are no longer affected by the magnetic field. Penetration of the accelerating field into S is reduced to tolerable proportions by covering the face of S near to T with wire mesh.

It remains to accelerate the ions into a 90° sector magnet mass analyser. The electrode system described above is surrounded by an open-ended metal cylinder extending as far as T. This is held at ground potential as is the collision chamber. The whole is enclosed in the metal vacuum envelope held at the potential to which the ions are accelerated for the mass analysis (about $+500$ V).

The current of ions passing out of the mass analyser is measured by a ten-stage electron multiplier and vibrating-reed electrometer.

A check was made to examine whether the Wien filter exhibited energy discrimination. This was done by comparing the electron energy at which the cross-section is a maximum, as determined from total ion collection studies, with that obtained from the maximum value of $W_0^{1/2} F$, where W_0 is the peak ion energy and F the current transmitted through the filter (see (9.13)). The latter was found to be shifted to slightly lower energies, an effect which was reproduced when, instead of $W_0^{1/2} F$ the product $W_{1/2} F$, where $W_{1/2}$ is the full width of the distribution at half-height, is used, $W_{1/2}$ being proportional to $W_0^{1/2}$ (see (9.14)). The shift was consistent with the assumption that the detection efficiency of the filter for ions of energy W varies as W^{-n}, where $n = 1 \pm 0.2$. This will shift the peak of the observed ion distribution from its true value by $2n\beta kT$, where β is as in (9.14) and T is the gas temperature. At 300 °K this shift is 0.03 eV.

Measurements employing mass analysis but not energy analysis

In some cases, as for example in CO, more than one kind of negative ion may be produced and it is required to determine the absolute cross-sections for each separately. For this it is necessary to add a mass analyser to an apparatus which provides for total ion collection. However, when the cross-section for production of one kind of ion greatly exceeds that for the other, it is possible to dispense with mass analysis in a total collection experiment to obtain the larger cross-section. All that is then required is a measurement of the relative yield of the two ions as a function of electron energy.

Stamatovic and Schulz (1970) have carried out experiments of this kind in CO in which the cross-section for dissociative attach-

NEGATIVE ION FORMATION – DISSOCIATIVE PROCESSES 285

ment to produce O^- is more than three orders of magnitude greater than that for production of C^-. The electron gun used trochoidal motion of electrons in crossed electric and magnetic fields to provide a nearly monochromatic beam of electrons. Ions formed in the collision chamber were extracted and analysed by a quadrupole mass spectrometer and detected by a Channeltron multiplier. The big problem in this experiment was to observe the very weak C^- current against the background. This is discussed on p. 338.

Chantry (1969) has carried out experiments on the variation of dissociation cross-sections with temperature up to temperatures of 1000 °K, with mass analysis. The apparatus he used was simplified, as in that of Stamatovic and Schulz, by dropping the requirement for measuring absolute cross-sections but it was necessary that the variation with electron energy at each temperature should be correct. As in the experiments of Stamatovic and Schulz the ions were analysed by a quadrupole mass spectrometer and detected by a Channeltron multiplier. However, the ion extraction was different and was specially suitable for the high-temperature experiments. The collision chamber was in the form of an iridium tube, the section of which parallel to the electron beam measured 9 mm × 3 mm along and perpendicular to the beam respectively. Ions produced in the chamber were sampled through a slit 6 mm long and 0.5 mm wide. The temperature of the chamber, which could be varied by electrical heating, was measured by an iridium/iridium–rhodium thermocouple welded to the wall of the chamber opposite to the ion extraction slit (see Fig. 9.8). Check observations were made in a separate experiment in which the ion exit slit could be viewed with an optical pyrometer, at temperatures near 1000 °C, agreement being obtained to 5% with the thermocouple readings. The radiation shield enclosing the collision chamber (see Fig. 9.8) served the dual purpose of reducing heat loss by radiation and of providing an ion extraction electrode.

Results obtained with this apparatus for the temperature variation of the rate of production of O^- from N_2O by dissociative attachment are discussed on p. 358, in comparison with data from other experiments.

A third technique for studying the variation of cross-sections for

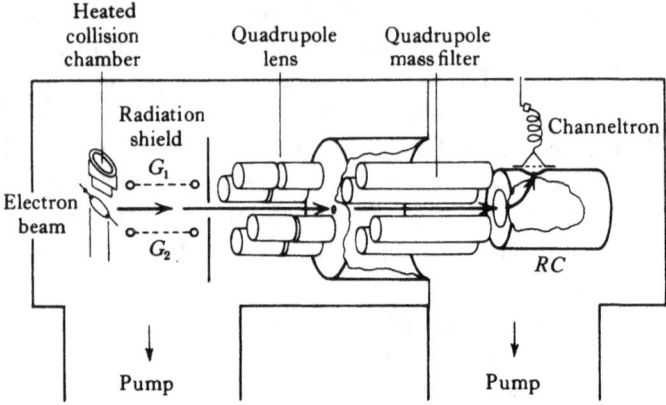

Fig. 9.8. Arrangement of the apparatus used by Chantry (1969) for studying dissociative attachment at high temperatures. G_1 and G_2 are planar grids.

dissociative attachment is the crossed-beam arrangement used by Henderson, Fite and Brackmann (1969) for O_2. In these experiments the electron beam intersects at right angles a molecular beam of hot O_2 molecules, the ions formed are extracted along the direction of the latter beam and are monitored by a quadrupole mass filter and electron multiplier.

Fig. 9.9 shows the general arrangement used. It was housed in a separate vacuum chamber from that enclosing the interaction and detection regions. Special care was taken in designing the iridium furnace source of hot molecules. A small aperture near the width of the furnace allowed for escape of the gas to form the molecular beam. Uniformity of temperature inside the furnace near the aperture was improved by iridium foils. The temperature was measured by an optical pyrometer viewing the inside of the furnace near the aperture and was monitored during the experiments by a photoelectric pyrometer. To prevent thermionic electrons from passing out through the aperture with the neutral gas and producing spurious signals through collisions with background gas, the furnace was biased positively with respect to the grounded walls of the vacuum chamber.

The O_2 beam was collimated by an aperture in the wall separating the two vacuum chambers and was modulated at 1440 Hz by a synchronously-driven toothed wheel. This made it possible

NEGATIVE ION FORMATION – DISSOCIATIVE PROCESSES

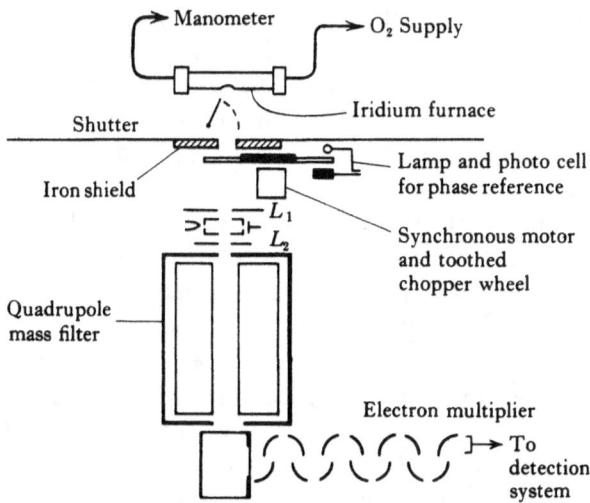

Fig. 9.9. Arrangement of the apparatus used by Henderson et al. (1969) to study dissociative attachment to hot O_2.

to use phase-sensitive detection (see Chapter 11, p. 432) to discriminate against background signals.

Electrons, emitted from a longitudinal filament, were confined by a magnetic field and collected by a tungsten surface heated dull red to reduce electron reflection. Although the ions produced were carried forward by the momentum of the neutral molecules from which they were formed, a weak extraction field was provided by the ion lens L_2.

The ions emerging from the mass filter were drawn out sideways by a high voltage on the first dynode of a 14-stage Cu–Be photomultiplier. In this way the background noise was greatly reduced as the photomultiplier was not directly exposed to light from the furnace and electron gun.

As in other experiments with mass analysis it is essential to check that the efficiency of ion collection does not depend on ion energy. Again such a check can be made by verifying that the observed variation of ion current with electron energy agrees with that observed in experiments in which total ion collection is assured. Having verified this, the absolute cross-section at room temperature can be calculated by comparison with the total ion collection results. The absolute values at higher temperatures T were obtained by multiplication of the signals per unit electron current by $(T/300)^{1/2}$, which is valid provided the total gas flow in the beam remains constant and the furnace operates under molecular flow conditions. Results obtained in these experiments are discussed on pp. 321–5.

A great number of experiments have been carried out in which ions produced by an electron beam passed through a gas or vapour are extracted and analysed by a mass spectrograph. Data were usually given in the form of the percentage probability that the ions produced by electrons of a given energy are of one or other type. However, in many cases, the sensitivity of the system depended strongly on the ion energy. Measurements were also made by carrying out observations at different electron energies of the appearance potentials of various ionic fragments. The interpretation of such data is complicated by temperature effects as well as by lack of uniformity of the detecting system as regards ion energy. In the case of diatomic or polyatomic ionic fragments, the internal energy of the fragment is also unknown. De Corpo and Franklin (1971) have attempted to overcome these difficulties in the manner described on p. 381.

Measurement of the angular distribution of the ionic momenta

Van Brunt and Kieffer (1970) have measured angular distributions of O^- ions produced by dissociative attachment in O_2, using equipment first employed to measure corresponding distributions of H^+ and D^+ ions produced by dissociative ionization of H_2 and D_2. Basically the experimental arrangement is similar to that used by Dunn and Kieffer (1963) who made the first observations of the latter distributions.

Fig. 9.10 illustrates the arrangement of the collision chamber and electron gun which were mounted within a stainless steel vacuum system. The electron beam produced ions in a field-free region which drifted through a pair of apertures into a 60° sector magnet tuned to select ions of a chosen momentum. These were then accelerated and focused into an electron multiplier. The electron gun could be rotated with respect to the ion detection system about

NEGATIVE ION FORMATION – DISSOCIATIVE PROCESSES

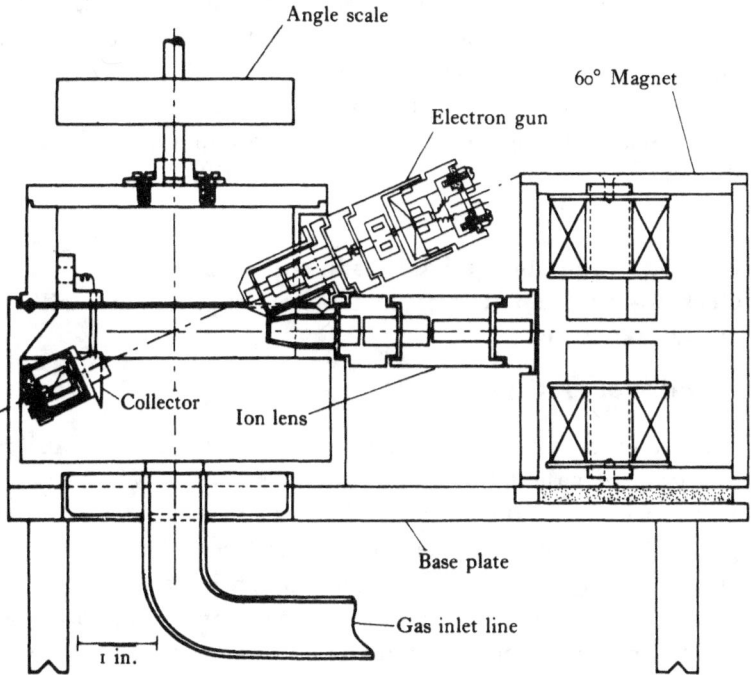

Fig. 9.10. Arrangement of the apparatus used by van Brunt and Kieffer (1970) for measuring the angular distribution of O⁻ produced by dissociative attachment in O_2.

an axis perpendicular to the plane of the paper, while the axis of the gun was oriented at an angle of 23° with respect to the same plane. With this geometry the angle θ of projection of the ions relative to the electron beam is given by

$$\cos \Theta = \cos 23° \cos \theta,$$

where Θ is the angle the plane perpendicular to the paper through the electron beam makes with the corresponding plane through the ion beam. Measurements could be made in the ranges $157° \geqslant \Theta \geqslant 23°$ and $337° \geqslant \Theta \geqslant 203°$.

The electron beam had a total divergence angle less than 2° at all energies in the range 5–300 eV and an energy spread at half-maximum of about 0.5 eV. The angular resolution was about ±1.2°.

As in other experiments described above, the surface of the

electron collector was coated with platinum black to reduce electron reflection while the inside surfaces of the scattering chamber, ion lens and magnet were coated with Aquadag for the same reason and also to reduce contact potentials.

The data obtained for the angular distribution of O^- ions from dissociative attachment in O_2 are discussed on p. 319.

9.4 Formation of negative ions in collisions of electrons with molecules – attachment experiments with electron swarms

Introductory remarks

The experimental study of the attachment of electrons drifting in a gas under the influence of a uniform electric field began at a very early stage in the investigation of the passage of ions through gases. While some of the techniques used in the early work have survived with comparatively small modifications, much of the data obtained in the earlier work are of doubtful value because of the profound influence of common impurities such as air and water vapour on the rate of attachment of slow electrons to gas molecules. In modern applications special precautions are taken to eliminate these effects by working under high or ultrahigh vacuum conditions.

Among the early techniques which have been employed in recent work are the electron-filter method originally introduced by Loeb and Cravath in 1929 and the diffusion method developed by Bailey (1925) four years earlier. Since the Second World War, techniques depending on the observation of the time development of electron pulses in passing through the gas under investigation have been used very effectively. Another technique which has also been applied during the same period is that of observing the time decay of electron and ion concentrations in discharge afterglows. Both of these methods have become available because of the possibility of making observations with high time resolution.

As contrasted with beam techniques, experiments with electron swarms provide information about attachment rates at much lower energies but the rates involved are averages over electron velocity distributions which depend on other transport properties

NEGATIVE ION FORMATION – DISSOCIATIVE PROCESSES 291

of the electrons. It is only in comparatively recent times that knowledge of these properties, together with the computational resources required, has made possible a complete analysis of swarm data.

General description and theory of swarm experiments

Consider a stream of electrons drifting through a gas containing N molecules cm^{-3} under the influence of a uniform electric field F. Let α_a be the chance that an electron of the stream in passing unit distance in the direction of the field will be attached to a gas molecule. The loss of electron current in a distance dx due to this cause is

$$dI = -I\alpha_a \, dx. \qquad (9.15)$$

The ratio of the electron current in the stream at two points x_1, x_2 will therefore be

$$I_2/I_1 = \exp\{-\alpha_a(x_2 - x_1)\}. \qquad (9.16)$$

Attachment experiments involving electron swarms have been devised to measure α_a, but it is necessary to obtain further information concerning the swarm before the mean attachment cross-section $\bar{Q}_a$ or probability of attachment per collision can be determined. In passing a distance δx in the direction of the field, an electron, because of its random motion, actually moves through a much greater distance than δx. If c and W are respectively the random and drift velocities of an electron, the actual distance traversed by the electron will be $c\delta x/W$, so that

$$\alpha_a = N\bar{Q}_a c/W. \qquad (9.17)$$

This may be rewritten in terms of the chance of attachment h per collision and l, the mean free path of the electrons in the gas, to give

$$\alpha_a = hc/lW. \qquad (9.18)$$

We may also define the *attachment frequency* ν_a by

$$\begin{aligned}\nu_a &= W\alpha_a \\ &= N\bar{Q}_a c, \end{aligned} \qquad (9.19)$$

so that the time rate of change of electron concentration is given by

$$\frac{dn_e}{dt} = -\nu_a n_e. \qquad (9.20)$$

In accordance with the usual notation, the *attachment rate coefficient k* is given by

$$kN = \nu_a, \qquad k = \overline{Q_a c}. \qquad (9.21)$$

There is here an unfortunate conflict of definition as α_a is referred to as the *attachment coefficient* by analogy with the long-established term ionization coefficient. To try to retain both of these associations we shall distinguish k by always including the extra adjective 'rate'.

The quantities c and W may be measured in other experiments as described for example in Massey and Burhop (1969).

Experimental methods

The method of the electron filter

This method, developed by Loeb and Cravath (1929) was used particularly by Bradbury in an extensive series of measurements of α_a in 1933–4.

In this method α_a is determined by direct measurement of the ratio I_1/I_2 of the electron current at two places at a distance x apart, perpendicular to the direction of the applied field. The electron content of the mixed negative-ion and electron current is measured by using a high-frequency filter in a manner suggested by Loeb. This filter consists of a grid between alternate wires of which a high-frequency alternating electric field is applied. This field can be adjusted to sweep the highly mobile electrons to the grid wires but leave the heavy negative ions unaffected. The decrease in current when the stream passes through this filter is then a measure of the contribution of the electrons to the total current.

Fig. 9.11 is a schematic illustration of the apparatus actually used. Electrons generated photoelectrically from the plate A diffused under the action of an electric field between plates P and A, maintained uniform by the guard rings B, C, D, E. One half of each of the guard rings E and C consisted of a grid of fine platinum

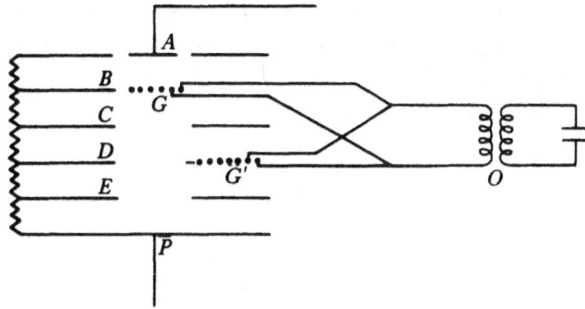

Fig. 9.11. Arrangement of the electron filter to study attachment.

wires (G, G') which could be slid into position to intercept the electron stream. (In the figure G is shown in position, G' not.) An alternating potential could be applied across the grid wires by an oscillator O.

The apparatus was enclosed in a pyrex tube and could be baked out at 200 °C and evacuated to a pressure as low as 10^{-6} torr. This is important in connexion with the removal of impurities.

In carrying out measurements, the current to P was measured first with one grid in place but no alternating field and then with the alternating field applied. Similar measurements were carried out at the second grid, and from these I_2/I_1 and hence α_a could be determined.

Pulse methods

In a typical experiment using a pulse technique, a pulse of electrons liberated from a photocathode is allowed to drift a distance d through the gas, in a uniform electric field F, towards a collector anode. During their passage some of the electrons will form negative ions in collision with gas molecules. We assume for present purposes that these ions will not suffer electron detachment before reaching the anode. Later, in Chapter 12, p. 530, we shall describe how it is possible to adapt the technique to allow for and measure the rate of detachment also.

The drift velocity of the negative ions is only about 10^{-3} of that of the electrons so that, after collection of the pulse due to free electrons, there will be an appreciable, measurable, delay before the ion current pulse arrives. Ions received at a time t after the electron pulse will have been produced at a distance $u_i t$ from the anode, where u_i is the ion drift velocity. At this distance the electron pulse will have been reduced in intensity by a factor $\exp(-\alpha_a x)$, where $x = d - u_i t$. The negative-ion current will therefore vary

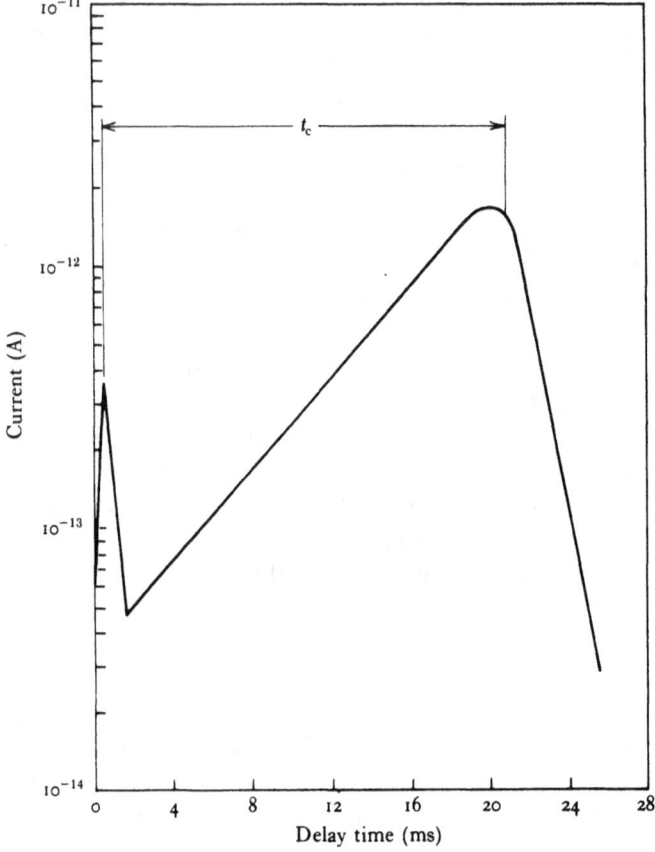

Fig. 9.12. General form of the current signal as a function of delay time in pulse experiments.

as $\exp(\alpha_a u_i t)$. Fig. 9.12 illustrates the general shape of the current signal as a function of delay time since generation of the electron pulse. The exponential rise is seen to cut off sharply after a time t_c. This will be the time required for negative ions to traverse the full distance d, so $u_i = d/t_c$. It follows that, if the shape of the current signal can be recorded, u_i and hence α_a may be determined.

An alternative procedure is to measure the ratio of the total flux of negative ions to that of electrons received at the anode. The number δn_i of negative ions produced at a distance between x and $x + \delta x$ from the cathode is given by

$$\delta n_i = \alpha_a n_e(x)\, \delta x$$
$$= \alpha_a n_e(0) \exp(-\alpha_a x)\, \delta x. \quad (9.22)$$

The total number of negative ions received at the anode is therefore

$$n_i = n_e(0)\, \alpha_a \int_0^d \exp(-\alpha_a x)\, \mathrm{d}x.$$

Hence

$$n_e(d)/n_i = \exp(-\alpha_a d)\{1 - \exp(-\alpha_a d)\}^{-1}. \quad (9.23)$$

In practice the transient current flow is measured by the transient potential differences developed across a resistor R. If an integrating circuit is introduced with time constant $RC \gg d/u_i$, the time t_i taken for negative ions to drift the distance d between the electrodes, the potential difference developed up to the drift time t_e ($\ll t_i$) for the electrodes will be

$$V(t_e) = \frac{n_e(0)e}{C\alpha_a d}\{1 - \exp(-\alpha_a d)\}. \quad (9.24)$$

The factor $\{1 - \exp(-\alpha_a d)\}/\alpha_a d$ represents the fraction of electrons which arrive unattached. At a time $t_1 > t_i$ but still $< RC$, all of the charge in the pulse will have been collected so

$$V(t_1) = n_e(0)e/C. \quad (9.25)$$

The ratio $V(t_1)/V(t_e)$ may be simply and accurately measured on an oscilloscope. Fig. 9.13 illustrates the typical appearance of an oscillogram. The height of the initial spike is proportional to $V(t_e)$. The voltage then grows to a flat plateau of height proportional to $V(t_i)$. From the ratio of these heights α_a may be determined.

The first application of pulse techniques was made by Doehring in 1952, depending on the observation of the time variation of the negative-ion signal. It was extensively developed by Chanin, Phelps and Biondi (1959, 1962) who applied it to study many important attachment processes.

Fig. 9.14 illustrates the general arrangement of the electrodes in

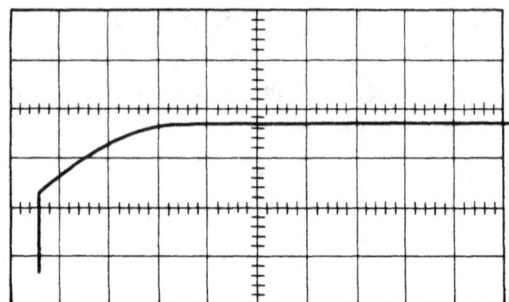

Fig. 9.13. General appearance of a typical oscillogram in Grünberg's experiment (1969) to measure attachment coefficient.

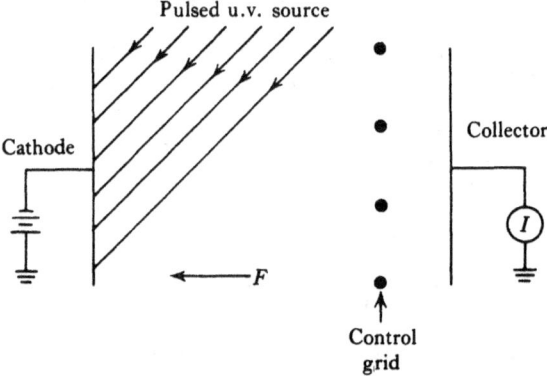

Fig. 9.14. Arrangement of the electrodes in the pulse experiments of Chanin et al. (1959, 1962).

their experiments. Electron pulses were produced by focusing ultraviolet light from a peaked mercury discharge lamp on to the cathode. To measure the negative-ion current as a function of time after the pulse, a control grid was used of basically the same type as in the electron filter experiments of Bradbury (see p. 293). Normally a bias was applied to the grid to prevent either electrons or negative ions from reaching it. To open this gate, rectangular voltage pulses were applied to each half of the grid so that the field between alternate wires was reduced to zero. By varying the time of opening of the grid relative to the pulse of light, the current to the collector could be measured at any desired time interval after the electron pulse. Measurements were made at two drift distances,

2.45 and 10.16 cm. Special care was taken to ensure good vacuum conditions, the drift tube being mounted on an ultrahigh vacuum handling system. The vacuum system was baked out at 300 °C for about 16 hours before each set of measurements. The residual gas pressure after bakeout was as low as 5×10^{-9} torr, rising at about 5×10^{-10} torr/min. To determine the conditions under which electron detachment was not important, the control grid could be operated as a filter (see p. 292) so it would collect electrons but not positive ions. If no delayed electrons were observed, detachment could be neglected. In Chapter 12, p. 530, we shall describe how the technique was extended to deal with conditions in which detachment is important.

The circuit integration technique was first introduced by Grünberg (1969). It requires a circuit with a time constant $\simeq 1$ s. As the drift tube has a capacitance of only about 20 pf, an input amplifier resistance of around 10^{11} Ω is required. Once these conditions are met, the method has the advantage of good accuracy and only requires that the electron pulse length t_p be such that $t_e + t_p \ll t_i$, where t_i and t_e are the ion and electron transit times. This makes it possible, when carrying out measurements under conditions in which α_a is large, to increase t_p to obtain an increased total flux of electrons.

Bortner and Hurst (1958) have introduced a pulse method depending on pulse height measurements which is especially suitable for studying electrons drifting in non-attaching gas containing a small partial pressure of gas which attaches readily.

Bailey's diffusion method
This is less direct than the methods we have described earlier.

Fig. 9.15 illustrates the principle of the method. A uniform electric field F is maintained between five parallel plates A, B, C, D, E contained in the gas under investigation. B, C, D contain parallel and superposed slits. Electrons generated photoelectrically from the plate A drift downwards under the influence of the electric field. By measuring the currents reaching the electrodes C, D, E with two sets of homologous values of plate displacement, gas pressure and electric field, the quantity α_a may be determined. We introduce the following symbols:

S, the ratio of the current passing through a slit to the total current arriving on a plane containing the opening. Suffixes 0, 1, 2, 3 refer to plates B, C, D, E respectively;

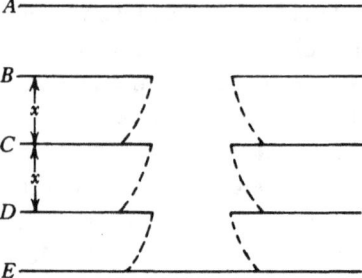

Fig. 9.15. Illustrating the principle of Bailey's method (1925) for measuring attachment coefficients.

R_e, the fraction of electrons falling on a plane which pass through the slit;

R, the corresponding fraction for ions;

r, the corresponding fraction for ions formed between plates B and C, or C and D;

p, the gas pressure;

k, the factor by which the mean energy of the random motion of the swarm electrons exceeds that of gas molecules in temperature equilibrium at NTP;

n_0, N_0, the number of electrons and ions respectively passing through the slit in plate B;

n_1, N_1 and n_2, N_2, corresponding quantities for plates D and E respectively;

x, the distance between the plates.

Then

$$S_1 = \frac{n_1 + N_1}{n_0 + N_0}, \quad S_1 S_2 = \frac{n_2 + N_2}{n_0 + N_0},$$

$$n_1 = n_0 R_e \exp(-\alpha_a x), \quad n_2 = n_1 R_e \exp(-\alpha_a x),$$

$$N_1 = N_0 R + r\{1 - \exp(-\alpha_a x)\} n_0,$$

$$N_2 = N_1 R + r\{1 - \exp(-\alpha_a x)\} n_1.$$

Eliminating r from these equations we obtain

$$R_e \exp(-\alpha_a x) = \frac{S_1(R - S_2)}{R - S_1} = a_1, \text{ say.} \quad (9.26)$$

Now the quantities R_e, R are known functions of F/kx, but k, assumed equal to unity for the ions, is not known for the electrons. To avoid this difficulty use is made of the fact that k is a function of F/p so the quantity a_n is obtained corresponding to (9.26) but with F, p, x

NEGATIVE ION FORMATION – DISSOCIATIVE PROCESSES 299

increased in the same ratio n. k is then unaltered and so is R_e, giving

$$R_e \exp(-n\alpha_n x) = a_n.$$

If it be assumed that the attachment probability h is independent of pressure (which may not always be justified in view of our introductory remarks), then $\alpha_n = n\alpha_a$ and

$$\alpha_a = \frac{1}{x(n^2-1)} \log\left(\frac{a_1}{a_n}\right), \quad R_e = \left(\frac{a_1^{n^2}}{a_n}\right)^{1/(n^2-1)}.$$

This gives α_a and R_e in terms of quantities which are all, with the exception of R, directly measurable. The calculation of R depends on diffusion theory and is not likely to be seriously in error. Additional support for the general theoretical basis of the method is also provided by measurements carried out in pure hydrogen which is known to have no attachment properties.

Modern modifications of the techniques have depended on choosing geometrical conditions which make simpler the analysis which gives S.

Attachment rates from measurements in static discharge afterglows

When the exciting source is cut off, the electron concentration, which has been generated in a gaseous discharge, will decay at a rate determined by the rates of diffusion to the walls of the containing vessel, recombination to positive ions and attachment to neutral molecules to produce negative ions. These three loss processes may be distinguished in principle by their variation with time and with gas pressure.

If diffusion is alone important the rate of decay will, at least at not too short times in the afterglow, vary as $\exp(-t/\tau)$, where τ is given by

$$\tau = 2\Lambda^2/D_a, \tag{9.27}$$

Λ being the diffusion length of the container and D_a the ambipolar diffusion coefficient (see Chapter 15, p. 645),

$$D_a = (D^+ \mu_e + D_e \mu^+)/(\mu_e + \mu^+). \tag{9.28}$$

D^+ and D_e are the diffusion coefficients and μ^+, μ_e the mobilities of the positive ions and electrons respectively. If the electrons and ions have come to equilibrium with the gas at temperature T,

$$D^+/\mu^+ = D_e/\mu_e = kT/e,$$

so

$$D_a \simeq 2D^+. \tag{9.29}$$

D^+, D_e and hence D_a are inversely proportional to the gas pressure.

The derivation of the formula (9.28) is briefly as follows. Immediately after cessation of the exciting source, electrons will diffuse much more rapidly to the walls than the massive positive ions. This charges the wall negatively so that an electric field is set up between the axis of the discharge and the walls in a sense to resist electron passage to the walls. This field will increase until a value F is reached for which the rate of flow of electrons to the walls exactly balances that of the positive ions. If n_e, n^+ are the respective concentrations of electrons and of positive ions and $\mathbf{j}_e$, $\mathbf{j}^+$ their corresponding particle current densities, we have (see also Chapter 15, p. 645)

$$\mathbf{j}_e = -D_e \operatorname{grad} n_e - \mu_e \mathbf{F} n_e, \qquad (9.30)$$

$$\mathbf{j}^+ = -D^+ \operatorname{grad} n^+ + \mu^+ \mathbf{F} n^+. \qquad (9.31)$$

To maintain the condition $|n^+ - n_e| \ll n_e$ for existence of a plasma, $\mathbf{j}_e = \mathbf{j}^+$ so that $\mathbf{F}$ may be eliminated from (9.30) and (9.31) to give

$$\mathbf{j} = -D_a \operatorname{grad} n, \qquad (9.32)$$

where

$$D_a = (D^+ \mu_e + D_e \mu^+)/(\mu_e + \mu^+).$$

In a decaying plasma in which there is no source of ionization

$$\operatorname{div} \mathbf{j} + \frac{\partial n}{\partial t} = 0, \qquad (9.33)$$

so

$$-D_a \nabla^2 n + \frac{\partial n}{\partial t} = 0. \qquad (9.34)$$

This equation is satisfied by $n = R(r)T(t)$ with

$$\frac{\partial T}{\partial t} + \tau^{-1} T = 0. \qquad (9.35)$$

$$-D_a \nabla^2 R - \tau^{-1} R = 0. \qquad (9.36)$$

The allowed values τ_s of τ are those for which the solution of (9.36) is a proper function of r throughout the container and vanishes at the container walls. The complete solution will then be

$$n = \sum_s A_s R_s(r) \exp(-t/\tau_s). \qquad (9.37)$$

At sufficiently large times the only important term will be that which has the largest value of τ_s and this is given by (9.27) with Λ^2 determined from the boundary conditions.

If recombination is the dominant loss process,

$$\frac{dn_e}{dt} = -\alpha_r n^2, \qquad (9.38)$$

where α_r is the recombination coefficient. This equation may readily be integrated to give

$$\{n_e(t)\}^{-1} = \{n_e(0)\}^{-1} + \alpha_r t, \qquad (9.39)$$

a linear instead of exponential relation between n_e^{-1} and t.

Finally, with attachment as the main process of electron loss

$$\frac{dn_e}{dt} = -\nu_a n_e, \qquad (9.40)$$

where ν_a is the attachment frequency. In this case we have again an exponential relation $n_e(t) = n_e(0) \exp(-\nu_a t)$ but, in contrast to diffusion loss, the exponent is not proportional to $1/p$ but increases either as p for a two-body attachment process, or p^2 for a three-body one.

On the basis of these considerations it would seem to be possible to determine, from observations of the time variation of the electron concentration during the afterglow stage, which of the loss processes is dominant. Furthermore, if conditions may be adjusted so that attachment is the only important process, the attachment coefficient may be obtained.

In practice complications arise from the existence of processes leading to electron production, such as detachment collisions or collisions between metastable atoms. Furthermore, there will usually be more than one ionic species present and ionic reactions may occur which complicate the interpretation of the results.

While observations of the rate of decay of electron concentrations in discharge afterglows has proved mainly of importance for the measurement of ambipolar diffusion and recombination coefficients, some applications have been made to measure attachment rates.

Most experiments have been carried out in afterglows of discharges produced by microwave breakdown or by photoionization, the mean electron concentration being determined as a function of time by measurement of the variation with time of the resonant frequency of the cavity within which the discharge is generated. We illustrate the procedure employed and the precautions taken in a recent experiment carried out by Truby (1968) to measure the rate of attachment of electrons in iodine vapour.

To reduce the effect of diffusion loss the experiments were carried out in a helium–iodine mixture, the helium acting as a buffer, inhibiting diffusion but being inert in other respects. This mixture was enclosed in a cavity with diffusion length $\Lambda = 0.484$ cm. Ionization was produced by a pulse of ultraviolet radiation, in the wavelength range 1150–1350 Å, emitted by a low-pressure hydrogen flash lamp.

The resonant frequency of a cavity may be determined by measurement of its reflection coefficient for microwaves as a function of microwave frequency. At resonance the reflection coefficient is a minimum. In the experiments concerned it is necessary to determine the cavity frequency as a function of time in the afterglow. In most experiments, which have been concerned primarily with measurement of diffusion and recombination coefficients, this was done by a repetitive process, the ionizing pulses being repeated at regular intervals. After each pulse a single frequency measurement is made at a time which is varied from pulse to pulse, so building up the relation between frequency and time in the afterglow. It was found, however, that when working with gases in which negative ions are formed, this repetitive procedure can lead to spurious results due to gradual build-up of negative-ion concentration. Truby (1968) therefore arranged to sweep the probing microwave frequency between suitable limits continually during the afterglow stage, the sweep frequency being high compared with the electron concentration decay times. In this way the variation of cavity frequency with time in the afterglow could be obtained using a single pulse of ionizing radiation. Of course such observations were repeated many times but at such long time intervals that there was no possibility of complication from build-up effects.

The mean electron concentration is proportional to the shift in resonance frequency from the value with no ionization present according to a well-known relation. Although initially the distri-

NEGATIVE ION FORMATION – DISSOCIATIVE PROCESSES

bution of ionization may be far from uniform within the cavity, marked departures from uniformity soon smooth out due to diffusion.

In Truby's experiments the microwave power used in the probing signals was about 10 μW and the duration of the ionizing pulse was varied between 10 and 50 μs. The light intensity fell by a factor of 100 about 7 μs after pulse termination.

Special precautions had to be taken to ensure purity in the gas. Fig. 9.16 shows the general arrangement of the gas control and

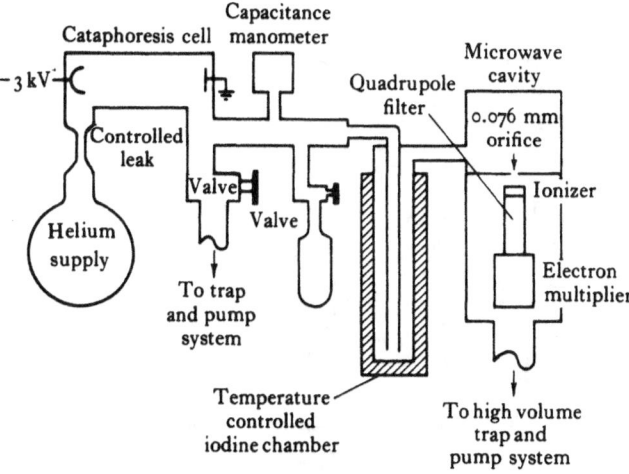

Fig. 9.16. Arrangement of the gas control and sampling apparatus in Truby's experiments (1968) on attachment in I_2.

sampling which arranged for the helium–iodine mixture to flow continuously through the cavity. Helium, purified by cataphoresis, entered the cavity through a tantalum chamber containing iodine crystals. The iodine vapour pressure in this chamber could be controlled by varying the chamber temperature. Mixing of the helium and iodine was carried out by two different methods. In one the iodine crystals were condensed in the chamber before introducing the helium while in the other the iodine was added at a vapour pressure of 0.2 torr to the helium upstream of the cavity. The iodine vapour pressure in the cavity was monitored by allowing the mixture to diffuse out of the cavity through an 0.076 mm hole into an ionizer and a quadrupole mass spectrometer. The iodine signal from the spectrometer was calibrated by measurement of the absorption of 4840 Å radiation by the mixture in the cell with the iodine vapour at a pressure of 0.2 torr, the absorption coefficient of I_2 for this radiation being known. The measured pressures agreed within about $\pm 5\%$ with those determined from the temperature of the iodine chamber.

The gas-handling system, iodine chamber, cavity and quadrupole analyser chamber were all baked out at 350 °C for 24 hours before measurements. Immediately after baking the pressure was 3×10^{-9}

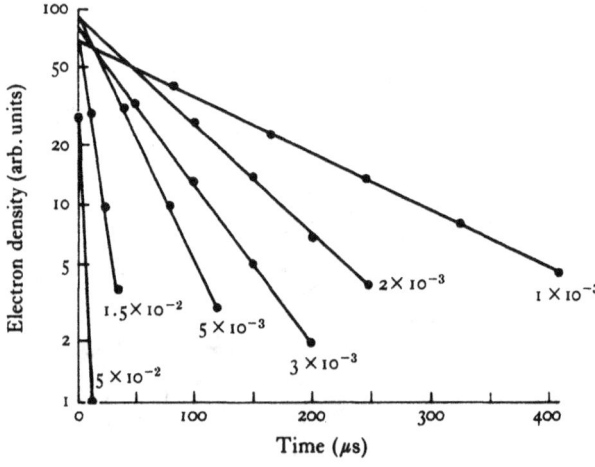

Fig. 9.17. Observed decay of electron concentration with time in the afterglow for different vapour pressures of I_2, as indicated in torr, in 20 torr of He. From Truby (1968).

torr and rose at about 10^{-9} torr s^{-1} with the system isolated. It was verified that elimination of the cataphoresis stage made no significant difference in the results while the quadrupole analyser showed no evidence of contamination of the iodine by either chlorine or bromine. The iodine vapour pressures used ranged from 10^{-3} to 5×10^{-2} torr. The buffer helium pressure was about 20 torr and the flow rate through the cavity about 1 cm^3 s^{-1}.

Fig. 9.17 shows the observed decay of electron concentration with time in the afterglow for different pressures of iodine, the initial values being of order 10^8 cm^3 s^{-1}. It will be seen that the decay rate increases with the iodine pressure. Assuming two-body attachment to be the sole cause of the decay, a plot of attachment frequency, derived from the results of Fig. 9.17, against iodine pressure should be a straight line with 45° slope. Fig. 9.18 shows that this is very well satisfied except for a small deviation at the lowest pressure of 10^{-3} torr which is outside the estimated experimental error of 5%. To account for this, experiments were carried out to determine the contribution due to diffusion which might

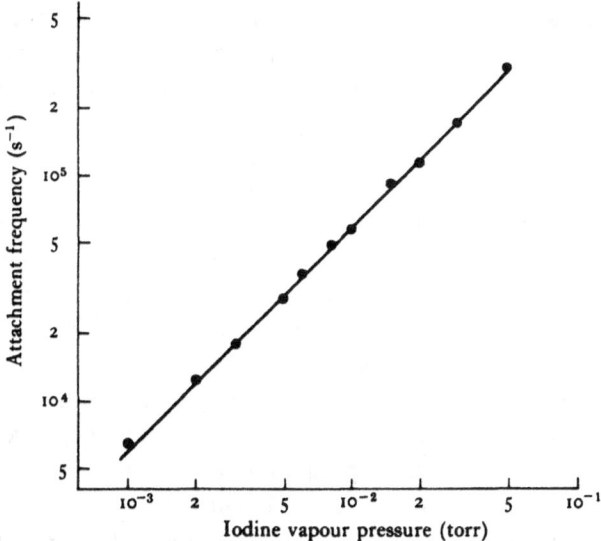

Fig. 9.18. Variation of electron attachment frequency with iodine vapour pressure, derived from data shown in Fig. 9.17.

not be negligible at these low pressures. Measurements were carried out with a very low iodine pressure and helium pressures ranging from 2.3 to 20 torr to determine the ambipolar diffusion coefficient. This can be done because under these conditions the decay frequency of the electrons is given by

$$\nu = \nu_a + D_a p/p\Lambda^2,$$

where $D_a p$ is a constant. A plot of ν against $1/p$ gives D_a. In this way $D_a p$ was found to be 725 cm² s⁻¹ torr⁻¹. This value yields a 10% correction to the observed ν_a for 10^{-3} torr of iodine but negligible modification at pressures greater than 2×10^{-3} torr.

The derived value of the attachment coefficient is 1.8×10^{-10} cm³ s⁻¹ at 295 °K corresponding to a mean attachment cross-section of 1.7×10^{-17} cm². Experiments were also carried out with pulses repeated at a frequency of 62.5 Hz and an iodine pressure of 10^{-3} torr. It was found that the rate of decay of electron concentration increased with the number of preceding pulses, reaching a maximum about six times greater than that obtained after a single ionizing pulse and remaining high for a large number of pulses. Build-up of negative-ion concentration is likely to be one important factor responsible for these high decay rates. The results obtained in these experiments and in similar experiments in bromine and nitric oxide are discussed in relation to results obtained in other experiments on p. 340–1.

A second method of obtaining attachment coefficients from afterglow observations has been developed by Puckett, Kregel and Teague (1971) which involves time-resolved measurements of the currents of positive and negative ions which diffuse through a small orifice in the wall of the cavity into a mass spectrometer. As described earlier, during the initial phase of the afterglow, negative ions are mainly prevented from reaching the walls by the repulsive field between the axis (or centre) of the discharge and the walls. This field is set up to balance the loss of negative charge to the walls through electron diffusion with that of positive charge which is assisted out by the field. Eventually, as the negative-ion concentration builds up relative to that of the electrons, a situation is reached in which electron loss no longer controls the field distribution. Under these circumstances equal rate of loss of positive and negative charge to the walls can only occur if the field collapses so that negative ions can diffuse out at an equal rate to the positive. The onset of this phase will be marked by a sudden increase in the negative-ion current diffusing into the mass spectrometer from an earlier relatively very small value. Puckett *et al.* (1971) show that the time in the afterglow at which this transition occurs may, under suitable conditions, be directly related to the attachment frequency.

Thus if n^+, n^-, n_e are the concentrations of positive and negative ions and of electrons respectively in the afterglow,

$$n^+ = n^- + n_e, \tag{9.41}$$

NEGATIVE ION FORMATION – DISSOCIATIVE PROCESSES 307

and in the first phase (electron-loss controlled)

$$\frac{dn^+}{dt} = -\nu_d n^+, \qquad (9.42)$$

where ν_d is the diffusion frequency D_a/Λ^2 if only one positive ion is present.

If detachment is negligible we also have, in the same phase,

$$\frac{dn^-}{dt} = \nu_a n_e, \qquad (9.43)$$

where ν_a is the attachment frequency.

It follows from (9.41), (9.42) and (9.43) that, in the first phase,

$$\frac{dn^-}{dt} + \nu_a n^- = \nu_a n^+(0) \exp(-\nu_d t), \qquad (9.44)$$

where $n^+(0)$ is the initial positive-ion concentration. Solving (9.44) we obtain

$$n^- = n^+ \{\nu_a/(\nu_a - \nu_d)\}\{1 - \exp(\nu_d - \nu_a)t\}. \qquad (9.45)$$

The transition between the two phases will occur at a time T such that $n_e \simeq 0$, so, substituting $n^- = n^+$ in (9.45),

$$T(\nu_a - \nu_d) = \log(\nu_a/\nu_d). \qquad (9.46)$$

As ν_d may be measured by the usual technique, ν_a may be obtained if T can be determined.

In practice more than one ion may be present, in which case the theory needs amplification though remaining basically the same.

Puckett and his colleagues (1971) carried out experiments on these lines in nitric oxide. The cavity was a stainless steel cylinder of 18 in. diameter and 36 in. length, which could be baked and maintained under ultrahigh vacuum conditions. Ionization was produced by photoionization with pulsed krypton resonance radiation (1165–1236 Å). Fig. 9.19 illustrates the observed time variation of currents of positive and of negative ions diffusing through the wall orifice during the afterglow period. The experimental conditions are indicated.

At times in the afterglow less than 170 ms, the currents of both positive ions NO^+ and $(NO)_2^+$ varied smoothly with time in the expected way. During this period the current of negative ions, exclusively NO_2^-, was small and fluctuated markedly. However, soon after 170 ms this current suddenly increased by two orders of magnitude and then decayed smoothly out to 500 ms. The positive-ion currents on the other hand fluctuated much more markedly at times beyond 170 ms. It seems clear that the change in phase occurred close to $T = 170$ ms. Puckett et al. chose T as the time at which the rate of change of positive-ion current was a maximum.

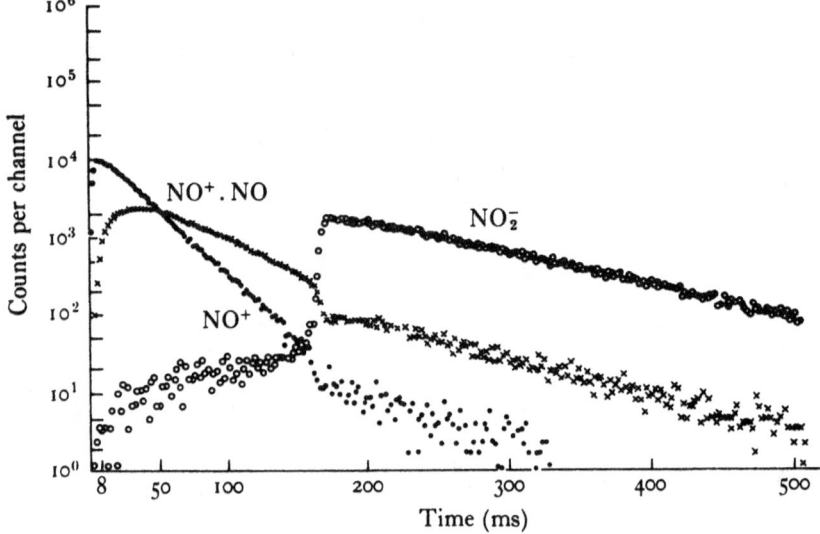

Fig. 9.19. Variation with time in the afterglow of currents of NO^+, $NO^+.NO$ and NO_2^- ions diffusing through an orifice in the cavity wall, observed by Puckett et al. (1971) in NO. Pressure 5×10^{-2} torr, pulse length 8 ms, pulse interval 1 s, orifice potential ± 50 mV.

This is based on the fact that, as the repelling field breaks down, any remaining electrons will quickly be lost to the walls and this will be apparent in an equally rapid loss of positive ions.

Experiments were carried out over a range of NO pressures from 50 to 700 torr with measurable traces of NO_2 ranging up to 10^{-3} torr also present. From these results it was possible to determine separately the contributions to the attachment frequency from NO and NO_2. It was found that the best fit to the data was given by assuming the attachment processes to be a two-body one to NO_2

NEGATIVE ION FORMATION – DISSOCIATIVE PROCESSES 309

with a rate constant of $(1.4 ^{+2.0}_{-0.9}) \times 10^{-11}$ cm^3 s^{-1} and a three-body one to NO with a rate constant of $(6.8 \pm 3.4) \times 10^{-32}$ cm^6 s^{-1}. Discussion of these results and their relation to measurements made using other techniques is given on pp. 340 and 365.

Attachment rates from measurements in flowing afterglows

The flowing afterglow technique, which has been so successfully applied to the measurement of the rates of ionic reactions under thermal and near-thermal conditions, has also been applied to the measurement of attachment rates. A description of the technique is given in Chapter 12, p. 526. In principle, buffer gas flowing at a velocity v down a tube is ionized at some stage and the ions are carried with the flow past a section of the tube at which reactant gas is injected into the tube at a measured flow rate. The gain or loss of ions through reaction with the injected gas is determined by a mass spectrometer at the end of the flow tube. From observations at different injected flow rates of reactant gas the rate of the particular reaction can be determined.

In particular, Fehsenfeld (1970) applied this technique to measure attachment coefficients of thermal electrons in SF_6. This is so high that it was not possible to inject pure SF_6 at a sufficiently small flow rate to avoid complete depletion of the electrons. Instead, analysed mixtures of traces (near 0.05%) of SF_6 in Ar and He were used. As a check on the data obtained from observation of the increase of SF_6^- current with increase in injected SF_6 flow, measurements were made of the reduction of electron density using Langmuir probes and found to correlate well. The attachment rate coefficient, at pressures between 0.1 and 1.5 torr of the buffer He gas, was found to be nearly independent of temperature between 300 and 500 °K. Discussion of these results in relation to other data for attachment to SF_6 is given on p. 372.

9.5 Discussion of results of experiments on attachment of electrons to diatomic molecules

H_2, HD and D_2

The cross-sections for dissociative attachment to H_2 and the isotopic molecules are very small and special precautions must be taken to ensure that observed effects do not arise from impurities, particularly water vapour, which give rise to H⁻ ions. Thus, in 1930, Lozier, applying the Tate–Lozier technique (see p. 274) to H_2,

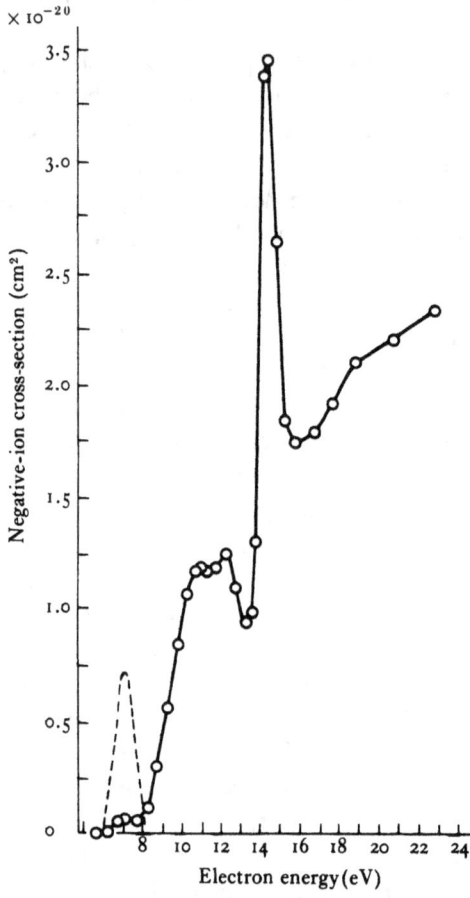

Fig. 9.20. Variation with electron energy of negative-ion production by electron impact in H_2, observed by Schulz (1959). ---- additional production observed when the reagent grade hydrogen was introduced without a liquid air trap.

observed two peaks at electron energies of 6.6 and 8.8 eV, both of which he ascribed to H_2O rather than to H_2. It was not until 1959 that Schulz applied ultrahigh vacuum techniques to investigate total ion production in H_2. With the whole apparatus baked at 400 °C, a background pressure as low as 2×10^{-10} torr was obtained without a liquid air trap, although in the actual experiments such a trap was employed to reduce the water vapour pressure. Fig. 9.20 illustrates the results obtained. Without the liquid air trap a further peak near 6.8 eV was obtained, presumably due to water vapour.

A few years after Schulz obtained these results, Rapp, Sharp and Briglia (1965) measured the total cross-section for negative-ion formation as a function of electron energy, not only in H_2 but also in

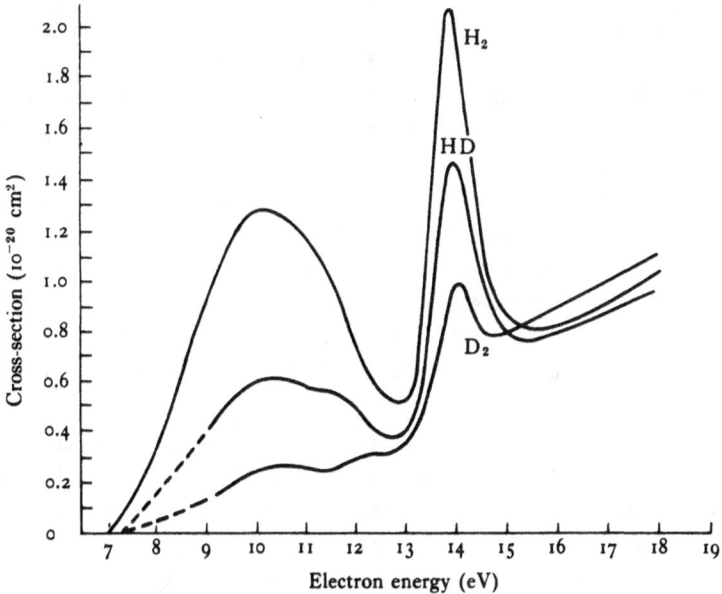

Fig. 9.21. Cross-sections for negative-ion formation in H_2, HD and D_2 as functions of electron energy, observed by Rapp et al. (1965).

the isotopic molecules HD and D_2. Their results, illustrated in Fig. 9.21, show a strong isotope effect. An even more pronounced effect of this kind was observed by Schulz and Asundi (1965), who were particularly concerned with checking the validity of some earlier evidence that a weak attachment process also occurs at an electron energy of about 3.7 eV. For this purpose they worked with pressures of H_2 up to 5×10^{-2} torr by proper differential pumping arrangements. Their apparatus included a mass spectrograph and they definitely observed in H_2 a dissociative attachment process with an appearance potential of 3.75 ± 0.07 V. The ions concerned were certainly H⁻, and the intensity of their production was proportional to the H_2 pressure. It was further found that this attachment process shows a very strong isotope effect as may be seen by reference to Fig. 9.22 which shows the attachment cross-sections observed in H_2, HD and D_2. Thus for HD the peak cross-section is about 0.06 of that for H_2 while for D_2 it is an order of magnitude smaller still.

These remarkable results may be interpreted in terms of the potential-energy curves for H_2 and H_2^- already discussed in Chapter 6, p. 176 and Chapter 7, p. 222. We reproduce in Fig. 9.23 the probable disposition of the two curves for states of H_2^- which dissociate into ground state H and H⁻, relative to the ground $^1\Sigma_g^+$ state of H_2.

According to the Franck–Condon principle the probable transitions by capture to the lowest vibrational level of H_2 lie within the vertical lines indicated. These cut the curve for the ground $(1s\sigma_g)^2(2p\sigma_u)^2 \Sigma_u^+$ state of H_2^- over a range of nuclear separations over which it is rendered very diffuse by autodetachment. A transition satisfying the Franck–Condon principle will lead to production of H⁻ ions of zero kinetic energy when the electron energy is equal to $D - E_a$, where D is the dissociative energy of H_2 (4.44 eV) and E_a the electron affinity (0.75 eV) of H. This agrees within experimental error with the threshold for the weak process observed by Asundi and Schulz.

The validity of this interpretation can be checked from the observed isotope effect. As described on p. 268, the attachment cross-section Q_{da} can be written

$$Q_{da} = Q_c \exp(-\bar{\Gamma}\tau/\hbar), \qquad (9.47)$$

where Q_c is the cross-section for the initial capture of the electron, $\bar{\Gamma}$ is the mean width of the level of H_2^- due to autodetachment and τ is the time taken for the atom and ion to separate over the range of nuclear separation from the initial capture to the separation R_c beyond which the negative ion is stable. For an atom and ion of masses M_1, M_2 respectively, τ will be proportional to $\{M_1 M_2/(M_1 + M_2)\}^{1/2}$, while the other quantities are effectively independent of the masses. Substituting the observed values of Asundi

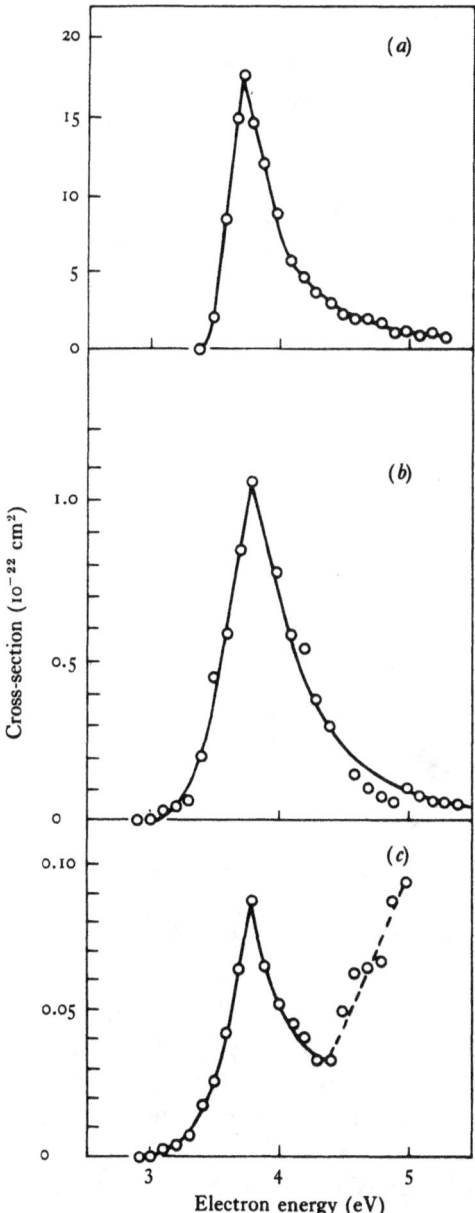

Fig. 9.22. Cross-sections for negative-ion formation in H_2, HD and D_2 by low-energy electrons, as observed by Schulz and Asundi (1965). (a) H^- from H_2, (b) D^- from HD, (c) D^- from D_2.

and Schulz we have

$$Q_{da}(H_2)/Q_{da}(HD) = \frac{1.6 \times 10^{-21}}{2.1 \times 10^{-22}} = \exp\left[\frac{\bar{\Gamma}\tau}{\hbar}\{(1.33)^{1/2} - 1\}\right] \quad (9.48)$$

$$Q_{da}(H_2)/Q_{da}(D_2) = \frac{1.6 \times 10^{-21}}{8 \times 10^{-24}} = \exp\left[\frac{\bar{\Gamma}\tau}{\hbar}\{2^{1/2} - 1\}\right]. \quad (9.49)$$

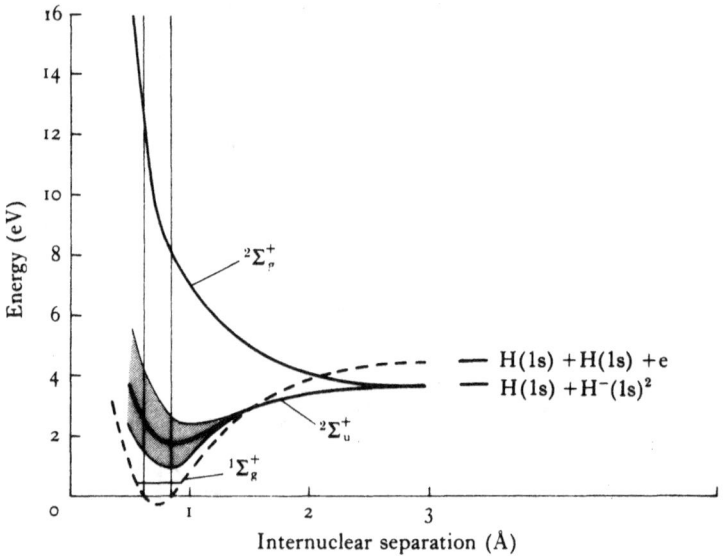

Fig. 9.23. Potential-energy curves of H_2^- responsible for dissociative attachment in H_2. The broken curve is that for the ground Σ_g^+ state of H_2.

From (9.48) we find $\bar{\Gamma}\tau = 8.6 \times 10^{-15}$ eV s and from (9.49) 8.4×10^{-15} eV s, showing a remarkable degree of consistency. Moreover, τ may be estimated as of the order 10^{-14} s giving $\bar{\Gamma}$ close to 1 eV. This very large width, corresponding to a lifetime not much greater than the time taken for an electron of 4 eV energy to traverse the molecule, is to be expected – the ground state of H_2^- near the equilibrium nuclear separation is of type II according to the definition outlined in Chapter 4. It is also quite close to that expected theoretically as may be seen by reference to Fig. 6.4 of

NEGATIVE ION FORMATION – DISSOCIATIVE PROCESSES   315

Chapter 6. Finally it is also to be noted that, on substitution of the derived value of $\bar{\Gamma}\tau$ in (9.4), we find

$$Q_c = 9 \times 10^{-16} \text{ cm}^2,$$

which is not unreasonable.

The isotope effect for the much higher peak found near 10 eV gives $\bar{\Gamma}\tau$ between 2.5×10^{-15} and 3×10^{-15} eV s so $\bar{\Gamma} \simeq 0.2$ to 0.3 eV. Although smaller than for the small peak at 3.7 eV, $\bar{\Gamma}$ is still so large that it seems probable that the H_2^- state concerned is also of type II. It is natural to suppose that this state is the $(1s6_g)(2p\sigma_u)^2\,{}^2\Sigma_u^+$ state which also dissociates to ground state H and H^-. If it is of type II it must lie a little above rather than a little below the $b^3\Sigma_u$ repulsive state of H_2 (see Fig. 6.3), at least over a considerable range of nuclear separation. This would be consistent with the fact that in the united atom limit this is certainly true, the H_2 state tending to the $1s2p2^3P$ state of He which certainly lies below the $1s2p^2\,{}^2D$ excited state of He^-. Although the theory cannot yield very reliable results for the real and imaginary parts of the potential-energy curve for this state, the discussion in Chapter 6, p. 178, shows that the conclusions derived from the observations are not inconsistent with the theory.

Further evidence is provided by use of the formula (9.4) for the attachment cross-section. To agree with the observed peak cross-section the width Γ_0 due to autodetachment to the ground state must be close to 0.004 eV. If this were the only autodetachment process possible, no isotope effect would be observed. It follows that a further state of H_2 must lie close below the H_2^- state.

The minimum kinetic energy of ions produced in this transition is about 2.5 eV, so the half-width of the energy spread due to the temperature motion is about 0.3 eV at the working temperature of 77 °K. Nevertheless the linear extrapolation technique (see p. 276) does give the correct value 3.7 eV for the appearance energy $D - E_a$ for ions of zero kinetic energy.

The isotope effect of the 14.2 eV peak gives $\bar{\Gamma}\tau$ as only a little smaller than for the peak at 10 eV, suggesting again that the upper state is of type II despite the fact that type I excited states of H_2^- have certainly been observed (see Chapter 7, p. 223).

The rise in the cross-section for negative-ion formation at

electron energies beyond 17 eV is probably due to polar dissociation into $H^+ + H^-$, the threshold energy for which, following (9.12), is 17.2 eV.

O_2

Fig. 9.24 illustrates the variation with electron energy of the cross-section for production of negative ions in O_2 as observed by Rapp and Briglia (1965) using the apparatus described on p. 278. The general form of the variation is essentially the same as observed in

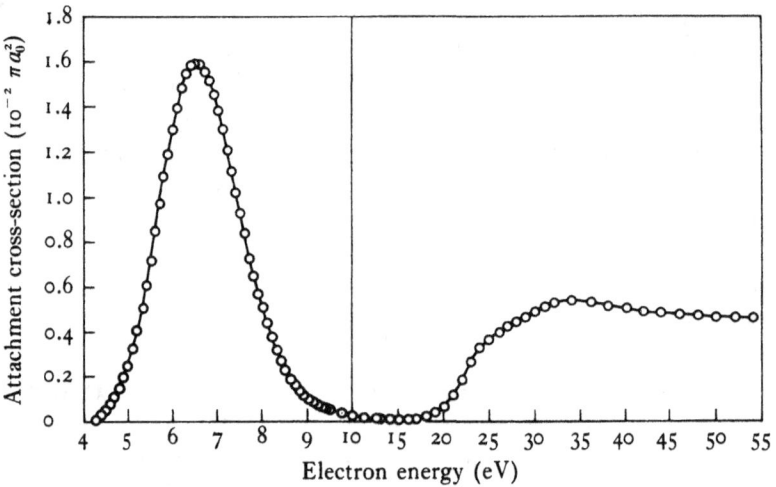

Fig. 9.24. Cross-sections for negative-ion production by electron impact in O_2, observed by Rapp and Briglia (1965).

the much earlier experiments of Lozier (1934) and of Hagstrum and Tate (1941).

The sharp peak at an energy of 6.5 eV is due to dissociative attachment, the cross-section for which is much larger than for attachment to H_2. Results obtained by different observers agree quite well as to the location, height and width of the peak.

Similar good agreement between different observers applies to the process with an onset potential near 16 eV. It is one of polar dissociation (p. 272)

$$e + O_2 \longrightarrow O^+ + O^-. \qquad (9.50)$$

With the dissociation energy $D(O)_2 = 5.11$ eV, the ionization energy

$E_i(O)$ of O 13.54 eV, the electron affinity $E_a(O)$ of O is 1.46_5 eV and the onset potential for ions of zero kinetic energy is 17.18 eV. Confirmation that the process is indeed (9.50) is provided by observations of the variation with electron energy of O^+ production in O_2, using mass analysis to distinguish between O_2^+ and O^+. A threshold for this production is observed close to 17.18_5 eV.

Chantry and Schulz (1967) have used the equipment described on p. 282 to measure the energy distribution of the ions produced in dissociative attachment in O_2, particularly with the aim of verifying

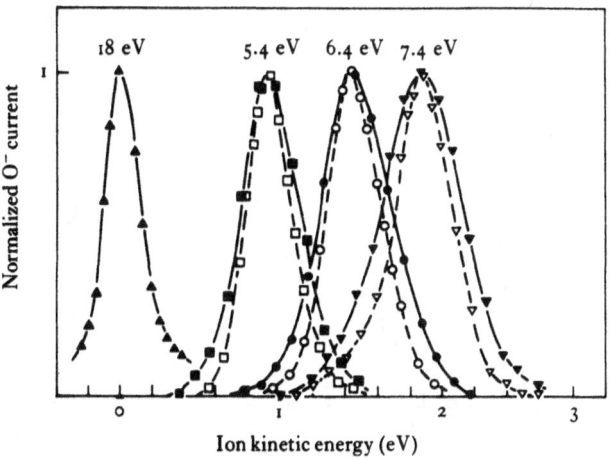

Fig. 9.25. Energy distributions of ions produced by impact in O_2 of electrons, of different kinetic energies as indicated, observed by Chantry and Schulz (1967). ▲, ■, ●, ▼ observed at room temperature, □, ○, ▽ observed with the gas inlet cooled by liquid nitrogen.

that temperature motion of the target molecules leads to the effects described on p. 276. Fig. 9.25 illustrates observed distributions for electrons of different kinetic energy measured at room temperature and also with the inlet gas cooled by liquid nitrogen. It will be seen that the widths of the distributions are less at the lower temperature. To calibrate the ion energy scale the distribution was measured for electrons of 18 eV energy, in which case the ions arise from polar dissociation. The sharpness of the rise from threshold for this process together with the measured value of the onset potential, the insensitivity of the distribution to the electron energy and the relative narrowness of the distribution all suggest strongly that the peak occurs at zero ion energy. Assuming this, the ion energy scale is as shown in Fig. 9.25.

The electron energy scale was calibrated by reference to a peak energy of the dissociation process, the onset threshold for the polar process and the threshold for positive-ion production. Agreement with all these was obtained by applying the same correction to the electron accelerating voltage.

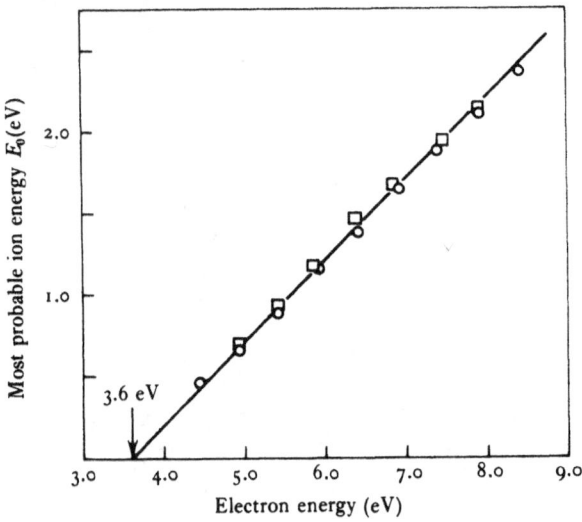

Fig. 9.26. Most probable ion energy as a function of electron energy for dissociative attachment in O_2, observed by Chantry and Schulz (1967). □, ○ two separate sets of observed results.

According to the theory outlined on p. 276, a plot of the most probable ion energy as a function of electron energy should have a slope of $\frac{1}{2}$ and extrapolate to zero at an electron energy equal to $D(O_2) - E_a(O)$. Fig. 9.26 shows such a plot obtained by Chantry and Schulz. It is indeed of slope $\frac{1}{2}$ and gives $D(O_2) - E_a(O) = 3.6$ eV which agrees within experimental error with the expected value 3.64_5 eV.

The theory of p. 276 predicts that the energy width $W_{1/2}$ of the ion energy distribution at half-maximum is given by

$$W_{1/2}^2 = 11\beta(1-\beta)kT\{E_e - (D - E_a)\}, \qquad (9.51)$$

where $\beta = \frac{1}{2}$ for O_2 and E_e is the electron energy. In practice allowance must be made for the width $W_{1/2,\,i}$ introduced by imperfect instrumental resolution, so we should write, for the observed half-width $\tilde{W}_{1/2}$,

$$\tilde{W}_{1/2}^2 = W_{1/2}^2 + W_{1/2,\,i}^2, \qquad (9.52)$$

where $W_{1/2}$ is given by (9.51). Fig. 9.27 shows plots of $\tilde{W}_{1/2}^2$ as a function of electron energy observed at room temperature and with the inlet gas cooled. As expected, in both cases a linear relation is

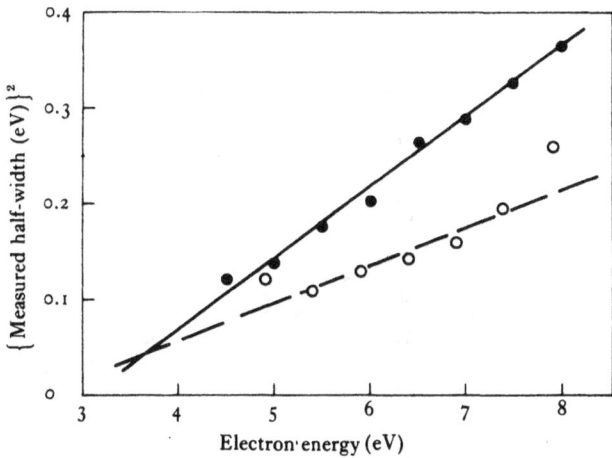

Fig. 9.27. Measured half-width of the ion-energy distribution as a function of electron energy for dissociative attachment in O_2, observed by Chantry and Schulz (1967). –●– observed at room temperature, –○– observed with the gas inlet cooled in liquid nitrogen.

found. The two straight lines intersect at an electron energy which is consistent with that, $D - E_a$, predicted from (9.51). The value of $\tilde{W}_{1/2}^2$ at this intersection should be equal to $W_{1/2,\,i}^2$. This gives $W_{1/2,\,i} \simeq 0.2$ eV which is consistent with the value found from observation of the energy distribution of O_2^+ ions which are almost certainly formed with negligible kinetic energy.

Turning now to the nature of the upper potential-energy curve of O_2^- which is involved in the dissociative attachment process, we see, by reference to Fig. 6.12, that there are many possibilities.

A Σ^- to Σ^+ transition is forbidden so that, since the ground state of O_2 is a $^3\Sigma_u^-$ state, we can exclude the upper Σ^+ state of O_2^-.

Nevertheless this leaves Π_u, Π_g, Δ_u, and Σ_u^- as possibilities. As discussed on p. 271, some further selection may be made if observations can be carried out of the angular distribution of the ions. Such experiments have been carried out by van Brunt and Kieffer (1970) using the equipment described on p. 288. Fig. 9.28 shows results which they obtained for electron energies ranging from 5.75

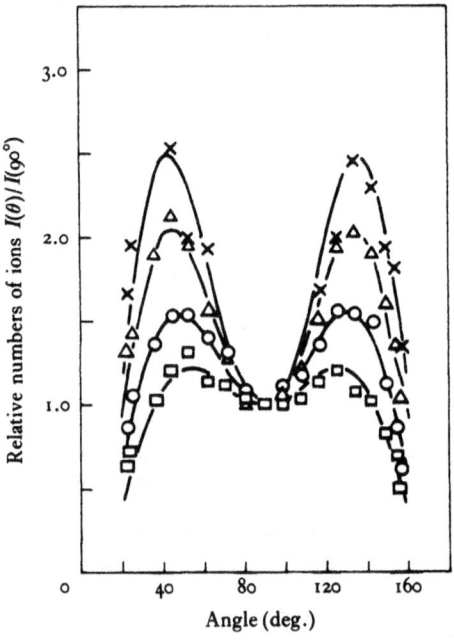

Fig. 9.28. Angular distributions of O⁻ ions produced by dissociative attachment in O_2, observed by van Brunt and Kieffer (1970). ×, △, ○, □ observed points for electron energies of 8.40, 7.80, 6.70 and 5.75 eV respectively; —— calculated assuming the form (9.53) with suitable choice of α.

to 8.40 eV. It will be seen that the angular distributions fall off very rapidly at angles θ near 0 and 180°. This excludes Σ_u^- which gives a distribution the dominant term of which varies as $\cos^2\theta$. The distribution was found to be well represented by

$$(\sin\theta + \alpha \sin\theta \cos^2\theta)^2, \tag{9.53}$$

where α decreases from 0.35 to 0.187 as the electron energy decreases from 8.40 to 5.75 eV. This is consistent with the assumption that the upper state is a Π_u state. For this, $\sin^2 \theta$ is the dominant term with the next term of the same form as in (9.53). The importance of the second term should decrease with decrease of electron energy, as observed. While possible combinations arising from other states cannot be excluded, they give less-satisfactory agreement with the observed data.

Evidence about the slope of the upper potential-energy curve at nuclear separations close to the equilibrium separation of the ground state is available from the somewhat unexpected observation of a strong temperature effect for the dissociative attachment process. This was first observed by Fite and Brackmann (1963), who were using a crossed-beam apparatus, similar in form but somewhat less refined than that shown in Fig. 9.9. The aim of the experiments was to try to observe radiative attachment of electrons to free oxygen atoms, produced in either a discharge or oven source. To avoid production of O^- by dissociative attachment from residual undissociated O_2 they proposed to work with electron energies below the threshold 3.7 eV for the dissociative process. Carrying out check experiments they found, however, that not only was the threshold reduced to almost zero electron energy with the discharge source but also was greatly reduced with the oven source heated to 2100 °C. The former result is readily understood for the discharge will produce metastable O_2 molecules in $^1\varDelta_u$ states. Dissociative attachment to such molecules can be expected to occur at much lower electron energies (see below).

The profound effect of temperature was much more surprising, so Henderson *et al.* (1969) carried out experiments to investigate the matter further, using the arrangement described on p. 286 and illustrated in Fig. 9.9, specially designed for the purpose. Fig. 9.29(a) shows results which they obtained for the variation of the dissociative attachment cross-section with electron energy for four different temperatures ranging from 300 to 1930 °K. It will be seen that, not only does the threshold decrease from near 4 to about 1.2 eV over this temperature range but there is also a considerable increase in the peak value as well as a decrease of the electron energy at the peak.

Henderson *et al.* (1969) also carried out experiments to determine whether rotational excitation played a vital part in these effects. This

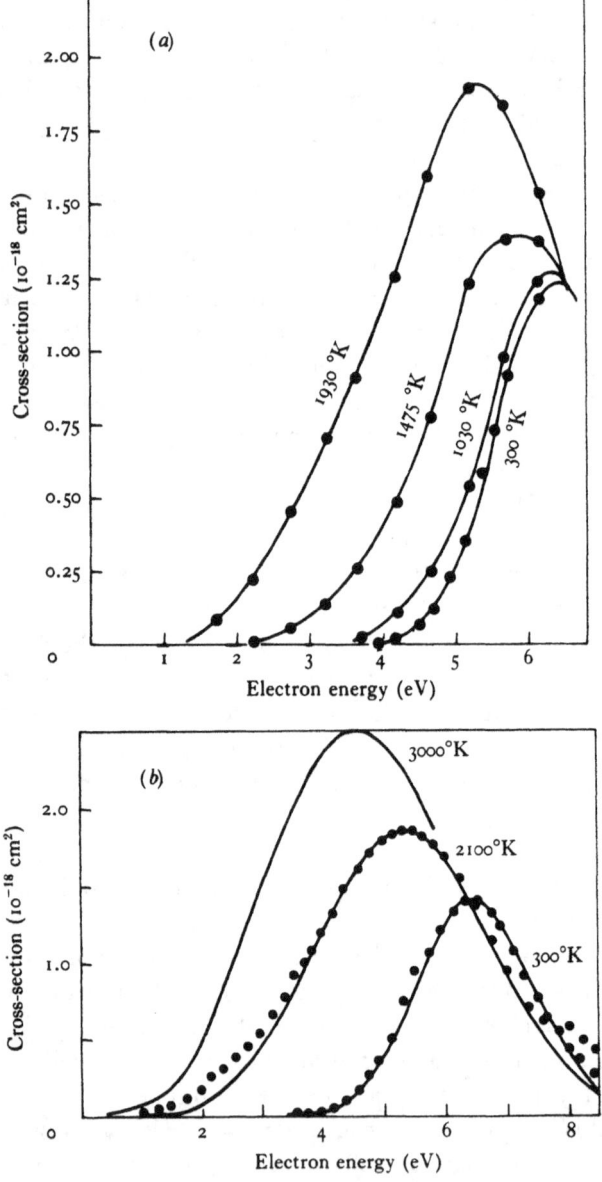

Fig. 9.29. Total cross-sections for dissociative attachment of electrons to O_2 as function of electron energy for different gas temperatures as indicated. (a) –●–●– observed by Henderson et al. (1969); (b) —— calculated by O'Malley (1967), ● observed by Fite and Brackmann (1963).

they were able to do by comparing results obtained with different exit apertures and oven pressures. With low pressures and a wide aperture, so that effusive conditions apply, vibrational and rotational temperatures are high and equal to the furnace temperature. On the other hand with a high pressure p and a small aperture diameter d ($pd \simeq 5$ torr mm), the vibrational temperature remains high but the rotational is greatly reduced, to about 600 °K, for a furnace temperature of 2000 °K. No change occurred in the observed results with the small aperture, indicating that rotational excitation plays no important role.

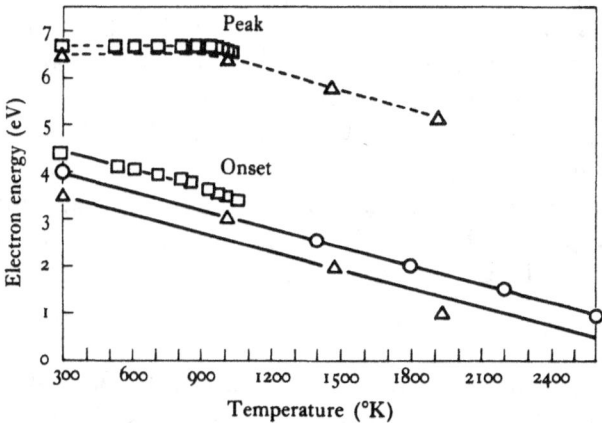

Fig. 9.30. Variation with temperature of the electron energy at the onset and peak of the dissociative attachment cross-section in O_2. –△– observed by Fite, Brackmann and Henderson, –□– observed by Spence and Schulz, –○– calculated by O'Malley (onset only).

Experiments to observe the temperature effect have also been carried out by Spence and Schulz (1969), using the equipment described on p. 281 (see Fig. 9.6). Fig. 9.30 shows the corresponding results for the electron energy at onset and at the peak.

O'Malley (1967) applied the theory outlined on p. 267 to interpret these results. Despite the fact that the fraction of molecules possessing 1 eV of excitation energy is not more than 2.5×10^{-3} at 2000 °K the presence of even the limited amount of vibrational excitation at this temperature can lead to a profound change in the survival factor $\exp\{-\bar{\Gamma}\tau/\hbar\}$ in (9.4).

Assuming that the vibrational and rotational states are in thermal equilibrium at the temperature T, the observed attachment cross-section for electrons of energy E is given by

$$\bar{Q}_{da}(T, E) = N\Sigma_v \Sigma_J \exp\{-(E_v + E_J)/kT\} Q_{da}^{v;J}(E), \quad (9.54)$$

where

$$N^{-1} = \Sigma_v \Sigma_J \exp\{-(E_v + E_J)/kT\}.$$

$Q_{da}^{v,J}(E)$ is the attachment cross-section for O_2 molecules in vibrational and rotational states of energies E_v, E_J respectively specified by the quantum numbers v, J. It can be calculated when the potential energy V_i in the initial state of O_2 and the potential energy V_f and width Γ_a of the final, autodetaching, state of O_2^- are known as functions of nuclear separation. V_i is well known but V_f and Γ_a have to be determined to give good agreement with observation. For this purpose O'Malley used a parametric representation for V_f in terms of an expansion in powers of ΔR, the separation from the equilibrium position for O_2. Instead of the same procedure for Γ_a, he worked in terms of the survival factor ρ of (9.3) from which Γ_a can be derived (see (9.3)). ρ was written in the form

$$\rho = (R_c - R)^n \sigma, \quad R \leqslant R_c = 0, \quad R > R_c, \quad (9.55)$$

where R_c is the separation beyond which the upper state is stable towards autodetachment. σ and n are constants to be determined.

The best fit to the observed results of Fite and Brackmann is shown in Fig. 9.29(b). The potential curve V_f determined in this way is shown in Fig. 6.12. It will be seen that it clearly indicates that the upper state possesses a minimum below the dissociation limit, probably around 1 eV in depth. As discussed above, the upper state is probably a $^2\Pi_u$ state.

R_c is found to be 1.44 Å and the level width is approximately given by $4(R_c - R)$ eV, where R_c and R are in Å. The partial level width $\Gamma_{a,\chi}$ for autodetachment to the ground state of O_2 may be determined, using (9.4), from the measured absolute cross-section, and is found to be $0.034E$, where E, the electron energy, is in eV.

The effect of rotational excitation is small, as may be seen from Fig. 9.31 in which contributions from different factors to the cross-section at 2100 °K are shown. It will be seen that the variation of the survival factor with temperature is solely responsible for the shift in energy of the peak. Without a significant variation of ρ with T, as is the case for many molecules, increase of temperature will produce a broadening of the cross-section curve but no shift of the peak.

Burrow (1973) has measured the cross-section for dissociative attachment to $O_2(a\,^1\Delta_g)$ metastable molecules. The experiment was carried out by crossing the electron beam with a stream of

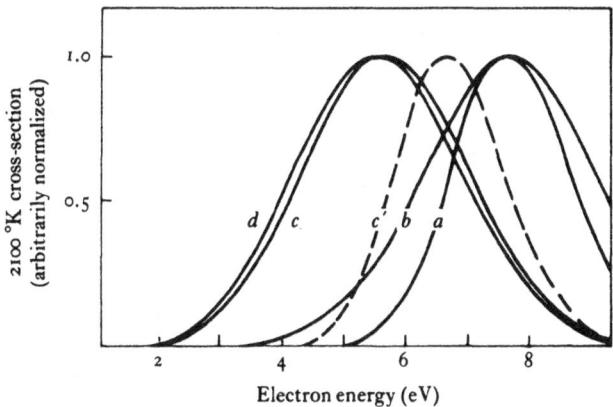

Fig. 9.31. Dissociative attachment cross-section for electrons in O_2 at 2100 °K calculated with different approximations (from O'Malley, 1967). Curves a, no excited vibration or rotation and survival factor $\rho = 0$; b, includes excited vibrational states; c, as for b but including a finite survival factor; c', curve a modified to include the survival factor; d, as for c but including rotation. The absolute values are normalized to unity at the peak. Curves a and b represent cross-sections nearly two orders of magnitude greater than for the other cases.

oxygen which has been passed through a microwave discharge. This discharge produced not only a $^1\Delta_g$ molecules but also molecules in vibrationally excited ground states and in the upper metastable b $^1\Sigma_g^+$ state. By locating the interaction region sufficiently far from the discharge, the proportion of the undesired excited species was reduced to negligible proportions, advantage being taken of the fact that a $^1\Delta_g$ molecules are remarkably stable both against deactivation by impact with normal O_2 (Becker, Groth and Schurath, 1971) and with Pyrex glass walls (Clark and Wayne, 1969).

The proportion of a $^1\Delta_g$ molecules present was determined by observing the fractional reduction in the intensity of excitation of a particular state of O_2 from the ground state by electron excitation in a trapped-electron experiment (Schulz and Dowell, 1962), when the discharge was turned on. The presence of a $^1\Delta_g$ molecules not only leads to production of O^- ions below the threshold for production from ground state O_2 but also to positive ionization at energies below the ionization energy of O_2. It was verified that the ratio of the negative- and positive-ion currents at chosen electron

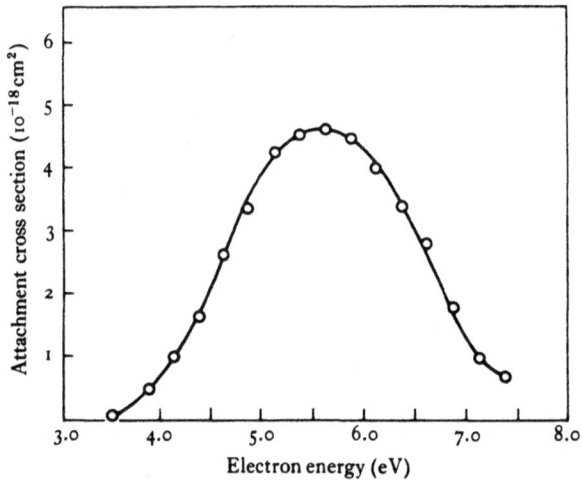

Fig. 9.32. Total cross-section for dissociative attachment of electrons to O_2 (a $^1\Delta_g$) metastable molecules, observed by Burrow (1973).

energies below the respective thresholds remained constant as experimental parameters affecting the a $^1\Delta_g$ fraction were changed.

Fig. 9.32 shows the observed detachment cross-section for $O_2(a\,^1\Delta_g)$. It has nearly the same shape as for ground state O_2 if the electron energy scale for the latter were reduced by 0.98 eV, the excitation energy of a $^1\Delta_g$. Analysis of the data in terms of the same theory as used by O'Malley for $O_2(^3\Sigma_g^-)$ shows that the decay constant for autodetachment from the $^2\Pi_u$ state of O_2^- to the a $^1\Delta_g$ state is approximately the same as to the ground state.

So far we have been discussing results obtained from beam experiments which did not provide evidence of the occurrence of processes leading to negative-ion formation at electron energies less then 3.7 eV, at least at room temperature. However, swarm experiments carried out by Bradbury in 1933 using the filter method (see p. 292) gave results for the probability h per collision of negative-ion formation in O_2 as a function of mean electron energy which are illustrated in Fig. 9.33. It will be seen that, in addition to a peak at a mean energy close to 2 eV, the curve, after passing through a minimum near 1.2 eV, begins to rise again steadily as the mean energy is further reduced. Qualitatively similar results were obtained a few years later by Healey and Kirkpatrick (1941) using Bailey's diffusion method (p. 297).

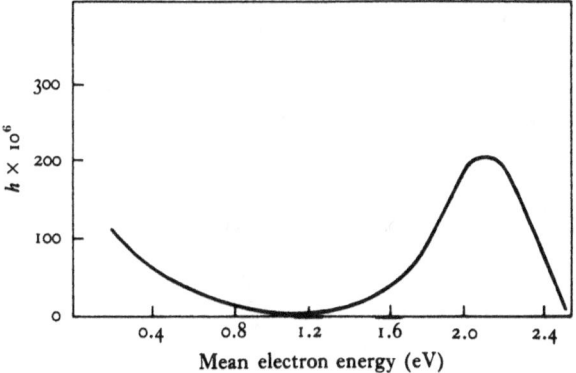

Fig. 9.33. Variation of attachment probability h per collision with mean electron energy, observed by Bradbury (1933).

While it seems likely that the peak near 2 eV is due to dissociative attachment observed in the beam experiments, the low energy behaviour suggests that a new process of formation sets in at a lower mean energy. Bloch and Bradbury (1935) proposed that this process is of a kind similar to that described as (c) on p. 270, leading to production of O_2^-. This process is a three-body one. The apparent absence of a pressure dependence, according to experiments at the time, could only be explained by assuming that, at the pressures concerned, the stabilization rate of the excited O_2 initially formed was so high as to lead to saturation (see (9.6)). This required rates of vibrational deactivation on collision which were far too large.

The introduction of further techniques of the type described on p. 293 resolved the difficulty. Thus Chanin et al. (1962) found that in fact the attachment coefficient for low-energy electrons is proportional to the square of the pressure p over the working pressure range, as required for a three-body process rate coefficient proportional to p.

Fig. 9.34 shows results obtained by Grünberg (1969), using the technique described on p. 297, for the variation of α_a/p with F/p, α_a being the attachment coefficient and F the electric field

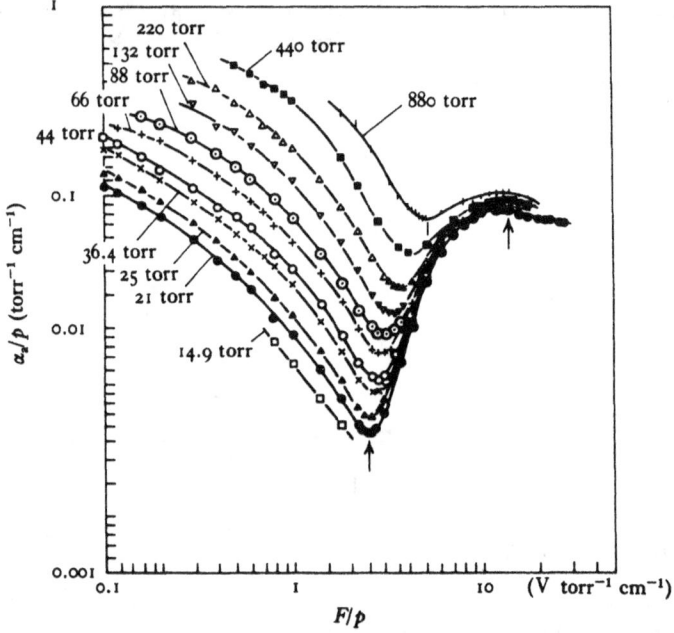

Fig. 9.34. Variation of α_a/p with F/p for electrons in O_2, observed by Grünberg (1969), at different pressures p, as indicated.

strength. For $F/p < 2.5$ V cm^{-1} torr^{-1}, α_a/p increases with p. As F/p increases beyond 2.5 V cm^{-1} torr^{-1} the curves for different processes converge to the same envelope. This is what would be expected if, for given p, the low-energy process for which α_a/p increases with p is dominated essentially by one for which α_a/p is independent of p. To examine the variation of α_a/p with p at low F/p we show in Fig. 9.35 a plot of α_a/p^2 as a function of F/p for $F/p < 3$ V cm^{-1} torr^{-1}. The curves for different p are indistinguishable for $p < 44$ torr, showing that under these conditions α_a

is proportional to p^2. At higher pressures α_a/p^2 decreases with the pressure, which would be expected if saturation is being approached. At atmospheric pressure, about 10^{10} gas kinetic collisions occur per second. Only a fraction of these will be effective in producing vibrational deactivation of the initially formed vibrationally excited O_2^-. It follows that the lifetime of this complex must be at least as long as 10^{-10} s.

The nature of the three-body attachment process described on p. 270 differs a little from that proposed initially by Bloch and Bradbury (1935) whose suggestions were made at a time when the electron affinity of O_2 was unknown.

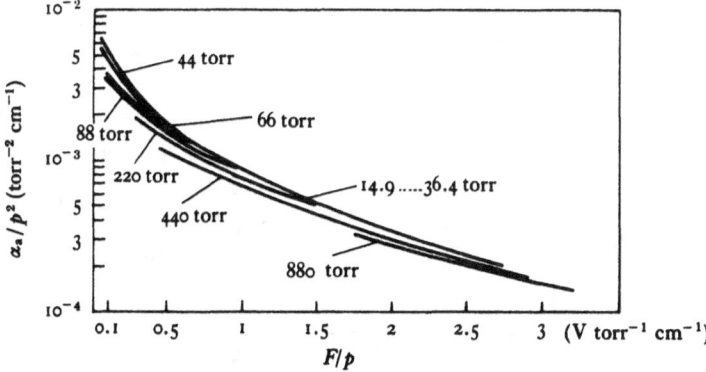

Fig. 9.35. Plot of α_a/p^2 against F/p for electrons in O_2 at different pressures p, taken from the data of Fig. 9.34.

The mechanism proposed on the basis of present knowledge of $E_a(O_2)$ is that an electron undergoes resonant capture into a vibrational state of O_2^- which lies above the ground level of O_2. If left to itself the O_2^- state will revert through autodetachment to the initial state of O_2 + free electron. However, stabilization can occur through collision with a third body, occurring before autodetachment takes place. The known properties (see Chapter 6, p. 190) of the potential-energy curves for O_2^- near to the equilibrium separation for O_2 show that the vibrational states of O_2^- with $v > 3$ lie above the ground level of O_2.

According to this picture, just as for vibrational excitation of O_2 (see Chapter 6, p. 186) we should expect the cross-section for the process to exhibit resonance maxima as a function of the electron energy E whenever $E = E_s$, where E_s is the energy excess of a vibrational state of O_2^- above the ground state of O_2. Schulz and Spence (1972) verified this expectation, using the apparatus described on p. 281.

Fig. 9.36 shows the variation of the effective cross-section with electron energy at room temperature. The maxima may be correlated directly with the energies of the vibrational states of O_2^- con-

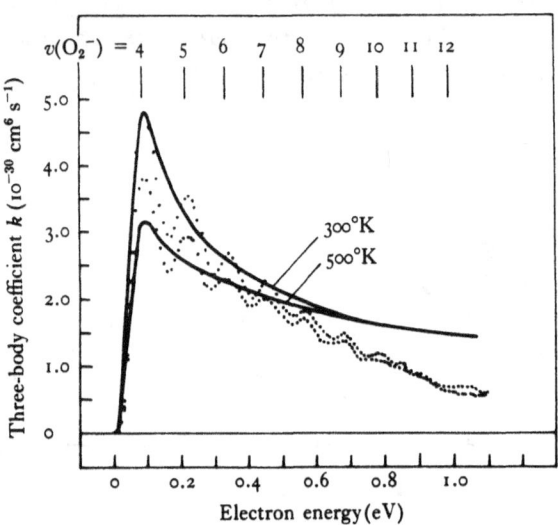

Fig. 9.36. Rate constant k for production of O_2^- from O_2 in three-body collisions with electrons at 300 and 500 °K. ——— observed by Pack and Phelps (1966), ····· observed by Spence and Schulz (1972) (normalized to agree with Pack and Phelps at the peak of their curve at 300 °K). Locations of the peaks in terms of vibrational quantum numbers v of states of O_2^- are indicated.

cerned. Comparison with results obtained by Pack and Phelps (1966), using pulse techniques, is shown in the figure, in which the three-body rate coefficients are given for temperatures of 500 as well as 300 °K. Since the swarm experiments refer to average values as functions of mean electron energy, the agreement is remarkably good. Discrepancies at high electron energies can probably be attributed to the averaging involved in the swarm data.

The absolute value of the rate coefficient at the peak, as found by comparison with the height of the peak for the dissociative attach-

NEGATIVE ION FORMATION – DISSOCIATIVE PROCESSES 331

ment process at 6.7 eV, is $(5.6 \pm 1.3) \times 10^{-30}$ cm^6 s^{-1} which agrees well with the value 4.8×10^{-30} cm^6 s^{-1} obtained from the swarm experiments. There seems little doubt that, in O_2, the low-energy attachment process is of the type (c) described on p. 278. The long lifetime of the vibrationally excited O_2^- is consistent with the analysis of resonance effects in vibrational excitation of O_2 by electron impact (see p. 188).

Swarm experiments of the pulse type have been carried out to determine how the attachment rate coefficient varies with the nature of the third body which produces collisional stabilization. For a mixture of O_2 with some other gas of molecules X, the rate of decrease of electron concentration is given by

$$\frac{dn_e}{dt} = -k_0 \{n(O_2)\}^2 n_e - k_1 n(O_2) n(X) n_e, \qquad (9.56)$$

where k_0 and k_1 are three-body rate coefficients for the processes in which O_2 and X act respectively as the third bodies, $n(O_2)$, $n(X)$, n_e being respectively the concentration of O_2, X and of electrons. The attachment frequency ν_a is then

$$\nu_a = k\{n(O_2)\}^2 + k_1 n(O_2) n(X). \qquad (9.57)$$

ν_a may be measured by a pulse technique, such as that described on p. 294, as a function of $n(X)$, $n(O_2)$ being fixed. From such observations k_1 may be obtained, k_0 being known from experiments in pure O_2 or derived by extrapolation of results to $n(X) = 0$.

Fig. 9.37 shows three-body attachment coefficients measured in this way (as a function of mean electron energy) for He, N_2, O_2, CO_2 and H_2O as third bodies. It will be seen that these range over several orders of magnitude. The low value for He is expected, both because of its small mass and its lack of internal degrees of freedom. Of special interest is the very high value found for H_2O. This means that quite small fractional concentrations of water vapour may increase considerably the rate of attachment of thermal electrons.

Using known data on the attachment rates for the three-body and dissociative attachment processes it is possible to calculate the expected attachment coefficient α_a for electron swarms as a function of F/p up to values which correspond to mean electron energies of 1–3 eV. This requires knowledge of the momentum transfer, as well as electronic excitation and vibrational excitation cross-sections, as functions of electron energy. Hake and Phelps (1966) carried out calculations of this kind using the best available information about the various cross-sections. Their results for α_a

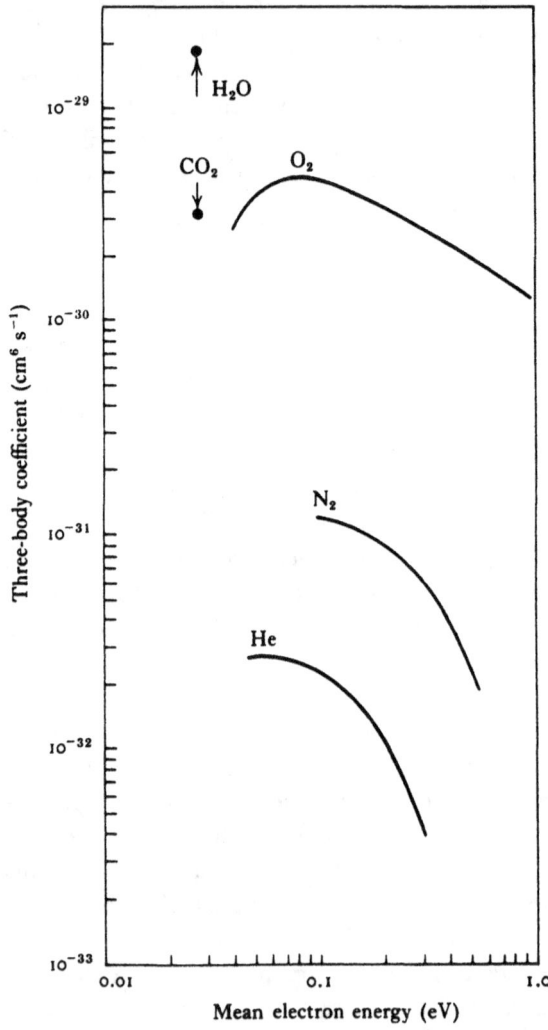

Fig. 9.37. Three-body attachment coefficients of electrons in O_2 for different third bodies at a gas temperature of 300 °K. —— observed by Chanin *et al.* (1962).

compare well with directly observed values obtained in different experiments (see Massey, 1969, p. 1019), although at the higher values of F/p there is considerable scatter among the experimental results.

There seems little doubt that negative-ion formation in O_2 is now well known and well understood although it is only in comparatively recent years that the problems of interpretation arising from earlier experiments have been thoroughly cleared up.

CO

Fig. 9.38 shows the cross-section for negative-ion production in CO as a function of electron energy, as observed by Rapp and Briglia (1965) using the apparatus described on p. 278. Again the variation with electron energy agrees at least qualitatively with that observed in earlier experiments by Lozier (1934) and by Hagstrum and Tate (1941).

The production process, which has an onset near 9 eV and peak near 10 eV, arises from dissociative attachment, while that which has an onset near 20 eV is due to polar dissociation. Although both O^- and C^- ions are stable, in both processes the ions produced are almost exclusively O^-.

Further detailed information about the dissociative attachment processes has been obtained from experiments of Chantry (1968), using the equipment described on p. 282, which were particularly concerned with the measurement of the energy distributions of the O^- ions at different electron energies, and those of Stamatovic and Schulz (1970), using the equipment referred to on p. 285, which were directed towards the study of processes leading to C^- production.

Fig. 9.39 compares the variation with electron energy of the cross-section for O^- production observed by Chantry and by Stamatovic and Schulz with that observed by Rapp and Briglia (1965) which is shown in Fig. 9.38. The differences can be ascribed to the fact that the electron energy distribution was broader in the last-named results because the retarding potential-difference technique was not used as in the other experiments. Measurements were also made by Rapp and Briglia (1965) using this technique and the results are indistinguishable from those of Chantry. The electron beam used by Stamatovic and Schulz was even more nearly homogeneous in energy so that the rise from threshold is sharper and structure is apparent in the shape of the cross-section curve.

In particular there is evidence of some new process beginning at an electron energy close to 11 eV. The existence of a second dissociative process with threshold near this energy was already established in Chantry's experiments from his observations of the O⁻ energy distribution as functions of electron energy. Fig. 9.40

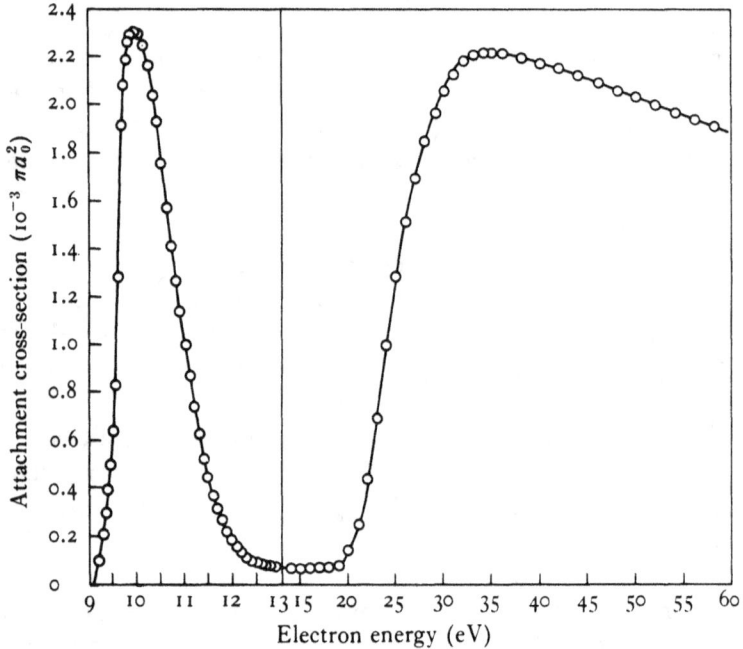

Fig. 9.38. Cross-sections for negative-ion production in CO, observed by Rapp and Briglia (1965).

shows the observed distributions for electron energies of 9.60, 9.95, 10.45 and 11.00 eV. For the first three of these energies a single peak is observed at an ion energy increasing with the electron energy. If the peak energy is plotted against the electron energy as in Fig. 9.26 the points fall on a straight line of slope 12/28 which cuts the electron energy axis at 9.70 ± 0.1 eV. The slope is exactly as expected from the theory of p. 276, according to which it is given by $M_1/(M_1 + M_2)$, where M_1, M_2 are the masses of the

NEGATIVE ION FORMATION – DISSOCIATIVE PROCESSES 335

ion and atom which result from the dissociation. Also 9.70 eV agrees within experimental error with the expected value $D(CO) - E_a(O)$ for dissociation into ground state atoms and ions since $D(CO) = 11.09$ eV and $E_a(O) = 1.465$ eV.

At the highest electron energy, 11.00 eV, in Fig. 9.40, the ion energy distribution shows two peaks, one near 0.6 eV, which is as expected for dissociation into ground state atoms and ions, and the other close to zero. If observations are extended to higher electron

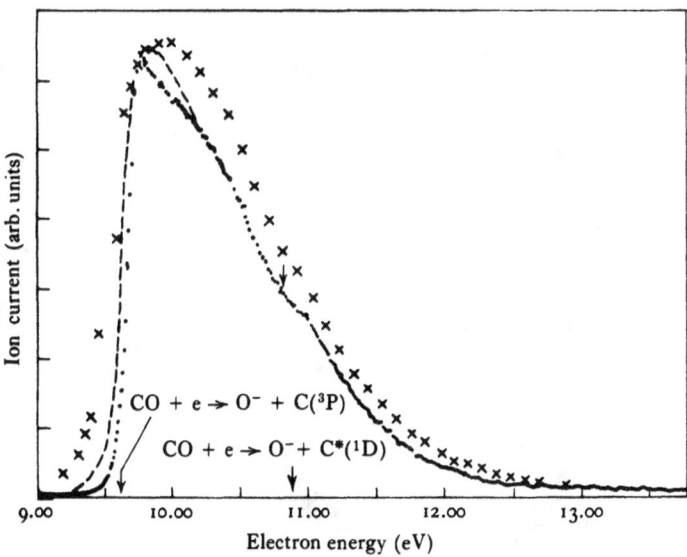

Fig. 9.39. Comparison of observed cross-sections for dissociative attachment in CO. Observed × Rapp and Briglia (1965), ----- Chantry (1968), · Stamatovic and Schulz (1960). Scales are adjusted so all agree at the peak.

energies, the second peak moves to higher energies giving again a linear plot with the same slope 12/28 as for the main process, but the threshold energy for production of ions of zero energy is at 10.95 ± 0.10 eV, 1.25 eV higher. Since this agrees within experimental error with the excitation energy (1.26 eV) of the ^1D level of C, it seems that we are dealing with the dissociative process

$$CO + e \longrightarrow O^-(^2P) + C(^1D). \tag{9.58}$$

The cross-section for this latter process was determined by observing the O^- current as a function of electron energy with the

Wien filter tuned to accept only ions of zero kinetic energy. Two peaks were obtained separated by the excitation energy of C(^{1}D). The ratio of the peak height gives the ratio of the cross-

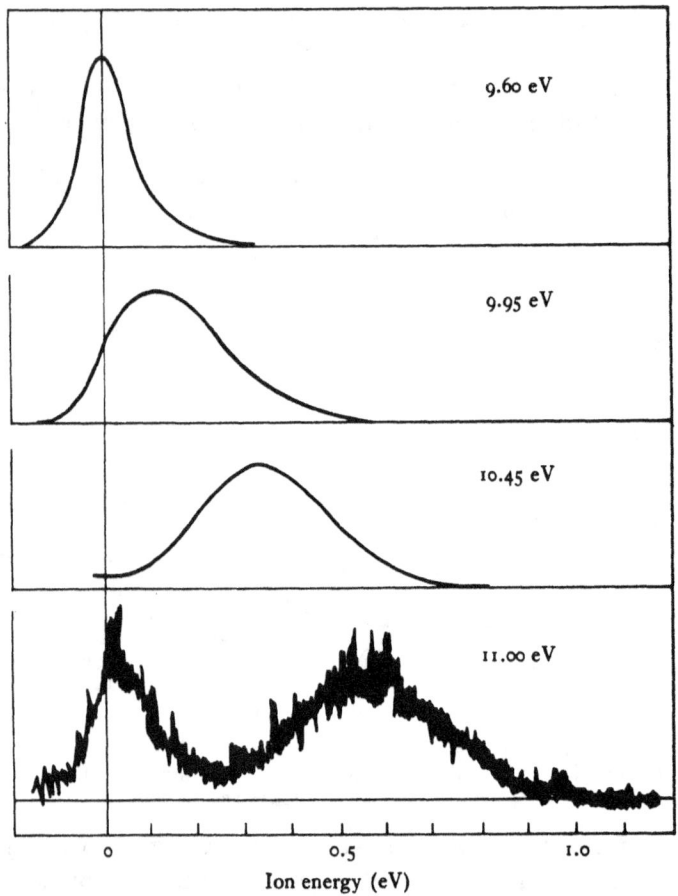

Fig. 9.40. Energy distributions of ions produced by electron impact in CO at different electron energies, as indicated. From Chantry (1968).

sections for dissociation into normal ^{3}P and excited ^{1}D C atoms. This was found to be 21 so that, taking the absolute value for the main process at its peak, following Rapp and Briglia (see Fig. 9.38), as 2.0×10^{-19} cm², that for the process (9.58) is 9.5×10^{-21} cm². This is not inconsistent with the structure observed in the experi-

NEGATIVE ION FORMATION – DISSOCIATIVE PROCESSES 337

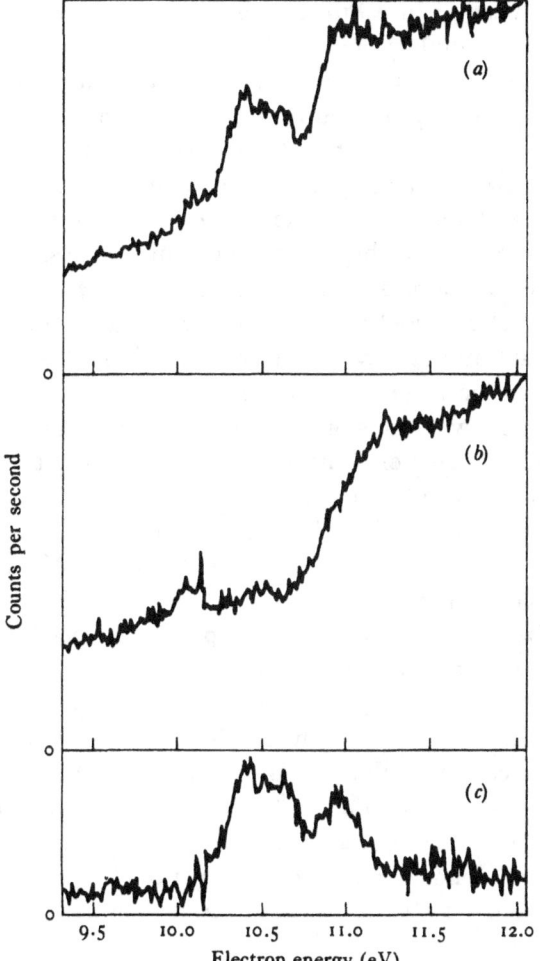

Fig. 9.41. Evidence for the formation of C⁻ from CO by electron impact, as observed by Stamatovic and Schulz (1970). (a) Total signal from C⁻ plus background from u.v. photons; (b) signal from u.v. photons only; (c) residual signal from C⁻.

ments of Stamatovic and Schulz (1970) in which the electron energy resolution was higher.

The occurrence of a dissociative attachment process in CO yielding C⁻ ions was first observed by Lagergren (1955) at an appearance potential of 10.07 ± 0.1 eV. Stamatovic and Schulz (1970), with their mass analyser tuned to receive C^- ions, found that the cross-section for C^- production was very low. In fact it was observed, with the mass analyser detuned, that the background signal due to ultraviolet photons produced in the collision chamber by electron impact was comparable with that due to C^- with the analyser tuned. However, by accumulating the signals for about 15 h, first with the analyser tuned and then detuned, it was possible to subtract the photon background to obtain the wanted integrated signal from C^-. It was verified that no significant change in the accumulated background signal occurred if, instead of detuning the mass analyser, the C^- current was prevented from reaching the analyser by reversing the ion extraction potential.

Fig. 9.41 shows results obtained in this way revealing clear evidence of C^- production. The appearance energy was determined, by comparison with that for O^-, as 10.20 ± 0.04 eV which is compatible with that found by Lagergren (1955). This threshold for production of ions with zero kinetic energy is 9.84 eV since $E_a(C) = 1.25$ eV (see Chapter 3, p. 44).

It will be noted that two peaks are present in the C^- production curve. The first almost certainly corresponds to production of normal $O(^3P)$ and $C^-(^4S)$. The onset for the second is 10.8 eV, only 0.6 eV higher than for the first. Since $C^-(^2D)$, while probably stable, lies nearly 1.25 eV above the ground $C^-(^4S)$ state, the second peak must arise, like the first, from dissociation into normal O and C^-. Its origin is still not clear.

The cross-section for C^- production at the first peak is found to be as low as $(6 \pm 1.5) \times 10^{-23}$ cm², over 3000 times smaller than the peak value for O^-.

There is no evidence of any low-energy attachment process as in O_2 – a result which is not surprising in that CO^-, like N_2^-, is not stable.

NO

In this gas the processes of negative-ion formation by electron impact are similar to those in O_2, involving dissociative attachment and polar dissociation for electrons of energy a few eV or higher and a three-body attachment process for thermal and near-thermal electrons.

Fig. 9.42 shows the cross-section for dissociative attachment as a function of electron energy as measured by Rapp and Briglia (1965), using the apparatus described on p. 278. In the same figure comparison is made with the results of Chantry (1968), using the equipment described on p. 282 and normalized to agree with Rapp and Briglia's results at an electron energy of 8e V. The agreement in shape is very satisfactory. Again, the variation of the cross-section with electron energy agrees at least qualitatively with the early

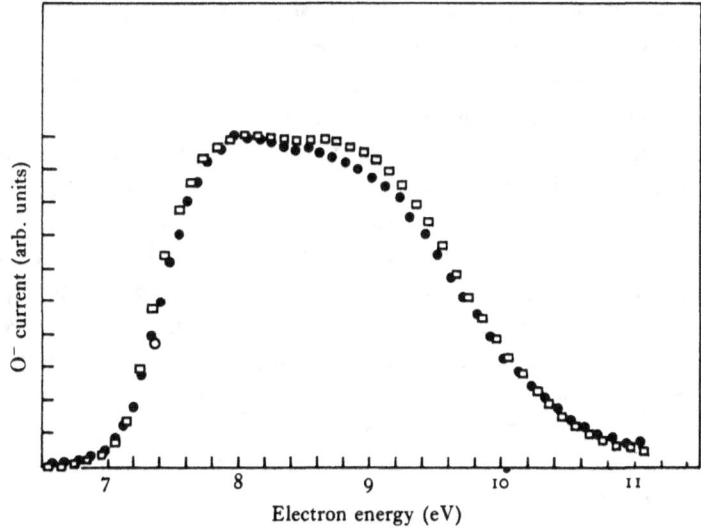

Fig. 9.42. Cross-sections for negative-ion production in NO. □ observed by Rapp and Briglia (1965), ● observed by Chantry (1968), normalized to agree with □ at the peak.

observations of Hanson (1937), using the Tate–Lozier method, and Hagstrum and Tate (1941), using a mass spectrograph.

Chantry (1968) was especially concerned with the measurement of the energy distribution of the ions produced. In contrast to CO he found, over the whole energy range for which the dissociative process is significant, that the distribution possessed only a single peak. The variation of the peak ion energy with electron energy followed a single straight line of slope 14/30 which intersected the electron energy axis at 7.5 ± 0.1 eV. The slope is exactly what would be expected for a single process. If the process leads to ground state

N and O⁻, the appearance energy for ions of zero kinetic energy would be $D(NO) - E_a(O) = 6.49 - 1.46 = 5.03$ eV. This is 2.45 eV below the observed value but this agrees within experimental error with the excitation energy of the ^2D term of the ground configuration of N. The dissociative process therefore seems to be

$$NO + e \longrightarrow N(^2D) + O^-(^2P).$$

The two separate peaks in the shape of the cross-section curve may arise from transitions to two different states of NO⁻ which tend to the same limit at large nuclear separations, but this is undecided.

Turning now to the results of swarm experiments, early measurements for NO, made by the filter method, already obtained evidence of attachment coefficients proportional to the square of the pressure, in contrast to early work in O_2. Further measurements have been made by afterglow techniques. Gunton and Shaw (1965) and later, Weller and Biondi (1968), obtained attachment rates from observations of electron decay times in afterglows in NO photoionized by pulsed ultraviolet radiation, the method being essentially as described on p. 299. In both experiments, which were operated at room temperature, attachment was found to proceed through a three-body reaction. Gunton and Shaw measured a rate coefficient of $(2.2 \pm 0.2) \times 10^{-31}$ cm⁶ s⁻¹ over an NO pressure range from 3 to 10 torr, but found an anomalous increase above this pressure to about 3.5×10^{-31} cm⁶ s⁻¹ near 15 torr. Weller and Biondi obtained $(1.3 \pm 0.1) \times 10^{-31}$ cm⁶ s⁻¹ over a pressure range from 1 to 5 torr.

Later measurements have since been made by Puckett, Kregel and Teague (1971) using the technique based on the observation of the transition time between electron and negative-ion dominated ambipolar diffusion as described on p. 306. They found a three-body rate constant for attachment to NO of $(6.8 \pm 3.4) \times 10^{-32}$ cm⁶ s⁻¹, considerably smaller than the values obtained in the earlier experiments.

However, the pressure in the afterglow experiments was less than 1 torr and there is evidence of an increase in the apparent attachment rate with pressure. This was confirmed by the drift-tube experiments of Parkes and Sugden (1972) who found that the rate increased from 2×10^{-31} to 8×10^{-31} cm⁶ s⁻¹ as the pressure increased from between 1 and 2 torr to 100 torr.

NEGATIVE ION FORMATION – DISSOCIATIVE PROCESSES 341

In all the experiments NO_2^- was identified as the dominant ion, no evidence of NO^- being found. NO has a very low electron affinity (0.024 eV, see Chapter 11, p. 482) so NO^- will lose its electron rather easily in collisions. Parkes and Sugden (1972) have analysed the reactions taking place in NO allowing for detachment and in this way have been able to interpret the pressure variation of the apparent attachment rate and derive the true attachment rate coefficient. As this discussion involves knowledge of the rates of detachment and other reactions it is deferred to Chapter 12, p. 561.

The mechanism of the three-body attachment process in NO could be expected to resemble that in O_2. Electrons are captured with high probability when they possess a kinetic energy close to $E_v(NO^-) - E_0(NO)$, E_v being the energy of NO^- in the vibrational state of quantum number v and E_0 that of the ground state of NO. Because of its low electron affinity, in all states of NO^-, $E_v(NO^-) > E_0(NO)$ for $v \geqslant 1$. (See Chapter 6, p. 191.) Stabilization can occur if the vibrationally-excited NO^- loses its excitation in a collision of the kind with a third body. In view of the fact that NO^- has not been observed as a direct product, the process of stabilization may be more complicated but no clear alternative has been suggested. While the existence of resonance peaks due to capture of electrons to form vibrationally-excited NO^- has been observed in experiments in elastic scattering and vibrational excitation of NO by electron impact, no experiments have yet been carried out to observe three-body attachment to NO with electron beams of homogeneous energy.

Halogen molecules

Experiments to measure rates of attachment to halogen molecules are difficult to perform because of their high chemical reactivity. This makes it very difficult to avoid effects due to impurities.

Of the different halogens, I_2 has been the most extensively studied. On p. 302 we described the most recent experiments up to the time of writing, those of Truby (1968) who observed the decay of electron concentration in an afterglow in helium-buffered iodine which was ionized by a pulse of ultraviolet radiation. The special precautions taken to minimize the effect of impurities and of the build-up of negative-ion concentration are described on pp. 302–5.

The measured attachment rate coefficient at 295 °K is 1.8×10^{-10} cm^3 s^{-1} corresponding to a mean attachment cross-section for thermal electrons of 1.7×10^{-17} cm^2. Earlier experiments by Biondi (1958), using essentially the same technique, obtained a much larger mean attachment cross-section 3.9×10^{-16} cm^2. The source of the larger value may have been build-up of negative-ion concentration (see p. 302) but the situation is still not clear.

At about the same time as Biondi was making the afterglow measurements, complementary experiments were carried out by Fox (1958) using the same two instruments as in the experiments of Hickam and Fox (1956) on attachment in SF$_6$ (p. 370), a 90° sector mass spectrometer and a total ion collection tube. In both of these the electron energy was controlled by the retarding-potential difference method and the vacuum chambers could be baked out at 400 °C before introducing the iodine. After bake-out the mass spectrum of positive ions showed traces only of H$_2$O$^+$ and CO$^+$. Although, when iodine was admitted at a pressure of about 10^{-5} torr, no HI$^+$ ions were observed, their concentration built up to give an observable peak after 15 min or so and continued to increase thereafter. As HI produces I$^-$ ions very readily by dissociative attachment, it was most important to work under conditions in which its concentration was known to be small.

Working with the total ion collector under these conditions, Fox (1958) obtained the results shown in Fig. 9.43 for the variation of the total collected negative ion current with electron energy. From the observed mass spectra the only ion present was I$^-$. To calibrate the energy scale, and at the same time determine the electron energy distribution, the total collected negative-ion current was measured with SF$_6$ substituted for the I$_2$. As an example of the problems encountered when working with the halogens, the energy scale as determined by the accelerating voltage was changed through introducing iodine, presumably because of surface changes on the electrodes. This effect was eliminated by admitting I$_2$ and SF$_6$ many times until the electrode surfaces were brought to a stabilized state.

The observed current from SF$_6$ is shown in Fig. 9.43, obtained after carrying out this procedure. After subtraction of the known fraction of SF_5^- ions produced from SF_6, the SF_6^- current was derived as shown.

On the low-energy side of the peak the SF_6^- and I^- currents nearly coincide in shape and position in the energy scale. As SF_6^- ions are produced with maximum probability by electrons of nearly

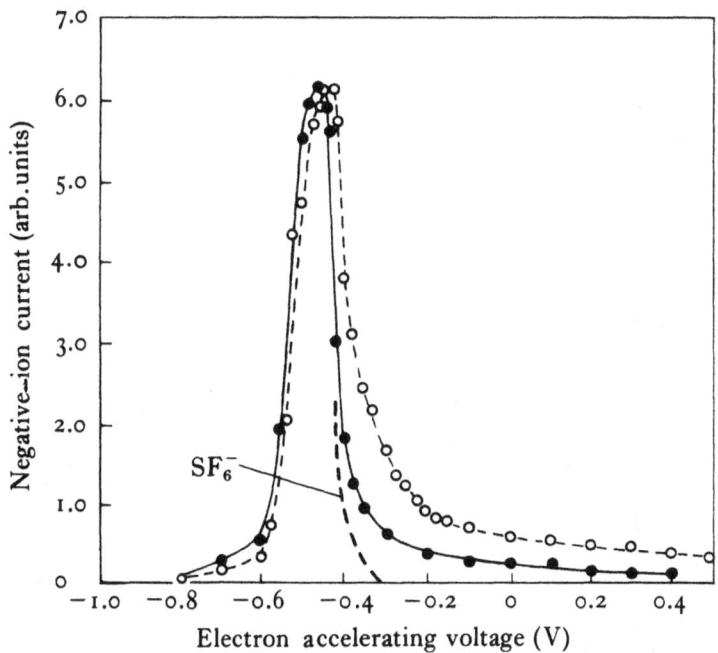

Fig. 9.43. Variation with electron energy of the total negative ion current arising from electron impact in I_2 compared with that in SF_6. ○ I^- from I_2, ● $SF_6^- + SF_5^-$ from SF_6, ---- SF_6^- derived from relative abundance measurements (from Fox, 1958).

zero energy (see p. 370) the same must apply to I^-. On the high-energy side, however, the I^- production persists to much higher energies than that of SF_6^-.

The mean attachment cross-section for $\bar{Q}_a(\bar{E})$ for electrons of mean energy $\bar{E}$ is given by

$$\bar{Q}_a(\bar{E}) = \int Q_a(E) f(E, \bar{E}) dE, \qquad (9.59)$$

where Q_a is the attachment cross-section for electrons of energy E and $f(E, \bar{E})$ is the fraction of electrons with energy between E and

$E + \mathrm{d}\bar{E}$ when the mean energy is $\bar{E}$. $f(E, \bar{E})$ is known from the shape of the SF_6^- production curve and $Q_a(E)$ from that of I^- so that, by inverting (9.59), we may obtain $Q_a(E)$, apart from a normalizing factor. At the time, this was chosen so that $Q_a(\bar{E})$ for thermal electrons agreed with the afterglow observations of Biondi (1958),

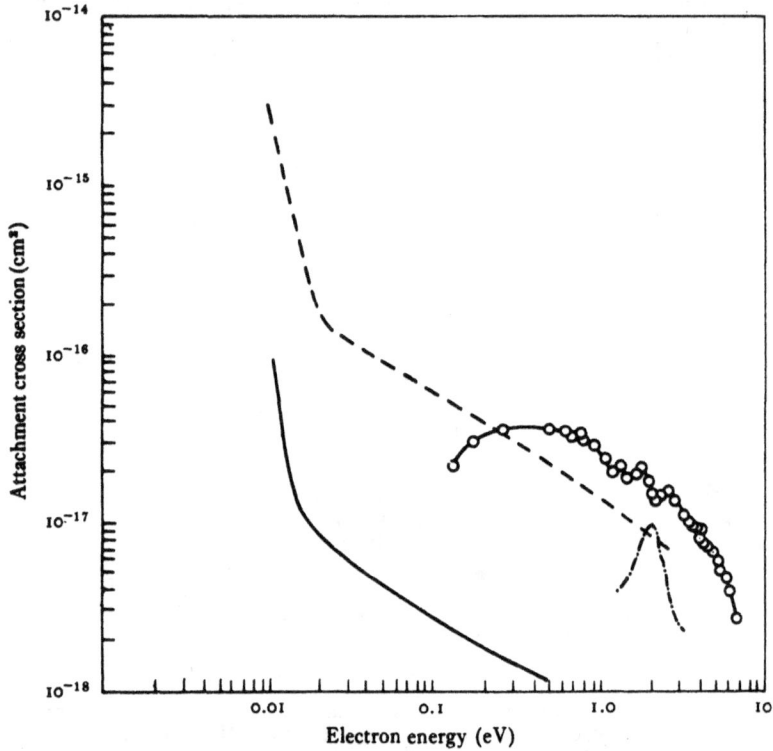

Fig. 9.44. Cross-sections for dissociative attachment of electrons to I_2. ---- derived by Biondi and Fox using microwave data of Biondi (1958), ——— as for ---- but normalized using microwave data of Truby (1968), –O–O– observed by Buchdahl (1941), –·–· observed by Healey (1938).

giving $Q_a(E)$ as shown in Fig. 9.44. If instead the results of Truby (1968) are used, the cross-sections also shown in Fig. 9.44 are over 20 times smaller.

Data obtained in earlier experiments are not sufficiently reliable to help in clarifying the position. Buchdahl (1941) measured the negative-ion production with equipment of Tate–Lozier type while Healey (1938) obtained attachment coefficients from diffusion experiments. Their results are shown in Fig. 9.44. The one thing which is clear is that we still are far from having reliable results, although it does seem very probable that the I$^-$ ions are formed with maximum probability by electrons of nearly zero energy.

For Br$_2$, Truby (1971) has measured the attachment rate coefficient for thermal electrons using the same method as in his experiments in I$_2$. He obtained a value of 0.8×10^{-12} cm^3 s^{-1} at room temperature corresponding to a mean cross-section of 0.8×10^{-19} cm^2, much smaller than for iodine.

Bailey, Makinson and Somerville (1937) measured the attachment coefficient for Br$_2$ using the diffusion method. Their experiments covered a range of mean electron energy from 1.2 to 3.0 eV. A peak corresponding to a mean cross-section of about 10^{-17} cm^2 was found at 1.75 eV. This is consistent with the maximum found in the beam experiments of Blewett (1936).

The only data for Cl$_2$ are the early swarm measurements of Bailey and Healey (1935) by the diffusion method and of Bradbury (1934) using the filter method. A maximum mean attachment cross-section of 5×10^{-18} cm^2 was observed at a mean electron energy near 1.5 eV.

9.6 Attachment to polyatomic molecules

A great number of experiments have been carried out to study attachment to polyatomic molecules, with several different aims – to determine electron affinities or dissociation energies of radicals, to assess the importance of attachment in determining flame-quenching or physiological behaviour of molecules, to interpret gas analysis data, to understand the factors which determined the effectiveness of different gases and vapours in increasing dielectric breakdown strength and so on. While it would be out of place here to attempt any comprehensive account of all of these investigations we shall draw attention to some of the problems of interpretation which arise, particularly in relation to the determination of electron affinities. A number of examples of special interest in this context and otherwise will then be chosen for more detailed discussion, while in Chapter 15 possible applications of attachment data will be briefly considered.

The appearance energy E_t for a reaction

$$XY + e \longrightarrow X + Y^-,$$

where X and Y are no longer atoms but more or less complex radicals, is given by

$$E_t = D(XY) - E_a(Y)$$
$$+ W + E_{int}(X) + E_{int}(Y^-). \qquad (9.60)$$

where $D(XY)$, $E_a(Y)$ and W have their usual significance, being respectively the dissociation energy of XY into normal X and Y, the electron affinity of Y and the translational energy of X and Y^-. The additional terms $E_{int}(X)$ and $E_{int}(Y^-)$ allow for the vibrational and rotational excitation of the product ions and neutral radicals.

It follows that, even if E_t, W and $D(XY)$ are known, it is only possible to derive a lower limit to $E_a(Y)$ since $E_{int}(X)$ and $E_{int}(Y^-)$ will not, in general, be known. De Corpo and Franklin (1971) have adopted a semi-empirical procedure, described on p. 381, which overcomes this difficulty in a certain limited series of cases.

However, apart from this problem, there is the further one that at any finite temperature the target molecule XY will possess internal energy. Because of the complexity of a polyatomic molecule the energy quanta, even of vibration, will be quite small and easily excited at ordinary temperatures. The result of this will be to blur the threshold, often to such an extent as to make it extremely difficult to determine unless measurements are made over a wide range of temperatures and extrapolated to zero temperature. Effects of this kind will be especially marked when the neutral molecule XY and ion XY^- have different geometrical configurations so that the dissociative attachment cross-section will depend very strongly on the presence of vibrational modes in the neutral molecule which lead to distortion in the direction of the geometry of XY^-. Such behaviour is already manifest in such simple molecules as CO_2 and N_2O as described on pp. 353 and 358 respectively. A comprehensive investigation of temperature effects in attachment to halogen-containing molecules is described on p. 379.

We now consider attachment to specific polyatomic molecules.

H_2O

The study of negative-ion formation in water vapour is of special interest, partly because of the frequent presence of water vapour as an impurity in experimental work and partly because three different negative ions H^-, O^- and OH^- could well result from dissociative attachment.

In fact all three ions have been observed with intensities in the decreasing order H^-, O^-, OH^-. The peak cross-section for production of H^- is two orders of magnitude larger than the peak for production from H_2 so that a discharge in water vapour is a more efficient source of H^- ions than a similar discharge in H_2.

Fig. 9.45(a) shows the total negative-ion current as a function of electron energy observed by Schulz (1960). The dominant, H^-, component of this current, as observed by Compton and Christophorou (1967), is shown in Fig. 9.45(b). Many of the H^- ions possess considerable kinetic energy so that special arrangements had to be made to avoid their loss to the walls before entering the mass spectrograph. The electron current was pulsed and the ion-collecting potential applied as a pulse within 1 μs from the termination of the current pulse.

Buchel'nikova (1958) has also measured the cross-sections for H^- production, finding values of 4.8×10^{-18} and 1.3×10^{-18} cm^2 at the peaks near 6.4 and 8.8 eV.

For H^- ions from H_2O, the effect of temperature motion of the target molecules on the apparent appearance potential should be small, as noted on p. 277. The linear extrapolation method (p. 276) should therefore be applicable to the determination of threshold electron energies for production of ions of zero kinetic energy. Fig. 9.46 shows the application of the method in which the ion-retarding potential is plotted against the appearance potential, using data obtained by Schulz (1960) for the H^- production processes associated with the two peaks in Fig. 9.45(b). Both processes give points falling on the same straight line which extrapolates to give 4.4 eV as the appearance potential for zero energy ions. This would be expected to be equal to $D(OH–H) - E_a(H)$, where $D(OH–H)$ is the energy required to dissociate H_2O into normal OH and H, and $E_a(H)$ is the electron affinity of H.

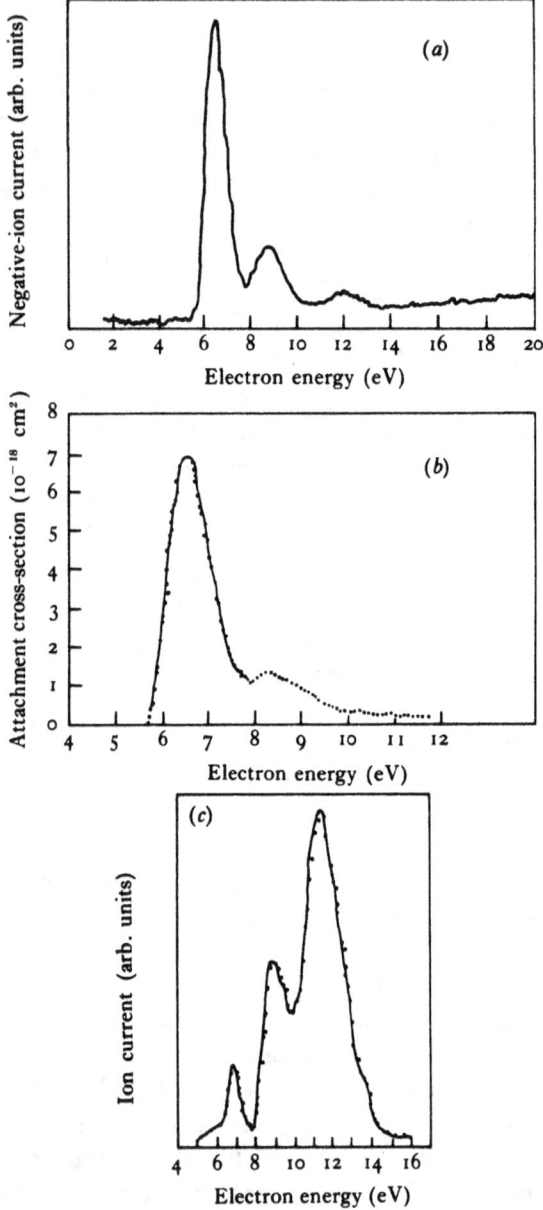

Fig. 9.45. Cross-sections for production of negative ions in H_2O as a function of electron energy. (a) Total ion production, observed by Schulz (1960); (b) H^- ions, observed by Compton and Christophorou (1967); (c) O^- ions, observed by Compton and Christophorou (1967).

Taking D as 5.11 eV, as given from thermochemical data, and E_a as 0.75 eV, $D - E_a = 4.36$ eV, in close agreement with expectation.

The variation of the O⁻ current with electron energy has also been observed by Compton and Christophorou (1967). It is in general less than 3% of the H⁻ current and shows three peaks as seen from Fig. 9.45(c).

OH⁻ ions were not observed as products of primary attachment

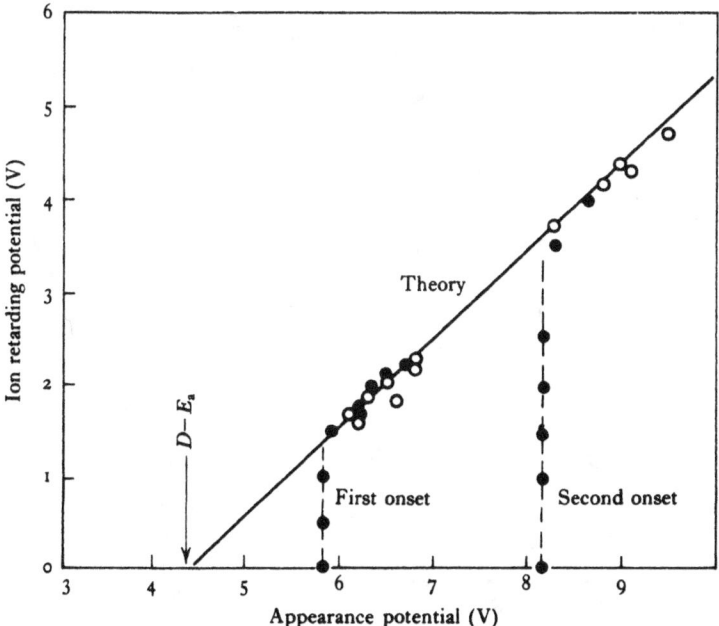

Fig. 9.46. Plot of retarding potential versus appearance potential for H⁻ ions from H_2O, as observed by Schulz (1960).

reactions until 1969, when Doumont, Henglein and Jäger found them to be produced with an intensity about 10 times smaller than that of O⁻.

With the same equipment Doumont et al. observed the O⁻ current, finding peaks at nearly the same electron energy as in the experiments of Compton and Christophorou (1967) and also in other experiments by Dorman (1966).

Measurements have also been carried out on negative-ion production from D_2O. Compton and Christophorou found the first D⁻ peak to be 0.3 eV narrower than for H⁻ from H_2O, probably because of the smaller spread of the ground state vibrational wave function of D_2O. For O⁻, pronounced isotope effects, also observed by Dorman (1966), were found.

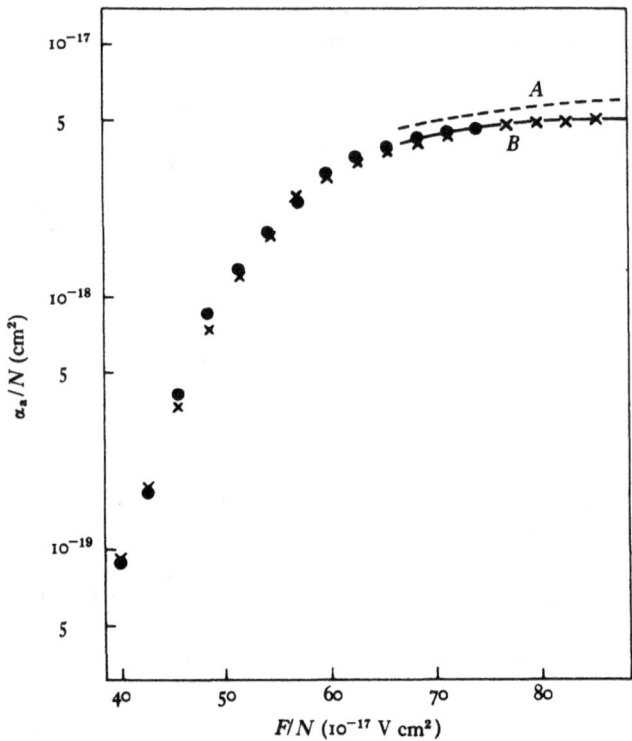

Fig. 9.47. Observed value of α_a/N as a function of F/N for electrons in water vapour, N being the number of molecules per cm³ and α_a the attachment coefficient. Curves A, Crompton et al. (1965); B, Ryzko (1966). Points ●, ×, Parr and Moruzzi (1972), measured at pressures of 10 and 5 torr respectively.

It seems that the general features of the negative-ion production by electrons with energies of a few eV and higher are quite well established although much remains to be done before precise results are available for the relative cross-sections – ion energy discrimination may be serious in some cases. The question remains as to whether any new attachment process occurs at thermal or near thermal energies.

Early measurements of attachment coefficients by swarm methods gave very inconsistent results. This may have been partly due to the presence of O_2 as an impurity – it has been pointed out above (p. 331) that three-body attachment to produce O_2^- is particularly effective with H_2O as third body. Even so, in many cases the O_2 concentration would have had to be impossibly high so that there must have been other major contributory factors.

In recent years more consistent results have been obtained. Fig. 9.47 compares results for the attachment coefficient as a function of F/N obtained by Parr and Moruzzi (1972) using Grünberg's method (p. 297) with those of Crompton, Rees and Jory (1965) using the diffusion method as described on p. 297 and of Ryzko (1966) using a pulse technique. In all of these experiments the attachment process was found to be a two-body one. No negative-ion formation was observed for $F/N < 3.9 \times 10^{-16}$ V cm^2 in agreement with observations by Pack, Voshall and Phelps (1962) using the pulse method described on p. 294. According to measurements by Crompton et al. (1965), for $F/N = 5.15 \times 10^{-16}$ V cm^2, the characteristic electron energy is 0.87 eV. Although a detailed check has not been carried through it seems probable that all of the attachment observed in these swarm experiments can be attributed to the dissociative processes observed with electron beams.

The excited states of H_2O^- which are concerned in the dissociative attachment processes have been discussed theoretically by Claydon, Segal and Taylor (1970).

NH_3

The situation for NH_3 is rather similar to that for H_2O. Beam experiments (Sharp and Dowell, 1969) show that dissociative attachment processes occur in two electron energy ranges, between 5 and 6 eV and between 10 and 11 eV. In the former region two reactions have been identified with appearance energies 5.3 ± 0.1 and 5.45 ± 0.05 eV, the first leading to H^- and the second to NH_2^- production. Both have nearly the same cross-section, 1.4×10^{-18} and 1.5×10^{-18} cm^2 respectively.

The most recent swarm measurements by Parr and Moruzzi (1972), using Grünberg's method, are in reasonable agreement with the much earlier results of Bailey and Duncanson (1930) using the diffusion method, but both are considerably smaller than the early results of Bradbury (1934) using the filter method. The attachment process was found to be a two-body one and there is no evidence that any new process exists at near thermal energies.

CO_2

Fig. 9.48 shows the cross-section for O^- production by dissociative attachment in CO_2 as a function of electron energy observed by

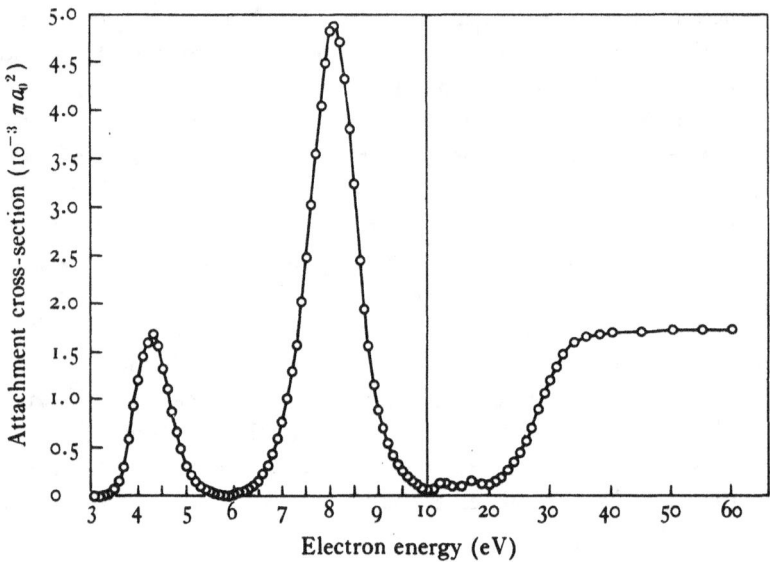

Fig. 9.48. Cross-sections for negative-ion production in CO_2, observed by Rapp and Briglia (1965).

Rapp and Briglia (1965) using the apparatus described on p. 278. The first two, major, peaks have been observed in a number of earlier experiments which agree quite well as to their location. Much fainter peaks are also observed near 13 and 17 eV which were not observed in earlier studies. Beyond 20 eV the ions are produced by polar dissociation.

Chantry (1972a) has used the apparatus described on p. 282 to investigate further the nature of the processes involved. He verified the existence of the peak at 13 eV and showed that its height varied linearly with the gas pressure, just as for the much higher

NEGATIVE ION FORMATION – DISSOCIATIVE PROCESSES 353

peaks at lower energies. This shows that the source of the peak is a primary process but it is difficult to measure the peak cross-section relative to that for the lower energy peaks. In total ion collection experiments such as those of Rapp and Briglia (1965) the background due to collection of scattered electrons may be serious when the negative-ion current is small, while with mass spectroscopic measurements there may be energy discrimination in the sampling of the ions. By comparing results obtained for the peak height ratios in the two types of experiment Chantry determined the ratio of the peak at 13 eV to that at 8.2 eV as 0.013 ± 0.003 at 300 °K.

Energy distributions of the ions produced were measured for all three processes. Typical results are shown in Figs. 9.49(a) and (b). For the process with peak at 4.4 eV the distributions show a peak near zero electron energy which does not vary in position with electron energy. However, as the energy increases the tail of the distribution extending to higher ion energies becomes increasingly prominent. According to (9.13) the peak ion energy for production by attachment of electrons of energy E_e is given by

$$W_{\text{in}} = (1 - \beta)\{E_e - D(\text{CO–O}) + E_a(\text{O}) + E^*\}, \quad (9.61)$$

where β is the ratio 16/44 of the masses of O and CO_2, $D(\text{CO–O})$ is the bond dissociation energy 5.45 eV, $E_a(\text{O})$ is the electron affinity 1.465 eV of O and E^* the excitation energy of the products. With $E^* = 0$ the appearance energy for ions with zero kinetic energy would therefore be 3.99 eV and for higher-energy electrons the peak ion energy will increase by $7(E_e - 3.99)/11$. The fact that the peak remains at zero energy must mean that this surplus energy is taken up in vibrational and rotational excitation of the CO fragment. A further problem remains in that the observed appearance energy for ions of zero kinetic energy, obtained by linear extrapolation, which is valid under these circumstances, is 3.85 eV and not 3.99 eV at 300 °K.

The resolution of this difficulty, which was noted prior to Chantry's work, came from experiments by Schulz and Spence (1969) who studied the effect of gas temperature on the onset energy using the apparatus described on p. 281. Fig. 9.50 shows their observed results in which a linear decrease of this energy with

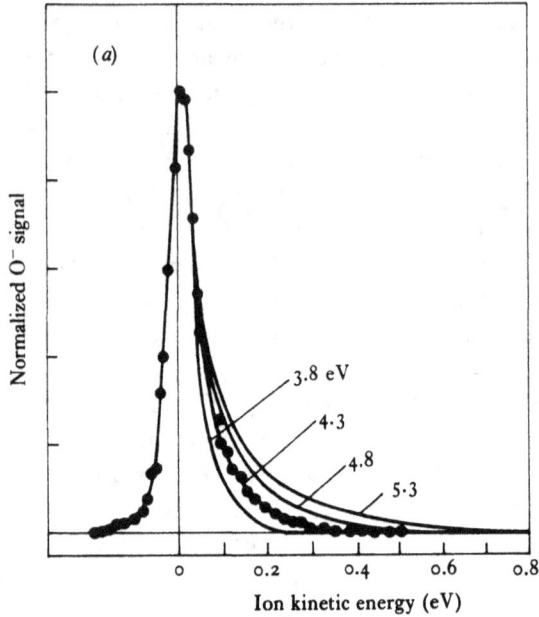

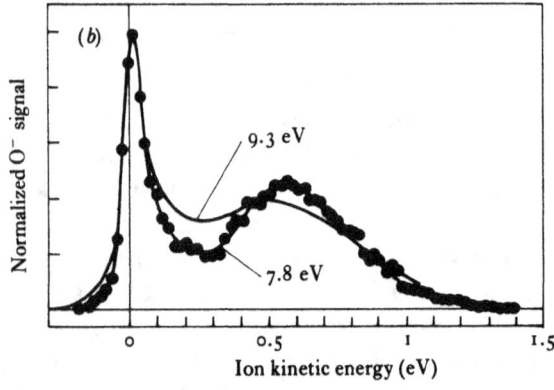

Fig. 9.49. Kinetic-energy distributions of O⁻ ions resulting from dissociative attachment in CO_2, observed by Chantry (1972a). (a) Ions produced by electrons with energies close to the peak at 4.4 eV (see Fig. 9.48). (b) As for (a), for ions produced by electrons with energies close to the peak at 8.2 eV (see Fig. 9.48).

temperature is observed between 300 and 950 °K. In a later experiment (Spence and Schulz, 1969), measurements were extended down to 200 °K. Extrapolation to 0 °K is not inconsistent with the expected value of 3.99 eV. In addition the peak cross-section was

NEGATIVE ION FORMATION – DISSOCIATIVE PROCESSES 355

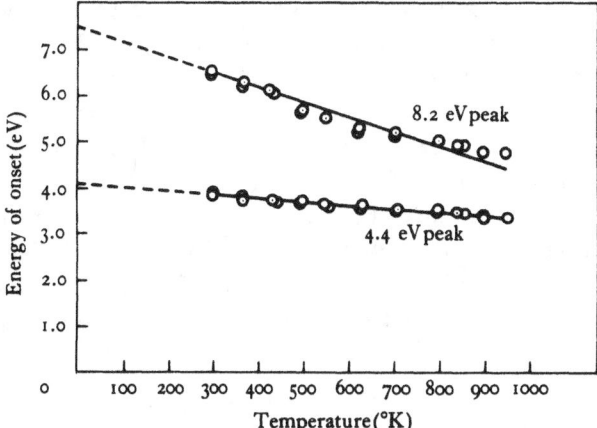

Fig. 9.50. Temperature dependence of the onset energy for production of O⁻ ions from CO_2 by dissociative attachment, as observed by Spence and Schulz (1969). ○ observed points. The two lines refer to the processes peaking at 4.4 and 8.2 eV respectively (see Fig. 9.48), as indicated.

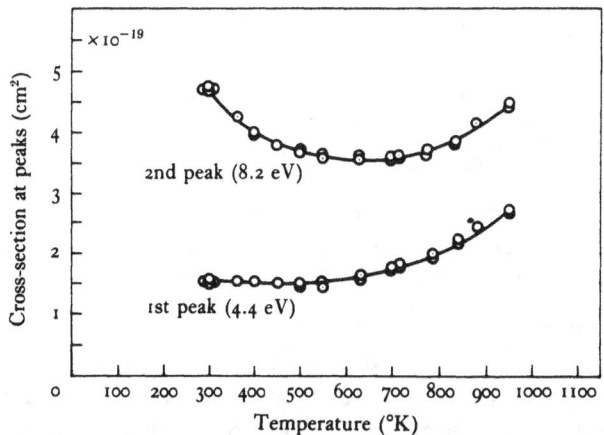

Fig. 9.51. Temperature dependence of the peak cross-sections for production of O⁻ ions from CO_2 by dissociative attachment, as observed by Spence and Schulz (1969). ○ observed points. The two curves refer to the processes peaking at 4.4 and 8.2 eV respectively (see Fig. 9.48), as indicated.

found to increase with temperature above 600 °K, as shown in Fig. 9.51.

Direct evidence of the production of vibrationally-excited CO was obtained by Stamatovic and Schulz (1973*a*) using the apparatus described on p. 285 in which the electron energy was very well defined by means of a trochoidal monochromator. Fig. 9.52 shows the variation of the attachment cross-section with electron energy covering the first peak, which they observed. This clearly shows structure. Mendas and Stamatovic (1973) have analysed these data, taking account of the energy distribution of the electron beam, and shown that they can be interpreted in terms of the production of CO with vibrational quantum numbers v up to 5.

Ion energy distributions observed by Chantry (1972a) for the process with a peak at 8.2 eV are shown in Fig. 9.49(b). Again they

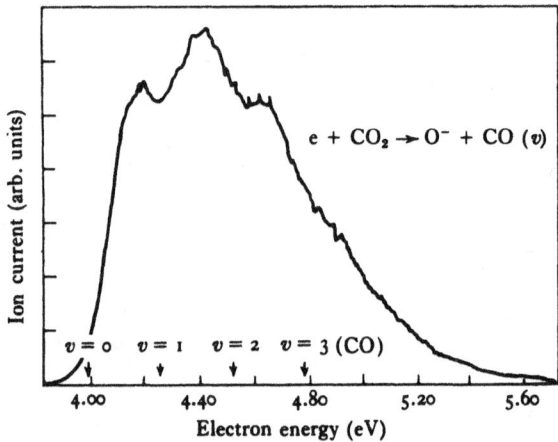

Fig. 9.52. Variation, with electron energy, of the dissociative attachment cross-section for electrons in CO_2, observed by Stamatovic and Schulz (1972).

show a sharp peak near zero energy which does not shift as the electron energy changes. In addition, there is a peak near 0.6 eV followed by a considerable tail at higher energies. The origins of the two peaks is not yet clear. Spence and Schulz (1969) found temperature effects for this process, of comparable magnitude to those for the lower-energy process, as may be seen from Figs. 9.50 and 9.51. There is a difference in the behaviour of the peak value which at first falls with temperature but the cross-section integrated over the peak behaves in much the same way as does the peak at 4.4 eV.

Spence and Schulz (1973b), using a mass analyser as in their experiments in CO, have observed both the weak C^- and O_2^- production from electron impact in CO_2. For the former, the cross-section has three peaks at electron energies of 16.0 ± 0.2, 17.0 ± 0.2 and 18.6 ± 0.1 eV, the last being the largest with a cross-section of 2.0×10^{-22} cm². The cross-section for O_2^- production is much smaller. It has two peaks of approximately 1.5×10^{-24} cm² at electron energies of 11.3 ± 0.2 and 12.9 ± 0.2 eV.

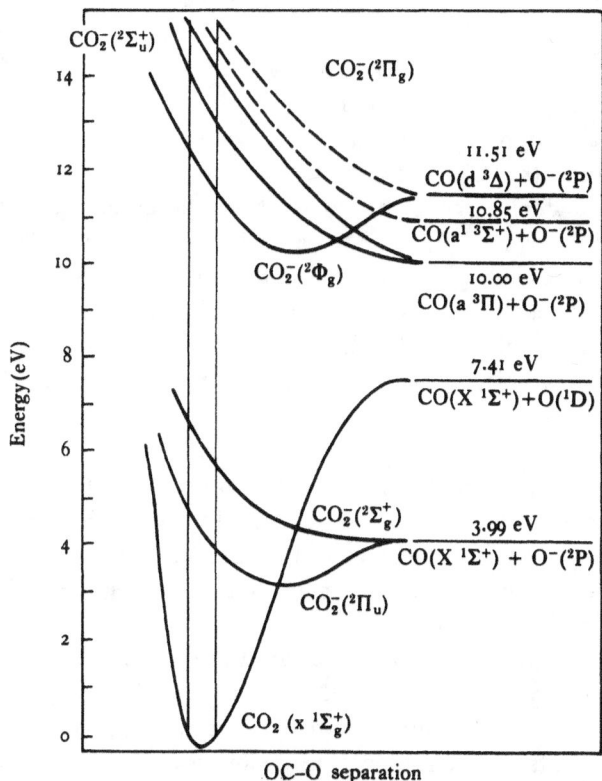

Fig. 9.53. Distribution of potential-energy curves for linear configurations of CO_2^- which are concerned in dissociative attachment to the linear $X^1\Sigma_g^+$ ground state of CO_2. From Claydon et al. (1970).

Although the energy of the triatomic systems depends on the geometry as well as the atomic separations, the ground state of neutral CO_2 is linear and according to the Franck–Condon principle the transitions occurring on electron capture will take place to CO_2^- states possessing nearly the same configuration. The best simplified way of considering the energies of the states of CO_2^- involved in dissociative attachment in relation to onset and peak energies is to examine them as a function of CO–O separation with the bond angle = 180°. This is done in Fig. 9.53. In this diagram the CO_2^- states responsible for the three observed attachment processes are the $^2\Pi_u$, $^2\Sigma_g^+$ and $^2\Phi_g$ states, the first of these being also responsible for vibrational excitation. The disposition of the curves shown in Fig. 9.53 is in qualitative accord with theoretical expectation.

Swarm experiments in CO_2 have produced no evidence of any new attachment process occurring with thermal or near-thermal electrons although CO_2^- ions have been observed as products of ionic reactions (see Chapter 6, p. 218).

Hake and Phelps (1966) have used the total attachment cross-section data shown in Fig. 9.48 to calculate attachment coefficients α_a for electrons in CO_2 as a function of the ratio F/p of electric field to gas pressure. Very good agreement is found between observed and calculated values of the difference $\alpha_i - \alpha_a$ of ionization and attachment coefficients, but is less satisfactory for α_a alone.

N_2O

Just as for CO_2, the lowest energy of the ion N_2O^- will be reached for a bent configuration whereas the neutral molecule is linear in its equilibrium state. This geometrical difference has a much more profound influence on the rate of dissociative attachment of slow electrons in N_2O than in CO_2.

Measurements of the total ion current as a function of electron energy made by Schulz in 1961 and by Rapp and Briglia in 1965, which for most gases agree quite well, gave very considerably different results at electron energies below 1.5 eV as may be seen from Fig. 9.54. At higher energies the disagreement disappeared. Mass-spectroscopic observations showed that O^- ions alone are produced in the observed attachment.

The threshold for dissociative attachment should provide an upper limit to $D(N_2-O) - E_a(O)$, where $D(N_2-O)$, the dissociation energy of the N_2-O bond, is taken from thermochemical arguments as 1.67 eV. On this basis the threshold should lie above 0.2 eV whereas Schulz (1961) found it to be very close to zero, a conclusion reached also by Curran and Fox (1961) from their mass spectroscopic observations.

It was suggested by Kaufman (1967) that the reason for this anomaly is that, at low electron energies, most of the attachment occurs to vibrationally-excited N_2O, so reducing the apparent threshold energy. At the same time the observed cross-section in the same energy region could be expected to vary rapidly with temperature. This could lead to discrepancies between results obtained by different observers with experiments in which no attempt was made to control the temperature. The fact that the excitation of

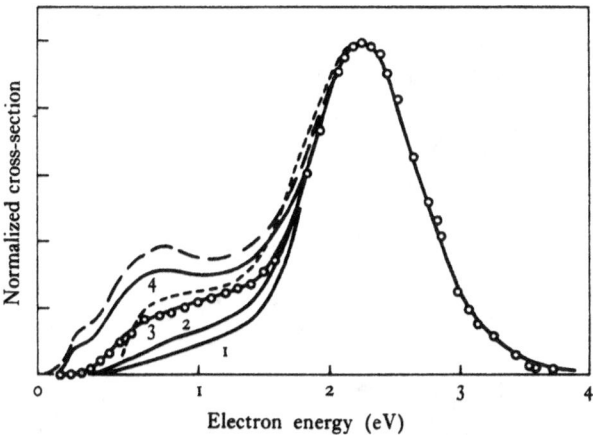

Fig. 9.54. Variation with electron energy of the cross-section for O^- production from N_2O by electron impact, observed in different experiments. -- observed by Schulz (1961), ---- observed by Rapp and Briglia (1965), —— observed by Chantry (1969) with the temperature baths as follows: (1) liquid N_2, (2) solid CO_2 in acetone, (3) room temperature, (4) hot oil. All curves are normalized to agree at the peak. From Chantry (1969).

bending vibrations would tend to modify the geometry of the neutral molecule towards that of the negative ion adds to the plausibility of Kaufman's suggestion.

Chantry, in 1969, carried out a comprehensive series of experiments over a wide range of temperatures. Between nominal temperatures of 77 and 420 °K he used the apparatus described on p. 282 (see Fig. 9.7), in which the energy distributions of the negative ions and the variation of total ion current could both be measured. At higher temperatures, up to 1040 °K, measurements were made, using the apparatus (see Fig. 9.8) described on p. 285, of the variation of total ion current with electron energy, but the energy distributions of the ions could not be measured.

Fig. 9.54 compares results obtained for the variation of total ion current with electron energy, using the first apparatus, with the

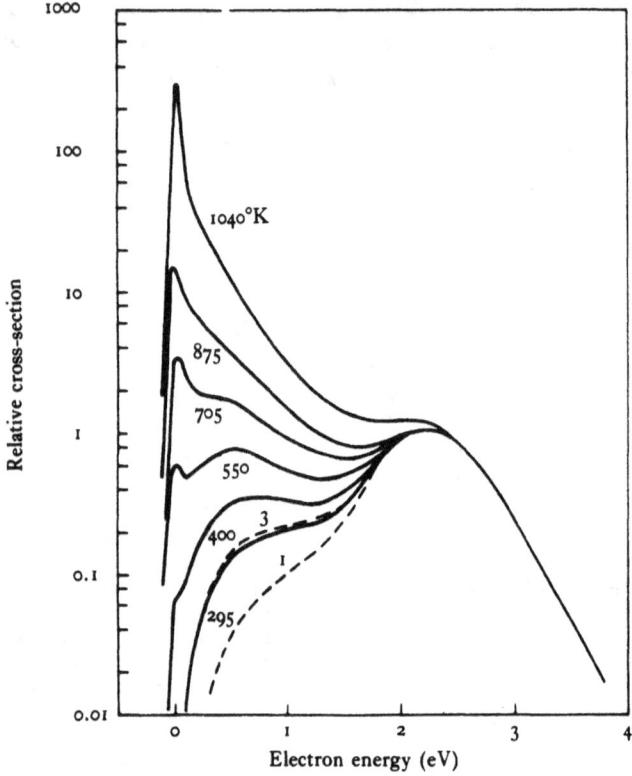

Fig. 9.55. Variation with electron energy of the cross-section for O⁻ production from N₂O by electron impact, observed by Chantry (1969) at high temperatures, as indicated. All curves are normalized to the same peak height at 2.25 eV and coincide at higher energies. Broken curves (1) and (3) are the curves (1) and (3) of Fig. 9.54 replotted.

measurements of Schulz (1961) and of Rapp and Briglia (1965) obtained nominally at room temperature. In all cases the curves were normalized to agree at the peak at 2.25 eV. The sensitivity of the attachment cross-section to temperature at electron energies below 1.5 eV is apparent. It is even more clearly seen in the data obtained with the high temperature equipment which are illus-

NEGATIVE ION FORMATION – DISSOCIATIVE PROCESSES

trated in Fig. 9.55, all but the highest curves being normalized to agree at the peak at 2.25 eV and coincide thereafter. For the curve at 1040 °K, normalization was such as to produce coincidence with the other curves at energies beyond 2.25 eV.

The room temperature curves obtained in both experiments

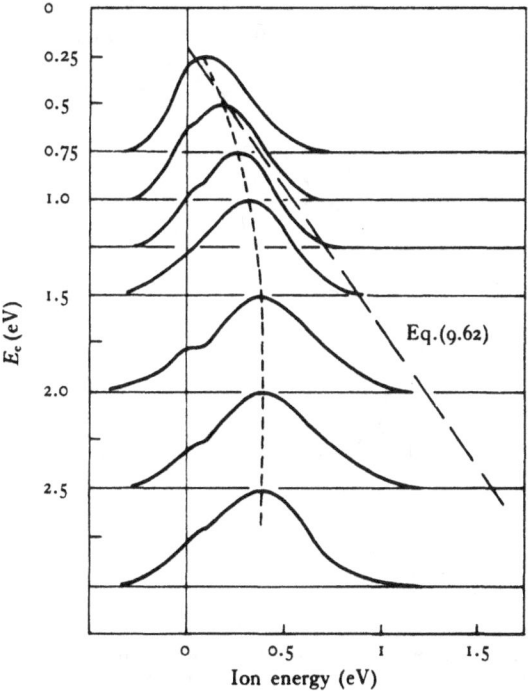

Fig. 9.56. Kinetic-energy distributions of O$^-$ ions produced by electron impact in N$_2$O at different electron energies, as indicated on the side. From Chantry (1969). The broken line is the locus of peak energy according to equation (9.62).

agree very well (see Fig. 9.55), showing that there is no ion energy discrimination affecting the high temperature results.

The precise values of the temperature for the data shown in Fig. 9.54 are not known, but from experiments with the same equipment in O$_2$, in which the widths of the O$^-$ energy distributions were measured, it appears that, for curve 1 the temperature is near 160 °K. Curve 4 nearly coincides with that observed at 370 °K with the high-temperature equipment.

Returning to the high-temperature observations, observed results at very low electron energies are consistent with a cross-section which is a maximum at zero energy and falls very steeply between zero and 0.1 eV. The peak value at a temperature of 1000 °K is at least as high as 10^{-15} cm^2 which is very large compared with measured values for other dissociative attachment processes. As pointed out by Chantry (1969), the impact of very slow electrons on hot N_2O should provide an intense source of O^- ions.

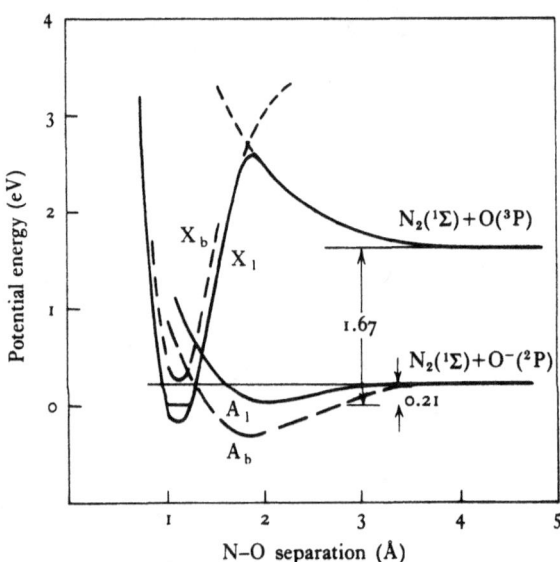

Fig. 9.57. Potential-energy curves (A) for the N_2O^- ground state relative to those (X) for the ground state of N_2O, as functions of the N–O separation, keeping that between the two N atoms as fixed. Suffices l and b refer to linear and bent configurations respectively. From Chantry (1969).

Further information about the nature of the processes occurring can be obtained from the energy distributions of the ions, which were observed by Chantry (1969) using his first apparatus. The zero of the ion energy scale was obtained by working with mixtures of N_2O and CO or NO and observing the O^- kinetic-energy distribution at electron energies just below the established thresholds for production of O^- from the mixture. A typical set of results covering an electron energy range from 0.25 to 2.5 eV is shown in Fig. 9.56.

If attachment occurs only to N_2O in its ground vibrational state, producing N_2 also without vibrational excitation, the peak ion energy would be given by equations (9.13) and (9.14) as

$$W_0 = (1 - \beta)\{E_e - (D - E_a)\}, \qquad (9.62)$$

where β is the ratio $16/44$ of the mass of O^- to that of N_2O, E_e is the electron energy and $D - E_a = 0.2$ eV. This would give the locus of peak energy indicated in Fig. 9.56.

It will be seen that, at electron energies about 0.5 eV, the peak occurs at a lower ion energy, whereas at 0.25 eV it is at a little higher energy. This latter result is evidence that attachment is occurring to molecules possessing vibrational and/or rotational excitation, while the former suggests that the excess energy is taken up in internal excitation of the neutral, N_2, product.

A qualitative understanding of the situation can be gained by ignoring any dependence of the energy of N_2O or N_2O^- on the N–N separation which is taken to be as in N_2. We then need only consider plots of the energy of the system as a function of the N–O separation for different bond angles. Such plots are shown in Fig. 9.57, following Chantry (1969). The suffixes l and b refer to linear and bent systems respectively. X denotes the ground electronic state of N_2O and A that of N_2O^-. The curve X_l has the deepest minimum for N_2O, A_b for N_2O^-. Allowance has been made for the observation of stable N_2O^- ions resulting from certain reactions (see Chapter 12, p. 578) so that the minimum of A_b falls appreciably below the dissociation limit. Account has also been taken of the failure to observe the associative detachment reaction

$$N_2 + O^- \longrightarrow N_2O + e.$$

This requires that all intersections between A and X curves occur at an energy which lies above that for dissociation into N_2 and O^-. From X_l, in the absence of vibrational excitation, energy close to 1 eV is required to produce a transition, in accordance with the Franck–Condon principle, to the A_l curve. On the other hand, as bending vibrations of neutral N_2O are more and more strongly excited, the initial curve will tend to follow X_b. From this curve, a transition to A_b requires little if any energy. Furthermore, the time taken before the N_2O^- complex so formed enters a region in which autodetachment is no longer possible, is very short. The survival factor $\exp(-\rho)$ in the simple formula (9.3) for the dissociative attachment cross-section will therefore be close to unity, while the cross-section Q_c for formation of the complex will be large because of the low electron energy.

Swarm experiments to study the attachment of electrons with thermal or near thermal energies in N_2O have been carried out by Phelps and Voshall (1968) using the pulse technique described on p. 293. They worked at considerably higher pressures (29–191 torr) than in the beam experiments and found attachment to occur through a three-body process with a rate coefficient at room temperature of $(6 \pm 1) \times 10^{-33}$ cm^6 s^{-1}, provided the ratio F/N of electron field strength to molecule concentration was less than 2×10^{-17} V cm^{-2}. At higher F/N the mean electron energy is substantially above thermal and the attachment process was primarily of two-body type. Warman and Fessenden (1968) measured attachment rate coefficients for thermal electrons by observing the rate of decay of electron concentration in an afterglow in N_2O as described on p. 299. They also found attachment to be of three-body type with a rate coefficient 5.6×10^{-33} cm^6 s^{-1}, agreeing quite well with the pulse experiments.

These experiments could not be extended to low enough pressures to measure the two-body rate coefficients but Warman, Fessenden and Bakale (1972) in a series of measurements, by the afterglow technique, were able to measure the coefficient not only for pure N_2O but also for mixtures of N_2O with a number of other non-attaching gases. Especial care was taken to reduce the concentration of O_2, as an impurity, to as low a value as possible. This is necessary because the measured rate coefficient for thermal electrons in N_2O is so small that even a small admixture of an attaching impurity would vitiate the results. With the procedure adopted, the O_2 fraction, as measured mass-spectrometrically with O_2 enriched in the isotope ^{17}O, was below 3×10^{-7}. The fractions of NO and NO_2 were comparable and that of CO_2 less than 10^{-4}.

The two-body rate coefficient deduced by extrapolation to zero N_2O pressure was found to be 1×10^{-15} cm^3 s^{-1}. Very nearly the same value was obtained from observations in mixtures of N_2 and N_2O but with hydrocarbons as diluents a rate constant 6 times

NEGATIVE ION FORMATION – DISSOCIATIVE PROCESSES 365

larger was found. The suggested explanation of this result is that, in pure N_2O and in mixtures with N_2, detachment occurs through the processes

$$O^- + N_2O \longrightarrow NO + NO^-, \quad (9.63a)$$

$$NO^- + N_2O \longrightarrow NO + N_2O + e. \quad (9.63b)$$

While N_2 is inert as far as these reactions are concerned, hydrocarbons RH react with O^- according to

$$O^- + RH \longrightarrow R + OH^-, \quad (9.64)$$

so reducing the chance of O^- reacting as in (9.63a).

In order that associative detachment of N_2 and O^- should not occur, N_2O^- cannot be formed below the dissociation limit by attachment of a low energy electron to N_2O. This is in contrast with the situation for O_2 and NO and means that the lifetime of any N_2O^- formed by direct capture will be very short. To stabilize it by collision, the third body would need to be in appreciable interaction with the N_2O at the instant of the initial attachment. The process could then be best regarded as one of dissociative attachment to a transient dimer.

NO_2

Attachment of thermal electrons in NO_2 has been investigated by static afterglow techniques. Mahan and Walker (1967) studied attachment in an afterglow in a mixture of NO, NO_2 and different rare gases by measurement of the electron concentration as a function of time (see p. 299). The initial plasma was produced by photoionization of the NO (at a pressure of 0.085 torr) with $Ly\alpha$ light. It was verified that no contribution came from ionization of NO_2. Measurements were made over a total pressure range from 3 to 70 torr, the NO_2 partial pressure being varied from 0.002 to 0.018 torr. It was found that the apparent attachment rate coefficient was independent of NO_2 and rare gas pressure but depended on the nature of the inert gas. Thus for He, Ne, Ar, Kr, Xe and N_2 it was found to have the values, in 10^{-11} cm^3 s^{-1}, 2.0, 3.1, 4.5, 3.0, 2.5 and 4.0 respectively.

Mahan and Walker suggested that this behaviour is due to the series of reactions

$$NO_2 + e \longrightarrow (NO_2^-)^*,$$

$$(NO_2^-)^* \longrightarrow NO_2 + e,$$

$$(NO_2^-)^* + M \longrightarrow NO_2^- + M,$$

$$(NO_2^-)^* + M \longrightarrow NO_2 + M + e.$$

If the rate coefficients for the respective reactions are k_1, k_2, k_3 and k_4 then, when the $(NO_2^-)^*$ concentration is in equilibrium, we have

$$\frac{dn_e}{dt} = -\frac{k_1 k_3 [NO_2][M] n_e}{k_2 + (k_3 + k_4)M}, \qquad (9.65)$$

n_e being the electron concentration and [X] that of the molecule X. If $k_2 \ll (k_3 + k_4)M$, (9.65) reduces to

$$\frac{dn_e}{dt} = \frac{k_1 k_3 [NO_2] n_e}{k_3 + k_4}, \qquad (9.66)$$

in which case the attachment is apparently of first order with rate coefficient $(k_1 k_3)/(k_3 + k_4)$. If k_4 is not negligible compared with k_3 this will depend on the nature of the rare gas, as observed.

As described on p. 306, Puckett et al. (1971) studied the transition from an electron-dominated diffusion regime to one dominated by negative-ion diffusion, in an afterglow produced by photo-ionization of 0.050–0.5 torr of NO containing up to 0.001 torr of NO_2. By analysing their data they found an apparent first order rate coefficient of $(1.4 ^{+2.0}_{-0.9}) \times 10^{-11}$ cm^3 s^{-1}, consistent with that observed by Mahan and Walker.

O_3

A major difficulty in laboratory measurements of attachment rates in ozone, as with the halogens, is that of avoiding the effects of impurities arising from chemical reactions between the ozone and the experimental chamber. Stelman, Moruzzi and Phelps (1972) have carried out measurements of rate coefficients for attachment of low-energy electrons in O_3 in which they have taken especial care to avoid serious contamination by impurities.

The drift tube they used is essentially the same as that described on p. 296. Measurements could be made not only at room temperature but also at 200 °K by enclosing the drift tube in a plastic box filled with dry ice. Fig. 9.58 shows the arrangement of the ozone preparation and gas-handling system, especially designed to reduce the rate of destruction of ozone in the experimental chamber to an acceptable value. A major source of destruction was found to be back-diffusion of pump oil and other contaminants. This was minimized by including the traps B and C, cooled in liquid N_2. The ozone was produced by passing O_2 through a microwave discharge and condensing about 0.1 cm³ of liquid O_3 in the trap A cooled in liquid N_2. O_2 was then removed by pumping, after which

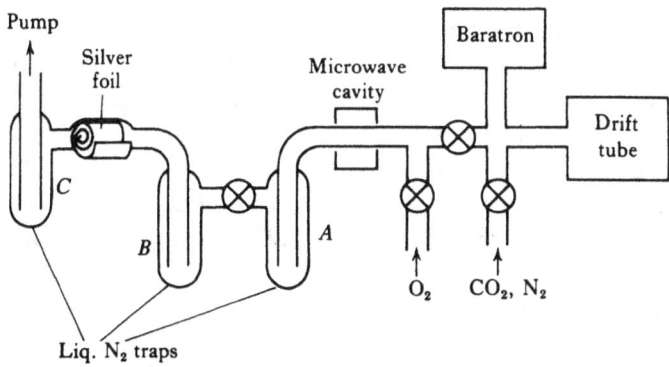

Fig. 9.58. Arrangement of the ozone preparation and gas-handling system in the experiments of Stelman et al. (1972). The Ag foil converts O_3 to O_2 on pump-out.

the desired amount of O_3 was admitted to the drift tube to the required density which was measured from the absorption of a selected ultraviolet mercury line. In many experiments CO_2 was added to the O_3 in the drift tube, in which case the total pressure was measured with a Baratron pressure gauge.

All valves exposed to the ozone were of stainless steel with stainless-steel-to-glass seals, and all brazed joints of the original tube were replaced with Heliarc welds. Spot welds were made using aluminium instead of copper electrodes, and all Kovar-glass feedthroughs were treated with H_2O_2. It was an advantage to keep the system filled with ozone as often as possible as this reduced the ozone destruction rate during the experiments.

In pure O_3 the attachment coefficient α_a was found to be proportional to the ozone density for densities between 1.5×10^{16} an 7×10^{16} cm^{-3}, suggesting that the process concerned is one of dissociative attachment

$$O_3 + e \longrightarrow O_2^- + O,$$

a process observed by Curran (1961) in mass spectrometer experiments with electron beams. As no data are available about the drift velocities of electrons in pure O_3 as a function of F/p, it is not possible to convert the measured values of α_a into rate coefficients. However, in CO_2 containing less than 10% of O_3, it was possible to use previously measured data on drift velocities and characteristic energies in CO_2, it being assumed that the small amount of O_3 present had proportionally small effect.

At low CO_2 pressures (less than 20 torr) the apparent two-body rate coefficient k_{app} was found to be effectively independent of O_3 pressure as for a two-body attachment process. At higher CO_2 pressures, k_{app} was found to increase with decreasing pressure of O_3, an effect ascribed to the presence, at a constant partial pressure, of an impurity with a large attachment coefficient, despite the stringent precautions taken.

Fig. 9.59 shows the two-body rate coefficient, taken from the observations at low CO_2 pressure, as a function of the characteristic electron energy. For comparison, the rate coefficient calculated from attachment cross-sections measured in beam experiments by Chantry (1972b), is also included. While agreeing reasonably well at characteristic energies of 0.2 eV and higher, the drift tube gives considerably smaller rate coefficients at low energies. Assuming a Maxwellian velocity distribution, the drift-tube data are consistent with an attachment cross-section

$$Q_a = (1.5 \pm 0.3) \times 10^{-17} E^{0.96} \text{ cm}^2,$$

where E is the electron energy in eV.

SF_6

A great deal of attention has been paid to the experimental study of electron attachment in halogen-containing compounds because of their dielectric strengths and flame-inhibiting properties (see

NEGATIVE ION FORMATION – DISSOCIATIVE PROCESSES

Chapter 15, p. 660). Of these, sulphur hexafluoride has proved to be of exceptional interest because, in it, attachment reactions involving slow electrons take place very readily. Over a wide range of pressures, SF_6^- is the main ion produced by impact of electrons

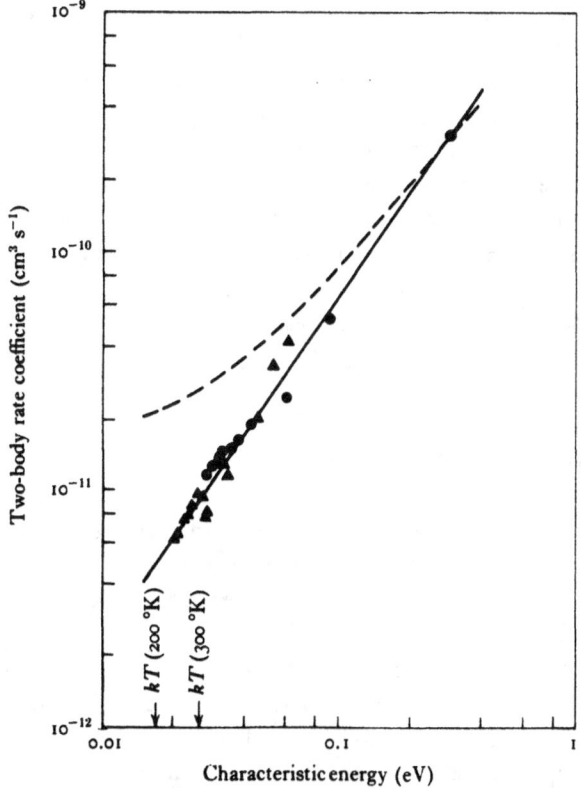

Fig. 9.59. Variation, with characteristic energy of the electrons, of the two-body rate coefficient for attachment to O_3 dilute in CO_2. —— observed by Stelman et al. (1972), ▲ at 200 °K, ● at 300 °K. ---- derived from attachment cross-sections, observed by Chantry (1972b), using electron beams.

with nearly zero energy. The rate of production of this ion is so fast that SF_6 is used in experiments as a 'scavenger' for slow electrons. In this capture process the SF_6^- ion first produced will be unstable towards autodetachment but its lifetime is so long that it may be observed in a mass spectrometer with normal path length. Furthermore, at pressures typical of swarm experiments there is a high probability of collision stabilization before autodetachment can occur. In fact, under these conditions, the formation of SF_6^- appears to occur as a two-body process because the chance of stabilization is effectively unity (see p. 269), no matter what atoms or molecules act as third bodies.

It is not surprising that a complex ion such as SF_6^- should have a long lifetime against autodetachment. The energy brought in by the captured electron will be rapidly distributed among a number of internal modes of motion so that, on the average, a considerable time will elapse before autodetachment.

The long lifetime means that the capture process will only occur if the electrons have a sharply defined energy – the resonance width will be very small (for a lifetime of 10 μs the width is only 6×10^{-11} eV). It follows that, in any experiments, the observed variation of SF_6^- production with electron energy will simply be determined by the shape of the electron-energy distribution. This provides a convenient means of determining this distribution in certain circumstances.

A still further use of SF_6^- production is to calibrate the electron-energy scale in experiments on the production of negative ions from other substances. This is because the resonance energy for capture by SF_6 is very close to zero. As SF_6 is relatively inert chemically, it can be mixed, at any rate at low pressures, with most other substances without reaction.

The main features of negative-ion production in SF_6 outlined above were established in the experiments of Hickam and Fox in 1956. They used both a total ion collector and a 90°-sectored-field mass spectrometer. Both the electrode systems used to produce the electron beam made it possible to use the retarding potential difference method to obtain results for electrons with energy defined to about 0.1 eV. Fig. 9.60 shows results they obtained with the mass spectrometer for the variation with electron energy of the currents of SF_6^- and SF_5^- ions. The variation of the total ion current agreed well with that observed in the total ion collection instrument.

Fig. 9.61 gives a comparison between the shape of SF_6^- current peak and the energy distribution of the electrons obtained by a retarding potential analysis. It can be seen that the shapes agree

NEGATIVE ION FORMATION – DISSOCIATIVE PROCESSES

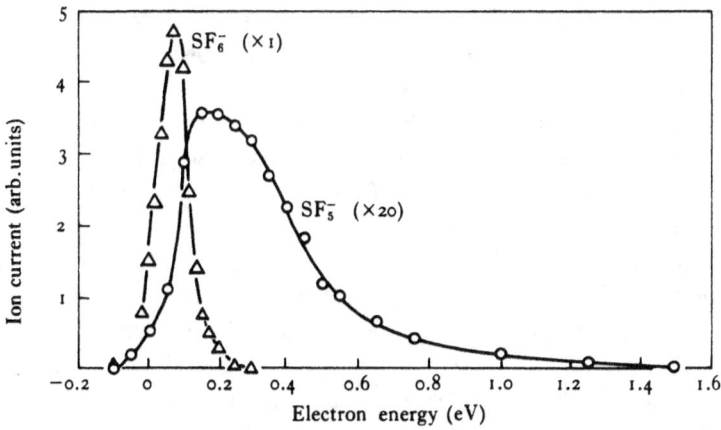

Fig. 9.60. Variation with electron energy of the yield of SF_6^- and SF_5^- ions by electron impact in SF_6, observed by Hickam and Fox (1956).

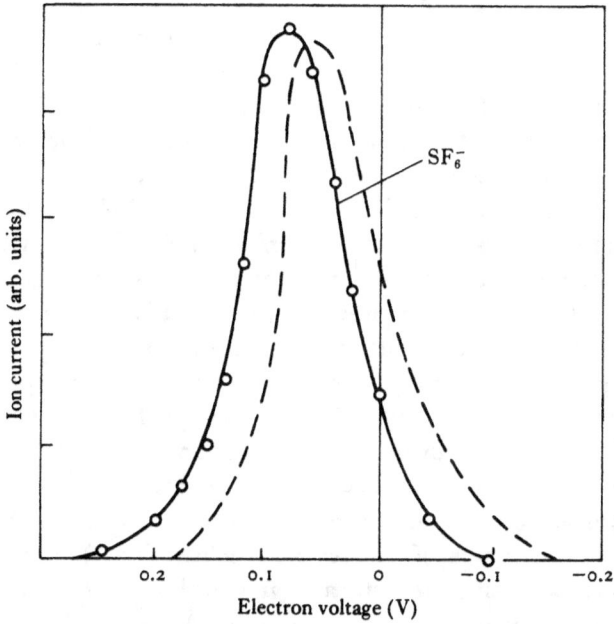

Fig. 9.61. Comparison of the electron energy distribution (---) determined by a retarding potential analysis with the shape of the SF_6^- peak (-○-).

very well although the peak of SF_6^- production is shifted by about 0.03 eV to a higher energy than that of the electron-energy distribution.

The SF_6^- pressure in the mass spectrometer was as low as 10^{-5} torr so that the SF_6^- ions observed could not have been stabilized by collision. The fact that the observed rate of production was proportional to the pressure then shows that the ions initially formed by electron capture must have a lifetime at least of some μs in agreement with the results obtained in later time-of-flight experiments (see p. 373).

From the total ion-collection measurements, the cross-section for capture of an electron by SF_6 at the peak must be of order 10^{-15} cm². This is consistent with later measurements carried out by Rapp and Briglia (1965), using the apparatus described on p. 278, and by Buchel'nikova (1958) (see p. 280), who obtained peak values of 2.1×10^{-16} and 5.1×10^{-16} cm² respectively. The resonance peak is so sharp that the results obtained depend very much on the electron-energy distribution.

Spence and Schulz (1973), using the equipment described on p. 281, found no variation with temperature of the rate of attachment of slow electrons to SF_6.

Further information is available from observations at higher pressures where collision stabilization of the initially formed ion becomes important. As described on p. 309, Fehsenfeld (1970) measured the attachment rate coefficients for SF_6 in a flowing afterglow buffered with helium over the pressure range 0.1–1.5 torr and temperatures between 293 and 523 °K. As in the low-pressure beam experiments, the rate coefficient was found to be independent of pressure, but in this pressure range this could only be because collision stabilization occurred so fast that almost all the SF_6^- ions initially formed were stabilized before autodetachment. The observed coefficient 2.2×10^{-7} cm³ s⁻¹ was also independent of temperature in the experimental range in agreement with the beam measurements of Spence and Schulz (1973). Mahan and Young (1966), using the static afterglow technique in the electron-diffusion dominated regime (see p. 299), obtained 3.1×10^{-7} cm³ s⁻¹.

If, at the pressures of helium concerned, collision stabilization is almost certain, then we would expect that it would be equally certain for other third bodies. In other words, the apparent two-body rate coefficient for production of negative ions in mixtures of

NEGATIVE ION FORMATION – DISSOCIATIVE PROCESSES 373

small concentrations of SF_6 in different gases should be independent of the nature of the gas. Experiments carried out by Davis and Nelson (1969, 1970) using a pulse technique confirm that this is indeed true. Table 9.1 gives a summary of their measured rate coefficients for SF_6 in small concentration in a variety of gases.

The result for helium agrees well with that measured by Fehsenfeld (1970). Apart from the measurements which have been made in helium using the static afterglow technique the rate has been measured in C_2H_4 by a drift-tube technique (Compton et al., 1966). While the absolute value is about $\frac{1}{4}$ of that measured by Nelson and Davis, the rate was found to be independent of temperature as in the experiments of Fehsenfeld (1970) in helium.

The magnitude of the rate coefficient observed in the swarm experiments corresponds to a mean attachment cross-section for electrons at room temperature of 1.4×10^{-14} cm^2, somewhat larger than the peak cross-section estimated from the beam experiments. This is not surprising, for the latter values were limited by the energy resolution of the electron source used.

It seems well-established that stabilization is almost certain at pressures of 0.1 torr in all gases. Since the time between collisions is of order 1 μs, this is consistent with the lifetimes observed in the time-of-flight experiments which we now describe.

The lifetime of the SF_6^- ion initially formed by capture of a slow electron has been investigated by time-of-flight (cf. Chapter 4, p. 130) and by ion cyclotron resonance techniques. Using the former technique Edelson, Griffiths and McAfee (1962) obtained a lifetime of 10 μs while a little later Compton, Christophorou, Hurst and Reinhardt (1966) obtained a value of 25 μs by much the same technique. A very different result was obtained by Henis and Mabie (1970), who measured the ion cyclotron resonance line width (see Chapter 11, p. 445) and derived a lifetime $\geqslant 500$ μs. A probable explanation of the large discrepancy between these results is forthcoming from the experiments of Odom, Smith and Futrell (1974). These authors used the ion cyclotron resonance technique in a different way.

SF_6^- and SF_5^- ions, and free scattered electrons, produced from a pulsed electron beam passing through SF_6, drifted with the same speed under the action of crossed electric and magnetic fields,

TABLE 9.1 *Apparent two-body rate coefficients for SF_6 in small concentrations in different gases and gas mixtures, measured by Davis and Nelson (1970)*

Main gas	He	H_2	N_2	CO	CO_2	CH_4	CF_4	C_2H_2	C_2H_4	C_2F_6	1% CH_4 +99% Ar	11% CH_4 +89% Ar
Range of SF_6 concentration (ppm)	2.1 5.0	23.9	2.5 2.7	5.2 6.1	1.1 2.5	11.7 24.2	17.7	16.7	6.4 24.4	18.7	2.9	6.3
Rate coefficient (10^{-7} cm^3 s^{-1})	2.12	2.16	2.075	2.22	2.06	2.10	2.21	2.14	2.12	2.04	2.10	2.15

towards a collector plate. With an SF_6 pressure of 3×10^{-7} torr many of the electrons were captured after a time interval of 400 µs. Free electrons could be removed selectively from the mixed swarm by application of an r.f. electric field of appropriate frequency to one of the trapping plates.

The number of heavy ions remaining at any time was obtained by measuring the current to the collector plate immediately after a pulse of an r.f. field had been applied to remove the electrons.

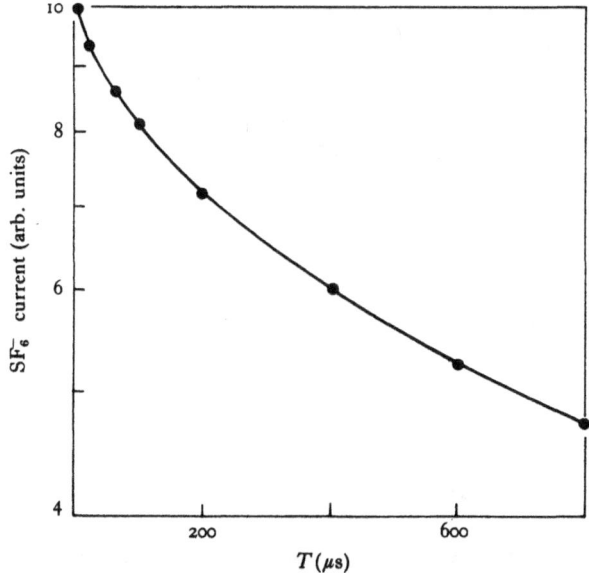

Fig. 9.62. Variation with time T after electron removal of the SF_6^- current, in the experiments of Odom et al. (1974). Note the logarithmic scale.

Owing to autodetachment, the SF_6^- current so measured should decrease as the duration T of the electron ejection pulse is increased. Fig. 9.62 shows that this is indeed so. However, if the ions possessed a single definite lifetime τ, the slope of the plot in Fig. 9.62 should be a straight line with slope equal to $1/\tau$. In fact, the apparent lifetime at any time T determined by the slope of the tangent to the curve in Fig. 9.62 at T, increases with T as shown in Fig. 9.63. It follows that the lifetime measured in any particular experiment will depend on the time since formation of the SF_6^-.

Although these preliminary experiments do not give information about the lifetime of SF_6^- within tens of μs of formation, they do suggest that the SF_6^- ions may be produced in a number of different initial states which differ in lifetime over a wide range.

Other halogen-containing molecules

Stockdale, Compton and Schweinler (1970) have studied negative-ion formation in a number of other hexafluorides, including SeF_6,

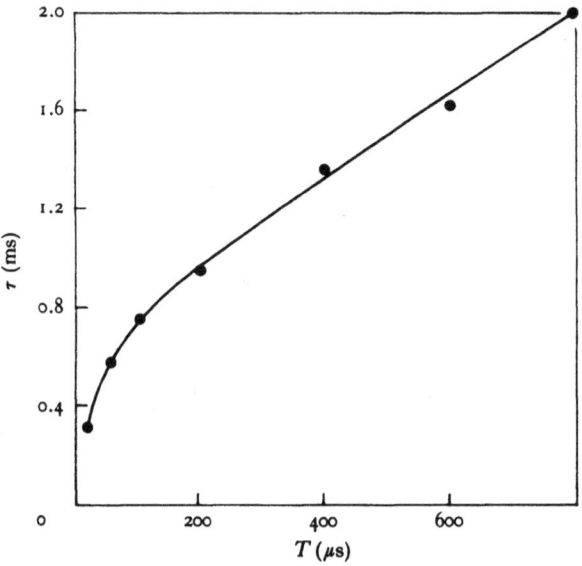

Fig. 9.63. Apparent autoionization lifetime τ as a function of time T derived from the curve of Fig. 9.62.

TeF_6, MoF_6, ReF_6 and UF_6. Of these, hexafluoride ions were formed only by MoF_6 and ReF_6, at least in the electron energy range 0 to about 10 eV. On the other hand SeF_6^-, TeF_6^- and MoF_6^- are all formed in charge transfer reactions with SF_6^{-*} at thermal energies. Begum and Compton (1969) also found that XeF_6 does not form XeF_6^-, although a variety of dissociated ions F^-, F_2^-, XeF^-, XeF_2^-, XeF_3^- and XeF_4^- were found.

A number of tetrahalides have been studied including CCl_4 (Reese, Dibeler and Mohler, 1956; Buchel'nikova, 1958; Fox and Curran, 1961; MacNeil and Thynne, 1968), CF_4 (Craggs and

McDowell, 1955), SiCl$_4$ (Vought, 1947; Jäger and Henglein, 1968), TiCl$_4$ (Marriott, Thorburn and Craggs, 1954) and XeF$_4$ (Begum and Compton, 1969). Among these, the only case in which there is clear evidence of the formation of an MX$_4^-$ ion at low electron energies is that of SiCl$_4$ observed by Jäger and Henglein. Many cases of mixed halogen tetrahalides have also been studied.

A considerable number of measurements have been made of attachment coefficients α_a as functions of F/p, for a wide variety of halogen-substituted aliphatic as well as aromatic hydrocarbons. A number of measurements have also been made for antiknock compounds such as tetraethyl lead. The interpretation of these measurements in terms of attachment rate coefficients as functions of characteristic electron energies requires knowledge of the electron drift velocity u and characteristic energy ϵ. Few such measurements are available in the vapours of the pure compounds but in some experiments this has been overcome by carrying out the measurements with the vapour under study diluted in nitrogen, for which gas not only are u and ϵ known but also the electron velocity distribution as functions of F/p (Engelhardt, Phelps and Risk, 1964).

One particularly comprehensive study on these lines has been carried out by Lee (1963) who was particularly concerned with flame-inhibitors and antiknock compounds, Thus he measured α_a/p as a function of F/p, over a range which corresponds to characteristic energies of 0.05–1.15 eV in N$_2$, for CF$_2$Br$_2$, CCl$_4$, CH$_2$Br$_2$, CH$_3$I, CF$_3$Br, CHCl$_3$, C$_2$H$_5$I, CH$_2$ClBr, CH$_3$Br, CF$_2$Cl$_2$, CHF$_3$, CF$_3$Cl, C$_2$H$_5$Br, CH$_2$Cl$_2$, CHF$_2$Cl, CH$_3$Cl, BF$_3$, CF$_4$ and the antiknock compounds Pb(C$_2$H$_5$)$_4$, Fe(CO)$_5$ and (CH$_3$C$_5$H$_4$)Mn-(CO)$_3$. A very wide spread in magnitude of α_a/p was found for the different compounds.

These and other results were analysed by Christophorou and Stockdale (1968) and extended to a number of other halogenated aliphatic compounds by Blaunstein and Christophorou (1968). Table 9.2 gives a number of attachment rate coefficients for thermal electrons which they derived.

Mahan and Young (1966) have used the static afterglow method, in the electron-diffusion dominated regime, to measure the rate coefficients for attachment of thermal electrons to prefluoromethyl

TABLE 9.2 *Attachment rate coefficients for thermal electrons in halogen-substituted hydrocarbons and antiknock compounds. Energy-integrated cross-sections are also given in many cases*

Compound	CH$_3$Cl	CH$_2$Cl$_2$	CHCl$_3$	CCl$_4$	C$_2$HCl$_3$	1,1,1-C$_2$H$_3$Cl$_3$	1,1,2-C$_2$H$_3$Cl$_3$
Rate coefficient (10^{-8} cm^3 s^{-1})	5.4 × 10^{-3}	1.4 × 10^{-3}	0.34	24	0.17	1.4	0.013
Energy-integrated cross-section (10^{-16} cm^2 eV)	1.5 × 10^{-3}	0.016	16.5	3.1	1.5	16	0.54

Compound	CF$_3$Cl	CF$_2$Cl$_2$	CFCl$_3$	C$_2$F$_3$Cl$_3$	CHF$_2$Cl	CHFCl$_2$	CHF$_3$	CF$_4$
Rate coefficient (10^{-8} cm^3 s^{-1})	0.040	—	9.8	3	5.4 × 10^{-3}	0.11	0.10	7 × 10^{-5}
Energy-integrated cross-section (10^{-16} cm^2 eV)	2.9 × 10^{-3}	—	14	8.2	—	—	0.16	—

Compound	CH$_3$Br	CH$_2$Br$_2$	C$_2$H$_2$Br$_4$	CF$_3$Br	CF$_2$Br$_2$	CH$_3$I	C$_2$H$_5$I
Rate coefficient (10^{-8} cm^3 s^{-1})	0.29	2.8	1.5	1.1	23	2.2	0.42
Energy-integrated cross-section (10^{-16} cm^2 eV)	0.37	5.9	0.52	1.5	26.9	4.3	1.4

Compound	C$_6$H$_5$Cl	C$_6$H$_5$Br	C$_6$D$_5$Br	o-C$_6$H$_4$Cl$_2$	o-C$_6$H$_4$CH$_3$Cl	o-C$_6$H$_4$CH$_3$Br
Rate coefficient (10^{-8} cm^3 s^{-1})	—	—	—	—	—	—
Energy-integrated cross-section (10^{-16} cm^2 eV)	0.09	0.71	0.71	2.3	0.15	0.48

Compound	Pb(C$_2$H$_5$)$_4$	Fe(CO)$_5$	(CH$_3$C$_5$H$_4$)Mn(CO)$_3$
Rate coefficient (10^{-8} cm^3 s^{-1})	9 × 10^{-3}	0.1	0.04

NEGATIVE ION FORMATION – DISSOCIATIVE PROCESSES

cyclohexane C_7F_{14} (see also p. 688). They found the high rate coefficient of 9.8×10^{-8} cm^3 s^{-1}.

It is clear from these data that the thermal attachment rate coefficient varies greatly between the different compounds but in many cases is as large as 10^{-8} cm^3 s^{-1} or higher. Advantage has been

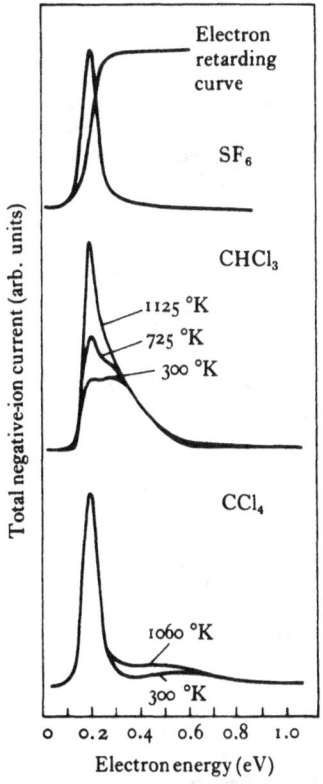

Fig. 9.64. Energy variation of cross-sections for attachment of electrons to SF$_6$, CHCl$_3$ and CCl$_4$ at different temperatures, observed by Spence and Schulz (1973).

taken of this for detecting traces of these strongly attaching compounds in a large excess of non-attaching gas (see Chapter 15, p. 682).

Spence and Schulz (1973a) have studied the temperature dependence of cross-sections for attachment of electrons, with energies extending down to less than 0.2 eV, to a number of polyatomic halides (SF$_6$, CCl$_4$, CFCl$_3$, CH$_2$Br$_2$, CH$_3$I, CHCl$_3$, CF$_3$Br and CH$_3$Br) using the apparatus described on p. 281. Fig. 9.64 shows the temperature dependence of the total cross-section for negative-ion production in SF$_6$, CHCl$_3$ and CCl$_4$. In SF$_6$ there is no observable dependence, while in CHCl$_3$ there is a marked increase

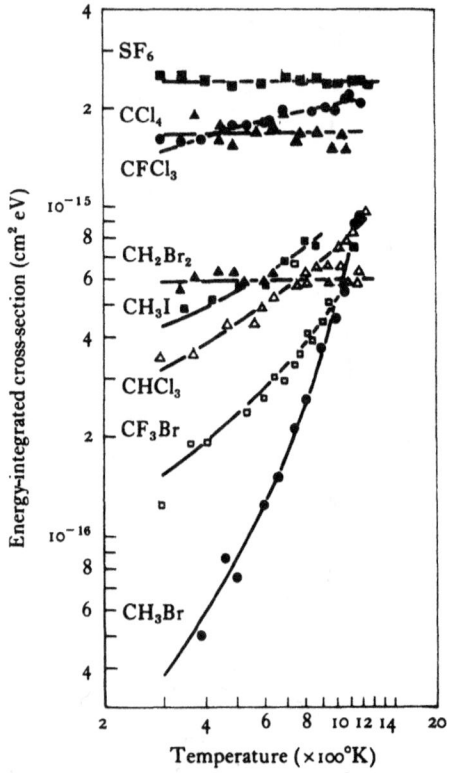

Fig. 9.65. Dependence on temperature of the energy-integrated cross-sections for attachment of electrons to SF$_6$, and a number of halogenated methanes, observed by Spence and Schulz (1973a).

of the low energy peak with temperature. In CCl$_4$, on the other hand, this peak does not vary but the much broader peak at higher energies increases with temperature.

Fig. 9.65 shows the variation with temperature of the total cross-sections for negative-ion production, integrated over the observed electron energy range, for all the molecules investigated. Absolute values for the cross-sections were obtained by normalizing to attachment measurements in swarm experiments (Blaunstein and Christophorou, 1968). It will be seen that as the temperature increases the integrated cross-sections, on the whole, tend to equality although at room temperature there is a very wide spread in magnitude.

As explained on p. 346, the problem of determining electron affinities of radicals from appearance potential observations with polyatomic molecules is complicated by the uncertainty about the amount of internal energy carried away by the reaction products. De Corpo and Franklin (1971) have attempted to overcome this difficulty by a semi-empirical procedure based on the analysis of a number of cases for which the electron affinities and ionization potentials of the products are known. They found that the excess internal energy can be written

$$E_{\text{int}} = \alpha N E_{\text{tr}},$$

where E_{tr} is the translation energy carried away by the products, N is the number of vibrational degrees of freedom and α is a constant equal to 0.42. Under these circumstances, if E_{tr} can be measured, it is possible to obtain from the threshold for the dissociative attachment,

$$e + XY \longrightarrow X + Y^-,$$

$D(XY) + E_i(X) - E_a(Y)$, in the usual notation.

In applying this to determine the electron affinities of halogen molecules de Corpo and Franklin studied the reactions

$$CCl_4 + e \longrightarrow Cl_2^- + CCl_2, \qquad (9.67a)$$
$$CBr_4 + e \longrightarrow Br_2^- + CBr_2, \qquad (9.67b)$$
$$CI_4 + e \longrightarrow I_2^- + CI_2, \qquad (9.67c)$$
$$CHI_3 + e \longrightarrow I_2^- + CHI, \qquad (9.67d)$$
$$BF_3 + e \longrightarrow F_2^- + BF. \qquad (9.67e)$$

The translational energies of the fragment ions were determined from measurement of the peak width at half-height of the mass spectrum line (Franklin, Hierl and Whan, 1967). From reaction (9.67a) they found $E_a(Cl_2) = 2.52$ eV, from (9.67b) $E_a(Br_2) = 2.87$ eV, from (9.67c) and (9.67d) $E_a(I_2) > 2.0$ eV and $E_a(I_2) = 2.6$ eV respectively and from (9.67e) $E_a(F_2) = 2.9$ eV. These values are compared with those obtained using other methods in Chapter 6, p. 211.

CHAPTER 10

Formation of Negative Ions by Capture of Bound Electrons

Negative ions may be formed in collisions between neutral systems through reactions of the type

$$X + Y \longrightarrow X^- + Y^+, \qquad (10.1)$$

in which X captures an electron from Y. A further possibility is that, in a collision with a neutral system Y, a positive ion X^+ may capture two electrons to form a negative ion

$$X^+ + Y \longrightarrow X^- + Y^{2+}. \qquad (10.2)$$

A number of reactions of both of these types have been studied experimentally and we now discuss the techniques employed, the results obtained and their interpretation.

10.1 Formation of negative ions in collisions between neutral systems

The threshold energy for (10.1) is $E_i(Y) - E_a(X)$, where $E_i(Y)$ is the ionization energy of Y and $E_a(X)$ the electron affinity of X. The atom with the lowest ionization energy is Cs with $E_i(\text{Cs}) = 3.89$ eV, and that with highest electron affinity is Cl with $E_a(\text{Cl}) = 3.6$ eV so that, if X and Y are atoms, the reaction (10.1) is always endothermic. This will still be true in most cases if X and Y are diatomic or polyatomic although there may well be radicals for which the electron affinity > 3.89 eV.

If the threshold energy can be measured, a further method for determining electron affinities becomes available. In general, however, this is only possible if $\Delta E = E_i(Y) - E_a(X)$ is small. Otherwise the cross-section rises so gradually from the threshold as the impact energy increases that no precise location of the threshold is possible. Even if ΔE is small, a number of problems arise which must be solved before reliable results for $E_a(X)$ can be obtained.

We first discuss the variation with impact energy of the cross-section for reactions such as (10.1). Consider a collision between two systems X and Y which result in transitions to states X', Y'', involving energy transfer $\Delta E(\infty)$. We can regard the process as one involving a transition between an initial and a final molecular state of XY. In very general terms we can say that the probability of the transition occurring in a collision will be small if the time τ of collision is long compared with the time T required for a transition to occur between the two states. Under these conditions the collision will be nearly adiabatic. Also, if $\tau \ll T$, there will not be time for a transition to occur and again the probability per collision will be small. We could expect the probability to be a maximum when $\tau \simeq T$. The problem is to estimate τ and T in any particular case.

As a crude first approximation we take $\tau \simeq a/v$, where a is the range of interaction between the systems and v their initial relative impact velocity. If in the same spirit we take $T = h/\Delta E(\infty)$, then the conditions are nearly adiabatic if

$$\lambda_0 = a\Delta E(\infty)/hv \gg 1. \tag{10.3}$$

If we take a as 5×10^{-8} cm, then for collisions between H atoms leading to production of H$^+$ and H$^-$ ions, $\Delta E = 12.8$ eV and λ_0 is > 1 for $E < 5$ keV. This would mean that the cross-section would be extremely small in the neighbourhood of the threshold. Even for Cs–Cl collisions leading to Cs$^+$ and Cl$^-$ for which $\Delta E = 0.29$ eV and a will be larger, λ_0 is > 1 for $E < 1$ keV if we take $a = 10^{-7}$ cm, and it would not appear that there would be much chance of accurate determination of the threshold. However, the criterion (10.3) is very crude and is likely to be particularly misleading for reactions of the type (10.1), where ΔE is small.

To see this we need only take account of the fact that while at infinite separation the energy difference between the initial and final molecular states is $\Delta E(\infty)$ this difference will vary with nuclear separation. Over a range Δa of separation, $\Delta E(R)$ may be much smaller than $\Delta E(\infty)$. The criterion (10.3) should then become

$$\lambda_1 = \Delta a \Delta E(R)/hv(R) \gg 1, \tag{10.4}$$

where $v(R)$ is the velocity of relative motion at separation R. This condition may be much less severe than (10.3). An important case in the present context to which these considerations apply is that of pseudo-curve crossing which we now discuss.

ION FORMATION – BOUND ELECTRON CAPTURE

Fig. 10.1 illustrates possible forms of the potential-energy curves of the initial and final molecular states as functions of the nuclear separation. If the transition probability between them vanished at all nuclear separations and impact energies the two curves would cross at the point R_c as shown on Fig. 10.1(a). However, if the two states do interact so that at finite velocities of impact the transition probability remains finite, the curves cannot intersect and the situation is as shown in Fig. 10.1(b). If X and Y approach infinitely slowly the interaction will follow the curve I_a. Equally well the

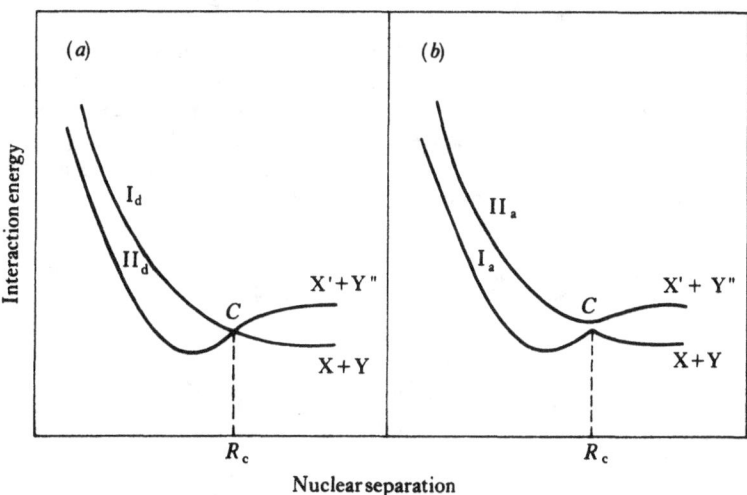

Fig. 10.1. Illustrating the pseudo-crossing of potential-energy curves.

interaction between the excited systems X' and Y" will follow the curve II_a if they approach adiabatically. On the other hand, if X and Y approach at finite velocity, sufficient for them to approach closer than R_c, a transition between I_a and II_a may occur with a probability P. If this happens the systems will proceed along II_a to a closest approach R_1 say, and then separate again. On the way out there will be a probability $1 - P$ that they will continue along II_a ending up at infinite separation as the excited systems X' and Y". The probability of this sequence of events will be $P(1 - P)$. Equally well X and Y may approach to R_c and continue on along I_a to a closest distance of approach R_2, the probability of this being

$1 - P$. If on the way out a transition takes place, with probability P to II_a, the systems will again appear at infinite separation in the excited states X', Y". This doubles the probability of the transition to these states occurring on collision.

Referring back to Fig. 10.1 it will be seen that, when a transition occurs at the near crossing, the systems follow very nearly the crossing curves I_d, II_d which are often referred to as the *diabatic* as distinct from the *adiabatic* interactions. We have then

$$I_d \simeq I_a, \quad II_d \simeq II_a, \quad R > R_c,$$
$$I_d \simeq II_a, \quad II_d \simeq I_a, \quad R < R_c.$$

The criterion (10.4) may be invoked to determine the conditions under which we would expect P to be small. Let $U(R_c)$ be the separation of the curves at the near crossing. Then we may take Δa to be such that at $R_c \pm \Delta a$ the separation between the curves is $\mu U(R_c)$, where μ is a number > 1 but of order unity. If V_d^I, V_d^{II} are the diabatic interactions, the separation between the curves at $R_c \pm \Delta a$ is given approximately by

$$\Delta a \left\{ \frac{d}{dR}(V_d^I - V_d^{II}) \right\}_{R=R_c}, \tag{10.5}$$

so

$$\Delta a \simeq \mu U(R_c) \Big/ \left\{ \frac{d}{dR}(V_d^I - V_d^{II}) \right\}_{R=R_c} \tag{10.6}$$

and (10.4) becomes

$$\mu \{U(R_c)\}^2 \Big/ hv(R_c) \left\{ \frac{d}{dR}(V_d^I - V_d^{II}) \right\}_{R=R_c} \gg 1. \tag{10.7}$$

Under these conditions we would expect P to be small.

The velocity $v(R_c)$ will depend on the relative angular momentum J of the colliding systems. Thus if M is the reduced mass of the colliding systems

$$\tfrac{1}{2} M v_J(R_c)^2 = \tfrac{1}{2} M v^2 - V_d^I - J^2/2MR_c^2. \tag{10.8}$$

J will of course be quantized so

$$J^2 = l(l+1)\hbar^2, \quad l = 0, 1, 2,$$

and we note the dependence on l by writing P_l for the probability instead of just P.

Since when $v_l(R_c)^2 < 0$ the closest distance of approach will be greater than R_c, we can assume that P_l will be negligible under these conditions.

An approximate formula for P_l, which is valid when the region of interaction is strictly confined to the neighbourhood of $R = R_c$, is (Mott and Massey, 1965, p. 804)

$$P_l = \exp\left\{-\frac{4\pi^2}{hv_l}U(R_c)^2 \bigg/ \frac{d}{dR}(V_d^I - V_d^{II})_{R=R_c}\right\}, \quad (10.9)$$

which is consistent with (10.7), v_l being written for $v_J(R_c)$.

To derive the cross-section for a collision in which the reaction

$$X + Y \longrightarrow X^+ + Y^- \quad (10.10)$$

occurs, we first note that the maximum possible cross-section for collisions in which the angular momentum quantum number is l, is

$$\pi(2l+1)/k^2, \quad (10.11)$$

where $k = Mv/\hbar$ is the wave number of the initial relative motion. Since $2P_l(1 - P_l)$ is the probability that the transition involved will take place, we may write for the contribution to the cross-section from collisions with the particular quantized angular momentum

$$\pi(2l+1)\,2P_l(1 - P_l)\,k^{-2}. \quad (10.12)$$

The required total cross-section is then

$$Q = 2\pi k^{-2} \sum_{l=0}^{l_1}(2l+1)P_l(1 - P_l), \quad (10.13)$$

where l_1 is the largest value of l for which v_l is > 0. As in collisions between systems of atomic mass $l \gg 1$ under most circumstances of importance,

$$Q \simeq 4\pi k^{-2}\int_0^{l_1} l P_l(1 - P_l)\,dl. \quad (10.14)$$

For collisions which lead to the reaction (10.10), at very large separations R

$$V_d^{II} = V_a^{II} = -e^2/R,$$

while $V_d^I = V_a^{II} \to 0$ at least as fast as R^{-6}. Hence if $\Delta E(\infty)$ is small, R_c is approximately given by

$$R_c = e^2/|\Delta E|. \quad (10.15)$$

On substitution in (10.14) we now have, using (10.9),

$$Q = 4\pi R_c^2\, I(\eta), \quad (10.16)$$

where

$$I(\eta) = \int_1^\infty \exp(-\eta x)\{1 - \exp(-\eta x)\} x^{-3}\, dx \qquad (10.17)$$

and

$$\eta = 4\pi^2 e^2 M^{1/2} \{U(R_c)\}^2 / 2^{1/2} h\, E^{1/2}\, \Delta E^2. \qquad (10.18)$$

$I(\eta)$ is illustrated as a function of η in Fig. 10.2. It has a maximum value of 0.113 when $\eta = 0.424$. Using (10.16) and (10.18), this means that the cross-section will rise with increasing energy E from the threshold to a

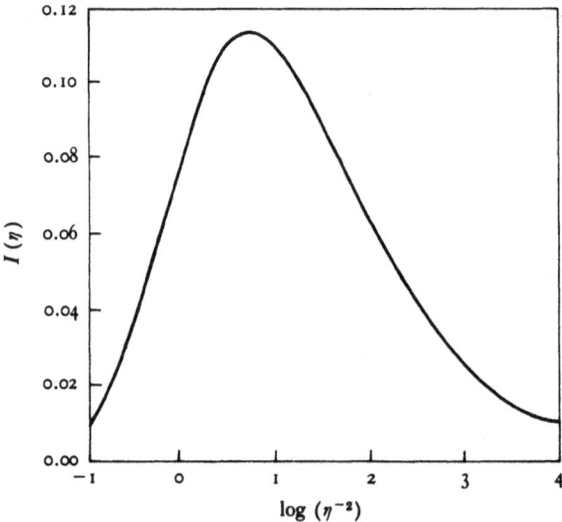

Fig. 10.2 The function $I(\eta)$ of Equation (10.17).

maximum when the initial impact velocity v $(=(2E/M)^{1/2})$ is given by

$$v_m = 4\pi^2 e^2 \{U(R_c)\}^2 / 0.424 h \Delta E^2. \qquad (10.19)$$

Before this formula may be applied to locate the maximum cross-section it is necessary to evaluate $U(R_c)$. For collisions of H atoms with alkali metal atoms in which H$^-$ is produced, Bates and Boyd (1956) estimate that $U(R_c)$ is close to 0.3 eV. Thus for Na–H collisions in which $\Delta E = 4.4$ eV we would have from (10.19) that $v_m = 3.6 \times 10^6$ cm s^{-1} corresponding to a relative impact energy of around 100 eV.

For Cs–Cl collisions, with $\Delta E = 0.29$ eV, R_c is as large as $80 a_0$.

The separation of the potential-energy curves with decreasing R will then be so gradual that it will no longer be a good approximation to regard the transition as taking place mainly for $R \simeq R_c$. Indeed, for such large values of R_c, $U(R_c)$ will be so small that P_t given by (10.9) will not be small but close to unity. In fact the transition will take place over a range of smaller values of R for which $U(R)$ will be considerably larger. Prediction of the cross-section in these cases requires a more elaborate theory.

The opposite extreme arises for, say, H–He collisions in which $\Delta E = 23.5$ eV. R_c as given by (10.15) is now of atomic dimensions and the assumptions that $V_d^{II} = -e^2/R$ and $V_d^I \to 0$ are certainly no longer valid. Under these circumstances in which curve crossing, if at all, occurs at atomic separations, accurate prediction again requires elaborate analysis but as a rough guide to the variation with collision energy the simple form (10.3) of the adiabatic condition may be used. This indicates that the maximum cross-section will occur at energies of a few keV.

We shall describe first the experiments which have been carried out to measure the capture cross-sections at energies of impact below 100 eV for cases in which $|\Delta E|$ is of the order of a few eV. In such cases there is hope of determining the threshold energy with reasonable accuracy. For collisions in which $|\Delta E|$ is considerably larger, experimental work has been confined to measurements using incident neutral beams of a few keV energy. The methods used and results obtained are discussed on pp. 394–414.

10.2 Low-energy experiments on the measurement of cross-sections and threshold energies

All experiments at low energies are confined to the study of collisions in which one of the colliding partners has a low ionization potential, which means in practice alkali metal atoms. Two types of source of beams of such atoms have been used. One, the sputter source, uses atoms sputtered from an alkali metal target by impact of Ar^+ ions. These atoms are collimated by suitable slits and velocity-selected by a mechanical selector. The second type of source uses charge transfer from a beam of ions, the only technique available for preparation of high-energy neutral beams. However, in order that the charge transfer cross-section should be high enough at the low energies with which we are now dealing, the ions and atoms must be of the same kind – for symmetrical charge transfer the cross-section increases as the impact energy decreases. In some experiments a device for increasing the energy resolution of the beam, such as a mechanical velocity selector, has been introduced with this type of source.

The neutral alkali metal beam either passes through a collision chamber containing the other neutral species as a gas or crosses a molecular beam of these molecules. The yield of negative ions is then measured as a function of the relative impact energy down to the threshold energy. It is easy to see that the accurate determination of this energy will be difficult even though the cross-section rises quite quickly above it. The thermal velocity of the target molecules and the velocity spread in the alkali atom beam will both blur the onset and their effects can only be unfolded by a detailed analysis.

The usual procedure is to assume simple plausible laws of variation of the cross-section with $E - E_t$, the energy above threshold, and then calculate what would be the observed variation when thermal and beam energy spread are both included. Comparison with the observed results gives the best values of E_t obtainable with the different assumptions. If there is not too-large a dependence on the assumed law, or if one form gives much better agreement with observation than others, a reasonably definite value of E_t is obtained.

From E_t the electron affinity of Y follows immediately if Y is an atom. If, however, Y is molecular the true threshold must refer to an initial state of Y and final state of Y⁻ which involve no vibrational or rotational excitation. That is to say, the electron affinity should be the adiabatic one. If this differs markedly from the vertical detachment energy the probability of transitions occurring on collision in which the final ion is in its lowest vibrational state will be very small. This will make it difficult to ensure that the threshold derived from the experiments, as described above, will be the true threshold and not one determined by the sensitivity of experimental detection of negative ion production. In some cases there is evidence of the occurrence of a pronounced tail in the cross-section extending to quite low energies.

Usually the observed variation of the cross-section with impact energy includes an extensive linear region which extrapolates to zero at an energy somewhat greater than the true threshold value. This apparent threshold may be effectively that for a collision in which the vertical detachment energy, as distinct from the adiabatic electron affinity, is involved.

Some typical experimental arrangements

Using sputter sources

Fig. 10.3 illustrates the experimental arrangement used by Baede, Auerbach and Los (1973) in which a sputter source provided the

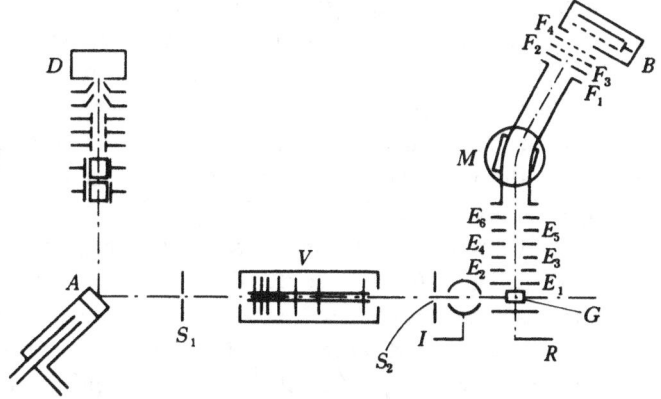

Fig. 10.3. Arrangement of the apparatus used in the experiments of Baede *et al.* (1973).

alkali metal atoms and a crossed-beam arrangement was employed.

Argon ions produced in the discharge chamber D were accelerated to 400 eV and collimated by passage through the electrode system so as to impinge on the alkali metal atom surface A. Sputtered atoms were collimated by passage through the slits S_1 and S_2 between which was placed the velocity selector V. The flux of atoms with chosen velocity passing through S_2 was measured by a surface ionization detector I and then crossed at right angles the target gas beam G produced from a standard type multichannel molecular beam source. The latter beam was collected on a liquid-air-cooled surface.

Negative ions formed in the interaction region between the beams were repelled by a negative potential on the plate R and accelerated through an extraction system of electrodes E_1 to E_6 into a 30° sector magnet mass spectrometer M. The electrodes E_1 to E_3 served to apply a uniform electric field of 450 V cm^{-1} to accelerate the ions from the interaction region which was held at -1.85 kV with respect to the earthed spectrometer. Electrodes E_4 to E_6 acted as an einzel lens to compensate for defocusing of the ion beam by the transition between the electric field and field-free regions. E_4 and E_6 were at earth potential and E_5 at a high voltage with respect to earth.

As the extraction voltage was kept constant the ions were selected in the spectrometer by varying the magnetic field. After acceleration from 2.2 kV to 4 kV through the electrode system E_1–E_4, the ions leaving the spectrometer were detected by a Bendix M 306 multiplier B because it is less sensitive to poisoning by halogens or exposure to air. Although no absolute measurements of cross-sections were made, relative measurements of cross-sections for production of atomic negative and molecular halogen ions were made. This involved checking that the counting efficiency was independent of the nature of the ions. For this an indirect procedure was necessary using a Johnston multiplier. By comparison of the output with this as detector with the d.c. current measured on the first dynode used as a Faraday cage, it was verified that the counting rate was independent of the mass of the ion. This procedure was only possible with a high ion signal obtained by removing the velocity analyser from the path of the ion beam. It was not possible with the Bendix multiplier because of its low saturation level. Having checked the independence of response of the Johnston multiplier, the Bendix performance in this respect could be determined by direct comparison.

It was also important that the ratio of the production cross-sections for different ions should be proportional to the peak heights in the mass spectrum. Since ions of different mass acquire different momenta in the collision, there is a danger that there may be discrimination because of this in the extraction stage. To minimize this the extraction voltage was increased until the collected ion current was saturated. A further check was made by examining whether the peak position in the mass spectrum was effectively

ION FORMATION – BOUND ELECTRON CAPTURE

independent of the energy of the primary beam. Discrimination may also occur in passage through, and exit from, the mass analyser. This was minimized by the focusing action of the einzel lens and by adjusting the exit slit so as to collect all ions of the chosen mass which entered the spectrometer.

Discussion of results obtained in these experiments is given on p. 394. In earlier experiments using a sputter source the target gas filled the collision chamber and the ions formed were collected by a condenser plate method rather similar to that used in measurement of cross-sections for dissociative attachment (see Chapter 9, p. 278).

Using charge transfer sources

The main difficulty in producing neutral beams by charge transfer in the low-energy range we are considering is the spreading of the initial ion beam by space charge. Helbing and Rothe (1969) developed a source of neutral Cs atoms in which this effect was reduced by using short ion paths and providing partial neutralization of space charge.

Fig. 10.4 illustrates the arrangement of this source, the electrode structure of which is confined within a box containing Cs vapour,

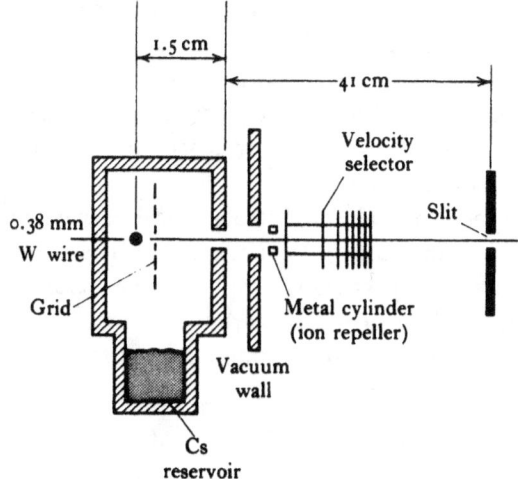

Fig. 10.4. Arrangement of the Cs atom source developed by Helbing and Rothe (1969).

operated between 110 and 120 °K. Cs⁺ ions are formed by surface ionization on a heated filament at positive potential. All other components are grounded so that ions are accelerated through a nickel grid about 0.25 mm from the filament. After passage through the grids, ions are converted to neutral atoms by charge exchange over a distance of 15 mm. The atom and ion beam emerges from the box to pass through a velocity selector after the ions have been removed electrostatically. In this arrangement, photoelectrons formed on the Cs-coated surfaces help to neutralize the space charge. By fixing the selected velocity and studying the transmitted flux of neutral atoms as a function of accelerating voltage it was verified that no appreciable fraction of Cs_2 or more complex molecules were present.

In later applications of this source the velocity selector has been replaced by a time-of-flight selector. The ion accelerating voltage in the source is pulsed. After a fixed delay time a signal opens an electronic gate. This remains open for a selected time and any particles arriving at the detector in this interval are counted. Otherwise they are rejected. It is a simple matter to relate the delay times to the particle velocities.

Results obtained in low-energy experiments

Collisions between alkali metal atoms and halogen molecules
Baede *et al.* (1973), using the apparatus described above (see Fig. 10.3), were able to measure relative cross-sections for production of atomic and molecular negative ions. Fig. 10.5 illustrates some of the results they obtained for these cross-sections as functions of the relative impact energy.

It will be seen that, although we are dealing with more complicated cases than the atom–atom collisions discussed on pp. 383–9, the variation with impact energy follows the general form expected when the threshold energy ΔE is around a few eV.

The threshold energy for production of atomic negative ions is given by

$$E_t^a = E_i + D(X_2) - E_a(X),$$

where E_i is the ionization energy of the alkali metal atom, D the dissociation energy of the halogen molecule X_2 and E_a the electron

ION FORMATION – BOUND ELECTRON CAPTURE

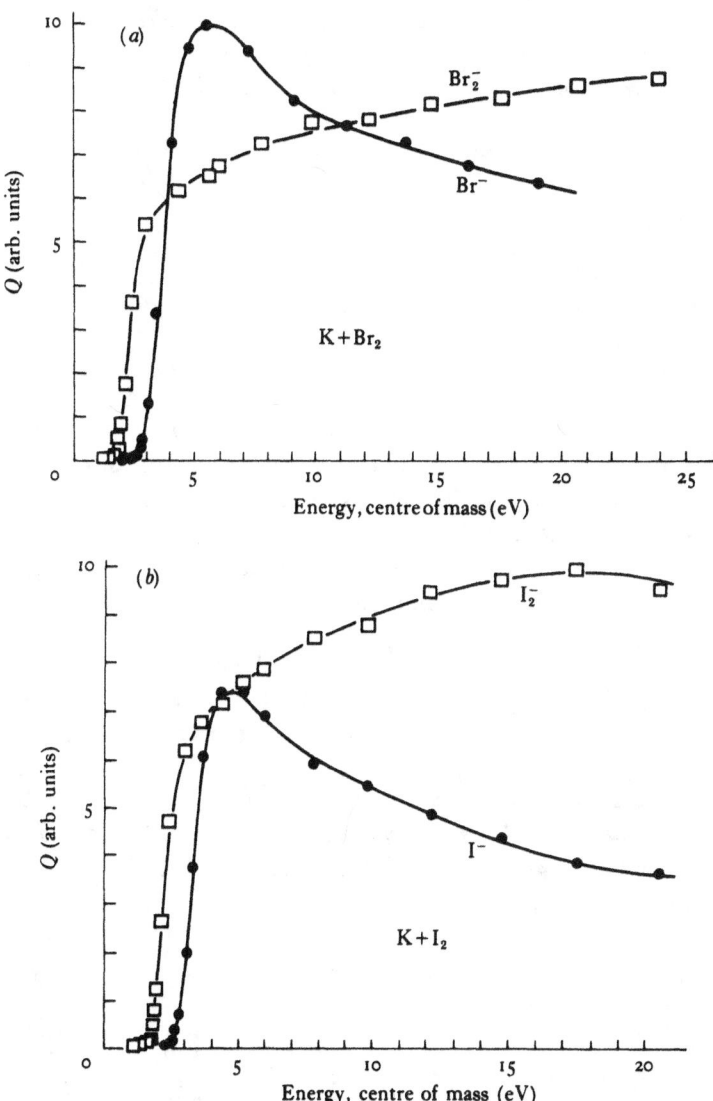

Fig. 10.5. Relative cross-sections for production of halogen atomic and molecular negative ions in collisions between K atoms and halogen molecules, observed by Baede *et al.* (1973). (*a*) K–Br$_2$, (*b*) K–I$_2$.

affinity of X. Since all of these quantities are well known, E_t^a may be accurately predicted so that a comparison with the threshold derived from the observed results for atomic ion production

TABLE 10.1 *Comparison of observed and calculated values for threshold energies for production of atomic negative ions by impact of alkali metal atoms with halogen molecules*

	Threshold energy (eV)	
Collision pair	Observed	Calculated
$K + Br_2$	2.95	2.95
$Na + Br_2$	3.75	3.75
$K + I_2$	2.87	2.82
$Na + I_2$	3.62	3.62

provides a valuable check on the validity of the analysis involved. Table 10.1 compares calculated and measured values for collisions of K and Na atoms with Br_2 and I_2 molecules. It will be seen that very good agreement is obtained, lending support to the validity of the method used to determine thresholds. For production of molecular ions X_2^- the threshold energy is

$$E_t^m = E_i - E_a(X_2),$$

where $E_a(X_2)$ is the affinity of X_2. Table 10.2 gives results obtained for $E_a(X_2)$ from the thresholds derived from the observation of

TABLE 10.2 *Adiabatic electron affinities of Cl_2, Br_2, I_2 and IBr, in eV, derived from observed thresholds for molecular ion production in collisions of alkali metal atoms with halogen molecules*

	Molecule			
Collision partner	Cl_2	Br_2	I_2	IBr
K (a)	–	2.49	2.54	2.55
(b)	2.30	2.60	2.50	–
Na (a)	–	2.64	2.55	2.06
(b)	2.50	2.55	2.55	–
Li (b)	2.45	2.55	2.30	–
Cs (c)	–	2.23	–	–
Mean value	2.45	2.55	2.55	2.55

(a) Sputter source, crossed beam, Baede et al. (1973).
(b) Sputter source, static target { Baede and Los (1971). Baede (1972).
(c) Charge transfer source, Helbing and Rothe (1969).

ION FORMATION – BOUND ELECTRON CAPTURE 397

Baede *et al.*, including results for IBr. Results obtained using sputter sources with static gas target are also included. Although no mass analyser was used in these experiments the observed threshold must refer to molecular ion production. The only value derived from experiments with a charge transfer source is that obtained by Helbing and Rothe (1969) from Cs–Br$_2$ collisions. It is substantially smaller than the other values given and has been ignored in calculating the mean values.

Comparison of these results with values obtained from the thresholds for other collision processes (see Chapter 12, p. 566) is not unsatisfactory and has been discussed in Chapter 6, p. 213.

Collisions between alkali metal atoms and other molecules

Fig. 10.6 shows the variation of the cross-section with impact energy for production of positive ions by impact of K atoms on O$_2$, observed by Lacmann and Herschbach (1970) using a charge transfer source (see p. 393). For this case the electron affinities of both products are known accurately (see Chapter 3, p. 44, and Chapter 6, p. 184). Using these, the threshold energies for production of O$_2^-$ and of O$^-$ are indicated. It will be seen that the former agrees well with that for K$^+$ production while at the latter there is a

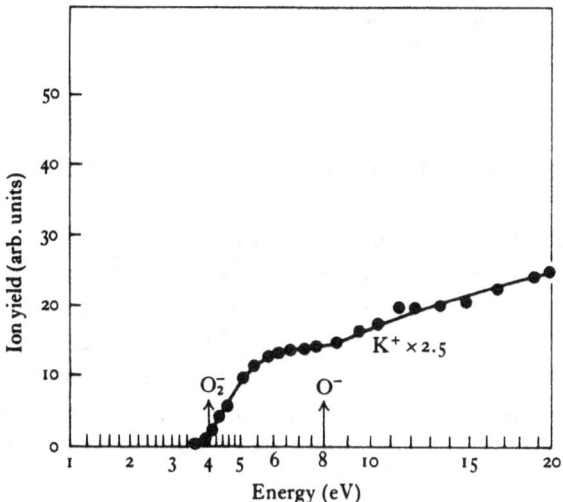

Fig. 10.6. Variation with impact energy of the cross-section for production of K$^+$ ions by impact of K atoms on O$_2$, observed by Lacmann and Herschbach (1970).

sudden rise in the cross-section, indicating the onset of a new process.

Measurements have also been made of the threshold for O_2^- production using a static gas target and sputter source (Baede, 1972). These give 0.6 and 0.4 eV respectively for the electron affinity, derived from experiments with K and Na beams respectively. These are to be compared with the known value 0.44 eV (see Chapter 6, p. 183).

Lacmann and Herschbach (1970) also measured the cross-section for K^+ production in K–NO collisions using a sputter source, and obtained a result consistent with an electron affinity $\simeq 0.0$ eV for NO. Similar results are obtained by Nalley, Compton, Schweinler and Anderson (1973) for NO^- production in Cs–NO collisions, using a charge transfer source and static target.

TABLE 10.3 *Adiabatic electron affinity (in eV) of NO_2 derived from observed thresholds for negative-ion production in collisions of alkali metal atoms with NO_2*

Method	Collision partner		
	Na	K	Cs
Charge transfer source – crossed beam (Leffert et al., 1973)	–	–	2.50
– static target (Nalley et al., 1973)	–	–	2.5
Sputter source – crossed beam (Baede, 1972)	2.55	2.45	–
– static target (Baede, 1972)	–	2.45	–

In view of its atmospheric interest considerable attention has been devoted to the study of collisions of alkali metal atoms with NO_2. Although the threshold for NO_2^- production seems to be poorly defined experimentally, the observed cross-sections having long tails near the threshold in most cases, consistent results for the electron affinity have been obtained. These include measurements made using the charge transfer source described on p. 393, with the time-of-flight technique and crossed-beam target, and using the sputter source with both static gas and crossed-beam targets.

These results compare well with those obtained from the observed thresholds for charge transfer (Chapter 12, p. 570). Com-

ION FORMATION – BOUND ELECTRON CAPTURE

parison with these and other results is discussed in Chapter 6, p. 216.

Compton and Cooper (1973), using the charge transfer source with static gas target, also observed the thresholds for production of SF_6^- and TeF_6^- ions in collision with Cs atoms. From these they obtained $E_a(SF_6) = 0.54 {}^{+0.1}_{-0.17}$ eV and $E_a(TeF_6) = 3.34 {}^{+0.1}_{-0.17}$ eV. Production of SF_5^- was also observed in these collisions and from the threshold value it was deduced that $E_a(SF_5) \geqslant 2.8 \pm 0.2$ eV, agreeing very well with the result obtained from charge transfer experiments (see Chapter 12, p. 221).

10.3 High-energy experiments – experimental methods

In general, in carrying out experiments on charge-changing collisions of energetic beams of particles in gases, account must be taken of the fact that a wide variety of such collisions can arise. We denote by Q_{if} the cross-section for a collision in which the charge of the incident particle is changed from ie to fe. Thus the cross-section for a collision in which a fast neutral atom captures an electron to form a negative ion will be Q_{0-1} usually written as $Q_{0\bar{1}}$.

Electron capture by H atoms

Consider now a beam of energetic hydrogen atoms entering a gas. After passage through a finite distance l in the gas there will also be present H^+ and H^- ions formed by electron loss from, and capture to, the neutral atoms in collisions with the gas. Let n_0, n_1, n_{-1} be the concentrations of H, H^+ and H^- respectively in the beam at some point. Then we have

$$\frac{dn_0}{d\mu} = -n_0(Q_{0\bar{1}} + Q_{01}) + n_1 Q_{10} + n_{-1} Q_{\bar{1}0}, \quad (10.20)$$

$$\frac{dn_{-1}}{d\mu} = -n_{-1} Q_{\bar{1}0} + n_0 Q_{0\bar{1}}, \quad (10.21)$$

$$\frac{dn_1}{d\mu} = -n_1 Q_{10} + n_0 Q_{01}, \quad (10.22)$$

where $\mu = Nl$, N being the concentration of molecules in the target gas. These equations will be valid provided we may neglect $Q_{1\bar{1}}$ and $Q_{\bar{1}1}$ which involve double change of charge.

If the initial rate of production of H⁻ ions from a neutral beam is measured so that on the right-hand side of (10.21) we may ignore all terms except those involving n_0, then

$$\frac{dn_{-1}}{d\mu} = n_0 Q_{0\bar{1}}. \tag{10.23}$$

This determines $Q_{0\bar{1}}$ directly provided n_0 can be measured.

An alternative procedure may be used as follows. If an experimental arrangement is set up which removes the beam particles if they change charge, then we remove the coupling terms in (10.20)–(10.22) and we have

$$n_0 = n_0(0) \exp\{-\mu(Q_{01} + Q_{0\bar{1}})\}, \tag{10.24}$$

$$n_{-1} = n_{-1}(0) \exp\{-\mu Q_{\bar{1}0}\}, \tag{10.25}$$

$$n_1 = n_1(0) \exp\{-\mu Q_{10}\}, \tag{10.26}$$

$n_0(0)$, $n_{-1}(0)$, $n_1(0)$ being the values of the separate concentrations at $\mu = 0$. Now consider the situation as μ increases so that an equilibrium is reached between the concentrations of the three species. Under these conditions let F_i be the fraction of particles on the beam with charge ie. Putting the left-hand terms in (10.20)–(10.22) equal to 0 we find

$$F_{-1} = Q_{0\bar{1}} Q_{10}/D, \; F_0 = Q_{\bar{1}0} Q_{10}/D, \; F_1 = Q_{\bar{1}0} Q_{01}/D, \tag{10.27}$$

where

$$D = Q_{\bar{1}0}(Q_{01} + Q_{10}) + Q_{0\bar{1}} Q_{10}. \tag{10.28}$$

Hence

$$F_{-1}/F_0 = Q_{0\bar{1}}/Q_{\bar{1}0}, \; F_0/F_1 = Q_{10}/Q_{01}. \tag{10.29}$$

It follows that if these ratios can be measured as well as the quantities n_0, n_{-1}, n_1, as functions of μ, all of the cross-sections can be separately determined.

Further possibilities arise if arrangements are included to measure not only the fluxes of fast particles but also of the charged slow products. Thus Curran and Donahue (1960) measured currents

ION FORMATION – BOUND ELECTRON CAPTURE 401

of slow H^+ and H^- ions produced when a beam of neutral H atoms passed through H_2 gas. The slow positive ions arose either from the reactions

$$H + H_2 \longrightarrow H^- + H_2^+, \qquad (10.30)$$

$$\longrightarrow H^- + H^+ + H, \qquad (10.31)$$

or

$$H + H_2 \longrightarrow H + H^+ + H^-. \qquad (10.32)$$

Their rate of production is therefore proportional to $Q_{0\bar{1}} + Q_p$, where Q_p is the cross-section for (10.32). Similarly the rate of production of slow negative ions will be proportional to $Q_{01} + Q_p$. By measuring both rates absolutely it is possible to determine $Q_{01} - Q_{0\bar{1}}$. Also, from the attenuation of the neutral atom beam when all charge-changing atoms are removed from the beam as it passes through the gas, $Q_{01} + Q_{0\bar{1}}$ could be measured from (10.24). In this way both Q_{01} and $Q_{0\bar{1}}$ could be determined separately.

In all experiments of this kind it is necessary to use detectors for fast neutral atoms which can be calibrated. To do this a thermal detector may be used. This consists of a sensitive thermocouple connected to a thin foil on which the beam impinges. To avoid a dependence of the signal on the position of the disc hit by the beam, a thin metal disc may be soldered to the centre of the foil so as effectively to increase the foil thickness over the target area. For calibration the thermal signal from a proton beam can be compared with an absolute measurement of the proton flux using a Faraday cage. With this available the thermal signal from a neutral beam may be converted to a measurement of absolute neutral particle flux. It is then possible to use secondary emission detectors as well as scintillation, proportional, or solid state counters to measure the flux of neutral particles in the keV energy range or above.

Fast H atom beams may be produced by charge transfer from a beam of protons traversing a suitable target gas such as H_2. Residual charged particles may be removed by electrostatic deflexion. In the charge transfer process, excited as well as normal H atoms will be produced. Many of the excited atoms will make radiative transitions to the ground state before entering the collision chamber. While the lifetimes of highly excited atoms and of metastable H(2s) atoms would be long enough to penetrate into this region they may be removed by application of a comparatively small electrostatic field. This quenches H(2s) by producing mixture with H(2p) and ionizes the highly excited atoms so they are removed together with residual charged particles.

Measurements of cross-sections $Q_{0\bar{1}}$ for capture of electrons by H atoms in passing through various gases have been carried out in a number of experiments. Stier and Barnett (1956) have obtained

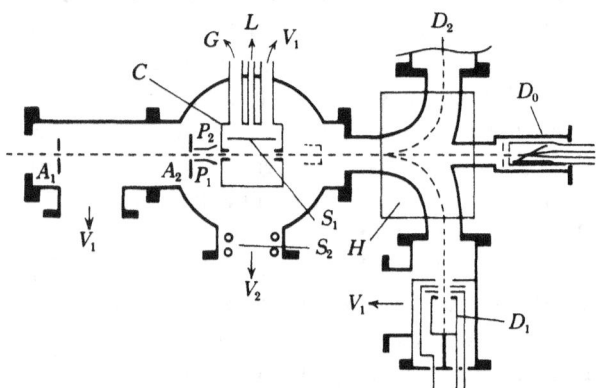

Fig. 10.7. Arrangement of the collision chamber and collecting system in the experiments of Williams and Dunbar (1966) and Williams (1967). C, collision chamber; V_1, V_2, diffusion pumps; G, ionization gauge; A_1, A_2, beam-defining apertures; S_1, S_2, liquid N_2-cooled surfaces; P_1, P_2, electrostatic deflexion plates, D_0, D_1, D_2, secondary electron type detector and Faraday cups; L, variable gas leak; H, magnetic field for beam analysis.

$Q_{0\bar{1}}$ by measuring the equilibrium ratios F_{-1}/F_1 and F_0 together with the attenuation of the neutral atom beam. A similar method has been used by Fogel, Ankudinov, Pilipenko and Topolia (1958) while Curran and Donahue (1960), confining their measurements to capture in H_2, obtained $Q_{0\bar{1}}$ from measurements of the slow negative- and positive-ion currents as well as of the attenuation of the neutral beam. McClure (1964) and, later, Williams (1967), used the rate of growth of the fast negative-ion current with gas pressure as given by (10.23) to determine $Q_{0\bar{1}}$ directly. Thus in the experiments of Williams (1967), using the apparatus developed by

Williams and Dunbar (1966), the H atom beam with flux density of order 10^{11} atoms s^{-1} cm^{-2} was formed by electron capture by fast protons on passage through H_2. After passage through the target gas the fast H$^-$ and H$^+$ formed through electron capture and loss respectively were deflected in opposite sense by a magnetic field so as to be collected in two Faraday cups. The neutral beam passed through undeflected and was measured with a calibrated secondary emission detector. Fig. 10.7 shows the general arrangement of the collision chamber and detecting system. Advantage was taken of the

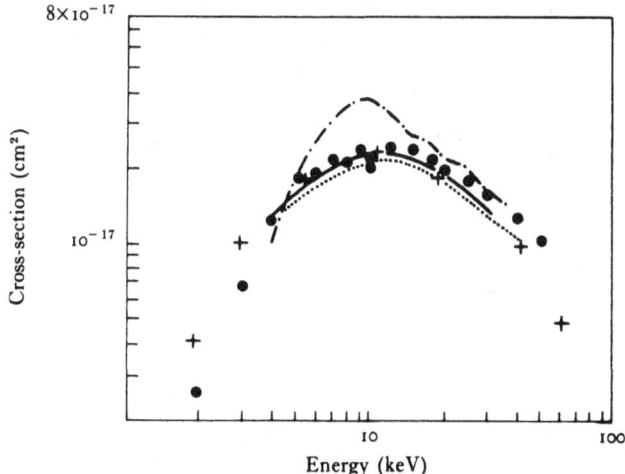

Fig. 10.8. Observed cross-sections $Q_{0\bar{1}}$ for one-electron capture by H atoms in H_2. ● Williams (1967), —— Stier and Barnett (1956), ··· Fogel et al. (1958), —·—·— Curran and Donahue (1960), +++ McClure (1964).

fact that Q_{10} for 10 keV protons has been measured with good accuracy for collisions with a number of gases. $Q_{0\bar{1}}$ was then obtained from measurements of $Q_{0\bar{1}}/Q_{10}$.

Fig. 10.8 shows the observed capture cross-section $Q_{0\bar{1}}$ for H atoms of energy between 2 and 50 keV in H_2. As discussed on p. 389, the cross-section has a maximum at an energy of several keV, a result found in all experiments. In other respects the agreement is moderately good but the structure observed by Curran and Donahue (1960) is not found by other observers.

Figs. 10.9(*a*), (*b*), (*c*), (*d*) give observed results for collisions

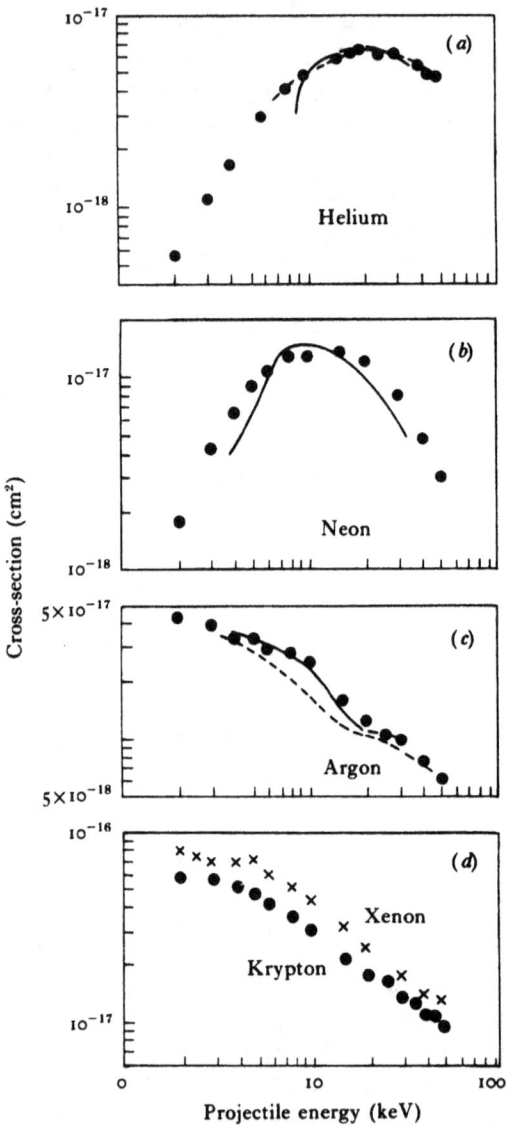

Fig. 10.9. Observed cross-sections $Q_{0\bar{1}}$ for one-electron capture by H atoms in rare gases. ●●● observed by Williams (1967) in He, Ne, Ar and Kr, ××× observed by Williams (1967) in Xe, —— observed by Stier and Barnett (1956), ---- observed by Fogel et al. (1958, 1960). (a) He, (b) Ne, (c) Ar, (d) Xe and Kr.

with the rare gases. The energy ΔE required for capture to occur decreases in the order 23.83, 20.8, 15.0, 13.2, 11.4 eV in going from He to Xe. Correspondingly we note that the location of the maximum moves to lower energies as the mass of the target increases as would be expected from the qualitative arguments of p. 389. At the same time the maximum cross-section increases as ΔE decreases.

As far as absolute magnitude is concerned, at around 10 keV the capture cross-sections are about one-tenth of the corresponding cross-sections for electron loss.

Production of He⁻

The measurement of cross-sections for charge-changing collisions of fast He atoms and ions is complicated by the fact that, in producing neutral He beams by charge transfer, allowance must be made for the presence of metastable 2^3S and 2^1S atoms as well as atoms in the ground state. Unlike metastable H(2s) these atoms cannot be removed by field quenching. Furthermore, the only He⁻ ions which are likely to be observable will be those in metastable $1s\,2s\,2p\,^4P$ states in which all three electrons have parallel spins (see Chapter 5, p. 126). Such ions cannot be formed with appreciable probability in single capture collisions of normal He atoms with other atoms. Thus, in addition to capturing an electron with the appropriate spin, it would be necessary to reverse the spin of one of the He electrons – a most unlikely event because of the weakness of the spin–orbit interaction. To produce He⁻(^4P) an exchange of an He electron with one of opposite spin in the target atom would have to take place at the same time as the capture. On the other hand, He⁻(^4P) could be produced by single electron capture of an electron by a 2^3S metastable He atom. We would therefore expect that 2^3S atoms in a neutral He beam would make a disproportionally large contribution to He⁻(^4P) production.

The first evidence for this came in 1960, when Fogel, Ankudinov and Pilipenko found that the yield of He⁻ ions formed in single collisions of fast He atoms through thin targets of rare gases depended strongly on the conditions of formation and hence, presumably, on the metastable atom content. Thus the chance of production of such an atom in charge transfer depends on the nature of the neutralizing gas. In particular, for alkali metal atoms the ionization energy is nearly the same as that required to ionize He(2^3S) so that, following the arguments of p. 389, with $|\Delta E|$ very small we would expect preferential production of the metastable atoms. Collins and Stroud (1967) found that comparatively high yields of He$^-$ were obtained when He$^+$ beams were fired through alkali metal vapour, presumably by a two stage process in which He(2^3S) is first produced by charge transfer and then captures an electron from a second alkali metal atom. In other experiments with He$^+$ and He beams it was shown that the observed yield of He$^-$ ions from different target gases approaches an equilibrium value as the target thickness increases in such a way that production through metastable atoms seems to be the only simple explanation.

A detailed study of the formation of He$^-$ ions, which confirmed that they are mainly produced through capture by He(2^3S) and obtained lower bounds to the capture cross-sections for a number of targets, has been carried out by Gilbody, Browning, Dunn and McIntosh (1969). Their study is based on the obvious generalization of (10.24) for a beam containing a fraction f of metastable atoms. If Q_{0*i} is the cross-section for a charge-changing collision in which such an atom acquires a charge ie, then the intensity I of the beam after traversing a target of thickness l will be given by

$$I = (1-f)I_0 \exp\{-\mu(Q_{01} + Q_{02})\} + fI_0 \exp\{-\mu(Q_{0*1} + Q_{0*2})\},$$

(10.33)

where $\mu = Nl$, N being the atomic concentration and I_0 is the initial intensity. It will be reasonable to assume that $Q_{01} \gg Q_{02}$ and $Q_{0*1} \gg Q_{0*2}$. If further $Q_{0*1} \gg Q_{01}$ it is possible, by observing I/I_0 as a function of μ over an extensive range, to analyse it to obtain the two terms separately. This gives the metastable atom fraction f as well as Q_{01} and Q_{0*1}.

By using different neutralizer gases it is possible to obtain beams with different values of f determined as above. The next step is to measure fluxes $I^\pm$ of fast He$^\pm$ ions produced in single charge-changing collisions of the beam with the target gas under investigation. These are given by

$$I^+/I_0 = f\mu(Q_{0*1} - Q_{01}) + \mu Q_{01}.$$

(10.34)

ION FORMATION – BOUND ELECTRON CAPTURE 407

$$I^-/I_0 = f\mu(Q_{0*\bar{1}} - Q_{0\bar{1}}) + \mu Q_{0\bar{1}}. \quad (10.35)$$

A plot of I^+/I_0 against f should therefore give a straight line, A say, of slope $\mu(Q_{0*1} - Q_{01})$ and intercept on the I^+/I_0 axis of μQ_{01}. Similarly a second straight line, B say, should be obtained by plotting I^-/I_0 as a function of f. If, as found, $Q_{0\bar{1}} \ll Q_{0*\bar{1}}$

$$\frac{\text{Slope of } B}{\text{Slope of } A} = \frac{Q_{0*\bar{1}}}{Q_{0*1} - Q_{01}} \quad (10.36)$$

and

$$\frac{\text{Slope of } B}{\text{Intercept of } A} = \frac{Q_{0*\bar{1}}}{Q_{01}}. \quad (10.37)$$

There should be consistency between the values of Q_{01} and Q_{0*1} obtained from the attenuation analysis using (10.33) and from the slope and intercept of the line A and also between the values of $Q_{0*\bar{1}}$ obtained from (10.36) and from (10.37).

Fig. 10.10 shows a schematic diagram of the apparatus used by Gilbody et al. (1969) to carry out the measurement programme outlined. A beam of He^+ ions from a van der Graaf generator was partially neutralized in the gas-filled canal B. Residual charged components were removed by an electrostatic field between the plates C leaving a neutral He beam which entered the target cell T via the 1 mm diameter aperture D. Although the region between B and D was maintained at a pressure less than 2×10^{-5} torr, some charged particles could be produced by collision of the He atom with residual gas. These were removed by a second pair of electrostatic deflecting plates H immediately before D. It can then be assumed that the beam entering D contains only normal and metastable helium atoms. It must be remembered, however, that both 2^3S and 2^1S will be present.

To fulfil the conditions under which (10.33) is applicable, charged particles formed on collision of the neutral beam with the target gas must be removed soon after formation. This was done again by electrostatic deflexion using a field of about 6 kV cm^{-1} between the plates E_1 and E_2.

The beam emerging from the target passed through the canal F into a high vacuum region where it was analysed into He, He$^+$ and

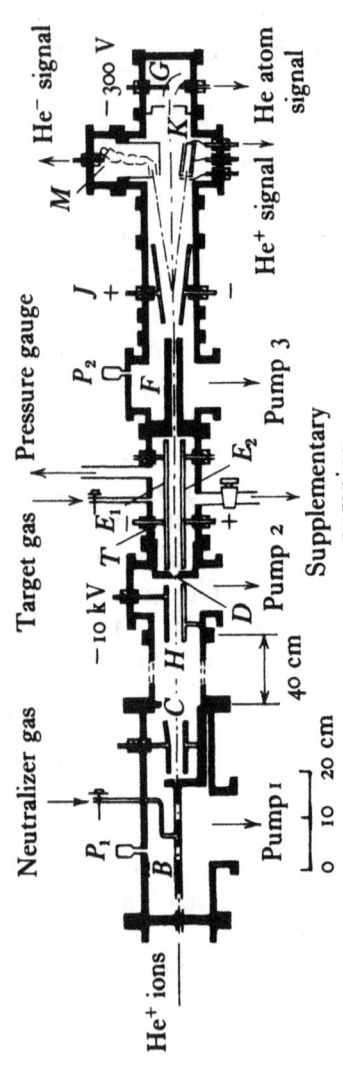

Fig. 10.10. Arrangement of the apparatus used by Gilbody et al. (1969) in their experiments on the production of He⁻ ions.

ION FORMATION – BOUND ELECTRON CAPTURE

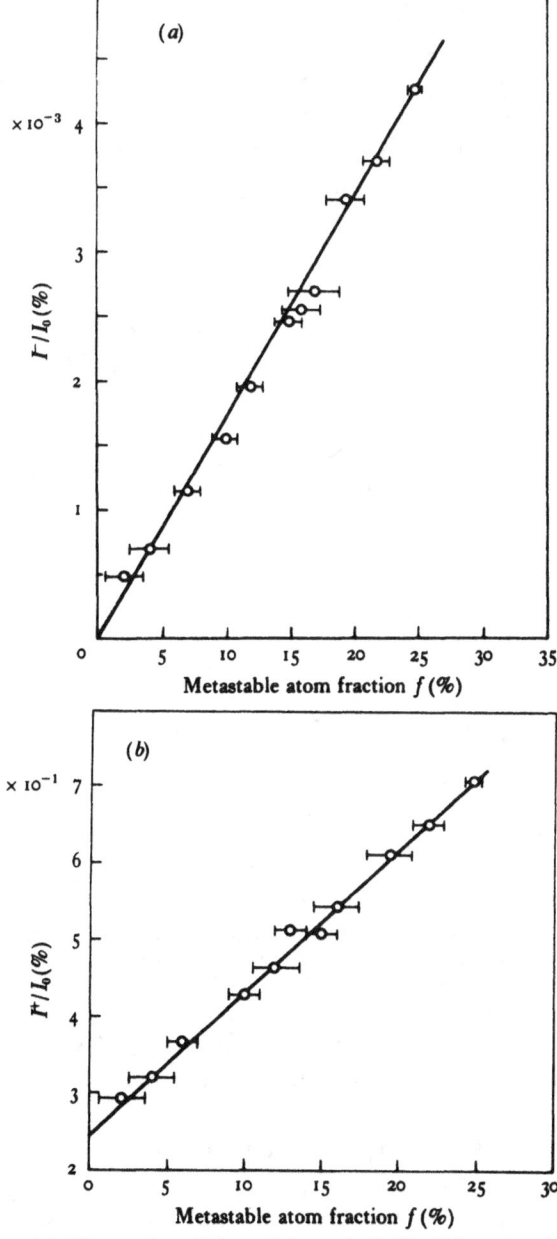

Fig. 10.11. (a) Observed variation of the ratio I^-/I_0 with metastable atom fraction f in the experiments of Gilbody, Dunn and Browning (1970). (b) Corresponding variation of I^+/I_0.

He⁻ components again by application of an electrostatic field at J. The He⁺ ions were detected by a Faraday cage and the neutral atoms by a secondary emission detector. It was assumed that, at the kinetic energies involved, there was no significant difference in the response to normal and metastable atoms. When the metastable fraction f was high the He⁻ current was large enough to be detected

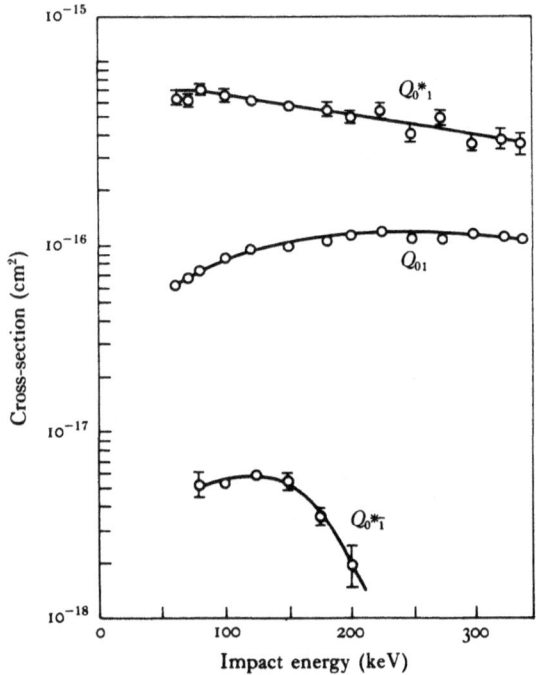

Fig. 10.12. Cross-sections $Q_{0*\bar{1}}$ for capture of electrons by metastable helium atoms in passage through hydrogen, as observed by Gilbody et al. (1970). Electron-loss cross-sections Q_{01} and Q_{0*1} are also shown for comparison.

with a Faraday cage, but otherwise an eleven stage Be/Cu multiplier was used. This was calibrated against the cage with the He⁻ current as large as possible.

Fig. 10.11 illustrates typical results obtained for the variation of I^+/I_0 and I^-/I_0 with f for an He beam of 100 keV energy in H_2. For this case the cross-sections Q_{0*1} and Q_{01} were found to be 66×10^{-17} and 8.6×10^{-17} cm² respectively. The line for I^-/I_0 in Fig. 10.11(a) passes through the origin showing that $Q_{0\bar{1}}$ is

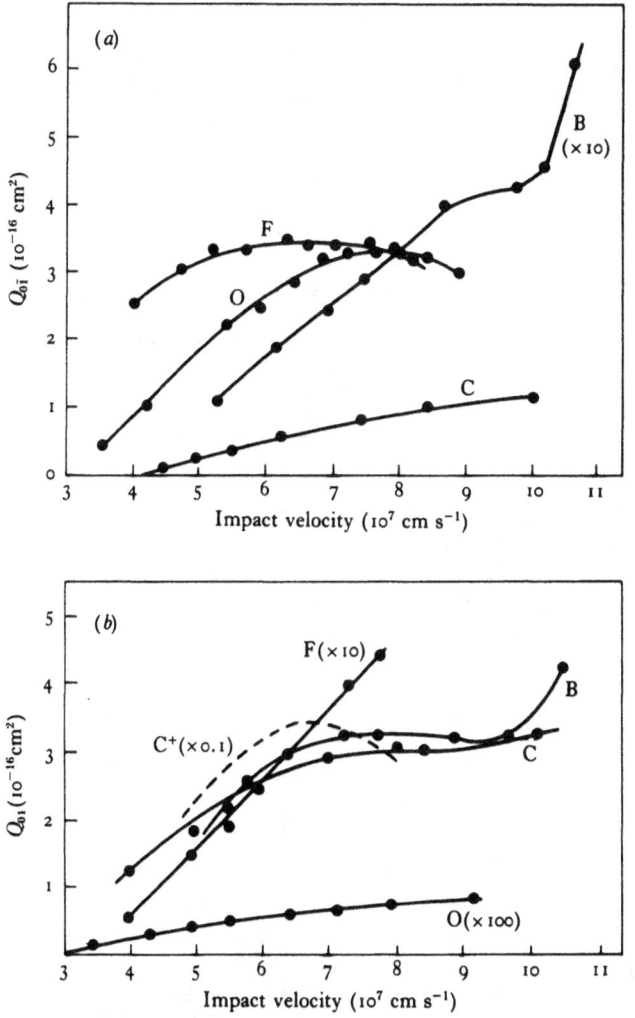

Fig. 10.13. Cross-sections $Q_{0\bar{1}}$ and Q_{01} for electron capture and loss respectively by B, C, O and F atoms in passing through krypton, observed by Fogel et al. (1959, 1960). (a) $Q_{0\bar{1}}$, (b) Q_{01}.

negligible. $Q_{0*\bar{1}}$ was found to be 4.97×10^{-18} and 5.44×10^{-18} cm² according to (10.36) and (10.37) respectively.

Fig. 10.12 shows the variation of $Q_{0*\bar{1}}$ with the kinetic energy of the metastable atoms (Gilbody et al., 1970). The electron-loss cross-sections Q_{0*1} and Q_{01} are also shown for comparison.

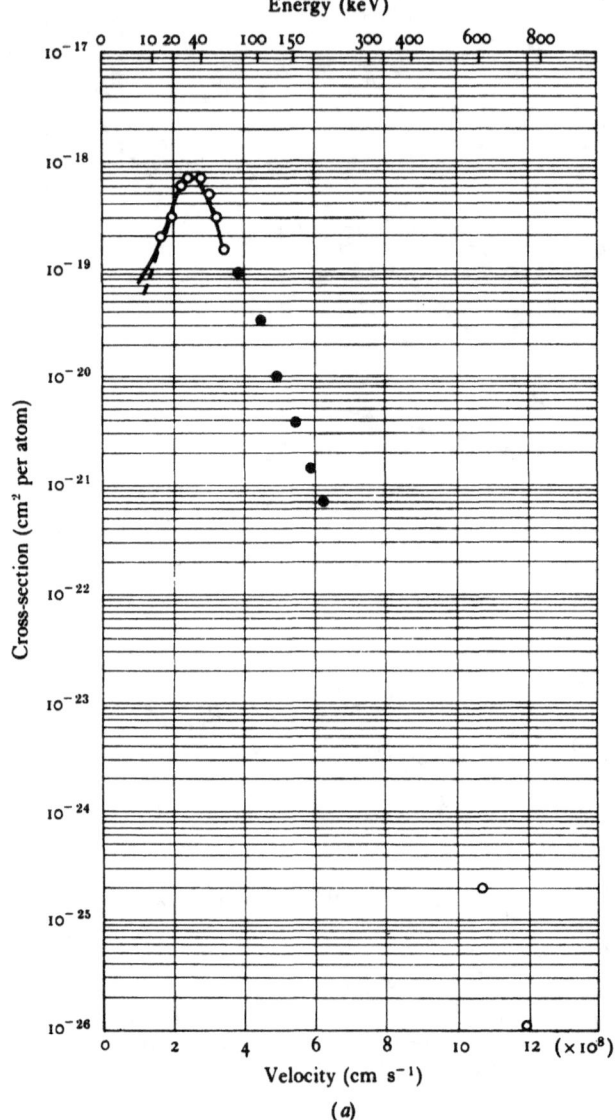

Fig. 10.14(a).

Fig. 10.14. Cross-sections $Q_{1\bar{1}}$ for double capture of electrons by protons in (a) helium, (b) argon. Observed by —— Williams (1966), ----- Fogel et al. (1958), ○ Schryber (1966), ● Toburen and Nakai (1969).

ION FORMATION – BOUND ELECTRON CAPTURE

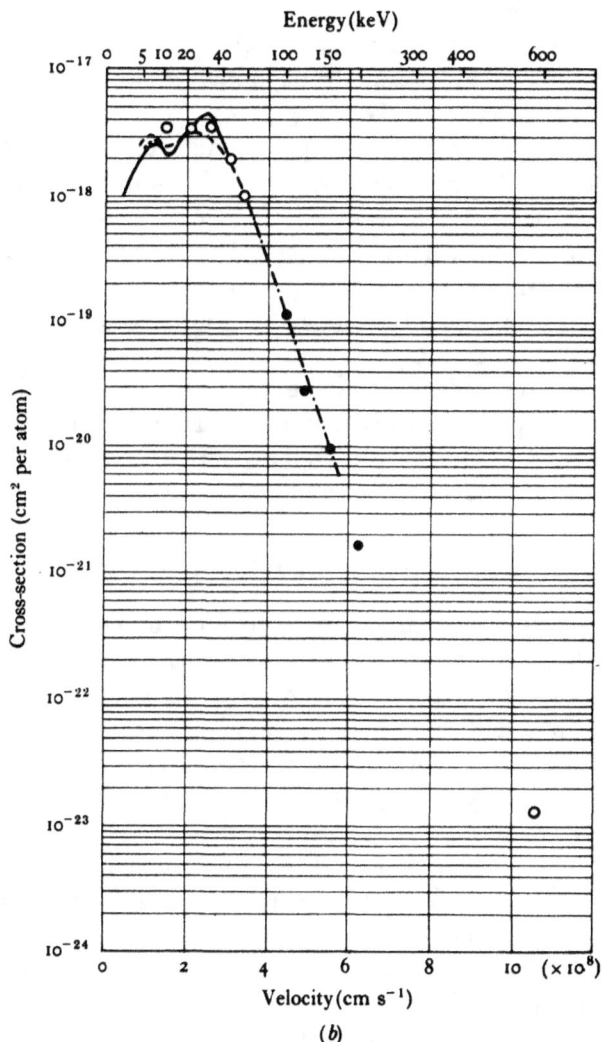

Fig. 10.14(b).

Both 2^1S and 2^3S metastable atoms in unknown proportions were present in the beam. As $He^-(^4P)$ cannot be formed by single electron capture to 2^1S, the cross-section $Q_{0*\bar{1}}$, as shown in Fig. 10.12, must be regarded as a lower bound only.

Capture by other atoms

Capture and loss cross-sections $Q_{0\bar{1}}$ and Q_{01} have been measured for B, C, O and F atoms passing through inert gases, H_2 and N_2, with energy in the range 10–60 keV (Fogel *et al.*, 1959, 1960). Results for Kr are shown in Fig. 10.13. In proceeding from B to F, $Q_{0\bar{1}}$ increases and Q_{01} decreases so that, for F at an impact energy of 50 keV, $Q_{0\bar{1}}$ is nearly ten times larger than Q_{01}, in sharp contrast to the behaviour for H. This is associated with the much larger electron affinity and high ionization potential of F. For C and B which have electron affinities comparable with that of H, $Q_{0\bar{1}}$ is considerably smaller than Q_{01}.

The same trend is noted, even more markedly, for Li which has an electron affinity even smaller than that of H and a much lower ionization potential. For 100 keV Li atoms in H_2, $Q_{0\bar{1}}$ is only about 10^{-18} cm² as compared with 2×10^{-16} cm² for Q_{01}.

10.4 Negative-ion formation by double capture of electrons

In the preceding discussion of the formation of negative ions through single capture of electrons by fast neutral atoms we have assumed that the cross-sections $Q_{1\bar{1}}$ for the simultaneous capture of two electrons by a positive ion are small enough to be negligible. Measurements of $Q_{1\bar{1}}$ for protons in different gases have, however, been made by a number of investigators, covering an energy range from several keV to 1 MeV.

A typical experimental method is that used by Williams (1966) and by Toburen and Nakai (1969) which depends on application of (10.23). A beam of fast protons is fired into a target gas and the rate of growth with gas pressure of the fast negative-ion current produced is measured.

Fig. 10.14 shows observed results for double capture by protons in He and in Ar. It will be seen, by comparison with Fig. 10.9, that $Q_{1\bar{1}}$ is indeed considerably smaller than $Q_{0\bar{1}}$, at least at impact energies less than 100 keV. This is also true (Fogel *et al.*, 1959, 1960) for C^+ ions in Kr at energies below 30 keV, $Q_{1\bar{1}}$ being about $0.1\ Q_{0\bar{1}}$. It appears to be less marked for O^+ ions in Kr. At the opposite extreme, for Li^+ in H_2, the peak value of $Q_{1\bar{1}}$ near 30 keV is only 2×10^{-20} cm² compared with 10^{-18} cm² for $Q_{0\bar{1}}$.

Oparin, Il'in, Serenkov, Solov'yev and Fedorenko (1971) have used electrostatic field detachment to investigate the formation of weakly-bound negative ions through double capture from positive ions. They used the formula (11.83) of Chapter 11 for the rate of ionization of an ion of binding energy ϵ by an electric field of strength F. This may be used to discriminate between ions with different values of ϵ. In particular they studied the C^- ions formed from C^+ ions of kinetic energy 100 keV in air. It was found that the intensity of the C^- beam fell by about one-third of its initial value when the electric field was increased from 60 to 200 kV cm^{-1}. This may be interpreted as evidence for the presence in the initial beam of C^- ions with binding energy about 0.03 eV. In view of the considerations of Chapter 4, p. 66, these were probably ions in bound metastable ^2D states.

Similar experiments with Si^- gave similar results, 17% of the ions being detached by an increase of the field from 30 to 160 kV cm^{-1}. The binding energy of these ions must be comparable with that of the metastable C^- and it is likely that they are in a metastable ^2P state (see Chapter 4, p. 66).

Conformation of the reliability of this technique is provided by experiments with He^-, in which case little detachment was suffered until the field exceeded 200 kV cm^{-1} but *all* the ions were destroyed by a field twice as large. This is to be expected because the only ions present will be those in the metastable state. The binding energy using (11.83) came out to be 0.076 eV, very close to that measured accurately by photodetachment (see Chapter 11, p. 450).

CHAPTER 11

Detachment of Electrons from Negative Ions – Photodetachment, Field Detachment and Detachment by Electron Impact

11.1 Introduction

To each of the processes of negative-ion formation which we discussed in Chapters 8–10 there is a corresponding inverse reaction which involves destruction of a negative ion. Thus we have the following possibilities:

(a) Photodetachment: $A^- + h\nu \to A + e$. This is the inverse of radiative attachment (see Chapter 8, p. 242).

(b) Detachment by electron impact: $A^- + e \to A + e + e$. This is the inverse of three-body attachment with an electron as third body (see Chapter 9, p. 264).

(c) Associative detachment: $A^- + B \to AB + e$. This is the inverse of dissociative attachment (Chapter 9).

(d) Detachment by ion or atom impact: $A^- + B \to A + B + e$. This is the inverse of three-body attachment with an atom or ion as third body (see Chapter 9, p. 265).

(e) Mutual neutralization: $A^- + B^+ \to A + B$. This is the inverse of ion pair formation by impact of two neutral systems (see Chapter 10, p. 383).

In addition to these we should include:

(f) Multiphoton detachment: $A^- + nh\nu \to A + e$. This possibility has become of practical significance through the development of lasers.

(g) Detachment by static electromagnetic fields. This is in a sense a limiting case of (f) for a large number of very low frequency photons.

It is possible to relate the rate of a process to that of its inverse by detailed balance considerations but such relations are not very useful in practice for relating rates of detachment reactions (a) to (e), as observed under experimental conditions, to those of the corresponding formation reactions. Thus, for (c), the inverse attachment reaction is observed under conditions in which the molecule is in

DETACHMENT BY PHOTON AND ELECTRON IMPACT

the ground electronic state in a low vibrational level. On the other hand the detachment process may produce molecules not only in highly excited vibrational levels but even in electronically excited states. No detailed balanced relations can be used to derive these rates from that observed for attachment to the ground state only. Similar remarks apply to (e). Again for (d) most experiments give rates of detachment for collisions in which the relative kinetic energy of A^- and B is far above thermal energies whereas for three-body attachment the major interest is concerned with rates under thermal or near-thermal conditions.

In fact the only cases in which it is useful to apply detailed balance relations come under (a) and even then are somewhat limited as we shall see below.

As it is rarely possible to derive the rate of the detachment reactions under experimental conditions from observed rates of the inverse attachment reactions and vice versa, there is separate need for experimental study of both types of reaction. At first this proved difficult for detachment but this situation has completely changed since the first major breakthrough was made by Branscomb and Fite in 1954, in the measurement of photodetachment cross-sections. All the detachment processes listed above are now the subject of extensive quantitative experimental study. Indeed it is through photodetachment measurements that by far the most precise values of electron affinities have been obtained as well as of the structural constants of molecular negative ions.

In this chapter we discuss photodetachment as well as detachment by static electromagnetic fields and by electron impact. Other detachment processes will be discussed in the succeeding three chapters.

11.2 Photodetachment – introductory theoretical considerations

In Chapter 8, p. 243, we defined the cross-section Q_a for radiative capture of an electron of initial velocity v by a neutral atom to form a negative ion. The cross-section Q_d for the inverse process of photodetachment in which an electron is ejected with velocity v from the negative ion, leaving an atom in its ground state, is related to Q_a by the formula (8.8) so that using (8.1) and (8.2),

$$Q_d = (32\pi^4 m^2 e^2 v\nu/3hc^3)|\mathbf{M}|^2, \qquad (11.1)$$

where $\mathbf{M}$ is the matrix element

$$\mathbf{M} = \iint \psi_i(\mathbf{r}, \mathbf{r}_a)\, \mathbf{r}\psi_c(\mathbf{r}, \mathbf{r}_a)\, d\mathbf{r}\, d\mathbf{r}_a. \qquad (11.2)$$

Here $\mathbf{r}$ is the coordinate of the ejected electron while $\mathbf{r}_a$ represents the aggregate of the coordinates of the remaining electrons. m is the electron mass, and ν the frequency of the radiation. ψ_i is the wave function of the ground state of the negative ion and ψ_c that of the continuum state in which the ejected electron moves under the influence of the field of the residual atom. ψ_i is normalized to unity as usual and ψ_c so as to have the asymptotic form, for large r,

$$\psi_c \sim \phi_0(\mathbf{r}_a)\, F(\mathbf{r}), \qquad (11.3)$$

where, in terms of the electron wave number $k\, (= mv/\hbar)$,

$$F(\mathbf{r}) \sim \{\exp(ikr\cos\theta) + \text{outgoing scattered waves}\} \qquad (11.4)$$

and $\phi_0(r_a)$ is the ground state atomic wave function.

The behaviour of the matrix element (11.2) when $v \to 0$, i.e. near the threshold for photodetachment, has been discussed in Chapter 8. Referring to pp. 244–5 we see from (8.5) that, under these conditions, when an electron is detached from an s orbital, and so is finally in a p orbital,

$$Q_d \propto vk^3 \qquad (11.5)$$
$$= \nu(\nu-\nu_0)^{3/2} \qquad (11.6)$$
$$= (A/\lambda)\{(\lambda_0-\lambda)/\lambda\lambda_0\}^{3/2}, \quad \text{s} \longrightarrow \text{p}, \qquad (11.7)$$

where ν_0 and λ_0 are the threshold frequencies and wavelengths respectively. On the other hand, when a electron is detached from a p orbital it may enter an s or d orbital. In the latter case the cross-section near threshold will vary as k^4 and so may be ignored giving

$$Q_d \propto vk \qquad (11.8)$$
$$= \nu(\nu-\nu_0)^{1/2} \qquad (11.9)$$
$$= (B/\lambda)\{(\lambda_0-\lambda)/\lambda\lambda_0\}^{1/2}, \quad \text{p} \longrightarrow \text{s}. \qquad (11.10)$$

The range of validity of these approximations may be very small as has been confirmed experimentally (see p. 456). When allowance is made for the effect on ψ_c of long range forces due to the polarizability α of the atom, it is found that (O'Malley, 1965), for detachment into an orbital of angular momentum quantum number L,

$$Q_d \propto vk^{2L+1} [1 - \{4 \alpha k^2 \log(k/a_0)(2L+3) \\ (2L+1)(2L-1)\} + O(k^2)]. \quad (11.11)$$

In the above considerations we have treated the problem as one in which essentially a single electron above is involved. To generalize it to apply to ions containing N electrons with coordinates $\mathbf{r}_1, \ldots, \mathbf{r}_N$, it is merely necessary to replace (11.2) by

$$\mathbf{M} = \int \Psi_i(\mathbf{r}_1, \ldots, \mathbf{r}_N) \left(\sum_{s=1}^{N} \mathbf{r}_s\right) \Psi_c(\mathbf{r}_1, \ldots, \mathbf{r}_n) \, d\mathbf{r}_1 \ldots d\mathbf{r}_N. \quad (11.12)$$

Ψ_i and Ψ_c are now properly symmetrized wave functions of $\mathbf{r}_1$ to $\mathbf{r}_N$. Thus, for photodetachment from H$^-$, Ψ_i is the symmetric ground state function $\Psi_i(\mathbf{r}_1, \mathbf{r}_2)$ and Ψ_c has the asymptotic form

$$2^{-1/2}\{\phi(r_1) F(\mathbf{r}_2) + \phi(r_2) F(\mathbf{r}_1)\}, \quad (11.13)$$

where $F(\mathbf{r})$ has the asymptotic form (11.4) with the scattered waves which arise when the two electrons have opposite spin. Owing to electron exchange effects these waves are of different amplitude from those arising when the electron spins are parallel.

In dealing with complex atoms and ions the initial and final states may be degenerate. If g_i and g_0 are the statistical weights of the ionic and neutral states respectively, then we replace Q_d given by (11.1) and (11.2) by

$$Q_d = (32\pi^4 m^2 e^2 v\nu/3hc^3 g_i) \sum |\mathbf{M}|^2, \quad (11.14)$$

the sum being taken over all $g_i g_0$ pairs of initial and final states. None of these modifications affect the threshold formulae (11.7), (11.10) and (11.11).

In all cases of photodetachment the wave functions Ψ_i and Ψ_c are only known approximately and it is desirable to have some check on the accuracy of the calculated Q_d. Usually the approximate wave functions will have been calculated by a variational method which can only be relied upon to yield good wave functions for values of the electron coordinates close to those which are most important in determining the total energy of the system, i.e. for values at which the wave function is close to its maximum. The matrix element (11.12) attaches greater weight to larger electronic coordinates.

However, if the functions Ψ_i, Ψ_c are exact, it can be shown that (11.2) may be transformed into alternative forms $\mathbf{M}_V$, $\mathbf{M}_A$, in which $\mathbf{r}$ is replaced by $(h/4\pi^2 \nu m)$ grad and $-(e^2/4\pi^2\nu^2 mr^3)\mathbf{r}$ respectively (Chandrasekhar, 1945). Since $(-i\hbar/m)$ grad and $e^2 \mathbf{r}/r^3$ are the velocity and acceleration operators, $\mathbf{M}_V$ and $\mathbf{M}_A$ are usually referred to as the *dipole velocity* and *dipole acceleration* matrix elements as distinct from the *dipole length* element (11.12) which we shall designate henceforth as $\mathbf{M}_L$. While $\mathbf{M}_V$, $\mathbf{M}_A$ and $\mathbf{M}_L$ would all be exactly equal if the wave functions Ψ_i, Ψ_c were exact, the main contribution to $\mathbf{M}_V$ comes from smaller r than for $\mathbf{M}_L$ and to $\mathbf{M}_A$ from even smaller r. If the wave functions are only approximate the three matrix elements calculated from them will differ and the magnitude of the difference will be an indication of the accuracy of the wave functions. In general $\mathbf{M}_V$ should give the best results because it depends on roughly the same values of r as are important for determining the total energy.

A further test of the accuracy of calculated cross-sections is provided by so-called sum rules which must be satisfied by the exact cross-sections as functions of frequency. If there are no bound excited states for the negative ion these take the integral form

$$\left(\frac{e^2}{ha_0}\right)^s \frac{mc}{\pi e^2} \int_{\nu_0}^{\infty} Q_d(\nu)\nu^{-s}\,d\nu = A_s, \qquad (11.15)$$

in which A_s is, in general, a function of the ground state wave function of the negative ion. For H⁻, the case to which the rules have been applied in detail, we have

$$A_0 = 2, \qquad A_1 = (2/3a_0^2)\langle(\mathbf{r}_1+\mathbf{r}_2)^2\rangle_{00}, \qquad A_2 = \alpha/a_0^3,$$
$$A_{-1} = \tfrac{2}{3}\{(E_T/E_H) - a_0^2\langle \mathrm{grad}_1 \cdot \mathrm{grad}_2\rangle_{00}\},$$
$$A_{-2} = (4\pi a_0^3/3)\langle\delta(\mathbf{r}_1)+\delta(\mathbf{r}_2)\rangle_{00}, \qquad (11.16)$$

where α is the polarizability, E_T the total energy of H⁻ and E_H that of H, and $\langle f(\mathbf{r}_1,\mathbf{r}_2)\rangle_{00}$ denotes

$$\iint \Psi_i f(\mathbf{r}_1, \mathbf{r}_2)\Psi_i\, d\mathbf{r}_1\, d\mathbf{r}_2,$$

where Ψ_i is the ground state wave function of H⁻. As this function is known with very high accuracy from variation methods (see Chapter

1, pp. 6–15) the A_s can be calculated to comparable accuracy. In applying these checks it must be remembered that $Q_d(\nu)$ should include not only the cross-section for detachment leaving the atom in its ground state but also in any excited state. If these latter contributions are omitted, the $=$ sign in (11.15) should be replaced by $\not=$.

The angular distribution of the electrons produced by linearly polarized light, in detachment from an orbital of quantum numbers n, l, is given in terms of the differential cross-section

$$I_d^{nl}(\theta)\,d\omega = (4\pi)^{-1}\,Q_d^{nl}\{1+\beta P_2(\cos\theta)\}d\omega, \qquad (11.17)$$

where Q_d^{nl} is the total cross-section for detachment from the orbital and θ is the angle between the direction of motion of the detached electron and the polarization of the incident light. The asymmetry factor β depends on the quantum numbers n and l and on the energy of the photoelectron. However, for $l = 0$, $\beta = 2$ independent of this energy. In the general case Cooper and Zare (1968) have given formulae for β in terms of HF wave functions, which represent a good approximation provided the magnetic substates of the initial state are equally populated and the atom is left in the ground state. These formulae have been generalized by Lipsky (1967) to cases in which the atom is left in an excited state.

For unpolarized light the angular factor in (11.17) is changed to

$$1 - \tfrac{1}{2}\beta P_2(\cos\theta). \qquad (11.18)$$

11.3 Calculated photodetachment cross-sections – H⁻

A very great deal of attention has been directed towards the accurate calculation of cross-sections for photodetachment from H⁻ because of the importance of this process in the solar photosphere (see Chapter 15, p. 678).

The calculated cross-section depends very much on the approximation assumed for the ground-state function for H⁻. As was shown by Massey and Bates (1940), who carried out the first calculation of Q_d using only $\mathbf{M}_L$, the continuum function Ψ_c is given to quite a high accuracy by

$$\Psi_c = 2^{-1/2}\{\phi(r_1)\exp(ikr_2\cos\theta_2) + \phi(r_2)\exp(ikr_1\cos\theta_1)\}. \qquad (11.19)$$

This is because the ejected electron must occupy a p orbital which is very little affected by interaction with the residual atom. In terms of the notation of Chapter 4, p. 71, the p phase shift η_1 for elastic scattering of slow electrons by H atoms is very small. Nevertheless, this will not necessarily apply to the calculation of $\mathbf{M}_A$ which depends on the wave function at small r for which (11.17) will certainly be less accurate.

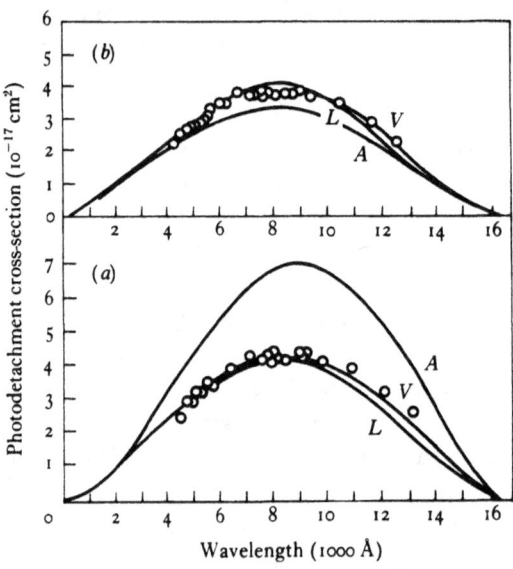

Fig. 11.1. Cross-sections for photodetachment from H$^-$, —— calculated using the 70-parameter Schwartz wave function for H$^-$ and Ψ_c taken as (a) a plane wave, (b) given by the 1s-2s-2p close-coupling approximation. L, V, A denote calculations using length, velocity and acceleration matrix elements respectively. ○ observed by Smith and Burch (1959), normalized to agree with V at 5280 Å.

This has been borne out by subsequent calculations which have used more and more sophisticated approximations for both Ψ_i and Ψ_c. Fig. 11.1 shows the cross-sections calculated using all three matrix elements and a very accurate 70-parameter wave function for H$^-$ which is an extension of the Hart–Herzberg 20-parameter function (see (1.10)) obtained by Schwartz (1961). Results are shown (Geltman, 1962) with Ψ_c given by (11.19) and also by the 1s-2s-2p close-coupling approximation (Doughty, Fraser and McEacharn, 1966) which allows for the interaction between the electron and H atom to quite a high accuracy.

It will be seen that the results obtained using $\mathbf{M}_L$ and $\mathbf{M}_V$ do not depend very much on the form assumed for Ψ_c but $\mathbf{M}_A$ is much more sensitive, as expected. With the more sophisticated approximation for Ψ_c, results obtained using $\mathbf{M}_A$ are very much closer to those from $\mathbf{M}_L$ and $\mathbf{M}_V$.

TABLE 11.1 *Values of A_s (see (11.15)) for photodetachment from H^-*

	s	3	2	1	0	−1	−2
A_s	Calculated directly	4000	212	15.1	2	1.01	2.76
	Derived from sum rules using calculated Q_d from ground state	3670	200	14.1	1.76	0.61	1.18

Table 11.1 gives a comparison of the quantities A_s given in (11.16) as calculated directly and as derived from the calculated Q_d using the sum rules. In the direct calculation, 70-parameter wave functions of the Pekeris form (see Chapter 1, p. 8) were used for A_{-2} and A_{-1} and the 70-parameter function of Schwartz (1961) for A_1, A_2 and A_3. For Q_d, values derived from $\mathbf{M}_V$ with the most elaborate wave functions have been used for $s = 3$, 2 and 1.

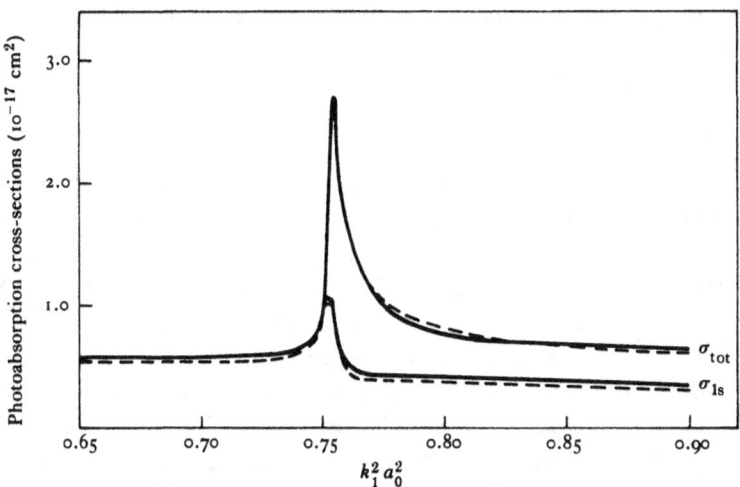

Fig. 11.2. Cross-sections for photodetachment by H^-, calculated by Macek (1967), in the neighbourhood of the threshold for the process $H^- + h\nu \rightarrow H(2s, 2p) + e$. The cross-sections are given as functions of $k_1^2 a_0^2$, where k_1 is the wave number of the ejected electron, so the threshold occurs at $k_1^2 a_0^2 = 0.75$. —— using length matrix elements, - - - - using velocity matrix elements.

It will be seen that in all cases the directly calculated values exceed those calculated from the sum rules, which is to be expected when no allowance is made for photodetachment which leaves the H atom in an excited state.

Macek (1967) has calculated cross-sections for detachment by radiation in the wavelength range from 1400 to 250 Å including the contribution from cases in which the H atom is left in an excited state with $n = 2$, the threshold for which is at 1131 Å. Elaborate close-coupling wave functions were used for the continuum states. Over the wavelength range covered beyond the $n = 2$ threshold, detachment leaving the atom in an excited state contributes nearly 50% of the total. Very close to the threshold a resonance effect was found, associated with the continuum wave function. At the peak, the cross-section is more than four times the background value (see Fig. 11.2).

Macek checked these cross-sections by using them in conjunction with those calculated earlier for the ground state, as shown in Fig. 11.2, to evaluate A_0. Using the velocity and length matrix elements he obtained the respective values 1.94 and 2.06 which are in reasonable agreement with the exact value 2.

Comparison of calculated with experimental cross-sections is discussed on p. 449.

Other atomic negative ions

Useful semi-empirical calculations of photodetachment cross-sections for atomic negative ions have been carried out by Cooper and Martin (1962) and by Robinson and Geltman (1967). While not attempting precise evaluation of the cross-sections, results may be obtained to useful accuracy for a number of atoms so that the dependence on the main atomic properties is brought out.

Both procedures start from the assumption that it is a sufficient approximation to treat the problem as a single electron one in which the active electron moves in the same effective atomic field in the initial and final states. Cooper and Martin take this field to be given in terms of the observed polarizability α by

$$V(r) - \tfrac{1}{2}\alpha e^2(r^2 + r_p^2)^{-2}, \qquad (11.20)$$

where $V(r)$ is the interaction calculated from the HF field of the neutral atom. The polarizability cut-off parameter r_p was chosen arbitrarily to be the average distance from the nucleus of the outer nl electrons of the neutral atom while the effective polarizability was adjusted so that the total interaction (11.20) was just strong enough to bind an electron with an energy E_a equal to the electron affinity. Robinson and Geltman used a very similar procedure although different in detail. Thus for $V(r)$ they used the form

$$V_{HS}(r) + e^2 r^{-1}\{1 - \exp(-r/r_0)\}, \quad (11.21)$$

where V_{HS} is obtained from a so-called Hartree–Fock–Slater potential (Herman and Skillman, 1963), in which a simplified approximation is made for the effect of electron exchange. This approximation is invalid for large r and must be corrected by adding the screened Coulomb term in which r_0 is treated as an adjustable parameter. The polarization term included an additional factor $\{1 - \exp(-r/r_p)\}$ which vanishes as $r \to 0$ and α was taken to be the observed polarizability. r_p was arbitrarily chosen as $1.5a_0$, $2.5a_0$, $3.5a_0$ and $4.5a_0$ for all atoms in successive rows of the periodic table. The parameter r_0 was then adjusted so as to give the correct binding energy for the active electron.

Results obtained using these procedures are compared with experimental data in Figs. 11.15 and 11.18. Robinson and Geltman have used their wave functions to calculate transition probabilities for two-quantum detachment and their results for I$^-$ are discussed on p. 493.

A number of other calculations have been carried out, but the most interesting are those of Moores and Norcross (1974) for photodetachment from the alkali metal ions. The special feature of these calculations is that they exhibit the structure which appears in the cross-section at and near the threshold for detachment which leaves the atom in the ^{2}P excited state. For the ground state wave functions they used the multiconfiguration functions of Weiss (1968) (see Chapter 2, p. 28), and for the continuum state a function obtained by using a close-coupling expansion (see Chapter 4, p. 94) in terms of up to four target states, the lowest two s states and the lowest p and d states. The problem is reduced to a two-electron one by supposing the valence electron to move in the field of the nucleus and closed shell core described by the same approximation as used by Norcross (1974) in the calculations described in Chapter 2, p. 29.

Fig. 11.3 shows the results obtained for Li$^-$, Na$^-$ and K$^-$ using

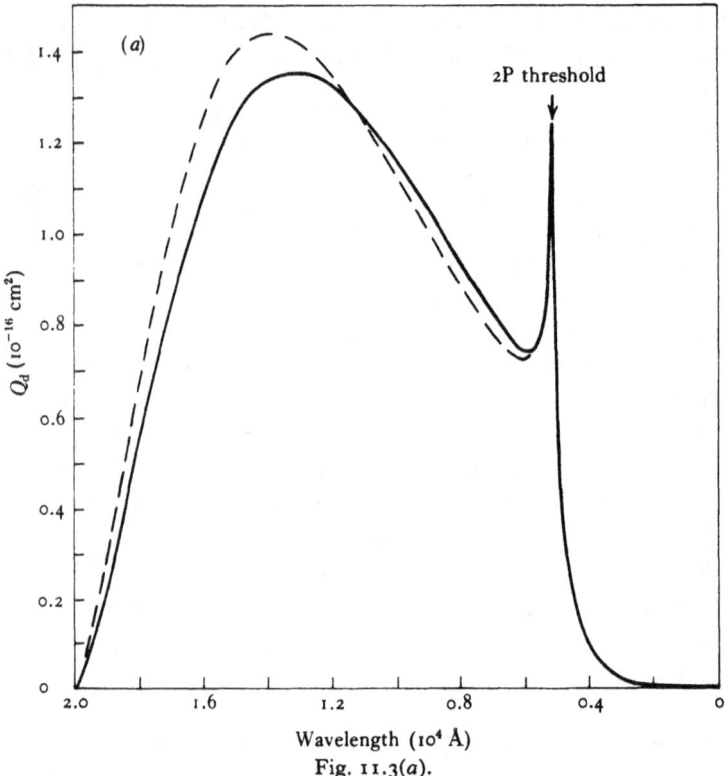

Fig. 11.3(a).

Fig. 11.3. Cross-sections for photodetachment from (a) Li$^-$, (b) Na$^-$ and (c) K$^-$ ions as functions of wavelength, calculated by Moores and Norcross (1974) using dipole length (----) and dipole velocity (———) matrix elements.

the dipole velocity and dipole length matrix elements. In all these cases a pronounced feature appears at the threshold for detachment which leaves the atom in the first excited P state. For Li$^-$ and Na$^-$ this takes the form of a pronounced cusp but is somewhat different for K$^-$, where the threshold is marked rather more by a step in the cross-section. These effects may be understood in quite general terms as arising from the opening of a new reaction channel and are also found in the elastic scattering (Chapter 5, p. 148) (Wigner, 1948; Mott and Massey, 1965, p. 380). Comparison of the calculated with observed behaviour (shown in Figs. 11.26 and 11.27 for K⁻ and Na⁻ respectively) shows that the main features indeed

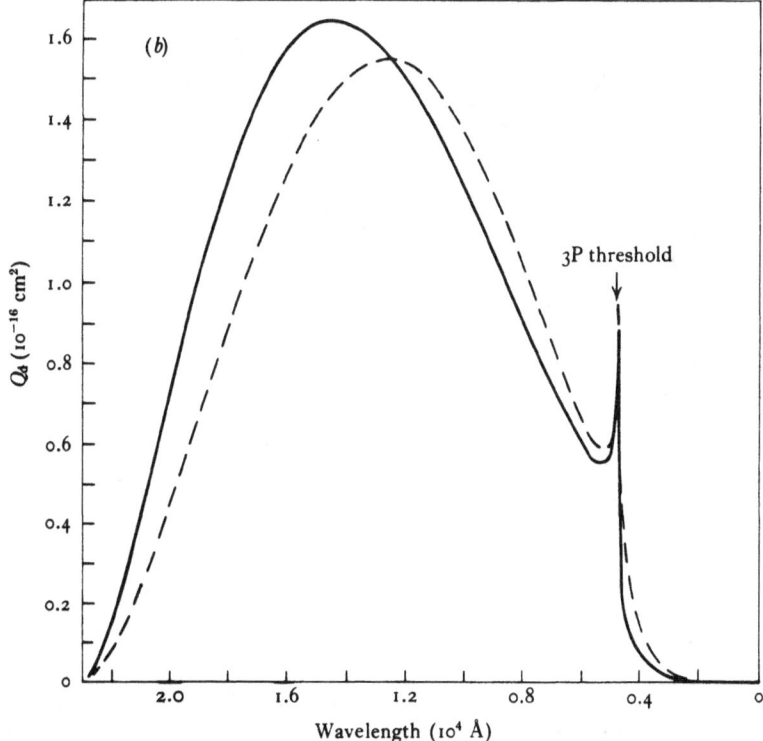

Fig. 11.3(b). For legend see p. 426.

appear experimentally and may be used to determine the threshold energy, and hence the electron affinity, with some precision.

Although the dipole velocity and dipole length matrix elements do not agree quantitatively, they nevertheless behave in a very similar way at the ²P threshold. However, Moores and Norcross (1974) have obtained closer agreement between the results obtained from the two matrix elements for K⁻ by using a close-coupling type wave function for the ground states (see Chapter 4, p. 94).

This calculation preserves the main structural features near the threshold (see p. 468 and Fig. 11.26).

Moores and Norcross (1974) have also calculated the asymmetry parameter β of (11.17) at frequencies just above the ^2P threshold. Their results are shown in Fig. 11.4 for Li^-, Na^- and K^-. The observed value for K^- (see p. 469) is included and is quite close to

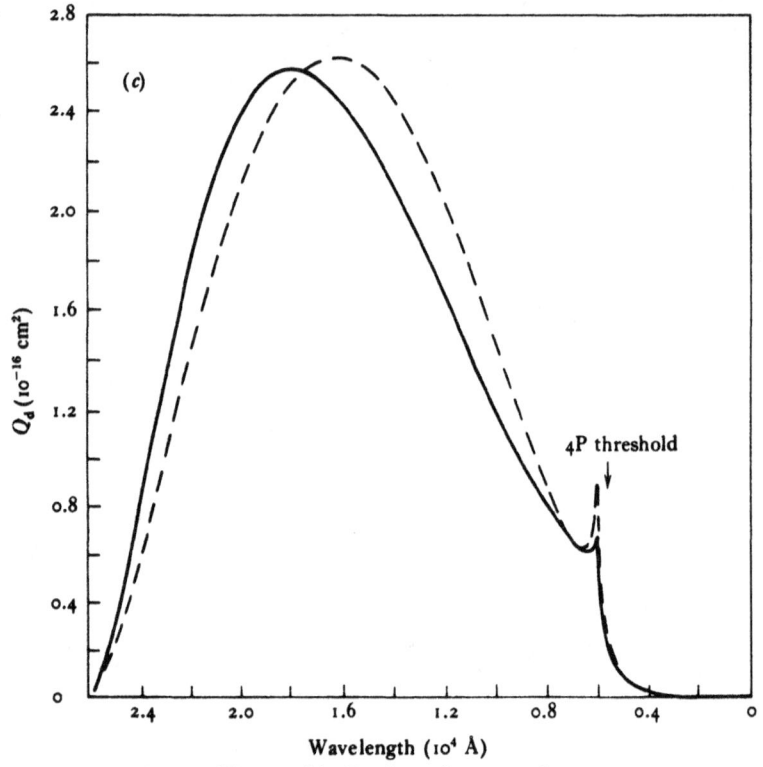

Fig. 11.3(c). For legend see p. 426.

that calculated with the dipole velocity matrix element, using the Weiss function for the ground state of K^-. Even better agreement, with either matrix element, is obtained using the close-coupling ground state function referred to above.

Rau and Fano (1971) have considered the threshold behaviour in photodetachment when a number of transitions between fine structure levels are involved. This is a matter of some importance in

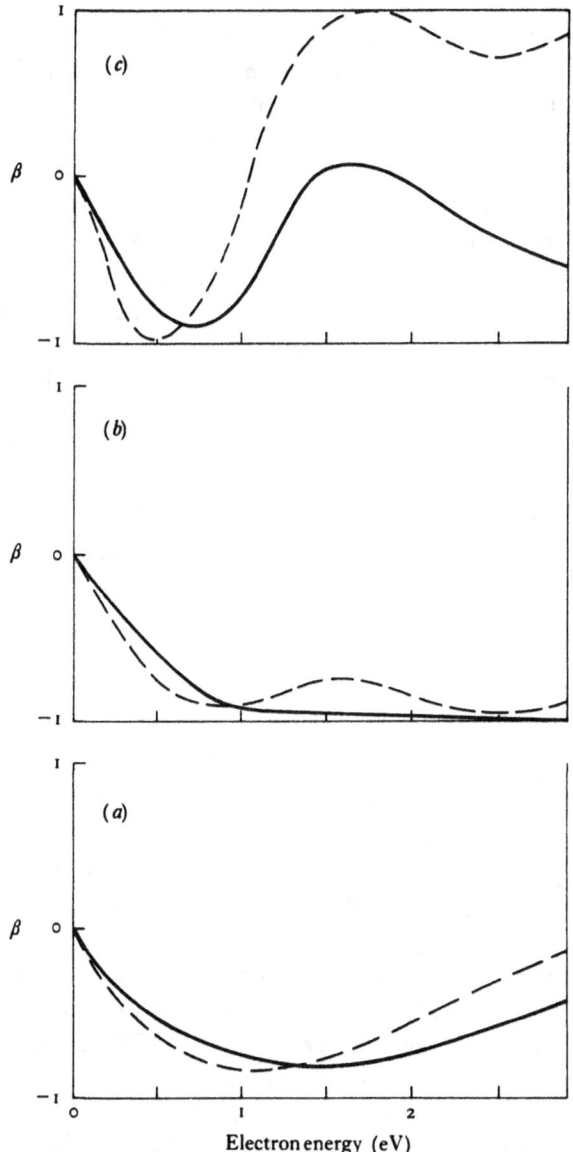

Fig. 11.4. Asymmetry parameter β for photoelectrons detached from (a) Li$^-$, (b) Na$^-$ and (c) K$^-$ ions, as a function of the energy of the ejected electron, calculated by Moores and Norcross (1974) using dipole length (-----) and dipole velocity (———) matrix elements.

the analysis of observed data to obtain electron affinities. It would appear, at first sight, that at each threshold for a transition between fine structure levels the cross-section should rise according to the laws (11.5) or (11.8), whichever is appropriate, with relative magnitudes in the ratio of the statistical weights of the corresponding levels. However, the final state should be considered as a single system of electron plus atom in *LS* coupling, giving rise to a number of states which then break up with equal probability to yield the final fine structure states. This leads, in general, to a different intensity distribution between the different transitions. This point of view has been followed through in detail by Rau and Fano for photodetachment of an initial p electron and applied to the data on S$^-$ observed by Lineberger and Woodward (1970) (see p. 457). Semi-quantitative agreement was obtained between the relative intensities of the fine structure transitions near threshold. The later experiments on Se$^-$ (see p. 458) provided a more definite test and confirm the correctness of the viewpoint adopted by Rau and Fano.

11.4 Photodetachment – experimental methods

Introduction

The first measurements of photodetachment cross-sections were made using a crossed-beam technique in which a negative-ion beam was crossed at 90° by a photon beam and the flux of electrons produced by photodetachment measured. In all of this work the problem of distinguishing the wanted signal from a much stronger background is a serious one. Although this was at first aggravated by the difficulty of obtaining monochromatic light sources of sufficient intensity, the use of modulated phase-sensitive detection made it possible to obtain accurate data. In later work use has been, and is being made, of laser sources which are now available with precisely defined wavelengths in the range of interest, and with high intensity. Using tunable dye-lasers it is possible to measure the photodetachment cross-section as a function of photon frequency over a considerable range extending from the threshold. A second method, using a fixed frequency laser, such as an argon ion laser, is to measure the energy distribution of the detached electrons.

If ν is the laser frequency and E_a the electron affinity, a sharp peak will be observed in the electron energy spectrum at an energy E_e given by

$$E_0 = h\nu - E_a. \qquad (11.22)$$

This makes it possible to determine E_a with precision. By working with polarized laser radiation it is also possible to measure the angular distribution of the photoelectrons. This can be of value in providing information about the nature of the initial and final states concerned.

Measurements of the electron energy spectrum are of particular value in studying photodetachment from molecular negative ions as the contributions from different vibrational transitions may be separately identified.

Some time after the first crossed-beam measurements were made, Berry, Reimann and Spokes (1961) made the first measurements of light absorption by a bulk concentration of negative ions produced by shock-heating alkali halides. This technique has proved very successful for measurement of the electron affinities of the halogen atoms and of the fine structure separations in the ground states of the corresponding negative ions. However, it cannot be applied very widely.

A second quite different method, in which light absorption by a bulk concentration of ions is observed, is one which uses ion cyclotron resonance to trap ions of a particular species in a suitable absorption chamber. The ion concentration may be monitored by the absorption of power at the resonance frequency. Photodetachment is then observed by the reduction of ion concentration when the ions are irradiated. This technique, using laser radiation sources, has been applied successfully to measure photodetachment thresholds and hence electron affinities.

An intermediate type of experiment was carried out by Woo, Branscomb and Beaty (1969) in which the ions were drifting in a buffer gas at pressure p under the action of an electric field F small enough for the mean energy of the ions to be not far above thermal. The aim of this work was to measure photodetachment from molecular ions which had undergone so many collisions ($>10^6$) that they were vibrationally deactivated. This is no longer so important as individual vibrational transitions may be separated in photo-electron energy spectra as mentioned above.

We now describe typical experimental arrangements using these techniques.

Crossed-beam methods

In a crossed-beam experiment in which the photon flux in the wavelength range between λ and $\lambda + d\lambda$, between points, measured along the ion beam, which are distant between x and $x + dx$ from some fixed reference point, is $n(\lambda,x)\,d\lambda\,dx$, the probability that an ion will suffer photodetachment is given by

$$P = v^{-1} \iint Q_d(\lambda)\, n(\lambda, x)\, d\lambda\, dx, \qquad (11.23)$$

where $Q_d(\lambda)$ is the cross-section for photodetachment by radiation of wavelength λ and v is the ion velocity. With ions of energy 300 eV and $Q_d \sim 10^{-18}$ cm^2, we find that the probability per cm path is only 2.5×10^{-7} per W cm^{-2} photon flux density. When crossed-beam experiments were first begun it was possible to illuminate about 2 cm of path with $\frac{1}{4}$ W cm^{-2} giving $P \simeq 10^{-7}$. At the same time the background pressure in the chamber through which the beam passed was around 10^{-6} torr, sufficient to give a chance of detachment through collisions with background gas as high as 10^{-5} per cm of path.

The problem of discriminating between the wanted signal of electrons produced by photodetachment from the much greater signal due to stripping collisions with background gas was solved by Branscomb and Fite (1954) using modulated phase-sensitive detection. The photon beam was chopped at 450 Hz. By means of a selective amplifier, detection could be limited to signals modulated at this frequency or close to it. This already greatly reduced the noise background but there still remained a noise component at the modulation frequency. This was largely eliminated by recording only signals which possessed the right phase relative to the chopper pulses.

Fig. 11.5 shows a typical arrangement of ion source and detector as used by Smith and Branscomb (1960). The ions were produced in a glow discharge at pressures between 25×10^{-6} and 75×10^{-6} torr and emitted through a hole in the anode with energy between

DETACHMENT BY PHOTON AND ELECTRON IMPACT 433

300 and 500 eV and energy spread of 25–50 eV. These ions were then accelerated by about 3000 V applied at electrode 1 and focused into a parallel beam by the decelerating lens formed by electrodes 1 and 2. They were then accelerated in a parallel beam by the

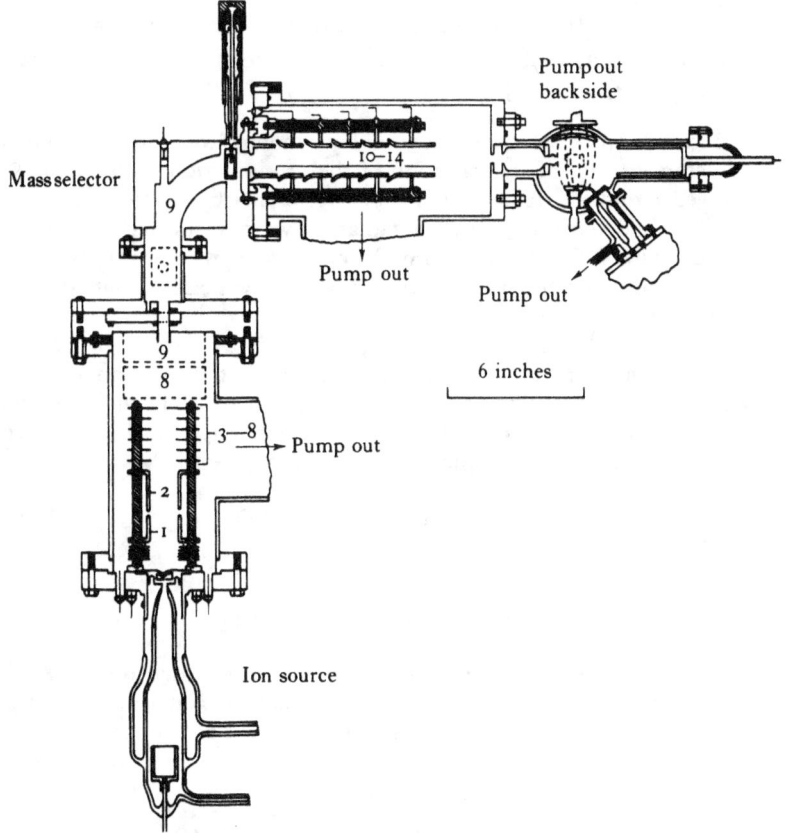

Fig. 11.5. Arrangement of the ion source and detector system in the photodetachment experiments of Smith and Branscomb (1960).

electrode lens system 3–8 to enter the 90° sector mass analyser 9. After passing through 9 the beam entered a further electrode lens system 10–14 which refocused it into the reaction chamber 15 with a kinetic energy of a few hundred eV.

A light beam from a carbon, or in later experiments, a xenon arc was focused on the ion beam in the reaction chamber, after passage through interference filters and, when necessary, absorption cells. In relation to Fig. 11.5 the optical beam passed normally to the plane of the paper.

Electrons produced in the reaction chamber were trapped by a magnetic field of 40 G and extracted by an electric field of 15 V cm^{-1}, both normal to the ion and photon beams. They were then accelerated to an energy of about 500 eV and detected with a ten-stage Ag–Mg electron multiplier.

In this early work, which did not employ a monochromatic light source, the determination of the variation of the cross-section with frequency involved complicated manipulation of interference filters. Great care was necessary in determining the total radiant power transmitted through the different filters. To eliminate any dependence of the sensitivity of the bolometer used on the wavelength, calibration was carried out against a calorimeter.

For absolute measurements of Q_d, the arc source was replaced by a very much more stable tungsten lamp. This irradiated a section of the ion beam uniformly so that the detachment probability P, as given by (11.23), was the same for all ions. Under these conditions the ratio of the current j_e of detached electrons to that j_i of the ions is given by

$$j_e/j_i = v^{-1} \int Q_d(\lambda) \{ \int n(\lambda, x)\, dx \}\, d\lambda, \qquad (11.24a)$$

with

$$\int n(\lambda, x)\, dx = \phi(\lambda)\, T(\lambda)\, (\lambda/hc) \int w(x)\, dx. \qquad (11.24b)$$

Here $w(x)$ is the incident power density at the ion trajectory, $T(\lambda)$ the transmission of the lamp envelope and other elements of the optical system and $\phi(\lambda)$ is the spectral distribution of the source normalized so

$$\int \phi(\lambda)\, T(\lambda)\, d\lambda = 1. \qquad (11.25)$$

$w(x)\, dx$ was measured with an absolute radiometer, $\phi(\lambda)$ from the colour temperature of the lamp and the emissivity of tungsten and $T(\lambda)$ with a spectrophotometer. Knowing these, $Q_d(\lambda)$ could be obtained in principle from observed values of j_e/j_i and v by inversion of (11.24a). This was carried out in practice by some iterative procedure.

With steady improvement in attainable vacua, in the intensity of light sources and the sensitivity of detectors, the range and accuracy of the crossed-beam technique was extended quite a long way beyond that attainable at the time of its introduction. One of the more remarkable achievements was the precise determination of the electron affinity of O which we discuss below on p. 451.

A big new step forward was taken in 1970 when Lineberger and Woodward published the results which they obtained for S⁻ using continuously tunable dye-laser sources. In particular, they improved the energy resolution to better than 1 meV. With different dyes it is possible to obtain lasing systems with tunable wavelengths in the wavelength range 5180–5700, 5400–5840, 4700–6200, 6000–6500 and 6400–7000 Å, so that a photon energy range from 1.75 to 2.4 eV is available. This is very convenient for the study

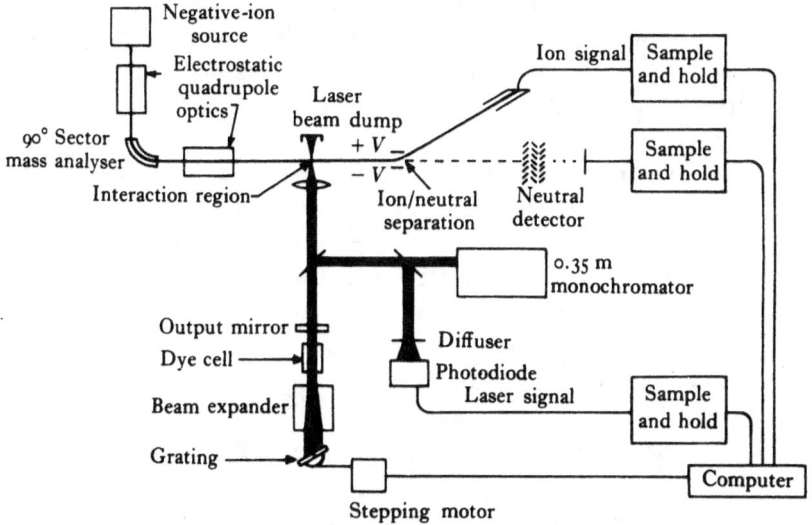

Fig. 11.6. Arrangement of the apparatus used by Lineberger and Woodward (1970) for studying photodetachment with a tunable dye-laser as light source.

of photodetachment, at photon energies close to the threshold, for a great number of atomic and molecular ions.

Fig. 11.6 illustrates a typical arrangement of an experiment of this kind. It differs from the earlier experiments described above, not only in the use of a laser beam, but also in determining the photodetachment cross-section by measurement of the flux of neutral particles produced instead of the current of electrons. For this purpose the ion beam is accelerated to a higher energy, 2 keV, so that the neutral, 'stripped', ions can be detected by a suitable second emission multiplier.

Referring to Fig. 11.6, the ions are extracted from a hot-cathode discharge source, accelerated to 2 keV and focused by an electrostatic quadrupole doublet on the entrance aperture of a 90° sector mass analyser. The emergent mass-selected beam is then focused by a second electrostatic quadrupole doublet, on to the first stage of the multiplier which is normally used to detect the neutral beam produced by detachment. Just before the ion beam enters the interaction region it is deflected through about 5°. This largely eliminates the contribution to the background due to stripping of the ion beam by background gas before reaching the interaction region, a contribution which is relatively large because the background pressure up to the deflector is about 100 times higher than in the interaction and neutral detection chambers, where it is only about 10^{-8} torr.

The dye cell of the laser, which is placed at the focus of an elliptical cavity, is pumped by a xenon flash lamp located at the other focus. The optical cavity of the laser is defined by a partially transmitting mirror and a grating with 1800 times per mm blazed at 5000 Å in first order. To tune the laser the grating is suitably rotated. The laser power is measured by deviating a portion of the beam to a solid state photodiode. Normally the laser pulse repetition rate is 5 s^{-1}, of duration about 0.3 μs and power output a few mJ/pulse. The envelope of several hundred pulses has a full width at half-maximum of 1–2 Å.

After crossing the interaction region the remaining ions are deflected by an electrostatic field into a Faraday cup, while the neutral atoms or molecules pass straight on to the first dynode of a 15 stage Cu–Be multiplier.

To discriminate against background due largely to stripping by background gas, the multiplier signal is monitored using a matched pair of fast sample holds. The first samples the signal after a delay (about 6 μs) since the laser pulse, closely equal to the time taken for the neutral beam to pass from the interaction region to the detector, while the second samples the signal about 1 μs earlier. This provides a background correction, obtained in a time which is short compared with the coherence time of the ion beam. In addition to this correction a further correction is included for laser-induced electrical noise in the atom channel. This is measured periodically by turning off the beam and averaging the difference between the signals on the two-sample holds over 100 laser pulses.

A single measurement of the cross-section is obtained as the average over about 300 laser pulses.

Checks are made to verify that the measured cross-section is independent of the laser power in the working range. If multi-photon processes were important it would increase with power, while if saturation of photodetachment were occurring it would decrease.

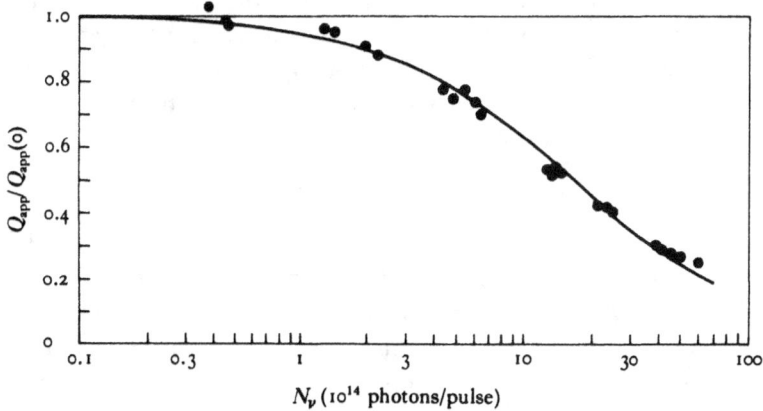

Fig. 11.7. Dependence of apparent photodetachment cross-sections Q_{app} for Ag⁻ on laser flux at a photon energy of 16 900 cm⁻¹ (2.095 eV). ● observed by Hotop and Lineberger (1973), —— calculated for average laser spot diameter 0.18 mm. $Q_{app}(0)$ is the limit for zero flux.

Use may be made of partial saturation to compare photodetachment cross-sections Q_d. Let N_ν be the number of photons in the laser pulse of duration τ, v the velocity of the ion beam and ab the assumed rectangular area illuminated uniformly by the laser, b being measured parallel to the ion beam. The number of neutral atoms produced by photodetachment for each shot of the laser is then given by

$$N = BI\{1 - \exp(-N_\nu\, Q_d b/ab\tau v)\}, \qquad (11.26)$$

where I is the ion current and B a constant. The apparent cross-section is given by

$$Q_{app} = CN/N_\nu I. \qquad (11.27)$$

A plot of Q_{app} against N_ν therefore takes the form shown in Fig. 11.7 which in fact refers to observed results for detachment from Ag⁻ (Hotop and Lineberger, 1973) (see p. 463). It follows that comparison of these saturation plots for two different negative ions under the same conditions can give directly the ratio of the true detachment cross-sections for those ions, independent of the efficiency of the detector for the different neutral atoms or of any errors in the absolute detection of the light flux.

Applications of this technique are described on pp. 454 *et seq.*

Measurement of the energy and angular distribution of the detached electrons

Consider a beam of negative ions of kinetic energy E_- and mass $M+m$, where m is the electron mass. Electrons ejected from the beam through photodetachment by photons of frequency ν, in a direction making an angle ϕ with the normal to the beam, will have an energy $E(\nu, \phi)$ in the laboratory system given by

$$h\nu - E_a = E(\nu, \phi) + (m/M)E_- - 2\{mE_- E(\nu, \phi)/M\}^{1/2}\sin\phi. \quad (11.28)$$

This is obtained by transformation from the centre of mass to the laboratory system allowing for the velocity of the ions. If measurements are made, as is usual, of the energy of electrons ejected normally to the beam in the laboratory system, $\phi = 0$, and we have

$$E(\nu, \pi/2) = h\nu - E_a - (m/M)E_-. \quad (11.29)$$

If ϕ is not accurately zero the third term on the right-hand side of (11.28) provides the appropriate correction.

As discussed on p. 421, the angular distribution of the electrons detached by linearly polarized light varies as $\{1 + \beta P_2(\cos\theta)\}$, where θ is the angle between the electric vector of the radiation and the direction of motion of the electron and β is a function of the initial and final electronic states. The distribution may therefore be measured by rotating the plane of polarization of light beam while keeping the direction of collection of the electrons fixed at 90°.

A convenient fixed frequency source for use in the measurement of the energy and angular distribution of the detached electrons is an argon ion laser which may be operated at wavelengths of 4880 and 5145 Å (2.5 and 2.4 eV). The earliest application was to the measurement by Brehm, Gusinow and Hall (1967) of the detachment energy of metastable He⁻ ions.

Fig. 11.8 shows a typical arrangement used in the study of photodetachment from NO^- and O_2^- which is very similar to that used for He^-.

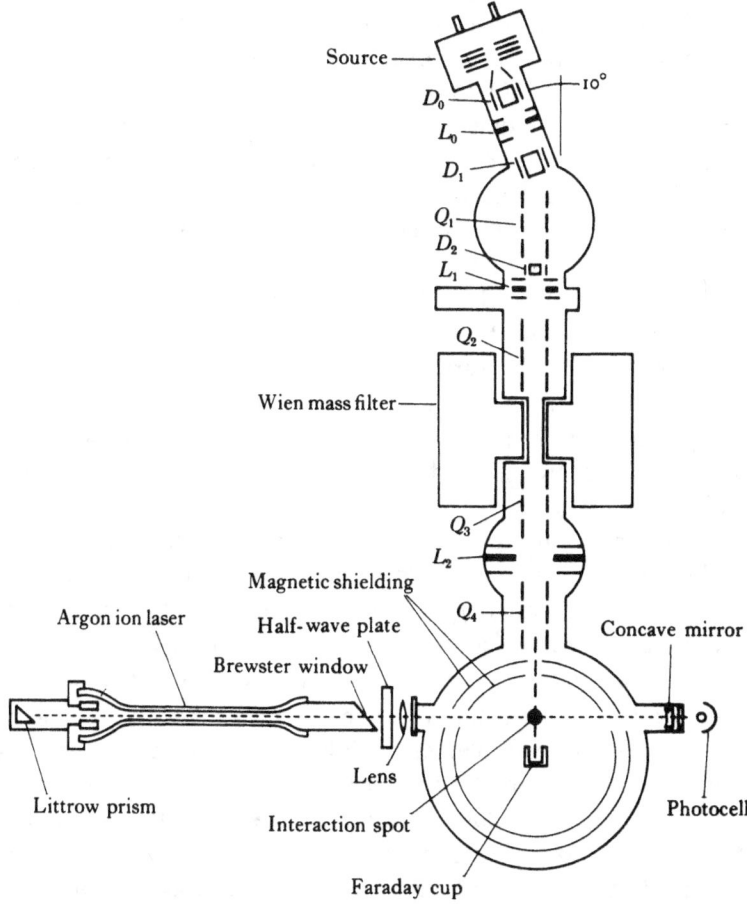

Fig. 11.8. Arrangement of the apparatus used for measurement of the angular and energy distribution of electrons resulting from photodetachment by laser light. From Siegel *et al.* (1972). D_0, D_1, D_2, vertical and horizontal deflectors; L_0, L_1, L_2 einzel lenses; Q_1, Q_2, Q_3, Q_4, twelve-element symmetrical quadrupole lenses.

A beam of negative ions, contaminated by neutral atoms and molecules, photons and electrons, extracted directly from a glow discharge source was passed through an einzel lens L_0 before deflection through 10° to remove the neutral and photon contaminants. Electrons were removed by a weak magnetic field. The deflected ion beam was then periodically focused through a quadrupole–einzel–quadrupole lens system and collimated before passing into a Wien velocity filter (see Chapter 9, p. 283) for mass selection. The filter could be set to transmit a single chosen mass, or at zero resolution to transmit all ionic species, or switched rapidly among up to four masses so that relative measurements could be made in a time so short that surface conditions in the interaction chamber could be assumed constant. After leaving the filter the beam was focused into the interaction region to an elliptical spot of area about 1 mm^2 and finally collected in a Faraday cup.

A linearly polarized laser beam was produced by a combination of Littrow prism and Brewster-angle window optics. The plane of polarization could be varied by rotation of a low-loss half-wave plate inside the laser cavity. By means of a lens, also inside the cavity, the laser beam was focused to a 0.1 mm diameter spot producing a photon flux of 200–500 kW cm^{-2} in the interaction region.

Electrons produced by photodetachment moving within a solid angle of $4\pi/2000$ sr about the direction normal to both intersecting beams were accelerated by a lens system and injected into a hemispherical analyser as shown diagrammatically in Fig. 11.9. The aperture defining the angular acceptance was located in a field-free area within the interaction region so that the acceptance angle was independent of electron energy. With the geometry used, the theroetical width of the energy peak at half-height was about 40 meV.

At the output end of the analyser, electrons which had been transmitted through were accelerated to 112 eV and passed through an exit slit to enter a high-gain, low-noise electron multiplier.

To improve stability against contact potential changes the interaction chamber was made of vacuum-fired graphite while the input and output optical components and the hemispherical analyser were of heavily gold-plated copper. As shown in Fig. 11.9 the interaction region analyser and particle multiplier were enclosed in two Permalloy magnetic shields by means of which the residual internal magnetic field was reduced to around 1 mG.

Peaks in electron energy spectra due to photodetachment

DETACHMENT BY PHOTON AND ELECTRON IMPACT 441

are normally so marked while the background is slowly variable that there is little need to determine a background correction if the aim is to measure the electron affinity. For the angular distribution measurements, however, it is necessary to distinguish an isotropic component from the background. This was done by storing signals obtained with and without the laser beam present, so that all data runs include an integral number of complete

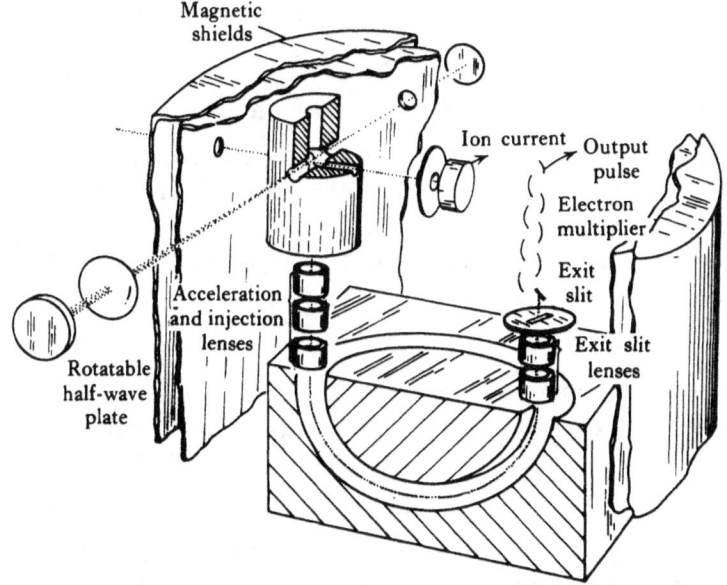

Fig. 11.9. Arrangement of the interaction chamber and the hemispherical electron monochromator in experiments using the apparatus shown in Fig. 11.8.

laser on and off cycles. The background correction is then obtained by subtraction. Small corrections are necessary because the angles measured are in the laboratory system.

To check whether the angle ϕ of collection of the electrons differs appreciably from zero, measurements were made simultaneously for H⁻ and D⁻. As the electron affinities are effectively the same the angle ϕ is given from (11.28) by

$$\sin\phi = \frac{(E_v^D - E_v^H) + mE_-(M_D^{-1} - M_H^{-1})}{2(mE_-)^{1/2}\{(E_v^D/M_D)^{1/2} - (E_v^H/M_H)^{1/2}\}}, \quad (11.30)$$

where E_v^H, E_v^D are the observed peak energies of the electron distributions arising from H⁻ and D⁻ respectively and M_H, M_D are the respective masses of H and D. In typical experiments ϕ was found to be < 0.02 rad.

The energy scale of the electron-energy analyser was calibrated against the difference between the electron affinities of S and O, measured precisely in other experiments, and the vibrational spacings in photodetachment from NO⁻ and O_2^-. In all cases a 3% correction factor was found. Applications of this technique are described on pp. 463 *et seq.*

Absorption by shock-heated alkali halides

It is possible by shock-wave heating of alkali halides to produce an atmosphere of negative atomic halogen ions of sufficient concentration and extent to produce measurable absorption of light by photodetachment. Thus for CsI the equilibrium constant K for the reaction

$$\text{CsI} \longrightarrow \text{Cs}^+ + \text{I}^-, \qquad (11.31)$$

calculated in cm³ from the partition functions in the usual way is 5×10^{17} at 3000 °K and 2.6×10^{19} at 4000 °K. Writing $n(X)$ for the concentration of the species X we have if $n(\text{Cs}^+) = n(\text{I}^-)$ that

$$K = n(\text{I}^-)^2/n(\text{CsI}). \qquad (11.32)$$

Hence if $n(\text{CsI}) = 10^{17}$ cm⁻³, $n(\text{I}^-)$ is between 10^{17} and 10^{18} cm⁻³ over the temperature range concerned. In fact this will be modified by the reactions

$$\text{Cs}^+ + \text{I}^- \longrightarrow \text{Cs} + \text{I}, \qquad (11.33a)$$

$$\text{I} + e \rightleftharpoons \text{I}^-, \qquad (11.33b)$$

$$\text{I} + \text{I} \rightleftharpoons \text{I}_2. \qquad (11.33c)$$

From the calculated equilibrium constants for reactions (11.33b) and (11.33c) respectively it is found that $n(\text{I}_2)$ is unimportant and $n(e)$ only becomes important at the highest temperature. In considering the effect of (11.33b) we may then take $n(\text{Cs}) = n(\text{I})$, in which case the calculated equilibrium constant shows that $n(\text{I})/n(\text{I}^-)$ varies from about 13 at 3000 °K to 9 at 4000 °K. This would still leave $n(\text{I}^-)$ between 10^{16} and 10^{17} cm⁻³. As it is possible to produce temperatures in the range concerned over distances of order 10 cm, perpendicular to a shock wave, we would expect substantial absorption to occur for radiation for which the photodetachment cross-section is between 10^{-17} and 10^{-18} cm^2.

This absorption has been observed for all four halogen ions by Berry and his collaborators. The precision with which the absorption edges have been located yields accurate values both of the electron affinity and of the fine structure splitting of the ground states of F^-, Cl^- and Br^-.

Fig. 11.10 shows the general arrangement used in these experiments. As usual the shock wave was generated by collapse of a

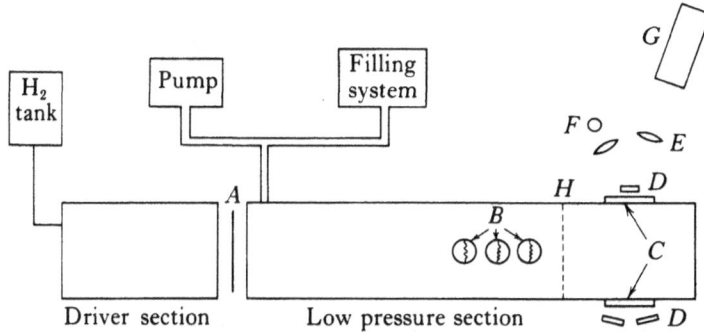

Fig. 11.10. Arrangement of apparatus used by Berry et al. for observing absorption of light by the vapour released through shock-heating of alkali halides. A, diaphragm; B, resistance gauges; C, fused silica windows; D, mirrors; E, collimating lenses; F, flash lamp; G, medium quartz spectrograph; H, sample.

diaphragm separating the high-pressure driver section from the low-pressure section. The driver gas was hydrogen initially at a pressure between 12 and 25 atmospheres while the low-pressure section was initially filled with argon at pressures between 15 and 25 torr. The salt specimen H weighing between 0.1 and 1 g was deposited on thin perforated aluminium foil or cellulose tissue. About 90 cm downstream, fused silica windows C were mounted on the top and bottom sides of the shock tube and three front-surfaced mirrors D formed a multiple reflection system.

Passage of the shock front was recorded by each of three resistance gauges B as a voltage change which initiated a signal to a variable delay circuit from which the flash lamp F was triggered.

After four traverses of the multiple reflection system, corresponding to a total path length of 34 cm in the shock tube, the light was dispersed by a Bausch and Lamb medium quartz spectrograph and recorded photographically. Comparison spectra were taken without any salt present.

Fig. 11.11 illustrates a typical absorption spectrum obtained from the vapour produced by shock heating CsCl. A spectrum

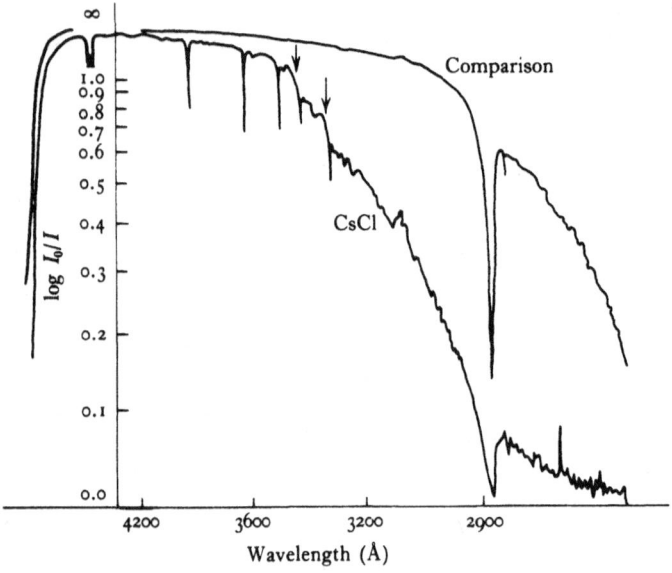

Fig. 11.11. Microdensitometer tracing of absorption spectrum of the vapour from CsCl evaporated by shock heating. A comparison spectrum taken without any CsCl present is also shown. The arrows indicate photodetachment thresholds for Cl^-.

taken without any CsCl present is shown for comparison. Many absorption lines due to neutral Cs atoms are present, superposed on a strong continuous absorption. This continuum has the same shape for a given halogen irrespective of the alkali metal. The extinction increases with shock temperature and depends on the size of the salt sample in a way which correlates well with the intensity and width of the atomic lines. As one can rule out the possibility of absorption by neutral or ionized molecules and by the neutral or positively-ionized halogen atoms which certainly falls in the ultra-

DETACHMENT BY PHOTON AND ELECTRON IMPACT 445

violet, it seems clear that the continuum can only be due to photodetachment from halogen negative ions.

The photodetachment thresholds for Cl⁻ are indicated in Fig. 11.11. Since the transition involved is from a p to an s orbital, the cross-section rises from threshold as $(\nu - \nu_0)^{1/2}$ (see (11.9)) and has an infinite slope at the threshold as a function of frequency. This is consistent with the rapid increase of extinction at the thresholds indicated. The two thresholds arise from transitions to the two fine-structure levels of the ^2P state of Cl, so that from observations such as are shown in Fig. 11.11 the fine structure separation can be determined. Comparison with values determined spectroscopically provides a further check on the interpretation of the shock wave results. Further details of the application of the shock wave technique to the determination of the electron affinities of the halogen atoms are discussed on p. 461. At the highest shock pressures sufficiently large concentrations of free electrons are produced to give rise to an observable affinity emission continuum which has already been discussed in Chapter 8.

Photodetachment studies using an ion cyclotron resonance spectrometer

By means of an ion cyclotron resonance spectrometer it is possible to trap negative ions for periods of the order of seconds under high vacuum conditions and also monitor their concentration with high sensitivity of detection. It is then possible to observe changes in concentration produced by irradiation with detaching radiation and hence to determine the variation of the photodetachment cross-section with the frequency of the radiation. There is the additional advantage that the concentrations of different ions in a mixture can be separately monitored.

The principle of the method depends on the fact that, in a uniform magnetic field H, an ion of mass M and charge e moving in a plane perpendicular to H with velocity v will describe a circle of radius Mcv/eH with an angular frequency

$$\omega_c = eH/Mc. \qquad (11.34)$$

ω_c is known as the cyclotron frequency, it is independent of v and in a given magnetic field, for singly-charged ions, will be characteristic of the mass of the ion. If an alternating electric field of angular frequency ω is applied in a direction normal to H, resonance acceleration will occur when $\omega = \omega_c$, as in a cyclotron. This will result in increase in the orbital radius until the ions reach the containing walls. The energy required for the acceleration must come from the electric power source and can be detected by the change in load.

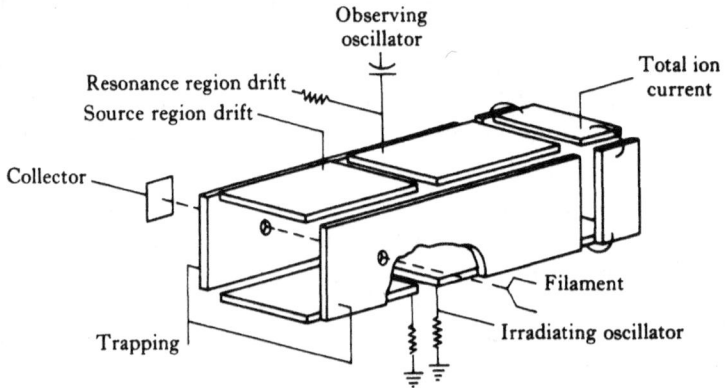

Fig. 11.12. Illustrating the typical arrangement of an ion cyclotron spectrometer.

The mean power absorption is given by

$$\bar{P} = \tfrac{1}{4} n_- e^2 E^2 (\nu/M) \{\nu^2 + (\omega_H - \omega)^2\}^{-1}, \qquad (11.35)$$

where n_- is the concentration of the negative ions of mass M, E the amplitude of the alternating field applied and ν the momentum transfer collision frequency. Hence, by observing $\bar{P}$ as a function of ω it is possible to determine both n_- and ν for the particular negative ions.

Fig. 11.12 illustrates the arrangement in a typical cyclotron resonance cell. It consists essentially of a rectangular box divided into two chambers and bathed by a uniform magnetic field in a direction parallel to the end section. The first chamber contains the gas or vapour from which the ions are produced by impact of electrons accelerated from a hot filament along the direction of the magnetic field. Motion of the ions parallel to the magnetic field is constrained by voltage applied to the side plates of the cell so that, apart from motion normal to the field, they are trapped. However, by application of a steady electrostatic field E_d applied to the upper and lower plates of the source chamber, the ions are given a sideways drift velocity HE_d/c which transfers them to the second, analyser, chamber. In the absence of any corresponding field E_d, after entering this chamber they will be trapped until they reach the walls through collisions with residual gas.

The concentration of ions is then monitored by a marginal oscillator detector. This is an oscillator operated under conditions which are just marginal for continuous oscillation so that the level of the oscillator is very sensitive to small impedance changes.

For application to photodetachment an optical glass window is fitted to the end of the resonance cell to allow a light beam from a suitable source to enter the cell in the direction of the longitudinal axis, parallel to the drift velocity of the ions. In the first experiments the light source was a xenon arc lamp and a monochromator or a Chromatix Model 1000 E Nd:YAG tunable laser. The photon flux was monitored with either an Optical Power Meter using a silicon solar cell or a Perkin–Elmer thermocouple.

In typical experiments first carried out by Brauman and Smyth (1969), the cell dimensions were $2.54 \times 2.54 \times 16.1$ cm. Typical electrical fields used were in V cm^{-1}, 0.70–0.85, 0.08–0.28 and less than 0.04 for trapping, drift and in the analysing chamber respectively. The pressure of the gas or vapour from the ions produced was kept as low as possible (10^{-7}–10^{-6} torr) consistent with usable negative-ion signals. Under these conditions ions could be trapped for several hundred ms.

In steady state experiments of this kind, Smyth, McIver, Brauman and Wallace (1971) were able to produce detachment from 50% of the trapped population of PH_2^- using a laser at an average power level of 30 mW.

A pulse technique has also been used in which the negative ions are formed by a 30 ms pulse of the electron beam and trapped within the cell for a time around 200 ms. During this time they could be irradiated with a number of laser pulses.

Results obtained using these techniques are discussed on p. 486.

Photodetachment from ions aged in a drift tube

One of the difficulties associated with the study of molecular negative ions is the uncertainty in the distribution of initial vibrational states. This can be overcome by measurement of photoelectron energy spectra with a laser source but otherwise is a serious complication. Woo, Branscomb and Beaty (1969) carried out a remarkable experiment in which the ions were allowed to age by drifting in a gas until reaching vibrational equilibrium at the gas temperature before photodetachment. The work

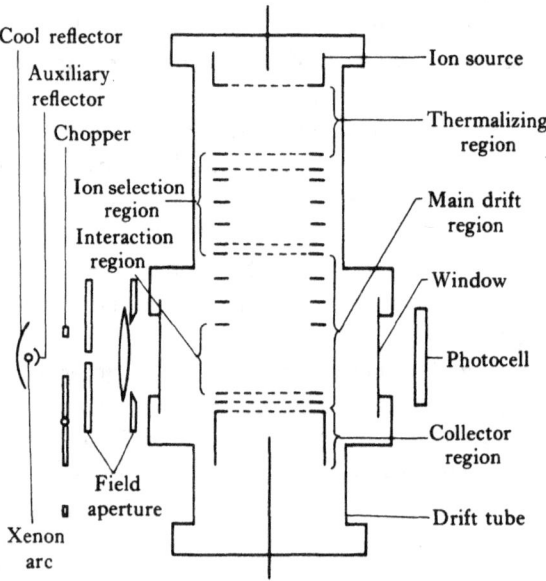

Fig. 11.13. Arrangement of the apparatus used by Woo *et al.* (1969) in their photodetachment experiments.

was particularly directed towards obtaining the rate of photodetachment of O_2^- ions by sunlight in the E region of the ionosphere.

Fig. 11.13 illustrates the general arrangement of the experiment. Negative ions from a pulsed discharge drifted under the action of an electric field F through O_2 at a pressure p, around 4 torr, in the thermalizing and ion-selection region. Before entering the region of interaction with the photons an ion would make 10^7 collisions at thermal energy with the gas molecules. When the ions arrived in the interaction region the collector was opened to receive the electrons produced by photodetachment.

Ions were selected by their mobility, using the time-of-flight across the ion-selection region. In the particular case of O_2^- this is not very satisfactory as O_4^- ions, which will be formed by ionic reactions within the gas, have nearly the same mobility as O_2^- (see Chapter 12, p. 541).

The light source, chosen because its spectrum resembles that of sunlight, was a high-pressure xenon arc. The pulse sequence required for preparing the ions was generated, in relation to the phase of the light chopping, by a photoelectric trigger. To avoid the need for absolute measurements of the light intensity within the interaction region, measurements were made for O_2^- relative to O^- which could be selected for study in the same experiment and for which the photodetachment cross-section is known quite accurately as a function of frequency.

A similar technique has been applied by Burt (1972, 1973) to study photodetachment from O_4^-, CO_3^- and $CO_3^- \cdot H_2O$. These results are discussed on p. 489, those for O_2^- on p. 482.

11.5 Application to different atomic negative ions
H⁻

Following the measurement of the relative cross-section Q_d for H⁻ by Branscomb and Fite (1954), absolute values were obtained a little later by Branscomb and Smith (1955) for wavelengths greater than 4000 Å using the technique described on pp. 432–4. At the standard wavelength, 5280 Å, they found $Q_d = (3.28 \pm 0.3) \times 10^{-17}$ cm². The accuracy of measured cross-sections relative to this value was improved by Smith and Burch (1959), who obtained results with a probable error of only 2%.

Fig. 11.1 illustrates the best observed cross-sections as functions of wavelength in comparison with elaborate theoretical values calculated as described on p. 421. It will be seen that the agreement with the most reliable theoretical results, those obtained using the velocity matrix element, is very good and there is little doubt that these results can be used with confidence in the important application to the absorption by the solar atmosphere which we discuss in Chapter 15, p. 678.

Hall and Siegel (1968) have measured the angular distributions of the electrons ejected from H⁻ by argon ion laser radiation using the technique described on p. 438. The results obtained agree closely with those given by the theoretical distribution which in this case, involving s–p transitions, is of the simple form $\cos^2 \theta$, where θ is the angle the direction of motion of the electron makes with the electric vector of the detaching radiation, independent of the electron energy.

He⁻

Photodetachment of the metastable He⁻(^{4}P) ion (Chapter 5, p. 126) was the first to be studied by observation of the energy spectrum of the detached electrons. Brehm *et al.* (1967) used for this purpose equipment similar to that described on pp. 438–42, except that a double-charge transfer ion source was used. Positive ions of helium, deuterium and hydrogen were extracted from a hot-cathode arc discharge, accelerated to 2.5 keV and focused through an oven containing potassium vapour at a pressure of a few times 10^{-3} torr. A few per cent of the ions picked up two electrons in charge transfer collisions with the K atoms (see Chapter 10, p. 414). This produced a well-defined beam of nearly monoenergetic He⁻ ions which were electrostatically separated from the positive and neutral beams.

Contact potentials were determined by measurement of the energy of the electrons produced by photodetachment from H⁻ or D⁻ for which the electron affinity is known accurately. The effective collection angle of the analyser (see p. 441) was determined by measuring the energy separation between the 2^3S and 2^3P final states.

From these measurements the threshold energy for the reaction

$$\text{He}^-(^4\text{P}) + h\nu \longrightarrow \text{He}(2^3\text{S}) + e, \qquad (11.36)$$

was found to be 80.0 meV with a probable error of ±2 meV. Comparison with theoretical values has already been discussed in Chapter 5, p. 127.

The cross-section for the process in which the atom is left in the 2^3P state could be measured relative to that for H⁻ because in both cases an s–p transition is involved so that the ejected electrons have the same angular distributions (see p. 421). For $\lambda = 5145$ Å the cross-section was found to be 7×10^{-19} cm². It is not possible to determine the cross-section for the process (11.36) in the same way because, as p–s and p–d transitions are involved, the angular distribution of the ejected electrons will be different from those detached from H⁻.

O⁻

The experimental study of photodetachment from O⁻ is now a classic in the subject. Soon after the crossed-beam method was developed it was natural to turn towards O^-, a negative ion of possible interest in the earth's ionosphere. At the time of the first experiments by Branscomb and Smith (1955) the accepted value of the electron affinity was 2.2 eV, derived from observed threshold potentials for production of O^- ions by dissociative attachment in O_2, CO and NO. It was a great surprise when the crossed-beam method yielded the considerably lower value 1.45 eV so that special attention was concentrated on checking it by very careful measurements near the photodetachment threshold. These confirmed the lower value and gave it as

$$E_a(O) = 1.465 \pm 0.005 \text{ eV}, \quad (11.37)$$

the probable error being so much smaller than in other determinations of electron affinities that (11.37) is used as a standard, having been confirmed from measurements using dye lasers. The reasons why the value derived from threshold potentials for dissociative attachment was erroneously high have been discussed in Chapter 5, p. 278.

The problem of determining the threshold wavelength is complicated by the presence of fine structure in the initial $O^-(^2P)$ and final $O(^3P)$ levels. Fig. 11.14 gives an energy level diagram, showing in particular the threshold energies for detachment from the $P_{3/2}$ and $P_{1/2}$ states, the former, although larger, corresponding to the electron affinity. To analyse this situation, Branscomb, Burch, Smith and Geltman (1958) took as a good approximation to the behaviour near the thresholds

$$Q_d(\lambda) = (\gamma B/\lambda)\{(\lambda_1 - \lambda)/\lambda_1 \lambda\}^{1/2}, \quad \lambda_0 < \lambda < \lambda_1$$
$$= (\gamma/\lambda)\{(\lambda_0 - \lambda)/\lambda_0 \lambda\}^{1/2} + (\gamma A/\lambda)\{(\lambda_0 - \lambda)/\lambda_0 \lambda\}^{1/2}, \quad \lambda < \lambda_0.$$
$$(11.38)$$

A and B are adjustable parameters, γ a constant and λ_0, λ_1 are the threshold wavelengths for detachment from the $^2P_{1/2}$ and $^2P_{3/2}$ states. In choosing (11.38) account was taken of the form of the threshold law for a p–s transition (see 11.10).

By using a set of seven filters with short wave cut-off not less than 7000 Å, the threshold being near 8500 Å, advantage was taken of the fact that the carbon arc light source gives a nearly uniform spectral distribution over the wavelength range. Then for a particular filter combination m, the detachment probability may be written

$$P_m = (CW/v) \int Q_d(\lambda) \, T_m(\lambda) \, d\lambda, \tag{11.39}$$

where W is the power incident on the filter of transmission $T(\lambda)$, v is the ion velocity and C a geometrical constant. By measuring P_m for each of the 7 filter combinations and substituting the forms (11.38) for $Q_d(\lambda)$, it was found that satisfactory solutions of the

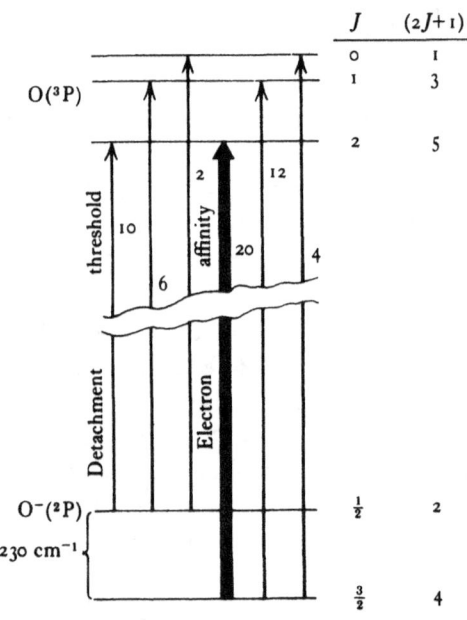

Fig. 11.14. Energy-level diagram showing photodetachment transitions between fine structure levels of O⁻ and O.

integral equations could be found by appropriate choice of the undetermined constant in (11.38). In particular λ_0 was found to be 8460 ± 30 Å yielding the value (11.37) for the electron affinity.

The absolute value of Q_d was measured, as described on p. 434, for light of wavelength 5280 Å, as 6.3×10^{-18} cm². Relative cross-sections were measured from threshold to 5200 Å with a probable error of 2% and extended to 3100 Å by using a scanning monochromator. Fig. 11.15 illustrates the final form of Q_d as a function of photon energy.

It will be seen that a new threshold appears at a quantum energy of 3.43 eV which agrees exactly with that expected for a detachment process in which the neutral atom is left in the first excited (^1D) state which lies 1.96 eV above the ground state. For this the threshold should be $1.96 + 1.46_5$ eV $= 3.42_5$ eV.

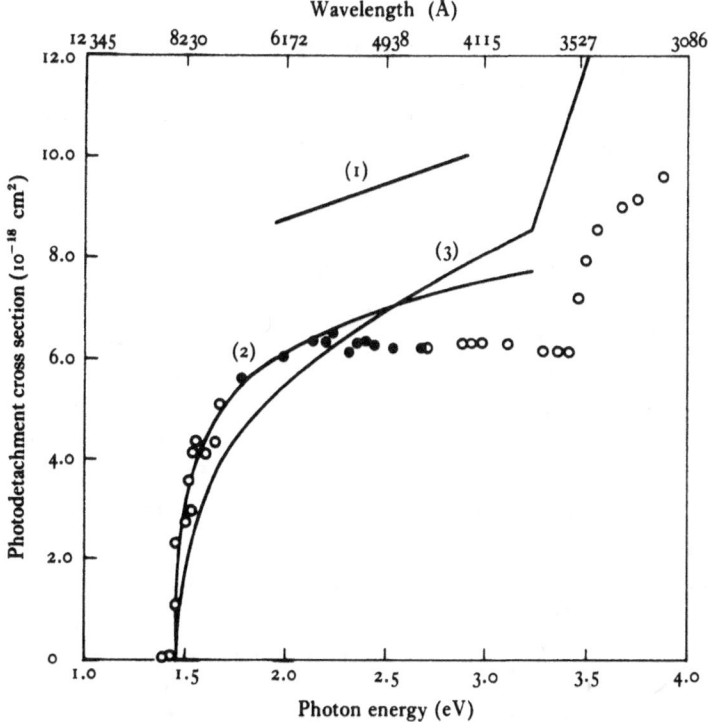

Fig. 11.15. Photodetachment cross-sections for O⁻. ●, ○, observed by Branscomb et al. (1965), (1) derived from analysis of arc emission spectra by Boldt (1959) (see Chapter 8, p. 253), (2) calculated by Cooper and Martin (1962), (3) calculated by Robinson and Geltman (1967).

In Fig. 11.15 comparison is made with the theoretical calculations of Robinson and Geltman (1967) and of Cooper and Martin (1962) (see p. 424). Further comparison with theory is provided by the measurement of the angular distribution of the photoelectrons produced by detachment using the 4880 Å argon ion laser line. Hall and Seigel (1968), using the method described on p. 438, find that the distribution is given by (11.17) with $\beta = -0.885 \pm 0.015$. Cooper and Zare (1968) using matrix elements calculated with the Robinson–Geltman model (p. 425) find $\beta = -0.95$, which is in reasonably good agreement.

According to the sum rules (see p. 420) the integral

$$(mc/\pi e^2) \int_{\nu}^{\infty} Q_d(\nu) \, d\nu \qquad (11.40)$$

should equal the total number, 9, of electrons in O^-, ν_0 being the threshold frequency. From threshold to 4000 Å the contribution is only 0.02 so that the greater part of the integral comes from much higher frequencies.

S^- and Se^-

The first studies of photodetachment from S^- were made by Branscomb and Smith (1956) using a similar technique to that for O^-. They determined the electron affinity as 2.07 ± 0.07 eV. Since then detachment from both S^- and Se^- has been studied using tunable dye-laser light as described on p. 435. In fact S^- was the first ion to be studied in this way. As a result very precise values of the electron affinity are now known for both S and Se as well as of the fine structure separation in the ground ^2P state of both ions.

Figs. 11.16(a) and (b) illustrate the observed variation of the detachment cross-section with photon energy. The energy level diagrams for both S^- and Se^- are similar to that for O^- shown in Fig. 11.14 except that the fine structure splittings are higher for S^- and much higher still for Se^-. In fact, in the latter case, the separation is so large that most of the Se^- ions in the experimental beam can be expected to be in the lowest $^2P_{3/2}$ level. This assists identification of observed thresholds, so adding to the advantage of studying a system with relatively widely spaced sublevels. Because of this we first discuss observations for Se^-, although S^- was studied earlier.

From the energy-level diagram we expect six thresholds. In fact, as may be seen from Fig. 11.16(b), 4 are clearly visible. The strongest transitions could be expected to arise from the ground $^2P_{3/2}$ level to the three ^3P levels of Se and the separation between their thresholds should then be as given from the known fine structure separ-

DETACHMENT BY PHOTON AND ELECTRON IMPACT 455

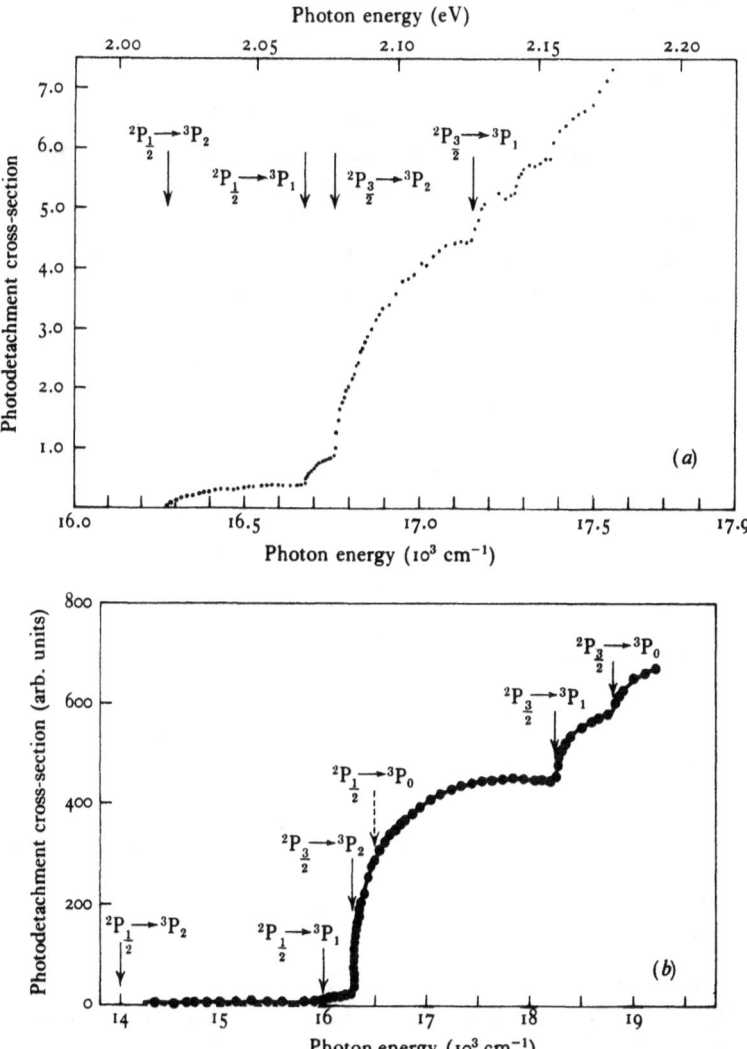

Fig. 11.16. Photodetachment cross-sections for (a) S⁻ and (b) Se⁻, measured with a tunable dye-laser as light source, by Lineberger and Woodward (1970) and by Hotop et al. (1973) respectively. The identifications of the transitions at thresholds are indicated.

ation in the atom. This leads to the identifications shown in Fig. 11.16(b). Of the much weaker transitions from the $^2P_{1/2}$ level there is clear evidence of a threshold near 16 000 cm⁻¹ which must be interpreted as associated with a transition to the $^3P_1$ level. A final $^3P_0$ level is ruled out because we should then see a $^2P_{1/2} \to {}^3P_1$ threshold around 15 450 cm^{-1} which is not observed. Again, if the transitions were to a $^3P_2$ level this should be the lowest threshold, yet there is clear evidence of photodetachment at lower photon energies. Also there is no evidence of what would then be the $^3P_1$ threshold at 18 000 cm^{-1}.

Accepting the interpretation of the threshold near 16 000 cm^{-1}, that for the $^2P_{1/2} \to {}^3P_2$ transition would occur near 14 000 cm^{-1},

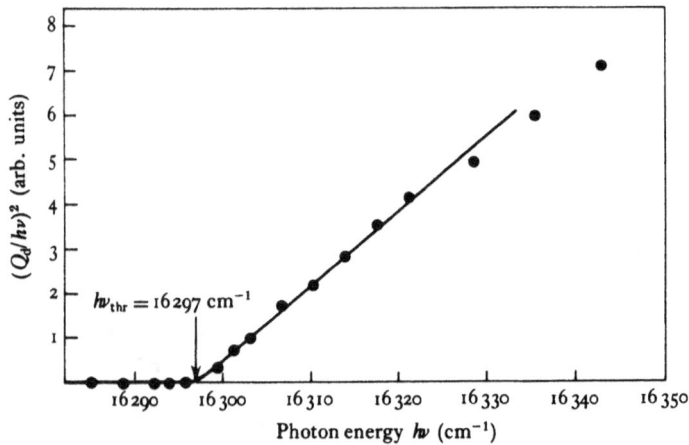

Fig. 11.17. Plot of observed value of $(Q_d/h\nu)^2$ as a function of photon energy near the threshold for the $^2P_{3/2}$–$^3P_2$ transition from Se$^-$. ● observed results, —— calculated from the law (11.9) with allowance for laser bandwidth. From Hotop et al. (1973).

beyond the range attainable with the dye-lasers, while that for $^2P_{1/2} \to {}^3P_0$ is apparently unresolved in the rising part of the cross-section for the $^2P_{3/2} \to {}^3P_2$ transition.

From the three most accurately determined thresholds ($^2P_{3/2} \to {}^3P_2$, $^2P_{3/2} \to {}^3P_1$, $^2P_{3/2} \to {}^3P_0$) the fine structure separation in Se$^-$ was determined as 2279 ± 2 cm^{-1} = 282.6 ± 0.3 meV.

To determine the electron affinity, plots were made of $(Q_d/h\nu)^2$ against photon energy. For transitions from p orbitals such plots should rise linearly from the threshold (see (11.9)). This was found to be true up to about 5 meV only above the threshold. Fig. 11.17

illustrates a typical plot for the $^2P_{3/2}$–$^3P_2$ threshold which gives the electron affinity directly. The full line is that calculated on the assumption of a linear threshold law but with allowance made for the finite band width ($\simeq 0.3$ meV) of the laser source. By choosing the threshold value to give the best fit it may be determined to ± 2 cm^{-1}. This gives for the electron affinity

$$E_a(\text{Se}) = 2.0206 \pm 0.0003 \text{ eV}. \qquad (11.41)$$

Comparison with values estimated empirically is given in Chapter 3, Table 3.6.

By comparison with measurements for O$^-$ the cross-section Q_d for the $^2P_{3/2}$–$^3P_2$ transition was found, at 18 000 cm^{-1}, to be $(7.5 \pm 2) \times 10^{-18}$ cm^2.

The relative intensities of the different transitions between fine structure levels which could be identified were also measured and will be discussed below in relation to measurements for S.

The earlier observations of Lineberger and Woodward (1970) for sulphur, clearly revealed four identifiable thresholds as indicated in Fig. 11.16(a). In this case we would not expect, and do not find, much preponderance of transitions from $^2P_{3/2}$ states. The electron affinity was found to be

$$E_a(\text{S}) = 2.0772 \pm 0.005 \text{ eV}. \qquad (11.42)$$

which agrees with the earlier measurement of Branscomb and Smith (1956) to within their quoted experimental error. Comparison with other estimates is given in Chapter 3, p. 58.

The fine structure separation in S$^-$ was determined as 482 ± 2 cm^{-1}. Hotop, Patterson and Lineberger obtained values for these separations for all the sulphur-like ions, O$^-$, S$^-$, Se$^-$ and Te$^-$ by the extrapolation procedure described in Chapter 3, p. 54. Their results for Se$^-$ and S$^-$, 2290 ± 50 and 488 ± 11 cm^{-1} agreed very well indeed with those measured from photodetachment as described above. For O$^-$ they obtain 181 ± 4 cm^{-1} which is somewhat smaller than that derived from shock wave emission measurements as described in Chapter 8, p. 253.

Relative intensities of the different transitions between fine structure levels were also measured for S$^-$ by Lineberger and Woodward (1970). Both their results and those of Hotop *et al.* for

Se⁻ do not agree with what would be expected on simple statistical considerations. The results for Se⁻ do suggest however that the theory of Rau and Fano (1971) (see p. 428) is basically correct.

C⁻

Photodetachment from C⁻ was first studied by Seman and Branscomb (1962) using the technique described on p. 432. The chief difficulty at the time was that of obtaining a sufficiently intense source of C⁻ ions. Thus from a hot-cathode discharge in carbon monoxide, currents of between 5×10^{-10} and 10^{-9} A were obtained,

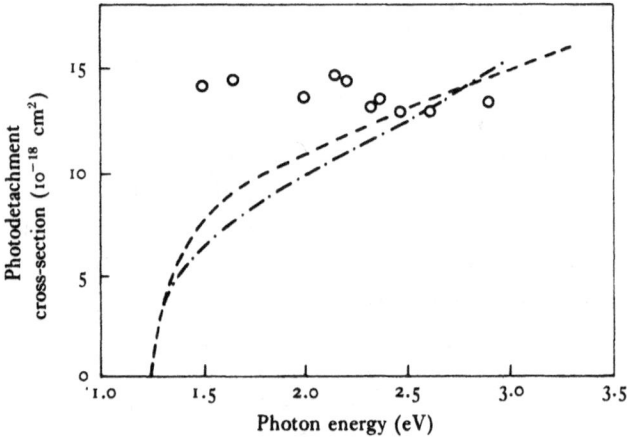

Fig. 11.18. Photodetachment cross-sections for C⁻. ○ observed by Seman and Branscomb (1962), —·—· calculated by Cooper and Martin (1962), – – – – calculated by Robinson and Geltman (1967).

about two orders of magnitude smaller than for H⁻ and O⁻. This is not surprising in view of the small cross-sections for C⁻ production by dissociative attachment (see Chapter 9, p. 338). Nevertheless, by improving the background vacuum and introducing an electron multiplier as detector, these difficulties were overcome.

Fig. 11.18 shows the observed detachment cross-section as a function of frequency. It is of interest to note that, up to photon energies of 3.0 eV, no second threshold is observed. This strongly supports the identification of the ground state of C⁻ as a ⁴S state, as for the isoelectronic atom N. If this is so then the only possible transition is to C ^{3}P, whereas two transitions would be possible if the ground state were ^{2}D, namely to ^{3}P and ^{1}D levels separated by 1.24 eV.

On the other hand, evidence was obtained of photodetachment occurring down to very long wavelengths, beyond 2 μm. The signals at these wavelengths were found to be proportional to ion beam currents and photon flux and were quite reproducible. They probably arise from photodetachment from the ^{2}D state of C⁻ which may well lie just below the ground ^{3}P state of C (see Chapter 4, p. 66).

The presence of this long wavelength tail complicates the determination of the threshold frequency for detachment from the ground state but Seman and Branscomb obtained the value 9900 ± 200 Å giving for the electron affinity of carbon

$$E_a(C) = 1.25 \pm 0.03 \text{ eV}. \tag{11.43}$$

Hall and Siegel (1968) have used the equipment described on p. 438 to measure the angular distribution of the photoelectrons from C⁻. In the course of these experiments they obtained, from measurements relative to O⁻ for which they assumed the electron affinity (11.37),

$$E_a(C) = 1.270 \pm 0.010 \text{ eV}. \tag{11.44}$$

The observed angular distributions for 4880 and 5145 Å detaching radiation were found to have the form (11.17) with $\beta =$ −0.715 and −0.805 respectively. Cooper and Zare (1968) found, using the model of Robinson and Geltman to calculate the matrix elements, the corresponding values −0.65 and −0.73 which are in reasonable agreement. The same model gives rather too large cross-sections at low photon energies as may be seen from Fig. 11.18.

Halogen negative ions

The only halogen negative ion to be studied by the crossed-beam method is I⁻. Steiner, Seman and Branscomb (1962) were able to measure photodetachment from an I⁻ beam produced from a hot-cathode discharge through a mixture of iodine vapour and ammonia, using a monochromator in conjunction with a 2500 W xenon lamp source of radiation. Over the wavelength region from 4000–3000 Å the wavelength resolution was 33 Å while for scanning between 4100 and 3900 Å near the threshold, it was improved to 18 Å.

Fig. 11.19 shows the observed variation of Q_d with wavelength. The existence of a second threshold is clear but could not be located with precision. However, particular attention was paid to determination of the first threshold and hence the electron affinity. By fitting the observed cross-section with the empirical form

$$Q_d = a(k + bk^2), \qquad (11.45)$$

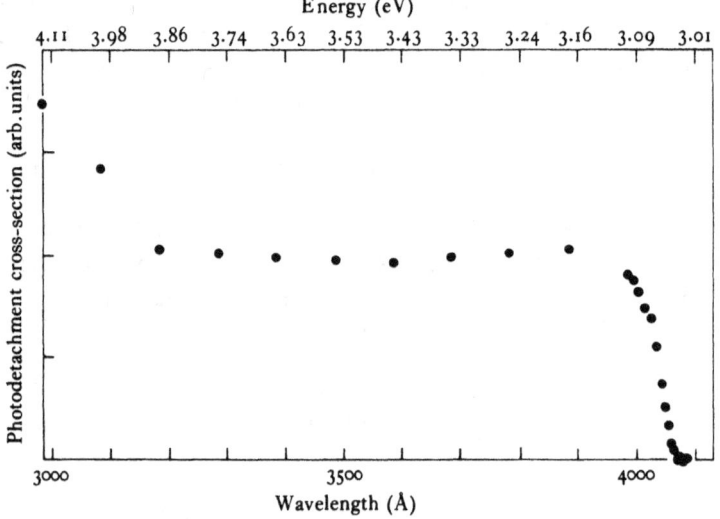

Fig. 11.19. Variation with wavelength of the photodetachment cross-section for I$^-$, observed by Steiner et al. (1962).

where k is the wave number of the attached electron, which agrees with the threshold law (11.8) for p–s transitions, and allowing for the slit function of the monochromator, the first threshold was located at a wavelength of 4051 ± 3 Å, giving for the electron affinity

$$E_a(\mathrm{I}) = 3.059 \pm 0.002 \text{ eV}. \qquad (11.46)$$

By comparison with the known Q_d for H$^-$ between 3600 and 4900 Å the magnitude of Q_d for I$^-$, assumed to behave as a step function at the threshold, was determined as $(2.1 \pm 0.5) \times 10^{-17}$ cm^2 but no attempt was made to obtain accurate absolute values.

Mandl and Hyman (1973) have observed the photodetachment cross-section to smaller wavelengths using the shock wave absorption technique applied to CsI. They find a peak at 2250 Å which they interpret as due to excitation of an autodetaching state of I^-.

Berry et al. (1961), from observation of the absorption of RbI evaporated by shock-wave heating (see p. 443), found

$$E_a(I) = 3.063 \text{ eV},$$

which agrees quite closely with the crossed-beam measurements. Comparison with other measurements of $E_a(I)$ is discussed in Table 3.3.

Similar measurements for the other halogens gave

$$E_a(F) = 3.448 \text{ eV}, \ E_a(Cl) = 3.613 \text{ eV}, \ E_a(Br) = 3.063 \text{ eV},$$

the first being due to Berry and Reimann (1963) and the latter two to Berry et al. (1961). For Cl and Br these values agree well with those determined from the threshold wavelength for emission of the affinity continuum, namely $E_a(Cl) = 3.610$ eV and $E_a(Br) = 3.366$ eV (see Chapter 8, p. 252). Comparison with values derived by other methods is given in Table 3.3.

In the absorption measurements with RbI only one threshold was observed but, as shown in Fig. 11.11, two thresholds were observed for alkali chlorides and bromides. The second threshold, also observed for alkali fluorides, arises from detachment which leaves the neutral atom in the $^2P_{1/2}$ state. This is confirmed by comparison of the separation between the two thresholds with the fine structure separation of the ground ^2P term of the neutral atoms, as measured spectroscopically. Thus for F, Cl and Br, the observed threshold separations are 0.052 ± 0.006, 0.108 ± 0.007 and 0.457 ± 0.007 eV respectively, in good agreement with the spectroscopic values 0.0501, 0.1092 and 0.4568 eV.

Negative ions of the noble metals

Photodetachment from Au^-, Ag^- and Pt^- has been studied by Hotop and Lineberger (1973), using tunable dye-lasers, as described on p. 435, and a sputter source of the negative ions (see Chapter 10, p. 389).

Using a similar source, Hotop, Bennett and Lineberger (1973) have applied the technique described on p. 438 to measure the energy spectrum of electrons produced by photodetachment from Cu⁻ and Ag⁻ by argon ion laser radiation.

For Cu⁻, Ag⁻ and Au⁻, the ground state is a $^1S_0$ state and the first threshold corresponds to detachment leaving neutral atoms in their ground $^2S_{1/2}$ states. The first excited state is about 1 eV higher, beyond the range of the tunable dye-lasers.

Fig. 11.20 shows Q_d observed for Au⁻ in the photon energy range

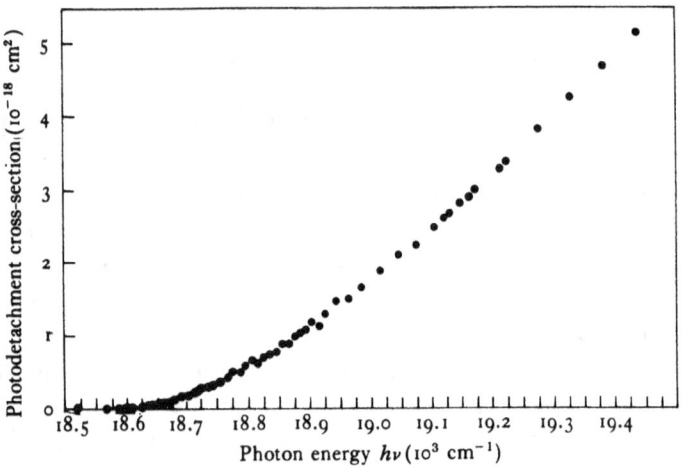

Fig. 11.20. Photodetachment cross-section for Au⁻, observed by Hotop and Lineberger (1973).

18 500–19 500 cm⁻¹ (2.29–2.42 eV). The threshold close to 18 600 cm⁻¹ (2.30 eV) was determined with high precision by plotting $(Q_d/h\nu)^{2/3}$ against the photon energy. For s–p transitions this should give a straight line, close to the threshold, intersecting the horizontal axis at the threshold energy. Fig. 11.21 shows that a straight line is indeed obtained out to photon energies as much as 50 meV above the threshold which is determined as 18 620 ± 5 cm⁻¹. This gives

$$E_a(\mathrm{Au}) = 2.3086 \pm 0.0007 \text{ eV}. \quad (11.47)$$

By comparison with O⁻, Q_d for Au⁻ was determined at a photon

DETACHMENT BY PHOTON AND ELECTRON IMPACT

energy of 19 120 cm^{-1} as

$$Q_d(Au^-) = 2.6 \times 10^{-18} \text{ cm}^2 \pm 30\%. \qquad (11.48)$$

For Ag$^-$, Q_d was found to be nearly constant in the range 6200–5700 Å at the value $(65 \pm 10) \times 10^{-18}$ cm^2 determined by comparison with O$^-$. Care had to be taken that the photon flux was not too high, for it was readily possible to obtain partial saturation of the photodetachment. Over the wavelength range attainable with the tunable dye-lasers it was not possible to observe the first threshold

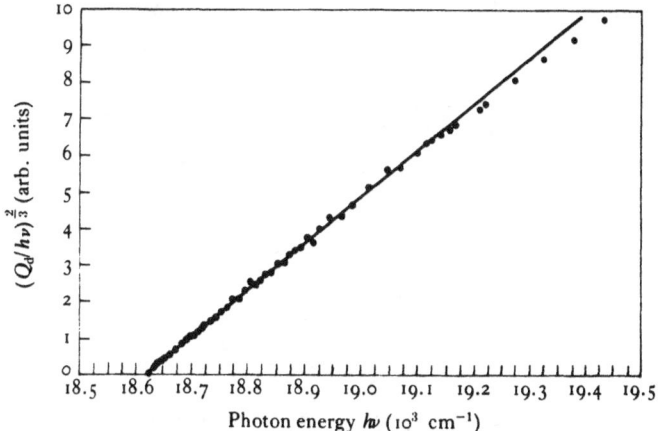

Fig. 11.21. Plot of observed value of $(Q_d/h\nu)^{2/3}$ as a function of photon energy near the threshold for photodetachment from Au$^-$. ● observed by Hotop and Lineberger (1973), —— calculated from the law (11.6) with allowance for the laser bandwidth.

for Q_d but it was estimated to be at a photon energy 1.5 eV. However, from observations of the photoelectron energy spectrum, calibrated by comparison with results for O$^-$ and OH$^-$ using electrical switching of the Wien filter (see p. 283), it was found that

$$E_d(Ag) = 1.303^{+0.007}_{-0.011} \text{ eV}. \qquad (11.49)$$

Similar measurements for Cu gave

$$E_a(Cu) = 1.226 \pm 0.010 \text{ eV}. \qquad (11.50)$$

Comparison with other determinations of these electron affinities is discussed in Chapter 3, p. 43.

Measurement of the angular distribution of the photoelectrons found that, for both Ag and Cu, the distribution is of the form (11.17) with the asymmetry parameter $\beta = 2.00 \pm 0.03$, as would be expected for s–p transitions (see p. 421).

For Pt⁻ the situation is more complicated, the energy level diagram being as shown in Fig. 11.22. As discussed in Chapter 2, p. 19, the ground state, in LS coupling, is expected to be

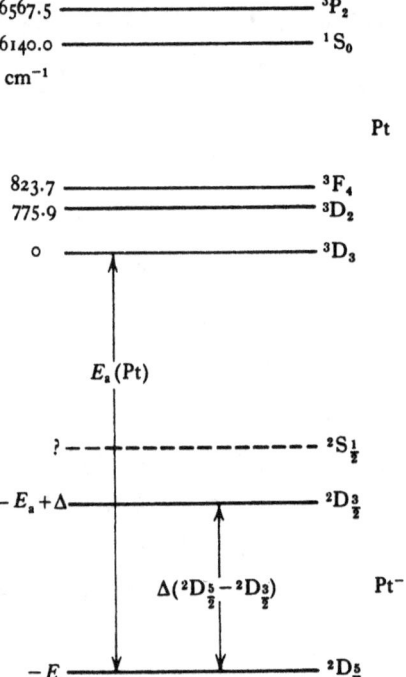

Fig. 11.22. Energy-level diagram showing photodetachment transitions between fine structure levels of Pt⁻ and Pt.

$5d^9 6s^2\, ^2D_{5/2}$. The estimated $^2D_{5/2}$–$^2D_{3/2}$ separation is around 10 000 cm⁻¹ (1.24 eV), which is less than the estimated electron affinity (2.1 eV, see Chapter 3). According to this, $^2D_{3/2}$ will be stable toward autodetachment while its radiative lifetime is likely to be as long as 10^{-2} s. A second excited state, the $5d^{10} 6s\, ^2S_{1/2}$ is also probably stable towards autodetachment and would have a long radiative lifetime. A Pt⁻ beam is likely to contain an appreciable fraction of these excited ions.

DETACHMENT BY PHOTON AND ELECTRON IMPACT   465

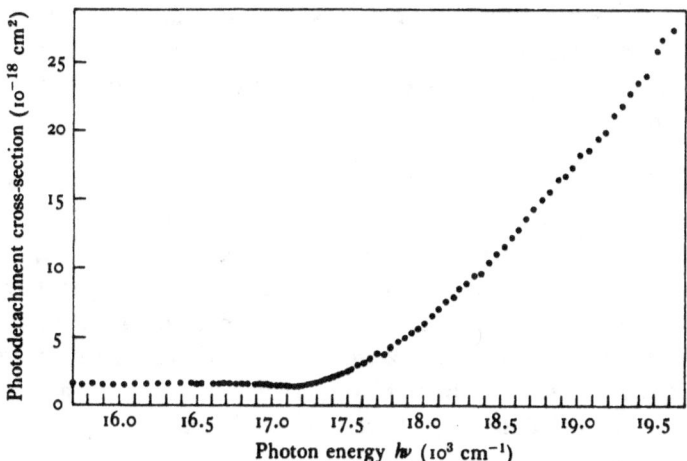

Fig. 11.23. Apparent photodetachment cross-section $Q_{d,0}$ for Pt$^-$, observed by Hotop and Lineberger (1973).

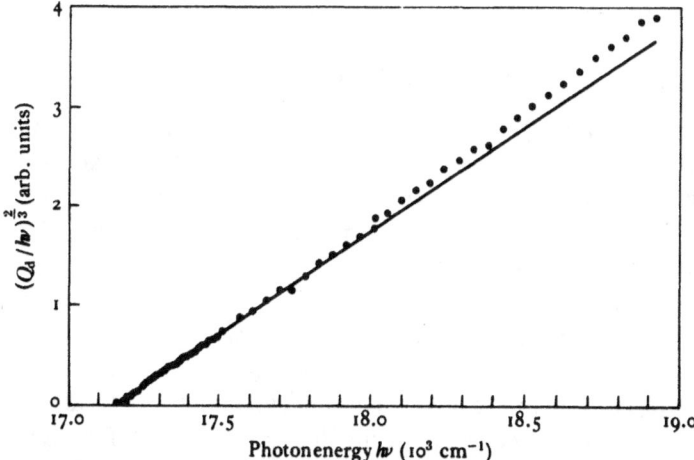

Fig. 11.24. Plot of $(Q_d/h\nu)^{2/3}$ as a function of photon energy near the threshold for photodetachment from Pt$^-$. ● Q_d obtained by subtracting the constant low-energy tail from $Q_{d,0}$, as in Fig. 11.23. —— calculated from the law (11.6) with allowance for the laser bandwidth.

Fig. 11.23 shows the observed cross-section $Q_{d,0}$ as a function of photon energy using tunable dye-lasers as photon sources. While there is clear evidence of a threshold near 17 200 cm^{-1}, the cross-section remains finite down to the lowest accessible energy. If the nearly constant 'tail' is subtracted from $Q_{d,0}$ to give Q_d, then a plot of

$(Q_d/h\nu)^{2/3}$ against photon energy, shown in Fig. 11.24, is of the correct form for an s–p transition and gives a threshold at $17\,160 \pm 16$ cm^{-1}. If this is interpreted as associated with transitions from the ground $^2D_{5/2}$ state we have

$$E_a(\text{Pt}) = 2.128 \pm 0.002 \text{ eV}. \qquad (11.51)$$

This is quite close to the value estimated by extrapolation (using the accurate value given above for $E_a(\text{Au})$) (see Chapter 3, p. 64).

On this basis the tail could be ascribed to detachment from metastable Pt$^-$ ions. It was not possible to eliminate PtH$^-$ as a possible source because of inadequate resolution in mass analysis. However, no gases containing H were used in the sputter ion sources and the small fraction of OH$^-$ formed relative to O$^-$ suggests that PtH$^-$ is unlikely to be important.

The absolute cross-section in the 'tail' was determined as 47×10^{-18} cm^2 by comparison of the saturation curve for Ag$^-$ with that for a selected wavelength in the tail (see p. 437).

Negative ions of the alkali metal atoms

Some very interesting observations have been made of photodetachment from the negative ions Li$^-$, Na$^-$, K$^-$, Rb$^-$ and Cs$^-$. The threshold energy for all these cases is close to 0.5 eV, well below the photon energy which may be reached with a tunable dye-laser. Nevertheless, the electron affinities may all be measured from observations of the energy distribution of electrons detached by the radiation from an argon ion laser. Furthermore, although the first lowest energy threshold cannot be observed with tunable dye-lasers, it is possible to observe the behaviour of the photodetachment cross-section in the neighbourhood of the thresholds for detachment which leave the atoms in the first excited $^2P_{1/2}$ and $^2P_{3/2}$ states. For Li$^-$, Na$^-$ and K$^-$ the calculations of Moores and Norcross, discussed on p. 425, suggest that the $^2P_{1/2}$ threshold should exhibit certain clearly marked features which could be used to determine the threshold energy quite accurately, and hence the electron affinity, since the excitation energies of the $^2P_{1/2}$ states are known very well from spectroscopic data for all these atoms. These predictions are completely confirmed from data obtained with tunable dye-lasers. Observations, with high-frequency resolution, also reveal remarkably clear window-type resonance effects (see Chapter 4, p. 92) very close to both the ^2P thresholds for Cs⁻ and to the $^2P_{1/2}$ for Rb⁻.

Fig. 11.25 shows the variation with frequency of the photodetachment cross-section for K⁻ observed by Patterson, Hotop, Kasdan, Norcross and Lineberger (1974). It will be seen that two quite sudden falls occur at the energy interval which is closely

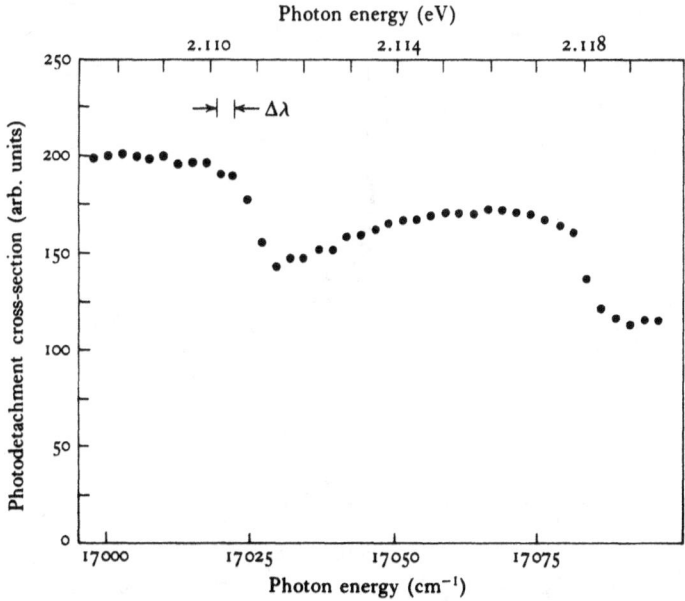

Fig. 11.25. Cross-sections for photodetachment from K⁻ observed by Patterson et al. (1974).

equal to that between the $^2P_{1/2}$ and $^2P_{3/2}$ levels of the neutral atom. Fig. 11.26 shows the detail at the first sharp fall on an expanded energy scale compared with the calculations of Moores and Norcross (1974) described on p. 428. For this comparison the energy scale of the calculated cross-sections (full line) has been displaced so that the peak just before the fall occurs at the same energy as observed. Also, the magnitudes of the observed cross-sections, which are not absolute, have been normalized to agree with the calculated (full line) at 4750 Å. It will be seen that the agreement in shape of the observed cross-section with that calculated with the dipole velocity matrix element is remarkably good over the range in which the cross-section is falling rapidly. The corresponding dipole length calculations (broken line) do not agree so well, apart from being considerably larger in absolute magnitude. Included in Fig. 11.26

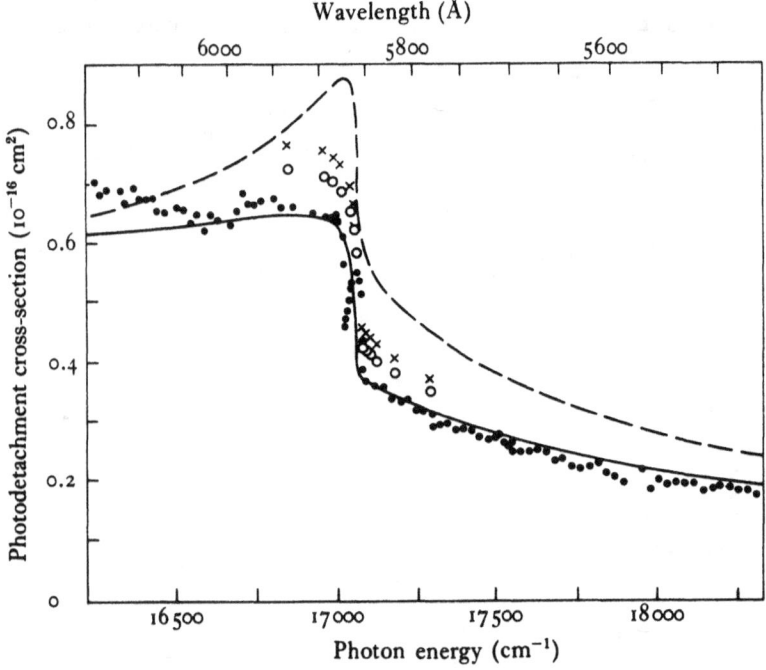

Fig. 11.26. Cross-sections for photodetachment from K^- near the threshold for the process which leaves the K atom in the $4^2P_{1/2}$ state. ● observed by Patterson et al. Calculated by Moores and Norcross: using the Weiss function for $K^-(4s^2)$, —— dipole velocity. ---- dipole length; using the close-coupling function for $K^-(4s^2)$, × dipole velocity, ○ dipole length. The normalization for comparison of theory and experiment is as described in the text.

are also some further calculations in which both the ground and continuum states of K^- are represented by the same close-coupling type of wave function (Chapter 4, p. 94, and p. 425 of this chapter). For these the dipole length and velocity calculations give results which are much closer and, when normalized in the same way as for the earlier dipole velocity results, give again very good agreement with the observed shape of the falling cross-section. It seems very safe to assume that, as according to the calculations, the threshold for the $^2P_{1/2}$ process is at the minimum reached in the sharp fall. This gives a value 0.5012 ± 0.0015 eV for the electron affinity of K, which is quite close to the theoretical value (0.492 eV) of Weiss (see Chapter 2, p. 29).

Assuming the electron affinity of K, values for the other alkali metal atoms were obtained relative to it by comparison of the energies of the peaks observed in the photoelectron spectra. The values obtained are given in Table 11.2.

Turning again to the threshold phenomena in photodetachment thresholds, Fig. 11.27 compares observed cross-sections for Na⁻ near the $^2P_{1/2}$ threshold with those calculated by Moores and Nor-

TABLE 11.2 *Electron affinities (in eV) of alkali metal atoms measured from photoelectron spectra and from the thresholds for photodetachment leaving the atom in the $^2P_{1/2}$ state*

Atom	Photoelectron spectra	Photodetachment threshold
Li	0.620 ± 0.007	—
Na	0.548 ± 0.004	0.543 ± 0.010
K	Comparison ion	0.5012 ± 0.0015
Rb	0.486 ± 0.003	0.4859 ± 0.005
Cs	0.470 ± 0.003	0.472 ± 0.003

cross using the same dipole length approximation as for K⁻ (full line in Fig. 11.26). Normalization of the energy and cross-section magnitude has been carried out as for K⁻. It will be seen that the agreement is again very good and confirms the theoretical prediction that the behaviour of the cross-section near the threshold is rather different from that for K⁻. From the threshold determined from the location of the peak, the electron affinity for Na given in the second column of Table 11.2 is obtained. It agrees, within the expected error range, with that obtained from photoelectron spectra.

Fig. 11.28 shows observed results for Cs⁻. In this case the cross-section exhibits two window-type resonances (see Chapter 4, p. 92), the first of which falls to a minimum differing from zero by an amount ascribable to the finite line width of the laser. From the measurements on the photoelectron spectra the locations of the $P_{1/2}$ and $P_{3/2}$ thresholds are shown, with the associated energy uncertainty, as the cross-hatched regions in the figure. It seems that the window resonances arise from the existence of autodetaching states which can make optically allowed transitions to the ground state, and which lie closely below the $P_{1/2}$ and $P_{3/2}$ levels of the neutral atom. The fact that the lower energy window drops to

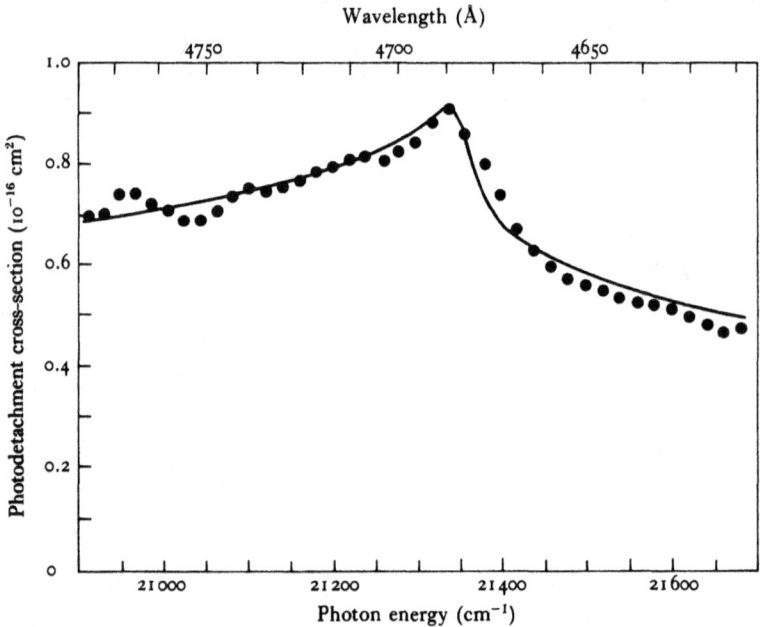

Fig. 11.27. Cross-sections for photodetachment from Na⁻ near the threshold for the process which leaves the Na atom in the $3^2P_{1/2}$ state. ● observed by Patterson et al. (1974). Calculated by Moores and Norcross (1974) using the Weiss function for Na⁻($3s^2$), —— dipole velocity.

zero while the second does not can be understood because the first occurs below the threshold for a second channel to open. On the other hand for the higher energy at which the second window occurs, two channels are open and the second supplies a background.

The inset of Fig. 11.28 shows an expanded view of the region near the $^2P_{1/2}$ threshold. The sharp peak about 2.5 meV above the minimum was taken to occur at the actual threshold and yields the electron affinity given in the second column of Table 11.2. This again agrees within error limits with that derived from the photoelectron spectrum.

Fig. 11.29 shows the remarkable window resonance observed for Rb⁻ near the $5\,^2P_{1/2}$ threshold, which has a width of approximately 0.15 meV. The corresponding resonance at the $^2P_{3/2}$ threshold is considerably wider, about 1 meV. As judged from the photoelectron spectrum, the threshold actually occurs 0.5 meV above the resonance minimum but the error limits of ± 3 meV stretch beyond the wavelength scale of Fig. 11.29.

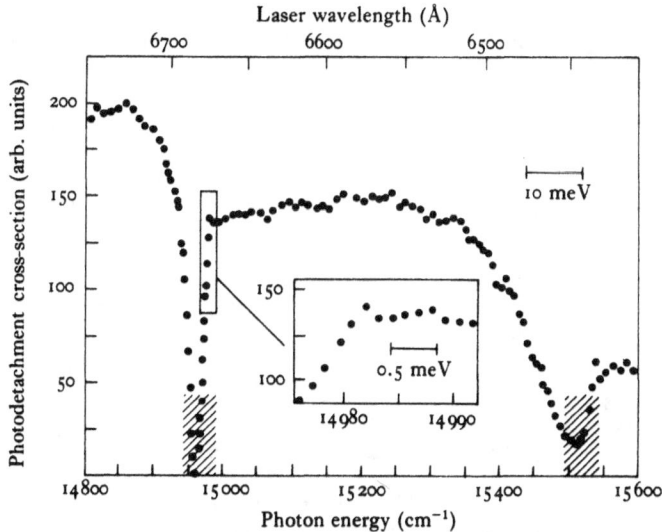

Fig. 11.28. Cross-sections for photodetachment from Cs⁻, observed by Patterson *et al.* (1974). The $6^2P_{1/2}$ and $6^2P_{3/2}$ thresholds, as determined from the photoelectron spectrum, lie within the cross-hatched regions.

The identification of the autodetaching states involved in these window resonances is not definite at the time of writing but preliminary calculations suggest that for Cs⁻ it may well be of 6p 7s configuration with a strong admixture of 6p 5d. These resonances are the narrowest so far observed.

The values of the asymmetry parameter β determining the angular distributions of the photoelectrons calculated by Moores and Norcross (1974) as a function of frequency have been shown in Fig. 11.4. The observed value for K at the argon ion laser frequency -0.64 ± 0.02 is quite close (see Fig. 11.4) to that calculated with the dipole velocity matrix element as in the full-line curve of Fig. 11.26. Agreement within experimental error is obtained with either of the matrix elements and the close-coupling calculations shown in Fig. 11.26.

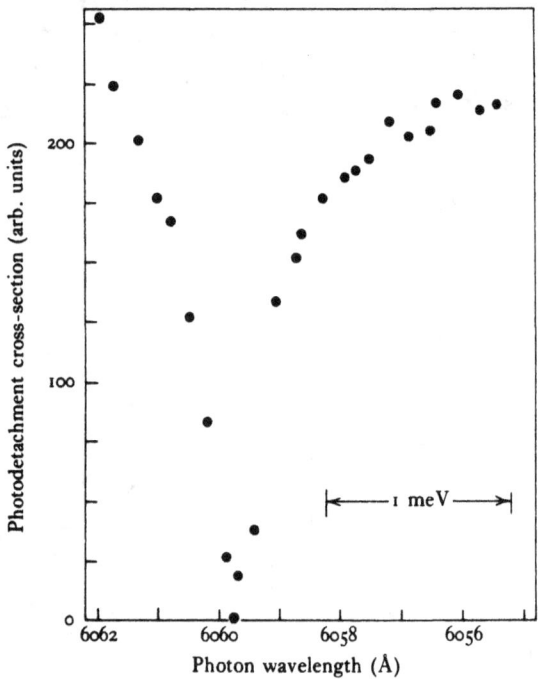

Fig. 11.29. The 'window' resonance in the photodetachment cross-section of Rb⁻, observed by Patterson *et al.* (1974).

Other atomic ions

Feldmann, Rackwitz, Heinicke and Kaiser (1973) have measured the following electron affinities by determining the threshold frequency for photodetachment with a high-pressure xenon arc source and prism monochromator.

$E_a(\text{Ge}) = 1.20 \pm 0.1$ eV, $E_a(\text{Sn}) = 1.25 \pm 0.1$ eV,
$E_a(\text{Cr}) = 0.66 \pm 0.05$ eV, $E_a(\text{P}) = 0.77 \pm 0.05$ eV,
$E_a(\text{As}) = 0.80 \pm 0.05$ eV, $E_a(\text{Sb}) = 1.05 \pm 0.05$ eV,
$E_a(\text{Bi}) = 0.9$ to 1.2 eV, $E_a(\text{Te}) = 1.9 \pm 0.15$ eV.

DETACHMENT BY PHOTON AND ELECTRON IMPACT

In addition they found no threshold for photodetachment from B^-, Ga^-, In^- and Tl^- at photon energies >0.5 eV, the lowest energy available from their light source.

These results are included in the discussion of electron affinities given in Chapter 3.

11.6 Application to different molecular negative ions

As pointed out earlier, the interpretation of data on photodetachment from molecular ions, particularly at photon energies close to the threshold, is complicated by the existence of states of vibration and rotation which may be excited in the initial ion and in the final neutral molecule. Thus it is difficult to prepare molecular-ion beams which do not possess vibrational excitation – the actual vibrational distribution will usually not even be known. When the potential-energy curves of the ground states of the ion and neutral molecule have nearly the same equilibrium position and are very similar in shape about equilibrium, no problems arise as the strongest transitions will normally be between the lowest vibrational states and there will be a definite threshold corresponding to this. This situation prevails for OH^- and SH^-. In other cases, however, the Franck–Condon principle shows that, in general, these will not be the strongest transitions and it will not be possible to interpret the data to obtain the adiabatic electron affinity unless the separate vibrational transitions can be disentangled. The only effective way of doing this which has so far been applied successfully is to measure the photoelectron energy spectrum with sufficiently high resolution to separate individual vibrational transitions. This has proved very effective for NO^- and O_2^-. Otherwise interpretations of observed thresholds in terms of adiabatic electron affinities must be regarded as highly provisional.

In the discussion below we shall use (v', v'') to denote a transition between ionic and neutral states with respective vibrational quantum numbers v'', v'.

OH^-, OD^- and SH^-

For these molecules the equilibrium separations in the ion and corresponding neutral molecule are nearly the same, so that the electron affinities and structural properties of the ions can be determined from measurements of the photodetachment cross-sections Q_d as functions of frequency.

The first experiments were carried out (Smith and Branscomb, 1955) for OH⁻ using the filter method to determine the frequency variation of Q_d, as in the experiments on O⁻ described on p. 451. Some time later this work was repeated (Branscomb, 1966) using a more intense hot-cathode ion source with a monochromator as in the experiments on I⁻ (see p. 459). This yielded a resolution of about 100 Å which corresponds to about 0.015 eV at 7000 Å, an energy spread much less than the vibrational quanta (0.4 eV).

Finally, Hotop and Lineberger (1973) have carried out measurements close to the threshold using dye-laser sources and so attaining much higher resolution (~1 Å). We first discuss the conclusions arrived at by Branscomb which have been confirmed and extended from the laser experiments.

Results obtained for OH⁻ and OD⁻ are shown in Fig. 11.30, including those from the earlier experiments. It will be seen that there is no vibrational structure visible near the threshold, while the background at lower frequencies is $< 10^{-3}$ of the signal at 6000 Å. From this evidence the threshold must refer to a single vibrational transition. Furthermore, if it can be assumed that the fraction of OH⁻ in excited vibrational states is negligible, the transition can only be the (0, 0) one as otherwise there would be observable detachment at longer wavelengths.

Again, there is no evidence of the (1, 0) threshold which should occur near 5830 Å for OD⁻ and 5500 Å for OH⁻. In fact it can be said that the intensity of the (1, 0) transition is < 0.02 of that for (0, 0). By using simple harmonic approximations to the vibrational wave functions for the ions and neutral atoms, Branscomb was able to show that this small intensity-ratio requires that the difference between the equilibrium separations is < 0.0020 Å for OH and 0.0017 Å for OD.

Analysis of the observed threshold behaviour to obtain the electron affinity is complicated by the need to allow for rotational excitation of the OH⁻ ions and also for the fine structure of the ground state of OH which includes $^2\Pi_{3/2}$ and $^2\Pi_{1/2}$ substates. Assuming a rotational temperature of 1000 °K and step-function behaviour of Q_d at the threshold, as well as allowing for the slit function of the monochromator, Branscomb found that the observed behaviour is well reproduced if the electron affinity of OH is given by

$$E_a(\text{OH}) = 1.83 \pm 0.4 \text{ eV}. \quad (11.52)$$

It was also found that $E_a(\text{OD})$ is the same within ± 0.01 eV.

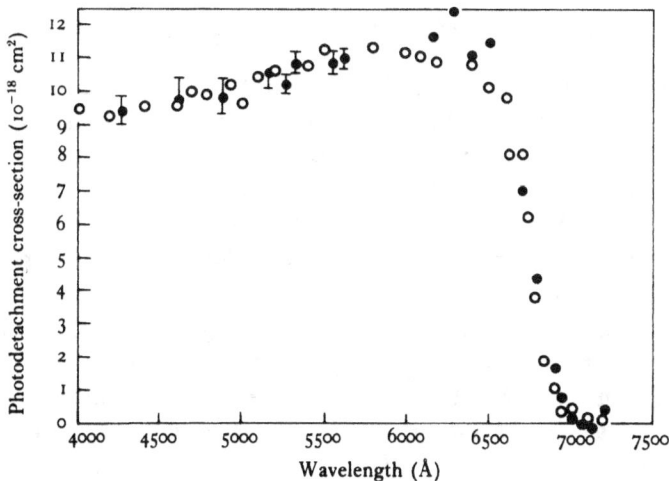

Fig. 11.30. Photodetachment cross-sections for OH⁻ and OD⁻, observed ● for OH⁻, ○ for OD⁻, by Branscomb (1966) using a monochromator with 100 Å resolution. ⊥ observed by Smith and Branscomb (1955) using band pass filters.

Branscomb was then able to show that the fundamental oscillation frequency of the ion and corresponding neutral molecule must also be very nearly equal. Thus the difference ΔE_a of the electron affinities of OH and OD is given in terms of the fundamental frequencies $\nu(\text{AB})$ of the different molecules concerned by

$$E_a = \tfrac{1}{2}h[\{\nu(\text{OH}) - \nu(\text{OD})\} - \{\nu(\text{OH}^-) - \nu(\text{OD}^-)\}]. \quad (11.53)$$

If the potential-energy curves of the electronic states of the isotopic molecules are effectively the same, ν is proportional to $\mu^{-1/2}$, where

μ is the reduced mass. We then have

$$\nu(\text{OH}^-) - \nu(\text{OD}^-) = 0.272\nu(\text{OH}^-), \qquad (11.54a)$$

$$\nu(\text{OH}) - \nu(\text{OD}) = 0.272\nu(\text{OH}), \qquad (11.54b)$$

so

$$\Delta E_\text{a} = 0 \pm 0.01 \text{ eV} = 0.136h\{\nu(\text{OH}) - \nu(\text{OH}^-)\}, \qquad (11.55)$$

from which it follows that

$$h\{\nu(\text{OH}) - \nu(\text{OH}^-)\} \leqslant 0.07 \text{ eV } (560 \text{ cm}^{-1}). \qquad (11.56)$$

Branscomb's conclusion that the equilibrium separations of the ions are very closely the same as for the neutral atoms was confirmed and extended by the observations of the photoelectron spectrum made by Celotta, Bennett and Hall (1974). They find that the (1, 0) transition is $< 6 \times 10^{-4}$ times as strong as the (0, 0), which means that the equilibrium separation for OH⁻ differs from that of OH by less than 0.001 Å.

Fig. 11.31 shows the variation with energy near the threshold of the cross-section for photodetachment from OH⁻ measured by Hotop and Lineberger (1973) with the highly monochromatic dye-laser source. A rather sharp onset, observed also for OD⁻, is apparent at 14 700 cm⁻¹. This was not observed in the earlier experiments with their lower frequency resolution. Hotop and Lineberger ascribe the onset to the excitation of a certain rotational branch associated with the $^2\Pi_{1/2}$ state of OH and have carried out a detailed analysis of the threshold behaviour in terms of the known structures of the molecules. As a result they obtain the electron affinities

$$E_\text{a}(\text{OH}) = 1.825_4 \pm 0.002 \text{ eV}, \qquad (11.57)$$
$$E_\text{a}(\text{OD}) = 1.823_0 \pm 0.02 \text{ eV}. \qquad (11.58)$$

which are more precise than the values (11.52) determined by Branscomb. The difference in the electron affinities

$$E_\text{a}(\text{OH}) - E_\text{a}(\text{OD}) = 20 \text{ cm}^{-1} = 2.5 \text{ meV} \qquad (11.59)$$

is considered to be real, and more accurate than the individual values (11.57) and (11.58).

These results have already been discussed in Chapter 6, p. 180, in relation to the theoretical work of Cade (1967), with which they are in good agreement.

SH⁻ and SD⁻ have been studied in a similar way by Steiner (1968), the ions being produced in a hot-cathode arc discharge through a mixture of NH_3 and SF_6 at a pressure of 0.06 torr. Again only one threshold is found, which may be taken as that for the (0,0) transition. Analysis of the results shows that the equilibrium

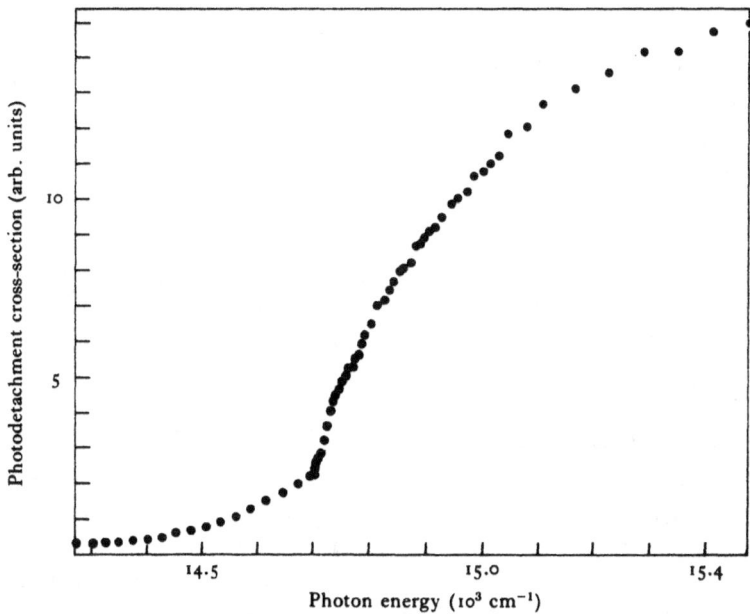

Fig. 11.31. Photodetachment cross-section for OH⁻ near the threshold, measured by Hotop et al. (1973) using a tunable dye-laser light source.

separation of the ion differs from that of the neutral molecule by < 0.02 Å and the fundamental vibrational quanta by < 0.035 eV (300 cm⁻¹). The electron affinity of SH is found to be

$$E_a(SH) = 2.319 \pm 0.010 \text{ eV}. \qquad (11.60)$$

CH⁻

Photodetachment from CH⁻ has been observed by Feldmann (1970) using a high-pressure xenon lamp and monochromator as light source. Two thresholds were observed at 0.74 and 1.94 eV respectively. The energy separation is about what would be expected if the processes involved refer to photodetachment from the ground $X^3\Sigma^-$ state of CH^- to the $X^2\Pi$ ground state and a $^4\Sigma^+$ excited state of CH respectively. However, the electron affinity, on this basis, is considerably smaller than that derived by Cade (1967) using the semi-empirical procedure discussed in Chapter 6, p. 183.

NH^-

Photodetachment from NH^- has been studied by Celotta *et al.* (1974), who measured the energy spectrum of the electrons detached from an ion provisionally identified as NH^-, by argon ion laser light. This gave

$$E_a(NH) = 0.38 \pm 0.03 \text{ eV}.$$

which agrees quite well with that derived semi-empirically by Cade (1967) (Chapter 6, p. 183).

O_2^-

Special interest was attached to the application of photodetachment techniques to the study of O_2^-, not only because of its importance in the upper atmosphere, but also because of the wide range of values obtained for the electron affinity of O_2 by different methods (see Chapter 6, p. 184). The first experiments, by Burch, Smith and Branscomb (1958, 1959), using the technique described on p. 432, gave disappointing results as no threshold was observed down to photon energies of 0.7 eV. This was not surprising when account is taken of the unknown distribution of vibrational levels in the O_2^- beam as well as of the likelihood that metastable species may also be present. The good fortune attendant on the study of OH^- and SH^- no longer prevails – the molecular constants of the ground states of O_2 and O_2^- are quite different.

These difficulties were overcome when it became possible to observe the energy spectra of electrons arising in photodetachment from O_2^- by highly-monochromatic laser light. Using the technique described on p. 438, Celotta, Bennett, Hall, Siegel and Levine

(1972) have made a very thorough study of O_2^- which has yielded precise values of the electron affinity of O_2 as well as of the molecular constants R_e and B_e for O^-.

Fig. 11.32 shows a typical photoelectron spectrum which they observed, using argon ion laser light at 4880 Å. It will be seen that there is evidence of two series of peaks, that the fourth and ninth

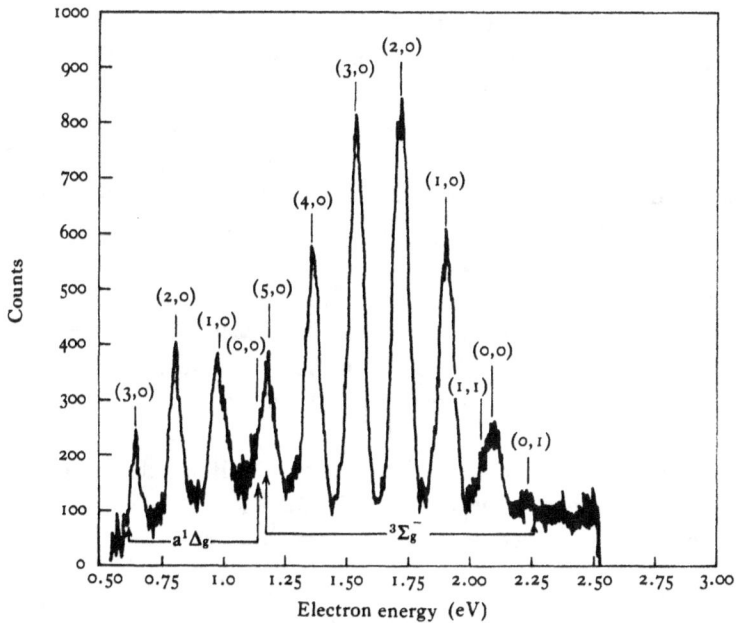

Fig. 11.32. Energy spectrum of the photoelectrons arising through photodetachment from O_2^- by argon ion laser light, as observed by Celotta et al. (1972).

peaks, counted from the left, are anomalously broad and that there is evidence of a further peak beyond the ninth.

To anticipate the results of the analysis, the identifications of the transitions involved are indicated in Fig. 11.32. These include the $(v',0)$ series, with v' ranging from 0 to 5, associated with transitions to the ground $^3\Sigma_g^-$ states of O_2 and with v' ranging from 0 to 3, associated with transitions to the excited $a\,^1\Delta_g$ state of O_2. The justification for these identifications may be outlined as follows.

(a) The separations between the successive (v',0) peaks for the $^3\Sigma_g^-$ final state are nearly equal and closely the same as the accurately known separations of the vibrational levels of $O_2(^3\Sigma_g^-)$. This shows that they all arise from the same initial vibrational level of O_2^-.

(b) Although nearly equal, the separations decrease slowly but steadily on moving to the left. This shows that v' increases by one unit in moving to the next peak on the left.

(c) Similar considerations, though with less accuracy, apply to the peaks associated with the a $^1\Delta_g$ final state.

(d) The fourth peak is broadened because of overlap between the $^3\Sigma_g^-$ and a $^1\Delta_g$ series.

(e) To determine the initial vibrational quantum number v'' we note that the envelope of the intensity distribution in each series has a single maximum. According to the Franck–Condon principle this can only occur if $v'' = 0$ so that the initial vibrational wave function has a single peak.

(f) It remains to determine the final vibrational wave number v' for one peak in each progression. This was done in three ways.

(i) By measuring the separations with high enough accuracy to determine v' by comparison with the accurately known values for O_2.

(ii) By measurement of the isotope shift in the separations. The difference $\Delta\nu$ between the frequencies of the transitions for two isotopic molecules is given, for $v'' = 0$, by

$$c\Delta\nu = (1-\rho)\omega_e'(v'+\tfrac{1}{2}) - (1-\rho^2)\omega_e' x_e'(v'+\tfrac{1}{2})^2 \\ - \{(1-\rho)(\tfrac{1}{2}\omega_e'') - (1-\rho^2)(\tfrac{1}{4}\omega_e'' x_e'')\},$$

(11.61)

where $\rho^2 = M_1/M_2$, the ratio of the reduced masses of the respective isotopic molecules. The constants ω_e'', x_e'' for O_2^- are available with sufficient accuracy from analysis of the experimental observation of resonances in the transmission of electrons through O_2, as described in Chapter 6, p. 189.

(iii) By analysing the intensity variation of the bands in relation to the Franck–Condon principle. Accurate determination of these relative intensities involves measurement of the angular distributions of the ejected electrons, by the method described on p. 438, as the asymmetry factors defining these

DETACHMENT BY PHOTON AND ELECTRON IMPACT 481

distributions (see p. 421) vary with electron energy. Such measurements were carried out for all electron energies corresponding to the peaks in Fig. 11.32.

To apply the Franck–Condon principle it is assumed that the intensity of any particular transition is proportional to the square of the overlap integral between initial and final vibrational wave functions. To obtain these functions for the initial state the potential-energy curve of this O_2^- slit was assumed to be of the Morse form (see Chapter 6, p. 163) with ω_e'' and $\omega_e'' x_e''$ chosen as in (ii) above and the equilibrium separation R_e'' regarded as adjustable.

It is also necessary to take account of rotational excitation. From the relative proportion of O_2^- ions with $v'' = 0$ and 1 (see below) respectively, the effective source temperature was found to be 630 °K. At this temperature the most probable initial rotational quantum number J'' of the ion is 13.

Using the overlap factors derived from these data, R_e'' and v' were adjusted to give the best agreement with the observed intensity ratios. The best fit was obtained with $R_e'' = 1.341 \pm 0.010$ Å, somewhat smaller than that 1.374 Å, predicted by Badger's rule (1934). The rotational constant B_e'' is given by

$$B_e'' = B_e'(R_e'/R_e'')^2$$
$$= 1.17 \pm 0.02 \text{ cm}^{-1}, \qquad (11.62)$$

if we take the spectroscopic values for B_e' and R_e'.

All these methods gave consistently the results shown in Fig. 11.32 for all $(v', 0)$ transitions. There remains the weak peak beyond the $(0, 0)$ peak. This may be ascribed to a $(0, 1)$ transition, the separation from $(0, 0)$ being consistent with R_e'' in (ii) above. The $(1, 1)$ maximum then falls as indicated in Fig. 11.32 and would explain the fact that the $(0, 0)$ peak appears broadened on the left-hand side.

Having identified the transitions, the threshold energy for the $(0, 0)$ transition was found to be 0.438 ± 0.001 eV. To obtain the electron affinity from this, allowance must be made for the fine structure of the O_2^- ground state and for rotational excitation.

To correct for the latter it is first noted that, from considerations of conservation of angular momentum, it can be shown that the final distribution of rotational states should be symmetrically distributed about J''. In that case the correction to be applied is to subtract the excess of rotational energy of O_2 as against O_2^- when the rotational quantum number is J'', i.e.

$$-(B'_e - B''_e)J''(J''+1), \qquad (11.63)$$

with $J'' = 13$ and B''_e given by (11.62). This amounts to -51.7 cm^{-1} (-6.4 meV).

The second correction arises because the ground state of O_2^- includes $^2\Pi_{1/2}$ and $^2\Pi_{3/2}$ substates. Extrapolation suggests that the doublet is inverted with estimated separation 150 ± 30 cm^{-1}. Assuming that each substate contributes in proportion to its statistical weight, it is then possible to estimate the shift from the position of the peak for the transition from the lowest $^2\Pi_{3/2}$ substate due to the occurrence of transitions from $^2\Pi_{1/2}$. This is found to involve a correction $+62.3$ cm^{-1} (7.7 meV).

When these corrections are included, the electron affinity of O_2 is found to be

$$E_a(O_2) = 0.440 \pm 0.008 \text{ eV}. \qquad (11.64)$$

Comparison of this result with values obtained by other methods is discussed in Chapter 6, p. 185. The implications of the derived value of R''_e are also considered in that section.

Using the method described on p. 448, Woo *et al.* (1969) obtained, for the rate of photodetachment from O_2^- by sunlight, 0.3 ± 0.15^{-1}. As pointed out on p. 449, interpretation of this result is possibly complicated by the presence of O_4^- ions.

NO$^-$

Photodetachment from NO$^-$ has been studied in a similar way to O_2^- by Siegel, Celotta, Hall, Levine and Bennett (1972).

Fig. 11.33 shows a typical observed energy spectrum of photodetached electrons due to argon ion laser radiation at 4880 Å. The identification of the transitions (v', v'') between initial and final vibrational levels with quantum numbers v'', v' respectively, shown in Fig. 11.33, was arrived at in exactly the same way as for O_2^-.

It is of interest to note that, in contrast to O_2^-, there is no evidence of any transitions from initial states with $v'' > 0$. This remained true no matter how the pressure and discharge current conditions in the source were changed. It is consistent with the fact, which follows from the derived electron affinity, that only the level with $v'' = 0$ falls below the $v' = 0$ level of NO, so all ions with $v'' > 0$ are unstable towards autodetachment (see p. 191).

In carrying out the intensity distribution analysis as in (*f*) (iii) above, the molecular constant ω_e'' and R_e'' were both regarded as variable. The best fit was obtained with

$$R_e'' = 1.258 \pm 0.010 \text{ Å}, \quad \omega_e'' = 1470 \pm 200 \text{ cm}^{-1} \text{ (0.182} \pm 0.023 \text{ eV)},$$
(11.65)

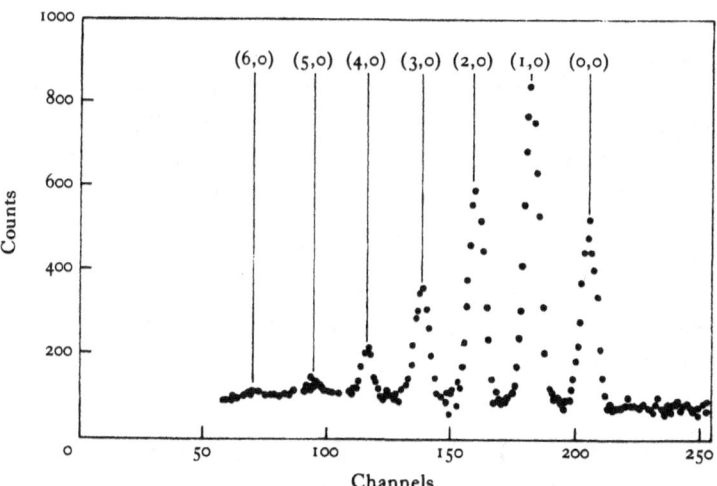

Fig. 11.33. Energy spectrum of the photoelectrons arising through photodetachment from NO⁻ by argon ion laser light, as observed by Siegel *et al.* (1972).

giving, as in (11.62),

$$B_e'' = 1.427 \pm 0.02 \text{ cm}^{-1} \text{ (0.177} \pm 0.003 \text{ meV)}. \quad (11.66)$$

The value for ω_e'' compares with 0.170 eV obtained by Spence and Schulz from analysis of their data on the location of the resonance in the transmission of slow electrons through NO (see Chapter 6, p. 193).

In deriving the electron affinity from the energy peak for the (0,0) transition, corrections had to be made both for rotational excitation and for the fine structure in the final NO($^2\Pi$) state. The former was calculated on the assumption that the effective source temperature for NO⁻ is 630 °K, determined from the O_2^-

data as described above, a reasonable assumption since both ions were extracted from the same source. The net correction for both effects was −12.5 meV. Including this the electron affinity is found to be

$$E_a(NO) = 0.024^{+0.010}_{-0.050} \text{ eV}. \qquad (11.67)$$

The relation of these results to the structure of NO⁻ is discussed in Chapter 6, p. 184.

C_2^-

Fig. 11.34 shows the photodetachment cross-section for C_2^- near the threshold observed by Feldmann (1970), using the same technique as in his experiments with CH⁻ and SO⁻. There is a very clear threshold which, when allowance is made for accurate voltage calibration, gives

$$E_a(C_2^-) = 3.54 \text{ eV}. \qquad (11.68)$$

The relation of this to theoretical expectation has been discussed in Chapter 6, p. 207, while its relevance for two-photon detachment experiments is considered on p. 494.

S_2^-

Celotta, Bennett and Hall (1974) have also measured the energy spectrum of electrons from S_2^- obtained from a hot-cathode discharge in SO_2 at a pressure of 2.3×10^{-2} torr containing traces of COS and CS_2. SO_2^- ions were also produced strongly by this source but the electron energy spectrum arising from these ions did not seriously overlap that from S_2^- which was very similar to that from O_2^- and SO⁻ and was analysed in the same way. Without allowing for rotational and fine structure effects it was found that

$$E_a(S_2) = 1.663 \text{ eV}, \qquad (11.69)$$

with an error estimate of ±0.04 eV.

SO⁻

The cross-section for photodetachment from SO⁻ has been investigated by Feldmann (1970) using a xenon lamp and monochromator as light source (see also CH⁻ and C_2^-). He found a sharp threshold at 1.09 eV which he associated with the adiabatic electron affinity.

DETACHMENT BY PHOTON AND ELECTRON IMPACT 485

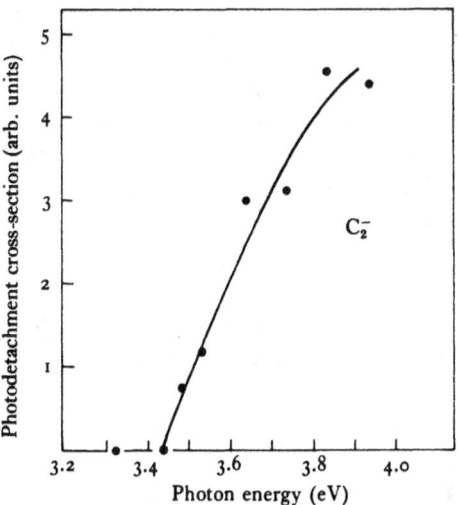

Fig. 11.34. Photodetachment cross-section for C_2^- near the threshold, as observed by Feldmann (1970).

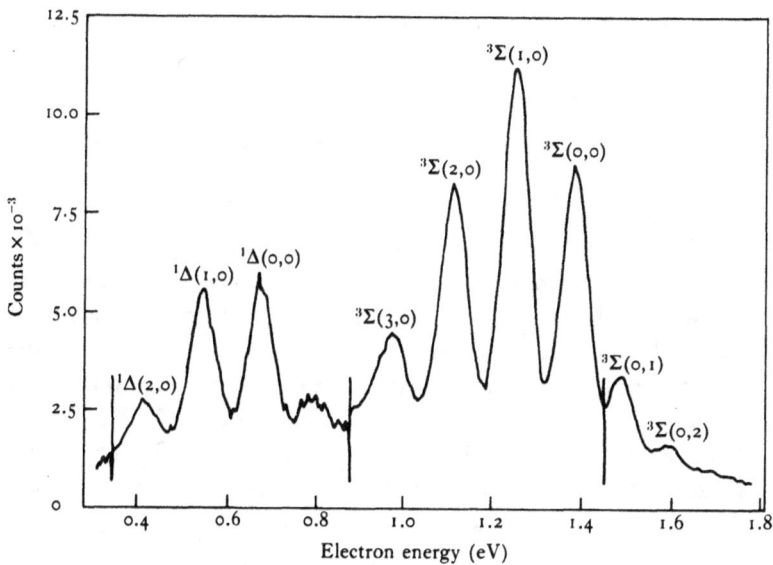

Fig. 11.35. Energy spectrum of the photoelectrons arising through photodetachment from SO^- by argon ion laser light, observed by Bennett (1972).

The photoelectron spectrum from SO⁻ has been observed by Bennett (1972) and is reproduced in Fig. 11.35. The identification of the transitions was carried out in the same way as for O_2^-. From the energy of the peak for the (0,0) transition to the ground $^3\Sigma$ state, the electron affinity is found to be 1.14 eV, somewhat higher than determined by Feldmann.

NH_2^-, PH_2^-, AsH_2^-

Photodetachment from these three radicals has been studied (Smyth and Brauman, 1972a, b) using the ion cyclotron resonance technique (see p. 445).

In these experiments two light sources were used, a 1000 W xenon arc lamp with a grating monochromator and a Chromatix Model 1000 E Nd: YAG laser, tunable over a useful frequency range. The negative ions were produced by dissociative attachment in NH_3, PH_3 and AsH_3 respectively (see Chapter 9, p. 351).

In all these cases Q_d rose sharply from a well-defined threshold so that arguments similar to those used in the analysis of data on OH^- and SH^- may be applied, suggesting strongly that the threshold corresponds to transitions between states of the ion and neutral radical which involve no vibrational excitation. This being so the electron affinities of the three radicals are found to be

$$E_a(NH_2) = 0.744 \pm 0.022 \text{ eV}, \quad E_a(PH_2) = 1.25 \pm 0.03 \text{ eV},$$
$$E_a(AsH_2) = 1.27 \pm 0.03 \text{ eV}.$$

Celotta et al. (1974) have observed the energy spectrum of electrons detached from NH_2^- by argon ion laser light, obtaining a single peak, the energy of which was calibrated from the position of the single peak observed for O^-. The detachment energy required to produce the electrons with the peak energy was found to be 0.779 eV. It was not possible to determine how far this differs from the adiabatic electron affinity but it is improbable, from the nature of the observed single peak, that it could be by more than ±0.020 eV. Allowing for other sources of error this gives

$$E_a(NH_2) = 0.779 \pm 0.037 \text{ eV}, \qquad (11.70)$$

which is consistent with the measurements of Smyth and Brauman

(1972b) and also those of Feldmann (1970), using the same technique as for CH^-, C_2^- and SO^-, which gave 0.76 ± 0.04 eV.

SO_2^-

This case is of special interest as it is the first triatomic molecule to which the photoelectron technique has been applied and the results analysed in detail to obtain the adiabatic electron affinity.

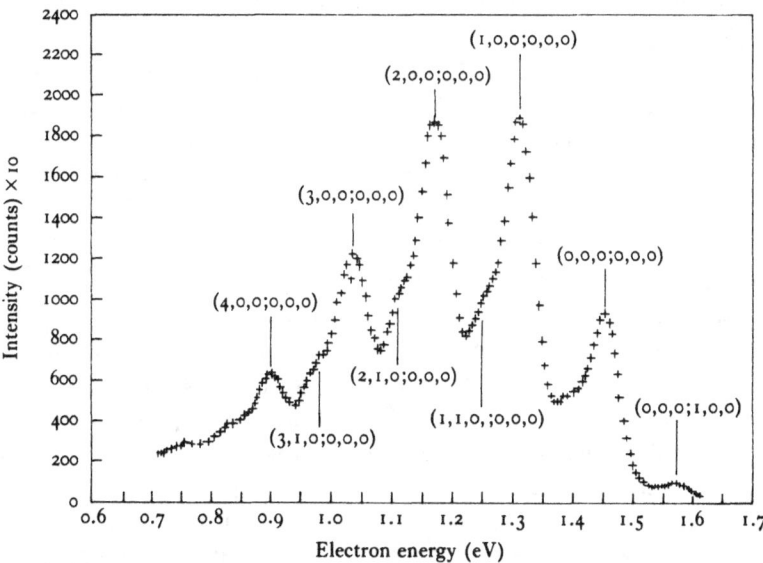

Fig. 11.36. Energy spectrum of photoelectrons arising through detachment from SO_2^- by argon ion laser light, as observed by Celotta et al. (1974).

SO_2 and SO_2^- have the same triangular structure (see Chapter 6, p. 215). For SO_2 the fundamental frequencies, in wave numbers, are 1151, 519 and 1360 cm^{-1}. The vibrational quantum numbers associated with these modes will be designated v_1, v_2, v_3 respectively, the neutral molecule and negative ion being distinguished by ' and " respectively.

Fig. 11.36 shows the photodetachment spectrum observed by Celotta et al. (1974). It contains five main peaks with a smaller peak at 1.57 eV. On the low-energy side of each of the main peaks there is evidence of a subsidiary peak. The separations between the main peaks agree closely with those between successive levels of symmetrical stretching vibration of SO_2, as determined spectroscopically. At the same time the separations between the main peaks and the associated subsidiary peaks are equal to those between successive levels of bending vibration. It is natural therefore to adopt the identification of the main and subsidiary peaks indicated in Fig. 11.36. The small peak at 1.57 eV can then be tentatively ascribed to detachment from SO_2^- molecules possessing one quantum v'' of the first ('symmetrical stretching') mode of vibration.

Accepting these identifications, the energy required to produce the (000;000) transition is found to be 1.097 eV, which should be the adiabatic electron affinity within ± 0.036 eV when allowance is made for rotational effects and for other estimated errors. This is to be compared with 1.0 ± 0.05 eV obtained by Feldmann from determination of the detachment threshold using a conventional light source as in his experiments on CH^-, C_2^-, SO^- and NH_2^-. The fact that an appreciable contribution arises from the (000;100) transition, as shown from Fig. 11.36, means that Feldmann would tend to observe a threshold at a smaller energy than the true threshold for the (000;000) transition. It is not surprising that his threshold value is about 0.1 eV smaller than that derived from the photoelectron spectrum.

From the separation between the (000;000) and (000;010) peaks the frequency of the symmetrical mode for SO_2^- is determined as 989.4 cm^{-1} (122.5 meV).

These results for SO_2^- are encouraging and suggest that the measurement of the energy spectra of electrons detached from polyatomic negative ions by laser light will prove to be very useful for studying the properties of these ions as well as for determining the adiabatic electron affinity.

NO_2^-

Warneck (1969) has measured cross-sections for photodetachment from NO_2^- using a technique similar to that described on p. 432. Absolute values were obtained by comparison with H^-.

For this ion the threshold is not well defined (see also Chapter 10,

DETACHMENT BY PHOTON AND ELECTRON IMPACT

p. 398 and Chapter 12, p. 570) so that it is difficult to derive the adiabatic electron affinity from the observations. The value estimated from his data by Warneck, 3.1 eV, seems to be considerably too high as judged by comparison with values derived by other methods (see Chapter 6, p. 217).

Other negative ions of atmospheric interest

The technique described on p. 448 for observing photodetachment from a swarm of drifting negative ions has been applied by Burt (1972*a*, *b*) to observe effective rates for photodetachment by sunlight from CO_3^-, $CO_3^-.H_2O$ and O_4^- ions, all of which are important in the lower ionosphere (see Chapter 15, p. 667). The rates were measured relative to that for O^-, derived from the measured cross-section Q_d, which is 1.448 s^{-1}. Burt found the corresponding values 1.4, 1.1 and 1.9 s^{-1} for CO_3^-, $CO_3^-.H_2O$ and O_4^- respectively. Only rough values were obtained for the threshold energies, 1.8, 2.1 and 1.9 eV respectively, and it is not possible to decide how they are related to the adiabatic electron affinity (see Chapter 6).

Sinnott and Beaty (1971) have applied a similar technique, but with the xenon arc source replaced by a tunable dye-laser, to observe photodetachment from O_3^-. They obtained evidence of a threshold at 2.09 eV which agrees within experimental error with that found by Burt (1972*a*) using the xenon arc source. It is difficult to say, however, whether the threshold energy is effectively the adiabatic electron affinity (see Chapter 6, p. 215).

11.7 Multiphoton detachment

The advent of lasers has made it possible to observe processes which can only occur through simultaneous interaction with more than one photon. In particular, photodetachment from I^- by ruby laser radiation has been not only observed but studied in some detail by Hall, Robinson and Branscomb (1965). This is despite the fact that the detachment energy is 3.063 eV, whereas the quantum energy of the laser radiation is only 1.785 eV.

Fig. 11.37 shows the experimental arrangement used. It was of the usual crossed-beam type except that the light source was a 20 MW Q-switched ruby laser. The laser beam was focused to a spot of area 1 mm^2 on to the focus of the I$^-$ beam which covered an area of 9 mm^2.

The power of the laser beam was monitored as a function of time, by reflecting it from a silvered mirror onto a MgO screen and then to a biplanar photodiode. Comparison with a liquid-cell calorimeter provided absolute calibration.

As usual the detached electrons were extracted in a direction

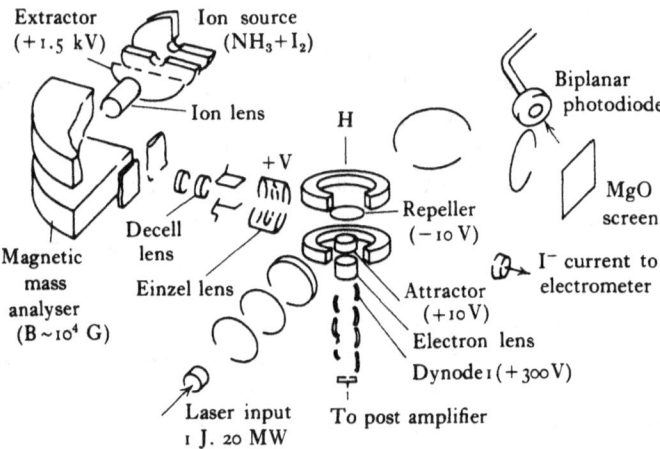

Fig. 11.37. Arrangement of apparatus used by Hall *et al.* (1965) for investigating two-quantum photodetachment.

perpendicular to both beams, along which a uniform magnetic field was applied, and focused on the first dynode of an electron multiplier.

To calibrate the multiplier and electron collection system the pulse-height distribution and total signal due to a single electron incident on the first dynode were first measured. The collection efficiency was then determined in an ingenious way using a weak H$^-$ beam from which photodetachment can occur in single collisions with a ruby laser photon. With the powerful light pulse available it was readily possible to produce complete detachment from the H$^-$ beam, so that the absolute value of the detached electron current

DETACHMENT BY PHOTON AND ELECTRON IMPACT 491

was equal to the initial H⁻ current. Comparison with the signal given by the collection system under these circumstances gave the necessary calibration. As a further check the absolute value of the H⁻ single photon detachment cross-section was measured using an attenuated laser beam and checked against earlier measurements.

It was necessary to show that the observed signals from the I⁻ beam did not arise from ultraviolet photons generated in the optical system as second harmonics of the laser photons. To check this, tests were made in which a red absorbing filter, transparent in the ultraviolet, was placed in front of the laser beam just after passage through the lens system. No residual signal was observed. It was also verified, by inserting a duplicate window in the beam, that no ultraviolet photons were introduced by the vacuum window.

For a 2-quantum process the transition probability per ion is given by

$$W = \delta F^2, \qquad (11.71)$$

where F is the photon flux and δ depends on the photon frequency. Fig. 11.38 shows the relation on a log–log scale between the number of detached electrons per ion and the time integral of F^2 observed by Hall et al. (1965). Although there is a considerable scatter among the experimental points, they are consistent with the relation (11.71) although the slope of the best-fitting straight line is a little higher than unity. δ is found to be given by

$$\delta = (350 \pm 140) \times 10^{-51} \text{ cm}^4 \text{ s}, \qquad (11.72)$$

but there are some difficulties of interpretation as the distribution of intensity of the laser light throughout the interaction region was probably not uniform. Allowance for this would certainly reduce the true value of δ below (11.72).

According to second-order perturbation theory the two-quantum process occurs through virtual transitions to intermediate states. The amplitude for such a transition from the initial state o of energy E_0 to an intermediate state j of energy E_j is proportional to the dipole matrix element $\langle \mathbf{e} \cdot \mathbf{r} \rangle_{j0}$ between the two states in the direction $\mathbf{e}$ of polarization of the incident radiation. Similarly for a transition from j to the final state f of energy E_f, the amplitude is proportional to $\langle \mathbf{e}\cdot\mathbf{r}\rangle_{fj}$. Summing over all intermediate states, the cross-section $Q_d^{(2)}$ per unit light intensity is then given by

$$Q_d^{(2)} = \frac{\alpha a_0^2}{4\pi} \frac{\mathcal{J}}{\mathcal{J}_0} \int \left| \sum_j \frac{\langle \mathbf{e}\cdot\mathbf{r}\rangle_{fj}\langle \mathbf{e}\cdot\mathbf{r}\rangle_{j0}}{E_0 - E_j + h\nu} \right|^2 h\nu k\, d\omega. \quad (11.73)$$

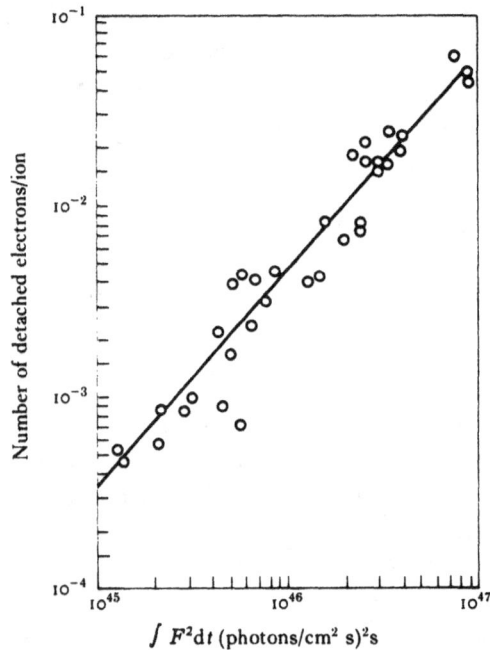

Fig. 11.38. Relation between photodetachment probability and $\int F^2 dt$, where F is the laser flux, observed by Hall et al. (1965).

ν is the frequency of the radiation and k is the wave number of the ejected electron given by

$$k^2 = (2m/\hbar^2)(2h\nu + E_0), \quad (11.74)$$

α is the fine structure constant. $\mathcal{J}$ is the light intensity and $\mathcal{J}_0$ the atomic unit of intensity which is such that the field strength in the radiation field is e/a_0^2 (5.14×10^9 V cm^{-1}). It is given by

$$\mathcal{J}_0 = 7.02 \times 10^{-16}\,\text{W cm}^{-2}. \quad (11.75)$$

The angular integration (dω) is over the direction of motion of the ejected electron. The matrix elements, the energies and the wave number k are all expressed in atomic units. For detachment from negative ions the intermediate states form a continuum, in which case the sum over j is replaced by an integral and care must be taken that the wave functions appearing in the matrix elements are properly normalized.

The magnitude of $\mathcal{J}_0$ gives some indication of the flux of photons required to produce two-quantum processes at rates comparable with those for single quanta. For photons of quantum energy 1 eV the flux corresponding to $\mathcal{J}_0$ is 5×10^{35} photons cm^{-2} s^{-1}. Actually, more detailed calculations show that the critical flux is considerably less than this. The quantity δ in (11.71) is given by

$$\begin{aligned} \delta &= W/F^2 \\ &= FQ_2^{(d)}/F^2 \\ &= (Q_2^{(2)}/\mathcal{J})(\mathcal{J}/F) \\ &= (Q_2^{(2)}/\mathcal{J}) \times 1.6 \times 10^{-19} \, V \, \text{cm}^4 \, \text{s}^{-1}, \end{aligned} \qquad (11.76)$$

where V is the photon energy in eV and $\mathcal{J}$ is in W cm^{-2}.

Robinson and Geltman (1967) calculated δ for conditions appropriate to the experiments of Hall *et al*. They used the model described on p. 425 to calculate the matrix elements and allowed for distortion of the waves representing the unbound motion of the detached electron in the intermediate states by the atomic field. Their result for δ was 180×10^{-51} cm^4 s which is quite close to the observed value, particularly when allowance is made for uncertainty in the latter for the reasons mentioned. It seems very well established that two-quantum detachment was in fact observed.

It is also possible to achieve photodetachment through a two-step process in which the first stage involves excitation, by near-resonance single-photon absorption, to a real bound excited state of the ion, if one exists. This is then followed by absorption of a second photon which produces detachment from the excited state. Such a process will be energetically possible with photons of frequency ν provided

$$h\nu > E_{\text{ex}} \text{ and } > E_{\text{a}} - E_{\text{ex}},$$

where E_{ex} is the excitation energy of the real bound excited state and E_{a} the electron affinity.

For dye-laser radiation these conditions are satisfied for the C_2^- ion, for which the electron affinity is near 3.5 eV (see p. 208) and there is strong evidence that a bound excited state exists at an excitation energy near 2.3 eV, which combines optically with the ground state (Chapter 6, p. 208). Lineberger and Patterson (1972) therefore carried out detachment experiments with C_2^-. Provided the laser power is not too large, photodetachment will only occur

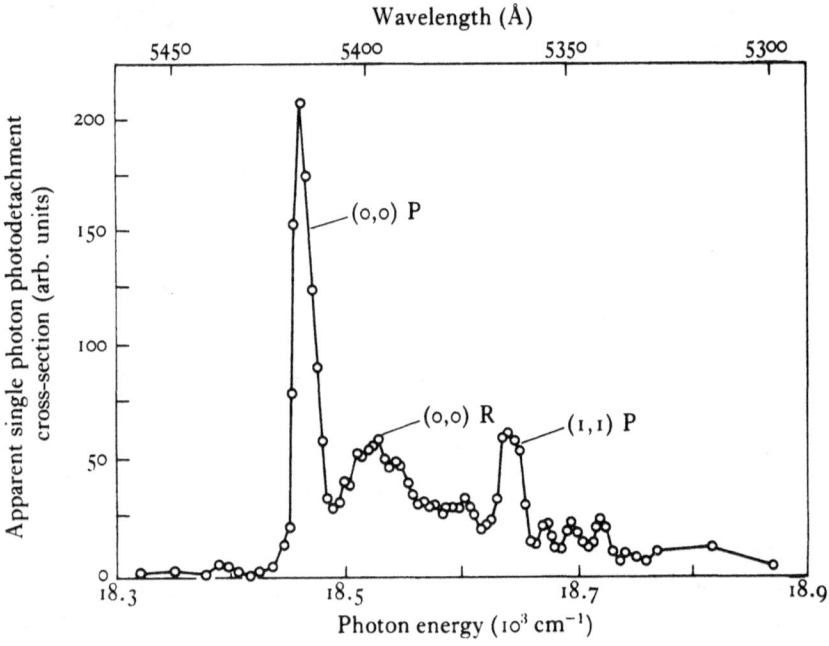

Fig. 11.39. Variation with photon energy of the apparent photodetachment cross-section for C_2^-, observed by Lineberger and Patterson (1972).

through near-resonant absorption of the first quantum, but it is possible to work with sufficient power for this first stage to be saturated so that the observed photodetachment will be proportional to the laser flux F rather than to F^2. We may then define an apparent photodetachment cross-section Q_d^{app} as if the process were a single-photon one.

Fig. 11.39 shows the variation with photon energy of the observed apparent photodetachment cross-section. The peaks occur at energies which correspond closely to the frequencies (0,0)P,

(0,0)R and (1,1)P bands derived by Herzberg and Lagerquist (1968) from their analysis of the C_2^- spectrum (Chapter 6, p. 208). This provides strong evidence that the observed detachment is really due to the two-step process described above. There is even evidence of a further electronic transition which could be identified as the $^1\Sigma_u^+ - ^1\Pi_g$ which would be expected by analogy with the isoelectronic molecule N_2^+ (see Chapter 6, p. 210).

Fig. 11.40 shows the observed variation of Q_d^{app} with laser flux. The saturation at sufficiently high fluxes, corresponding to constant

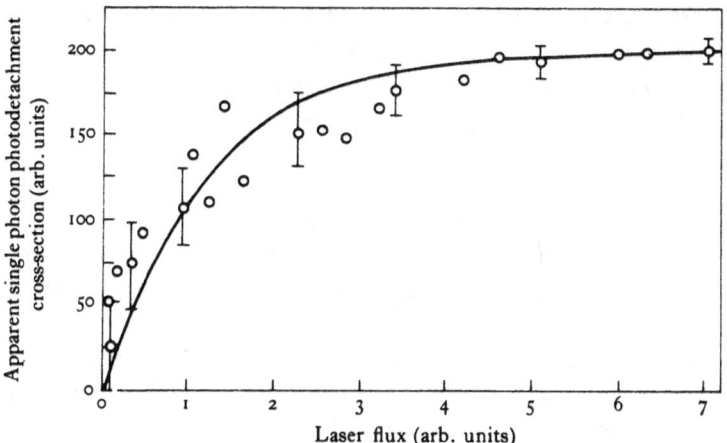

Fig. 11.40. Variation with laser flux of the apparent photodetachment cross-section for C_2^- due to 5415 Å radiation (at the (0,0)P peak in Fig. 11.39), observed by Lineberger and Patterson (1972).

Q_d^{app}, is clearly seen. At low fluxes however, Q_d^{app} is proportional to the flux as is the case for an unsaturated two-photon process.

11.8 Detachment from negative ions in electrostatic fields

Theoretical considerations

The application of a uniform electric field F to an atom modifies the potential field in which the electron moves so that it is confined by a finite potential barrier through which it has a finite penetration probability P. Thus for a neutral H atom the potential has the form

$$V(r, \theta) = -e^2 r^{-1} - Fe\, r \cos\theta, \tag{11.77}$$

the polar axis being taken along the field direction. Then, for $\theta = 0$, an electron possessing the ground state energy $e^2/2a_0$ encounters a barrier of finite height $2e^{3/2}F^{1/2}$ and thickness $r_t = e/2a_0 F$ provided $r_t \gg a_0$.

The rate of leakage of a particle through a spherical potential barrier (as in Chapter 4, Fig. 4.3), i.e. one in which the potential is a function $V(r)$ of r only, can be written in the form

$$w = w_0 P,$$

where w_0 is the number of times per second that the particle impinges on the barrier and P is the chance that it will penetrate when it does so impinge. If E is the energy of the particle, P is given to a good approximation by the semi-classical formula

$$\bar{P} = \exp\left[-\int_{r_1}^{r_2} \left\{\frac{2m}{\hbar^2}(-E+V)\right\}^{1/2} dr\right], \qquad (11.78)$$

where m is the mass of the particle and r_1, r_2 are the values of r for which $V(r) = E$. This is valid provided the wavelength of the motion of the particle is small compared with the barrier thickness $r_2 - r_1$.

This may be applied to a case such as the H atom in a uniform field provided proper account is taken of the directional dependence of V. Thus approximately we replace (11.78) by

$$\bar{P} = \tfrac{1}{2}\int_0^\pi P(r,\theta)\sin\theta\, d\theta. \qquad (11.79)$$

Oppenheimer (1928) applied these considerations to the ionization of H obtaining the formula

$$w = \frac{e^2}{2ha_0}\frac{F_H}{F}\exp\left(-\frac{2}{3}\frac{F_H}{F}\right), \qquad (11.80)$$

with $F_H = e/a_0^2$. In this analysis w_0 is simply given by $|E_H|/h$ where E_H is the energy $-e^2/2a_0$ of the ground state of H.

Interest in the extension of this work to detachment from H⁻ developed particularly in relation to the design of circular accelerators of H⁻ ions which have certain advantages (see Chapter 15, p. 690). Because of their motion with velocity v in a uniform magnetic field H the ions experience an electric field in their rest frames which is given by

$$F = 0.3\beta\gamma H \times 10^6 \text{ V cm}^{-1}, \qquad (11.81)$$

where $\beta = v/c$, $\gamma = (1-\beta^2)^{-1/2}$ and H is in kG. Such a field is by no means negligible from the point of view of detachment.

As a rough approximation the problem for H⁻ may be treated as a single-body one by using

$$E = -E_a(\text{H}), \quad V(r) = V_H(r) - Fer\cos\theta, \qquad (11.82a)$$

where E_a is the electron affinity of H and V_H the mean static field of the hydrogen atom,

$$V_H(r) = -e^2(r^{-1} + a_0^{-1})\exp(-2r/a_0). \qquad (11.82b)$$

For w_0, in the same vein, we would take $\frac{1}{2}|E_H|/h$ since the two electrons share the total energy which is close to that of H.

Calculations were carried out on these lines by Darewych and Neamtam (1963) which give values for the lifetime $\tau = w^{-1}$ as a function of F not far from those obtained in more elaborate calculations by Hiskes (1962) and by Mullen and Vogt (1968) (see Fig. 11.43) in which polarization of the atom by the departing electron is taken into account.

The measurement of the rate of field detachment from H⁻

A number of experiments have been carried out to measure the mean lifetime of H⁻ ions in the presence of an electric field as seen by the ions. In early experiments Kaplan, Paulikas and Pyle (1963) observed that D⁻ ions were completely dissociated at an electric field strength of 4.2 MV cm⁻¹. Somewhat later Cahill, Richardson and Verba (1966) measured the lifetimes of H⁻ ions for field strengths between 2.2 and 2.4 MV cm⁻¹. A later detailed study was carried out by Stinson, Olsen, McDonald, Ford, Axen and Blackmore (1969) covering a range from 1.8 to 4.4 MV cm⁻¹. The arrangement used in these experiments is illustrated in Fig. 11.41. The H⁻ beam was obtained by passing protons of 500 keV energy from a Cockcroft–Walton injector through a canal 25 cm long containing water vapour at a pressure of 0.2 torr. A considerable fraction of the protons captured two electrons during this passage. The mixed beam then entered a 50 MeV proton linear accelerator, after passage through which the accelerated H⁺ and H⁻ ions were separated by an analysing magnet. The H⁻ ions passed for 10 m along a beam line,

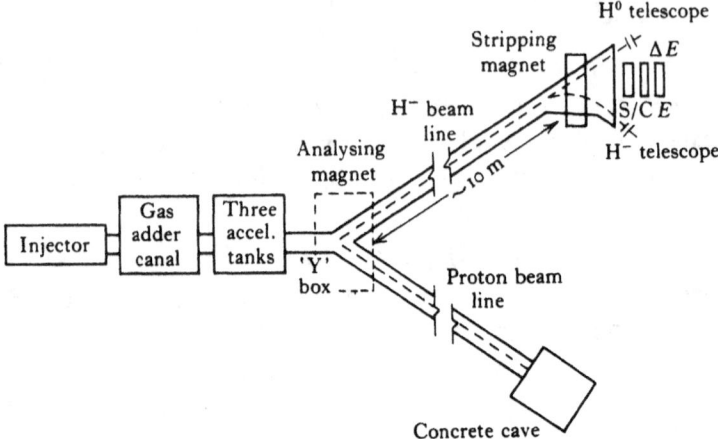

Fig. 11.41. General arrangement of the experiments of Stinson *et al.* (1969) to measure the rate of detachment from H⁻ by a steady electrostatic field.

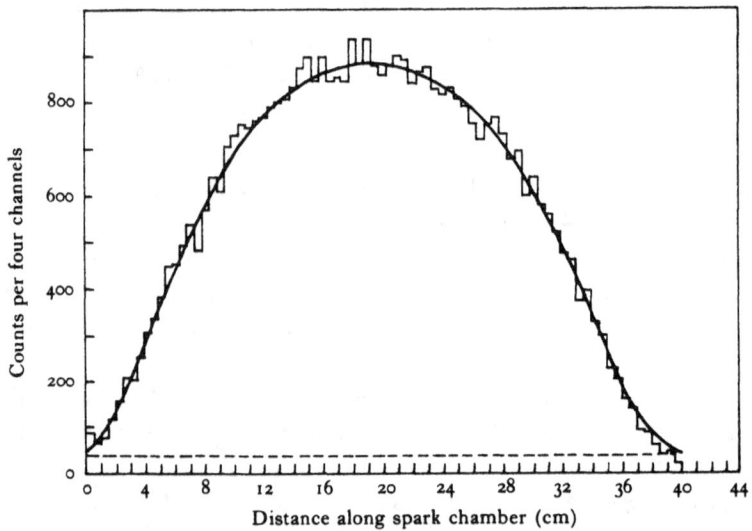

Fig. 11.42. Distribution of stripped H⁻ ions along the spark chamber, observed in the experiments of Stinson *et al.* (1969), the stripping field being 21.69 kG. The broken line is the background due to stripping in collisions with residual gas while the full line is the best fit obtained to the distribution, in the analysis to determine the lifetime towards field detachment.

maintained at a pressure of 6×10^{-6} torr, before entering the stripping magnet to produce field detachment. H⁻ ions deflected by this magnet were detected by a fast counter telescope while neutral

H atoms present in the beam at the entrance to the field were detected by a similar system. H atoms produced through detachment either by the magnetic field or in collisions with residual gas in passage through the field were detected by a position-sensitive acoustic spark chamber triggered by a particle identification system (Swales, 1968).

Fig. 11.42 shows a typical distribution of stripped H^- ions along the spark chamber with a stripping field of 21.69 kG. The back-

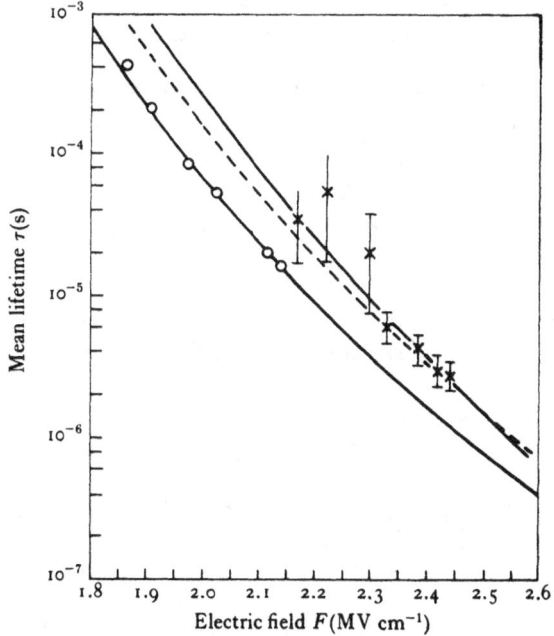

Fig. 11.43. Mean lifetime of H^- ions as a function of the applied steady electrostatic field. Observed –O– Stinson et al. (1969), × Cahill et al. (1966). Calculated —— Hiskes (1962), - - - - Mullen and Vogt (1968).

ground due to stripping by collisions with the residual gas was determined by measuring the distribution at a field of 16.97 kG, so small that field detachment was negligible. It is indicated by the broken line in Fig. 11.42. Analysis of the data proceeded essentially by an iterative procedure in which a particular form for the variation of lifetime with field strength was assumed and the distribution calculated from the trajectories of ions in the stripping field.

Fig. 11.43 shows the variation of mean lifetime with electric field determined in this way. These results are compared with the earlier observations of Cahill *et al.* and the theoretical calculations of Hiskes (1962) and of Mullen and Vogt (1968). It is not surprising that the theory overestimates the lifetime as it assumes the electron moves radially in the field. The measurements of Stinson *et al.* were particularly concerned with the design of the magnet for the 500 MeV variable energy sector focused H⁻ cyclotron (see Chapter 15, p. 691) at the University of British Columbia and led to a reduction in the maximum magnetic field.

Field detachment from excited negative ions

Smirnov and Chibisov (1965) have made approximate calculations of the rate of detachment of electrons from weakly-bound states of negative ions by uniform electric fields F. They write the asymptotic form of the wave function for the state concerned in the form

$$\Psi \sim Br^{Z/\gamma - 1} e^{-\gamma r},$$

where the binding energy is given in atomic units by $-\tfrac{1}{2}\gamma^2$, Z is the effective charge on the atomic core and B is determined by the inner form of the wave function. They then find that the rate of detachment W is given approximately, for an electron in an orbital with angular momentum quantum numbers l and m, by

$$W = B^2 \frac{(2l+1)}{2\gamma^m} \frac{m!(l+m)!}{(l-m)!} \left(\frac{2\gamma^2 e}{a_0^2 F}\right)^{\frac{2Z}{\gamma} - m - 1} \exp(-2\gamma^3 e/3a_0^2 F).$$

(11.83)

Oparin, Ill'in, Serenkov, Solov'yev and Fedorenko (1970) have used field detachment to distinguish weakly-bound negative ions in a mixed beam as described in Chapter 10. p. 415.

11.9 Detachment by electron impact

Theoretical considerations

The differential cross-section for an inelastic collision in which the nth state of the atom is excited from the ground state by electron impact is given to a high accuracy by Born's approximation, provided the velocity v of the electron is large compared with the orbital velocities u of the atomic electrons concerned. Thus for a collision in which the electron velocity changes from **v** to $\mathbf{v}_n$ and its direction of motion from that of the unit vector $\mathbf{n}_0$ to $\mathbf{n}_1$, where $\mathbf{n}_0 \cdot \mathbf{n}_1 = \cos\theta$, we have for the differential cross-section (Mott and Massey, 1965, p. 327)

$$I_{0n}(\theta)\,d\omega = (4\pi^2 m^2/h^4)(k_n/k) \times$$
$$\times |\iint \psi_0(\mathbf{r}_a)\exp\{i(\mathbf{k}-\mathbf{k}_n)\cdot\mathbf{r}\}\,V(\mathbf{r},\mathbf{r}_a)\,\psi_n^*(\mathbf{r}_a)\,d\mathbf{r}\,d\mathbf{r}_a|^2\,d\omega, \quad (11.84)$$

where $\mathbf{k} = (mv/\hbar)\mathbf{n}_0$, $\mathbf{k}_n = (mv_n/\hbar)\mathbf{n}_1$ and ψ_0, ψ_n are the initial and final atomic wave functions. The total cross-section for the particular inelastic collision is then

$$Q_{0n} = \int_0^{2\pi}\int_0^{\pi} I_{0n}(\theta)\sin\theta\,d\theta\,d\phi. \quad (11.85)$$

There is no difficulty in extending (11.84) to collisions in which the final atomic state is one of a continuum as in ionization. Thus let $\psi_\kappa(\mathbf{r}_a)$ be the wave function for such a state, in which an electron moves with wave number κ in the field of the core, normalized so as to have the asymptotic form

$$\psi_\kappa(\mathbf{r}_a) = \psi_c(\mathbf{r}_c)\{\exp(i\boldsymbol{\kappa}\cdot\mathbf{r}_1) + g(\theta_1)r_1^{-1}\exp(i\kappa r_1)\}. \quad (11.86)$$

Here $\mathbf{r}_c$ represents the aggregate of coordinates of the core electrons, $\mathbf{r}_1$ the coordinates of the unbound (ejected) electron while $\psi_c(\mathbf{r}_c)$ is the core wave function. The differential cross-section $I_{0\kappa}(\theta)\,d\omega\,d\kappa$, for a collision in which an electron is ejected with wave numbers between κ and $\kappa + d\kappa$ leaving the core in a state with wave function ψ_c, is then given by

$$I_{0\kappa}\,d\kappa\,d\omega = (4\pi^2 m^2 \kappa^2 k'/kh^4)|\iint \psi_0(\mathbf{r}_a)\exp\{i(\mathbf{k}-\mathbf{k}')\cdot\mathbf{r}\}$$
$$\times V(\mathbf{r},\mathbf{r}_a)\,\psi_\kappa(\mathbf{r}_a)\,d\mathbf{r}\,d\mathbf{r}_a|^2\,d\kappa\,d\omega. \quad (11.87)$$

This may be applied in principle to detachment from H$^-$ by electron impact. However, for collisions at high energies, a simplified but accurate asymptotic form may be obtained which makes it possible to integrate the cross-section over all allowed values of κ and sum over all final states of the atom.

Born's first approximation is certainly applicable under these conditions. Then it has been shown by Inokuti, Kim and Platzman (1967) that the total cross-section for all inelastic collisions with a two-electron atom, initially in its ground state, takes the asymptotic form

$$Q_{in}^t = 4\pi a_0^2\,\alpha(v)\,M_t^2\log\{4c_t/\alpha(v)\}. \quad (11.88)$$

Here $\alpha(v) = (2\pi e^2/hv)^2$, v being the impact velocity, while M_t and c_t are quantities which depend only on the ground state wave function $\psi_0(\mathbf{r}_1,\mathbf{r}_2)$ of the target atom.

For H⁻, all inelastic collisions result in detachment, as there are no bound excited states, so that $Q_{in}^t = Q_{de}$, the cross-section for detachment by electron impact leaving the H atom in the ground or any excited, including ionized, state.

M_t is then given by

$$M_t^2 = \langle (\mathbf{r}_1 + \mathbf{r}_2)^2 \rangle / 3a_0^2, \qquad (11.89)$$

where $\langle \ \rangle$ denotes the average over the ground-state density distribution $\psi_0^2(\mathbf{r}_1, \mathbf{r}_2)$. For c_t we have

$$M_t^2 \ln c_t = 2L_{-1} + I_1 - I_2, \qquad (11.90)$$

where

$$L_{-1} = \int_{E_a}^{\infty} \frac{d\mathscr{F}}{dE} \frac{R}{E} \log\left(\frac{E}{R}\right) dE, \qquad (11.91)$$

$$I_1 = \int_1^{\infty} 2S_{in}(K)(Ka_0)^{-4} \, d(Ka_0)^2, \qquad (11.92)$$

$$I_2 = \int_0^1 \{M_t^2 - 2S_{in}(K)/K^2 a_0^2\}(Ka_0)^{-2} \, d(Ka_0)^2, \qquad (11.93)$$

$$S_{in}(K) = \tfrac{1}{2}|\langle |\exp(i\mathbf{K}\cdot\mathbf{r}_1) + \exp(i\mathbf{K}\cdot\mathbf{r}_2)|^2\rangle - \\ - |\langle \exp(i\mathbf{K}\cdot\mathbf{r}_1) + \exp(i\mathbf{K}\cdot\mathbf{r}_2)\rangle|^2, \qquad (11.94)$$

R being the Rydberg energy $e^2/2a_0$. All these quantities can be calculated given an accurate wave function $\psi_0(\mathbf{r}_1,\mathbf{r}_2)$ except the differential oscillator strength $d\mathscr{F}/dE$ which is related to the total photodetachment cross-section Q_d by

$$Q_d(\nu) = (\pi e^2 h/mc) \frac{d\mathscr{F}}{dE}. \qquad (11.95)$$

Because of this, the determination of L_{-1} is the least accurate and it is not worthwhile to calculate the other quantities to greater accuracy. Inokuti and Kim (1968) proceeded as follows.

M_t^2 was calculated from (11.89) using the Pekeris wave function (Chapter 1 (1.12)) as well as a number of simpler, less-accurate functions including the 20-parameter function (1.10). $I_1 - I_2$ was then calculated only for these simpler functions and extrapolated as a function of M_t^2 to the value 7.484 of M_t^2 given by the Pekeris function.

It is not possible to determine $d\mathscr{F}/d\nu$ and hence L_{-1} from (11.91) using either observed or calculated cross-sections $Q_d(\nu)$ because the former do not cover a sufficiently large frequency range and the latter do not allow for detachment which leaves the H atom in an excited state. Kim and Inokuti therefore adopted the procedure of correcting an initial

DETACHMENT BY PHOTON AND ELECTRON IMPACT

approximation to $d\mathscr{F}/d\nu$ so that when substituted for $(mc/\pi e^2 h)Q_d$ in the formula (11.15) it gave good agreement with the accurately calculated values for s ranging from -2 to $+2$ (see Table 11.1) which were obtained using accurate ground state wave functions.

In this way Kim and Inokuti found that

$$Q_{de} \sim 4\pi a_0^2 \, \alpha(v) \left[-7.484 \log\{\alpha(v)\} + 25.3 \pm 1.5\right]. \quad (11.96)$$

This formula should be valid at sufficiently high but non-relativistic impact velocities. It is compared with experiment in Fig. 11.49.

The calculation of Q_{de} at lower impact energies is much more difficult and has only been carried out for collisions which leave the H atom in its ground state. The simple Born approximation (11.84) ignores the fact that the incident electron moves in a repulsive field which has the asymptotic form e^2/r. At high impact energies this neglect is unimportant and it is reasonable to represent the initial and final wave functions of the colliding electron in terms of undisturbed plane waves $\exp(i\mathbf{k} \cdot \mathbf{r})$, $\exp(i\mathbf{k}' \cdot \mathbf{r})$ respectively. In an attempt to extend the applicability of the approximation to lower impact energies Bely and Schwartz (1969) replaced the plane waves in (11.84) by waves distorted by the Coulomb repulsion e^2/r. This is a good approximation for the initial wave function, representing motion in the field of H^-. For the final wave function it is not so obvious but it was assumed that the ejected electron in fact continues to maintain an average repulsive interaction approximately of the same Coulomb form. For the H^- ground state an HF wave function (see Chapter 1 (1.17)) was used and for the continuum state the ejected electron was represented by a wave function for motion in the field of the nucleus and of the remaining electron, the wave function for which was taken to be unaltered from that in the HF ground state function. In the most successful calculations, as judged by comparison with experimental results (see Fig. 11.50), the wave function for the ejected electron was constrained to be orthogonal to the initial HF function.

Measurement of detachment cross-sections

Cross-sections for detachment from H^- by electron impact have been measured over an energy range from 12 to 1000 eV using crossed beams of ions and electrons intersecting at 90°, and from the threshold to 30 eV using the so-called inclined-beam technique in which the beams intersect at angles of 10 or 20° (see Chapter 14, p. 634). Experiments of this kind, in which both interacting particles are charged, are in principle more difficult than corresponding experiments in which one beam is neutral. This is largely because space-charge effects in the interaction region may alter the trajectories of the charged particles, thereby giving rise to spurious signals which cannot be removed by modulation techniques. As the magnitude of this effect will depend on the velocity of the ions it is possible to check whether it is important by checking whether the observed cross-section is dependent on the ion velocity.

Despite this and other difficulties good measurements are now available. The first experiments were carried out using somewhat different techniques to eliminate background effects, by Tisone and Branscomb (1966) and by Dance, Harrison and Rundel (1967). These experiments, of the crossed-beam type, were repeated somewhat later by Peart, Walton and Dolder (1970) with improved signal-to-background ratio. The same authors also measured cross-sections $Q_d^e(2)$ for the two-electron detachment

$$H^- + e \longrightarrow H^+ + 3e. \qquad (11.97)$$

To improve the accuracy of measurement at electron energies below 20 eV they carried out further experiments by the inclined-beam technique which yielded evidence of a resonance effect at electron energies of 14.53 eV.

Fig. 11.44 shows the general arrangement of the crossed-beam experiments by Peart et al. H$^-$ ions, produced from a hot-cathode discharge in ammonia, were accelerated and focused and then passed through five differentially pumped chambers before deflexion through 60° by a magnetic field M_1 to enter the interaction chamber. After passing through a defining aperture they were deflected by an electrostatic field back through 60° so as to be moving parallel to their original direction. The double deflexion, by reducing the length of ion beam viewed directly by the neutral atom detector D, reduced the contribution to the background due to H atoms produced by detachment collisions of the ions with residual gas.

DETACHMENT BY PHOTON AND ELECTRON IMPACT

After the second deflexion the ions passed through the interaction region in which they were crossed by the electron beam from the gun G which was collected in Ce. Those ions which were unaffected were then deflected by a second magnet M_2 to a screened collector C while the neutral atoms passed straight through to the particle-multiplier detector D. The efficiency of this detector was determined by comparing its count rate with the weak current of H^+ and H^- ions collected by a Faraday cup and measured with a vibrating

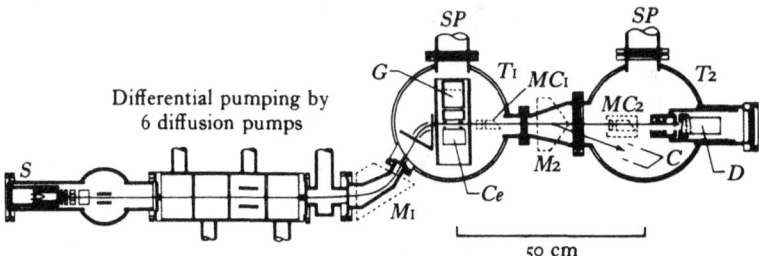

Fig. 11.44. General arrangement of the apparatus used by Peart et al. (1970) for measurement of cross-sections for detachment of electrons from H^- by electron impact.

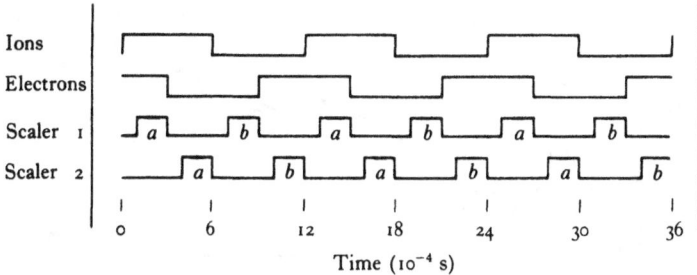

Fig. 11.45. Wave forms of ion and electron beams and scaler gates in the experiments of Peart et al. (1970).

capacitor electrometer. Little difference was found for the oppositely charged ions so the efficiency for neutral atoms was taken as the mean.

To subtract the background signal both the electron and ion beams were pulsed at a frequency of 1.67 kHz but out of phase by 90°. The output pulses from the detector were fed in parallel to two scalers which were gated so that counting takes place over two portions a, b of the pulse and the wave forms of the ion and electron beams and scaler gates are as shown in Fig. 11.45.

If R is the true signal, and B_i, B_e, B_n are the backgrounds associated with the ion beam, the electron beam and background noise respectively, the following signals will be recorded.

(scaler 1)$_{\text{gate } a}$ $R + B_i + B_e + B_n$,

(scaler 1)$_{\text{gate } b}$ B_n,

(scaler 2)$_{\text{gate } a}$ $B_i + B_n$,

(scaler 2)$_{\text{gate } b}$ $B_e + B_n$.

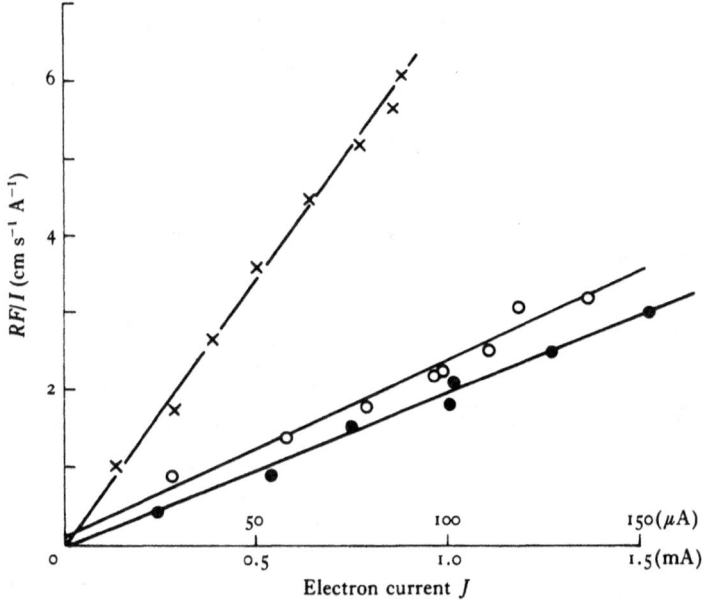

Fig. 11.46. Plots of observed values of RF/I (see p. 507) against electron current, for different electron beam energies E. From Peart *et al.* (1970). ● $E = 19.4$ eV, × $E = 152$ eV, ○ $E = 602$ eV.

The true signal R is then given by

(scaler 1 − scaler 2)$_{\text{gate } a}$ − (scaler 2 − scaler 1)$_{\text{gate } b}$.

In a typical experiment the ratio of signal-to-background was about 0.1 as compared with about 0.02 and 0.01 in the earlier experiments of Dance *et al.* and of Tisone and Branscomb respectively. The improvement in the later work was largely due to the reduced pressure in the neighbourhood of the electron gun (3×10^{-10} torr as against 3×10^{-9} and 5×10^{-8} torr respectively).

The detachment cross-section is given by

$$Q_d^e(1) = \frac{R}{I\mathcal{J}} \frac{vV}{(V^2 + v^2)^{1/2}} \frac{e^2 F}{D}, \qquad (11.98)$$

where R is the measured rate of production of H by electron impact, I and $\mathcal{J}$ are the currents of ions and electrons with velocities V and v respectively, D is the detector efficiency and F is an overlap factor given by

$$F = \int_{-\infty}^{\infty} i(z)\,dz \int_{-\infty}^{\infty} j(z)\,dz \bigg/ \int_{-\infty}^{\infty} i(z)j(z)\,dz, \qquad (11.99)$$

which allows for uneven distribution of intensity across the ion and electron beams. Thus with z measured in the mutually perpendicular direction, $i(z)dz$, $j(z)dz$ are the respective ion and electron currents passing between z and $z + dz$. It is important to measure F when both beams are present because when they interact, space-charge effects may change the beam profiles. The measurement was made by moving a shutter in the form of a right-angled bracket across the interaction region so that both profiles could be intercepted simultaneously.

To determine $Q_d^e(1)$ and obtain a valuable check on the validity of the results, RF/I was plotted as a function of the electron current $\mathcal{J}$. According to (11.98) this should give a straight line passing through the origin from the gradient of which $Q_d^e(1)$ can be derived. Fig. 11.46 shows typical plots for three electron energies which conform well with expectation.

A further important check, because of the danger of spurious signals due to space-charge effects (see above), was made by measuring $Q_d^e(1)$ as a function of the energy of the ion beams, the energy of the electron beam being adjusted in each case to give a constant relative kinetic energy. Fig. 11.47 shows typical results which show no dependence on the ion energy and hence provide strong evidence that space-charge effects were unimportant. By making measurements at different residual gas pressures it was also verified that the derived values of Q_d^e were independent of the ratio of signal-to-background over the working region.

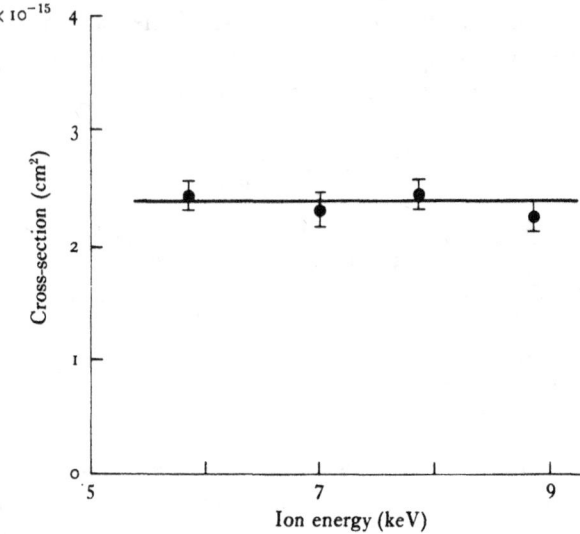

Fig. 11.47. Detachment cross-sections Q_d^e measured at different ion energies in the experiments of Peart *et al.* (1970), the electron beam energy being 52.4 eV.

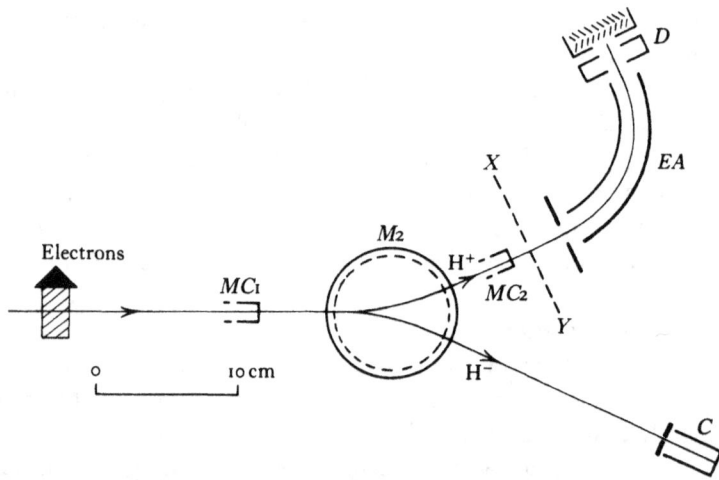

Fig. 11.48. General arrangement of the apparatus used by Peart, Walton and Dolder (1971) to measure cross-sections for double detachment of electrons from H⁻ by electron impact. MC_1, MC_2 are movable collectors, C is a fixed collector.

To measure the cross-section for double detachment, the detecting system was changed as shown in Fig. 11.48. By means of a magnet M_2 the H⁺ ions produced in this process were deflected vertically away from the main H⁻ beam by an electric field between the electrodes EA to impinge on the first dynode of a particle multiplier D. By comparison with direct current measurements with a Faraday cup the detection efficiency for protons with the experimental energy (8 keV) was $88 \pm 3\%$.

Finally, because it was not found to be practicable to make accurate measurements at electron energies below 10 eV with beams intersecting at 90°, further experiments were carried out in which the electron gun and collector assembly were modified so that the electron beam intersected the ion beam at 20°. This raised problems concerning the screening of the gun from strong magnetic fields, which was important because of the relatively low-energy electron beams involved. The magnet $M\text{1}$ in Fig. 11.44 was replaced by a much smaller magnet which deflected the H⁻ beam through 10° only, while the interaction region was shielded by two concentric mumetal boxes.

With this arrangement $V/(v^2 + V^2)^{1/2}$ in (11.98) is replaced by

$$V \sin\theta/(V^2 + v^2 - 2vV\cos\theta)^{1/2}, \qquad (11.100)$$

with $\theta = 20°$, and the cross-section again deduced from a plot of RF/I against J. Apart from the fact that this plot should be linear and pass through the origin, an additional check is available in that measurements may be made down to electron energies below the threshold 0.755 eV for the process. If the results are valid then the existence of this threshold should be apparent. These checks, and others carried out as in the crossed-beam experiments gave favourable results.

Results obtained for single detachment

Fig. 11.49 shows the results obtained for the single-detachment cross-section $Q_d^e(1)$ in the crossed-beam experiments, including the results obtained in the earlier experiments.

It will be seen that for energies greater than 50 eV there is very good agreement between the results obtained by Peart et al. (1970) and by Dance et al. (1967). These two sets of results are also consistent with the theoretical results of Inokuti and Kim (1968) which should be quite accurate at high electron energies. This is shown in Fig. 11.49. In contrast the results of Tisone and Branscomb (1966) are much too high at these energies. It seems from the

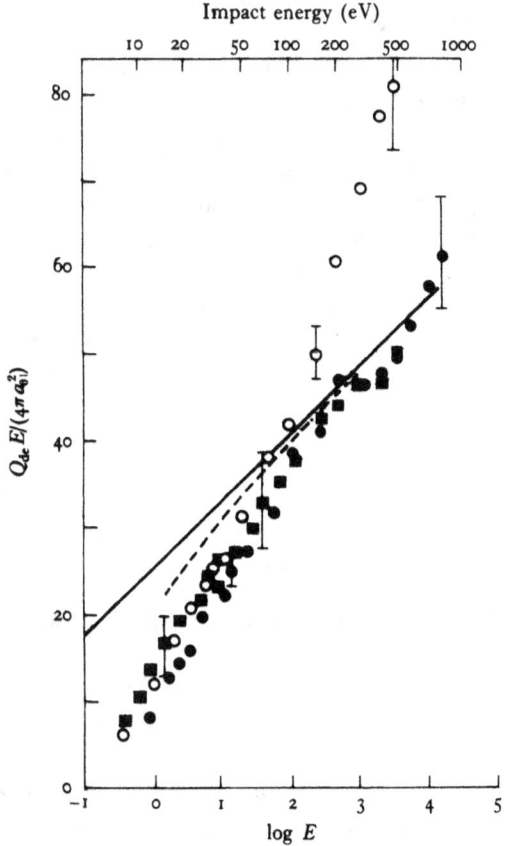

Fig. 11.49. Plots of $Q_{de}E/4\pi a_0^2$ against $\log E$ for impact of electrons of energy E (measured in Rydbergs) with H⁻. Observed ● Peart et al. (1970), ■ Dance et al. (1967), ○ Tisone and Branscomb (1966). Calculated —— Inokuti and Kim (1968), --- Born's approximation.

evidence of Fig. 11.49 that Born's approximation gives good results at energies greater than about 100 eV.

At lower energies it is difficult to obtain reliable theoretical results, but the calculations of Bely and Schwartz (see p. 503) give surprisingly good agreement with the results of Peart et al. down to 17 eV.

Fig. 11.50 shows the results obtained by Walton, Peart and Dolder (1971) at energies below 30 eV using the inclined-beam technique. It will be seen that these results are independent of ion energy between 7 and 10 keV and agree well with those obtained in the crossed-beam experiment by the same authors. Moreover there is clear evidence of a threshold near the correct value.

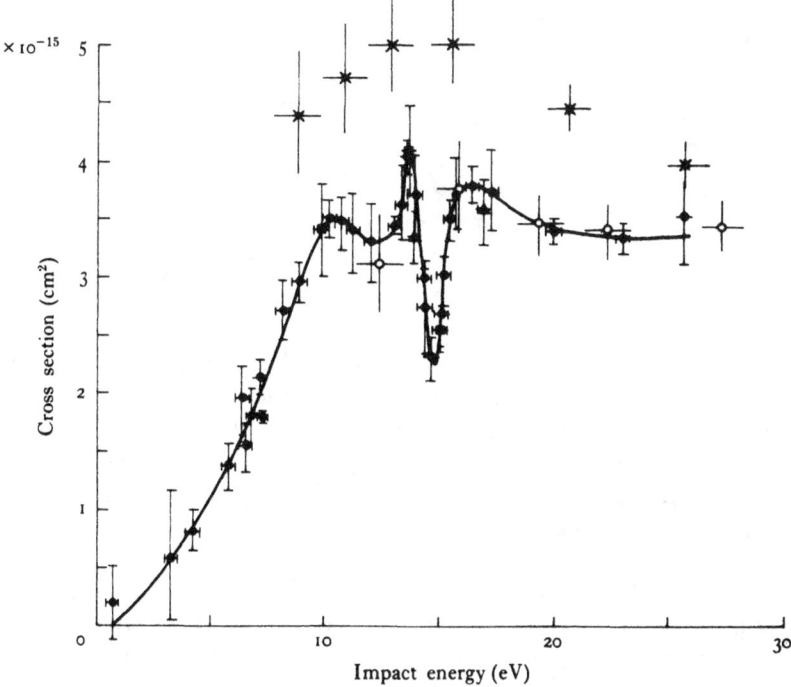

Fig. 11.50. Observed cross-sections $Q_{de}(1)$ for detachment of electrons from H⁻ by electron impact, at low impact energies. Observed; 90° intersection: ○ Peart et al. (1970), × Dance et al. (1967); 20° intersection: Walton et al. (1971) for different ion beam energies, ▲ 7 keV, ● 8 keV, ■ 10 keV.

A specially interesting feature is the resonance-like structure near 14.2 eV. Theoretical evidence of the existence of a $(2s)^2 2p\ ^2P^0$ resonance state of H⁻ at an energy close to 14.2 eV with a width $\simeq 1$ eV has been discussed in Chapter 5, p. 151, and it is likely that the observed structure is associated with this level. The same theoretical work suggested also that a less marked resonance, arising from a

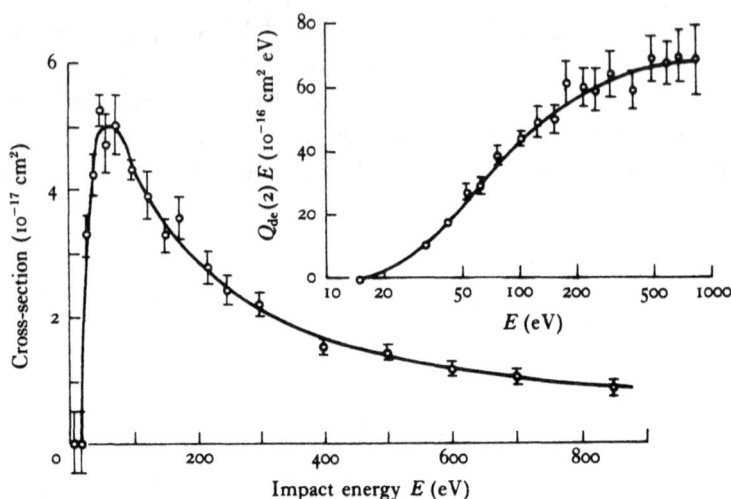

Fig. 11.51. Observed cross-sections $Q_{de}(2)$ for double detachment of electrons from H⁻ by electron impact. ⚬ observed by Peart et al. (1971), —— best fit to experimental points.

(2p)² ²P⁰ state, should occur near 17.2 eV. Evidence of this was obtained in a later experiment by Peart and Dolder (1973) using the inclined-beam technique, with the angle θ in (11.100) reduced to 10°, and improved experimental conditions which measured the signal-to-background ratio by more than an order of magnitude compared with the first inclined-beam experiments.

Double detachment

Fig. 11.51 illustrates the measured cross-section $Q_d^e(2)$ as a function of electron energy. Over a large part of the energy range (100–800 eV) the ratio $Q_d^e(1)/Q_d^e(2)$ has a nearly constant value close to 40. The inset shows the variation of $Q_d^e(2)$ with $\log E$.

CHAPTER 12

Detachment, Charge Transfer and Other Reactions between Negative Ions and Neutral Systems at Low and Intermediate Energies

In the preceding two chapters we have discussed processes which lead to detachment of electrons by negative ions through the action of electromagnetic fields, either static or at high frequency as in photodetachment, and through electron impact. As already explained in Chapter 11, p. 416, detachment can also occur through impact with neutral atoms or molecules. However, it is not appropriate to discuss such detachment reactions in isolation as a considerable variety of other reactions can take place and the experimental techniques which yield data on detachment rates are usually applicable also to the study of the other types of possible reaction. Furthermore, it is necessary to distinguish the different types of reactions which are observed. We therefore consider in this chapter the experimental and theoretical methods available for the study of all kinds of reactions between negative ions and neutral atoms or molecules, confining ourselves only to collisions at near thermal energies, or at intermediate energies in which attention is concentrated largely on behaviour near the threshold for the reaction. In the following chapter we shall discuss reactive impacts at higher energies.

12.1 Classification of types of ionic reaction

The following possibilities arise on impact of an atomic ion A^- with an atom B.

(a) Direct detachment

$$A^- + B \longrightarrow A + B + e. \qquad (12.1)$$

This reaction will necessarily be endothermic by an amount E_a, the electron affinity of A, if B is in its ground state. If B is in an excited state B^*,

$$A^- + B^* \longrightarrow A + B + e \qquad (12.2)$$

will be exothermic if $E_{ex}(B)$, the excitation energy of B* is greater than E_a. The process is in fact analogous to Penning ionization

$$A^* + B \longrightarrow A + B^+ + e,$$

A* being a metastable atom.

(b) Associative detachment

$$A^- + B \longrightarrow AB + e. \qquad (12.3)$$

This will be exothermic if $D(AB)$, the dissociation energy of AB, is greater than E_a.

(c) Charge transfer

$$A^- + B \longrightarrow A + B^-, \qquad (12.4)$$

exothermic if $E_a(B) > E_a(A)$.

(d) Ionic association. If the collision occurs in the presence of a third body M then the reaction

$$A^- + B + M \longrightarrow A^- \cdot B + M \qquad (12.5)$$

can take place, leading to production of a more complex ion.

If the colliding systems are not monatomic additional rearrangement collisions may occur of the type

$$AC^- + BD \longrightarrow AB^- + CD,$$
$$\longrightarrow AB^- + C + D,$$
$$\longrightarrow ABC^- + D,$$
$$\longrightarrow A + B + CD + e.$$

12.2 Orbiting collisions

Consider a classical collision between a particle of mass M and velocity v and an infinitely massive centre of force exerting an interaction energy $V(r)$, where r is measured radially outwards from the centre. If the impact parameter in the collision is p, so the particle possesses an angular momentum $J = Mvp$ about the centre, the radial component of the velocity of the particle when at a distance r from the centre is given by

$$\tfrac{1}{2}Mv_r^2 = \tfrac{1}{2}Mv^2 - V_{\text{eff}}(r), \qquad (12.6a)$$

where
$$V_{\text{eff}}(r) = V(r) + J^2/2Mr^2, \qquad (12.6b)$$

the term $J^2/2Mr^2$ representing the contribution from the centrifugal force.

If the conditions are such that v_r is small over an appreciable range of r, the angular motion will cause the particle to make more than one revolution about the centre of force before passing out of this region of r. This is known as *orbiting* and will certainly occur if, at the closest distance of approach r_c, where $v_r = 0$, $V_{\text{eff}}(r)$ has a maximum so v_r changes very slowly with r, i.e. if

$$\tfrac{1}{2}Mv^2 - V_{\text{eff}}(r_c) = 0, \qquad (12.7a)$$

$$\left\{ \frac{d}{dr} V_{\text{eff}}(r) \right\}_{r=r_c} = 0. \qquad (12.7b)$$

We now apply these considerations to collisions between an ion A⁻ and a neutral system B, the velocities now referring to relative motion and M being the reduced mass. The long-range interaction between the two systems at separation r will have the asymptotic form $-\tfrac{1}{2}\alpha e^2/r^4$, where α is the polarizability of B. For orbiting to occur the conditions (12.7a), (12.7b) require

$$E + \tfrac{1}{2}\alpha e^2 r^{-4} - (J^2/M)r^{-2} = 0. \qquad (12.8a)$$

$$-\frac{J^2}{2Mr^3} + \frac{\alpha e^2}{r^5} = 0, \qquad (12.8b)$$

where $E = \tfrac{1}{2}Mv^2$. Eliminating r by use of (12.8b) we find that

$$E = \tfrac{1}{8}J^4/\alpha e^2 M^2, \qquad (12.9a)$$

or, since $J = Mvp$,

$$E = 2\alpha e^2/p^4. \qquad (12.9b)$$

Thus, for collisions in which $p = p_c = (2\alpha e^2/E)^{1/4}$, orbiting will occur and the interacting systems will remain in close proximity for a relatively long time during which they can react. It is therefore useful to relate the effective cross-section for a particular reaction to the orbiting cross-section $Q_{\text{orb}} = \pi p_c^2$ which is given by

$$Q_{\text{orb}} = \pi(2\alpha e^2/E)^{1/2}. \qquad (12.10)$$

We would expect, for a particular reaction, that the cross-section Q_r will be $c_r Q_{\text{orb}}$, where $c_r \leqslant 1$ and in many cases will be comparable with unity.

The reaction rate constant $k = vQ_r$ will be given by

$$k = 2\pi(\alpha e^2/M)^{1/2} c_r, \tag{12.11}$$

which will be independent of relative kinetic energy if c_r is so independent. For collisions between O⁻ and O for example, with $\alpha \simeq 5.5 a_0^3$, we have

$$k = 7.5 \times 10^{-10} c_r \text{ cm}^3 \text{ s}^{-1}. \tag{12.12}$$

We shall refer to a reaction as *fast* if c_r is comparable with unity.

12.3 The theory of detachment reactions

It is possible to go a little further in developing a theory of detachment reactions as, for example, the reactions

$$\text{H} + \text{H}^- \longrightarrow \text{H}_2 + \text{e}, \tag{12.13a}$$

$$\longrightarrow \text{H} + \text{H} + \text{e}, \tag{12.13b}$$

the first of which is of the associative type, the second direct detachment.

Referring to Chapter 6, p. 175, we see that an H atom and H⁻ ion in their ground states can interact in two different ways corresponding to the $^2\Sigma_g^+$ and $^2\Sigma_u^+$ states of H_2^-, the first of which is attractive and the second repulsive. Consider first the case in which the interaction $V_1(R)$ is attractive and we have the situation shown in Fig. 12.1(a) (see also Fig. 6.2). For nuclear separations less than R_s the potential-energy curve for H_2^- lies above that for the ground state of H_2 and so for such separations the system is unstable towards autodetachment, the lifetime being $\hbar/\Gamma_1(R)$, where $\Gamma_1(R)$ is the line width of the $^2\Sigma_g^+$ state of H_2^- at separation R.

The relative motion of the atom and ion during the impact can be treated classically. In an impact in which the relative velocity is v, the reduced mass is M and the impact parameter is p, the nuclear separation will vary with time according to the equations

IONIC REACTIONS: LOW AND INTERMEDIATE ENERGIES

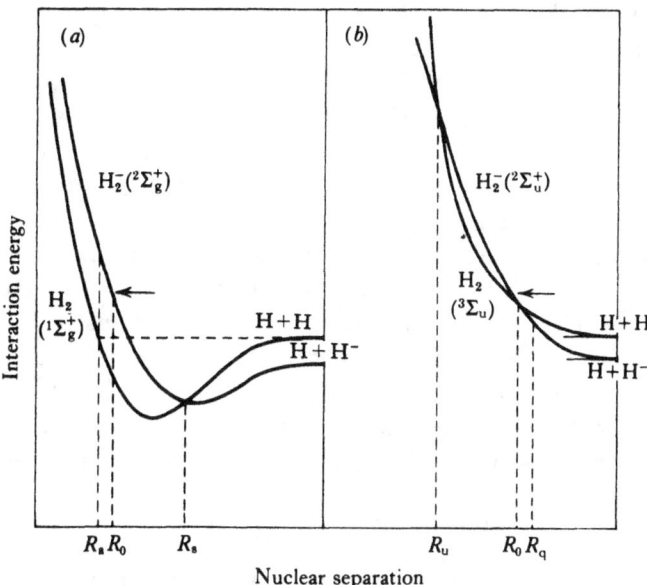

Fig. 12.1. Schematic potential-energy curves for H_2 and H_2^- relevant to the theory of detachment.

(12.6a) and (12.6b) so that

$$\frac{dR}{dt} = v_r = \{v^2 - p^2 v^2 R^2 - 2V_1(R)/M\}^{1/2}. \quad (12.14)$$

The probability of autodetachment during the collision will be given by

$$P_1(p) = 1 - \exp\left\{-\hbar^{-1} \int_{-\infty}^{\infty} \Gamma_1(R)\,dt\right\}$$

$$= 1 - \exp\left\{-2\hbar^{-1} \int_{R_c}^{R_0} \Gamma_1(R) \frac{dt}{dR}\,dR\right\}, \quad (12.15)$$

where R_0 is the closest distance of approach. Knowing $V_1(R)$ and $\Gamma_1(R)$, $P_1(p)$ may be calculated from (12.15) and (12.14). The cross-section for detachment when the atom and ion interact in the $^2\Sigma_g^+$ state will now be given by

$$Q_d(^2\Sigma_g^+) = 2\pi \int_0^\infty P_1(p)\,p\,dp.$$

Provided the closest distance of approach is $> R_a$ (see Fig. 12.1) autodetachment will lead to stable $H_2(X^1\Sigma_g^+)$ molecules, i.e. to associative detachment. If, however, $R_0 < R_a$, autodetachment occurring for $R_0 < R < R_a$ will lead to two H atoms, i.e. will contribute to process (12.13b). In that case the associative detachment cross-section will be obtained from (12.15) with the upper limit R_0 in the integral replaced by R_a.

We may carry out a similar calculation for the case (see Fig. 12.1(b)) in which the atom and ion interact in the $^2\Sigma_u^+$ state with energy $V_2(R)$. If the $^2\Sigma_u^+$ state lies above the repulsive $^3\Sigma_u$ state of H_2 for $R > R_q$ autodetachment will almost always occur with rate $\Gamma_2/\hbar$ to leave the neutral molecule in that state rather than in the ground state, and hence will lead to the process (12.13b) and not to associative detachment. The appropriate cross-section will be

$$Q_d(^2\Sigma_u^+) = 2\pi \int_0^\infty P_2(p)\,p\,\mathrm{d}p.$$

If for $R = R_u$ the repulsive curves intersect again then for $R > R_u$ autodetachment can only occur to the ground state and will be relatively improbable.

To represent experimental conditions we must allow for the fact that there is an equal probability, $\frac{1}{2}$, that the atom and ion will interact in either state. Hence, provided the impact energy is such that $R_0 > R_a$, we have that the cross-sections Q_d^a and Q_d^d for the processes (12.13a) and (12.13b) respectively will be given by

$$Q_d^a = \pi \int_0^\infty P_1(p)\,p\,\mathrm{d}p, \qquad (12.16a)$$

$$Q_d^d = \pi \int_0^\infty P_2(p)\,p\,\mathrm{d}p, \qquad (12.16b)$$

with P_1 given by (12.14) and (12.15) and P_2 by the corresponding expressions with V_1, Γ_1 replaced by V_2, Γ_2 respectively.

These theoretical considerations have been developed by Dalgarno and Browne (1967) and by Herzenberg (1967). They may readily be applied in principle to discuss more complicated cases. Using values for V_1 and Γ_1 calculated by Herzenberg and Mandl (see Chapter 6, p. 176) Dalgarno and Browne calculated for (12.13a) a rate as high as 1.9×10^{-9} cm^3 s^{-1} at 300 °K, which varies only slowly with temperature. This is quite close to the orbiting value,

IONIC REACTIONS: LOW AND INTERMEDIATE ENERGIES 519

3×10^{-9} cm^3 s^{-1}, which is independent of temperature, a result which is not surprising when it is noted that the lifetime towards autodetachment is not much larger than the time of collision when orbiting occurs.

The reaction (12.13b) is endothermic by 0.75 eV so its rate is very low at ordinary temperatures. However, at 16 000 °K Dalgarno and Browne obtain a rate constant of 4.9×10^{-11} cm^3 s^{-1} which falls to 1.8×10^{-12} at 4000 °K, 2.9×10^{-17} at 1000 °K and 1.1×10^{-21} cm^3 s^{-1} at 500 °K.

12.4 Ionic mobilities

A very great deal of experimental work has been concerned with the study of the mobilities of ions, both positive and negative, in different gases. A swarm of ions drifting in a gas at pressure p under the influence of an electric field F will acquire, through a balance between energy gained from the field and lost in collisions, a drift velocity v_d which will be a function of F/p. The mobility μ of the ions at the pressure p is defined by v_d/F so that μp is also a function of F/p.

At small F/p, where the mean energy of the ions acquired by the ions from the field is small compared with the thermal energy, the mobility is related to the diffusion coefficient D of the ions in the gas by

$$\mu = eD/kT. \qquad (12.17)$$

Since D is proportional to p^{-1}, μp is constant under these conditions, i.e. the drift velocity is proportional to F/p at small F/p.

For ions of mass M_1 drifting in a gas composed of molecules of mass M_2 the diffusion coefficient D is given by

$$D = kT \frac{M_1 + M_2}{M_1 M_2 \nu}, \qquad (12.18)$$

provided the collision frequency ν is independent of relative impact velocity v. ν is given by $nQ_d v$, where n is the concentration of gas molecules and Q_d is the diffusion, or momentum transfer, cross-section. If $I(\theta)\,d\omega$ is the differential cross-section for a collision in which the relative velocity vector is turned through an angle θ into the solid angle $d\omega$, Q_d is given by

$$Q_d = \int_0^{2\pi}\int_0^{\pi} (1 - \cos\theta) I(\theta) \sin\theta \, d\theta \, d\phi. \tag{12.19}$$

Under these conditions

$$\mu = e \frac{M_1 + M_2}{M_1 M_2 \nu}. \tag{12.20}$$

If ν is not constant it must be replaced by some mean $\bar{\nu}$ which will depend on the interaction between ion and atom.

For ions drifting in a gas of atoms of a different species a good approximation to the mobility is obtained by treating the relative motion of the ion and atom as classical with an interaction at a separation r

$$V(r) = -\tfrac{1}{2}\alpha e^2 r^{-4}, \quad r > r_0,$$
$$\longrightarrow \infty, \quad r < r_0, \tag{12.21}$$

α being the polarizability of the atom. Langevin (1905) thereby derived the formulae

$$\mu = (4\pi M)^{-1/2} \alpha^{-1/2} n g(\lambda), \tag{12.22}$$

where M is the reduced mass of ion and atom, n the atom concentration,

$$\lambda = (2kTr_0^4/\alpha e^2)^{1/2} \tag{12.23}$$

and $g(\lambda)$ is a function which tends to 0.510 as $\lambda \to 0$. In this limit the polarization energy at the hard sphere radius r_0 is large compared with the mean kinetic energy of relative motion of ion and atom.

The reduced mobility is defined as the mobility when n is equal to Loschmidt's number. In the limit of small λ the reduced mobility μ' is given simply by

$$\mu' = 35.9/(\alpha M)^{1/2} \text{ cm}^2 \text{ V}^{-1} \text{ s}^{-1}, \tag{12.24}$$

where α is measured in atomic units (a_0^3) and M is measured in multiples of the proton mass.

When the ions and the gas atoms are of the same species, symmetrical charge transfer is important and reduces the mobility below the value given by (12.22). The same general conclusions

IONIC REACTIONS: LOW AND INTERMEDIATE ENERGIES 521

apply to molecular ions and molecular gases but the formulae are now only semi-quantitative.

The problem of calculating μp at higher values of F/p is quite difficult and has not been solved in the general case. If the relative motion of the ions and neutral systems is determined mainly by the polarization interaction (12.21) Wannier (1951) showed that the mean energy of ions of mass M_1 drifting in a gas whose molecules are of mass M_2 is given by

$$\bar{\epsilon} = \tfrac{1}{2}M_1 u^2 + \tfrac{1}{2}M_2 u^2 + \tfrac{3}{2}kT, \qquad (12.25)$$

where u is the drift velocity. At high F/p the drift velocity is proportional to $(F/p)^{1/2}$ at least for collisions dominated by charge transfer or in which the colliding systems behave like rigid spheres.

Because of the anisotropy impressed on the system by the direction of the electric field for high F/p, the ionic diffusion can no longer be expressed in terms of a single diffusion coefficient but becomes strictly a tensor quantity which involves both a longitudinal and transverse diffusion coefficient D_L and D_T respectively.

Much of the earlier work directed towards the measurement of ionic mobilities was vitiated by failure to allow for reactions occurring between the ions and the molecules of the main gas or of impurities. The rates of such reactions are often fast enough to change profoundly the nature of the ions in a swarm drifting through a gas.

In recent years, experiments have been designed which determine the composition of the ions in a drifting swarm as a function of time so that it is possible to measure not only the mobilities of the different ions but also the rates of the reactions which take place between them and the neutral gas.

12.5 Experimental methods for measuring mobilities and/or reaction rates for negative ions in gases

We shall confine our description to some of the most recent experimental methods. Results obtained by earlier methods will, however, be referred to in the discussion of individual mobilities and reaction rates.

The combined drift tube and mass-spectrometer technique

As a typical sophisticated example of this technique we describe the equipment used by McDaniel and his collaborators (McDaniel, Martin and Barnes, 1962) which has been applied to the study of negative as well as positive ions.

In principle, the method employed is to produce a pulse of ions, from a suitable source, which drifts a distance l under the action of a uniform electric field F in the chosen gas at pressure p. Ions reaching an orifice at the bottom of the drift space pass out in an emergent jet from which the core is extracted by a suitable skimmer and passed into a quadrupole mass spectrometer. Ions of a selected species are then detected individually by an electron multiplier. By repeated pulse operation using a 256-channel time-of-flight analyser the time spectrum of arrival of different ions at the exit aperture is measured for different drift distances l which may be varied.

If we measure time t from the beginning of the input pulse and z along the direction of the electric field from the entrance aperture, the flux of primary ions leaving through the exit aperture at time t is given by

$$\mathcal{J}(z, t) = \tfrac{1}{4}As \exp(-\alpha t)(u + zt^{-1}) \times$$
$$\times \{1 - \exp(-r_0^2/4D_T t)\}\exp\{-(z - ut)^2/D_L t\}(\pi D_L t)^{-1/2}. \quad (12.26)$$

Here A is the area of the exit aperture, r_0 the initial radius of the circular section of the entering pulse, s the initial surface density of the ions, u their drift velocity, D_L, D_T their longitudinal and transverse diffusion coefficients and α the rate at which ions are lost through reactions with the gas. It is assumed that no reverse reactions occur which replenish the primary ions.

$\mathcal{J}(z, t)$ has the characteristic shape shown in Fig. 12.2 which reproduces data obtained for N^+ ions in N_2. Although in principle four quantities u, D_L, D_T and α are to be determined from observations of $\mathcal{J}(z, t)$ as a function of z and t, this can often be done very accurately without much difficulty. Thus u is given to quite a good approximation by $(z_1 - z_2)/(t_1 - t_2)$, where t_1, t_2 are the mean drift times for drift distances z_1, z_2 respectively. D_L is easily derived from the width and shape of the spectrum for fixed z.

To derive D_T and α it is best to work, not with $\mathcal{J}(z,t)$, but with the total flux through the exit aperture.

$$K(z) = \int_0^\infty \mathcal{J}(z,t)\,dt. \qquad (12.27)$$

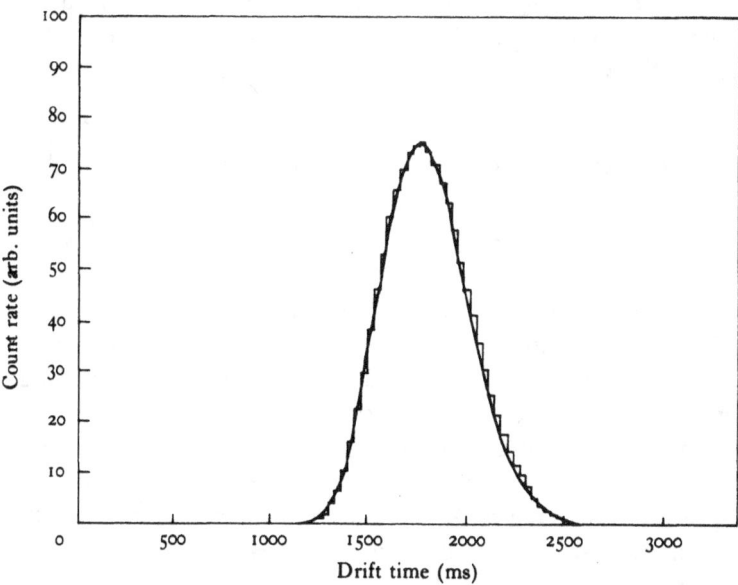

Fig. 12.2. The function $\mathcal{J}(z,t)$ of eq. (12.26) of the text, for N^+ ions in N_2 with $F/N = 1.22 \times 10^{-16}$ V cm², $p = 0.078$ torr, drift distance = 6.247 cm. The histogram gives the observed results while the solid line is the best analytical fit to the data. From Moseley, Snuggs, Martin and McDaniel (1968).

The decrease of $K(z)$ with z is determined by both D_T and α. If $\alpha = 0$ then D_T may be obtained without difficulty. Again, if F/p is small, D_T is given by (12.17) so that α may then be obtained readily. When F/p is not small, advantage may be taken of the fact that, while D_T varies as p^{-1}, α varies as p. Hence by making measurements at relatively low pressures, α is small and a good approximation to D_T may be obtained by using the value of α obtained at small F/p. This may then be used at relatively high pressures to obtain a better approximation to α and so on.

In some cases an alternative procedure may be used to determine α based on observations of the time spectrum of secondary ions produced through the same reaction as that which depletes the primary ions. This is possible provided the secondary ions once formed do not further react and the condition

$$(u - u_s)t^{1/2} \gg 2(D_L + D_{Ls}) \qquad (12.28)$$

is satisfied, u_s and D_{Ls} being the drift velocity and longitudinal diffusion coefficient of the secondary ions. If the transverse diffusion coefficient D_{Ts} of the secondary ions satisfies

$$D_{Ts}/D_T = u_s/u, \qquad (12.29)$$

as it will do at low F/p because of (12.17), then it may be shown that

$$\frac{\partial \log n_s(z, t)}{\partial t} = \frac{\alpha u_s}{u - u_s}, \qquad (12.30)$$

where n_s is the number density of the secondary ions collected at time t and drift distance z. As u_s may be determined from arrival time spectra observed for the secondary ions, measurements of the current density of secondary ions as a function of t will give α. It is possible to extend this analysis even when (12.17) is not satisfied but if the secondary ions are also depleted through reactions at a rate α_s it is not possible to determine α and α_s unless they depend on different powers of the gas pressure.

Fig. 12.3 illustrates a typical arrangement of equipment for carrying out experiments on these lines. The pressure in the drift tube may be varied between 0.02 and 1.0 torr and the drift distance between 1 and 44 cm. The main vacuum chamber, of stainless steel, is baked out at 200 °C prior to measurements being made. At the same time the drift chamber is heated to 300 °C so that a background pressure not greater than 2×10^{-8} torr can be maintained in the chamber after isolation.

The exit aperture and that of the skimmer are 0.079 cm in diameter. No accelerating voltage need be applied to the ions until, after passing through the skimmer, they enter the analysis region which is at a pressure less than 10^{-6} torr. This greatly reduces the chance that any weakly-bound ions will be broken up in comparatively energetic collisions with gas molecules.

IONIC REACTIONS: LOW AND INTERMEDIATE ENERGIES 525

Results obtained in experiments of this kind are discussed on pp. 541 and 548.

Fig. 12.4 illustrates a second type of combined drift tube and mass spectrometer first introduced by Moruzzi aud Phelps (1966) particularly for the study of reactions involving negative ions.

The drift tube consists of an electron source and an anode, 2.35 cm apart, between which a uniform electric field may be

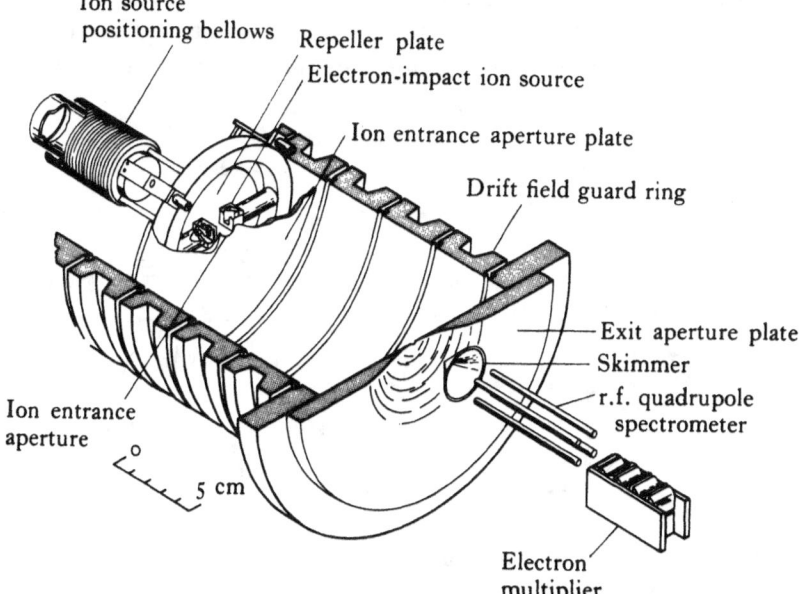

Fig. 12.3. Arrangement of the ion drift tube used by McDaniel and his collaborators (1962) to measure mobilities and reaction rates of positive and negative ions in gases.

applied. Ions and electrons passing through an aperture of 0.01 cm diameter in the anode enter the mass spectrometer chamber. They are then accelerated and focused into the r.f. mass spectrometer by means of a cylindrical lens. Ions of the selected mass pass through the spectrometer and are further accelerated to the dynode of an electron multiplier. About 1 in 10^6 of the ions reaching the anode eventually strike the dynode. The mass spectrum is scanned by sweeping the radiofrequency of the spectrometer.

Primary ions are formed in the drift tube by electron impact with the gas molecules. A photoelectric source was found to be most convenient for the electrons. For this a thin layer of gold or platinum was deposited on a quartz window and illuminated on the rear surface by an ultraviolet lamp operated at 10–20 W. The great advantage of this source is its lateral extent. It can be assumed to provide a uniform current over an area large compared with that over which the current from a point source would be spread by lateral diffusion in travelling to the anode. Under these circumstances lateral diffusion may be neglected in analysing the data. As longitudinal diffusion leads only to minor modifications in reaction rates

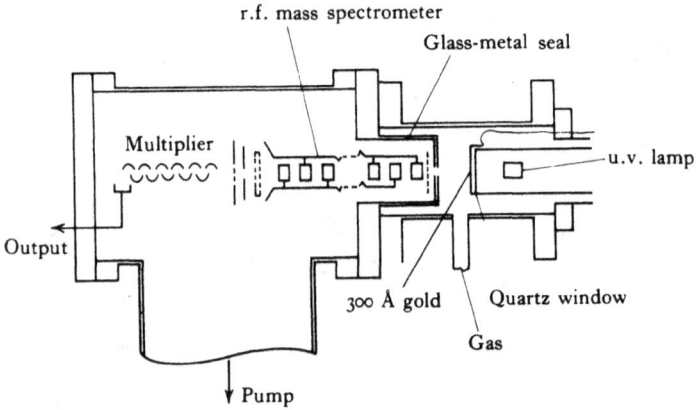

Fig. 12.4. Arrangement of the drift tube and mass spectrometer used by Moruzzi and Phelps (1966).

determined essentially from observations of the variation of different ion currents with pressure p at fixed F/p, it may also be ignored in the analysis.

Applications of this equipment are described on pp. 540 and 549.

The flowing afterglow method

This is a very versatile method for measuring the rates of ionic reactions of all kinds. Essentially the method depends on the production of the ions under study at some point in a tube through which a buffer gas is pumped at a flow rate of order 10^4 cm s^{-1}.

Once produced, the ions are carried downstream by the buffer gas past a nozzle through which the neutral atoms or molecules, whose reactions with the ions are under study, are injected into the stream at a measured rate. The ions reaching the end of the tube are sampled through a fine aperture by a quadrupole mass spectrometer, the buffer gas stream being pumped out at the side. The reaction rate is determined from the rate of variation of the ion current passing through the mass spectrometer with the flow rate of injected reactant molecules. Thus the ion current will vary as $\exp(-\alpha n)$, where n is the concentration of reactant molecules and $\alpha = kl/v$, where l is the distance from the nozzle to the exit aperture, v is the flow velocity of the buffer gas and k is the required rate constant for the reaction.

With flow rates of 10^4 cm s^{-1} and tube lengths of some tens of centimetres, the reaction time l/v is some ms which is suitable for gas pressures around 0.25 torr.

One of the advantages of the method is that it is possible to study reactions of ions with radicals such as N or O atoms. For example, N atoms may be produced in a subsidiary flow tube by subjecting N_2 to a microwave discharge and then introducing it through a nozzle in the usual way. The flow rate of N atoms may be measured by a titration method using the reaction

$$N + NO \longrightarrow N_2 + O. \tag{12.31}$$

Completion of this reaction may be observed visually so, at completion, the NO flow rate then gives a measure of that of N.

The reaction (12.31) may also be used to provide a supply of O atoms with flow rate the same as that of the NO.

Checks have been made to determine whether any important error is introduced in the injection system due to the high-speed flow of the buffer gas. Results obtained using different inlet nozzles with apertures ranging from pinhole size to 1 cm diameter have agreed within 30%. Errors from this source may be further reduced from theoretical estimates of the importance of diffusion and mixing effects.

The determination of the effective flow velocity for use in the determination of the reaction rate coefficient presents some difficulties. Experiments, in which plasma was produced upstream and its passage down the tube observed, gave values for the transport velocity of the plasma which were greater than those for the flow velocity determined from absolute flow rate and density measurements. This reflects the fact that the velocity distribution across a section of the flow tube is parabolic. The afterglow plasma tends to radiate most strongly from the region near the axis so its measured transport velocity gives the axial flow rate which exceeds the mean rate. As the ions are sampled along the axis of the tube it is the former rate which is appropriate.

In many cases the ions will react with the neutral molecules in more than one way and it will be important at least to determine which of these possibilities is the most probable, even if accurate information on the relative rates cannot be obtained. Useful information to this end may be obtained by relating the rate of decrease of the mass-analysed current of the reacting ions with the rate of growth of that of the various product ions.

Some ingenuity is necessary in order to introduce the desired negative ions into the flow tube. For O^- there is no difficulty. Small fractional concentrations of O_2 or CO_2 may be added to the buffer gas and the mixture subjected at some point to a pulsed discharge. O_2^-, on the other hand, is conveniently produced through the exothermic charge transfer reaction

$$NO^- + O_2 \longrightarrow NO + O_2^-, \qquad (12.32)$$

the NO^- ions being produced by introducing NO_2 into the buffer gas upstream of the discharge. Other examples are given in connexion with the discussion of results obtained, on pp. 547 *et seq.*

Experiments in static afterglows

We have already described in Chapter 9, p. 299, how attachment rates for electrons in static afterglows in various gases have been measured. Before the introduction of the flowing afterglow technique by Ferguson, Fehsenfeld, Dunkin, Schmeltekopf and Schiff, in 1964, a number of attempts had been made to determine the rates of ionic reactions from observations of the rate of decay, with

IONIC REACTIONS: LOW AND INTERMEDIATE ENERGIES 529

time in the afterglow, of currents of different ionic species, sampled through a small orifice in the side of the afterglow cavity and analysed by high time-resolution mass spectrometry. Such a procedure has the disadvantage, overcome by use of the flowing afterglow technique, that the components of the afterglow plasma remain mixed throughout the measurements and it is hard to be sure that allowance is made, in the analysis of the data, for all the possible reactions which may occur. For the study of negative-ion reactions the problem is more complicated because, as described in Chapter 9, p. 306, in the early stages of the afterglow the negative ions are largely prevented from reaching the cavity walls by the repulsive potential set up through electron-controlled ambipolar diffusion. At a later stage, if the electron concentration has been reduced sufficiently through attachment, a transition occurs to a new regime in which the repulsive potential breaks down and negative ions diffuse freely to the walls and hence may be sampled by a mass spectrometer.

Puckett and Lineberger (1970), in the experiments referred to on p. 307, have attempted to determine rate constants for negative-ion reactions produced in the afterglow resulting from photoionization of mixtures of NO and H_2O in which they concentrated attention only on the later regime, dominated by negative ions and not electrons. In pure NO, as described in Chapter 9, p. 307, the main negative ion is NO_2^-. The presence of water vapour leads to hydration of these ions. Assuming this to occur through the three-body process

$$NO_2^- + H_2O + NO \longrightarrow NO_2^-.H_2O + NO, \qquad (12.33)$$

the rate of decay of the NO_2^- concentration with time is given by $e^{-\nu t}$, where

$$\nu = \nu_D + k[NO][H_2O], \qquad (12.34)$$

ν_D being the contribution from diffusion loss and k the rate coefficient for the reaction (12.33). ν_D will not be appreciably affected by the addition of small fractional concentrations of water vapour and so may be determined from observations in pure NO. Results obtained in this way for the rate coefficient k are discussed on p. 560.

Extension of the pulse method

In Chapter 9, p. 293, we described the pulse method introduced by Chanin, Phelps and Biondi for the study of the rate of attachment of electrons in different gases. This method may be extended to determine rates of detachment, when the conditions are such that there is an appreciable chance that, after attachment, the electron will be detached again in a further collision before reaching the collector electrode (Phelps and Pack, 1961).

The general arrangement of the electrodes for this purpose is the same as shown in Fig. 9.14. Electron pulses are produced by focusing pulses of ultraviolet light on the cathode. The current to the anode can be measured, as a function of time from initiation of the electron pulse, by means of a control grid. By applying a short voltage pulse at a chosen time to the grid, electrons and negative ions are collected by it, so reducing the current to the anode plate. Application of the grid pulse at different times then enables the time variation of the total current of electrons and negative ions to be obtained. If a pulse of high frequency (~ 1 MHz) voltage is applied to the grid, negative ions are unable to follow it and are unaffected. This makes it possible to determine the electron component of the anode current alone.

The transit time t_e for electrons which remain free throughout their passage is much shorter than that t_i for negative ions. Occurrence of detachment will show up in the arrival of electrons at the anode at times $> t_e$ but $< t_i$ from the initiation of the electron pulse.

If n_e, n_i are the respective concentrations of electrons and of negative ions in the drift region at a time t after initiation of the pulse, and at a section which is at a distance x from the source, then we have

$$\frac{\partial n_e}{\partial t} + u_e \frac{\partial n_e}{\partial x} = -\nu_a n_e + \nu_d n_i, \tag{12.35}$$

$$\frac{\partial n_i}{\partial t} + u_i \frac{\partial n_i}{\partial x} = \nu_a n_e - \nu_d n_i, \tag{12.36}$$

where u_e and u_i are the respective drift velocities of the electrons and negative ions, ν_a is the attachment frequency per electron and ν_d the detachment frequency per negative ion.

If Q_e is the charge emitted in the electron pulse per unit area of cathode and it is assumed that the pulse duration is infinitesimal, solution of the equations (12.35), (12.36), consistent with the initial conditions, gives for the electron current density $\mathcal{J}_e$ received at time t on the grid,

$$\mathcal{J}_e/Q_e = \delta(t - t_e)\exp(-a) + \exp(-a)\,z(t - t_e)^{-1}\exp\{\tau(a - d)\}\,I_1(2z). \tag{12.37}$$

t_e is the transit time L/u_e for a free electron, L being the total drift path. τ and z are given by

$$\tau = (t - t_e)/(t_i - t_e), \qquad z = \{ad(1 - \tau)\tau\}^{1/2}, \tag{12.38}$$

with t_i the transit time L/u_i for negative ions, $a = \nu_a L/u_e$ the probability that an electron will form a negative ion during its transit time, while $d = \nu_d L/u_i$ is the probability that a negative ion will lose its electron during its transit time. $I_1(2z)$ is the usual Bessel function and δ the usual δ-function normalized so $\int \delta(t)\,dt = 1$.

To derive convenient methods for obtaining information about ν_d as well as ν_a, it is convenient to consider two limiting cases, one at low pressures in which the effect of detachment is small and the other at high pressures in which electrons and negative ions will take part in many attachment and detachment reactions along the drift path.

Considering first the case of low pressures we may take d as small. At times $t \ll t_i$ so $\tau \ll 1$, we then find

$$\mathcal{J}_e/Q_e \simeq \delta(t - t_e)e^{-a} + (ad/t_i)e^{-a}. \tag{12.39}$$

In practice the initial pulse will have a finite duration Δt. We then find, if $q_e = Q_e/\Delta t$,

$$\mathcal{J}_e/q_e = e^{-a} + \{ad(t - t_e)/t_i\}e^{-a}, \quad t_e < t < t_e + \Delta t, \tag{12.40}$$

$$\mathcal{J}_e/q_e = (ad\Delta t/t_i)e^{-a}, \qquad\qquad t > t_e + \Delta t. \tag{12.41}$$

In (12.40), e^{-a} represents the constant contribution $\mathcal{J}_0/q_e$ from electrons which have passed across the drift space without attaching, while the second term, increasing linearly with t, is due to electrons which have detached after first suffering attachment. Such electrons will appear if they suffer detachment in a time $t - t_e$, the probability of which is $d(t - t_e)/t_i$. Similarly, in (12.41), the only contribution, now constant, is from electrons arising from detachment which must have occurred in the time Δt. If we denote this by $\mathcal{J}_d/q_0$, then we have

$$\mathcal{J}_d/\mathcal{J}_0 = ad\Delta t/t_1,$$
$$= (\nu_a \nu_d/u_e) L\Delta t. \qquad (12.42)$$

Hence, as $\mathcal{J}_d$ and $\mathcal{J}_0$ may be measured, $\nu_a \nu_d/u_e$ may be obtained.

To determine ν_a/u_e and hence ν_d under the same conditions, use is made of the relation between the current $\mathcal{J}_{es}$, collected by the grid when all electrons which cross the drift tube without attachment are collected by it, and the total current $\mathcal{J}_t$ of electrons and negative ions, namely

$$\mathcal{J}_{es} = \mathcal{J}_t \exp(-\nu_a L/u_e). \qquad (12.43)$$

$\mathcal{J}_{es}$ may be measured by increasing the voltage of the r.f. pulse until the electron current collected by the grid is saturated, while $\mathcal{J}_t$ is the current received by the collector plate when no voltages are applied to the grid. In Chapter 9, p. 294, it was described how ν_a/u_e may be obtained from the shape of the total current pulse by a procedure valid in the absence of detachment. It was verified that under these conditions the relation (12.43) gave results for ν_a/u_e which agreed within experimental error.

Fig. 12.5 shows the variation with time of the current collected under typical experimental conditions, calculated according to the full formula (12.37) for conditions applying to the experiments of Pack and Phelps (1966a, 1966b) in O_2 at low pressures.

A striking example of the change of shape of the collected electron current pulse through detachment is shown in Fig. 12.16, which compares the current observed in pure O_2 at 5 torr pressure, which shows no contribution from delayed electrons, and that observed when a pressure of 0.001 torr of H_2 is added. The delayed electrons in the latter case arise from the exothermic associative detachment reaction

$$H_2 + O^- \longrightarrow H_2O + e.$$

Further details of this and the applications of the low-pressure technique are discussed on pp. 549–54.

Turning now to the analysis of the data at high pressures, in which attachment and detachment occur frequently during the passage from cathode to grid, we note that the ratio n_e/n_i of the concentrations of negative ions and electrons will remain unchanged throughout the drift space. The mixed swarm of ions and electrons will then

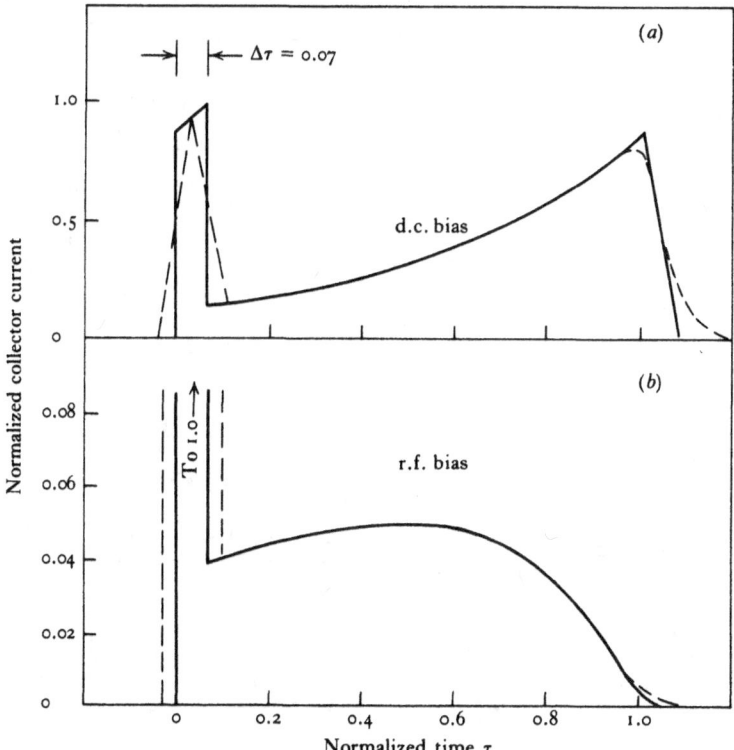

Fig. 12.5. Collector current as a function of normalized time τ calculated using the formula (12.37) applied to conditions in the pulse experiments of Pack and Phelps (1966a, b). Values assumed for the quantities involved are $a = 2$, $d = 0.2$, an electron emission pulse width of 0.07, $F/N = 3.1 \times 10^{-17}$ V cm^2, $N = 10^{18}$ cm^{-3}, temperature $T = 410$ °K, (a) with d.c. voltage applied to the grid so that both electrons and negative ions are collected, (b) with r.f. voltage applied to the grid so that only electrons are collected.

drift with a velocity given by

$$u = (u_e n_e + u_i n_i)/(n_e + n_i). \tag{12.44}$$

Under typical experimental conditions $u < 10\, u_i$ so, since $u_e \gg u_i$, $n_e \ll n_i$ and hence

$$u = u_i + u_e n_e/n_i. \tag{12.45}$$

Also, because of the dynamical equilibrium between attachment and detachment,

$$n_i/n_e = \nu_e/\nu_d. \tag{12.46}$$

If both attachment and detachment occur in two-body collisions, ν_a/ν_d will be independent of gas pressure p for fixed F/p. If, however, as for O_2, attachment arises from a three-body process, ν_a/ν_d will be proportional to p. Fig. 12.6 shows observed time variations of

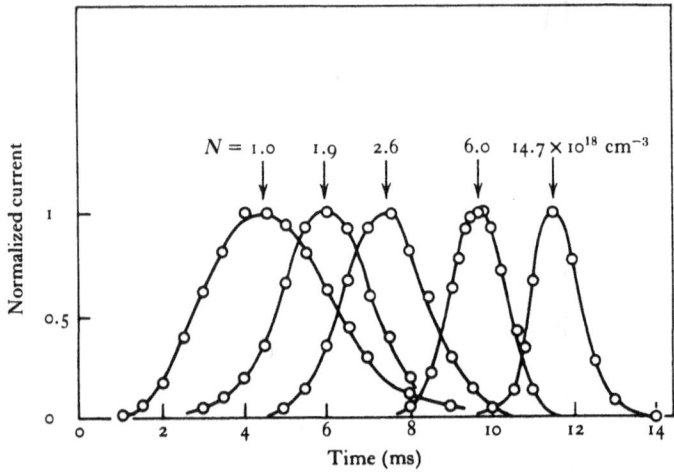

Fig. 12.6. Observed change in collector current with time in the pulse experiments of Pack and Phelps (1966a, b) in O_2 at high pressures, $T = 477\ °K$, $F/N = 3.1 \times 10^{-17}\ V\ cm^2$ and different O_2 concentrations N as indicated, showing the apparent increase of mean mobility with time as N increases.

collector current pulses, for different high gas pressures in O_2. It will be seen that the mean drift velocity decreases with the gas pressure in agreement with (12.45) and (12.46) for ν_a proportional to p^2 and ν_d to p.

Further discussion of the application of this technique to ions in O_2 is given on p. 544.

Experiments with low-energy ion beams

Mauer and Schulz (1973) have measured relative total cross-sections for, and the energy distributions of the electrons produced in, associative detachment collisions of O^- ions with CO, H_2 and O_2 molecules, using a beam method in which the ion energy ranged from 0.1 to 10 eV.

Fig. 12.7 illustrates the experimental arrangement used. The whole apparatus is bathed in a uniform magnetic field of about 100 G in the plane of the paper in the direction shown. O^- ions

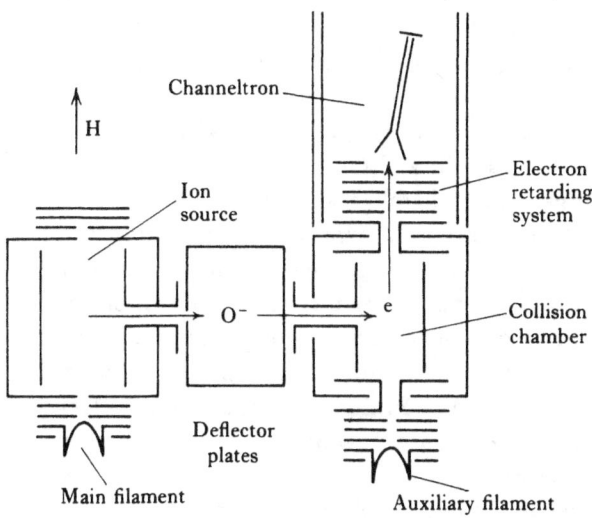

Fig. 12.7. Arrangement of the apparatus used by Mauer and Schulz (1973) for the measurement of the rates of associative detachment reactions.

produced from a suitable source are collimated and passed into a collision chamber containing the reactant gas. Electrons produced by detachment travel along the direction of the magnetic field through an electrode system with a channeltron electron multiplier.

The O^- ions are produced by dissociative attachment in O_2 at a pressure of about 10^{-2} torr due to an electron beam of around 10 μA. They are extracted through a channel attached to a plate. A second parallel plate near the other wall of the source chamber is maintained at a negative potential so as to repel negative ions towards the channel. Secondary ions such as O_2^- produced by charge transfer are prevented from leaving the source by the magnetic field. In any case, it is verified that the O^- current from the source varies linearly with the gas pressure. The ion energy is determined by the potential difference between the collision chamber and electron-retarding system, considered as a unit, with respect to the ion source and deflector plates. At the higher energies used ($\simeq 10$ eV) the ion current is about 2×10^{-12} A falling to 10^{-14} A at the lowest energies, about three times larger than the fluctuations associated with the beam and its measurement.

On leaving the source, the ions enter a highly evacuated region containing deflector plates which collimate the beam and improve its energy resolution. The beam then passes into the field-free collision chamber and is collected on a suitable electrode. The pressure of reactant gas in the chamber is about 10^{-2} torr.

Detached electrons moving along the magnetic field pass out via an exit channel through the 0.5 mm apertures in a guard plate and a system of retarding electrodes. Before entering the multiplier they pass through two further guard plates which screen off the high potential of the multiplier. The multiplier gain is about 1000 and the maximum counting rate about 10 s^{-1}. Background noise gives 0.1–0.3 counts s^{-1} which is comparable with the lowest signal observed.

The relatively large background signal at low ion energy arises not only from noise independent of the ion beam but from electrons produced in detachment collisions of the O^- ions with residual gas. As the latter will be dependent on ion energy it is necessary to determine the total background at every energy when measuring relative total detachment cross-sections. This is done by extrapolating the observed signals at different gas pressures to zero pressure, leaving a residual due to background.

The energy distribution of the detached electrons is measured by making a retarding potential analysis, sweeping the potential on the retarding plates. Although the distributions so obtained are distorted by the magnetic field the maxima should correspond closely to the real energies of the electrons.

The data are averaged by long-term accumulation using a PDP/8 computer. Results obtained are discussed on pp. 546 and 552.

12.6 The determination of threshold energies for endothermic charge transfer reactions

If the threshold energy for an endothermic reaction

$$X^- + Y \longrightarrow X + Y^-$$

can be measured, the excess of the electron affinity $E_a(X)$ of X over that $E_a(Y)$ of Y may be obtained. Just as in a corresponding case (see Chapter 10, p. 390) of the threshold for

$$X + Y \longrightarrow X^+ + Y^-,$$

problems arise in practice if Y is a molecule because the probability of a reaction producing Y^- in its ground state of internal motion may be very small. However, if a fairly sharply-defined threshold is

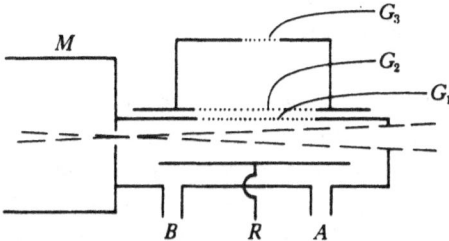

Fig. 12.8. General arrangement of the apparatus used by Chupka et al. (1971) for studying endothermic charge transfer reactions of halogen atomic ions X^- with gas molecules.

discerned, there is a high probability that it is the true threshold given by $E_a(X) - E_a(Y)$. Conversely, if the rate of the reaction as a function of relative energy exhibits a gradual tailing off at low energies, there is little hope of obtaining reliable information about the adiabatic electron affinity $E_a(Y)$.

Chupka, Berkowitz and Gutman (1971) have carried out a number of experiments in which the ions X^- are atomic halogen ions produced by polar photo-dissociation of halogen molecules (see Chapter 8, p. 255). As a result they have obtained quite good results for the electron affinities of a number of molecules Y, including the halogen molecules, O_2, O_3 and NO. For NO_2 less satisfactory results were obtained.

Fig. 12.8 illustrates the general arrangement of the apparatus used. Vacuum ultraviolet radiation forming the many-line continuum of H_2 is dispersed by a monochromator and enters a ionization chamber containing the chosen halogen vapour X_2 as well as gas molecules Y, at a total pressure between 10^{-4} and 10^{-3} torr. X^+ and X^- ions are formed selectively by polar photodissociation using radiation of the appropriate frequency. The X^- ions are repelled towards the grid G_1, maintained at a potential between 0.1 and 0.2 V above that of the repeller electrode R. After passage through G_1, the ions are accelerated to the desired kinetic energy by a potential applied between G_1 and G_2. They then enter the field-free region between grids G_2 and G_3 which constitutes the main reaction region. Primary X^- and secondary Y^- ions which pass through G_3 are accelerated and focused into a mass spectrometer. With slits of width 300 μm in the optical system and a resolution of 2.5 Å, typical intensities of X^- ions are 10^6 s^{-1} and of the product Y^- ions around 30 s^{-1} at energies about 1 eV above threshold.

Measurements are made first of the intensity of X^- ions as a function of the accelerating voltage between G_1 and G_2 and then, after retuning the mass spectrometer, of the product Y^- ions. The data analysed are the ratios of the Y^- to X^- intensity at different values of the centre of mass energy, the zero of the energy scale being taken as the energy at which the intensity in the steepest part of the initial rise is halfway to the maximum. Fig. 12.9 illustrates observed results for the reaction

$$I^- + Cl_2 \longrightarrow Cl_2^- + I$$

which are typical of many of the other cases investigated.

The problem of determining the threshold energy from data such as these is still quite difficult. Chupka *et al.* follow a similar procedure to that discussed in Chapter 10, p. 390, in connection with the corresponding problem for neutral–neutral collisions. Different relations are assumed for the energy variation of the reaction cross-section close to the threshold and averaged over the energy distribution arising from the thermal motion of the target molecules (see Chapter 9, p. 276). Comparison with the observed results enables a choice to be made among the assumed threshold laws and hence provides a value for the threshold energy.

For experiments of this kind, designed to determine threshold energies, I_2 is the most convenient halogen to use as a source of the reactant ions. Thus polar photodissociation occurs with considerable intensity in the wavelength range within the hydrogen continuum (see Chapter 8, p. 255) and its threshold is below that for ionization so it is possible to work under conditions in which no electrons are produced. Also, as the electron affinity of I is the

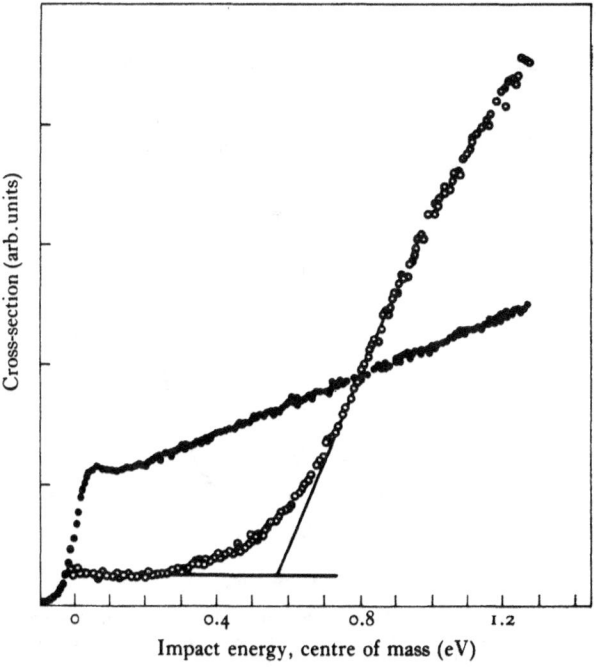

Fig. 12.9. Variation with impact energy near the threshold of the yield of Cl_2^- ions in the experiments of Chupka et al. (1971). ○ observed current of Cl_2^- ions, ● current of primary I^- ions.

lowest of all the halogen atoms, the threshold energy for an endothermic reaction with a given neutral molecule is lower. This leads to steeper thresholds and higher intensity. While the large mass of I atoms is a favourable factor in converting to centre of mass coordinates it is a disadvantage in that the thermal spread is greater.

Discussion of results obtained with this technique is given on p. 566.

12.7 Discussion of observed results

Mobilities and reactions of oxygen ions in oxygen

Fig. 12.10 shows results obtained by Moruzzi and Phelps (1966), using the apparatus shown in Fig. 12.4, for the relative intensity of different negative ions in oxygen as functions of the ratio F/p of electric field to gas pressure. It is clear from these data why, without

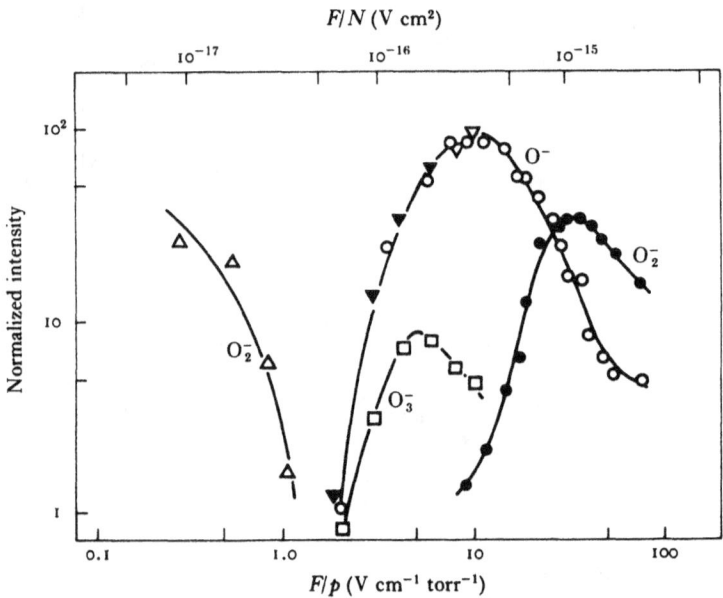

Fig. 12.10. Relative intensity of different negative ions in O_2 as a function of F/p, as observed by Moruzzi and Phelps (1966). ●, ○, $p = 1.76$ torr; □, △, ▼, $p = 3.37$ torr.

means of identification of the different ions, the early experiments on the mobilities of ions in oxygen gave somewhat confusing results. Nevertheless, three different ions, referred to as of types A, B and C were distinguished. Type C, observed at low F/p can be identified as O_2^-, type A, relatively abundant at low pressures and low to intermediate F/p as O^-, and type B, mainly observed at high pressures and intermediate F/p as O_3^-.

In these experiments Moruzzi and Phelps failed to observe O_4^- ions but Conway and Nesbitt (1968) (see Chapter 6, p. 220), as

IONIC REACTIONS: LOW AND INTERMEDIATE ENERGIES 541

described on p. 220, not only observed the ions but were able to determine their relative abundance as a function of temperature. Fig. 12.11 shows a typical mass spectrum which they obtained from O_2 passed over a tritium source of 2 Ci strength. It shows peaks due to O_2^-, O_3^-, $O_2^-.H_2O$, O_4^- and $^{16}O_3{}^{18}O^-$ ions. In addition, to assist in identification, a trace of D_2O was added giving rise to further peaks for $HOD.O_2^-$ and $DOD.O_2^-$. From data such as these they determined the equilibrium constant for the reactions

$$O_2^- + O_2 \rightleftharpoons O_4^- \qquad (12.47)$$

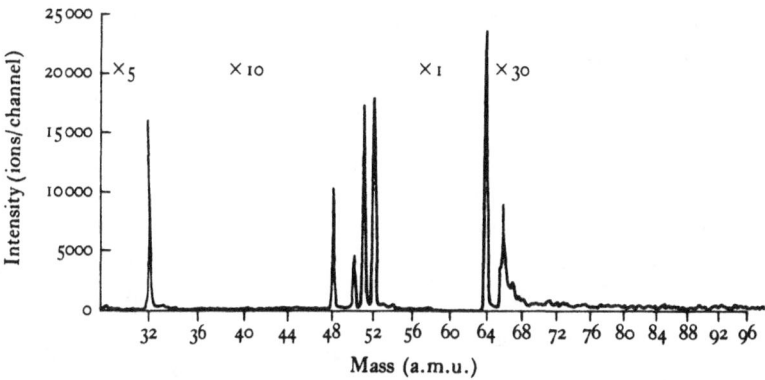

Fig. 12.11. Negative-ion mass spectrum, observed by Conway and Nesbitt (1968) for ions in oxygen. The relative sensitivity is shown to the left of each mass range.

as a function of temperature over the range 273.2–352.8 °K, at three different pressures (6, 9 and 12 torr). They were then able to determine the enthalpy change associated with the reaction as -13.55 ± 0.16 kcal mol^{-1} (0.588 eV) consistent with the theoretical considerations referred to on p. 220.

Since then the mobility of O_4^- has been measured in drift tube experiments while the rates of various reactions which lead to its formation and transformation have been determined by drift tube and flowing afterglow techniques.

The zero field mobilities of O^-, O_2^-, O_3^- and O_4^- ions in O_2 have been measured by Snuggs, Volz, Schummers, Martin and McDaniel (1971) using the apparatus described on p. 524. They find, for the respective ions, mobilities in cm^2 V^{-1} s^{-1} of 3.20 ± 0.09,

2.16 ± 0.07, 2.55 ± 0.08 and 2.14 ± 0.08 respectively. The relatively low mobility of O_2^- is due to the possibility of resonant charge transfer

$$O_2^- + O_2 \longrightarrow O_2 + O_2^-, \quad (12.48)$$

which increases the probability of large-angle scattering of the ions and hence reduces the mobility. In fact, O_4^- has a mobility very close to that of O_2^- and this was certainly a source of confusion in early experiments which lacked the means for mass analysis of the ions. Except for O^-, the results obtained by Snuggs et al. agree quite well with those of McKnight (1970) also obtained using a combination of drift tube and mass spectrometer. In both experiments measurements have been extended up to values of F/p as high as 50 V cm^{-1} torr^{-1} for O^- and O_2^- and 10 V cm^{-1} torr^{-1} for O_3^-. The agreement is good for O_2^- over the entire range of F/p but less satisfactory for O^- and O_3^-.

The rates of the three-body associative reactions which lead to the production of O_3^- ions from O^- and of O_4^- from O_2^- have been measured by the drift tube plus mass spectrometer technique. Up to values of F/N of 3×10^{-16} V cm^2 the rate coefficient for

$$O^- + O_2 + O_2 \longrightarrow O_3^- + O_2, \quad (12.49)$$

at 300 °K appears to be nearly constant at 1.0 and 0.9 × 10^{-30} cm^6 s^{-1} according to Snuggs et al. (1971) and to McKnight (1970) respectively. For the reaction

$$O_2^- + O_2 + O_2 \longrightarrow O_4^- + O_2 \quad (12.50)$$

McKnight and Sawina (1971) find rate coefficients for the forward and reverse reactions of 3×10^{-31} cm^6 s^{-1} and 2×10^{-14} cm^3 s^{-1} respectively at 300 °K and $F/N = 2 \times 10^{-16}$ V cm^2.

We list in Table 12.1 the rates of various reactions of O^-, O_2^-, O_3^- and O_4^- ions with neutral oxygen allotropes which have been observed under thermal or near-thermal conditions. The table includes values of the energy released or absorbed in the reaction, when this is known, as well as the rate constant for orbiting collisions when this is relevant. We now discuss any special features associated with these reactions, the methods used to measure their rates and any additional information available about them from theoretical or other methods.

TABLE 12.1   *Rate constants for reactions of oxygen negative ions in oxygen at* 300 °K

Reaction	Energy release (eV)	Rate constant	
		Obs.	Calc. from orbiting/radius
Three-body association		10^{-30} cm^6 s^{-1}	
(1) $O^- + O_2 + O_2 \to O_3^- + O_2$		1.0, 0.8	
(2) $O_2^- + O_2 + O_2 \to O_4^- + O_2$		0.3	
Direct detachment		10^{-10} cm^3 s^{-1}	
(3) $O^- + O_2 \to O + O_2 + e$	-1.46_5		
(4) $O_2^- + O_2 \to O_2 + O_2 + e$	-0.44		
(5) $O_2^- + O_2(^1\Delta_g) \to O_2 + O_2 + e$	0.6	2	
Associative detachment		10^{-10} cm^3 s^{-1}	
(6) $O^- + O \to O_2 + e$	3.6	1.9	7.5
(7) $O^- + O_2 \to O_3 + e$	-0.4	0.01	
(8) $O^- + O_2\,(^1\Delta_g) \to O_3 + e$	0.6	3	
(9) $O_2^- + O \to O_3 + e$	0.6	3.3	6
Charge transfer		10^{-10} cm^3 s^{-1}	
(10) $O^- + O_2 \to O + O_2^-$	-1.025	3×10^{-5}	
(11) $O^- + O_3 \to O + O_3^-$	0.5	5.3	
(12) $O_2^- + O_2 \to O_2 + O_2^-$	0		
(13) $O_2^- + O_3 \to O_2 + O_3^-$	1.5	4.0	
Rearrangement		10^{-10} cm^3 s^{-1}	
(14a) $O_4^- + O \to O_3^- + O_2$		} 4.0	7
(14b) $O_4^- + O \to O^- + 2O_2$			

For references see text.

Of the direct detachment reactions, the first two listed in the table occur through transfer of kinetic energy from relative motion of the ion and O_2 molecule to the attached electron. Detachment from O_2^- was investigated by Pack and Phelps (1966a, b), using the pulse technique described on pp. 530–4. They worked at low values of F/N for which O_2^- is the dominant ion (see Fig. 12.10). At low pressures they found that the quantity $\nu_a \nu_d / u_e$, derived from their observations as described on p. 532, varies, as expected, as N^3. Thus in Fig. 12.12 typical plots of $\nu_a \nu_d / u_e$ against N^3 are shown, for different gas temperatures. As explained on p. 532, ν_a / u_e may be separately measured so that ν_d / N, the detachment rate constant, may be derived. Fig. 12.13 illustrates typical results showing how the rate constant varies with temperature for $F/N = 3.1 \times 10^{-17}$ V cm^2.

At high pressures in the same range of F/p, the observed results for the mean drift velocity were found to be consistent with the relation (12.45) so that, from known values of the electron drift

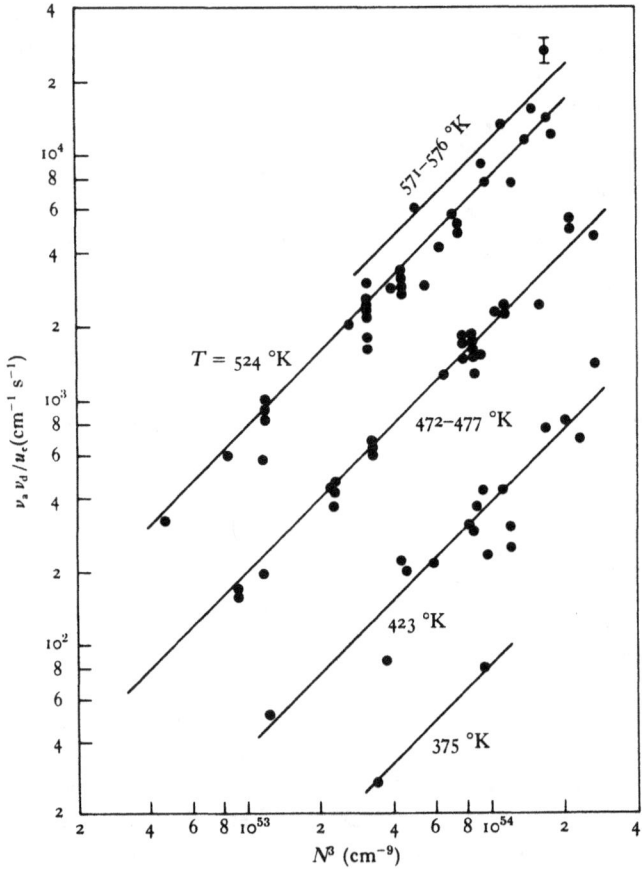

Fig. 12.12. Variation of $\nu_a \nu_d/u_e$ with N^3 (see p. 532), observed in pulse experiments in O_2 by Pack and Phelps (1966a, b), for $F/N = 3.1 \times 10^{-17}$ V cm^2 and different gas temperatures, as indicated.

velocities u_e, the ratio n_e/n_i of the electron to negative-ion concentrations could be obtained. Since, at these pressures, this ratio is determined from the balance between the forward and reverse reactions

$$2O_2 + e \rightleftharpoons O_2^- + O_2, \qquad (12.51)$$

the equilibrium constant $K = n_e N/n_i$ for these reactions can be obtained as a function of gas temperature over the experimental range. Using the statistical formula for K in terms of the partition

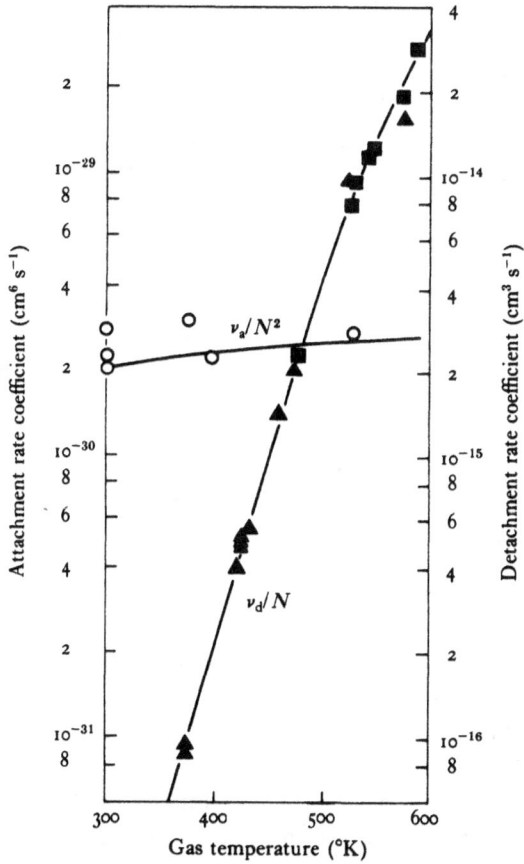

Fig. 12.13. Rate coefficients v_d/N, v_a/N^2 for detachment and attachment of electrons in O_2, for $F/N = 3.1 \times 10^{-17}$ V cm^2, measured as a function of temperature by Pack and Phelps (1966a, b). ▲ v_d/N at low pressures, ■ v_d/N at high pressures, ○ v_a/N^2.

functions for the ions, electrons and neutral molecules, the electron affinity of O_2 can be derived, provided the molecular constants of the ion are known. At the time there was no direct information about the relevant constants but Pack and Phelps (1966b) assumed that they are the same for O_2^- as for O_2 and obtained for the electron affinity of O_2

$$E_a(O_2) = 0.43 \pm 0.02 \text{ eV}.$$

This has proved to be the first accurate determination, agreeing within experimental error with the precise photodetachment measurements (0.440 ± 0.008 eV) discussed on p. 478.

From the measured equilibrium constants, values for the detachment rate coefficient may be obtained from observed values for the corresponding attachment rate coefficient and are shown in

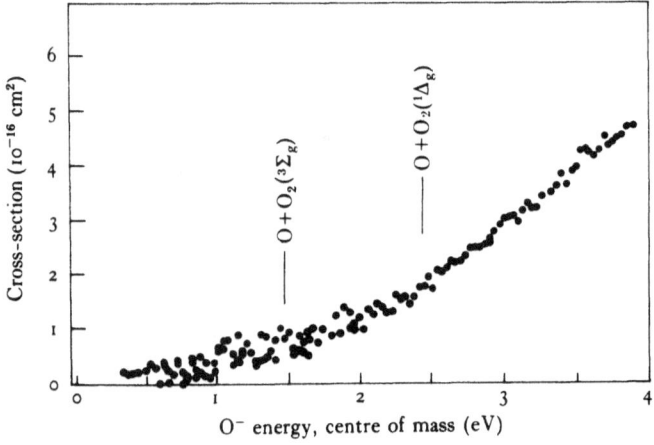

Fig. 12.14. Observed total cross-sections for detachment reactions of O^- in O_2, observed by Mauer and Schulz (1973).

Fig. 12.13. It will be seen that these values, obtained from measurements at high pressures, are consistent with measurements at low pressure.

Detachment from O^- in collisions with O_2 has also been studied by Mauer and Schulz (1973) using their beam technique (see p. 535). The observed cross-section for all detachment processes is shown as a function of the relative kinetic energy of the O^- ions in Fig. 12.14. The absolute values have been normalized by comparison with observed results for detachment in CO (see p. 552 and Fig. 12.18a). While it is not possible to distinguish contributions from associative detachment (reaction (7) of Table 12.1), which has a threshold at 0.4 eV, from direct detachment with a threshold at 1.46_5 eV, it seems likely that the major fraction arises from the former. Assuming this, the beam data are consistent with a linear extrapolation at zero at the threshold.

Mauer and Schulz (1973) measured the energy distribution of the electrons produced by detachment collisions at relative kinetic energies of 2.61 eV. Peaks were found at electron energies of 0.450, 0.286 and 0.179 eV, but their origin is obscure.

The reaction (5) of Table 12.1 is of particular interest because it is the first of its kind to be observed (Fehsenfeld, Albritton, Burt and Schiff, 1969), in which the electron is detached through transfer of internal excitation energy. In principle, the process is similar to that of Penning ionization of atoms and molecules by metastable helium and other rare gas atoms. The excitation energy available (1.0 eV) is more than adequate to detach the electron from O_2^-, the excess being carried off as kinetic energy of that electron. The rate measurement was made using the flowing afterglow technique. O_2^- and O^- ions were produced by flowing the buffer helium gas containing an admixture of O_2 through an electron beam fired from an electron gun. $O_2(^1\Delta_g)$ molecules were introduced downstream at a measurable flow rate and the variation of the O_2^- and O^- concentrations with this flow rate determined in the usual way. The excited molecules were produced by a microwave discharge in O_2 after which the products were passed through a glass-wool plug containing mercuric oxide to remove atomic oxygen. To measure the concentration of $O_2(^1\Delta_g)$ in the mixture with normal O_2, the intensity of the 1.27 μ emission from the excited molecules was observed with a PbS photometer. For this purpose the mixture was passed through a cell of length 20 cm and depth 6 cm.

The measured rate coefficient for the reaction (5) is quite high as would be expected for an exothermic reaction in which no transfer from kinetic energy of relative motion to electron energy is involved.

We turn now to the associative detachment reactions. Reaction (6) has been studied by flowing afterglow techniques (Fehsenfeld, Ferguson and Schmeltekopf, 1966). It is an exothermic reaction which is of particular interest in connexion with the behaviour of the E region of the ionosphere at night. The rate is so high that it prevents any build-up of negative ions in that region even when there is no loss by photodetachment (see Chapter 15, p. 663).

The endothermic reaction (7) (Fehsenfeld et al., 1966) has a low rate constant at ordinary temperature which has been estimated only very roughly from flowing afterglow experiments. On the other hand the reaction (8), the rate of which has been measured by the technique described above in connexion with reaction (5), is exothermic and the rate constant is quite high.

Of the charge transfer reactions (10)–(13), (12) is a symmetrical process in which no change of internal energy is involved if the product molecules are in their ground states. The rate constant for this process has not been measured at thermal energies but some evidence as to its magnitude can be obtained from the measured mobility of O_2^- ions in O_2.

For the reactions (11) and (13), the rate coefficients, measured by the flowing afterglow technique (Ferguson, 1967), are very high, showing that both reactions are exothermic. This requires that the electron affinity of O_3 exceed 1.46_5 eV, consistent with results from observations of charge transfer between I^- and O_3 (see p. 570) and of photodetachment from O_3^- (see p. 489).

On the other hand the reaction (10) is certainly endothermic. The rate coefficient has been measured over a range of F/N from 5.5×10^{-16} to 50×10^{-16} V cm^2 by drift tube techniques (McKnight, 1970; Snuggs et al., 1971). There is quite good agreement between the results obtained in different experiments, at least up to $F/N = 10^{-15}$ V cm^2. The rapid rise of the rate coefficient with F/N is exactly what would be expected for an endothermic reaction of this kind. According to the expression (12.25) the mean energy of O^- ions when $F/N = 6.3 \times 10^{-16}$ V cm^2 is only 0.3 eV, considerably below the threshold for the reaction (10). However, at the highest values of F/N studied, the mean energy of the ions will certainly exceed the threshold. Measurements of the cross-section for the reaction (10) at high energies carried out using beam techniques are discussed in Chapter 13, p. 603, but are not accurate enough to relate to the drift tube data.

Finally, in Table 12.1, we include the rearrangement collisions of O_4^- with O atoms which lead to two alternative sets of products. The rates of these reactions are important for the understanding

IONIC REACTIONS: LOW AND INTERMEDIATE ENERGIES 549

of the ionic composition of the lower ionosphere as O_4^- probably plays an important role in the reaction chain (see Chapter 15, p. 668). The flowing afterglow method (Fehsenfeld, Ferguson and Bohme, 1969) has been applied to measure the total rate coefficient for reactions (14) between O_4^- and O which lead to destruction of O_4^- and it is found to be high, approximately 60% of the orbiting cross-section. It was not possible to determine quantitatively the relative probabilities of the two alternative reaction paths but it was established that (14a) is the more probable.

Reactions of oxygen negative ions with atoms and molecules of other species

We next consider the information available about the rate constants for reaction of O^-, O_2^-, O_3^- and O_4^- with neutral atoms and molecules of other species.

In Table 12.2 we list, in a similar way to Table 12.1, the reactions studied, classified under the same four headings, as well as the energy released in each reaction, the observed reaction rates and the calculated rate for orbiting collisions. The choice of reactions studied has been largely determined by their relevance to the understanding of the composition of the lower ionosphere of the earth or of the ionospheres of Venus and Mars in which CO_2 is especially important.

Of the three-body association reactions, involving He as third body, the rates of which have been measured by the flowing afterglow technique (Adams, Bohme, Dunkin, Fehsenfeld and Ferguson, 1970), the reactions with CO_2 are very much faster than those with O_2 and N_2. Moruzzi and Phelps (1966), using their drift-tube technique described on p. 525, also found high rate coefficients for association of O_2^- and CO_2 both with CO_2 and with O_2 as third bodies. It seems in fact that the binding of O^- to CO_2 is higher than that to O_2. Evidence for this was obtained from experiments carried out by Pack and Phelps using the pulse technique described on p. 530.

Fig. 12.15 shows the observed variation of collected electron current with time from initiation of the pulse, observed in pure O_2 and in O_2 containing 0.23 per cent of CO_2, at low O_2 pressures and low F/N. In the former case the delayed electron current due to direct detachment from O_2^- is about 0.25 of the peak current. Addition of the small concentration of CO_2 reduced the delayed

TABLE 12.2 *Rate constants for reactions, at 300 °K of oxygen negative ions with atoms and molecules of other species*

Reaction	Energy release (eV)	Rate constant	
		Obs.	Calc. from orbiting/radius
Three-body association		10^{-30} cm^6 s^{-1}	
(1) $O_2^- + O_2 + He \rightarrow O_4^- + He$	–	0.34	
(2) $O^- + CO_2 + He \rightarrow CO_3^- + He$	–	260	
(3) $O^- + N_2 + He \rightarrow N_2O^- + He$	–	0.04	
(4) $O_2^- + CO_2 + He \rightarrow CO_4^- + He$	–	47	
(5) $O_2^- + CO_2 + CO_2 \rightarrow CO_4^- + CO_2$	–	9	
(6) $O_2^- + CO_2 + O_2 \rightarrow CO_4^- + O_2$	–	20	
(7) $O_2^- + N_2 + He \rightarrow N_2O_2^- + He$	–	0.04	
Associative detachment		10^{-10} cm^3 s^{-1}	
(8) $O^- + N \rightarrow NO + e$	5.1	2.2	7
(9) $O^- + N_2 \rightarrow N_2O + e$	0.15	0.1	
(10) $O^- + NO \rightarrow NO_2 + e$	1.6	5	10
(11) $O^- + H_2 \rightarrow H_2O + e$	3.5	15	16
(12) $O^- + CO \rightarrow CO_2 + e$	4.0	5	10
(13a) $O_2^- + N \rightarrow NO_2 + e$	4.1	5	8
(13b) $\rightarrow NO + O + e$	1.0	–	–
Charge transfer		10^{-10} cm^3 s^{-1}	
(14) $O^- + NO_2 \rightarrow NO_2^- + O$	0.8	12	
(15) $O_2^- + NO_2 \rightarrow NO_2^- + O_2$	1.85	8	
Rearrangement		10^{-10} cm^3 s^{-1}	
(16) $O^- + N_2O \rightarrow NO^- + NO$	0.14	2.2, 2.5	
(17) $O_2^- + N \rightarrow O^- + NO$	2.5	–	8
(18) $O_3^- + NO \rightarrow NO_3^- + O$		0.1	10
(19) $O_3^- + CO_2 \rightarrow CO_3^- + O_2$		0.4	
(20) $O_4^- + NO \rightarrow NO_3^- + O_2$		2.5	10
(21) $O_4^- + CO_2 \rightarrow CO_4^- + O_2$	0.2	4.3	
(22) $O_4^- + H_2O \rightarrow O_2^-.H_2O + O_2$		10	

For references see text.

current by a factor greater than 2. This suggests that in the presence of the CO_2 a more stable negative ion, X^- say, is produced. If this ion does not suffer detachment and an equilibrium is set up between O_2^- and X^- the ratio of the delayed currents shown as II and I in

IONIC REACTIONS: LOW AND INTERMEDIATE ENERGIES

Fig. 12.15 would be

$$n(O_2^-)/\{n(O_2^-) + n(X^-)\}, \qquad (12.52)$$

where $n(O_2^-)$, $n(X^-)$ are the concentrations of O_2^- and X^- respectively. It was found from observations at different partial pressures of CO_2, at a fixed temperature, that the ratio $n(X^-)/n(O_2^-)$ is pro-

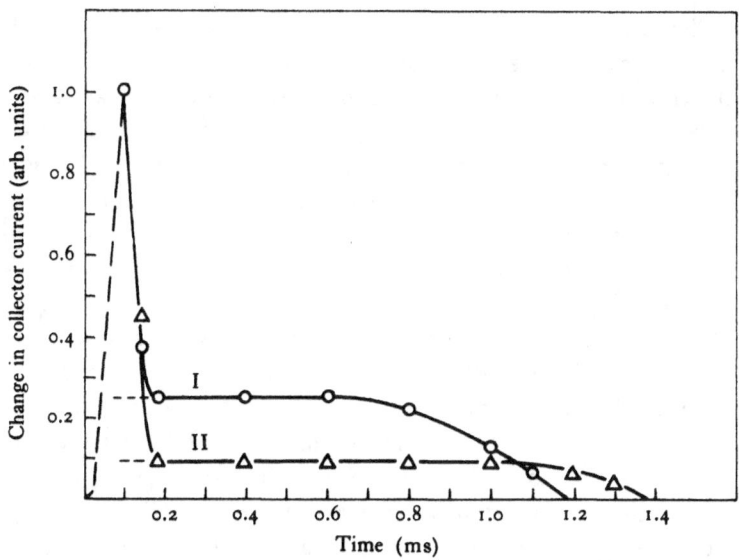

Fig. 12.15. Observed change of collector electron current with time in the pulse experiments of Pack and Phelps (1966b), at 461 °K and $F/N = 3.1 \times 10^{-17}$ V cm². I for pure O_2 at a concentration of 1.1×10^{18} cm^{-3}, II for O_2 containing 0.23 per cent of CO_2.

portional to $n(CO_2)$, suggesting that X^- is CO_4^-, the equilibrium being

$$CO_2 + O_2^- \rightleftharpoons CO_4^-. \qquad (12.53)$$

From observations at high pressures and low F/N it is found, from the same argument as outlined on p. 534, that

$$n_e/n(X^-) \propto 1/n(O_2)n(CO_2),$$

as would be the case if the new ion were CO_4^-, formed under these conditions through

$$e + O_2 + CO_2 \rightleftharpoons CO_4^- + \Delta E. \qquad (12.54)$$

Accepting this, the equilibrium constant for (12.54) is obtained from (12.52) and thence, from statistical mechanics assuming complete freedom of internal motion in CO_4^- but no vibrational activation of O_2 or CO_2, it is found that $\Delta E = 1.2 \pm 0.1$ eV. This would give the energy of dissociation of CO_4^- into O_2^- and CO_2 as 0.8 ± 0.1 eV, higher than that for O_4^- into O_2^- and O_2 (0.6 eV, see p. 541).

The associative detachment reactions (8) and (13) are highly exothermic and the rate coefficients measured by the flowing afterglow method are large, being in both cases substantial fractions of the orbiting rate coefficient. The reactions (10), (11) and (12) of O^- ions with NO, H_2 and CO have been studied not only by the flowing afterglow method but also by the pulse (p. 530) and drift-tube (p. 522) methods. Furthermore, the cross-sections for the detachment reactions with H_2 and CO have been measured as a function of ion energy and the energy distributions of the ejected electrons observed at selected ion energies by Mauer and Schulz (1973) using the beam technique described on p. 535. Fig. 12.16 shows the dramatic effect on the delay time of the pulse in O_2, under conditions in which O^- is the dominant ion, of the addition of relatively small concentrations of H_2, due to the production of delayed electrons by associative detachment.

Fig. 12.17 compares the results obtained for the rate coefficients as functions of F/N for the reactions with CO, H_2 and NO. A mean ion energy scale determined from Wannier's formula (12.25) is added. On the whole there is consistency with the thermal energy results obtained by the flowing afterglow method. Over the range of ion energies studied for CO and H_2 the rate constant varies little, so that the detachment cross-section would be expected to vary inversely as the ion velocity. This is confirmed by the beam measurements for both cases, as may be seen from Fig. 12.18. In each case the absolute values of the cross-sections have been normalized to agree with the flowing afterglow results at thermal energies. At somewhat higher energy the cross-section attains a minimum, after which it rises to a broad maximum at an ion energy near 2 eV in the laboratory system before falling gradually at higher energies. At low ion energies detachment can probably take place only via the lowest attractive states of the compound ion, CO_2^- or

IONIC REACTIONS: LOW AND INTERMEDIATE ENERGIES

H_2O^- respectively. As the energy increases, however, an increasing contribution comes via a repulsive state or states (see p. 357) and this leads to the secondary maximum. In the centre of mass system

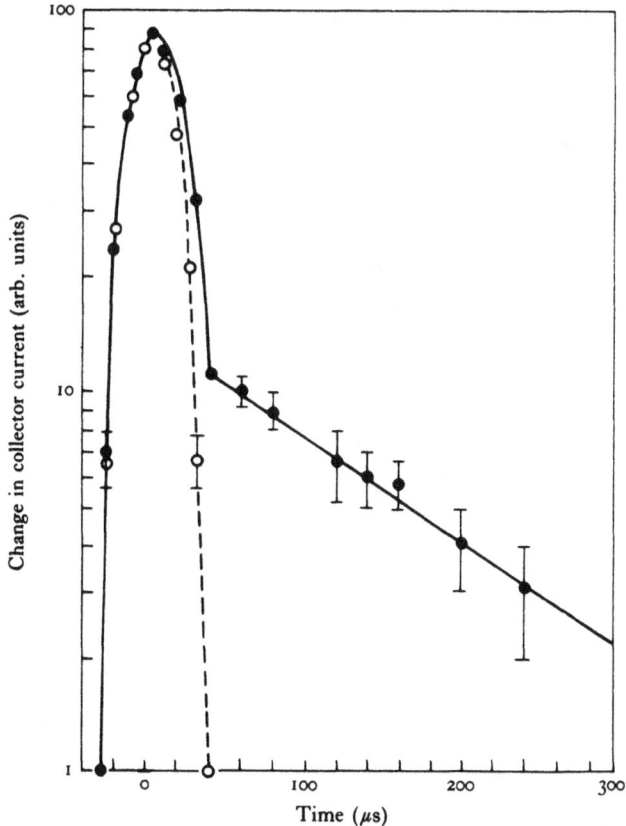

Fig. 12.16. Observed change of electron current to the collector with time delay in the pulse experiments of Moruzzi, Ekin and Phelps (1968) for $F/p = 5$ V cm^{-1} torr^{-1}. –○– in pure O_2 at 5 torr pressure, –●– for O_2 at 5 torr containing 0.001 torr partial pressure of H_2.

this occurs at a relative energy of 1.5 eV for CO but at a much lower energy 0.25 eV for H_2. This is not inconsistent with theoretical expectation.

Further light is thrown on the nature of the transitions involved by the observed energy distributions of the detached electrons

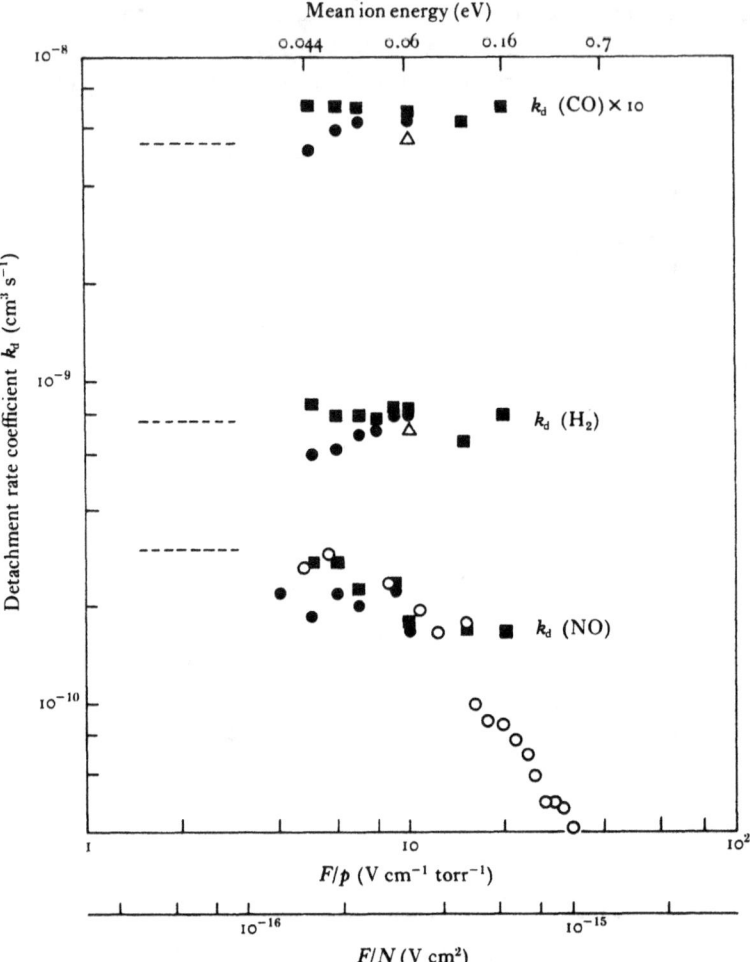

Fig. 12.17. Observed rate coefficients for associative detachment of electrons from O⁻ ions in collisions with CO, H₂ and NO. From pulse experiments (Pack and Phelps, 1966b) using attachment coefficient data, ■ from Huxley, Crompton and Bagot (1959), ● from Chanin, Phelps and Biondi (1962). From combined drift tube and mass spectrometer (Moruzzi and Phelps, 1966) ○. From flowing afterglow experiments (Fehsenfeld et al., 1966) -----. From beam experiments (Mauer and Schulz, 1973) △.

shown in Figs. 12.19 and 12.20. It was pointed out on p. 536 that, while the widths of the observed peaks are affected by the magnetic field applied to collect the electrons, the energies at which the peaks occur should be unmodified.

IONIC REACTIONS: LOW AND INTERMEDIATE ENERGIES

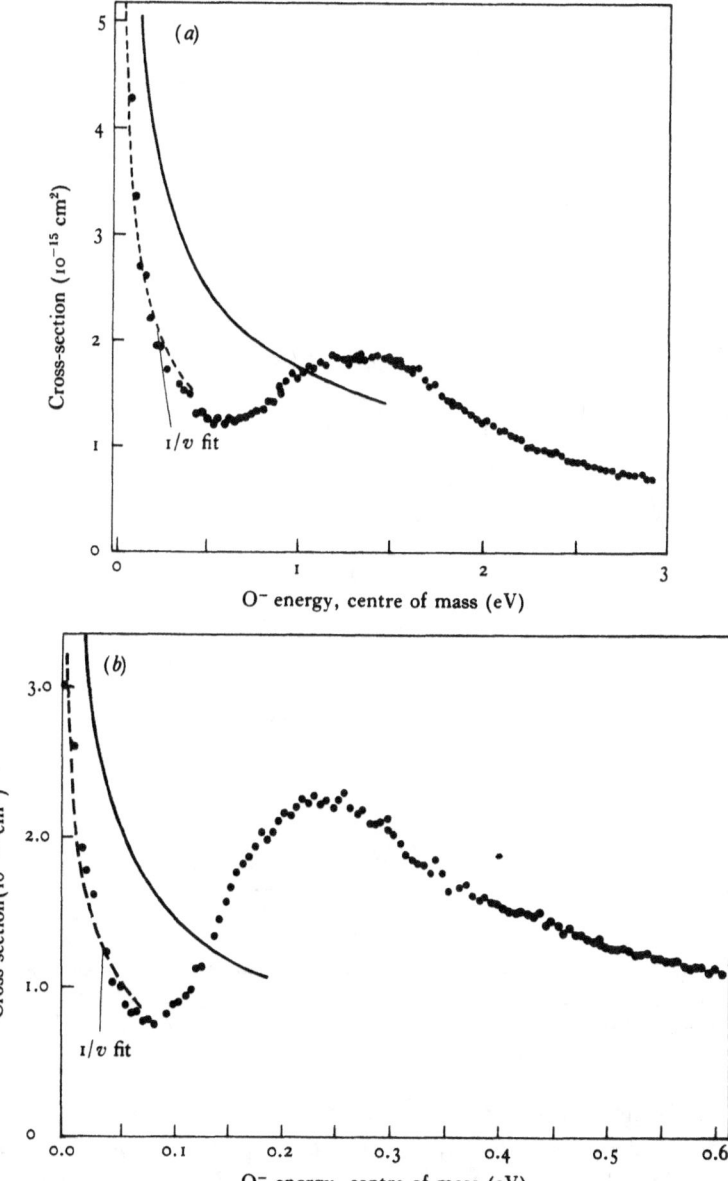

Fig. 12.18. Observed total cross-section, ● for detachment reactions of O^- in (a) CO and (b) H_2, observed by Mauer and Schulz (1973). The line ——— gives the orbiting cross-section (12.14), the line ---- represents a fit to the data assuming a variation as $1/v$, where v is the ion velocity.

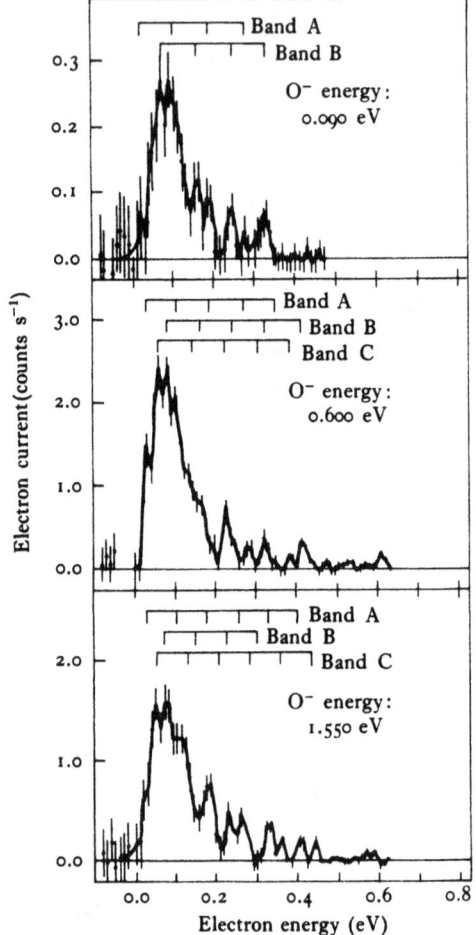

Fig. 12.19. Energy distribution of electrons detached from O$^-$ in collisions with CO, as observed by Mauer and Schulz (1973). The energies of the incident ion are as indicated.

It is first noteworthy that all the distributions involve only low energy electrons. This means that the resultant molecule is produced in a highly excited state of internal motion. An estimate of the rotational energy based on considerations of conservation of angular momentum shows that only about 0.05 eV can be ascribed to it. This leaves about 2.5 eV of vibrational energy in each case.

IONIC REACTIONS: LOW AND INTERMEDIATE ENERGIES 557

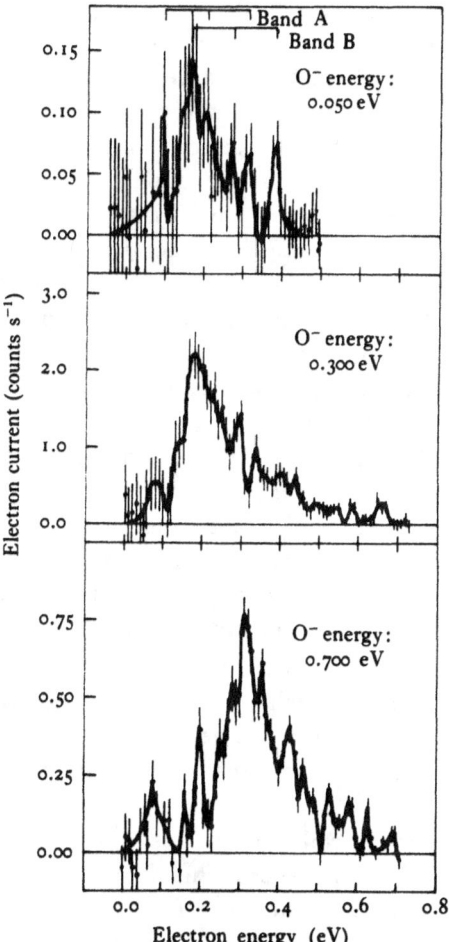

Fig. 12.20. Energy distribution of electrons detached from O⁻ in collisions with H₂, as observed by Mauer and Schulz (1973). The energies of the incident ion are as indicated.

For CO, at the low ion energy of 0.090 eV in the centre of mass system, the peaks can be analysed into two sets of bands, each with a spacing of about 0.08 eV displaced by 0.056 eV from each other. The vibrational spacing of the bending mode of CO_2 is about 0.067 eV, as compared with 0.248 and 0.165 eV for the asymmetric and symmetric stretching modes respectively, so that it appears that the detachment process leaves the CO_2 molecule excited mainly in the bending mode. The two displaced bands probably arise from separate attractive electronic states of CO_2^- which cross into the autodetaching region (see p. 357) at different configurations.

At a considerably higher ion energy, in the centre of mass system, of 0.600 eV a third band series is observed with the same spacing of 0.08 eV. This can be ascribed to the accessibility at these impact energies of the repulsive state. This structure persists in the electron energy distribution at higher energies.

The interpretation of the observed electron energy distributions for H_2 is more complicated because of the anharmonicity of the vibrations of the H_2O molecule. At the lowest ion energy shown in Fig. 12.20, the maxima in the electron energy distribution can be analysed into two bands, each with a spacing near 0.105 eV, displaced by 0.040 eV. Again it seems clear that bending vibrations are mainly excited as the other modes have spacings of 0.458 and 0.471 eV. Near the potential minimum of H_2O the bending vibrational quanta are of 0.200 eV energy but, because of the anharmonicity, they may be much smaller at the degree of excitation which occurs in the detachment process.

The charge transfer reaction (14) is of interest in that its high rate (Ferguson, 1967) shows that it must be exothermic and hence the electron affinity of NO_2 must be greater than that, 1.465 eV, of O (for further discussion of the electron affinity of NO_2, see p. 570).

Of the rearrangement collisions, (16) is of interest as a source of NO^- in flowing afterglow experiments. It has been studied by a variety of techniques including the flowing afterglow (see also p. 564), the ion cyclotron resonance spectrometer and the drift tube, and in low-energy beam experiments. As an example of the latter, Chantry (1969) used essentially the same equipment as in his experiments on dissociative attachment of electrons to N_2O (see Chapter 9, p. 358). While at low pressures, only O^- ions produced in this way were observed, at higher pressures a current of NO^- ions was also detected which increased as the square of the N_2O pressure p. This indicated that these secondary ions resulted from the successive reactions

$$e + N_2O \longrightarrow O^- + N_2, \qquad (12.55)$$

IONIC REACTIONS: LOW AND INTERMEDIATE ENERGIES 559

$$O^- + N_2O \longrightarrow NO^- + NO, \qquad (12.56)$$

(12.56) being exothermic by 0.14 eV. The fraction of O^- converted to NO^- is given by $[N_2O]Ql$, where $[N_2O]$ is the concentration of N_2O molecules, l the path length of O^- ions in the chamber and Q the cross-section for (12.56). From the observed ratio of NO^- to O^- currents and the chamber dimensions, Q was found to be 3×10^{-16} cm². As the O^- resulting from (12.55) have a peak energy of 0.38 eV (see Chapter 9, p. 361) the corresponding rate coefficient is 7×10^{-11} cm³ s⁻¹. This value is included in Fig. 12.21 which shows the results obtained for the rate coefficient as a function of the kinetic

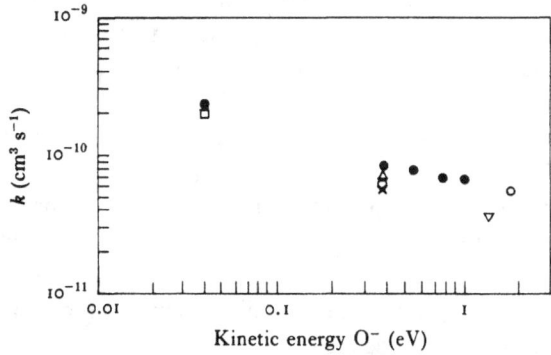

Fig. 12.21. Rate constants k for the reaction of O^- with N_2O as a function of the kinetic energy of the O^- observed in different experiments. ● Marx et al. (1973) (ion cyclotron resonance), × Stockdale et al. (1970), □ Parkes (1972a) (drift tube), ○ Paulson (1966), ▽ Paulson (1970), △ Chantry (1969).

energy of O^-, in the different experiments which have been carried out. The agreement is quite good. At room temperature the agreement is also good between the values obtained by ion cyclotron resonance (2.5×10^{-10} cm³ s⁻¹) and from the flowing afterglow technique (2.2×10^{-10} cm³ s⁻¹) both measurements being by Marx, Mauclaire, Fehsenfeld, Dunkin and Ferguson (1973).

There are a number of other possible exothermic reactions between O^- and N_2O namely

$$O^- + N_2O \longrightarrow O_2^- + N_2, \qquad (12.57)$$
$$\longrightarrow O_2 + N_2 + e, \qquad (12.58)$$
$$\longrightarrow NO + NO + e, \qquad (12.59)$$

in which the energy released is 2.46, 2.02 and 0.12 eV respectively. From the ion cyclotron resonance experiments (Marx et al., 1973) it was deduced that at thermal energies the rate constant for (12.57) is 1×10^{-12} cm^3 s^{-1} and the sum of those for (12.58) and (12.59) $< 10^{-11}$ cm^3 s^{-1}.

It is of interest to note that in the flowing afterglow experiments difficulties were encountered due to the rapid detachment of electrons from the NO$^-$ product. Thus, with argon as buffer gas and CO$_2$ as the source of O$^-$ ions, the only observed ionic product was NO$^-$ but with a rate of production less than that of loss of O$^-$. This is probably due to detachment from NO$^-$ in collisions with CO$_2$. This difficulty was overcome by using He as a buffer gas with O$_2$ as source gas for the O$^-$ ions. The O$_2$ reacts rapidly with the NO$^-$ product to produce O$_2^-$ which does not react further. It is then found that the rate of production of O$_2^-$ accurately balances the rate of loss of O$^-$. Further discussion of results obtained in studying detachment from NO$^-$ is given on p. 561.

The rearrangement reactions (17)–(22) (Ferguson, 1969; Fehsenfeld et al., 1969) are of interest in connexion with the interpretation of the negative-ion composition in the lower ionosphere. (22) is of special interest as the likely primary reaction which leads to hydration of O$_2^-$ (see also p. 578).

Reactions of negative ions other than those of oxygen

Three-body associative reactions

Puckett and Lineberger (1970), using their technique for studying static afterglows in electronegative gases in the later phase in which they are controlled by negative ion instead of electron diffusion (see p. 306), measured the rate constant for the primary hydration of NO$_2^-$ by the reaction

$$\mathrm{NO_2^- + H_2O + NO \longrightarrow NO_2^- \cdot H_2O + NO}. \quad (12.60)$$

They worked with pure NO and mixtures containing an admixture of water vapour as described on p. 307. The rate constant they obtained was $1.3 \pm 0.3 \times 10^{-28}$ cm^6 s^{-1}. Marx et al. (1973), using the ion-cyclotron resonance technique as in their experiments on

IONIC REACTIONS: LOW AND INTERMEDIATE ENERGIES

the reactions of O^- with NO, measured the rates of

$$NO^- + N_2O + Ar \longrightarrow NO^-.N_2O + Ar, \quad (12.61)$$

and

$$NO^- + CO_2 + Ar \longrightarrow NO^-.CO_2 + Ar, \quad (12.62)$$

as 7.1×10^{-30} and 36×10^{-30} cm^6 s^{-1} respectively. The rate of production of $NO^-.CO_2$ with CO_2 as the third body has been measured by Parkes (1972b) using the pulsed drift-tube technique as 75×10^{-30} cm^6 s^{-1}.

Associative detachment reactions
A number of these reactions which have been studied, by the flowing afterglow technique, are listed in Table 12.3, together with the energy released in the reaction and the measured rate constant at thermal energies.

TABLE 12.3 *Associative detachment reactions involving negative ions other than those of oxygen*

Reaction	Energy release (eV)	Rate constant (10^{-10} cm^3 s^{-1}) Obs.
$OH^- + O \rightarrow HO_2 + e$ (a)	1.0	2.0
$OH^- + N \rightarrow HNO + e$ (a)	2.4	0.1
$C^- + CO \rightarrow C_2O + e$ (b)	1.1	4.1
$C^- + CO_2 \rightarrow 2CO + e$ (b)	4.3	0.47
$C^- + N_2O \rightarrow ?$ (b)		9.0
$C^- + H_2 \rightarrow CH_2 + e$ (b)	2.0	<0.001
$Cl^- + H \rightarrow HCl + e$ (a)	0.8	Fast
$Cl^- + O \rightarrow ClO + e$ (a)	−0.9	0.1
$Cl^- + N \rightarrow ClN + e$ (a)	−0.9	0.1

References
(a) Fehsenfeld et al. (1966); (b) Fehsenfeld and Ferguson (1970).

Direct detachment reactions
NO^-. In Chapter 9, p. 340, we described the results obtained for the rates of attachment of electrons in NO at low mean energies which seem to show a systematic error depending on the pressure. It has been shown by Parkes and Sugden (1972) how these results

may be interpreted in terms of detachment, the reactions involved being

$$e + 2NO \longrightarrow NO^- + NO, \qquad (12.63_1)$$

$$NO^- + NO \longrightarrow e + 2NO, \qquad (12.63_2)$$

$$\left.\begin{array}{l} NO^- + 2NO \longrightarrow (NO)_2^- + NO, \\ (NO)_2^- + NO \longrightarrow NO_2^- + N_2O \end{array}\right\} \quad (12.63_3)$$

As evidence for this they studied attachment in a drift tube with NO containing a small admixture of O_2. Under these conditions the fast charge transfer reaction

$$NO^- + O_2 \longrightarrow O_2^- + NO, \qquad (12.63_4)$$

followed by

$$O_2^- + 2NO \longrightarrow NO_3^- + NO, \qquad (12.63_5)$$

overcomes the detachment.

This interpretation was tested by Parkes and Sugden (1972) in the following way. Let $\alpha_i \delta x$ be the chance that, in a pulsed drift-tube experiment, a charged particle, ion or electron, will undergo the reaction (12.63_i) with i ranging from 1 to 5. Let $j_e(x)$ be the electron current on the axis at a distance x from the cathode. The detachment reaction (12.63_2) is very fast so that a steady state relation is set up between $j_e(x)$ and $j_{NO}(x)$, the NO$^-$ current. This gives

$$j_{NO}(x) = \frac{\alpha_1}{\alpha_2 + \alpha_3} j_e(x). \qquad (12.64)$$

The number of stable ions collected per pulse in pure NO will then be

$$j_s = \frac{\alpha_3 \alpha_1}{\alpha_2 + \alpha_3} \int_0^L j_e(x)\,dx, \qquad (12.65)$$

where L is the length of the drift tube.

With O_2 present we must take account of the reaction (12.63_4) giving a number

$$j_s' = \frac{\alpha_1(\alpha_3 + \alpha_4)}{\alpha_2 + \alpha_3 + \alpha_4} \int_0^L j_e(x)\,dx. \qquad (12.66)$$

In the limit of high O_2 concentration

$$j_s'' \longrightarrow j' = \alpha_1 \int_0^L j_e(x)\,dx, \qquad (12.67)$$

so that

$$\frac{j_s''}{j' - j_s''} = \frac{\alpha_4 + \alpha_3}{\alpha_2}. \qquad (12.68)$$

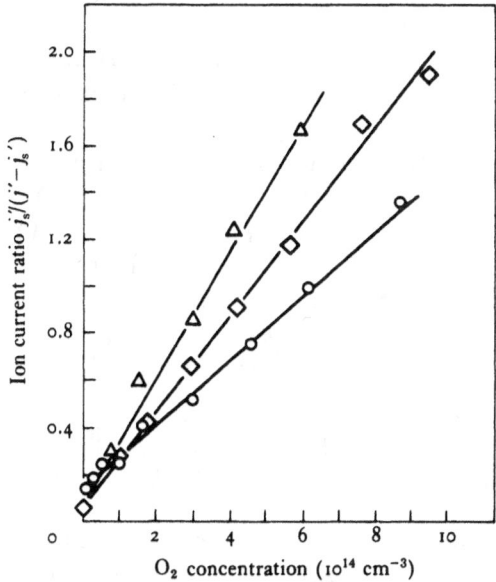

Fig. 12.22. Variation of negative ion current ratio (see eq. (12.68)) in NO with concentration of added O_2, observed by Parkes and Sugden (1972). ◇ [NO] = 4.1×10^{16} cm^{-3}, $F/N = 3.1 \times 10^{-17}$ V cm^2; △ [NO] = 4.1×10^{16} cm^{-3}, $F/N = 6.2 \times 10^{-17}$ V cm^2; ○ [NO] = 7.7×10^{16} cm^{-3}, $F/N = 3.1 \times 10^{-17}$ V cm^2.

In terms of the rate coefficients k_i for the various reactions

$$\frac{\alpha_4 + \alpha_3}{\alpha_2} = \frac{k_4[O_2] + k_3[NO]^2}{k_2[NO]}. \qquad (12.69)$$

Hence if $j_s''/(j' - j_s'')$ is plotted against $[O_2]$, a straight line should result with k_4/k_2 determined from the slope and k_3/k_2 from the intercept. Typical plots shown in Fig. 12.22 are consistent with this analysis. Taking for k_4 the value 5.0×10^{-10} cm^3 s^{-1} (see Table

12.5) it is found in this way that $k_2 = 5.0 \times 10^{-12}$ cm^3 s^{-1} and $k_3 = 7.6 \times 10^{-30}$ cm^6 s^{-1}.

The above analysis would be invalid if the added O_2 modified the rate of the initial attachment process as would be the case if the reaction

$$e + O_2 + NO \longrightarrow O_2^- + NO$$

were fast. It was verified by adding NO to O_2 in the drift tube that NO is not appreciably more effective than O_2 as a third body in O_2^- production.

j' is the number of ions which would be measured if the initial attachment process were irreversible. By comparison with the current of O_2^- obtained with O_2 in the drift tube and use of the measured three-body attachment rate coefficient in O_2 (see p. 328) the rate coefficient k_1 for (12.63$_1$) was found to be 10^{-30} cm^6 s^{-1}. At high pressures of NO, $j_s \to j'$ as α_3 is then $\gg \alpha_2$. The apparent attachment rate measured under these conditions (see Chapter 9, p. 340) should be equal to the true rate for (12.63$_1$). Parkes and Sugden (1972) measured an apparent attachment rate coefficient of 8×10^{-31} cm^6 s^{-1} in their high-pressure experiments, which is in good agreement with that derived from j'.

Knowing k_1 and k_2, the electron affinity $E_a(NO)$ of NO may be derived by application of statistical mechanics as in the derivation of the electron affinity of O_2 by Pack and Phelps (1966b). In this way $E_a(NO)$ is found to be 0.026 ± 0.018 eV, which is consistent with the precise value 0.024 eV derived from photodetachment experiments (see Chapter 11, p. 482).

Finally, Parkes and Sugden showed that with the observed values for the various rate coefficients the apparent attachment rate coefficients at lower pressures are consistent with the above analysis.

Direct measurements of the rates of detachment from NO$^-$ in collision with various molecules have been carried out by McFarland, Dunkin, Fehsenfeld, Schmeltekopf and Ferguson (1972) using the flowing afterglow technique. With argon as buffer gas NO$^-$ ions were produced through the reaction (12.56), N_2O being the source gas. The reactant was then added downstream at a measured inflow rate in the usual way and the reduction of NO$^-$ concentration measured as a function of this rate. Measurements of this kind were made over a temperature range from 193 to 506 °K for He, Ne,

IONIC REACTIONS: LOW AND INTERMEDIATE ENERGIES 565

TABLE 12.4 *Rate constants k_d in 10^{-11} cm^3 s^{-1} for direct detachment from NO^- collisions with neutral molecules, observed by McFarland et al.* (1972)

Neutral atom or molecule	Temperature			
	193 °K	285 °K	382 °K	506 °K
He	6.1×10^{-3}	2.4×10^{-2}	5.1×10^{-2}	0.1
Ne		2.9×10^{-3}	7.3×10^{-3}	1.6×10^{-2}
H_2	4.2×10^{-3}	2.3×10^{-2}	5.0×10^{-2}	9.4×10^{-2}
CO	1.0×10^{-2}	5.0×10^{-2}	8.3×10^{-2}	0.12
NO		0.50	1.4	3.0
CO_2	0.19	0.83	1.8	3.7
N_2O	6.8×10^{-2}	0.51	1.7	3.5
NH_3	0.64	2.0	3.8	5.7

H_2, NO, CO, CO_2, N_2O and NH_3 as reactants. Table 12.4 summarizes the observed detachment rate constants.

Parkes (1972b) has also measured detachment rates at room temperature in NO, N_2O and CO_2 using the drift-tube technique and obtains values in 10^{-11} cm^3 s^{-1} of 0.5, 0.6 and 1.0 respectively, which agree well with the flowing afterglow results. In the same experiments Parkes also measured the attachment rate for production of NO^- ions in CO_2. Under these conditions the ions once produced by the initial three-body process are very quickly stabilized by the fast clustering reaction (see p. 561)

$$NO^- + 2CO_2 \longrightarrow NO \cdot CO_2^- + CO_2,$$

for which the rate coefficient is as high as 7.5×10^{-29} cm^6 s^{-1}. The measured attachment rate should therefore be the true rate. It was found to be 17×10^{-31} cm^6 s^{-1}. Using this in conjunction with the measured detachment rate the equilibrium

$$e + NO + CO_2 \rightleftharpoons NO^- + CO_2$$

yields a further value for $E_a(NO)$ which comes out to be 0.036 ± 0.005, quite close to the precise value 0.024 eV.

There seems little doubt that the interpretation of the various results obtained for attachment and detachment involving NO^- outlined above is substantially correct.

F⁻ *and* Br⁻. Mandl, Evans and Kivel (1970), following earlier experiments by Berry *et al.* (1968) on Br⁻, have studied electron detachment from F⁻ in a shock-heated mixture of caesium fluoride and argon as described in Chapter 11, p. 442. The rate of change of the concentration [F⁻] with time after dissociation is given by

$$\frac{1}{[F^-]} \frac{d[F^-]}{dt} = -k_{Ar}[Ar] - k_{Cs}[Cs^+], \qquad (12.70)$$

where k_{Ar}, and k_{Cs} are the rate constants for destruction in collisions with Ar and Cs⁺ respectively. [Ar] is known from the initial pressures and the shock velocity and [F⁻] was measured by optical absorption at wavelengths of 3440 Å and between 2100 and 3100 °K. To determine [Cs⁺] measurements were made of the intensity of the recombination emission bremsstrahlung which gives [Cs⁺][e] and of the bremsstrahlung emission which gives [e].

It was found by measurement at different argon pressures that, at temperature T in °K,

$$k_{Ar} = 1.2 \times 10^{-11} \exp\{-4 \times 10^4/T\} \text{ cm}^3 \text{ s}^{-1},$$
$$k_{Cs} = 5.6 \times 10^{-9} \exp\{-4 \times 10^4/T\} \text{ cm}^3 \text{ s}^{-1}.$$

The activation energy is therefore 3.45 eV, which is equal to the electron affinity of F (3.45 eV).

For Br⁻, Berry found

$$k_{Ar} = 5.8 \times 10^{-11} \exp\{-3.9 \times 10^4/T\} \text{ cm}^3 \text{ s}^{-1},$$

the activation energy 3.36 eV being again close to the electron affinity of Br (3.37 eV).

Charge transfer reactions

The rates of some exothermic reactions are given in Table 12.5. The fast reaction (4) is the one made use of in the flowing afterglow experiments on the rate of the O⁻–NO reaction in order to convert the NO⁻ product into stable negative ions before suffering detachment. (2) is of interest in showing the electron affinity $E_a(NO_2)$ of NO_2 is greater than that, 1.83 eV, of OH. On the other hand, from similar experiments with F⁻ and Cl⁻, it was well established that in these cases the charge transfer reactions with NO_2 are endothermic showing that $E_a(NO_2) < E_a(F)$ (3.45 eV). Further evidence about $E_a(NO_2)$ is discussed on p. 570.

Although it is useful to obtain evidence about the otherwise unknown electron affinity of a stable molecule or radical by observing that certain reactions are exothermic, as judged by their high thermal rate coefficient, much more definite information can be obtained by determining the threshold energy for onset of an endothermic reaction. This technique has been applied especially by Chupka et al. (1971). They observed the threshold behaviour of the cross-sections for charge transfer reactions involving negative atomic halogen, and particularly I^-, ions using the apparatus and the method of analysis described on p. 537.

TABLE 12.5 *Rate constants at thermal energies for charge transfer reaction of negative ions, other than those of oxygen, with various neutral atoms and molecules*

Reaction	Energy release (eV)	Rate constant (10^{-10} cm^3 s^{-1}) Observed
(1) $H^- + NO_2 \to NO_2^- + H$	1.54	29 (a)
(2) $OH^- + NO_2 \to NO_2^- + OH$	~0.46	10 (a)
(3) $NH_2^- + NO_2 \to NO_2^- + NH_2$	~1.55	10 (a)
(4) $NO^- + O_2 \to O_2^- + NO$	0.42	9 (a), 5 (b)

References
(a) Ferguson (1967, 1969); (b) Parkes (1974) (see p. 361).

Similar experiments have been carried out by Hughes, Lifshitz and Tiernan (1973) (see also Lifshitz, Hughes and Tiernan, 1970) using a tandem mass spectrometer.

We begin first by considering the results obtained by this technique for two molecules whose electron affinities have been determined with precision by photodetachment techniques, namely O_2 (see Chapter 11, p. 478) and NO (see Chapter 11, p. 482). Fig. 12.23 shows observed results obtained by Berkowitz, Chupka and Gutman (1971) for the O_2^- ions produced by the reaction

$$I^- + O_2 \longrightarrow I + O_2^-. \qquad (12.71)$$

After subtraction of the background, the threshold behaviour was analysed using five different assumed laws of variation of the cross-section at energies just above the threshold – a step function, a function showing a linear rise with energy from zero at the threshold, a combined linear and step function and an exponential function rising from zero at the threshold. Using the full three-dimensional energy distribution function which allows for the thermal motion of the target molecules, it was found that the best fit with the observed behaviour was obtained either with an exponential function or a combined linear-plus-step function. The former gives

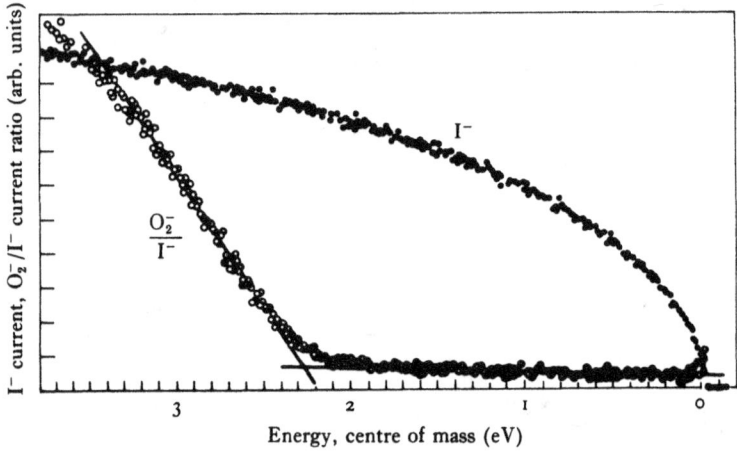

Fig. 12.23. Variation with energy in the centre of mass system of the yield of O_2^- ions near the threshold in the experiments of Berkowitz et al. (1971). ○ observed rates of O_2^- to I^- current, ● current of primary I^- ions.

a threshold energy which leads to an electron affinity $E_a(O_2)$ of 0.51 eV while the latter gives 0.61 eV. Allowing for about 0.03 eV of rotational excitation these become 0.48 and 0.58 eV respectively compared with the accurate value 0.44 eV from photodetachment.

For NO, the linear, linear-plus-step and exponential functions all give good fits to the observed data and give for $E_a(NO)$, 0.11, 0.09 and 0.11 eV respectively (allowing for 0.03 eV rotational energy), as compared with the accurately determined value of 0.024 eV (see Chapter 11, p. 482).

Hughes et al. (1973) have also studied charge transfer from I^- and from O^- ions in NO. From the thresholds they obtained

$E_a(\text{NO}) = 0.015 \pm 0.1$ and -0.14 ± 0.2 eV, which are also consistent with the photodetachment data.

It seems from these two cases that the method is capable of giving electron affinities correct to 0.1 eV but this cannot be taken as applying in general since it depends on the sharpness of the observed threshold. Even for O_2^- and NO^- production the tail of the observed distribution extends to about 0.3 eV beyond the threshold as determined.

Results have also been obtained by Hughes *et al.* (1973) for SO_2 from determinations of the thresholds for charge transfer reactions with I^- and S^-. The resulting values of $E_a(SO_2)$, namely 1.00 ± 0.1 and 0.97 ± 0.1 eV, are consistent with the value determined with some precision from photodetachment (1.097 ± 0.036 eV, see Chapter 11, p. 487), showing that the charge transfer threshold technique can give good results for triatomic molecules in suitable cases.

Turning now to cases for which no photodetachment data exist we refer first to the experiments of Chupka *et al.* (1971) on the electron affinities of the halogen molecules in which they determined the thresholds for charge transfer reactions of atomic halogen

TABLE 12.6 *Threshold energies for charge transfer reactions of atomic halogen ions* X^- *with halogen molecules* Y_2 *together with the derived electron affinities* $E_a(Y_2)$ *of the molecules*

Projectile ion	Target molecule	Threshold energy (eV)	Electron affinity (eV) Observed	Selected value
I^-	F_2	0	3.06	
Br^-	F_2	0.27	3.10	3.08 ± 0.10
F^-	F_2	0.38	3.07	
I^-	Cl_2	0.66	2.41	
Cl^-	Cl_2	[1.31]	[2.30]	2.38 ± 0.10
Br^-	Cl_2	1.01	2.35	$[2.32 \pm 0.1]$
I^-	Br_2	0.59	2.48	2.51 ± 0.10
Br^-	Br_2	0.84	2.53	$[2.62 \pm 0.2]$
I^-	I_2	0.49	2.57	2.58 ± 0.10
Br^-	I_2	0.77	2.59	$[2.42 \pm 0.2]$

Observed by Chupka *et al.* (1971) except for values in [ ] observed by Hughes *et al.* (1973).

ions X^- with halogen molecules Y_2. A typical set of observed data, for I^- and Cl_2 is shown in Fig. 12.9. Table 12.6 summarizes the results obtained.

Hughes et al. (1973) have also studied charge transfer in Cl_2, Br_2 and I_2. The selected values which they obtained are included in square brackets [ ] in Table 12.6. In all three cases they are consistent with those determined by Chupka et al.

These results are reasonably consistent with those obtained from the threshold for ion-pair production in collisions between alkali atoms and halogen molecules (see Chapter 10, p. 394). A comparative discussion of results on the electron affinities for these molecules obtained by different methods is given in Chapter 6, p. 210.

In many cases, in the experiments of Chupka et al. (1971), mixed halogen ions XY^- were observed, resulting from the rearrangement reaction.

$$X^- + Y_2 \longrightarrow XY^- + Y.$$

Quantitative threshold data for these reactions were obtained for IBr^-, found both in I^-–Br_2 and Br^-–I_2 collisions, giving $E_a(IBr) = 2.7 \pm 0.2$ eV.

The method has also been applied by Berkowitz et al. (1971) to O_3. With the best fit to the observed data, obtained with the linear-plus-step function, they found $E_a(O_3) = 1.96$ eV, allowing for about 0.04 eV of rotational excitation. Comparison with results obtained by other methods is given in Chapter 6, p. 215.

NO_2 has been investigated in both sets of experiments, by Berkowitz et al. (1971) and by Hughes et al. (1973). It was found in both experiments that the rate of production of NO_2^- showed an extensive tail. Thus Fig. 12.24 shows results for I^-–NO_2, obtained by Berkowitz et al., in which the tail extends to well over 0.5 eV beyond the linear extrapolation threshold and cannot be ascribed only to thermal effects. It may arise because of the change in shape between the equilibrium state of NO_2, with a bond angle of 134°, and NO_2^- for which it is 116.8°. Berkowitz et al. conclude that $E_a(NO_2) \geqslant 2.04$ eV (allowing for rotational motion).

Hughes et al. (1973) found values of 2.21 ± 0.1, 2.34 ± 0.1 and 2.17 ± 0.2 eV from the charge transfer reaction with I^-, Br^- and F^- respectively, leading to a selected mean value of 2.28 ± 0.1 eV. The

IONIC REACTIONS: LOW AND INTERMEDIATE ENERGIES

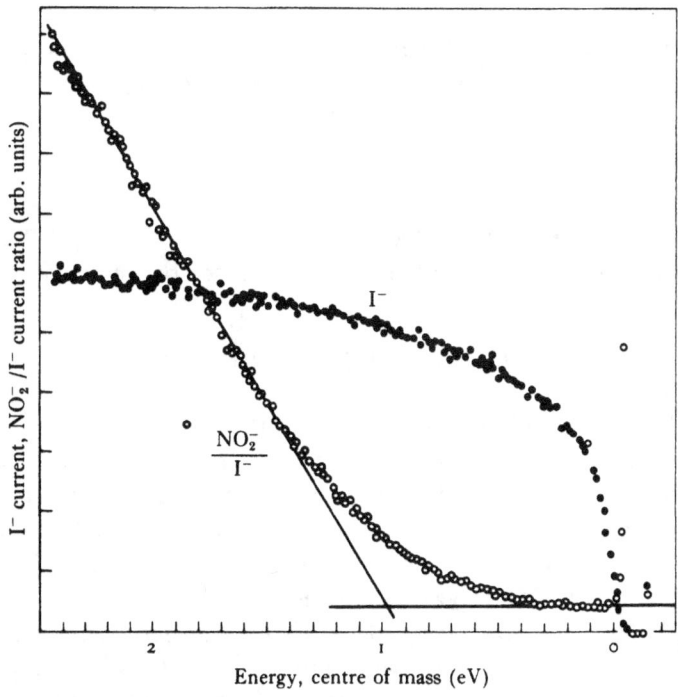

Fig. 12.24. Variation with energy in the centre of mass system of the yield of NO_2^- ions near the threshold, in the experiments of Chupka et al. (1971). ○ observed ratio of NO_2^- to I^- current, ● primary I^- current.

difference between the extreme values obtained in different reactions is somewhat larger than for the other cases they investigated.

A number of charge transfer reactions involving NO_2^- or NO_2 were also studied so that it was possible to bracket the electron affinity between quite narrow limits. Rate coefficients at impact energies in the centre of mass system around 0.3 eV were measured for charge transfer reactions of O^-, D^-, S^-, SH^-, SO_2^- and CS_2^- with NO_2 and found to lie between 0.4×10^{-9} and 1.2×10^{-9} cm³ s⁻¹. These large values indicate that all the reactions are exothermic, further evidence for which was the observed decrease of the cross-section with impact energy of the target ions concerned. SH has the largest electron affinity so

$$E_a(NO_2) \geqslant E_a(SH) = 2.32 \text{ eV}.$$

It is of interest to note that this conclusion is opposite to that obtained by Vogt (1969), who considered the reaction with SH⁻ to be endothermic. It does agree however with measurements made by the flowing afterglow technique (Dunkin et al., 1972).

Evidence was obtained that the reaction

$$NO_2^- + Cl_2 \longrightarrow Cl_2^- + NO_2$$

is exothermic, the rate coefficient being 2.2×10^{-10} cm³ s⁻¹. The inverse reaction on the other hand could not be detected. This means that

$$E_a(NO_2) < E_a(Cl_2) = 2.4 \text{ eV}.$$

Similar conclusions came from study of the reactions with I_2.

Despite the difficulties, it seems from these experiments that $E_a(NO_2)$ is fairly well determined. Evidence concerning it, obtained from experiments on ion-pair production in alkali metal atom collisions with NO_2, has been discussed in Chapter 10, p. 398, and all data on $E_a(NO_2)$ are compared in Chapter 6, p. 216.

Lifshitz et al. (1973) have studied charge transfer reactions of S⁻ and O⁻ ions with SF_6, from which they obtained $E_a(SF_6) = 0.6$ and 0.67 eV respectively. In agreement with these results it was found that the rate constant for the transfer reaction between O_2^- and SF_6 is as high as 3.6×10^{-11} cm³ s⁻¹, as would be expected for an exothermic reaction. However, it was found that the rate coefficient for charge transfer reactions between SF_6 and polyatomic negative ions, or SF_6^- and polyatomic neutral molecules, were in some cases abnormally low, even when the reaction was exothermic. Thus no charge transfer was observed either between SF_6^- and SO_2 or SO_2^- and SF_6, while the rate coefficient for the reaction of SF_6^- with NO_2 was found to be between one and two orders of magnitude lower than for the reaction between O⁻ and NO_2. On the other hand, fast charge transfer is found between SF_6^- and SeF_6 and UF_6 (Stockdale et al., 1970). Transfer to SF_6 also occurs at a fast rate from highly-excited rare gas atoms and is used to monitor the flux of such atoms (Freund, 1971). These results show that additional care must be taken in deciding whether a reaction is endothermic or exothermic, when dealing with polyatomic systems.

IONIC REACTIONS: LOW AND INTERMEDIATE ENERGIES 573

Lifshitz et al. (1973) have also measured relative cross-sections for dissociative charge transfer

$$X^- + SF_6 \longrightarrow SF_5^- + F + X,$$

where X refers to S^-, O^- and Br^-. Fig. 12.25 shows results obtained for O^-. The threshold for production of SF_5^- may be interpreted as

$$E_a(X) + D(SF_5-F) - E_a(SF_5)$$

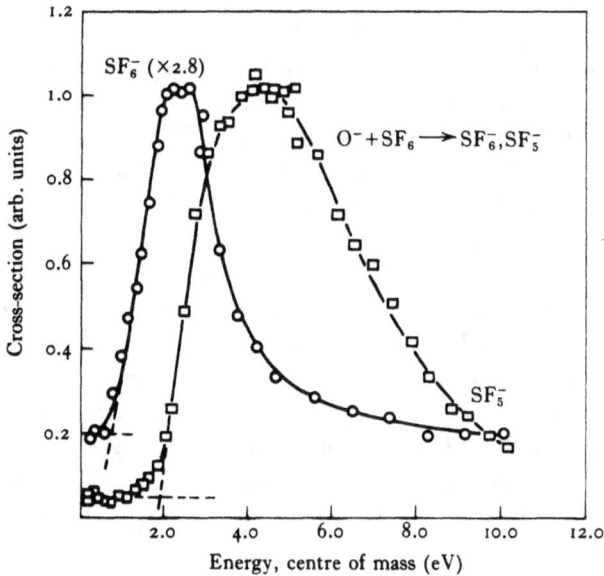

Fig. 12.25. Relative cross-sections for the dissociative charge transfer process $O^- + SF_6 \rightarrow SF_5^- + F + O$, observed by Lifshitz et al. (1973).

if there is no internal energy or excess kinetic energy carried away by the products. On this assumption, taking the dissociation energy of SF_6 into SF_5 and F as 3.3 eV (Cottrell, 1958) it is found that $E_a(SF_5) \geqslant 2.8 \pm 0.1$ eV. This is not inconsistent with the estimation of the threshold energy for

$$SF_5^- + NO_2 \longrightarrow SF_5 + NO_2^-,$$

which gives $E_a(SF_5) = 3.2$ eV.

While good results are clearly obtainable using the charge transfer technique it is essential to study as many reactions as possible because in some cases, for one reason or another, anomalous results are found, as for example with Cl^- and SO_2 and for other polyatomic cases referred to above.

Some interesting results have been obtained by Lifshitz et al. (1973) and by Hughes et al. (1973) using isotopic techniques. Thus for Br the relative proportion of Br_2 molecules involving $^{79}Br^{79}Br$, $^{79}Br^{81}Br$ and $^{81}Br^{81}Br$ nuclei is 0.255:0.499:0.244. It was found that the product ions from the reaction

$$Br^- + Br_2 \longrightarrow Br_2^- + Br$$

contain the isotopic nuclei in nearly these proportions when the relative impact energy is as high as 10.5 eV and this remains the same whether the incident ion is $^{79}Br^-$ or $^{81}Br^-$. This suggests that, under these conditions, electron transfer alone occurs, leaving the Br atoms in the target unchanged. At lower energies, however, the situation changes on both counts, indicating that atomic rearrangement is occurring. Similar results were found for Cl^- and Cl_2. Hughes et al. (1973) found that the data at energies of 1.9 eV were what would be expected if a linear symmetrical intermediate complex Br_3^- or Cl_3^- were formed in the collisions.

Isotopic experiments were also carried out to study symmetrical charge transfer reactions between polyatomic systems. It was found that no $^{34}SF_6^-$ ions resulted from collision of $^{32}SF_6^-$ ions in SF_6. In other words we have the surprising result that the symmetrical reaction

$$SF_6^- + SF_6 \longrightarrow SF_6 + SF_6^-$$

is too slow to be observed in these experiments. On the other hand, similar experiments with NO_2^-, SO_2^- and CS_2^- gave rate coefficients, in 10^{-10} cm^3 s^{-1}, of 4.2, 6.7 and 1.5 respectively for symmetrical charge transfer.

Rearrangement collisions

In Table 12.7 we give the rates of a number of rearrangement reactions which have been studied at thermal energies, usually by the flowing afterglow technique. Apart from the reaction (1), all are concerned with collisions involving one of the strongly bound ions NO_2^-, NO_3^- or CO_4^-, which are of interest in the interpretation

IONIC REACTIONS: LOW AND INTERMEDIATE ENERGIES 575

of the composition of the lower ionosphere (see Chapter 15, p. 670). The reactions chosen have been studied either because of their interest in this connexion or to obtain further information about the electron affinities or bond energies of the ions.

To produce the NO_3^- ions in the afterglow experiments, O^- ions are produced by electron impact in N_2O and converted to NO^- by the further reaction (12.56) with N_2O. By adding an excess of NO_2, NO_3^- ions are formed through the reaction

$$NO_2^- + NO_2 \longrightarrow NO_3^- + NO. \quad (12.72)$$

The observations show that the reactions of NO_2^- with HCl and HBr are clearly exothermic. It follows that $E_a(NO_2) < E_a(Cl) - D(HCl) + D(H-NO_2)$.

Unfortunately $D(H-NO_2)$, the dissociation energy of HNO_2 into H and NO_2, is not well known. Taking, however, the best estimate of 3.37 eV for this energy it is found that $E_a(NO_2) < 2.60$ eV.

TABLE 12.7 *Rate constants for rearrangement reactions at thermal energies involving negative ions other than those of oxygen*

Reaction	Rate constant (10^{-10} cm³ s⁻¹)
(1) $NO^- + HCl \rightarrow Cl^- + HNO$ (a)	0.16
(2) $NO_2^- + O_3 \rightarrow NO_3^- + O_2$ (b)	0.18
(3) $NO_2^- + NO_2 \rightarrow NO_3^- + NO$ (c)	0.04
(4) $NO_2^- + HCl \rightarrow Cl^- + HNO_2$ (a)	14
(5) $NO_2^- + HBr \rightarrow Br^- + HNO_2$ (a)	19
(6a) $NO_2^- + H \rightarrow OH^- + NO$ (d)	3
(6b) $\phantom{NO_2^- + H \rightarrow{}} \rightarrow HNO_2 + e$	
(7) $NO_3^- + O \rightarrow ?$ (a)	<0.1
(8) $NO_3^- + N \rightarrow ?$ (a)	<0.1
(9) $NO_3^- + NO \rightarrow ?$ (a)	<0.01
(10) $NO_3^- + HCl \rightarrow Cl^- + HNO_3$ (a)	<0.01
(11) $NO_3^- + HBr \rightarrow Br^- + HNO_3$ (a)	6.3
(12) $CO_3^- + O \rightarrow O_2^- + CO_2$ (d)	0.8
(13) $CO_3^- + NO \rightarrow NO_2^- + CO_2$ (e)	0.09
(14) $CO_3^- + NO_2 \rightarrow NO_3^- + CO_2$ (e)	0.8
(15) $CO_4^- + NO \rightarrow NO_3^- + CO_2$ (c)	0.48

References
(a) Ferguson, Dunkin and Fehsenfeld (1972), (b) Fehsenfeld and Ferguson (1968), (c) Fehsenfeld et al (1969), (d) Fehsenfeld and Ferguson (1972), (e) Ferguson (1967).

As far as NO_3 is concerned it seems that, although the rate constant is quite small, the reaction (3) is exothermic, in which case $E_a(NO_3)$ must exceed $E_a(NO_2)$ by more than 0.9 eV. More information is available from reactions (10) and (11). (10) may well be endothermic in which case $E_a(NO_3) > 3.5 \pm 0.2$ eV, the uncertainty arising from the heat of formation of NO_3. It was found that, when studying the reaction (11) by the flowing afterglow method, the

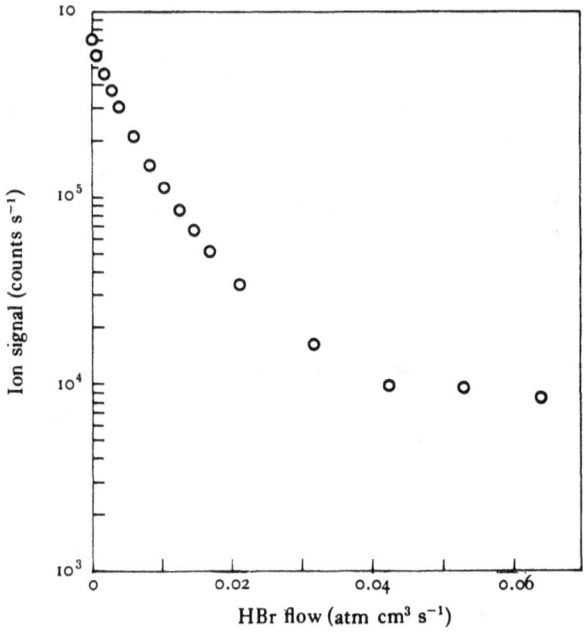

Fig. 12.26. Decrease of NO_3^- signal with increase in HBr flow rate, as observed in the experiments of Ferguson et al. (1972), at a helium buffer pressure of 0.42 torr and a temperature of 288 °K.

concentration of NO_3^- at first decreased rapidly, as would be expected for an exothermic reaction, but the loss rate then fell rapidly, as shown in Fig. 12.26, indicating that an equilibrium

$$NO_3^- + HBr \rightleftharpoons Br^- + HNO_3 \qquad (12.73)$$

had been reached. If this is so then

$$[Br^-]/[NO_3^-] = K[HBr]/[HNO_3], \qquad (12.74)$$

where K is the equilibrium constant and [X] denotes the concentration of X. All of the concentrations are available except that of HNO_3. If this is taken to be a constant fraction f of the concentration of NO_2 added to produce the NO_3^- ions by reaction with NO_2^-, it is found that indeed the condition (12.74) is satisfied as may be seen from Fig. 12.27. The fact that the equilibrium is attained

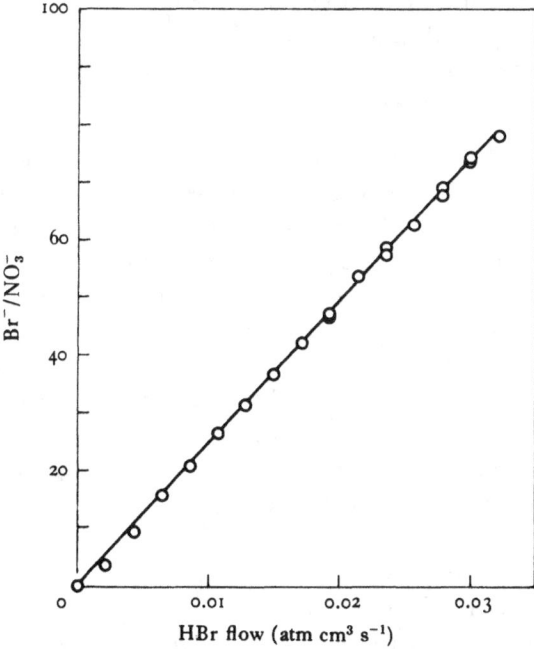

Fig. 12.27. Ratio of Br^- to NO_3^- ion signals as a function of the HBr flow rate in the experiments of Ferguson et al. (1972), at a helium buffer pressure of 1.03 torr and a temperature of 288 °K.

even when [HNO_3] is relatively so small indicates that the energy released in the forward reaction must be very small. Ferguson, Dunkin and Fehsenfeld analysed the situation in more detail in terms of measurements of K/f at three different temperatures and found that the energy released is 0.045 ± 0.015 eV. This leads to $E_a(NO_3) = 3.9 \pm 0.2$ eV, the major uncertainty being the energy required to dissociate HNO_3 into H and NO_3.

Finally, Berkowitz et al. (1971), using the same technique as in their experiments described on p. 569, determined the threshold energy for the reaction

$$\text{I}^- + \text{HNO}_3 \longrightarrow \text{NO}_3^- + \text{HI} \qquad (12.75)$$

as 1.817 eV. This gives, in conjunction with available theoretical data, including the uncertain heat of formation of NO_3, $E_a(\text{NO}_3) > 2.8$ eV which is not inconsistent with the results of Ferguson *et al.*

An interesting reaction which was observed by Chantry (1969), in his experiments on negative-ion production by electrons in N_2O (see pp. 359–63) as a function of gas pressure, is one which leads to N_2O^- production

$$\text{NO}^- + \text{N}_2\text{O} \longrightarrow \text{N}_2\text{O}^- + \text{NO}. \qquad (12.76)$$

N_2O^- ions were observed at a rate varying as the cube of the pressure of N_2O. This could well arise from the sequence of reactions beginning with (12.55) and (12.56) and then (12.76). If this is correct, the cross-section for (12.76) comes out to be about 10^{-16} cm² at the NO^- energy involved. An alternative possibility is that in some three-body reaction involving N_2O molecules, O^- ions are directly converted to N_2O^-. This is excluded because it requires a three-body rate coefficient of 2×10^{-27} cm⁶ s⁻¹ which is impossibly high. The difficulty in producing N_2O^- ions from N_2O because of the geometrical differences between the equilibrium configurations has been discussed in Chapter 9, p. 363.

Cluster exchange reactions

The rate coefficients for the reactions

$$\text{O}_2^- \cdot \text{H}_2\text{O} + \text{CO}_2 \longrightarrow \text{O}_2^- \cdot \text{CO}_2 + \text{H}_2\text{O},$$
$$\text{O}_2^- \cdot \text{H}_2\text{O} + \text{NO} \longrightarrow \text{O}_2^- \cdot \text{NO} + \text{H}_2\text{O},$$

which involve an exchange of molecules associated with the negative ion, have been measured by the flowing afterglow technique (Adams *et al.*, 1970). Both coefficients are large, being 5.8×10^{-10} and 3.1×10^{-10} cm³ s⁻¹ respectively. The reactions

$$\text{O}_4^- + \text{H}_2\text{O} \longrightarrow \text{O}_2^- \cdot \text{H}_2\text{O} + \text{O}_2,$$
$$\text{O}_2^- \cdot \text{CO}_2 + \text{NO} \longrightarrow \text{O}_2^- \cdot \text{NO} + \text{CO}_2,$$

are also both exothermic (see (22) of Table 12.2 and (15) of Table 12.7 respectively). It follows that the strength of binding to O_2^- increases in going from O_2 to H_2O to CO_2 to NO. This order is not the same as that for binding to O_2^+.

CHAPTER 13

Detachment, Charge Transfer and Other Reactions involving Negative Ions – Collisions at High Impact Energies

In the previous chapter we have discussed reactions involving negative ions which occur, in most cases, at thermal or near-thermal energies (up to a few eV). Because of this almost all of the experimental methods have depended on the observations of the behaviour in a gas of a swarm of ions with a distribution of velocities depending on the temperature and/or applied electric fields. In only a few cases have beam experiments, using a collimated beam of ions with narrow energy spread, been carried out. We now consider the study of reactions at higher energies extending from a few eV to many keV, using beam methods which are readily applicable at these energies.

In practice most of this work as yet has been confined to the study of charge transfer and direct detachment processes with little done about rearrangement reactions. We shall begin by discussing some of the theoretical considerations which are applicable within this energy range. In particular we consider symmetrical charge transfer processes which are difficult to study at thermal energies except indirectly through the mobility (see Chapter 12, p. 542 for the effect on the mobility of O_2^- in O_2).

13.1 Symmetrical charge transfer – theoretical considerations

The simplest charge transfer process involving negative ions is

$$H^- + H \longrightarrow H + H^-. \qquad (13.1)$$

To avoid unnecessary complication at the outset we first outline the theory for a charge transfer process in which only a single electron is involved, that between a proton and a hydrogen atom

$$H^+ + H \longrightarrow H + H^+. \qquad (13.2)$$

There is no difficulty then in extending the analysis to apply to the process (13.1) in which we are particularly interested here.

We distinguish the protons by the subscripts a, b, the latter applying to the initial ion. The coordinates of the electron relative to the respective protons we denote by r_a, r_b, the nuclear separation by R. A proton and a hydrogen atom in its ground state can interact in two different ways corresponding to the $^2\Sigma_g$ and

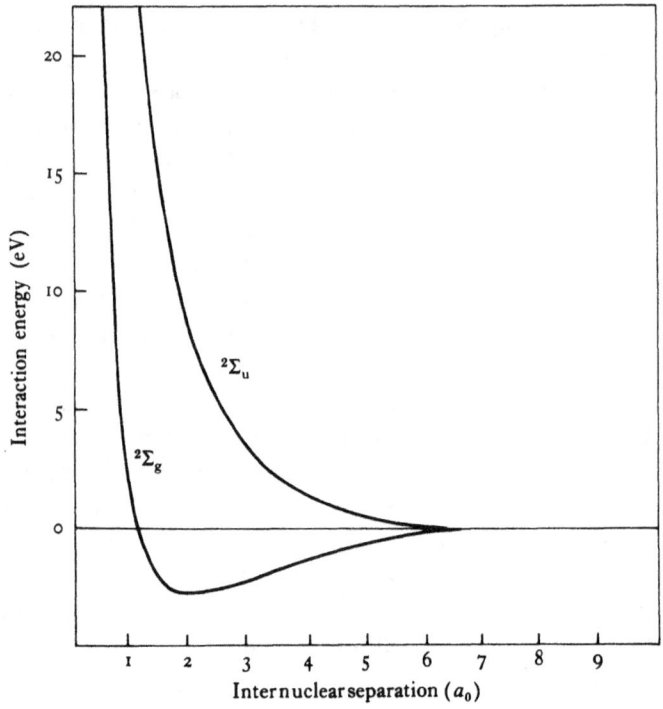

Fig. 13.1. The potential-energy curves for the $^2\Sigma_g$ and $^2\Sigma_u$ states of H_2^+.

$^2\Sigma_u$ states of H_2^+. The potential-energy curves for the two states, shown in Fig. 13.1, while tending to the same limit as $R \to \infty$, differ for finite R. We distinguish them respectively as $V_g(R)$, $V_u(R)$. As $R \to \infty$ the electronic wave functions for the respective states take the form

$$\psi_{g,u} \sim 2^{-1/2}\{\phi_0(r_a) \pm \phi_0(r_b)\}, \qquad (13.3)$$

where the ϕ_0 are ground state wave functions for atomic hydrogen.

IONIC REACTIONS AT HIGH ENERGIES 581

Consider now a collision between the proton and H atom for which the wave number is k. If the collision were to proceed under the interaction V_g then, following the same procedure as for electron–atom collisions described in Chapter 4, p. 68, we may represent the scattering in terms of a wave function F_g for the relative motion which has the asymptotic form

$$F_g \sim \exp(ikZ) + R^{-1}\exp(ikR)f_g(\Theta), \qquad (13.4)$$

where R, Θ, Φ are spherical polar coordinates with axis along the direction of initial relative motion. This function will be associated with the electronic wave function ψ_g so that the overall wave function is $F_g\psi_g$. However, for large R, this will have the asymptotic form

$$2^{-1/2}\exp(ikZ)\{\phi_0(r_a) + \phi_0(r_b)\}. \qquad (13.5)$$

This cannot be regarded as representative of a physical situation as initially we suppose the electron is attached to proton a so its wave function is $\phi_0(r_a)$ and the required asymptotic form is therefore

$$\exp(ikZ)\phi_0(r_a). \qquad (13.6)$$

To obtain the solution which gives this form we need only introduce the corresponding function $F_u\psi_u$ which is similar to $F_g\psi_g$ except that the interaction involved is now V_u. We then have

$$2^{-1/2}\{F_g\psi_g + F_u\psi_u\} \sim \exp(ikZ)\phi_0(r_a) + \tfrac{1}{2}\{(f_g+f_u)\phi_0(r_a) + \\ + (f_g-f_u)\phi_0(r_b)\}R^{-1}\exp(ikR). \qquad (13.7)$$

The incident plane wave is associated now with the electron on nucleus a, while the scattered spherical wave is associated with each possibility. Thus there is a scattered amplitude $\tfrac{1}{2}(f_g+f_u)$ associated with the electron remaining on nucleus a, i.e. with elastic scattering, and one $\tfrac{1}{2}(f_g-f_u)$ associated with the electron on nucleus b, i.e. with charge transfer. In terms of the differential cross-sections $I_{el}(\Theta)\,d\Omega$, $I_{tr}(\Theta)\,d\Omega$, for elastic scattering and charge transfer respectively, in which the direction of relative motion is turned through an angle Θ into the solid angle $d\Omega$, we therefore have

$$I_{el}(\Theta) = \tfrac{1}{4}|f_g+f_u|^2, \qquad (13.8)$$

$$I_{tr}(\Theta) = \tfrac{1}{4}|f_g-f_u|^2. \qquad (13.9)$$

Furthermore we may express f_g and f_u in terms of the respective phase shifts η_l^g, η_l^u for the scattering of waves of relative angular momentum $\{l(l+1)\}^{1/2}\hbar$ by the interactions $V_{g,u}$ just as in scattering by a centre of force, as outlined in Chapter 4, p. 71. Thus

$$f_{g,u}(\Theta) = (2ik)^{-1} \sum (2l+1) \exp(2i\eta_l^{g,u} - 1) P_l(\cos\Theta), \qquad (13.10)$$

giving

$$I_{tr}(\Theta) = (\tfrac{1}{16}k^2)| \sum(2l+1)\{\exp(2i\eta_l^g) - \exp(2i\eta_l^u)\} P_l(\cos\Theta)|^2. \qquad (13.11)$$

The probability of charge transfer at an angle Θ is given by

$$P_{tr}(\Theta) = I_{tr}/(I_{tr}+I_{el}), \qquad (13.12)$$

where

$$I_{tr}+I_{el} = \tfrac{1}{2}\{|f_g|^2 + |f_u|^2\}. \qquad (13.13)$$

The total cross-section for charge transfer Q_{tr} is now

$$Q_{tr} = 2\pi \int_0^\pi I_{tr}(\Theta)\sin\Theta\,d\Theta$$

$$= (\pi/k^2)\sum(2l+1)\sin^2(\eta_l^g - \eta_l^u). \qquad (13.14)$$

Thus Q_{tr} is determined by the *differences* between the phase shifts of the same order l, for scattering by the interactions V_g, V_u respectively.

To extend this analysis to the symmetrical negative-ion process (13.1) offers no difficulty provided we ignore detachment. Then, referring to Fig. 6.3, we see that an H⁻ ion and H atom in its ground state can also interact in 2 different ways corresponding to the $^2\Sigma_g$ and $^2\Sigma_u$ states. Without the possibility of autodetachment these curves are similar to the corresponding ones for the H⁺–H case and the charge transfer cross-section is given by the interactions V_g and V_u which correspond respectively to the $^2\Sigma_g$ and $^2\Sigma_u$ states in Fig. 6.3.

Q_{tr} has been calculated for H⁻–H collisions by Dalgarno and McDowell (1956) and by Davidovic and Janev (1969), detachment being ignored in both cases. The former authors used as the H⁻ wave function (1.17). In an attempt to improve the accuracy of the calculation of the difference $V_g - V_u$ at large R, Davidovic and Janev expressed the asymptotic form in terms of the parameters defining low-energy singlet scattering of electrons by H atoms, the polarizability of H and the binding energy of the active electron in H⁻. Their results differ only at energies below 30 eV in giving smaller cross-sections. Comparison between theory and experiment is shown in Fig. 13.8.

For more complicated cases a semi-empirical procedure is applicable as for positive ions. If, as is usually the case, $V_g - V_u$ can be represented with good accuracy at large R by

$$V_g - V_u = AR\exp(-\lambda R), \qquad (13.15)$$

then at relative ion energies greater than a few eV, for which semi-classical approximations are valid, Firsov (1951) and Demkov (1952) have shown that

$$Q_{tr} = \tfrac{1}{2}\pi p_0^2 + \pi^3 p_0/16\lambda, \qquad (13.16)$$

where

$$p_0^3 \exp(-2\lambda p_0) = \pi \lambda v^2 \hbar^2/8A^2, \qquad (13.17)$$

v being the velocity of relative motion. When v is small, but the relative energy remains well above thermal, we may write

$$p_0 = -(1/2\lambda)\log(\pi \lambda v^2 \hbar^2/8A^2 \bar{p}^3), \qquad (13.18)$$

where $\bar{p}$ varies very slowly with v. This gives

$$Q_{tr} = (\pi/2\lambda^2)\{\log v + \tfrac{1}{2}\log(\pi\lambda\hbar^2/8A^2\bar{p}^3)\}^2. \qquad (13.19)$$

For positive-ion charge transfer Rapp and Francis (1962) took

$$A = e^2/a_0^2, \quad \lambda = (E_i/E_H)^{1/2}/a_0, \qquad (13.20)$$

where E_i, E_H are the ionization energies of the atom concerned and of H respectively. This gives good results in comparison with experiment. Although it has not been tested to the same extent, the obvious extension to negative-ion cases is to replace E_i by E_a, the appropriate electron affinity.

The relation of charge transfer cross-sections to mobility follows because with the assumption of the form (13.15) the diffusion cross-section Q_d for collisions between the ions and atom is given to the same approximation as (13.19) by

$$Q_d = \pi p_0^2 + (\pi A/v\hbar\lambda)^2 p_0^4 \exp(-2\lambda p_0)$$
$$+ 3\pi^3 \alpha^2 e^4/8v^4 M^2 p_0^6, \quad (13.21)$$

where

$$(4A^2/\hbar^2 v^2 \lambda) p_0^3 \exp(-2\lambda p_0) + \frac{9}{8}\frac{\pi\alpha^2 e^4}{v^4 M^2 p_0^8} = \tfrac{1}{2}\pi, \quad (13.22)$$

α being the polarizability of the atom. If α is known, both Q_d and Q_{tr} may be specified in terms of the two parameters A and λ. Since Q_d determines the mobility, this affords the relation with charge transfer.

Effect of detachment on charge transfer cross-sections

It is possible, if the detachment cross-section Q_{de} is known, to correct the charge transfer cross-section calculated according to (13.14) to allow for detachment. Thus if Q'_{tr} is the corrected cross-section

$$Q'_{tr} \simeq Q_{tr} - \tfrac{1}{2}Q_{de}. \quad (13.23)$$

The analysis on which this is based (Bardsley, 1967) depends on the use of complex phase shifts. According to the discussion of the elastic scattering of particles by a centre of force given in Chapter 4, p. 71, the asymptotic form of the wave function for particles with angular momentum $\{l(l+1)\}^{1/2}\hbar$ about the centre is given by

$$rF_l(r) \sim [k^{-1} i^l (2l+1) \sin(kr - \tfrac{1}{2}l\pi)$$
$$+ \{\exp(2i\eta_l) - 1\}(2l+1)(2ik)^{-1} \exp(ikr)] P_l(\cos\theta). \quad (13.24)$$

The net outward particle flux at infinity is

$$j_r = (i\hbar/2m)\left(F_l \frac{\partial F_l^*}{\partial r} - F_l^* \frac{\partial F_l}{\partial r}\right). \quad (13.25)$$

If η_l is real, j_r vanishes, as it must if the scattering is purely elastic. However, if η_l were imaginary, say

$$\eta_l = \lambda_l + i\mu_l,$$

where λ_l and μ_l are real, then

$$j_r = (\hbar/4mr^2 k)(2l+1)^2 \{\exp(-4\mu_l) - 1\}\{P_l(\cos\theta)\}^2. \quad (13.26)$$

Hence, if μ_l is > 0, $j_r < 0$ which means that some of the ingoing particles are not scattered out again, i.e. they are absorbed by the scatterer. The cross-section for absorption of particles of angular momentum quantum number l is then given by

$$v Q_a^l = \int_0^\pi \int_0^{2\pi} j_r r^2 \sin\theta \, d\theta \, d\phi, \qquad (13.27)$$

so

$$Q_a^d = (\pi/k^2)(2l+1)\{1 - \exp(-4\mu_l)\}. \qquad (13.28)$$

The total absorption cross-section Q_a will be

$$Q_a = (\pi/k^2) \sum_l (2l+1)\{1 - \exp(-4\mu_l)\}. \qquad (13.29)$$

The corresponding elastic cross-section is given as before by

$$Q_{el} = (2\pi/k^2) \sum_l (2l+1)|\exp(2i\eta_l) - 1|^2,$$
$$= (2\pi/k^2) \sum_l (2l+1)(\cosh 2\mu_l - \cos 2\lambda_l) \exp(-2\mu_l), \qquad (13.30)$$

while the cross-section for all collisions

$$Q_{tot} = Q_a + Q_{el} = (2\pi/k^2) \sum_l (2l+1)\{1 - \exp(-2\mu_l)\cos 2\lambda_l\}. \qquad (13.31)$$

What significance can be attached to the absorption cross-section? It can be regarded as the cross-section for all processes which lead to loss of particles with the wave number k, i.e. the cross-section for all inelastic processes.

For collisions of negative ions, Q_a may often be interpreted as the total detachment cross-section as no discrete excitation of the ion can occur in most cases and excitation of the target atom or molecule will require much more energy.

All that is necessary now is to extend the concept of complex phase shifts to the case of symmetrical charge transfer where we now have in (13.14)

$$\eta_l^g = \lambda_l^g + i\mu_l^g, \quad \eta_l^u = \lambda_l^u + i\mu_l^u. \qquad (13.32)$$

The charge transfer cross-section then becomes

$$Q_{tr}' = (\pi/4k^2) \sum_l (2l+1)[\exp(-4\mu_l^g) + \exp(-4\mu_l^u)$$
$$- 2\cos\{2(\lambda_l^g - \lambda_l^u)\}\exp\{-2(\mu_l^g + \mu_l^u)\}], \qquad (13.33)$$

while the absorption cross-section, which is now identified as the detachment cross-section Q_{de} is given by

$$Q_{de} = (\pi/k^2) \sum (2l+1)\{1 - \tfrac{1}{2}\exp(-4\mu_l^g) - \tfrac{1}{2}\exp(4\mu_l^u)\}. \qquad (13.34)$$

For ion–atom collisions the wave number k, $=Mv/\hbar$, where M is the reduced mass, is very large compared with the reciprocal range of interaction so that a large number of terms are important in the sums over l. We may therefore convert the sums to integrals without serious error. If we assume that the real part of the phase shift is little affected by the absorption so that $\lambda_l^{g,\,u} \simeq \eta_{l0}^{g,\,u}$, where $\eta_{l0}^{g,\,u}$ are the phase shifts when detachment is ignored, then

$$Q'_{tr} - Q_{tr}^0 = -\tfrac{1}{2}Q_{de} + (\pi/k^2)\int_0^\infty l\cos\{2(\eta_{l0}^g - \eta_{l0}^u)\}$$
$$\{1 - \exp(-2\mu_l^g - 2\mu_l^u)\}\,dl. \qquad (13.35)$$

Q_{tr}^0 is the transfer cross-section calculated from (13.14) with η_{l0}^g, η_{l0}^u written for η_l^g, η_l^u.

The second term on the right-hand side of (13.35) will be very small because, for large l, $\mu_l^{g,\,u}$ are small, while for small l, η_{l0}^g and η_{l0}^u are large and oscillate rapidly with l. It follows that (13.23) is a good approximation.

Some evidence of the validity of this correction for detachment is discussed on p. 599 (see Fig. 13.8).

13.2 Unsymmetrical charge transfer – theoretical considerations

Unsymmetrical charge transfer is much more difficult to treat theoretically and it would be out of place here to attempt any detailed account of the methods available. For many purposes a semi-empirical treatment by Rapp and Francis (1962) provides useful guidance about the results to be expected for collisions involving atomic ions in collision with neutral atoms. This treatment is designed on the same lines as the semi-empirical theory of symmetrical charge transfer. According to it the cross-section for a process in which the magnitude of the difference of the electron affinities of the atoms involved is $\varDelta E$ should behave approximately as follows. It rises from threshold to a maximum at a relative velocity v such that (cf. Chapter 10, p. 384)

$$\varDelta E/\lambda\hbar v \simeq 1, \qquad (13.36)$$

where λ is given by (13.20) but with E_1 taken as the mean of the two electron affinities. At larger v, the cross-section behaves in very much the same way as that for symmetrical charge transfer with the same value of E_i. Application of this to some unsymmetrical cases is discussed on p. 602 (see Fig. 13.10).

13.3 Calculation of detachment cross-sections

At very high, but not relativistic, impact energies the cross-section for a detachment collision can be calculated using Born's approximation as in the corresponding case of detachment through electron impact. Thus, by an obvious generalization of (11.87), the cross-section for a detachment collision in which the electron is ejected with a wave number κ is given by

$$q_\kappa \, d\kappa = (8\pi^2 M^2 k'/kh^4) \, |\int V(R,\mathbf{r}_a,\mathbf{r})| \phi_0(r_a)|^2$$
$$\times \psi_0(r_i) \psi_\kappa(r_i) \exp\{i(\mathbf{k}-\mathbf{k}')\cdot \mathbf{R}\} \, d\mathbf{r}_a d\mathbf{r}_i d\mathbf{R}|^2 \, \kappa^2 d\kappa. \quad (13.37)$$

$\mathbf{k}$ and $\mathbf{k}'$ are the initial and final wave vectors of relative motion of the colliding ion and neutral atom, related to κ^2 by

$$\frac{\hbar^2}{M}\{k^2 - k'^2 - M\kappa^2/m\} = E_a, \quad (13.38)$$

where M is the reduced mass of the colliding systems and E_a the electron affinity. $\mathbf{r}_a, \mathbf{r}_i$ represent the aggregate of electron coordinates in the target atoms and negative ion respectively while $\mathbf{R}$ is the nuclear separation. $\phi_0(r_a)$ is the ground state wave function of the atom, $\psi_0(r_i)$ of the ion. $\psi_\kappa(\mathbf{r}_i)$ is the wave function for an unbound state of the ion which has the asymptotic form

$$\psi_\kappa(r) \sim \{e^{i\kappa \cdot r} + e^{i\kappa r} r^{-1} f(\theta)\} \chi(\mathbf{r}_{i-1}). \quad (13.39)$$

Here $\mathbf{r}$ denotes the coordinates of the active electron, $\mathbf{r}_{i-1}$ those of the remaining ionic electrons. $\chi(\mathbf{r}_{i-1})$ is the wave function of the neutral atom which results from the detachment. Finally, $V(R,\mathbf{r}_a,\mathbf{r}_i)$ is the interaction energy between the ion and atom.

Calculations on these lines have been carried out for detachment of electrons from H$^-$ in collisions with H and with He atoms but in both cases additional assumptions and approximations were made. Thus for detachment by He impact (Sida, 1955) the simple Hylleraas wave function (1.16) was used for the H$^-$ function $\psi_0(r_i)$ but a relatively good variational approximation was used for ψ_κ. The calculations for detachment in H impact (McDowell and Peach,

1959) were carried out, not only for detachment leaving the hydrogen atom unexcited, but also for cases in which this atom is left in any excited state. However, again the simple function (1.16) was used for $\psi_0(r_i)$ and in addition the continuum functions were taken to have the simple form

$$\psi_\kappa(r_i) = e^{i\kappa \cdot r} \phi_{nl}(r_{i-1}) \qquad (13.40)$$

where $\phi_{nl}(r_{i-1})$ is the wave function of the state nl in which the hydrogen atom is left after detachment occurs. In this case κ is given by (13.38) with E_a replaced by $E_a + E_{nl}$, where E_{nl} is the excitation energy of the nl state. Arguments were presented which suggest that the errors in the ground state and continuum functions tend to cancel.

The results obtained in these calculations are discussed on pp. 599 and 605.

An alternative semi-classical treatment of detachment collisions has been applied by Bates and Walker (1967) to calculate detachment cross-sections for impact of the H$^-$ ion with rare gases and with H$_2$. In this method the role of the neutral core to which the electron is initially attached is simply to determine the velocity of the electron. Relative to the target this velocity is given by $\mathbf{V} = \mathbf{v}_2 - \mathbf{v}_1$, where $-\mathbf{v}_1$ is the velocity of the incident hydrogen core relative to the nucleus of the target atom and $\mathbf{v}_2$ the velocity of the electron relative to the core. While $\mathbf{v}_1$ is fixed there will be a distribution of velocities v_2. Assuming this distribution to be the same as in a normal H atom the chance that v_2 lie between v_2 and $v_2 + dv_2$ is $f(v_2) dv_2$ where

$$f(v_2) = \frac{32 v_0^5 v_2^2}{(v_0^2 + v_2^2)^4}, \qquad (13.41)$$

$\frac{1}{2} m v_0^2$ being the ionization energy of an H atom with m the electron mass.

The target atom is then treated as a rigid spherical scattering centre of cross-section $Q_0(V)$ for electrons of velocity V relative to it. $Q_0(V)$ is taken to be the measured total cross-section of the target atom or molecule for free electrons of this velocity. Classical mechanics is then used to calculate the energy gained by the electron in the collision, relative to the projectile core. If this exceeds $E_a(H)$, the electron affinity of H, the electron will be detached.

It is then found, after a considerable amount of algebra, that the detachment cross-section, without allowance for the presence of two ionic electrons, is given by

$$Q_{de}^{(1)}(v_1) = (3\pi u^2)^{-1} \int_{E_a/4ut}^{\infty} Q_0\{(2\bar{T})^{1/2} v\} [\{8uv - (u-v)^2 - 1 - 2E_a/\bar{T}\}$$
$$\times \{1 + (u-v)^2\}^{-3} + \{1 + (u+v)^2 - E_a/\bar{T}\}^{-2}] \, dv, \quad (13.42)$$

where $u = v_1(2\bar{T}/m)^{-1/2}$, $v = V(2\bar{T}/m)^{-1/2}$ and $\bar{T}$ is the mean kinetic energy of the electron in its initial bound state according to (13.41).

To allow for the presence of two ionic electrons it is assumed that detachment occurs if either electron passes the target within a distance

$$p = \{Q_{de}^{(1)}/\pi\}^{1/2}. \quad (13.43)$$

If $2s$ is the mean value of the projection of the line joining the two electrons on a plane perpendicular to the direction of incidence and

$$\cos \alpha = s/p,$$

the full detachment cross-section Q_{de} is given by

$$Q_{de} = F Q_{de}^{(1)}, \quad (13.44)$$

where

$$F = 2 - (2\alpha - \sin 2\alpha)/\pi, \quad s < p,$$
$$F = 2, \quad s > p.$$

The assumption that the target behaves like a rigid scatterer implies that the angular distribution of electrons scattered by it is isotropic. At high velocities v_1,

$$Q_{de} \sim Q_0(v_1)\{1 - 7E_a/24mv_1^2\}, \quad (13.45)$$

which may be interpreted to mean that detachment occurs on the average for scattering outside the cone of semi-angle θ_c given by

$$\theta_c = (7E_a/6m)^{1/2} v_1^{-1}. \quad (13.46)$$

If the fraction of electrons scattered within this cone is small it is likely that the assumption of isotropic scattering does not lead to serious error.

Although this semi-empirical theory is far from precise it gives quite good results in comparison with experiment, as may be seen from the discussion on p. 605 (see, in particular, Figs. 13.12 and 13.13).

13.4 The measurement of charge transfer and detachment cross-sections

We first describe methods which are applicable with static gas targets and then consider the crossed-beam methods which must be applied when the targets are atomic or molecular radicals.

Some of the methods already discussed in Chapter 10, pp. 399–403, for the measurement of cross-sections for negative-ion formation by electron capture from neutral atoms and molecules are also applicable for the measurement of detachment cross-sections. Thus, if a fast beam of negative ions is passed through a gas and all beam particles are removed from the beam if they change charge, then the negative-ion current is reduced by the factor $\exp(-NlQ)$ after passage through the gas. Here N is the concentration of gas molecules, l the path length through the gas and Q the cross-section for a process in which the incident ion changes its charge. Under many circumstances Q is nearly equal to the detachment cross-section. For example, this will be so if the target gas molecules do not form stable negative ions, so excluding charge transfer, and at the same time the cross-section $Q_{\bar{1}1}$ for double detachment is negligible.

Equally well $Q_{\bar{1}0}$ can be obtained from measurements of the equilibrium fraction of different ions in a beam in which equilibrium has been attained between the charge states, as seen from (10.27) and (10.29).

In measurement of detachment and charge transfer cross-sections much use has been made of techniques which involve collection of the slow charged products and we now describe how these techniques may be used. We shall neglect the contribution from double detachment (see p. 615) in this discussion.

The condenser plate method

This is a method of wide generality which is applicable to the study of many ionic collision processes provided the ion energy is not too low. Fig. 13.2 illustrates a typical arrangement. The primary ion beam enters the collision chamber containing the target gas and passes between three pairs of parallel plate electrodes A_1 and A_2, B_1 and B_2, C_1 and C_2 before entering the collector D which will often be a Faraday cage. The electric field applied between the parallel plate electrodes is strong enough to collect on B_1 and B_2 all slow ions, positive and negative, as well as electrons formed by

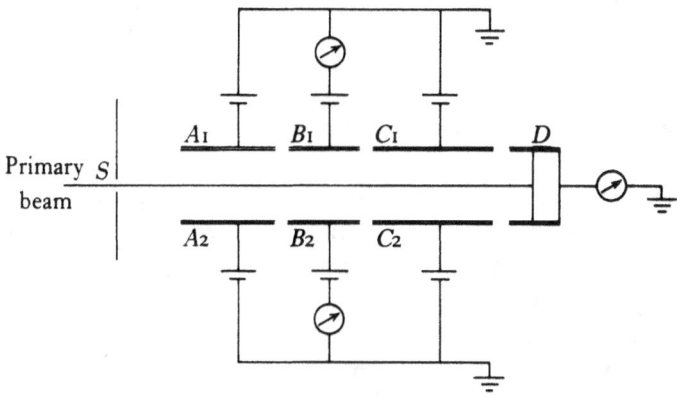

Fig. 13.2. Schematic diagram of a simple condenser plate system.

the beam in collisions with the gas molecules. The plates A_1 and A_2, C_1 and C_2 act as guard electrodes ensuring that the field between B_1 and B_2 is uniform and that these plates receive all the slow charged particles formed between them, and none formed elsewhere.

In general, with incident primary negative ions, there will result slow negative ions due to charge transfer, slow electrons due to detachment and pairs of slow electrons and positive ions due to ionization of the gas molecules. Let i_-, i_e, i_+ be the magnitudes of the currents to the plates arising from these respective sources. In addition we must allow for the possibility of secondary electron emission occurring from the negative plate, the magnitude of this current being i_s. We then have for the magnitude of the current collected by the negative electrode $B1$

$$i_1 = i_+ + i_s, \tag{13.47}$$

and for that collected by the positive electrode $B2$

$$i_2 = i_+ + i_e + i_- + i_s. \tag{13.48}$$

By subtraction of these two we have

$$i_2 - i_1 = i_e + i_-. \tag{13.49}$$

If charge transfer is excluded because the target gas forms no stable negative ions then $i_- = 0$ and (13.49) gives i_e directly. If the gas pressure is low enough for the negative ion in the beam to make at most a single collision in traversing the collision chamber then

$$(i_2 - i_1)/i_0 = nlQ_{de}, \tag{13.50}$$

where i_0 is the primary beam current, n the concentration of target molecules, l the length of the plates B and Q_{de} the detachment cross-section.

When charge transfer is possible some auxiliary device must be used to separate the slow negative-ion current from that of the electrons. One method which is often used is to apply a magnetic field of up to 300 G along the direction of the primary beam. Slow electrons, produced either by detachment or ionization, are held within the beam by the magnetic field and secondary electrons emitted from $B1$ are returned to it by the field. Hence, when the magnetic field is present, the current to $B1$ is

$$i_1' = i_+, \tag{13.51}$$

and to $B2$

$$i_2' = i_-. \tag{13.52}$$

Taken together with (13.49) this gives i_+, i_- and i_e separately and hence the cross-sections for charge transfer and ionization as well as for detachment.

The above analysis assumes that no secondary electron current reaches the electrodes other than that originating on the negative electrode $B1$. Some electrons may be produced at the edge of the beam-defining aperture S, with such energies and directions of motion as to reach the measuring electrodes. The presence of the guard electrodes reduces the risk of observing these electrons but it is desirable to include some means for suppressing the emission as, for example, by biasing the aperture positively with respect to a second slightly wider coaxial aperture.

It is also important that secondary electrons or reflected ions produced by impact of the fast primary ions on the detector are not permitted to escape into the collision chamber.

The effectiveness of the condenser plate method falls off at low primary ion energies because large-angle scattering and deflexion of the beam by the transverse electric fields become important.

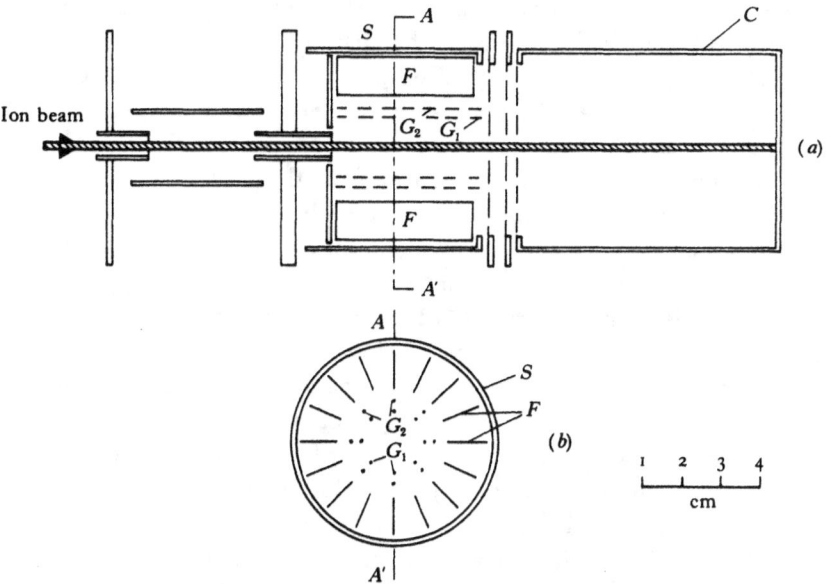

Fig. 13.3.(a) General arrangement of the apparatus developed by Bailey (1961) for studying detachment and charge transfer collisions of negative ions in gases. (b) Illustrating the ion–electron filter system.

Fig. 13.3 shows the arrangement used by Bailey (1961) and his collaborators to study detachment and charge transfer of negative ions in the energy range from 3.6 to 300 eV. The primary negative-ion beam passed through the collision chamber into the screened collector C. The collision region was enclosed within two cylindrical grids G_1 and G_2 coaxial with the beam. Slow negative ions and electrons passed through the grid G_1 to be accelerated by a d.c. potential on G_2 so as to enter a filter system F, a section of which is shown in Fig. 13.3(b) This operated in the same kind of way as the electron filters described in Chapter 9, pp. 292 and 297. Thus an r.f. potential applied between alternate vanes removed the electrons but had no effect on the negative ions which passed through to be collected on the cylindrical electrode S. Hence by measurement of the current to S with and without the applied r.f. voltage the slow negative-ion and electron currents could be measured separately and hence Q_{tr} and Q_{de} derived.

To correct the observed detached electron currents for secondary electrons and background scattering, the observed electron current $i_e(0)$, with residual gas only in the chamber, was subtracted from that i_e measured with gas present and the same primary ion current. This is valid if there is little chance of an ion being elastically scattered by the target gas in such a way as to produce secondary electrons which would not arise from the primary beam alone. For low-energy collisions however, for which the elastic scattering cross-section Q is relatively large and the detachment cross-section Q_d relatively small, a significant error may be introduced, particularly if it is desired to trace Q_d down to its threshold.

This was one of the aims of the experiment carried out by Wynn, Martin and Bailey (1970) to measure detachment collisions of various negative ions with rare gas atoms. Although no charge transfer can occur in these cases, the r.f. discriminator was used to correct the total scattered current for small contamination by elastically scattered negative ions. It was found, however, that near the threshold for Q_d the elastic scattering was large enough to give significant spurious signals due to the secondary electrons produced by the scattered as distinct from the primary ions. For measurements at these low energies they therefore used a modified apparatus, the general arrangement of which is shown in Fig. 13.4.

The collision chamber has cylindrical symmetry about the direction of the primary beam as axis. This beam, after passing through the equipotential defining region comprised by the electrodes DCL, DC and DCU, enters the collision chamber through a circular hole in the electrode DCU. The cylinder S and the plate SL are both maintained at earth potential and act as collector for the detached electrons. G_1, G_2 and G_3 are plane grids and the cylinder C is the collector for the transmitted ion beam. G_1 is earthed while G_2 is maintained about 2 V below earth. Negative ions scattered through angles less than α in Fig. 13.4 will not strike S or SL and will have sufficient energy to overcome the potential barrier presented by G_2 and enter the collector C. On the other hand, detached electrons which have energies less than 2 eV and are not directly collected by S or SL, reach the space between G_1 and G_2 and are reflected back by the barrier to be so collected. This method is applicable for collisions of ions with light atoms such as He for which the maximum scattering angle in the laboratory system is less than $\alpha(\simeq 30°)$.

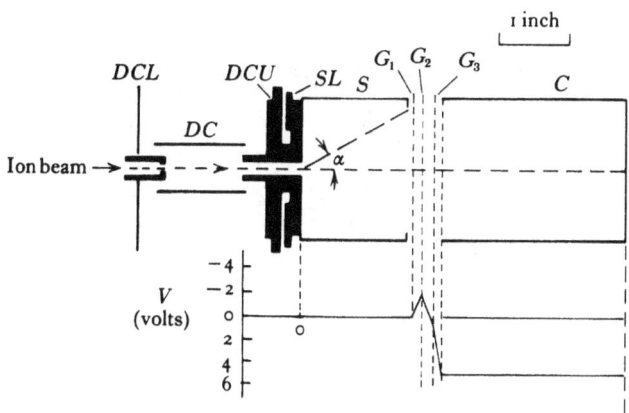

Fig. 13.4. Modified arrangement of the apparatus shown in Fig. 13.3, as used by Wynn et al. (1970) for studying detachment reactions near the threshold.

Another method was used by Roche and Goodyear (1969) which made use of a magnetic field in a somewhat different way from that used in the experiments at high energies described above.

Fig. 13.5(a) shows a section through the collision chamber which they used, in a plane containing the main beam. Separation of the current of slow detached electrons from that of negative ions was achieved by application of a uniform magnetic field along the axis of the chamber. This field was chosen so that the electron paths were so curved by it that a considerable fraction of the slow electrons were collected on the vanes v. The trajectories of the slow negative ions, on the other hand, were only slightly curved and by addition of the two guard electrodes g even those ions leaving at a small angle to the plane of the vanes would not reach the vanes. This situation is illustrated in Fig. 13.5(b).

To prevent secondary electrons arising from impact of the

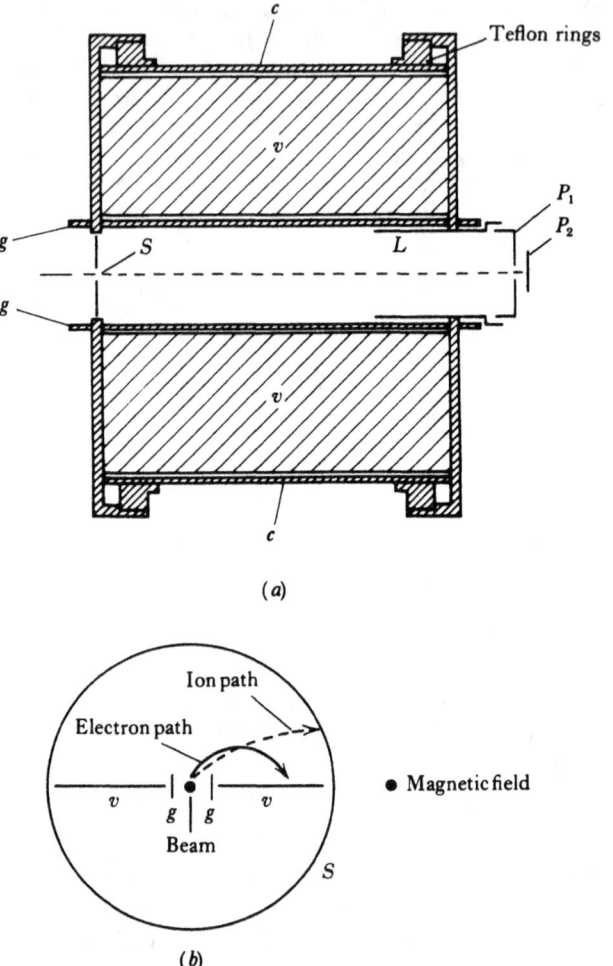

Fig. 13.5.(a) Arrangement of the collision chamber in the experiments of Roche and Goodyear (1969). v, vanes, g, guard electrodes, c, cylinder. (b) Illustrating the electron and ion paths in the collector system.

primary ion beam with the collector electrode P_1, the tube L (Fig. 13.5(b)) was inserted and P_1 was biased a few V positive with respect to the main chamber.

One difficulty about this method is that of calculating the collection efficiency of the vanes as this depends on the energy and angular distribution of the electrons. As far as the former is concerned it was found by a trajectory calculation that, if the magnetic field (a few G) is chosen so that the ratio of collected electron current to primary ion current is a maximum, the collection efficiency could be taken as 0.20 to about 15% accuracy, assuming the angular distribution to be isotropic. On the other hand, if the electrons were emitted in a direction normal to the beam with a uniform azimuthal distribution, the efficiency would rise to 0.36.

Results obtained using this method are discussed on p. 603.

Crossed-beam method

For measurement of the cross-section for the simplest charge transfer process (13.1) in which the target atoms are atomic hydrogen, it is necessary to resort to a crossed-beam technique.

Fig. 13.6 illustrates the general arrangement of the apparatus used by Hummer, Stebbings, Fite and Branscomb (1960) for this purpose. A beam of highly-dissociated hydrogen, issuing from a tungsten furnace source, was crossed at right angles by an H^- beam, the intersection being on the axis of the collision chamber. Before reaching the interaction region, any charged particles in the hydrogen beam were removed by electrostatic deflexion. The degree of dissociation D in the beam could be measured by electron impact ionization of the beam followed by mass analysis. Thus if q_1/q_2 is the ratio of the cross-sections for ionization of H and H_2 respectively, by electrons of the chosen energy, and P_1 and P_2 are the peak heights in the mass spectrum corresponding to H^+ and H_2^+ ions, then

$$1/D = 1 + \sqrt{2}\, \frac{q_2}{q_1}\frac{P_2}{P_1}. \qquad (13.53)$$

q_1/q_2 is known from earlier measurements.

The H^- beam was produced from a hot-cathode arc in water vapour and it was focused and collimated before entering the interaction region. Its intensity was measured by a Faraday cage.

Two plane electrodes were located symmetrically above and below the interaction region in planes parallel to the plane containing the two beams. A weak electric field could be applied across the plates while provision was also made for application of a magnetic field either parallel or perpendicular to the electrodes. In the former case, with a sufficiently strong electric field applied to produce saturation, the negative current consisted of slow negative ions only, whereas in the latter it consisted of slow negative ions and slow electrons.

To distinguish the wanted signals from the background, the hydrogen beam was chopped at a frequency of 100 Hz and the signals from the plates were passed through a high-gain narrow-band amplifier operating at this frequency, followed by a phase-sensitive detector using a signal from the chopper wheel as a reference.

Absolute values of the cross-sections were obtained by comparison with that for charge transfer by protons in atomic hydrogen,

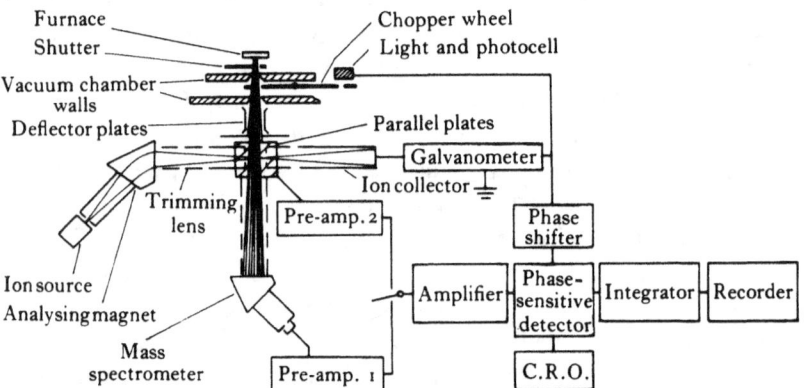

Fig. 13.6. General arrangement of the crossed-beam apparatus used by Hummer et al. (1960) to study H⁻–H collisions.

measured in an earlier experiment. This was carried out by reversing the interaction voltage on the ion source to obtain a beam of protons at a chosen energy, and the saturation current of positive ions to the collection plate measured with the same hydrogen beam as for the negative-ion experiments. Furnace conditions were chosen so that this beam was nearly completely dissociated.

The absolute measurement of the charge transfer cross-section for protons in atomic hydrogen was made by Fite, Brackmann and Snow (1958) by comparing the slow positive-ion signals S and S_0 obtained at two different oven temperatures T and T_0 respectively but with the same rate of inflow of H_2 gas to the oven. T_0 was such that no dissociation took place. We write

$$S = S_1 + S_2, \qquad (13.54)$$

where S_1 and S_2 arise from H and H_2 respectively in the partially dissociated beam. Also

$$S_2 = (1 - D) S_0 (T_0/T)^{1/2}, \qquad (13.55)$$
$$S_1 = 2^{1/2} D(Q_1/Q_2) S_0 (T_0/T)^{1/2}, \qquad (13.56)$$

the factor $(T_0/T)^{1/2}$ allowing for the variation of the intensity of the beam with temperature while the factor $2^{1/2}$ is the ratio of the mean velocity of the H atoms to that of the H_2 molecules at the same temperature. Q_1 and Q_2 are the charge transfer cross-sections for H and H_2 respectively. Eliminating S_2 and S_1 from (13.54) by means of (13.55) and (13.56) we have

$$Q_1 = (Q_2/2^{1/2} D) \left\{ \frac{S}{S_0} \left(\frac{T}{T_0} \right)^{1/2} + D - 1 \right\}. \qquad (13.57)$$

In the actual experiments, slow ions from charge transfer had to be distinguished from those from ionization by proton impact. This was done using magnetic fields as in the negative-ion experiments.

A similar technique was used by Snow, Rundel and Geballe (1969) to measure charge transfer and detachment cross-section for O^-–O collisions.

Results obtained are discussed below (see Figs. 13.7 and 13.10).

13.5 Observed results

H^-–H collisions

Fig. 13.7 illustrates the detachment cross-section for H^-–H collisions measured by Hummer *et al.* (1960) using the crossed-beam technique. Comparison is made with theoretical values calculated by McDowell and Peach (1959) using Born's approximation, as described on p. 588, by the semi-classical method of Bates and Walker (1967) (see p. 587) and by Bardsley (1967) from the impact parameter method which takes direct account of autodetachment (see p. 584). In applying the semi-classical method the total cross-section included for collisions of electrons with H atoms was taken to be the sum of the elastic cross-section, calculated theoretically, and measured cross-sections for ionization and for excitation of the 2p states.

It will be seen that the last two methods give better results than

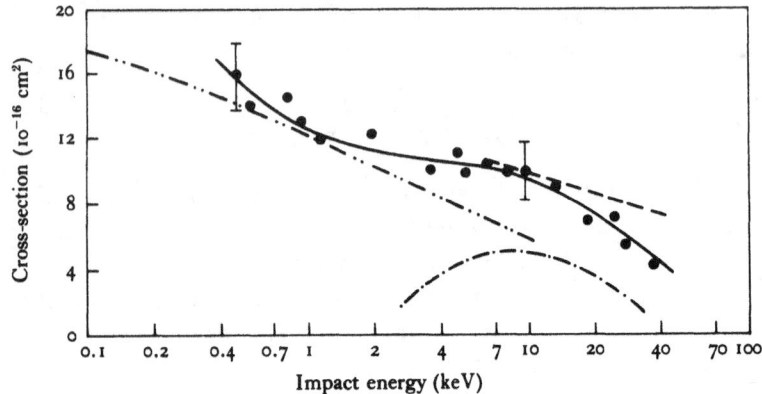

Fig. 13.7. Cross-sections for the electron detachment process $H^- + H \to H + H + e$. ● observed cross-section for the electron production by H^- (Hummer et al., 1960), —— best fit to observed data. Calculated - - - - by Bates and Walker, – · – · McDowell and Peach, – · · – · · Bardsley.

does Born's approximation but it must be remembered that further simplifications were made in carrying out the calculations with this approximation.

Fig. 13.8 shows $Q_{tr}^{1/2}$, where Q_{tr} is the charge transfer cross-

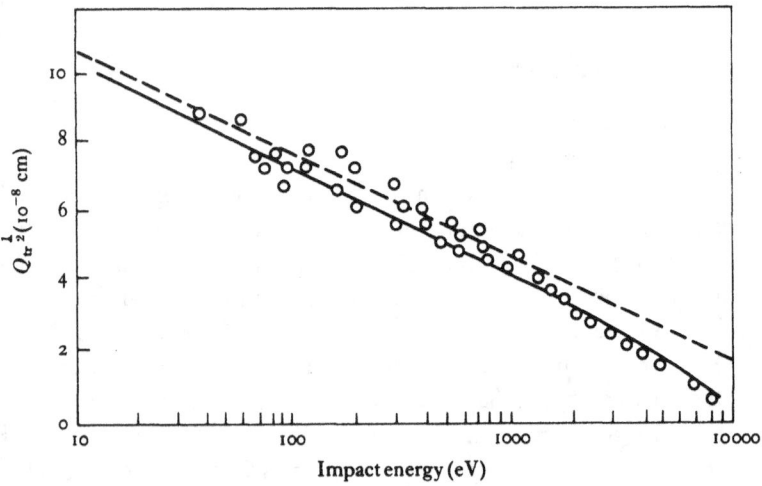

Fig. 13.8. Comparison of observed and calculated values of $Q_{tr}^{1/2}$, where Q_{tr} is the cross-section for the charge transfer process $H^- + H \to H + H^-$. ○ observed by Hummer et al. (1960). Calculated - - - - without allowance for detachment, —— with allowance for detachment.

section, as measured by Hummer et al. Results obtained theoretically by Dalgarno and McDowell (1956) (see p. 582), without allowance for autodetachment, are included for comparison. Also included are the results obtained if the correction (13.23) is included, Q_{de} being taken from the observed results in Fig. 13.7. Although there is a considerable scatter in the experimental points the inclusion of the correction definitely improves the agreement with theory at impact energies above 1 keV.

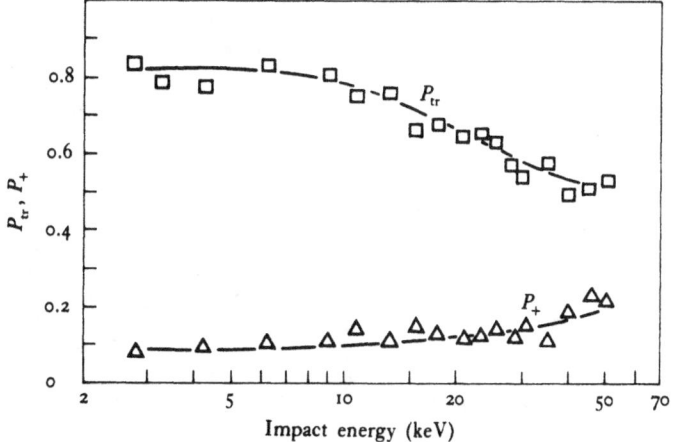

Fig. 13.9. Variation of the probability P_{tr} of charge transfer and P_+, of production of H$^+$ ions, in H$^-$–H collisions as a function of ion impact energy E, for scattering at 1.2° in the laboratory system, observed by Keever et al. (1968).

The probability P of charge transfer has been measured by Keever, Lockwood, Helbig and Everhart (1968) as a function of impact energy between 2 and 50 keV for collisions in which the scattering angle is 1.2° in the laboratory system. At the same time measurements were also made of P_+, the probability of production of a proton by either of the reactions

$$H^- + H \longrightarrow H^+ + H + 2e,$$
$$\longrightarrow H^+ + H^- + e.$$

Fig. 13.9 shows the results obtained.

Charge transfer reactions involving atomic negative ions and neutral atoms

The crossed-beam technique has been applied by Snow et al. (1969) to measure charge transfer cross-sections for O⁻–O, H⁻–O, O⁻–H, C⁻–H and C⁻–O collisions. They normalized their results, which are illustrated in Fig. 13.10, by comparison with the absolute measurements for H⁻–H made by Hummer et al. In Fig. 13.10 the cross-sections calculated for H⁻–H, O⁻–O and C⁻–C

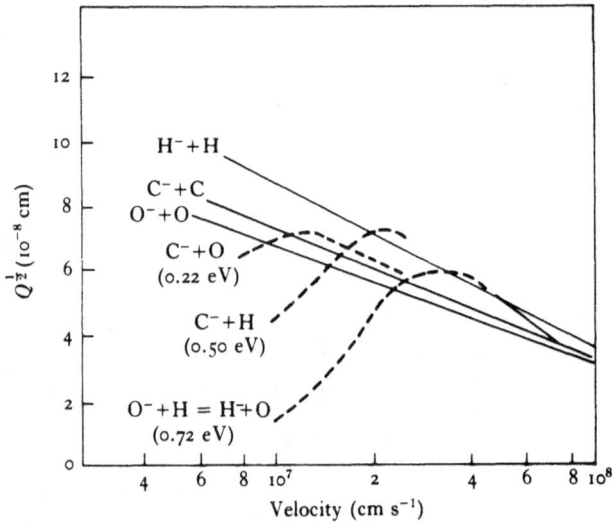

Fig. 13.10. Cross-sections for charge transfer reactions involving negative ions and atomic targets, observed by Snow et al. (1969). ---- observed curves for C⁻–O, C⁻–H and O⁻–H and H⁻–O reactions with electron affinity differences ΔE_a, as indicated. —— calculated cross-sections for symmetrical reactions H⁻–H, C⁻–C, and O⁻–O according to the semi-empirical theory of Firsov (1951) and of Rapp and Francis (1962).

collisions using the semi-empirical formulae (13.19) and (13.20) are also shown. In O⁻–O they are rather larger than the observed values.

According to the semi-empirical theory for unsymmetrical charge transfer given by Rapp and Francis (1962) (see p. 586) the results for H⁻–O, O⁻–H, C⁻–H and C⁻–O should rise with increasing impact velocity to a maximum at a velocity which increases with the magnitude of the difference between the electron affinities of the atoms concerned. At higher velocities the cross-section should lead to that which would be followed for symmetrical charge transfer between ions in which the electron affinity falls midway between that of the atoms actually involved. Reference to Fig. 13.10 shows these features are clearly present. The ratio of the cross-sections for the forward and backward reactions

$$H^-(^1S) + O(^3P) \rightleftharpoons H(^2S) + O^-(^2P)$$

should be that, 4/3, of the statistical weights. The observed ratio is 1.4 ± 0.4 which is consistent with this.

Charge transfer reactions involving molecules

Fig. 13.11 illustrates results obtained for a number of charge transfer reactions of H^- and O^- ions with neutral molecules. Of these $H^- - O_2$ and $O^- - O_2$ were studied by Bailey and Mahadevan (1970) using the apparatus described on p. 593 which enabled them to distinguish between charge transfer and detachment. The measurements of Snow et al. (1969), using the crossed-beam technique, were normalized to those of Bailey for $H^- - O_2$. While mass analysis of the product ions could be carried out, no means were available for distinguishing between charge transfer and detachment. The same applies to the measurements of Rutherford and Turner (1967). In all these cases, however, detachment was probably not important.

For the reactions of O^-, O_2^- and OH^- with NO_2, data at thermal energies is available from measurements using the flowing afterglow technique (Chapter 12, p. 526). The values obtained for the mean cross-section, 1.70×10^{-14}, 1.60×10^{-14} and 1.50×10^{-14} cm², are not incompatible with the beam results at the lowest energy 3 eV at which observations were made, namely 0.6×10^{-14}, 0.55×10^{-14} and 0.75×10^{-14} cm².

Cross-sections for charge-transfer collisions of Cl^-, Br^- and I^- ions with a variety of molecular gases, O_2, Cl_2, I_2, CO_2, NO_2, SF_6, C_6H_6 and CCl_4, have been measured by Dimov and Rosljakov (1971) over the impact energy range from 100 or so eV to a few keV. The condenser plate technique was used, discrimination between slow electrons and negative ions being carried out as described on

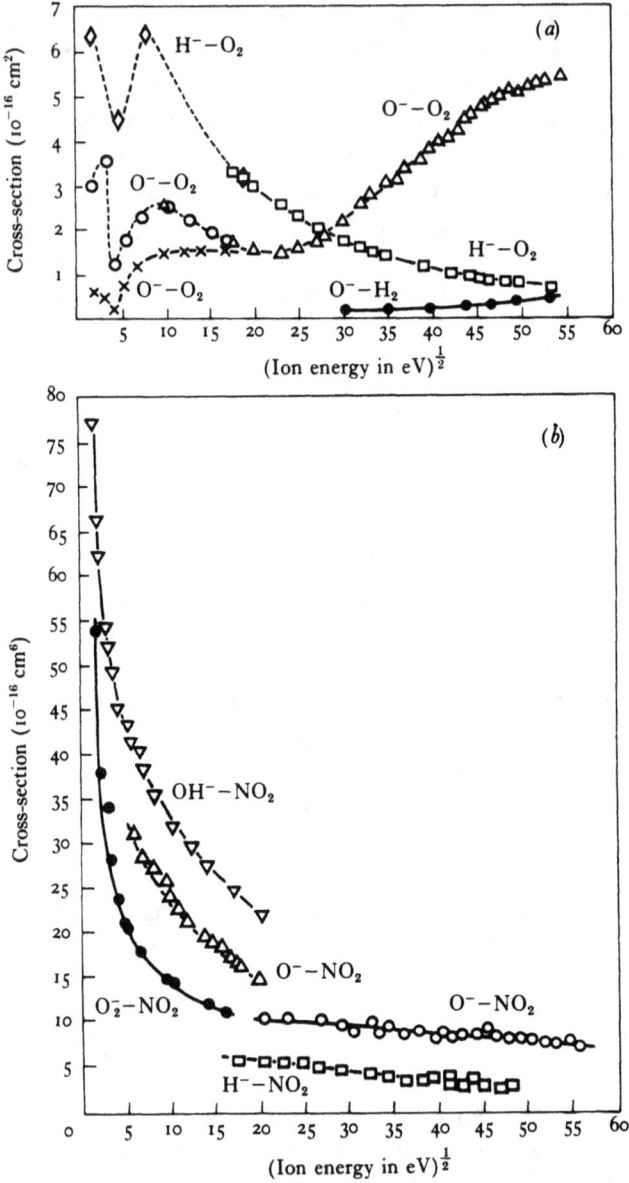

Fig. 13.11. Cross-sections for charge transfer reactions involving negative ions and molecular targets. (a) Reactions in O_2. $O^- – O_2$, observed by × Rutherford and Turner, ○ Bailey, △ Snow, Rundel and Geballe. $H^- – O_2$, observed by ◇ Bailey, □ Snow, Rundel and Geballe. $O^- – H_2$, observed by ● Snow, Rundel and Geballe. (b) Reactions in NO_2. $O^- – NO_2$, observed by △ Rutherford and Turner, ○ Snow,

p. 591. Provision was also made for mass analysis of the product ions. It was found that in O_2 and NO_2 simple charge transfer occurs, the ions produced being very largely O_2^- and NO_2^- respectively. With CO_2, SF_6, C_6F_6 and CCl_4 on the other hand, atomic negative ions, O^-, F^- and Cl^- were predominantly among the reaction products. With Cl_2 and I_2, molecular rather than atomic ions were generally more abundant but the relative probabilities were more comparable.

Detachment reactions involving H^- ions and other neutral species

Cross-sections for detachment of electrons from H^- ions in collision with rare gas atoms have been measured by Hasted and his collaborators (Hasted, 1952; Stedeford and Hasted, 1955; Hasted and Smith, 1956) and by Bydin (1966a), for impact energies in the range 0.1–40 keV, using the condenser plate method as described on on p. 592. Other measurements for H^- ions in collisions with rare gas atoms and with a number of molecular gases have been carried out by Stier and Barnett (1956), Whittier (1954), Williams (1967) and by Pilipenko, Gusev and Fogel (1966), all using the technique of charge analysis of a fast beam after passage through the target gas (see p. 400). In the experiment by the latter authors a condenser plate technique was used at the same time to verify that charge transfer collisions were unimportant in the cases investigated (O_2, NO and CO).

The results obtained in these different experiments do not agree well, as may be seen from Figs. 13.12 and 13.13. These figures also include theoretical results obtained by the semi-classical method of Bates and Walker (1967) (see p. 588). As far as can be judged this method gives quite good results. For H^-–He collisions it is considerably more satisfactory than Born's approximation as judged from the calculations of Sida (1955) (see p. 587), which, however,

---

Rundel and Geballe. H^-–NO_2, observed by □ Snow, Rundel and Geballe. O_2^-–NO_2, observed by ● Rutherford and Turner. OH^-–NO_2, observed by ▽ Rutherford and Turner. In both (a) and (b) the data of Snow et al. are normalized to agree with the absolute cross-sections observed by Bailey for O^-–O_2 at 300 eV.

used the simple wave function (1.16) for H⁻. Moreover, it does not include contributions from excitation or ionization of the target or of the resultant H atom. The semi-classical approximation includes allowance for target excitation and ionization but not for that of the H atom.

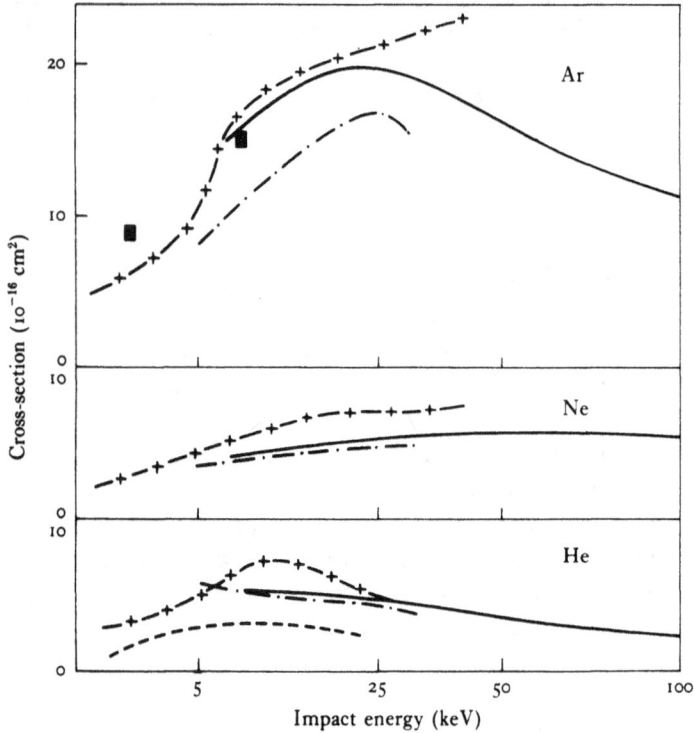

Fig. 13.12. Cross-sections for detachment of electrons from H⁻ in collisions with He, Ne and Ar atoms, as indicated. ——— calculated by Bates and Walker (1967), semi-classically, ---- calculated by Sida (1955) using Born's approximation, —·—·— observed by Stier and Barnett (1956), —+—+— observed by Stedeford and Hasted (1955), ■ observed by Bydin (1966a).

Andreev, Ankudinov, Dukel'skii and Orbeli (1969) have measured cross-sections for detachment from H⁻ leaving the H atom in a two-quantum excited state. This they achieved by measuring the intensity of the Lyα emission accompanying the passage of an H⁻ beam through the detaching gas. Experiments were carried out for H⁻ ions in the kinetic-energy range 5–40 keV in rare gases. Care was taken to establish that the observed emission was proportional to the gas pressure and so arises in primary collisions. The ratio $Q_{de}(2s)/Q_{de}(2p)$ of cross-sections for detachment, leaving the neutral H atom in the 2s and 2p states respectively, was measured by com-

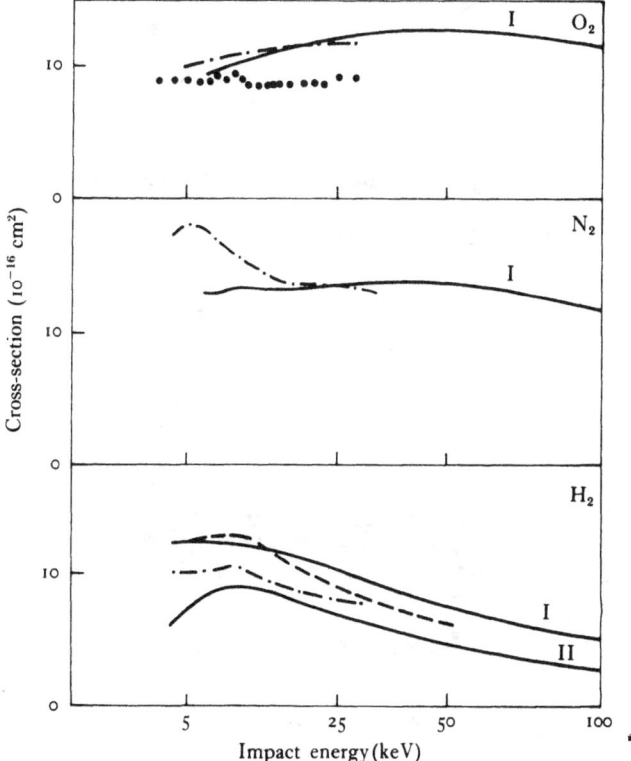

Fig. 13.13. Cross-sections for detachment of electrons from H⁻ in collision with H_2, N_2 and O_2. ___I___ calculated by Bates and Walker (1967) semi-classically, ___II___ calculated by McDowell and Peach (1959) for 2H, using Born's approximation, —·—· observed by Stier and Barnett (1956), ---- observed by Whittier (1954), ●●● observed by Pilipenko et al. (1966).

paring the intensity of the Ly α emission with and without application of a constant electric field of between 600 and 800 V cm⁻¹ to the collision chamber. Without this field emission came only from H(2p) atoms produced in primary collisions but with the field (H2s) atoms were quenched to H(2p) and also contributed to the emission. Absolute cross-sections were obtained by comparison of the Ly α emission with that obtained in electron capture by protons (Andreev, Ankudinov and Bobashev, 1966) for which absolute values were available. Fig. 13.14 illustrates the results obtained for $Q_{de}(2s)$ and $Q_{de}(2p)$ which are seen by comparison with Figs. 13.12

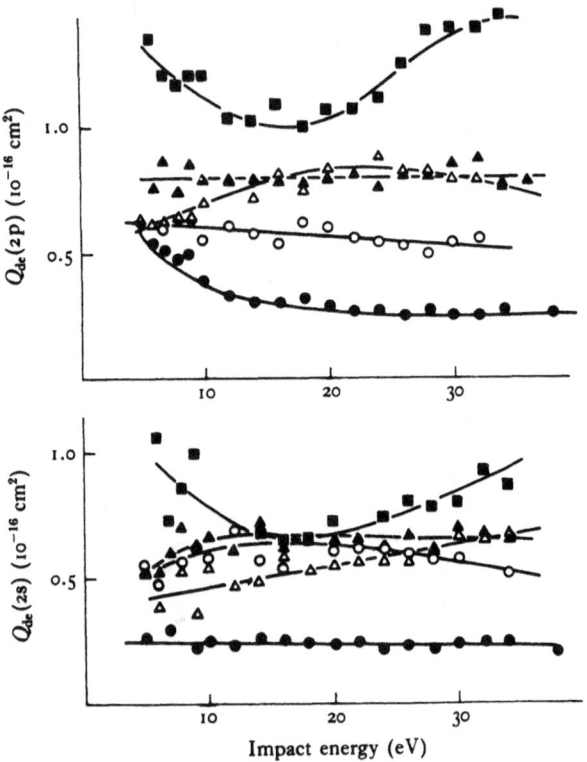

Fig. 13.14. Cross-sections, $Q_{de}(2s)$ and $Q_{de}(2p)$ for detachment of electrons from H⁻ leaving the atom in the 2s and 2p states respectively, in collisions with rare gas atoms as observed by Andreev et al. (1969). ● He, ○ Ne, ▲ Ar, △ Kr, ■ Xe.

and 13.13 to be small compared with the total detachment cross-section Q_{de}.

Absolute differential cross-sections for detachment collisions of H⁻ in He, Ar, H₂, N₂ and O₂ have been measured by Geballe and Risley (1973) for collision energies between 200 eV and 10 keV. This is an extension of earlier work by Edwards, Risley and Geballe

(1971) who measured the energy distribution of detached electrons and observed structure due to excitation of autodetaching states which, having lifetimes of order 10^{-14} s, break up before the ion can travel far from the point at which excitation occurs. The results of this work, as far as the autodetaching states are concerned, have already been discussed in Chapter 5, p. 119. Fig. 13.15 shows the observed differential cross-section $I(\epsilon, \theta)\,d\epsilon\,d\Omega$ for production

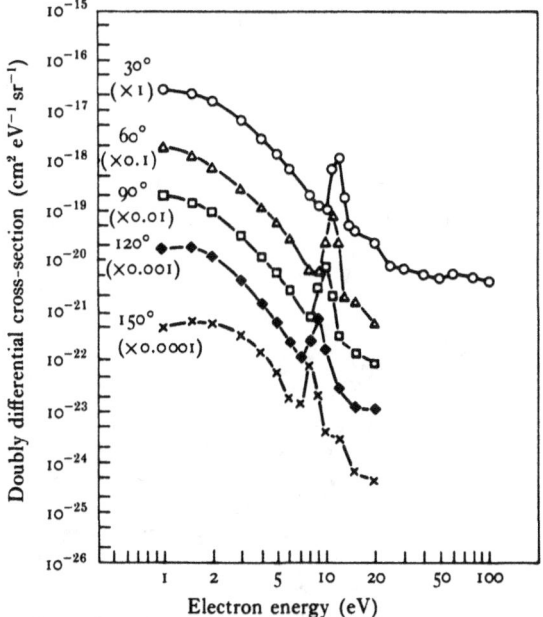

Fig. 13.15. Observed differential cross-sections $I(\epsilon, \theta)\,d\epsilon\,d\Omega$ for production of electrons by collisions of H$^-$ ions with 200 eV kinetic energy in helium, plotted as functions of ϵ for different angles θ. From Geballe and Risley (1973).

of electrons with energy between ϵ and $\epsilon + d\epsilon$ relative to the incident ion and moving in directions lying within the solid angle $d\Omega$ about a direction making an angle θ with that of the incident H$^-$ beam. These measurements relate to a beam of 200 eV energy in helium. The angular distribution of electrons with the same kinetic energy is not far from isotropic. In these experiments the separate autodetachment peaks shown in Fig. 5.1(b) are not separately resolved and appear in Fig. 13.15 as a single peak which moves to lower energies as the angle θ is increased, a purely kinematic effect.

Fig. 13.16 shows the energy distributions obtained by integrating $I(\epsilon, \theta)\,d\Omega$ over all scattering angles. The peak which occurs at lower energies as the ion energy decreases is located at an energy for which the electron velocity is equal to the velocity of the ion. This supports the basic approach to the semi-classical theory, described

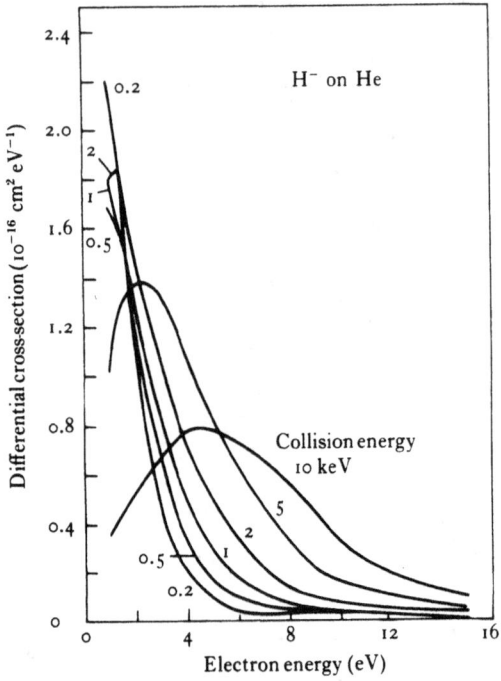

Fig. 13.16. Energy distributions of electrons detached from H⁻ ions in collisions with He, integrated over all angles of ejection, for different incident ion energies as indicated. From Geballe and Risley (1973).

on p. 588, in which the electron is detached as in an elastic collision with the target atom, with velocity equal to that of the ion.

Detachment from He⁻

The properties of the metastable He⁻ ion have been discussed in Chapter 5, p. 126, and experiments concerned with the formation of the ions have been described in Chapter 10, p. 405. There has been considerable interest in measuring cross-sections for detachment from these ions by impact with various atomic and molecular targets, particularly because of the use of the ions in tandem accelerators arranged to provide energetic beams of He^+ or He^{2+} ions (see Chapter 15, p. 690).

No serious complication arises, in making such measurements, for the finite lifetime of the ions. This loss mechanism may easily be allowed for from the data available about its rate (see Chapter 4, p. 130). Ryding, Wittkower and Rose (1968) have measured detachment cross-sections for ions with energies between 400 and 1500 keV in H_2, He and Ne. The method used was of the type described in Chapter 10, p. 400. A beam of He^- ions produced by charge transfer to He^+ was passed through a gas target. The

TABLE 13.1 *Detachment cross-sections $Q_{\bar{1}0}$ and $Q_{\bar{1}1}$ in 10^{-16} cm^2, for He^- ions in H_2, He and Ne, observed as a function of impact energy by Ryding et al. (1968)*

Energy (keV)	$Q_{\bar{1}0}$			$Q_{\bar{1}1}$		
	H_2	He	Ne	H_2	He	Ne
400	2.60	2.97	5.91	0.180	0.307	0.665
600	2.06	2.21	5.42	0.175	0.260	0.772
800	1.68	1.89	4.78	0.189	0.220	0.712
1000	1.46	1.65	4.49	0.166	0.235	0.843
1150	1.39	1.45	4.19	–	0.230	0.777
1350	1.17	1.23	3.82	0.103	0.221	0.675
1500	1.17	1.05	–	0.124	0.204	–

fractional composition of the beam issuing from the target was measured as a function of the effective target thickness (see Chapter 10, p. 400). For the He^0 fraction this growth rate is at first linear with a slope given by the desired cross-section $Q_{\bar{1}0}$ for single detachment. Similarly, from the initially linear rate of growth of the He^+ fraction the corresponding cross-section $Q_{\bar{1}1}$ for double detachment was obtained. Table 13.1 gives the results obtained in this way.

Measurements at lower impact energies (4–30 keV) have been made by Simpson and Gilbody (1972) in H_2, He and Ar using essentially the same method. It was found that for each target gas the cross-section $Q_{\bar{1}0}$ behaved in much the same way with impact energy as for detachment from H^- though the absolute magnitudes were somewhat greater, by a factor of about 2 for H_2 and He and a much smaller factor, near 1.1, for Ar.

Since metastable ions are produced essentially by attachment of an electron to an excited 2^3S helium atom (see Chapter 10, p. 405) it is to be expected that detachment of an electron, if it occurs directly, will leave the atom predominantly in this state. Pedersen and Hvelplund (1973) have investigated this by applying the technique used by Gilbody, Dunn, Browning and Latimer (1970) for determining the fractional concentration of metastable atoms in a neutral helium beam, to the neutral beam resulting from detachment. For 50 keV He^- in H_2 they found that in about 58% of the detachment collisions the atom is left in a metastable state. This is rather less than might be expected and suggests that detachment may occur with appreciable probability through some less direct process than first envisaged.

Detachment reactions involving other negative ions

Detachment cross-sections for Br^- and I^- ions with energies in the range 0.2–2 keV in rare gases have been measured by Bydin and Dukel'skii (1957) using the condenser plate method. The measured apparent threshold energies in the centre of mass system were substantially greater than the electron affinities of the halogen atom concerned.

Detachment cross-sections for atomic halogen ions have also been measured in O_2, Cl_2, I_2, CO_2, NO_2, SF_6, C_6H_6 and CCl_4, by Dimov and Rosljakov (1971) in conjunction with their charge transfer measurements (see p. 603). The results are generally similar to those in the rare gases. Energy distributions of electrons produced by detachment from Cl^- on collision with H_2 and He atoms have been measured by Cunningham and Edwards (1973). The results obtained, which are mainly concerned with the identification of autodetaching states of Cl^-, have been discussed in Chapter 5, p. 146.

Detachment from O^- has been studied extensively, particularly in the impact energy range below 100 eV. Typical results are shown in Fig. 13.17 for O^-–Ar collisions including data obtained by Hasted (1952, 1954) using the condenser plate technique, by Wynn et al.

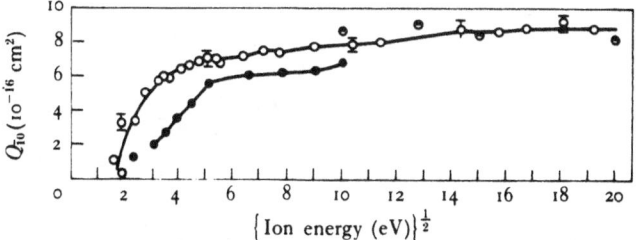

Fig. 13.17. Observed cross-sections $Q_{\bar{1}0}$ for detachment of electrons from O$^-$ in Ar. ○ Wynn et al. (1970), ● Roche and Goodyear (1969), ⊖ Hasted (1954).

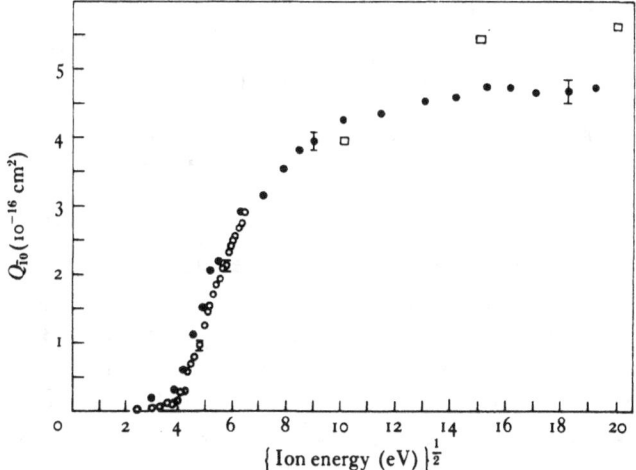

Fig. 13.18. Observed cross-sections $Q_{\bar{1}0}$ for detachment of electrons from O$^-$ ions in collision with He atoms. Wynn et al. (1970), ● original apparatus, ○ modified apparatus; Hasted (1954) □.

(1970), using the technique described on p. 594, and by Roche and Goodyear (1969), using that described on p. 595.

Edwards et al. (1971) have measured energy distributions of electrons arising from detachment collisions of O$^-$ in helium. Their results, which relate particularly to the identification of autodetaching states of O$^-$, have been discussed in Chapter 5, p. 143.

The behaviour of the cross-section near threshold has been investigated by Wynn et al. (1970) for O$^-$–He collisions, using the modified arrangement of their apparatus, especially suited to the

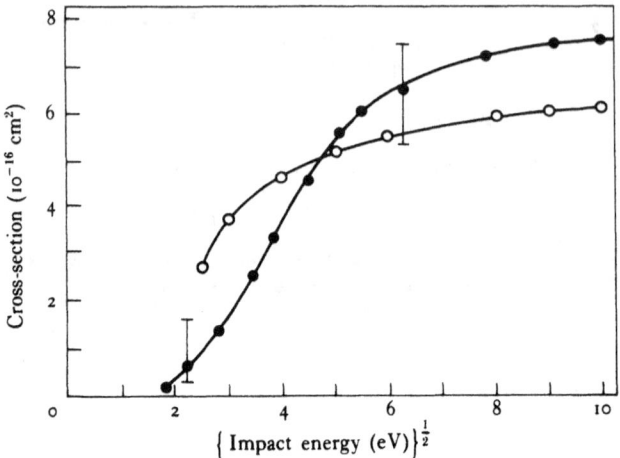

Fig. 13.19. Observed cross-sections for detachment of electrons from O^- in O_2. –O– observed by Bailey, –●– observed by Roche and Goodyear.

study of detachment collisions at low energy in which the mass of the ion is much greater than that of the target atom. Fig. 13.18 shows the results they obtained. These are compared in the figure with results measured using the unmodified apparatus (see Fig. 13.3) and the results of earlier measurements by Hasted and Smith (1954) using the condenser plate method. Similar measurements have been carried out by Wynn et al. for O_2^-–He and OH^-–He collisions.

Fig. 13.19 shows observed results for O^- and O_2. It is of interest to note that for the reaction between O_2^- and O_2 in which symmetrical charge transfer can occur, the cross-section falls rapidly as the impact energy decreases below 100 eV in a similar manner to the unsymmetrical case.

Cross-sections for detachment from negative ions of the alkali metal atoms on impact with rare gas atoms have been measured by Bydin (1966b) using the condenser plate technique.

Double detachment

The first measurements of cross-sections for detachment collisions in which two electrons are detached were made by Dukel'skii and

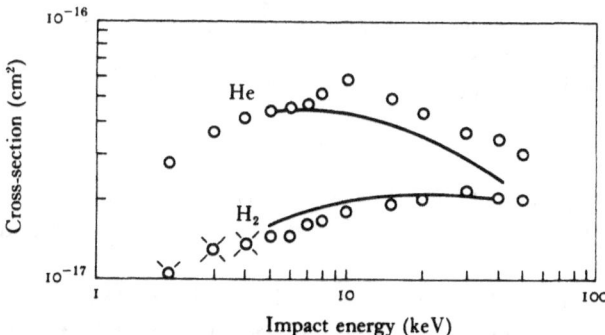

Fig. 13.20. Observed cross-sections $Q_{\bar{1}1}$ for double detachment of electrons from H⁻ in collisions with He and H₂. ○ Williams (1967), × Tisone and Branscomb (1964) normalized to agree with Williams at 4 keV, ——— Fogel et al. (1957).

Fedorenko (1956). They made observations for collisions of heavy negative ions Cl⁻, Br⁻, I⁻, Na⁻, Sb⁻, Bi⁻ and Sb₂⁻, with kinetic energy between 5 and 17.5 keV in He, Ar, H₂ and N₂. The cross-sections they obtained for double detachment were in the range 10^{-16}–10^{-17} cm². A little later Fogel, Ankudinov and Slabospitskii (1957) made the first measurements for detachment from H⁻. Further measurements have been made by Tisone and Branscomb (1964) and by Williams (1967).

These experiments make use of the fact that, when a beam of

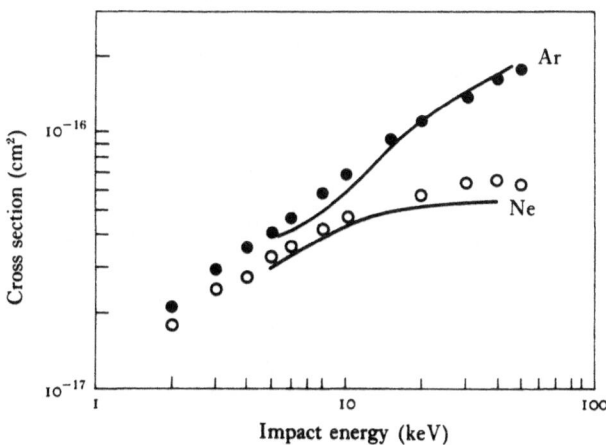

Fig. 13.21. Observed cross-sections $Q_{\bar{1}1}$ for double detachment of electrons from H⁻ in collisions with Ne and Ar. ● Ar, ○ Ne, Williams (1967), ——— Fogel et al. (1957).

negative ions of some keV energy is fired through a length l in gas at not too high pressure p, the ratio of the positive (I^+) and negative ion (I^-) currents in the beam emerging from the gas is given by

$$I^+/I^- = ap + bp^2,$$

where $a = lQ_{\bar{1}1}/kT$ and b involves a number of other cross-sections. By measuring I^+/I^- as a function of p, a may be obtained and hence $Q_{\bar{1}1}$.

Fig. 13.20 gives the results obtained in the different measurements for H^- in H_2 and He. There is quite good agreement for the former but, for He at energies above 10 keV, Williams finds considerably larger cross-sections than do Fogel *et al.* For H^- in Ar and Ne, however, the two experiments give results which agree well, as may be seen from Fig. 13.21. In all cases $Q_{\bar{1}1}$ is considerably smaller than $Q_{\bar{1}0}$, as may be seen by comparison with Figs. 13.12 and 13.13. Values of $Q_{\bar{1}1}$ for He^- ions in various gases, measured by Ryding *et al.* (1968), using the same technique, are given in Table 13.1, in which they may be directly compared with $Q_{\bar{1}0}$ measured in the same experiments.

CHAPTER 14

Recombination of Negative and Positive Ions – Mutual Neutralization

In the previous two chapters we have discussed reactions of negative ions with neutral atoms and molecules, including those which lead to detachment of the electron from the negative ion. We now consider reactions of negative with positive ions. If the ions are both atomic we have the following possibilities.

(a) Radiative recombination

$$X^- + Y^+ \longrightarrow X + Y + h\nu, \qquad (14.1)$$
$$\longrightarrow XY + h\nu. \qquad (14.2)$$

(b) Mutual neutralization

$$X^- + Y^+ \longrightarrow X' + Y''. \qquad (14.3)$$

(c) Three-body recombination

$$X^- + Y^+ + M \longrightarrow XY + M. \qquad (14.4)$$

More possibilities, involving atomic rearrangements, occur when either or both of the ions are molecular, as for example

$$X^- + YZ^+ \longrightarrow XY + Z, \qquad (14.5)$$

but we shall not attempt to list all the types of reaction of this kind which may occur.

We may dismiss radiative recombination rather briefly by noting that, as for radiative attachment of electrons, the chance per collision will be not greater than 10^{-6} in general. The reasons are the same as given in Chapter 8, p. 243. Even allowing for the long-range Coulomb interaction, which will increase the collision cross-section, the radiative recombination coefficient at thermal energies is unlikely to exceed 10^{-14} cm^3 s^{-1}. It is not surprising that this type of recombination has not been detected experimentally.

For the other two kinds of reaction a considerable body of experimental and theoretical information now exists and we proceed to discuss this. As usual we begin with a survey of the theory, then describe some of the experimental methods for measuring the reaction rates and finally discuss the results obtained in the light of the theory.

14.1 Theoretical considerations

Mutual neutralization

In Chapter 10, p. 383, we have outlined the theory of the inverse process to mutual neutralization in which the neutral systems X and Y are in their ground states. Apart from the need to include the possibility that these systems may be formed in excited states, little modification need be made.

Let ΔE be the energy released in the reaction (14.3) and M be the reduced mass of the ions. Then as in (10.16) we have for the mutual-neutralization cross-section when the relative kinetic energy of the colliding ions is E,

$$Q_\mathrm{m} = 4\pi R_\mathrm{c}^2 I(\eta), \qquad (14.6)$$

where

$$I(\eta) = \int_1^\infty \exp(-\eta x)\{1 - \exp(-\eta x)\}\, x^{-3}\, \mathrm{d}x, \qquad (14.7)$$

with

$$\eta = 4\pi e^2\, M^{1/2}\, \{U(R_\mathrm{c})\}^2 / 2^{1/2}\, \hbar E^{1/2}\, \Delta E^2. \qquad (14.8)$$

$R_\mathrm{c} = e^2/\Delta E$ is the nuclear separation at which the potential-energy curves for the state of $\mathrm{X^- Y^+}$ would intersect that for $\mathrm{X'Y''}$ if it were not for the interaction $U(R)$ between them (see Fig. 10.1). M is the reduced mass.

For the simplest mutual neutralization reaction

$$\mathrm{H^-} + \mathrm{H^+} \longrightarrow \mathrm{H'} + \mathrm{H''}, \qquad (14.9)$$

$U(R)$ is given to a sufficient approximation by

$$U(R) = e^2 \iint (r_{b1}^{-1} + r_{a2}^{-1} - r_{ab}^{-1} - r_{12}^{-1})\, \psi(r_{a1}, r_{a2})\, \phi_s(r_{a1}) \\ \times \phi_n(r_{b2})\, \mathrm{d}\tau_1\, \mathrm{d}\tau_2. \qquad (14.10)$$

Here the nuclei, initially of $\mathrm{H^-}$ and $\mathrm{H^+}$, are distinguished as a, b respectively, $\mathbf{r}_{a1}$ is the vector separation of electron 1 from nucleus

a, r_{ab} is the nuclear separation, r_{12} that between the electrons. $\psi(r_{a1}, r_{a2})$ is the wave function for H⁻, ϕ_s and ϕ_n those for the final atomic hydrogen states.

Bates and Lewis (1955) carried out detailed calculations for reactions in which H represents a ground-state atom. They found that the only important contributions to the mutual-neutralization cross-section due to long-range pseudo-crossing are from cases in which $n = 2$ or 3, n being interpreted as the principal quantum number. For $n = 1$ or > 5, no pseudo-intersection occurs, while for $n = 4$, R_c is so large that $U(R_c)$ is negligible. Account was taken in the calculations of the orbital degeneracy of the excited H states.

TABLE 14.1 *Calculated mutual-neutralization coefficients α_{2m}, α_{3m} for H⁻–H⁺ collisions in which one H atom is left unexcited and the other in a state with total quantum number = 2, 3 respectively*

Temperature (°K)	Mutual-neutralization coefficient	
	$\alpha_{2m}(10^{-9}$ cm³ s⁻¹)	$\alpha_{3m}(10^{-7}$ cm³ s⁻¹)
250	7	1.3
500	5	0.9
1000	3.5	0.7
2000	3	0.5
4000	2.5	0.4
8000	2.5	0.3
16 000	3.5	0.3
32 000	6	0.25

Table 14.1 gives the results obtained for the mutual-neutralization coefficient as a function of temperature. In terms of (14.6), (14.7) and (14.8) this is given in cm³ s⁻¹ for each transition by

$$\alpha_m = 0.51 M^{-1/2}(\Delta E)^{-2} T^{-3/2} \int_{\Delta E}^{\infty} \exp\{a(\Delta E - z)\}\, I(\xi z^{-1/2}) z\, dz \quad (14.11a)$$

where

$$a = 1.16 \times 10^4 T^{-1}, \quad \xi = 247 M^{1/2}\{U(R_c)\}^2/\Delta E^2, \quad (14.11b)$$

U and ΔE being in eV and M the reduced mass on the O¹⁶ mass scale.

Bates and Boyd (1956) carried out further detailed calculations

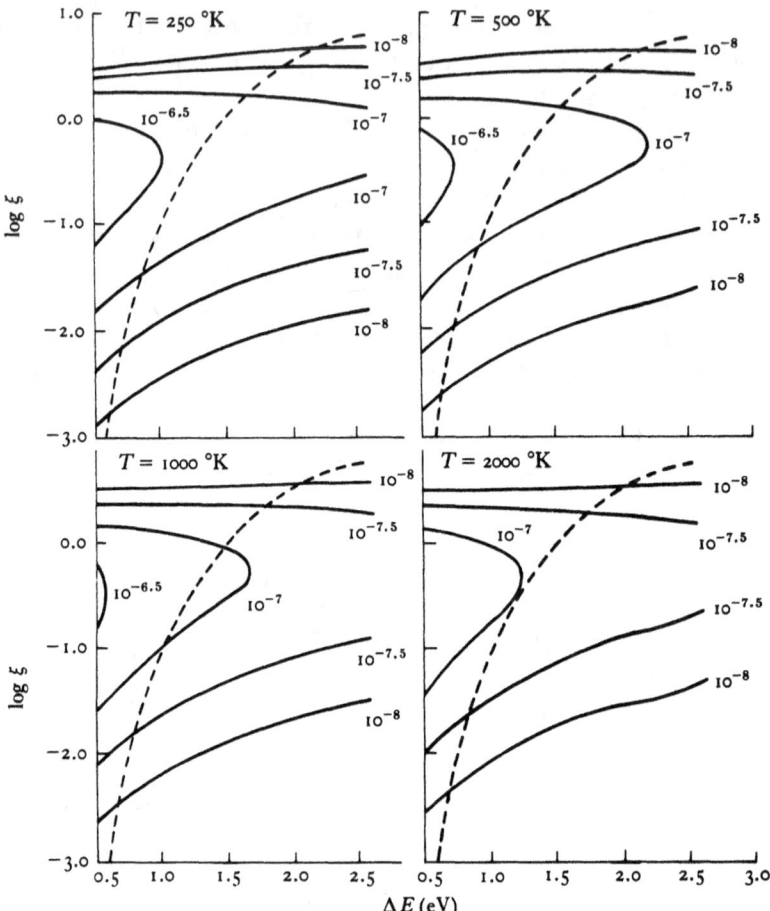

Fig. 14.1. Calculated variation of $\log \xi$ with ΔE for four different electron temperatures T, showing contour lines (———) of constant mutual-neutralization coefficient α_m for values of $M^{1/2} \alpha_m$ marked on each. The predicted variation of $\log \xi$ with ΔE is shown for each temperature by the broken lines. For the meaning of the symbols see the text.

for collisions of H⁻ ions with positive ions of the alkali metal atoms in which the electron is captured into the states with lowest and next lowest total quantum number which are available, the product H atom being unexcited. They found that to a good approximation for these cases $U(R_c)$ is a function $f(\Delta E)$ of ΔE only, a result which remained true when the H⁻–H⁺ cases were included. Thus ξ can be written as $247 \, M^{1/2} \, g \, (\Delta E)$, where $g(x) = x^{-2} f(x)$.

Contour lines of constant mutual-neutralization coefficient for fixed T and M may then be calculated from (14.11a) and plotted in a $\log \xi$, ΔE-plane as in Fig. 14.1. On the same diagram may be plotted the relation between $\log \xi$ and ΔE obtained from the calculation of $U(R_c)$. The intersections of this plot with the contour lines gives α_m for any ΔE at the temperature concerned.

Referring to Fig. 14.1, the first result which is apparent is that at any temperature the maximum value of α_m, considered as a function of ΔE, does not occur for $\Delta E = 0$ but at a value of around 1 eV. Also the maximum values are large, being 3×10^{-7} cm^3 s^{-1} at 250 °K (at $\Delta E = 1.2$ eV) and 10^{-7} cm^3 s^{-1} at 2000 °K (at $\Delta E = 1.3$ eV).

Results of the same order of magnitude were obtained by Bates and Massey (1947) and by Magee (1952) for $O^- - O^+$ collisions. Although the elementary pseudo-curve-crossing theory cannot be relied upon to give results of quantitative accuracy there is little doubt of the main conclusions. Mutual-neutralization coefficients at ordinary temperatures will in general be of order $10^{-7} - 10^{-8}$ cm^3 s^{-1}. Considered as far as dependence on ΔE is concerned, maximum values are obtained not for $\Delta E = 0$ but for $|\Delta E|$ close to 1 eV.

Recombination in three-body collisions

An elementary but instructive and remarkably successful theory of three-body recombination of ions at not too high pressures was developed as long ago as 1924 by J. J. Thomson. It gives good results at pressures below 100 torr. At much higher pressures an even earlier theory, due to Langevin (1903), gives good results. A theory which covers the entire pressure range has been given by Natanson (1959).

In recent years the subject has been analysed in much greater detail particularly by Bates and his associates (Bates and Moffett, 1966; Bates and Flannery, 1968). Their final results, which apply in the pressure range below 100 torr, do not differ greatly from those given by Thomson's theory but this is partly due to cancellation of rather larger errors.

Because of the insight they provide we shall first describe the theories of Thomson and Langevin in some detail and then say something of the more recent calculations.

Thomson's theory

This theory is based on the assumption that recombination occurs when either of two neighbouring ions of opposite sign loses sufficient kinetic energy by collision with a third body to describe a closed orbit relative to the other. For such an orbit to be described, the relative velocity v of two ions of charges $\pm e$ and masses M_1, M_2 at distances r apart must satisfy

$$\frac{1}{2} \frac{M_1 M_2}{M_1 + M_2} v^2 \leqslant \frac{e^2}{r}. \tag{14.12}$$

Assuming that each ion possesses the mean kinetic energy $\tfrac{3}{2}kT$, where T is the absolute temperature, this may be written

$$\frac{3}{2} kT \leqslant \frac{e^2}{r}. \tag{14.13}$$

The ions will not recombine unless they approach each other within a distance

$$r_0 = 2e^2/3kT. \tag{14.14}$$

If, then, there is some mechanism whereby two ions, approaching each other within a distance r_0, have their relative kinetic energy reduced to the mean value $\tfrac{3}{2}kT$, recombination will occur. Thomson assumes that collision with a third atom will provide such a mechanism. To calculate the recombination rate it is then only necessary to find the number of these collisions effected by either ion when within a distance r_0 of the other.

Let λ_1, λ_2 be the mean free paths of the positive and negative ions respectively. Then it may be shown that the chance of the positive ion colliding with a third body when within a distance r_0 of the negative ion is

$$s_1 = 1 + 2\{g_1^{-2} e^{-g_1} + g_1^{-1} e^{-g_1} - g_1^{-2}\}, \tag{14.15}$$

where $g_1 = 2r_0/\lambda_1$. Similarly we obtain s_2 for the negative ion. The number of times the ions come within a distance r_0 of each other is roughly

$$\pi r_0^2 (u_+^2 + u_-^2)^{1/2} n^+ n^-, \tag{14.16}$$

where u_+, u_- are the mean velocities of the ions and n^+, n^- their concentrations. This gives for the rate of recombination

$$\pi r_0^2 (u_+^2 + u_-^2)^{1/2}\, n^+ n^- (s_1 + s_2), \qquad (14.17)$$

i.e. an effective cross-section

$$Q_r = \pi r_0^2 (s_1 + s_2). \qquad (14.18)$$

For low pressures, $\lambda_1, \lambda_2 \gg r_0$ and we have

$$Q_r = \frac{4\pi r_0^2}{3}\left(\frac{r_0}{\lambda_1} + \frac{r_0}{\lambda_2}\right), \qquad (14.19)$$

while for high pressures

$$Q_r = 2\pi r_0^2. \qquad (14.20)$$

The mean free paths λ_1, λ_2 which appear are given to a good approximation by $1/n\bar{Q}_d^{(1)}$, $1/n\bar{Q}_d^{(2)}$ where $\bar{Q}_d^{(1)}$, $\bar{Q}_d^{(2)}$ are mean momentum transfer cross-sections for collisions between the respective ions and the gas atoms which are of concentration n.

At low pressures the recombination coefficient is given by

$$\alpha_r = \frac{32}{81}(3\pi^2)^{1/2}\frac{ne^6}{(kT)^{5/2}}\left(\frac{M_1 + M_2}{M_1 M_2}\right)^{1/2}(\bar{Q}_d^{(1)} + \bar{Q}_d^{(2)}). \qquad (14.21)$$

The variation of α_r with T will depend on the variation of $Q_d^{(1)}$ and $Q_d^{(2)}$ with T. If the interaction between the ions and atoms is determined mainly by polarization forces, $Q_d^{(1)}$ and $Q_d^{(2)}$ will vary approximately as $T^{-1/2}$ in which case α_r would vary as T^{-3}.

Langevin's theory

Thomson's theory becomes invalid when the pressure is so high that many-body collisions of order higher than 3 become important. Under these conditions it is appropriate to assume, following Langevin, that the ions drift towards each other under the influence of the Coulomb attraction between them. If μ_1 and μ_2 are the respective mobilities of these ions under the experimental conditions, the relative drift velocity when the ions are at a distance r apart will be

$$(\mu_1 + \mu_2)e^2/r^2. \qquad (14.22)$$

The number of positive ions drifting radially inwards per second across a spherical surface of radius r centred on the negative ion will therefore be

$$4\pi(\mu_1 + \mu_2) n^+ e^2, \qquad (14.23)$$

where n^+ is the concentration of positive ions. These ions will neutralize the negative unless deflected by collisions with other ions, a possibility which we ignore. The recombination coefficient is then given by

$$\alpha_r = 4\pi e^2 (\mu_1 + \mu_2). \qquad (14.24)$$

Since the mobilities are inversely proportional to the gas pressure p, so also will be the recombination coefficient.

This theory gives good results at pressures above about 2 atmospheres.

Improved theory in the low to intermediate pressure range

Consider the reaction

$$X^- + Y^+ + Z \longrightarrow XY + Z. \qquad (14.25)$$

In the analysis of Bates and Moffett (1966) the first step is to introduce rate coefficients for collisions in which energy exchange occurs between the translational energy of Z relative to the centre of mass of X^- and Y^+, and energy states of internal motion of X^- and Y^+. Ignoring the quantization of the levels of XY, let $K(E_i, E_f) dE_f$ be the rate coefficient for such collisions in which the internal energy of the ion pair is changed from E_i to between E_f and $E_f + \Delta E_f$.

The rate at which transitions take place between different energy states due to collisions is much faster than the rate at which the number density of free ions is changed, so a quasi-equilibrium distribution is set up between the internal states of the ions. Thus, if $n_i(E_i) dE_i$ is the number of ion pairs with energy between E_i and $E_i + dE_i$, we must have

$$n_i \int_\infty^{-E_b} K(E_i, E_f) dE_f = \int_\infty^{-E_b} n_f(E_f) K(E_f, E_i) dE_f, \qquad (14.26)$$

where E_b is the maximum binding energy possible for an ion pair. In terms of the n_i, the recombination coefficient may be calculated from

$$\alpha_r n^- n^+ = n_0 \int_{E_f \to \infty}^{E_s} \int_{E_i = E_s}^{-E_b} \{n_f K(n_f, n_i) - n_i K(n_i, n_f)\} dE_i dE_f. \qquad (14.27)$$

RECOMBINATION OF NEGATIVE AND POSITIVE IONS 625

where n^-, n^+ and n_0 are the concentrations of X^-, Y^+ and Z respectively and E_s is some arbitrarily chosen negative energy level. On the right-hand side the first term in the integral represents the rate at which the ion energy flows through E_s to lower energies while the second is that at which it flows to higher energies so the difference is the rate of recombination, which should be independent of E_s.

The problem of calculating α_r is formidable because it is necessary first to calculate the rate coefficients $K(E_f, E_i)$ and then to derive the n_i by solution of the integral equation (14.26).

Two special cases have been considered. Bates and Moffett (1966) dealt with that in which X, Y and Z are all of the same species so the reaction is

$$X^- + X^+ + X \longrightarrow X_2 + X. \qquad (14.28)$$

Under these conditions energy transfer to the third body takes place through the symmetrical charge transfer reactions

$$X^- + X \longrightarrow X + X^-,$$
$$X^+ + X \longrightarrow X + X^+.$$

For this case the Thomson theory gives very good results being in error by only a few per cent.

The second case, discussed by Bates and Flannery (1968), assumes that the interaction between an ion and a neutral atom at a distance r apart is given by (12.21) of Chapter 12. In this case, when the masses of X, Y and Z are all equal, the Thomson theory in the form (14.21) gives between 1.66 and 1.8 times the accurate value of α_r. When the masses are unequal it is not possible to express the results with such generality but it seems likely that there are few circumstances in which the Thomson theory gives seriously incorrect results.

14.2 The measurement of recombination and mutual-neutralization rates

The medium pressure range

We begin by describing experiments which are concerned with the measurement of recombination coefficients at pressures such that Thomson's theory should be valid, i.e. they are neither too low for mutual neutralization to be important nor so high that Langevin's theory should apply.

Experimental study of the recombination of ions at atmospheric pressures dates back to the end of the last century when Rutherford and Thomson (1897) showed that the rate of decay of ion density n_i after the ionizing agency was cut off followed the quadratic law

$$\frac{dn_i}{dt} = -\alpha n_i^2.$$

However, it was not until the work of Sayers in 1938 that reliable measurements were made in air at pressures between 100 and 1500 torr.

The principles on which these measurements were based were substantially the same as in the earlier experiments but certain serious sources of error were recognized and eliminated.

Gas between two collecting electrodes was ionized by a pulse of X-rays and the ionization still present at a time t after termination of the pulse measured by collecting the charges with a sufficiently large potential difference across the electrodes. The apparent recombination coefficient α_a is given by

$$\alpha_a = \frac{d}{dt}\left(\frac{1}{n_i}\right),$$

where n_i is the ion concentration between the electrodes at time t.

The value of α_a determined in this way was found to vary with the time after interruption of the ionization – that is to say the slope of a plot of $1/n_i$ against t varied with t. After a time of about 0.1 s, however, α_a remained constant. This effect is almost certainly due to lack of uniformity in the ionization produced by the pulse. It is only after about 0.1 s that the non-uniformities have been smoothed out.

A second complication was an observed dependence of α_a on the duration of the pulse. The steady value obtained by α_a decreased with increasing pulse length, probably because of the production by the X-ray beam of other chemical species, including complex molecules, which modify the recombination rate. This effect was particularly noticeable in pure O_2 but was absent in pure N_2 and He. To allow for it, the steady value of α_a, measured for different pulse lengths, was extrapolated to infinitely short pulses.

Finally, correction had to be made for loss of ions by diffusion to the plates. This limited the experiments to pressures greater than 100 torr. Thus the correction to the observed value of α is approximately $\frac{3}{2}D/nd^2$, where D is the diffusion coefficient of ions in the gas at the concentration n concerned and d is the plate separation. At s.t.p. D is 0.047 cm² s⁻¹ so that, under experimental conditions in which $d = 1.5$ cm, $n = 10^6$ cm⁻³, the correction at 100 torr is about 2.3×10^{-7} cm³ s⁻¹, as compared with a measured apparent value of α_a of a little over 10^{-6} cm³ s⁻¹.

Great care was taken in the design of the equipment to ensure good high-vacuum conditions, including provision for baking the whole electrode assembly at 500 °C for some hours.

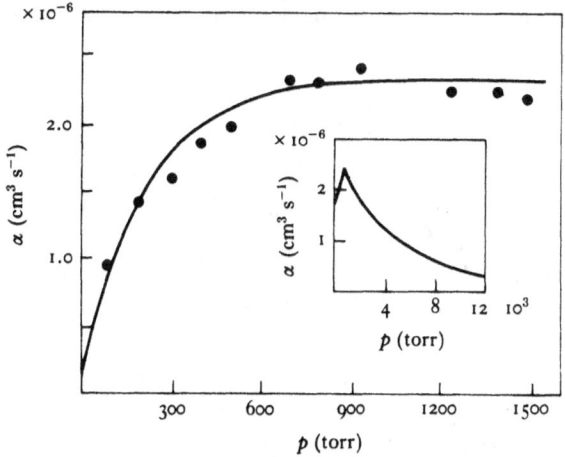

Fig. 14.2. Three-body recombination coefficient for ions in air as a function of pressure. —— calculated by Thomson's theory, ● observed by Sayers (1938). Inset: High-pressure region showing results of Mächler (1936).

Fig. 14.2 illustrates the results obtained which are seen to agree very well with the predictions of Thomson's theory.

Much more recently measurements, on the same basic lines, have been carried out by McGowan (1965, 1967), aimed at determining the effects of moisture and of organic impurities on recombination in air. The 180 keV X-ray beam producing the ionization was collimated to minimize the production of ions by irradiation of the electrodes. A further new feature was the availability of means for measuring the form of the initial ion distribution produced by a pulse. This was done by sweeping a plastic chamber with a narrow collecting electrode across the beam. If $n_i(x)$ is the observed ion concentration at a height x above the lower collecting plate and A is the area of each plate, the effective width $\bar{h}$ of the ionizing beam for recombination is then

$$\bar{h} = \{\int n_i(x)\,dx\}^2 / \int\{n_i(x)\}^2\,dx. \qquad (14.29)$$

At 300 °K for dry air at atmospheric pressure, McGowan found $\alpha_r = 2.2 \times 10^{-6}$ cm^3 s^{-1}, agreeing well with Sayers. For pure O_2 under the same conditions α_r was found to be 2.00×10^{-6} cm^3 s^{-1}, agreeing with earlier measurements by Gardner (1938). The value fell very markedly to between 1.65×10^{-6} and 1.45×10^{-6} cm^3 s^{-1} for moist air and even further with organic contaminants.

Measurements at high pressures

Measurements at pressures up to 2.2×10^4 torr have been carried out in air and CO_2 by Mächler (1936), using a radioactive ionization source (40 mg of Ra) and ion concentrations of 4×10^4 to 8×10^4 ions cm^{-3}. The results obtained in air shown in the inset of Fig. 14.2 are consistent with Langevin's theory in that at 18 °C the product $\alpha_r p$ is constant at pressures above 1×10^4 torr. At 52 °C, constancy is not reached till the pressure exceeds 1.5×10^4 torr. For CO_2 the situation is less clear as $\alpha_r p$ was found to fall off slowly with p at pressures above 8×10^3 torr.

Measurements at low pressures – rate of mutual neutralization

Mahan and Person (1964) measured rates of recombination of ions produced by photoionization of NO present to a partial pressure of 3×10^{-4} torr in a large excess of a rare gas (pressure of 100 torr or more), as a function of the rare gas pressure. They found by extrapolation to zero pressure that the limiting recombination rate coefficient was the same for all gases, which included not only the rare gases but also H_2, D_2 and N_2. This they interpreted as the rate coefficient for mutual neutralization between the negative and positive ions present. The value found, 2.5×10^{-7} cm^3 s^{-1}, is indeed of the order expected.

The only difficulty in accepting this interpretation is that no means were available for identifying the ions. It was assumed by Mahan and Person that the negative ions were produced by the sequence of processes involving fast three-body attachment of the initially-produced electrons to NO forming NO⁻ which then underwent charge transfer with NO_2 present as an impurity. However, the rate of detachment from NO⁻ by collisions with rare gas atoms is very fast (see Chapter 12, p. 561) and the partial pressure of NO_2 so small (of order 10^{-6} torr) that it is hard to see how this could lead to a preponderance of the negative charge carried by NO_2^- ions rather than by free electrons. Equally well, while the rate of direct dissociative attachment of electrons to NO_2 is fast, the mean time before a free electron would attach to the small impurity concentration of NO_2 would be as long as 0.1 s. Rather similar difficulties of interpretation occur in other experiments involving NO⁻ (see Chapter 9, p. 340, and Chapter 12, p. 561).

The pressure-dependent part of the observed recombination coefficient for the rare gases was found to agree quite well with the predictions of Thomson's theory for ions of mass 76 a.m.u., i.e. NO_2^-. NO and NO_2^+. NO, the ion–atom interaction being taken to be of the form (12.21). The results are nearly independent of the parameter r_0.

The first measurements of mutual-neutralization rate coefficients for mass-identified ions were made by Greaves (1964). To determine the ion concentration as a function of time he used the fact that the dielectric constant ϵ of a plasma, containing n ions cm⁻³, for electromagnetic radiation of angular frequency ω is given by

$$\epsilon = 1 - \frac{4\pi n e^2}{\omega^2}\left(\frac{1}{M^+} + \frac{1}{M^-}\right), \qquad (14.30)$$

where M^+, M^- are the respective masses of the positive and negative ions. It is assumed that $\omega \gg \nu$ the collision frequency of the ions in the plasma. If a condenser containing the plasma is included in an oscillating circuit, the frequency of the oscillation will depend on the ion concentration and, if the masses of the ions are known, the concentration can be obtained.

630 NEGATIVE IONS

The first application of this procedure was made by Yeung (1958) to measure mutual-neutralization rates in iodine. In this work no mass analysis of the ions could be made. Fig. 14.3 shows the arrangement used by Greaves in which such analysis was possible. The discharge tube was 8 cm in diameter and 27 cm long. A coaxial condenser was formed by an external copper sleeve

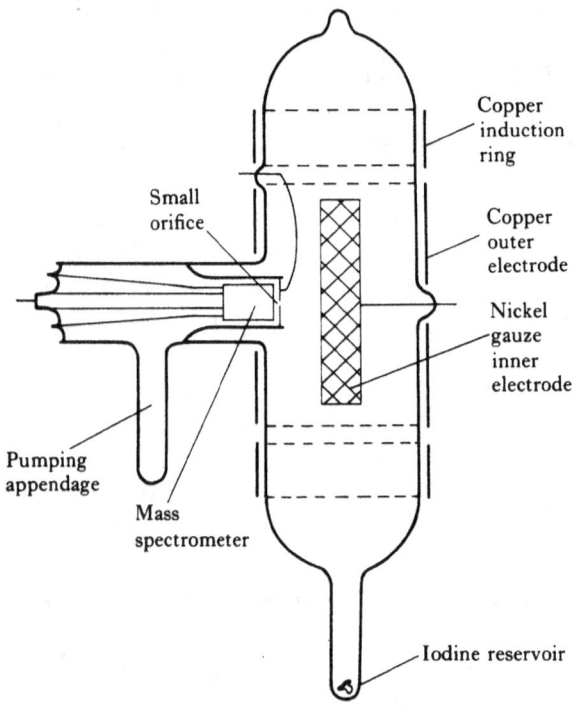

Fig. 14.3. Arrangement of the apparatus used by Greaves (1964) to measure mutual-neutralization coefficients for ions in iodine vapour.

enclosing an inner cylinder of nickel gauze. The iodine vapour pressure in the discharge tube could be varied between 0.03 and 1 torr by controlling the temperature of the iodine reservoir.

The vapour was ionized by pulses of 30 kW peak power, of duration 2 μs and repetition rate 25 s^{-1}. Free electrons present at the termination of the pulse formed I$^-$ ions very rapidly through dissociative attachment to I$_2$ (see Chapter 9, p. 341) and it was verified by means of the r.f. mass spectrometer that, in all the experiments, the only ions observed in the afterglow were I$_2^+$ and I$^-$.

Fig. 14.4 shows typical plots of reciprocal ion concentration $1/n_i$ against time in the afterglow at different iodine pressures. At pressures of 0.23 torr and above the plots are linear over a range of values wide enough to establish that ion loss is dominantly through recombination and not diffusion.

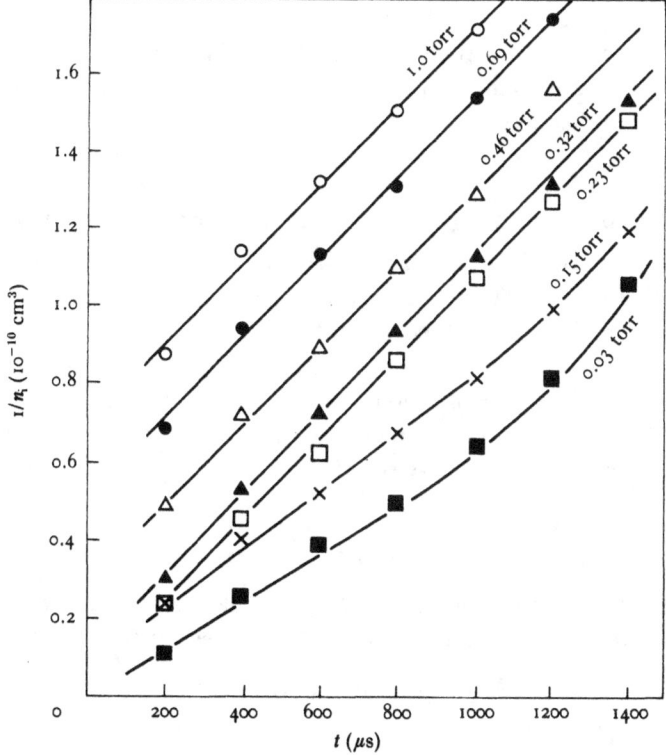

Fig. 14.4. Typical variation of the reciprocal ion concentration with time t in the afterglow for ions in iodine vapour at different iodine pressures, observed by Greaves (1964).

The mean values of the recombination coefficient determined from the slopes of the linear plots were found to be independent of pressure at all three ion temperatures studied (296, 328 and 338 °K), the respective values being 1.22×10^{-7}, 1.04×10^{-7} and 1.01×10^{-7} cm^3 s^{-1}. These are of the order of magnitude expected for mutual neutralization.

In a later experiment by a similar method Hirsch, Halpern and Wolf (1968) measured recombination rates between ions in O_2 contaminated by CO_2 and in airlike mixtures of O_2 and N_2. In the observed pressure range, 1–20 torr, ions were identified in the O_2–CO_2 mixture as CO_2^+ and CO_3^-, and in the airlike mixture as NO^+, NO_2^- and NO_3^-. Pressure-independent recombination coefficients were found with values $(5.9 \pm 1.2) \times 10^{-8}$ and $(4.4 \pm 1.0) \times 10^{-8}$ cm^3 s^{-1} for the O_2–CO_2 and O_2–N_2 mixtures respectively. These are again of the order of magnitude expected for mutual neutralization.

Measurements by merging and inclined-beam techniques

Many of the problems associated with the experimental study of ionic collisions at very low relative energies may be overcome by using the merging- or inclined-beam technique. In the latter, if A^+ and B^- are two ions, the collisions between which are to be studied, then a beam of B^- ions of kinetic energy E_2 and mass M_2 is fired along a beam of A^+ ions of kinetic energy E_1 and mass M_1. The relative energy of impact between the ions in these beams is given by

$$E = \tfrac{1}{2}M(v_2 - v_1)^2,$$

where M is the reduced mass $M_1 M_2/(M_1 + M_2)$ and v_1 and v_2 are the velocities of the A^+ and B^- ions in the respective beams. Thus

$$E = M\{(E_2/M_2)^{1/2} - (E_1/M_1)^{1/2}\}^2.$$

and, when $M_1 = M_2$,

$$E = \tfrac{1}{2}(E_2^{1/2} - E_1^{1/2})^2.$$

If $|E_2 - E_1|, = \Delta E$, is much smaller than $\tfrac{1}{2}(E_1 + E_2), = \bar{E}$, then

$$E \simeq \Delta E^2/8\bar{E}.$$

Thus if, for example, $E_1 = 5$ keV and $E_2 = 5.1$ keV, $E \simeq 0.25$ eV. It is thus possible to secure a low relative impact energy E using energetic beams which differ in energy, in the laboratory system, by much more than E.

A further useful feature is that the spread in relative energy produced by a finite energy resolution in the colliding beams is relatively much smaller. Thus a change δE in ΔE produces a change $\delta' E$ in E given by

$$\frac{\delta' E}{E} \simeq \frac{2 \delta E}{\Delta E}.$$

In the example cited, if the energy spread in each beam is 2 eV, the corresponding spread $\delta' E$ is only 0.01 eV.

In practice it is difficult to ensure parallelism of the merging beams and finite transverse velocity components modify the above considerations somewhat. Nevertheless it has been possible to

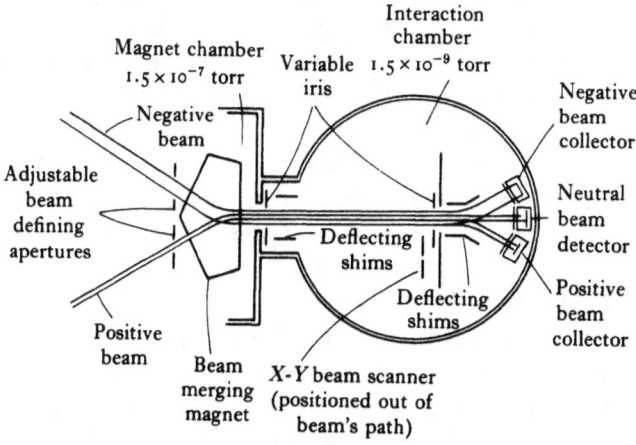

Fig. 14.5. General arrangement of the merging-beam apparatus used by Aberth et al. (1968) for measuring cross-sections for mutual neutralization.

develop the technique of using merging beams, the first successful experiments being those of Trujillo, Neynaber and Rothe (1966) who were concerned with low-energy collisions between ions and neutral atoms.

Fig. 14.5 shows the general arrangement of the apparatus developed by Aberth, Peterson, Lorents and Cook (1968) for application to the measurement of mutual-neutralization cross-sections. The positive- and negative-ion beams were ionized by magnetic deflexion and travelled about 30 cm during which collisions leading to mutual neutralization occurred. The beams were then separated magnetically, leaving the neutral products to continue to travel along the direction of the merged beams. The flux of these products was measured from the secondary electron emission produced from a suitable surface. To render such measurements absolute it is necessary to know the secondary emission coefficient γ for the neutral products. Since it was found that negative ions were only between 10 and 30 per cent more effective than the corresponding positive ions in producing secondary emission, γ for the neutral products was taken to be the mean of that for the positive and negative ions concerned in the experiment.

A strong background signal is always present due to electron stripping and capture reactions occurring between the ion beams and the background gas. To separate the wanted signal from this background, both ion beams were chopped at frequencies of 800 and 1000 Hz respectively and the beat frequency signal at 200 Hz was amplified by a phase-sensitive selective amplifier.

This technique has been applied to a number of mutual-neutralization reactions over a wide range of relative kinetic energy (0.1–400 eV). The results obtained are discussed below.

Inclined-beam method

A technique which can be regarded as a compromise between the merging-beam and the crossed-beam methods is the so-called inclined-beam method in which the two beams intersect at an angle of around 20°. This has the advantage of working with a well-defined collision region and the collision products remain well separated. On the other hand, it is not so suitable for studying collisions at very low relative energies.

Thus for ions of equal mass with kinetic energies E_1 and E_2 in the laboratory system colliding at an angle θ, the relative kinetic energy is given by

$$E = E_1 + E_2 - 2(E_1 E_2)^{1/2} \cos \theta,$$

so for $\theta = 20°$ and $E_1 = \bar{E}_2 = E$, $E \simeq \bar{E}/8$.

We have already described the use of the inclined-beam method for the measurement of cross-sections for detachment from H$^-$ by electron impact. A very similar technique has been applied by Harrison and his collaborators (Rundel, Aitken and Harrison, 1969; Gaily and Harrison, 1970a, b) to study mutual-neutralization collisions in the relative kinetic energy from 0.2 to 8 keV. The method of removing the background signals, which in these experiments largely arises from the capture of electrons from the background gas by the positive ions, was also essentially the same. Thus both beams were modulated according to the time scheme shown in Fig. 14.6, and output pulses from the neutral atom detector were corrected by two scalers gated as shown so that one recorded $S + B^+ + B^-$ and the other $B^+ + B^-$, S being the wanted signal, B^+ and B^- the background signals due to the positive and negative ions respectively. The required signal then was simply attained from the difference between the two recordings. This procedure

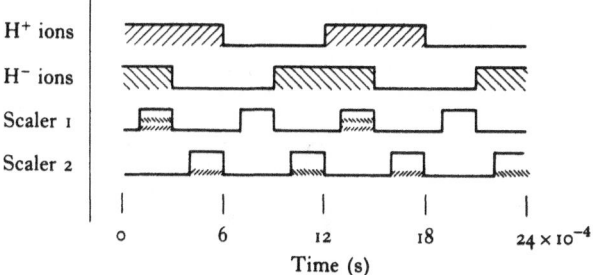

Fig. 14.6. Illustrating the pulsing system used in the inclined-beam experiments of Harrison and his collaborators for removing background signals.

proved effective even though B^+ was between 300 and 3000 times as large as the wanted signal S.

14.3 Results obtained for specific mutual-neutralization reactions

H^+–H^-

The rate of this simplest of all mutual-neutralization reactions has been measured using merging beams (Moseley, Aberth and Peterson, 1970) over a relative energy range from 0.15 to 3000 eV and using inclined beams (Rundel et al., 1969; Gaily and Harrison, 1970a) over the higher energy range 0.2–8 keV. Unfortunately there is only a small region of overlap between the observations taken by the different techniques. Fig. 14.7 shows the results obtained for the mutual-neutralization cross-section Q_m in the two inclined-beam experiments which are seen to agree within the rather large estimated experimental errors. For comparison the cross-sections measured in the merging-beam experiments at the higher energies

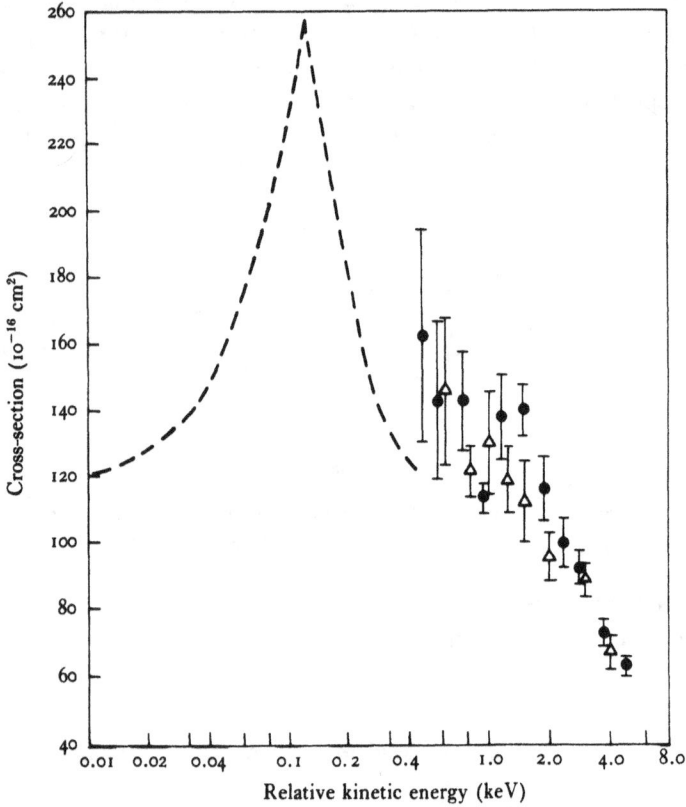

Fig. 14.7. Cross-sections for the mutual-neutralization reaction between H$^+$ and H$^-$ at relatively high energies. ● observed by Rundel *et al.* (1969), △ observed by Gaily and Harrison (1970*a*), ---- observed by Moseley *et al.* (1970).

are also shown. It seems that, while the trend of the cross-section with relative energy increase appears to be much the same, the cross-section at 300 eV obtained in the merging-beam experiments is smaller than those measured in the inclined-beam experiments. However, the experimental error in both sets of experiments is large

RECOMBINATION OF NEGATIVE AND POSITIVE IONS 637

near the upper and lower ends of their respective working energy ranges.

Fig. 14.8 shows the results of Moseley et al. at relative energies below 20 eV. The theoretical values calculated by Bates and Lewis, as described on p. 619, vary in much the same way with relative

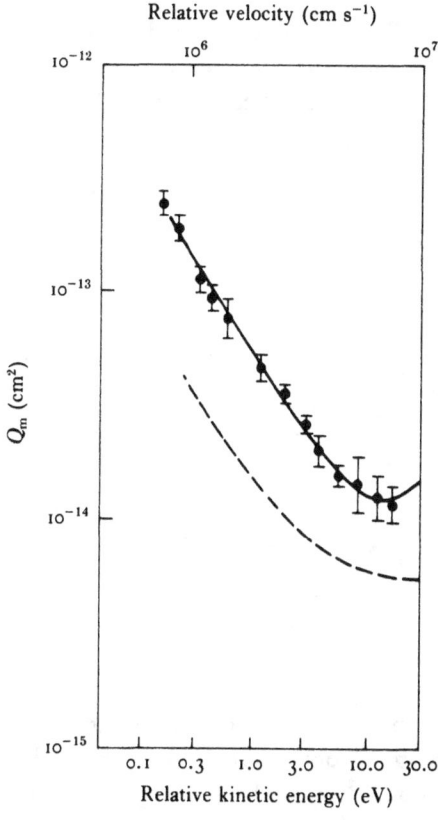

Fig. 14.8. Cross-sections Q_m for the mutual-neutralization reaction between H^+ and H^- at low energies. ⊥ observed by Moseley et al. (1970), ----- calculated by Bates and Lewis (1955).

velocity but are a factor of 3 or so too small. To extrapolate the observed results to 300 °K, Q_m was fitted over the obtained energy range by the form

$$Q_m = Av^{-2} + Bv^{-1} + C + Dv.$$

Assuming this form to remain valid at 300 °K, the mutual-neutralization coefficient at this temperature was found to be $(4.0 \pm 1.8) \times 10^{-7}$ cm³ s⁻¹, which is close to the magnitude anticipated.

N^+–O^-, N_2^+–O_2^-, O_2^+–O_2^-

The first measurements with the merging-beam technique, due to Aberth et al. (1968) were for the mutual-neutralization reaction between N^+ and O^- ions. Fig. 14.9 shows the results which they obtained for the mutual-neutralization rate $Q_m v$ where v is the relative velocity.

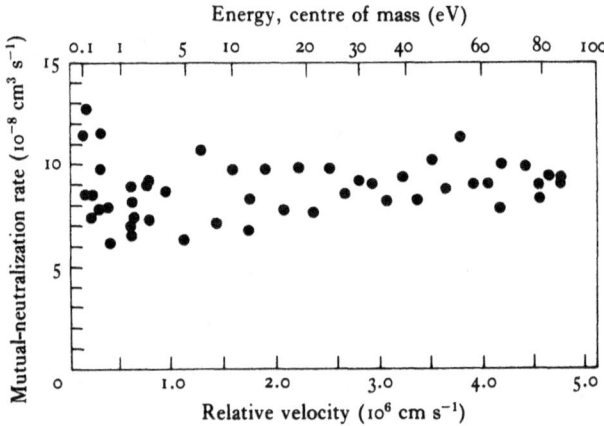

Fig. 14.9. Mutual-neutralization rate for N^+–O^- collisions observed by Aberth et al. (1968).

It will be seen that, for relative energies greater than 0.5 eV, the rate remains roughly constant at 9×10^{-8} cm³ s⁻¹ as the relative energy increases up to the highest energy investigated, 90 eV. However, as the energy falls below 0.5 eV there is evidence that the rate increases quite rapidly. While the constant rate above 0.5 eV is not far from the expected order of magnitude, the rise at low energies is not understood theoretically.

Rather similar behaviour was observed in later experiments of Aberth and Peterson (1970), for N_2^+–O_2^-, but for O_2^+–O_2^- there was no evidence of an increase in rate at very low relative impact energies.

He⁺–H⁻

If mutual neutralization takes place essentially through long range pseudo-crossing of potential-energy curves, it is to be expected that the rates for He⁺–H⁻ collisions would not differ greatly from those

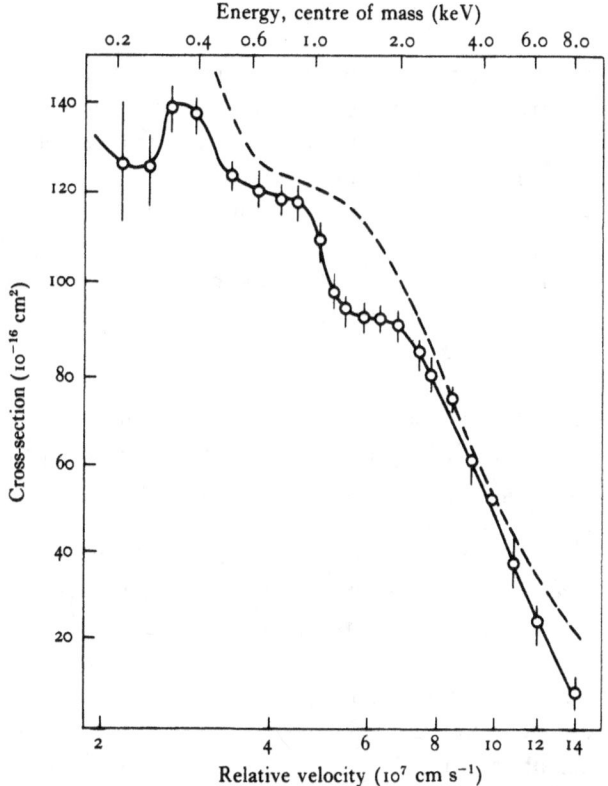

Fig. 14.10. Observed cross-sections for mutual-neutralization reactions: between He⁺ and H⁻ (○) observed by Gaily and Harrison (1970b), between H⁺ and H⁻ (----) observed by Gaily and Harrison (1970a). The upper scale is the relative energy for the He⁺–H⁻ collisions.

for H⁺–H⁻ at the same relative impact velocity. To check this Gaily and Harrison (1970b) applied the inclined-beam technique to measure cross-sections for He⁺–H⁻ to compare with their results for H⁺–H⁻, as shown in Fig. 14.10. It will be seen that, while the cross-sections are similar in magnitude and velocity variation, more structure is apparent for He⁺–H⁻.

CHAPTER 15

Negative Ions in Electric Discharges, Planetary and Stellar Atmospheres, Trace Analysis and Tandem Accelerators

It is to be expected that the nature of an electric discharge through an electronegative gas or vapour would be influenced by the formation of negative ions. As the oxygen in the earth's atmosphere is capable of forming such ions, it would also be anticipated that negative ions might be important constituents of the ionized layers in the upper atmosphere. It is perhaps not so obvious that the continuous emission spectrum of the sun and stars would be seriously influenced by the presence of negative ions in their outer atmospheres. Nevertheless, it has been established that the negative ion of hydrogen very largely determines the spectral distribution of the continuous solar emission in the observable wavelength region.

In this chapter we shall apply the considerations of the previous chapters, regarding negative-ion structure and modes of formation and destruction, to these phenomena. The natural order in which to discuss them is to deal first with glow discharges and with electrical breakdown in gases, then with the earth's upper atmosphere, which one might expect to resemble the plasma of a glow discharge in many ways, and finally with solar and stellar emission.

We shall also take the opportunity to discuss briefly how the production of negative ions has been made use of for detection of minute traces of certain substances in gases and for the design of tandem and other accelerators for charged particles.

15.1 Negative ions in glow discharges

General effects of negative ions in discharges

By way of illustration we first describe the effect of adding a trace of chlorine to a discharge in neon containing a little helium (Eméleus and Sayers, 1938).

At a pressure of 2 torr with an anode–cathode potential difference of 500 V, the pure neon discharge takes the form illustrated in Fig. 15.1(a). (1) is the primary dark space, (2) the cathode dark space, (3) the negative glow and (4) the Faraday dark space. Addition of a little chlorine changes the appearance to that shown in Fig. 15.1(b). The modifications are:

(a) Disappearance of the primary dark space.
(b) Shortening of the Faraday dark space.
(c) Appearance of a striated positive column (6) with strongly curved units.

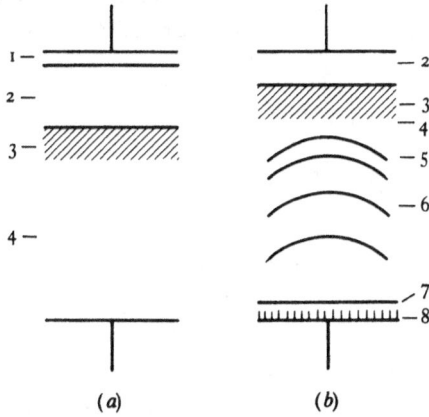

Fig. 15.1. Illustrating the effect of adding a trace of chlorine to a discharge in pure neon. (a) Before addition of chlorine, (b) after addition of chlorine.

(d) Separation of the positive column from the Faraday dark space by a dark sheath (5).
(e) Separation of the positive column from the anode by a dark sheath (7) bounded on the anode surface by a layer of neon light (8).

Similar, though more pronounced, effects accompanying glow discharges in pure iodine vapour and in helium containing a trace of iodine, were observed by Spencer-Smith (1935). In these cases it was found also that the positive column tends to turn into either a stationary or mobile ribbon. This contracted positive column ('ionenschlauch') is often found in discharges through electronegative vapours. In discharges of high current density more than one of these contracted columns may occur.

In discharges through gases containing less electronegative substances, such as oxygen, it is found that there are greater longitudinal fields in the positive column and a greater tendency for this to break into striations than with inert gases.

Analysis of a discharge in oxygen

Some of these general effects are almost certainly due to the presence of negative ions in considerable relative concentration in the discharge tube. A detailed experimental study of discharges in oxygen has been carried out by Thompson (1961a) in order to examine in as much detail as possible the effects arising from negative-ion production. For this purpose arrangements were made

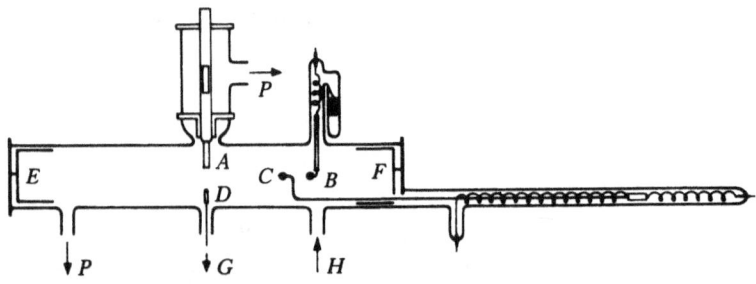

Fig. 15.2. Arrangement of apparatus used by Thompson to investigate glow discharges in oxygen. A, mass spectrometer; B, C, Langmuir probes; D, reference probe; E, F, electrodes; G, to pressure gauges; H, gas inlet; P, to pumps.

to measure as many as possible of the relevant quantities concerned, as may be seen from Fig. 15.2 which shows a section of the experimental tube, E and F being the electrodes.

Experimental measurements

The mass spectrometer probe A was of the r.f. type developed by Boyd and Morris (1955) and could be moved radially. This made it possible to determine the nature of the positive and negative ions and their radial distribution at the axial position of the probe. B and C were spherical Langmuir probes from the characteristics of which the electron concentration and energy distribution and the positive-ion concentration could be determined and, as it happened, also the negative-ion concentration. The probe C could be moved axially. A small plane probe in contact with the wall was also carried on the same glass stem as C so that wall current measurements could be correlated with electron energy distributions. Finally, D was a reference probe.

Measurements were carried out mainly in d.c. discharges at a pressure of 0.040 torr and currents between 1 and 100 mA. Under these conditions the negative glow and Faraday dark space extended over a third of the distance (100 cm) between cathode and anode.

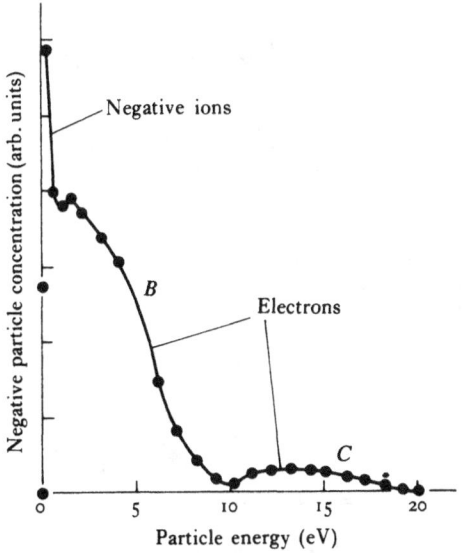

Fig. 15.3. Typical energy distribution of negative particles in oxygen.

In both pure oxygen and nitrogen the positive column was striated, the striations being quite flat in oxygen and markedly convex towards the cathode in nitrogen. As it is unlikely that negative ions played any important role in the latter case, the occurrence of striations is probably not dependent on the presence of such ions. Whereas the nitrogen discharge was very steady over a considerable range of tube current and gas pressure, it was not found possible to make useful measurements of electron energy distributions in oxygen, except under conditions in which the discharge current was close to 4 mA and the pressure to 0.040 torr. This inherent noisiness of the oxygen discharge is very likely to be due to the presence of considerable concentrations of negative ions.

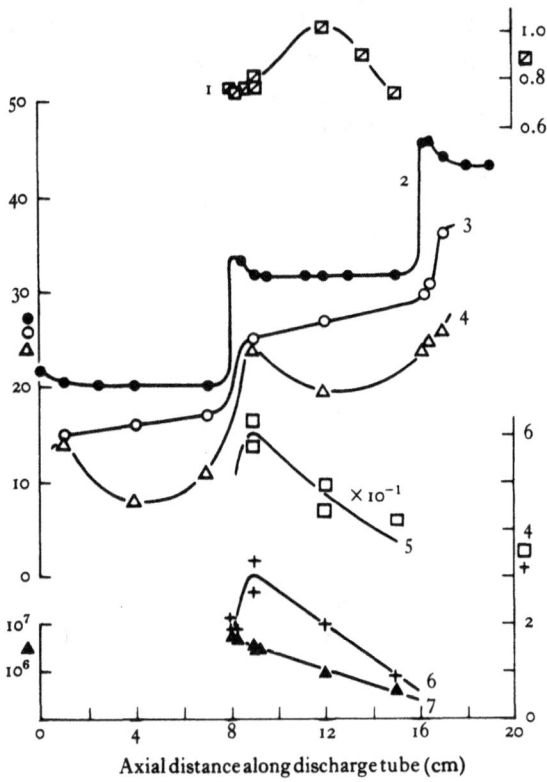

Fig. 15.4. Axial variation of currents, potentials and charged particle concentrations through striations in the positive columns of an oxygen discharge. Curve 1 (▨) wall current (μA cm^{-2}); 2 (●) space potential (V); 3 (○) floating potential (V); 4 (△) wall potential (V); 5 (□) negative-ion concentration (10^7 cm^{-3}); 6 (+) electron concentration group B (see Fig. 15.3) (10^7 cm^{-3}); 7 (▲) electron concentration group C (see Fig. 15.3) (cm^{-3}).

The analysis of Langmuir probe data obtained in a plasma containing a considerable concentration of negative ions was discussed earlier by Boyd and Thompson (1959).

Fig. 15.3 shows a typical energy distribution obtained for the negative particles at an axial position in the positive column in oxygen. The broad peaks which occur near 13 and 2 eV respectively also appear in corresponding distributions observed in nitrogen and hydrogen but the narrow sharp peak at low energies is found only in oxygen and is due to negative ions. From these measured distributions the relative concentrations of negative ions and electrons may be determined.

Fig. 15.4 shows the axial variation of wall currents, measured by the plane probe, of space potential, floating potential, wall potential, negative-ion concentration n^- and electron concentrations n_e in the energy groups, B and C respectively, shown in Fig. 15.3.

Ambipolar diffusion in the presence of negative ions

To correlate some of these data we first consider the effect of a high relative concentration of negative ions on the flow of ions to the wall. We have already discussed the case in which only positive ions and electrons are present, in Chapter 9, p. 299. Thus when negative ions are also present we need only add the appropriate third equation to (9.30) and (9.31) and we have for the current densities of $\mathbf{j}_e$, $\mathbf{j}^+$ and $\mathbf{j}^-$ of electrons, positive ions and negative ions respectively.

$$\mathbf{j}_e = -D_e \operatorname{grad} n_e - \mu_e \mathbf{F} n_e, \tag{15.1}$$

$$\mathbf{j}^+ = -D^+ \operatorname{grad} n^+ + \mu^+ \mathbf{F} n^+, \tag{15.2}$$

$$\mathbf{j}^- = -D^- \operatorname{grad} n^- - \mu^- \mathbf{F} n^-. \tag{15.3}$$

Here D_e, D^+ and D^- are the respective diffusion coefficients, μ_e, μ^- and μ^+ the mobilities and n_e, n^+, n^- the concentrations of the different charged particles, while $\mathbf{F}$ is the radial electric field strength. We write now

$$\mu_e = eD_e/kT_e, \quad \mu^+ = eD^+/kT^+, \quad \mu^- = eD^-/kT^-, \tag{15.4}$$

where T_e, T^+ and T^- are the respective temperatures, and take

$$T^+ = T^- = T_e/\gamma. \tag{15.5}$$

Within the plasma

$$n_e + n^- = n^+, \tag{15.6}$$

and at the boundary

$$\mathbf{j}^- + \mathbf{j}_e = \mathbf{j}^+.$$

Eliminating F from the equations (15.1) (15.2) and (15.3) gives

$$\mathbf{j}^+ = -D_a^+ \operatorname{grad} n^+, \quad (15.7)$$

$$\mathbf{j}^- = -D_a^- \operatorname{grad} n^-, \quad (15.8)$$

$$\mathbf{j}_e = -D_e^a \operatorname{grad} n_e, \quad (15.9)$$

where

$$D_a^+ = D^+ \left[\frac{(1+\gamma+2\lambda\gamma)(1+\lambda\mu^-/\mu_e)}{(1+\lambda\gamma)\{1+(\mu^+/\mu_e)(1+\lambda)+\lambda\mu^-/\mu_e\}} \right], \quad (15.10)$$

$$D_a^- = D^+ \left[\frac{1}{\gamma}\frac{\mu^-}{\mu_e} \frac{1+\gamma+2\lambda\gamma}{\{1+(\mu^+/\mu_e)(1+\lambda)+\lambda\mu^-/\mu_e\}} \right], \quad (15.11)$$

$$D_e^a = D^+ \left[\frac{1+\gamma+2\lambda\gamma}{1+(\mu^+/\mu_e)(1+\lambda)+\lambda\mu^-/\mu_e} \right], \quad (15.12)$$

$$\lambda = n^-/n_e. \quad (15.13)$$

The field F may now be obtained as a function of λ to give

$$\frac{F(\lambda)}{F(0)} = \frac{1 - D^+/D_a^+}{1+\lambda}\left(\frac{1+\gamma}{\gamma}\right). \quad (15.14)$$

To illustrate the dependence of the ambipolar diffusion coefficients and radial field on the negative-ion–electron ratio λ, numerical values (Thompson, 1959) have been obtained for oxygen for which $\mu^-/\mu_e = 0.0043$, $\mu^+/\mu_e = 0.0022$ and, in the observations of Thompson, $\gamma = 16$. Fig. 15.5 illustrates the ambipolar diffusion coefficients and Fig. 15.6 the field F. The behaviour is as expected and already described qualitatively in the discussion of Chapter 9, p. 300.

Thus for small λ, $D_e^a \simeq D_a^+$ and the behaviour is dominated by the electrons. For $\lambda > 1$, however, the importance of the negative ions grows rapidly as may be seen by the rapid fall of the ratio $F(\lambda)/F(0)$ and the tendency for D_a^- to reach equality with D_a^+.

To compare these predictions with experiment (Thompson, 1959) we show in Fig. 15.7 the radial variation of particle concentrations n_e, n^- and n^+ in the oxygen discharge, the first two being measured from distributions such as that shown in Fig. 15.3 and the third from the positive-ion current collected by the probe.

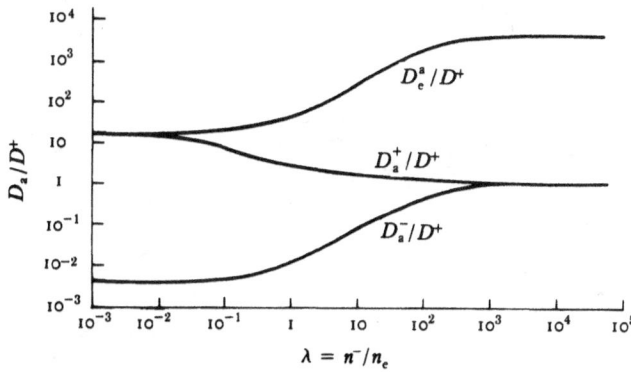

Fig. 15.5. Ambipolar diffusion coefficients as a function of $\lambda = n^-/n_e$, calculated for oxygen with $\gamma = 16$, as in the experiments of Thompson (1959).

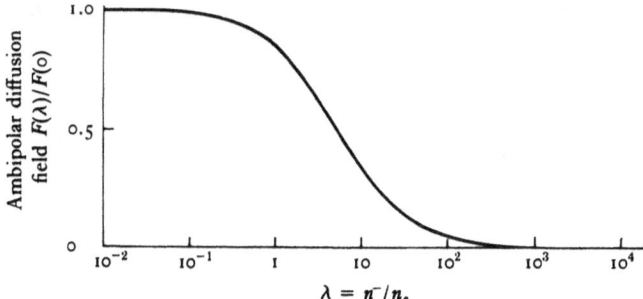

Fig. 15.6. Ambipolar diffusion field $F(\lambda)/F(0)$ as a function of λ calculated for the same conditions as in Fig. 15.5.

Under the discharge conditions, for which $\lambda \simeq 20$, the negative- and positive-ion concentrations should be nearly equal and it is seen that they do in fact vary in the same way across the radius whereas the electron concentration remains almost constant until very close to the wall. The negative-ion concentration varies nearly as $\exp\{V(r)/V^-\}$, where $V^- = 0.15$ eV and $V(r)$ is the space potential at a radius r. V^- was taken in subsequent analysis as giving the ion temperature.

Using the data of Fig. 15.7, a further check was applied by Thompson (1959). Fig. 15.8 shows the observed space potential for a discharge in nitrogen for which $\lambda = 0$. According to the formula (15.14) the expected curve for oxygen takes the form shown in Fig. 15.8. This agrees very well with that directly observed.

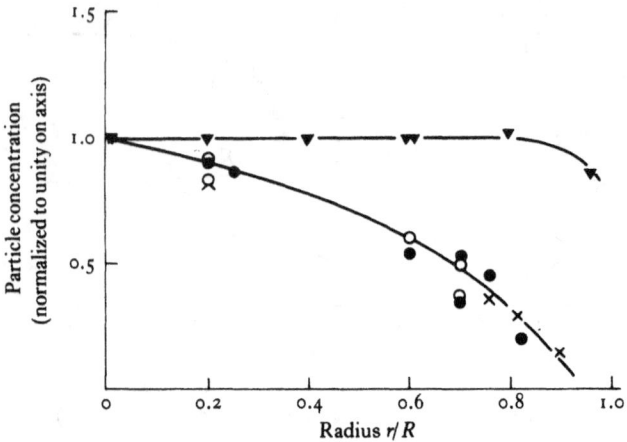

Fig. 15.7. Radial variation of charged particle concentrations, as observed by Thompson (1959) in an oxygen discharge. ▼ n_e, ○ n^+, ● n^-, determined from probe data. × $n^- = n^-(o)\exp\{V(r)/V^-\}$, where $V^- = 0.15$ eV and $V(r)$ is the space potential.

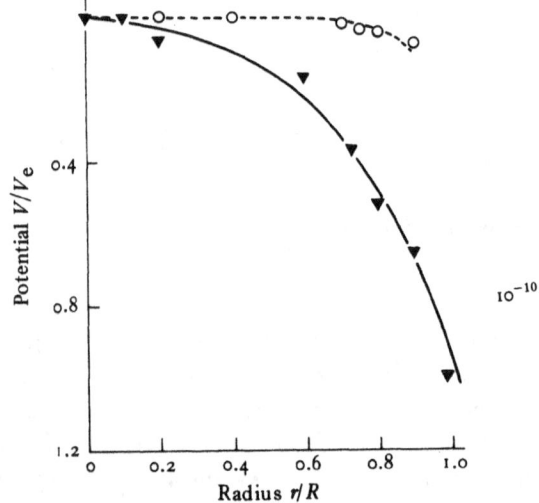

Fig. 15.8. Space potential observed as a function of radius in the positive column of discharges in oxygen and in nitrogen. –▼– observed by Thompson (1959) in nitrogen, ○ observed by Thompson (1959) in oxygen, ---- calculated for oxygen from –▼– using (15.14). V_e is written for kT_e.

The ion balance

With data of this kind combined with information on the relative abundance of different positive and negative ions it is possible to analyse the ion balance in some detail in relation to the rates of various processes of production and loss, the rates of many of which are known.

Thompson (1961a) found that in the negative glow and Faraday dark space n^- is $\simeq 0.01 n\phi$ but in the positive column

$$[O^-]/[O_2^+] = 0.9, \quad [O^+]/[O_2^+] = 0.014, \quad [O_2^-]/[O^-] = 0.1. \tag{15.15}$$

From the ratio $[O^+]/[O_2^+]$ it is possible to derive information about the degree of dissociation of the oxygen (Thompson, 1961b). Whereas O_2^+ ions are formed directly by impact ionization of O_2, O^+ can be formed not only directly from O but also from O_2, either by dissociative ionization or polar dissociation, both of which have threshold energies considerably higher than simple ionization. With the electron concentration and energy distribution as measured we find, approximately,

$$\text{rate of production of } O^+ = \left\{ 6 \times 10^{12} \frac{[O]}{[O_2]} + 4 \times 10^{10} \right\} \text{cm}^{-3} \text{ s}^{-1}. \tag{15.16}$$

The first term is due to direct ionization of O and the second to dissociative processes from O_2. For both of these processes cross-section data are available (Massey and Burhop, 1969; Massey, 1969).

O^+ ions are lost by charge transfer to O_2 (Massey, 1971) and to the walls at a rate determined from the measured values of $[O^+]$, $[O_2]$ and of the wall current. It is found then that

$$\text{rate of loss of } O^+ = 5.6 \times 10^{11} \text{ cm}^{-3} \text{ s}^{-1}. \tag{15.17}$$

Taken together with (15.16) it is seen that the degree of dissociation of O_2 is about 8%.

We are now in a position to consider the negative-ion balance in the positive column. Negative ions will mainly be formed in the gas phase through the dissociative attachment reaction

$$e + O_2 \longrightarrow O^- + O, \tag{15.18}$$

the cross-section for which is given as a function of electron velocity in Fig. 9.24. Using this in conjunction with the measured electron velocity distribution it is found that

$$\text{rate of production of O}^- \text{ ions} = 4 \times 10^{12} \text{ cm}^{-3} \text{ s}^{-1}. \quad (15.19)$$

Thompson (1961b) first examined what would be the situation if the negative ions suffered no detachment in the gas phase. In that case they could leave the plasma only by axial diffusion as the loss to the walls against the radial field is relatively small. From the measured values of the concentration gradient through the striations (see Fig. 15.4) it was found that n^- would build up to values between 0.5×10^{10} and 1.3×10^{10} cm^{-3}, whereas the measured values are at least an order of magnitude smaller. It follows that detachment in the gas must be the dominant loss process.

The rate coefficient for the associative detachment process

$$O + O^- \longrightarrow O_2 + e \quad (15.20)$$

is 1.9×10^{-10} cm^3 s^{-1} (see Chapter 12, p. 543). We then find, taking $[O] = 10^{14}$ cm^{-3} and $[O^-] = 4 \times 10^8$ cm^{-3}, that

rate of loss of O$^-$ ions due to associative detachment =

$$7.6 \times 10^{12} \text{ cm}^{-3} \text{ s}^{-1}. \quad (15.21)$$

This is sufficiently close to (15.19) to suggest that in fact (15.20) is the dominant reaction which leads to destruction of the negative ions. Some further evidence in support of this was adduced by Thompson (1961b) from measurement of the variation of plasma density with discharge current I_d. Thus O atoms are lost almost entirely by diffusion to the walls so that the [O] balance equation is simply

$$\text{rate of production of O} = \text{rate of wall loss of O} \quad (15.22)$$

The production rate comes almost entirely from the electrons of group C in the energy distribution (see Fig. 15.3), so we have

$$K_1[e]_C = K_2[O].$$

Similarly for the [O$^-$] balance,

$$K_3[e]_B = K_4[O][O^-]$$

since the major contribution of the dissociative attachment comes from the B group of electrons. From these we have

$$[O^-] = K[e]_B/[e]_C. \qquad (15.23)$$

Although $n^- \simeq 20n_e$, the discharge current is carried almost entirely by electrons because of their much greater mobility. In fact most of the current arises from the axial drift current of electrons of group B so that

$$I_D \propto [e]_B. \qquad (15.24)$$

However, the wall current of electrons will vary as $[e]_C$ for only these relatively energetic electrons will penetrate against the radial field. Hence for equilibrium the wall current of positive ions must vary as $[e]_C$. Probe measurements show that, under the working conditions the positive-ion current varies as $I_D^{1/3}$. It follows now from (15.23) that

$$[O^-] \propto I_D^{2/3}, \qquad (15.25)$$

a result which was consistent with the observed data.

If other possible detachment processes are examined, the only one which would give the same variation of O^- as (15.25) would be

$$O^- + O(^1D) \longrightarrow O + O + e \qquad (15.26)$$

and it seems unlikely that the concentration of $O(^1D)$ metastable atoms would be large enough to render this important.

In fact Thompson's analysis of the data on these lines was made before the rate of (15.20) had been measured directly and it provided the first estimate of this rate. The value he found was somewhat smaller than would appear from the numbers given above but this was because, at the time, the measured cross-sections for the attachment process (15.18) were too small by a factor of nearly 4.

While in the oxygen discharge $n^-/n_e \simeq 20$, in typical discharges in iodine $n^-/n_e \simeq 100$, as, for example, in a case investigated much earlier by Spencer-Smith (1935). It is not possible to analyse the negative-ion balance in such detail, as in the experiments of Thompson, partly because we have no information on the degree of dissociation of the I_2 and about the electron velocity distribution and also much less reliable information about the relevant reaction rates.

One important difference from the case of oxygen is clear however. The reaction

$$I^- + I \longrightarrow I_2 + e$$

is endothermic by 1.5 eV, in contrast to the exothermic reaction (15.20) for O^-, so the detachment rate constant due to it is likely to be considerably smaller, probably by a factor of at least 10. This would be more than sufficient to account for the increased ratio of negative-ion to electron concentration in the iodine discharge but account must be taken of the fact that the degree of dissociation of I_2 is likely to be much higher than that, 8%, found in the O_2 discharge. This is because the dissociation energy of I_2 is only 1.5 eV against 5.1 eV for O_2. Furthermore there is considerable uncertainty about the attachment rate in I_2. Referring to Chapter 9, p. 341, we see that the data for electrons of 4 eV energy are uncertain by a factor of 20 or so. The smallest value suggested from extrapolation of Truby's experiment gives a cross-section of only 2.5×10^{-19} cm², as compared with 1.4×10^{-18} cm² for oxygen (Fig. 9.24). It is clear that much remains to be done before the ion balance in a discharge in iodine vapour is understood. Meanwhile it seems clear that the ratio λ of negative-ion to electron concentration in a discharge through a gas which forms negative ions is not simply determined by the magnitudes of the electron affinities of the atoms and/or molecules concerned.

The investigation of discharges of this kind is complicated by the fact that they are usually very noisy electrically. Thus in oxygen there was only a very narrow range of tube currents and gas pressures for which the discharge was sufficiently noise-free for reliable energy distribution measurements to be made. Thompson (1961*b*) examined the conditions which need to be satisfied in order to maintain a double space charge layer at the walls, by extending the analysis of Langmuir (1929) which applied only to the case where negative ions are absent. His results suggest that it is not possible under many conditions to satisfy the sheath criterion and remove the negative ions at a sufficient rate in a steady discharge. This may well be the cause of the inherent instability.

An interesting possibility for investigating the importance of negative ions is to study the conditions in discharge afterglows in electronegative gases. In the later stages of the afterglow there is no source of free electrons, except from detachment, and the negative ion population builds up relative to the electrons. It is therefore possible in principle to follow the transition in behaviour from a plasma in which the negative-ion–electron ratio λ is very small to one in which it is large. Certain aspects of this transition have already been discussed in Chapter 9, in connexion with the use which was made of it to determine attachment rates. Further studies directed to understanding more about the changing ion balance in relation to the ambipolar diffusion fields would be of interest here.

15.2 The effect of negative ions on current build-up and electrical breakdown in gases

Consider the passage of electric currents between two plane parallel electrodes at a separation d, in a gas at pressure p, between which there is a uniform electric field F in a direction to accelerate electrons emitted from one of the electrodes. Let α_i be the ionization coefficient for electrons in the gas, that is to say the chance that an electron will undergo an ionizing collision when drifting through a distance δx in the direction of the field is $\alpha_i \delta x$. It then follows, as first derived by Townsend (1910) in his original theory of electrical breakdown, that the current i arriving at the anode is given by

$$i = i_0 \exp(\alpha_i d), \qquad (15.27)$$

where i_0 is the initial electron current leaving the cathode. This is valid provided no other collisions occur in the gas which can lead to gain or loss of charged particles. When this is so the conditions for electrical breakdown may be obtained as follows.

Let γ be the secondary emission coefficient for positive ions incident on the cathode surface. For a maintained discharge between the plates we must have

$$i = (i_0 + i_s) \exp(\alpha_i d), \qquad (15.28)$$

$$i_s = \gamma\{i - (i_0 + i_s)\}, \qquad (15.29)$$

where i_0 is the current of electrons released from the cathode by external means, i.e. irrespective of positive-ion bombardment, i the total electron current arriving at the anode and i_s the current released per second from the cathode by positive-ion bombardment. Eliminating i_s from (15.28) and (15.29) gives

$$i = i_0 \exp(\alpha_i d)/[1 - \gamma\{\exp(\alpha_i d) - 1\}]. \quad (15.30)$$

If the denominator vanishes, the current is self-sustaining and this gives the condition for electrical breakdown

$$1 - \gamma\{\exp(\alpha_i d) - 1\} = 0. \quad (15.31)$$

We have considered here only one secondary source of electrons but, in fact, if there are other such sources it is only really necessary to modify the interpretation of γ while retaining essentially the form (15.31).

The essential correctness of this picture of the development and onset of a spark discharge between plane parallel electrodes has been established by many experiments in gases in which negative ions are not formed, such as the rare gases. However, for other gases account must be taken of attachment and other ionic reactions.

Penning (1938) first derived the modified form of (15.27) taking account of attachment. If α_a is the attachment coefficient (see Chapter 9, p. 292) and n_e, n^+, n^- are the concentrations of electrons, positive and negative ions at time t at a point in a plane parallel to the electrodes at a distance d from the cathode, the equations of continuity take the form

$$\frac{\partial n_e}{\partial t} + u_e \frac{\partial n_e}{\partial x} = \alpha_i u_e n_e - \alpha_a u_e n_e, \quad (15.32a)$$

$$\frac{\partial n^+}{\partial t} + u^+ \frac{\partial n^+}{\partial x} = \alpha_i u_e n_e, \quad (15.32b)$$

$$\frac{\partial n^-}{\partial t} + u^- \frac{\partial n^-}{\partial x} = \alpha_a u_e n_e. \quad (15.32c)$$

u_e, u^+ and u^- are the drift velocities of the respective charged particles. We seek the steady-state solutions of these equations, subject to the condition that

NEGATIVE IONS – APPLICATIONS

$$\left.\begin{array}{l} u_e n_e = i_0 \\ u^- n^- = 0 \end{array}\right\} \quad x = 0. \qquad (15.33)$$

For the total negative current $u_e n_e + u^- n^-$ received at the anode we find

$$i = i_0 \left[\left\{ \left(\frac{\alpha_i}{\alpha_i - \alpha_a} \right) \exp(\alpha_i - \alpha_a) d \right\} - \frac{\alpha_a}{\alpha_i - \alpha_a} \right]. \qquad (15.34)$$

The presence of attachment makes itself felt by considering plots of $\log(i/i_0)$ as a function of electrode separation d for different values of F/p, it being remembered that both α_a/p and α_i/p are functions of F/p. When $\alpha_a = 0$ such plots are straight lines passing through the origin but when α_a is finite the curves will deviate from linearity to an extent which increases with increasing relative importance of α_a. This behaviour is illustrated in Fig. 15.9 which shows results obtained by Harrison and Geballe (1953) in their pioneering experiments in oxygen. It will be seen that, as F/p decreases so that the mean electron energy decreases, the curves show an increasing departure from linearity. Much more pronounced effects were observed in freon-12 and in CF_3SF_5.

In a paper which followed shortly afterwards, Geballe and Reeves (1953) considered how the breakdown condition (15.31) is modified when attachment occurs. They find that, in terms of the secondary coefficient γ, we have

$$\frac{i}{i_0} = \frac{\{\alpha_i/(\alpha_i - \alpha_a)\} \exp\{(\alpha_i - \alpha_a) d\} - \{\alpha_a/(\alpha_i - \alpha_a)\}}{1 - \gamma\{\alpha_i/(\alpha_i - \alpha_a)\} [\exp\{(\alpha_i - \alpha_a) d\} - 1]}, \qquad (15.35)$$

so that the breakdown condition becomes

$$\gamma\{\alpha_i/(\alpha_i - \alpha_a)\} [\exp\{(\alpha_i - \alpha_a) d\} - 1] = 1. \qquad (15.36)$$

When F/p is such that $\alpha_i/p \geqslant \alpha_a/p$, this condition can be satisfied for sufficiently large values of pd no matter what the values of α_i/p, α_a/p and γ may be. On the other hand, if F/p is such that $\alpha_i/p < \alpha_a/p$, the condition (15.36) tends in the limit of large pd to

$$\frac{\alpha_i}{p} = \frac{\alpha_a}{p}(1+\gamma)^{-1}. \tag{15.37}$$

This limiting condition will be satisfied for some value $(F/p)_c$ of F/p. It then follows that, for $F/p < (F/p)_c$ no breakdown will take place no matter how large pd may be. In practice $\gamma \ll 1$ so

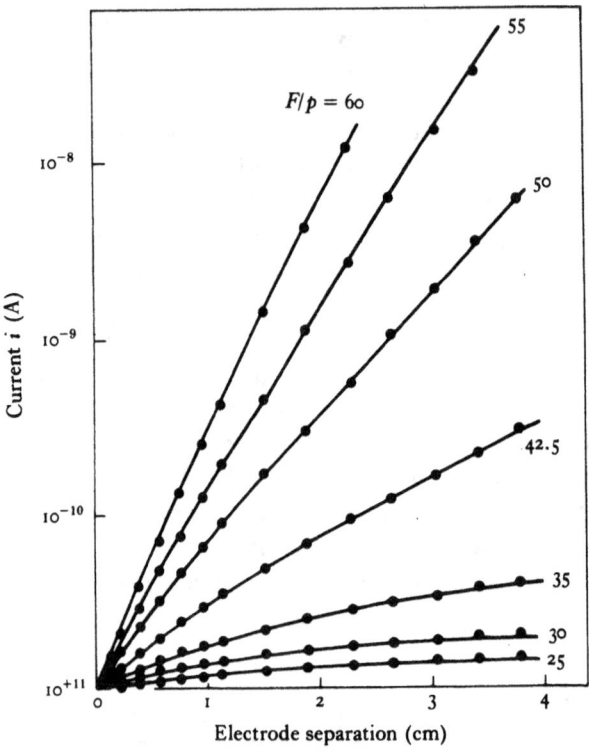

Fig. 15.9. Variation of $\log i$ with electrode separation d in oxygen, at a pressure of 11.2 torr, for various values of F/p as indicated in V cm^{-1} torr^{-1}, observed by Harrison and Geballe (1953).

that $(F/p)_c$ is simply given by the value of F/p for which

$$\alpha_i = \alpha_a.$$

It is also possible to study the growth of ionization in the gap between the electrodes with time after a pulse of electrons has been released from the cathode. Using the equations (15.32) it can be shown that, at a time t after the pulse $<T$, where T is the time

taken for an electron to drift across the gap, the number of electrons in the gap is given by

$$N_e(t) = N_e(0) \left[\exp\{(\alpha_i - \alpha_a) u_e t\} - 1\right]. \qquad (15.38)$$

Measurements of the electron, as distinct from the negative ion, current may be made by observing the intensity of the radiation emitted in the visible and near ultraviolet, excited by impact of the electrons with gas molecules. This may be done with a photomultiplier and oscilloscope.

For large F/p, $\alpha_i > \alpha_a$ and an exponential increase with time is observed, but for lower F/p with $\alpha_i < \alpha_a$ the current will actually fall with time.

Current build-up and breakdown in air and oxygen

A great deal of attention has naturally been paid to the study of current build-up and electrical breakdown in air and oxygen but a number of problems still remain. Results such as those obtained by Harrison and Geballe (1953) for oxygen at a pressure of 11.2 torr (Fig. 15.9) are of the expected form but it is difficult to analyse them to obtain accurate values for α_i/p and α_a/p. In particular, it is not easy to determine i_0 and if this is treated as an adjustable parameter the analysis is ambiguous.

To overcome this problem in O_2, Price, Lucas and Moruzzi (1972) took advantage of the fact that the associative detachment reaction between O^- and H_2 is fast (Chapter 12, p. 550) and therefore observed the effect on the current build-up of adding small fractional concentrations of hydrogen to the oxygen they were studying. It was found that, with only 2.5% of H_2 added, the mixture behaves like a non-attaching gas. This may be seen from the observed $\log i$–d curves shown in Fig. 15.10. Price et al. were then able to measure α_i/p for the O_2–H_2 mixture quite accurately and it can be assumed that this will differ by a few per cent at most from that for pure O_2. The results obtained were found to be 30% below earlier values derived from analysis of current build-up in pure O_2 (Prasad and Craggs, 1961).

With increased precision of measurement it has become neces-

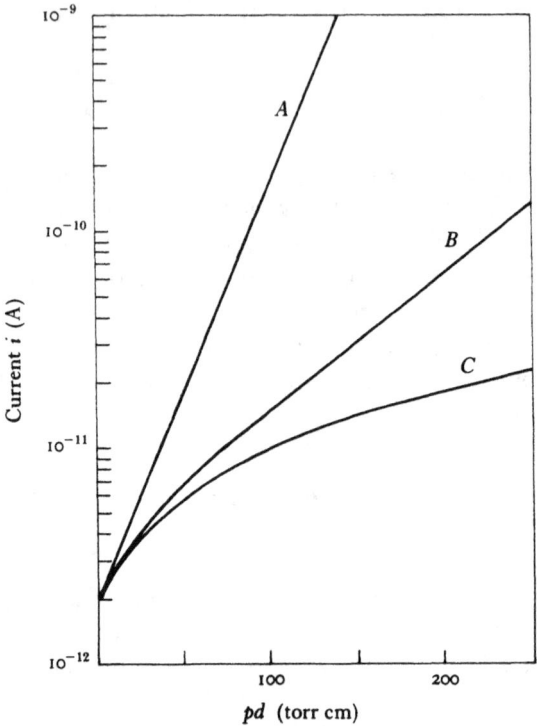

Fig. 15.10. Variation of $\log i$ with pd in O_2-H_2 mixtures for $F/p = 35$ V cm^{-1} torr, observed by Price et al. (1972). Curves A, 2.5% H_2; B, 1.0% H_2; C, 0% H_2.

sary to take into account not only attachment but also detachment and other ionic reactions. At pressures below 1000 torr, attachment takes place essentially through the dissociative process

$$e + O_2 \longrightarrow O + O^-, \qquad (15.39)$$

yielding primary O^- ions. These ions may suffer detachment through

$$O^- + O_2 \longrightarrow O + O_2 + e, \qquad (15.40)$$

and charge transfer

$$O^- + O_2 \longrightarrow O_2^- + O, \qquad (15.41)$$

as well as a number of other reactions involving production and loss of O_2^-, O_3^- and O_4^-. A completely satisfactory analysis of the situation, taking into account all these possibilities, is not yet available but some progress has been made by assuming that the processes (15.39) (15.40) and (15.41) alone are important. Thus, under these conditions Wagner (1971) has analysed data on the current build-up in oxygen and air. If the coefficients for the reactions (15.40) and (15.41) are denoted by α_d and α_t respectively the current build-up is given by

$$i/i_0 = A + B_1 \exp(a_1 d) + B_2 \exp(a_2 d), \quad (15.42)$$

where

$$a_{1,2} = \tfrac{1}{2}\{-U \pm (U^2 - 4V)^{1/2}\},$$
$$U = \alpha_t + \alpha_d + \alpha_a - \alpha_i,$$
$$V^2 = (\alpha_t + \alpha_d)(\alpha_a - \alpha_i) - \alpha_d \alpha_a,$$
$$A = \alpha_a \alpha_t / a_1 a_2,$$
$$B_{1,2} = -(\alpha_t + \alpha_d + \alpha_a + a_{1,2} + A a_{2,1})/(a_{2,1} - a_{1,2}). \quad (15.43)$$

A good fit to the data was obtained with α_i/p close to that measured by Price et al. (1972), α_a/p close to that calculated from measurements of dissociative attachment cross-sections (Chapter 9, p. 316) and α_d/p to measurements made by Frommhold (1964) using the pulse method (Chapter 12, p. 530), provided suitable values were also included for α_t/p. A similar fit was obtained for air in which the attachment and charge transfer coefficients were as for O_2, allowing for the partial pressures, while α_d/p was somewhat larger than for O_2. The same choice of coefficients was applied to the analysis of observed data on the temporal build-up of the electron current in air (Raether, 1964) and gave good agreement. No allowance was made for detachment from O_2^-.

At the time of writing, the interpretation of results at high F/p is still not fully clear and account may need to be taken of the role of O_3^- and, in some cases, O_4^-.

A further step towards obtaining more definite basic data was made by Price and Moruzzi (1973), by observing current build-up in O_2 containing small admixtures of CO_2. This acts in the opposite way to H_2 through the formation of cluster ions CO_3^- from O^- which are effectively stable towards detachment and charge transfer. Knowing α_i from the measurement with O_2–H_2 mixtures,

α_a may then be obtained to a few per cent from similar measurements with O_2–CO_2.

Current build-up and breakdown in halogen-containing substances

Gases and vapours such as SF_6, CCl_4, C_2F_6, etc. have long been known to have high dielectric strength and this may be ascribed to the ease with which electrons attach to these molecules (see Chapter 9, p. 368).

Geballe and Reeves (1953) made measurements of current build-up and breakdown in CCl_4, CF_2SF_5, CCl_2F_2 and SF_6. From the former measurements they derived α_i/p and α_a/p and hence the

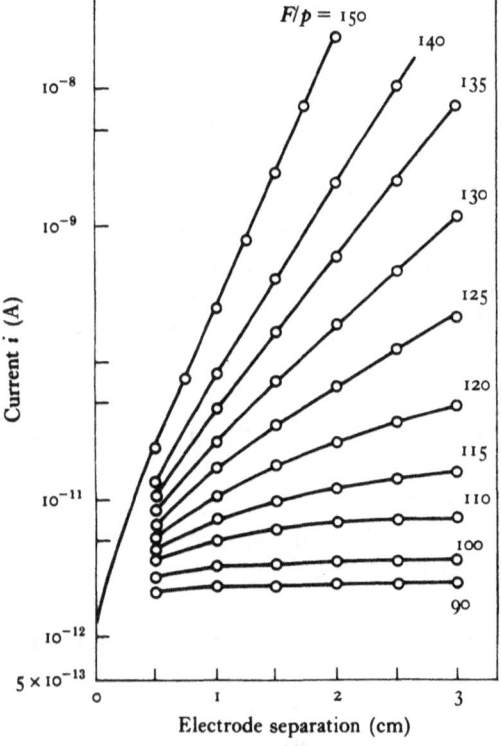

Fig. 15.11. Variation of $\log i$ with electrode separation d in SF_6 at a pressure of 5 torr, for various values of F/p, as indicated in V cm^{-1} torr^{-1}, observed by Bhalla and Craggs (1962).

value of $(F/p)_c$ at which $\alpha_i = \alpha_a$. According to (15.37), breakdown should not occur for $F/p < (F/p)_c$ and this they were able to verify directly.

In practice the situation may be complicated by local departures from the assumed conditions, as was found in a thorough study carried out with SF_6 by Bhalla and Craggs (1962). Fig. 15.11 shows the observed currents in SF_6, at a pressure of 5 torr, as functions of the electrode separation d for a number of values of F/p, the results being typical of an attaching gas. It was found from analysis of these data that $(F/p)_c = 117$ V cm^{-1} torr^{-1}, in good agreement with the earlier measurements of Geballe and Reeves (1953).

Fig. 15.12 shows the observed breakdown potential as a function of pd compared with the corresponding potentials for air. The greater breakdown strength of SF_6 is apparent. Finally, in Fig. 15.13, values of F/p obtained from the measured breakdown potentials are plotted as functions of pd. The broken curves represent breakdown occurring for $F/p < (F/p)_c$ which sets in at higher pd the higher the pressure. This type of breakdown is

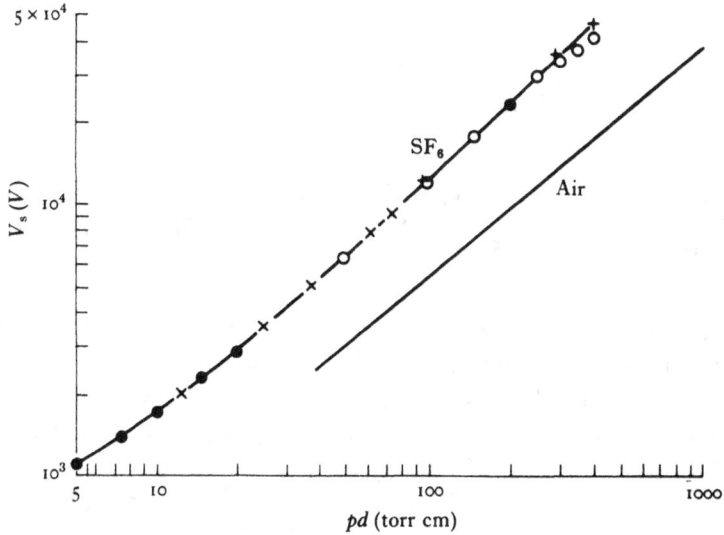

Fig. 15.12. Observed breakdown potential V_s as a function of pd for SF_6 (Bhalla and Craggs, 1962) compared with that for air. ● 5 torr, × 25 torr, ○ 100 torr, + 200 torr, SF_6 pressure.

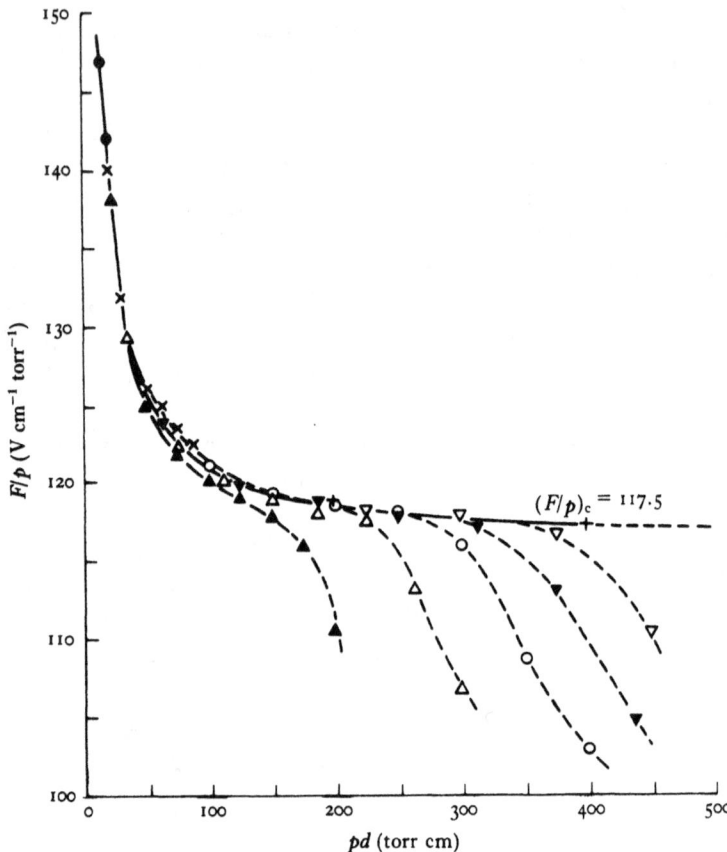

Fig. 15.13. Values of F/p at breakdown as a function of pd for SF_6 (Bhalla and Craggs, 1962). ● 5 torr, × 25 torr, ▲ 50 torr, △ 75 torr, ○ 100 torr, ▼ 125 torr, ▽ 150 torr, + 200 torr, SF_6 pressure.

associated with unstable currents and is observed to occur in the fringing field near the edges of the electrodes.

Observations have now been made of current build-up and breakdown in a number of halogen-containing molecules including C_2F_6 (Božin and Goodyear, 1968), C_4F_{10} (Razzak and Goodyear, 1968; Devins and Wolff, 1965), Cl_2 (Božin and Goodyear, 1967) and Br_2 (Razzak and Goodyear, 1969). Much remains to be done before the results can be thoroughly interpreted in terms of attachment and other reactions.

15.3 Negative ions in the terrestrial ionosphere

Introduction

The ionization present in the earth's atmosphere above an altitude of about 65 km arises from photoionization by solar radiation. At any particular altitude the concentration of electrons and of positive and negative ions of different species will depend not only on the rate of production by sunlight but on the rates of different processes which lead to neutralization, transfer of charge and chemical rearrangement. Intensive study of the ionized regions, the ionosphere, using ground-based techniques, began in the 1920s. It soon became apparent that the main regions capable of reflecting back radio waves, and hence making world-wide radio transmission possible, were located at quite high altitudes. Three separate regions were distinguished during the daytime, the E region at an altitude between 100 and 120 km, the F_1 near 180 km, and the main, F_2, region extending outwards from 220 km or so. Typical peak electron concentrations n_e observed in these regions were 1.5×10^5, 2.5×10^5 and 10^6 cm^{-3} respectively. Below the E region n_e fell off quite rapidly but could still be observed during the daytime down to 70 km or so and it became customary to refer to the sub-E ionization as constituting the D region.

At an early stage the nature of the processes giving rise to loss of electrons in the ionosphere was a matter of animated debate. From the point of view of their effectiveness in reflecting radio waves, an electron was lost equally well by attachment to a neutral molecule to form a negative ion as by recombination to a positive ion. In the absence of reliable information about reaction rates it was difficult to come to any conclusive decision as to the relative importance of these processes of electron loss. It was not recognized at first that, if attachment were dominant, it would be difficult to understand how the separate peaks of electron concentration associated with the E and F_1 regions could arise. A new and very important factor was first introduced in 1936 by Martyn and Pulley who had the prescience to suggest that the detachment reaction

$$O^- + O \longrightarrow O_2 + e$$

might well be important, although at the time they had no knowledge of the reaction rate. By this time more information about reaction rates gradually became available, largely from theoretical sources. Massey and Bates (1942) were able to show, from considerations of radiative attachment to O and photodetachment from O^-, that at least in daytime the concentrations of negative ions at E region heights and above must be much less than that of the electrons. They therefore sought for processes by which electrons in these regions were lost, ignoring the possibility of negative-ion formation, and suggested (Bates and Massey, 1947), for this role, dissociative recombination which has proved to be correct.

Quite soon after the Second World War, opportunity for detailed analysis of the properties and behaviour of the ionosphere became greatly expanded by the introduction of the techniques of space research, which made it possible to make *in situ* observations of the ionosphere and of the solar radiation at those levels, and of laboratory study of electronic and ionic reaction rates.

For a thoroughgoing theory of the ionosphere we need to have available the following data:

(*a*) the intensity and composition of the solar radiation,

(*b*) the composition of the neutral atmosphere as a function of altitude,

(*c*) the absorption and photoionization cross-sections of atmospheric atoms and molecules for solar radiation,

(*d*) the rates of recombination, attachment and other ionic reactions involving atmospheric ions and neutral constituents, as well as knowledge of atmospheric motion which may redistribute the ionization. Thanks to the availability of space vehicles we now have a great deal of information about (*a*) and (*b*) while laboratory experiments have yielded a remarkable, even if still not complete, amount of information about (*c*) and (*d*). Moreover, rocket-borne equipment has made measurements of positive-ion composition and is beginning to yield results about the negative ions. Taken together, there is sufficient material to check in some detail the analysis of ionospheric properties in the E and F regions, while extension to the lower, D, region has already proceeded some distance.

It is not difficult to show that negative ions must be quite unim-

TABLE 15.1 *Typical concentrations, in cm^{-3}, of the main neutral atmospheric constituents at 100 km and above*

Altitude (km)	O	N_2	O_2
100	5×10^{11}	8.1×10^{12}	2.0×10^{12}
130	3.3×10^{10}	1.2×10^{11}	2.1×10^{10}
160	9.4×10^{9}	2.0×10^{10}	2.8×10^{9}
220	2.6×10^{9}	2.5×10^{9}	2.7×10^{8}

portant in the E region and above. Thus typical concentrations of the main neutral constituents at altitudes of 100 km and above (CIRA, 1965) are given in Table 15.1. It will be seen that at 130 km and above the concentration of atomic oxygen exceeds that of O_2 by a factor which increases with altitude. Between 130 and 100 km the situation is reversed.

At the low pressures concerned, the only process which leads to negative-ion production is radiative attachment. For electrons of thermal energies, the rate of attachment to O, derived by detailed balance from the measured photodetachment cross-section of O^- (see Fig. 8.1) is approximately 1.2×10^{-15} [O] s^{-1}, while the rate of associative detachment from O^- in collisions with O is 1.9×10^{-10} [O] s^{-1}. It follows that, unless there is some additional attachment process, the ratio λ of the concentration n^- of O^- ions to that n_e of the electrons will be no greater than 10^{-5}. No significantly greater contribution can be expected to come from O_2^- which will suffer detachment through

$$O_2^- + O \longrightarrow O_3 + e, \qquad (15.44)$$

for which the rate coefficient is 3×10^{-10} cm^3 s^{-1} (see Table 12.1).

At lower heights the concentration of O falls and the main primary ion will be O_2^-. Detachment will continue to occur mainly through (15.44) so the ratio λ will be given approximately by 10^{-15} $[O_2]/3.0 \times 10^{-10}$ [O] $\simeq 3 \times 10^{-6}$ $[O_2]/[O]$. At a sufficiently low altitude, primary O_2^- ions will be formed at a significant rate through the three-body process

$$e + O_2 + O_2 \longrightarrow O_2^- + O_2, \qquad (15.45)$$

666   NEGATIVE IONS

for which the rate constant (Chapter 9, p. 327) is 1.6×10^{-30} cm^6 s^{-1}. For this to produce O$^-$ ions at the same rate as they are produced by radiative attachment,

$$1.6 \times 10^{-30} [O_2]^2 \simeq 10^{-15} [O_2], \qquad (15.46)$$

so $[O_2]$ must be about 6×10^{14} cm^{-3}.

The lower ionosphere

To examine the consequences of these numbers for the composition of the negatively-charged particles in the lower ionosphere it is necessary to trace the concentrations of O and O$_2$ to much lower

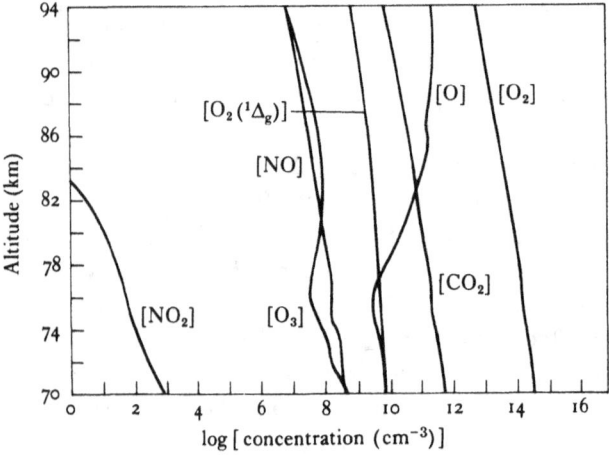

Fig. 15.14. Altitude profile of various neutral atmospheric constituents at altitudes between 70 and 94 km, as adopted by Reid (1970).

altitudes. For O$_2$ (CIRA, 1965) there is reasonably reliable information available from *in situ* pressure measurements with rocket-borne equipment because at these altitudes the O$_2$/N$_2$ ratio can be expected to be nearly the same as at ground. Much less reliable information is available for O and it is necessary to resort to a theoretical determination (Shimazaki and Laird, 1970) in which account was taken of molecular and eddy diffusion in modifying the distribution resulting from the equilibrium between photo-dissociation of O$_2$ and three-body recombination of O atoms. Fig. 15.14 shows the concentration between 94 and 70 km obtained in this way. Above 80 km it agrees quite well with two other theoretical determinations (Colegrove, Johnson and Hanson, 1966; Bowman, Thomas and Geisler, 1970) but there is marked disagreement below 80 km (Sechrist, 1972). Using the concentrations shown in Fig. 15.14, we find that, even at 80 km, λ will be still as small as 8×10^{-3} and only approaches unity a little below 70 km. Three-body attachment becomes as effective as radiative attachment at about the same altitude.

Once λ becomes of order unity we can expect that the lifetime of an O_2^- ion will be long enough for it to have an appreciable chance of undergoing some transformation in reaction with one of the neutral atoms and molecules present. In other words, the composition of the negative ions will become rapidly more complex, the dominant ion being the one which is most stable towards further reaction.

Sufficient information is available about reaction rates to make it practicable to calculate the composition of the negative ions, provided sufficiently reliable information is available about the neutral constituents which are likely to play a significant role. These include CO_2, O_3, NO, $O_2(^1\Delta_g)$ metastable molecules and H_2O. Altitude profiles of concentration for all of these except H_2O are shown in Fig. 15.14. That for CO_2 has been obtained by assuming the same mixing ratio as at ground and that for $O_2(^1\Delta_g)$ from airglow measurements (Evans, Hunten, Llewellyn and Jones, 1968). Theoretical calculations as for O were used to obtain the profiles for NO and O_3. The concentration of NO has been determined above 70 km by Meira (1971) from measurements of the intensity of the resonance fluorescence of the NO gamma bands. His results are quite close to the theoretical values shown in Fig. 15.14. In default of any reliable data on the concentration of water vapour, no estimates can be made of the degree of hydration of any ions which may be present.

Equations for the concentrations of the different species may be set up which include the various possible reactions which each species may undergo. For the reactions between electrons and various negative ions, no account has been taken, up to the time of writing, of the possibility of hydration, partly because there is little laboratory information about the various reaction rates and partly because of inadequate knowledge of the water vapour content. Eight negative ions remain to be considered, namely O^-, O_2^-, O_3^-,

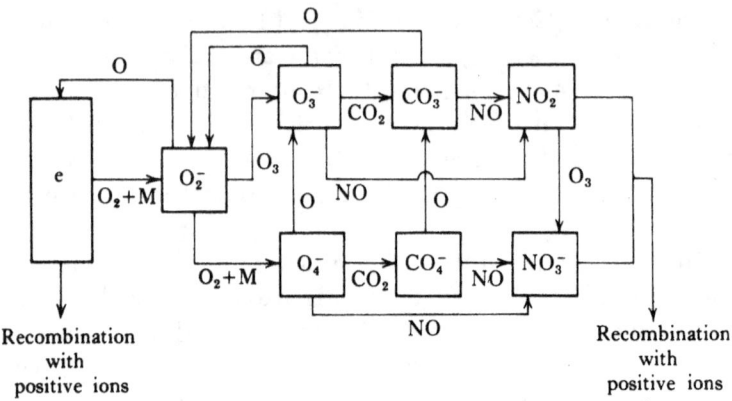

Fig. 15.15. Flow diagram illustrating reactions involving electrons and negative ions in the D region of the ionosphere. From Sechrist (1972).

O_4^-, CO_3^-, CO_4^-, NO_2^- and NO_3^-. It is usual to represent the reaction scheme in terms of a flow diagram, as shown in Fig. 15.15. The initial charged reactant and final charged product are specified by the boxes and the directed flow lines, while the neutral reactant is indicated on the flow line.

Basically the primary reaction is taken to be the three-body attachment

$$O_2 + O_2 + e \longrightarrow O_2^- + O_2. \qquad (15.47)$$

The O_2^- ions may then undergo three possible reactions

$$O_2^- + O \longrightarrow O_3 + e, \qquad (15.48a)$$

$$O_2^- + O_3 \longrightarrow O_3^- + O_2, \qquad (15.48b)$$

$$O_2^- + O_2 + M \longrightarrow O_4^- + M. \qquad (15.48c)$$

The O_3^- and O_4^- can undergo further reactions leading eventually to NO_3^- as the terminal stable ion which has the highest electron affinity. The rates of the relevant reactions are all given in Tables 12.1–12.3.

In addition to these reactions there must also be included those which lead to loss of electrons by recombination and of negative ions by mutual neutralization. For the latter a mean rate coefficient of 10^{-7} cm^3 s^{-1} is a reasonable assumption (see Chapter 14, p. 628).

Estimation of the rate of loss by recombination is more difficult. There is experimental evidence that the dissociative recombination coefficient increases with the complexity of the positive ion concerned. For NO^+ it is 4×10^{-7} cm^3 s^{-1} (Mehr and Biondi, 1969), while for hydrated protons $H^+(H_2O)_n$ it increases from 10^{-6} cm^3 s^{-1} for $n = 1$ to 10^{-5} cm^3 s^{-1} for $n = 7$ (Leu, Johnsen and Biondi, 1973). It is known from mass analysis carried out on rocket flights (Narcisi and Bailey, 1965; Goldberg and Blumle, 1970) that below about 82 km the major ions are hydrated protons, $H^+(H_2O)_2$ being the

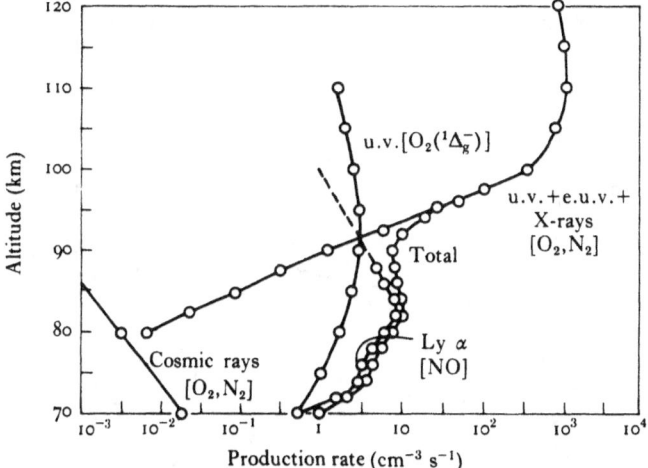

Fig. 15.16. Altitude profile of the rate of free electron production at altitudes between 70 and 120 km, calculated by Reid (1970). The contributions from photoionization of different constituents, as indicated, are shown separately as well as that due to ionization by cosmic rays.

most abundant. We would therefore expect a considerable increase in the recombination coefficient under these conditions to values possibly as high as 5×10^{-6} cm^3 s^{-1}.

The rate of electron production by solar radiation offers no particular difficulty once the atmospheric composition is known because the intensity of solar ionizing radiation, including Lyα which has penetrated through the O_2 absorption window, is fairly well known.

Fig. 15.16 shows the altitude profile of the rate of production of electrons from different sources calculated by Reid (1970) using the atmospheric composition in Fig. 15.14. Between 85 and 70 km it is almost entirely due to ionization of NO. In fact a serious problem arises here because it has not been possible to identify a loss process for NO^+ fast enough to balance the production rate. However, this is an incidental difficulty in the present context as we are mainly concerned with the negative ions.

Fig. 15.17 shows the negative-ion composition as a function of altitude also calculated by Reid (1970). This follows the general lines discussed above. O_2^- is dominant down to 85 km but at lower

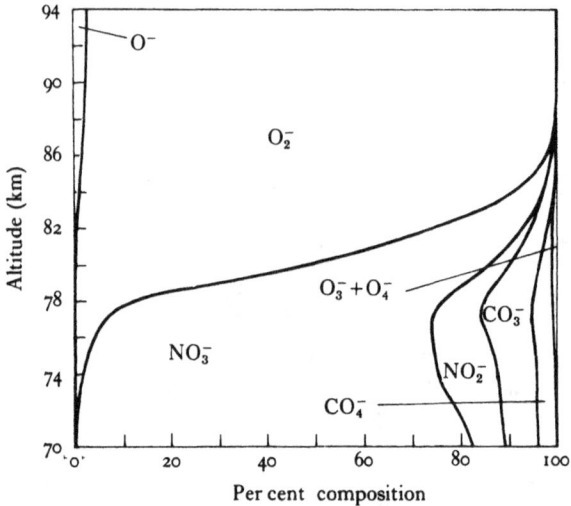

Fig. 15.17. Negative-ion composition as a function of altitude, as calculated by Reid (1970).

altitudes, where the total negative-ion concentration becomes comparable with that of the electrons, most of the ions are complex and predominantly NO_3^-.

The analysis of the ion composition at the relatively high pressures which prevail in the D region, using rocket-borne mass spectrometers, is difficult and especially so for the negative ions. This is because the rocket will normally be at a negative potential with respect to the surrounding atmosphere. If a positive potential is applied to the collector relative to the rocket this may draw in such a large electron current as to drive the main body of the rocket even more negative relative to the surroundings, thus rendering the applied positive voltage ineffective.

The first flights which produced definite results were made by Narcisi, Bailey, Della Lucca, Sherman and Thomas (1971). Below about 90 km the dominant species were found to be heavy ions with masses near 60, 62, 76, 80, 98, 116, 134 and 152 a.m.u. The last five, forming a sequence with a separation of 18 a.m.u. between successive members, together with that at mass 62, can probably be identified as $NO_3^-(H_2O)_n$ with n ranging from 0 to 5. The remaining ions with masses of 60 and 76 are probably CO_3^- and CO_4^- respectively.

Similar results were obtained in a flight during the solar eclipse of 7 March 1970 (Narcisi, Bailey, Wlodyka and Philbrick, 1972). Arnold, Kessel, Krankowsky, Wieder and Zähringer (1971) observed a rapid decrease in total negative-ion concentration above 78 km. The dominant ions were of mass 111 ± 1 and 125 ± 1, tentatively identified as $CO_4^-(H_2O)_2$ or $NO_2^-(HNO_2)H_2O$, and $NO_3^-(HNO_3)$, respectively. There was no sign of the hydrated ions observed by Narcisi and his collaborators but unclustered CO_3^-, NO_3^-, HCO_3^- and O_2^- as well as Cl^- were observed.

It is clear that much remains to be done before the difficult and complicated problem of determining the negative-ion composition and its variation is adequately solved, although there is some evidence that hydrated NO_3^- ions are dominant, at least in many conditions.

The electron concentration n_e may be calculated as a function of height once the effective recombination coefficient α_{eff} for electrons is known. Taking this to be 10^{-6} cm^3 s^{-1}, n_e comes out to be quite close to that observed at an altitude of 90 km but at lower altitudes becomes increasingly too large. It is possible then to derive the variation of α_{eff} with height which would provide agreement with the observed n_e. This was done by Reid (1970) who obtained the height profile of α_{eff} shown in Fig. 15.18. When the negative-ion concentration is negligible, α_{eff} is equal to the true electron recombination coefficient α_e but if λ, the ratio of negative-ion to electron concentration, is comparable with unity,

$$\alpha_{eff} = (1 + \lambda)(\alpha_e + \lambda \alpha_i), \qquad (15.49)$$

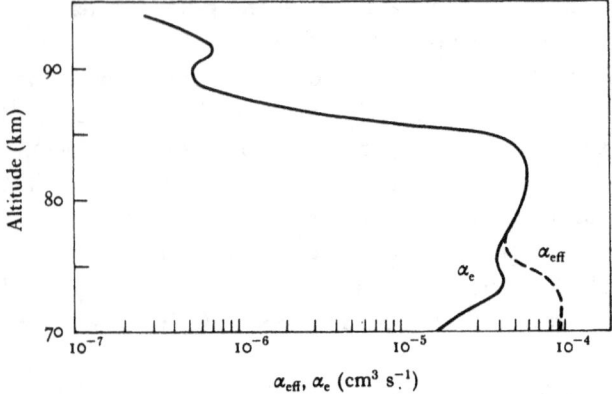

Fig. 15.18. Effective and true recombination coefficients α_{eff} and α_e, as a function of height, required to produce agreement between observed electron concentrations and those calculated by Reid (1970).

where α_i is the ion–ion mutual-neutralization coefficient. Thus if n^+ is the positive-ion concentration and q the rate of ion production,

$$\frac{dn^+}{dt} = q - \alpha_e n_e n^+ - \alpha_i n^- n^+. \tag{15.50}$$

Writing $n^+ = n^- + n_e$, $\lambda = n^-/n_e$, this gives

$$\frac{dn_e}{dt} = q(1+\lambda)^{-1} - (\alpha_e + \lambda\alpha_i) n_e^2, \tag{15.51}$$

it being assumed that λ remains effectively unchanged as recombination proceeds. Since, in equilibrium when $dn_e/dt = 0$, $n_e = (q/\alpha_{\text{eff}})^{1/2}$, (15.49) follows. Using the value of λ as a function of altitude calculated by Reid (1970), as described above, and taking α_i as 10^{-7} cm^3 s^{-1}, the required value of α_e remains equal to α_{eff} at altitudes above about 76 km and remains above 10^{-5} cm^3 s^{-1} below (see Fig. 15.18). It is very difficult to understand how such large values of α_e arise, even for heavily hydrated positive ions.

In this discussion we have followed closely the analysis of Reid (1970) which extended the earlier work on the same lines by le Levier and Branscomb (1968). Analyses have also been carried out by Sechrist (1972) and by Thomas, Gondhalikar and Bowman (1973) which differ in detail but lead to the same conclusions.

There is little doubt that the negative-ion–electron ratio is very small above 80 km and it seems likely that $NO_3^-\cdot(H_2O)_n$ are the dominant negative ions below that altitude. Nevertheless very serious problems remain, particularly about the loss mechanism which requires a quite sudden increase in the effective recombination coefficient α_{eff} to very large values as the altitude falls below 85 km. It is significant that rocket studies of positive-ion composition suggest that it is just at these altitudes that positive-ion clustering becomes important, but the required values of α_{eff} are large even for clustered ions. However, our knowledge of the neutral particle composition and of its variations is still rudimentary and too much should not be expected of a detailed theory at this stage.

15.4 Negative ions in the atmospheres of the sun and stars

H^- and the continuous emission spectrum of the sun

One of the most remarkable instances in which negative ions have played a decisive role in determining the nature of an observable phenomenon is that of the solar continuous spectrum. It has been shown beyond doubt that the absorption of negative hydrogen ions present in the solar photosphere determines the spectral distribution in the observable region. In order to appreciate fully how this has been established and why it arises it is necessary to summarize first the nature of the problem concerned and the method used to obtain a solution.

The emission spectrum in terms of atmospheric absorption coefficients

The problem is to calculate the amount of energy $I_\nu(\theta)\,d\nu\,d\omega$, with frequency between ν and $\nu + d\nu$, radiated per unit area per second by the sun within the solid angle $d\omega$ in a direction making an angle θ with the outward-drawn normal at any point of the emitting surface of the sun (the outer boundary of the photosphere).

We first define the emission and absorption coefficients j_ν, k_ν of the emitting material. j_ν is such that the total radiant energy with frequency between ν and $\nu + d\nu$ emitted per second from unit volume of the material of density ρ is $j_\nu \rho\, d\nu$. k_ν is such that the fractional loss of intensity of a beam of radiation with frequency between ν and $\nu + d\nu$ in passing a small distance dx through the material is $k_\nu \rho \, dx$. In terms of the absorption cross-section Q of the constituent atoms, as used in Chapter 11, $k_\nu = MQ$, where M is the mass of an atom. If the material is in thermal equilibrium at temperature T, then it is a well-known result of statistical thermodynamics that

$$j_\nu/k_\nu = 4\pi B_\nu, \qquad (15.52)$$

where B_ν is the Planck distribution function

$$B_\nu = \frac{2h\nu^3}{c^3} \bigg/ \left\{\exp\left(\frac{h\nu}{kT}\right) - 1\right\}. \qquad (15.53)$$

Use can be made of this relation even when thermal equilibrium is not strictly attained, T being regarded as an effective temperature to be determined.

To calculate $I_\nu(\theta)$ we consider the emission from an element of volume Q of solar material at a small depth x from the surface which is in such a direction as to pass out of the photosphere at P (Fig. 15.19) in a direction making an angle θ with the outward normal.

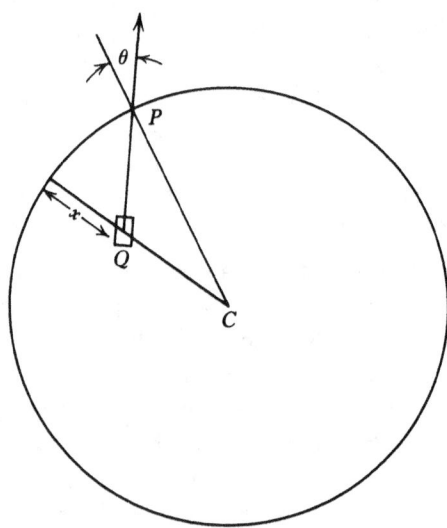

Fig. 15.19.

We choose the element to be a cylinder of unit cross-section, length ds and axis in the direction QP. The intensity of the radiation emitted by this element per unit solid angle in the direction QP is

$$\frac{1}{4\pi} j_\nu \rho \, ds = \frac{1}{4\pi} j_\nu \rho \, dx \sec\theta. \tag{15.54}$$

Owing to absorption in passing from Q to P only a fraction

$$\exp\left\{-\int_{-x}^{0} k_\nu \rho \, dx \sec\theta\right\} \tag{15.55}$$

will reach the surface at P and contribute to the solar emission. Adding the contributions from all volume elements Q we have

$$I_\nu(\theta) = \frac{1}{4\pi} \int_{-R}^{0} j_\nu \rho \sec\theta \exp\left\{-\int_{-x}^{0} k_\nu \rho \sec\theta \, dx\right\} dx, \tag{15.56}$$

where R is the radius of the sun.

To eliminate j_ν we use the relation (15.52), remembering, however, that the effective temperature T will depend on the depth x below the surface and on the frequency ν. It is also convenient to change the variable in (15.56) to τ, the optical depth, defined by

$$\tau_\nu = \int_{-x}^{0} k_\nu \rho \, dx, \tag{15.57}$$

so that

$$d\tau_\nu = k_\nu \rho \, dx. \tag{15.58}$$

We find then that

$$I_\nu(\theta) = \frac{2h\nu^3}{c^2} \sec\theta \int_0^\infty \exp\{-\tau_\nu \sec\theta\} \left\{\exp\left(\frac{h\nu}{kT_\nu}\right) - 1\right\}^{-1} d\tau_\nu. \tag{15.59}$$

In this way the problem becomes one of determining T_ν as a function of τ_ν. Instead of attempting this directly a simpler indirect procedure may be employed in which a suitable average over frequency is used.

We consider the flow of radiation through the photosphere which is assumed to be in radiative equilibrium. Referring again to the volume elements at a depth x we must then have:

(a) Radiation absorbed within the element = radiation emitted by the element, so that

$$k_\nu \rho \, ds \int I_\nu(\theta, x) \, d\omega = j_\nu \rho \, ds, \qquad (15.60)$$

where $I_\nu(\theta, x)$ is the value of I_ν at the depth x. Writing

$$J_\nu = \frac{1}{4\pi} \int I_\nu \, d\omega,$$

we have then

$$J_\nu = \frac{1}{4\pi} j_\nu / k_\nu$$

$$= B_\nu. \qquad (15.61)$$

(b) The increase in radiant intensity in direction θ, $dI_\nu(\theta, x)$, in passing through the element, is the excess of the intensity emitted in that direction by the element over that absorbed, i.e.

$$dI_\nu(\theta, x) = \frac{1}{4\pi} j_\nu \rho \, ds - k_\nu \rho I_\nu(\theta, x) \, ds, \qquad (15.62)$$

the emission from the element being supposed independent of direction. Using (15.52) and (15.61), and the relation $ds = dx \sec \theta$, gives

$$\cos \theta \frac{dI_\nu}{dx} = -k_\nu (I_\nu - J_\nu). \qquad (15.63)$$

We now integrate over all frequencies to give

$$\cos \theta \frac{dI}{dx} = -\bar{k}(I - J), \qquad (15.64)$$

where

$$I = \int I_\nu \, d\nu, \quad J = \int J_\nu \, d\nu,$$

and the mean absorption coefficient $\bar{k}$ is given by

$$\bar{k} I = \int k_\nu I_\nu \, d\nu \qquad (15.65)$$

$\bar{k}$, as so defined, depends on θ, but the approximation is now made of replacing it by an average over all angles in (15.64).

NEGATIVE IONS – APPLICATIONS

Introducing again the mean optical depth τ defined by

$$\tau = \int_{-x}^{0} \bar{k}\rho \, ds,$$

we have

$$\cos\theta \frac{dI}{d\tau} = I - J. \tag{15.66}$$

This equation must be integrated with the boundary condition that, at the surface $\tau = 0$, the inward flux of radiation is zero. Having done so, the mean temperature T at an optical depth τ is given by

$$J = \sigma T^4,$$

where σ is the Stefan–Boltzmann constant, a result obtained at once by integrating (15.61) over all frequencies.

A good approximate solution of (15.66) is obtained by assuming

$$I(\tau, \theta) = a_0(\tau) + a_1(\tau)\cos\theta,$$

which gives

$$T_\tau^4 = T_e^4(\tfrac{1}{2} + \tfrac{3}{4}\tau), \tag{15.67}$$

where T_e is the effective temperature. T_e is such that the total intensity emitted by the sun is the same as that which would be emitted if it were a black body at temperature T_e (5750 °K).

Returning now to the expression (15.59) for $I_\nu(\theta)$ we substitute the mean temperature T_τ given by (15.67) for T_ν and change the variable from τ_ν to τ, where

$$d\tau_\nu = \frac{k_\nu}{\bar{k}} d\tau.$$

This gives

$$I_\nu(\theta) = \sec\theta \frac{k_\nu}{\bar{k}} \int_0^\infty B_\nu(T\tau) \exp\left\{-\frac{k_\nu}{\bar{k}} \tau \sec\theta\right\} d\tau. \tag{15.68}$$

From the observed values of $I_\nu(0)$, $\int I_\nu(\theta)\,d\omega$ or $I_\nu(\theta)/I_\nu(0)$ it is now possible to determine $k_\nu/\bar{k}$ as a function of ν. A great deal of work has been carried out on these lines and good agreement between the different methods has been obtained. The resulting variation of $k_\nu/\bar{k}$ with wavelength λ is illustrated in Fig. 15.20.

The chief constituent of the sun and stars is hydrogen but, in order to explain the strong absorption at wavelengths > 5000 Å, it appeared necessary to suppose that atoms with low ionization potential, such as those of the metals, were present in sufficient proportion to provide this absorption. Grave difficulties were, however, associated with this interpretation. The main one concerns the strength of the metallic absorption lines. If the metals were present in sufficient proportion to provide the observed

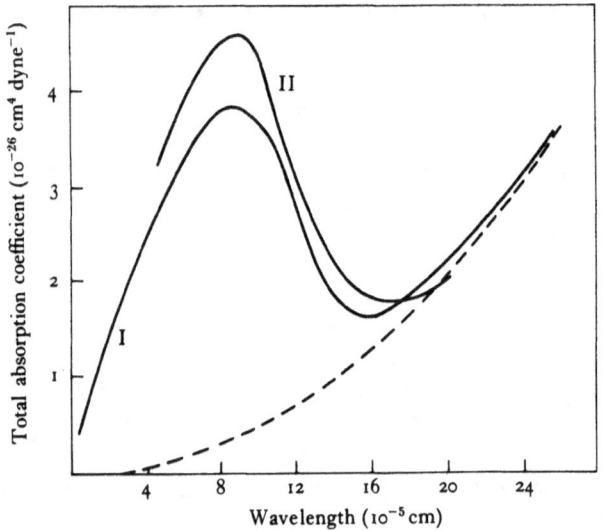

Fig. 15.20. The total absorption coefficient per hydrogen atom per unit electron pressure (see p. 681) at a temperature of 6300 °K. I calculated (see text), II derived from analysis of absorption in the solar atmosphere, - - - - contribution from free–free transitions.

continuous absorption they would produce much stronger absorption lines than were observed. Furthermore, they would give rise to a number of absorption edges which are not found (Stromgren, 1940; Deutsch, 1948).

The way out of this difficulty was first shown by Wildt (1939, 1941). He pointed out that, at the temperatures and electron pressures prevalent in the sun, a proportion of the atomic hydrogen would be present as negative ions which have a low threshold (0.755 eV) for continuous absorption.

The proportion of hydrogen present as H⁻ may be calculated in the following way. We consider the reaction

$$H + e \rightleftharpoons H^-$$

to be in thermal equilibrium at temperature T. Then, following the usual formula of statistical mechanics, the concentrations $n(H^-)$, $n(H)$, $n(e)$ per cm³ of H, H⁻ and e respectively are related by

$$\frac{n(H^-)}{n(e)n(H)} = \frac{f(H^-)}{f(H)f(e)},$$

where $f(H^-)$, $f(H)$, $f(e)$ are the corresponding partition functions. If A is the electron affinity of hydrogen

$$\frac{f(H^-)}{f(H)} = \tfrac{1}{2} e^{A/kT},$$

the factor $\tfrac{1}{2}$ being the ratio of the statistical weights of the ground states of H⁻ and H. Also

$$f(e) = 2(2\pi mkT)^{3/2}/h^3,$$

where m is the mass of the electron. Substituting numerical values and introducing the electron pressure p_e dynes/cm² $(= n_e kT)$, we find

$$\frac{n(H^-)}{n(H)} = p_e \phi(T), \qquad (15.69)$$

where

$$\log_{10} \phi(T) = 0.12 + \frac{5040}{T} A - 2.5 \log_{10} T, \qquad (15.70)$$

and the electron affinity A is measured in eV. The electron pressure p_e is determined by application of similar considerations to the ionization equilibrium of the metals, the abundances of which are determined from studies of the strength of metallic line absorption.

Estimates based on the rough absorption coefficients known at the time indicated that H⁻ would indeed provide the observed absorption and the first calculations using more accurate wave functions, those of Massey and Bates (1940), confirmed this.

Since then there has been a series of calculations using more and more accurate wave functions until we now have quite reliable results (see Chapter 11, pp. 421–4) which agree well with those obtained from photodetachment experiments (see Fig. 11.1). These absorption coefficients give quite good agreement with the solar absorption spectrum below 12 000 Å but at greater wavelengths become much too small (see Fig. 15.20). The reason for this discrepancy was soon traced to neglect of so-called free–free absorption in which an electron makes a transition between two continuum states in the field of an H atom.

The contribution from free–free absorption

The cross-section for free–free absorption of radiation of frequency ν in a medium in which there are $n(E)\,dE$ electrons per cm³ with energy between E and $E + dE$ is given by

$$Q_{\text{ff}}(\nu) = (8\pi^3 m^2 e^2 \nu/3ch^3) \iint v'\, n(E) |\mathbf{r}_{EE'}|^2 \, d\Omega\, dE. \quad (15.71)$$

Here $E' = E + h\nu$, $v' = (2E'/m)^{1/2}$ and

$$\mathbf{r}_{EE'} = \int \psi_E \mathbf{r} \psi_{E'}\, d\mathbf{r}. \quad (15.72)$$

ψ_E and $\psi_{E'}$ are the wave functions for electrons moving in the field of the neutral atom with energies E, E' respectively and normalized to have the asymptotic form

$$\psi_E \sim e^{i\mathbf{k}\cdot\mathbf{r}} + r^{-1} f_E(\theta)\, e^{ikr}, \quad (15.73a)$$

$$\psi_{E'} \sim e^{-i\mathbf{k}'\cdot\mathbf{r}} + r^{-1} f_{E'}(\pi - \theta')\, e^{ik'r}. \quad (15.73b)$$

Here $\mathbf{k}$ and $\mathbf{k}'$ are the wave vectors of the initial and final electron motions,

$$\cos\theta' = \cos\theta\cos\Theta + \sin\theta\sin\Theta\cos(\phi - \Phi), \quad (15.74)$$

where (Θ, Φ) are the polar angles of the direction of motion of the scattered electron relative to the incident and $d\Omega = \sin\Theta\, d\Theta\, d\Phi$. The similarity of these formulae to those for bound–free transitions may be seen by reference to Chapter 11, p. 419.

Equivalent formulae in terms of the dipole velocity and dipole acceleration matrix elements may be derived as in Chapter 11, p. 420.

The first calculations of free–free absorption coefficients for H⁻ were carried out by Wheeler and Wildt (1942) using the acceleration matrix element with undistorted plane waves for the free wave functions. This crude approximation failed to give adequate absorption in the infrared but when Chandrasekhar and Breen (1946) repeated the calculations, using wave functions distorted by the mean static field of the atom, they obtained absorption coefficients which were much too large.

A thorough study was carried out by Doughty and Fraser (1966) who took advantage of the fact that the major contribution to the free–free absorption coefficient comes from sp + ps transitions. They carried out their calculations using both length and velocity matrix elements. The wave functions for the s electrons were calculated using a 1s-2s close-coupling expansion including electron exchange (see Chapter 4, p. 94) while the p wave functions were taken as undistorted. It was confirmed by a number of checks at chosen energies that no important change in the results was produced by using more elaborate 1s-2s-2p close-coupling expansions for the wave functions. Very good agreement was obtained between the results obtained using dipole length and dipole velocity matrix elements.

It is convenient for application to the solar atmosphere to express the absorption coefficient as a function of electron temperature T in terms of the cross-section per H atom per unit electron pressure averaged over the Maxwellian velocity distribution of the electrons. We have from (15.69)

$$n(\mathrm{H}^-)/n(\mathrm{H}) = p_e \phi(T), \qquad (15.75)$$

where $\phi(T)$ is as in (15.70) and p_e is the electron pressure so that, if $\overline{Q}_a(T)$ is the mean absorption coefficient per H$^-$ ion at temperature T, the absorption coefficient per H atom per unit electron pressure is $\phi(T)\overline{Q}_a(T)$. Fig. 15.20 shows $\phi(T)\overline{Q}_a(T)$ calculated by Doughty and Fraser (1966) by the dipole length and dipole velocity matrix elements.

Combining the contributions from bound–free and free–free absorption we obtain the total absorption coefficient per H atom per unit electron pressure as a function of wavelength shown in Fig. 15.20. It is seen to agree well over a wide wavelength region with the curve derived from analysis of the solar atmosphere. In fact it is now true to say that the calculation of the absorption coefficient is more reliable than much of the analysis of other solar conditions.

It is of interest to note that, just as the discovery of helium as a solar constituent preceded its detection on earth, the absorption of light by negative ions in the gaseous phase has also been observed first in the solar atmosphere.

Negative ions in stellar atmospheres

The possibility of negative ions other than H^- playing a significant role in the absorption in stellar atmospheres has been examined by a number of authors but it seems likely in almost all cases that the abundance of the element concerned is not high enough. However, absorption by C^- may well be important as a factor in determining the opacity of the atmospheres of hydrogen-deficient carbon stars (Warner, 1967). An important contribution also comes in these stars from free–free transitions in the fields of He atoms, i.e. transitions between states of He^-. Such transitions are also considered to be a major source of opacity (Strittmatter and Wickramasinghe, 1971) in white dwarf stars classified as DB and DC by Greenstein.

15.5 The application of electron attachment to the qualitative and quantitative analysis of trace samples – the electron capture detector

We have already discussed the information available about the rate coefficients for dissociative attachment of thermal electrons to polyatomic molecules (Chapter 9, pp. 345–82). The values range widely for different molecules but for those which contain halogen atoms the rate coefficients k_a are often large, between 10^{-9} and 10^{-7} cm^3 s^{-1}. If a small fractional concentration of such molecules is being carried through a volume containing n_e electrons per cm^3 at such a rate that the residence time of a molecule in the volume is t_R, the chance that a molecule will undergo attachment on its way through is $k_a t_R n_e$. Taking $t_R \simeq 1$ s and n_e of order 10^7 cm^{-3} this will be between 10^{-2} and 1 for the molecules referred to above. Furthermore, if n_m is the concentration in cm^{-3} of the molecules, the chance that an electron will undergo attachment will be $k_a t_e n_m$, where t_e is the residence time of an electron in the chamber. Even for n_m as low as 10^9 cm^{-3} this ranges from t_e to $100 t_e$ for the molecules considered. With modern methods for detecting small changes in electron currents it would clearly be possible to detect the reduction of the electron current with t_e as small as 100 μs. For the more strongly attaching molecules it would seem practicable to detect concentrations as low as 10^7 cm^{-3} or even less.

Devices taking advantage of these possibilities have been developed for use as detectors in gas chromatography and have been very successful. Gas chromatography is a technique for separating components of a vapour mixture for analysis. The vapour is transported by an inert carrier gas through a chromatograph column during which it is exposed to reactions with either a stationary solid or liquid phase. During passage through the column, each component of the vapour mixture is retarded to an extent determined by its affinity for the stationary phase. Thus, if t_c is the time taken for the carrier gas to pass through the column, the corresponding time for a vapour component will be

$$t_c(1+k), \qquad (15.76)$$

where

$$k = \frac{C_s}{C_c}\frac{V_s}{V_c}.$$

C_c and C_s are the respective concentrations of the component in the carrier gas and in the stationary phase which occupy volumes V_c and V_s in the column.

It is necessary to detect the peak due to the emergence of the component from the column at its characteristic delay time. The electron capture detector is very suitable for this purpose when the component in question has a large attachment rate coefficient. In the direct current mode of operation, electrons are produced by ionization of the carrier gas by electrons from a β-ray source and extracted to an anode by application of a suitable potential. The current voltage characteristic will depend on the extent to which attachment takes place in the gas. Devices operating on these lines have been in use since 1960 when Lovelock and Lipsky first introduced them for this purpose. Using an ionization chamber with dimensions

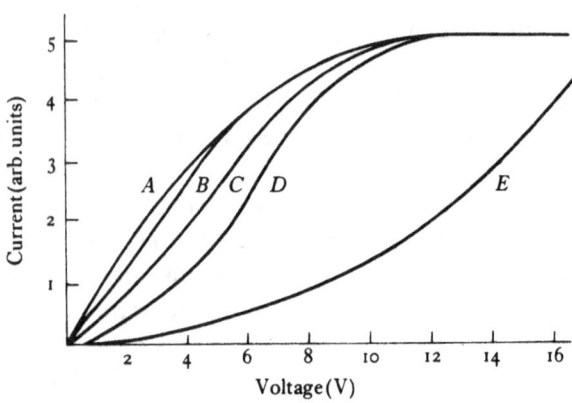

Fig. 15.21. Voltage–current characteristics of an electron capture detector containing pure nitrogen (curve A) and various organic molecules (B, hydrocarbon; C, ester; D, alcohol; E, halogenated hydrocarbon).

shown in Fig. 15.22 they obtained the current–voltage characteristics shown in Fig. 15.21 when the chamber contained pure nitrogen and when certain organic molecules were present. As expected, a higher voltage is necessary to produce saturation the more readily the molecules attach electrons. The voltage required to produce a current, which is a fixed chosen fraction of the saturation current, will depend on the nature and concentration of the attaching substance present (c.f. § 15.2).

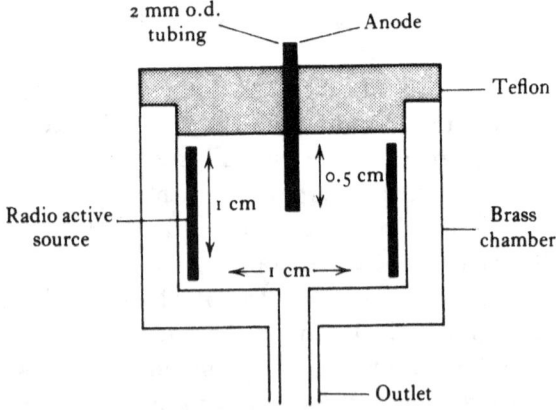

Fig. 15.22. Sectional view of an early electron capture detector.

An alternative procedure is to use a pulse technique (Lovelock, 1963). After lapse of a suitable interval during which a fraction of the electrons produced form negative ions by attachment, a short high-frequency voltage pulse is applied which collects the electrons but not the negative ions. The collected current will depend on the extent to which attachment has occurred during the period between pulses and so will vary with the nature and concentration of any attaching substance present. This method has the advantage that the interval between pulses may be chosen long enough for the electrons to have come to thermal equilibrium with the gas so that there is a greater possibility of interpreting the data in terms of basic attachment processes.

If N is the number of electrons produced per second within the detector, the electron concentration at time t after the electron injection begins, i.e. immediately after application of a collecting pulse, will be given by (Wentworth, Chen and Lovelock, 1966)

$$n_e(t) = (N/V\lambda)(1 - e^{-\lambda t}), \qquad (15.77)$$

where V is the volume of the cell and λ is given by

$$\lambda = \lambda_0 + k_a n_c. \qquad (15.78)$$

n_c is the concentration of attaching molecules, k_a the attachment rate coefficient and λ_0 includes the contributions from all other electron loss processes. This formula holds for $t \ll t_p$, the time between collecting pulses. In the limit $t \to \infty$ we have

$$n_e(\infty) \sim (N/V\lambda). \qquad (15.79)$$

Under the conditions in which $t_p \gg 1/\lambda$ the detector signal S will be proportional to $N/V\lambda$ so that, if S_0 is the signal when attaching molecules are absent,

$$n_c = \frac{N}{V k_a} \left(\frac{1}{S} - \frac{1}{S_0} \right). \qquad (15.80)$$

The limitation that $t_p \gg 1/\lambda$ is inconvenient in practice because it means that t_p is of the order of 1 ms. This requires very stringent conditions of cleanliness to ensure that $n_e(\infty)$ is large enough for satisfactory operation.

Maggs, Joynes, Davis and Lovelock (1971) have introduced a new mode of operation. According to (15.77) the mean electron current I_e flowing when the pulse interval is t_p is given by

$$\bar{I}_e = en_e V/t_p = \frac{eN}{\lambda t_p}\{1 - \exp(-\lambda t_p)\}. \qquad (15.81)$$

By varying t_p when λ varies so that λt_p is a constant, $\bar{I}_e$ can be maintained constant. Thus if the pulse-repetition frequency t_p^{-1} is f_0 when no attaching molecules are present, then the current will be unchanged when they are present if the frequency is changed to f, where

$$f - f_0 = Ck_a n_c, \qquad (15.82)$$

C being some constant. This provides a linear relationship between the signal, measured as the change in repetition frequency, and the concentration of attaching molecules.

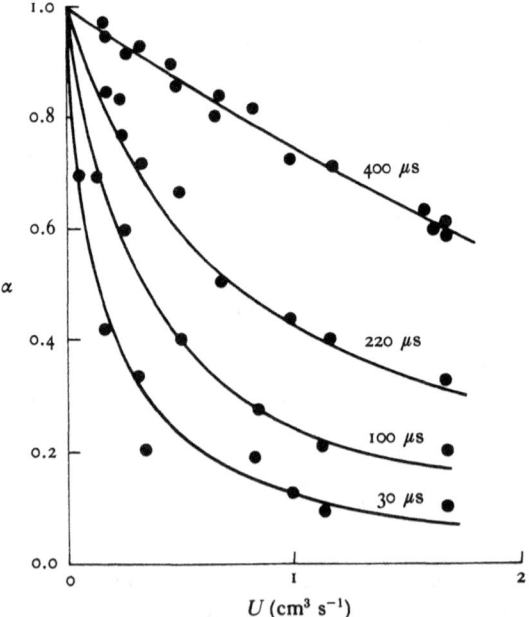

Fig. 15.23. Proportion α of SF_6 ionized in an electron capture detector at different flow rates, for different intervals between electron clearing pulses. —— calculated from known attachment coefficients, ● observed.

A quantitative absolute determination of the concentration of attaching molecules may be made if, for every electron which is permanently attached to form a negative ion during the period between pulses, one neutral molecule of the substance under study is lost from the stream. This use of the detector has been studied by Lovelock, Maggs and Adlard (1971).

In equilibrium, if U is the rate of flow of the carrier gas in cm^3 s^{-1}, the proportion of sample molecules lost by attachment in passage through the detector of volume V will be given by

$$\alpha = k_a n_e V / (k_a n_e V + U). \tag{15.83}$$

For SF$_6$, k_a is known and n_e may be determined from the current in the detector so that α may be calculated for different flow rates. Fig. 15.23 shows the calculated variation of α with U under experimental conditions using a detector with $V = 2$ cm^3 in which the electrons were produced through ionization by an Ni 63 source of 15 m Ci strength. Lovelock *et al.* (1971) compared the calculated values with those measured, in the following way. We consider two identical detectors connected in series and suppose that S_1 and S_2 are the signals, measured in terms of the loss of electrons by attachment, when a quantity N_c of molecules pass into the system. We then have

$$S_1 = \alpha N_c, \quad S_2 = \alpha(N_c - \alpha N_c),$$

so

$$\alpha = 1 - \frac{S_2}{S_1}, \quad N_c = \frac{S_1^2}{S_1 - S_2}. \tag{15.84}$$

It is seen from Fig. 15.23 that good agreement was obtained with the calculated curves.

As a further large scale check, a room of 85 cm^3 volume was provided with pressure ventilation at 47 l s^{-1}, with a fan to ensure mixing. 35 μl of SF$_6$ was introduced into the room and samples taken over a period of three hours. Fig. 15.24 shows the comparison between concentrations of SF$_6$ obtained from the detector measurements and those calculated from the known ventilation rate. There is good agreement, the scatter of points at short times being probably due to inadequate mixing.

Lovelock et al. consider that the detector will operate satisfactorily in this way if the attachment rate coefficient is greater than 3×10^{-9} cm^3 s^{-1} (see Table 9.2).

In earlier work Chen, George and Wentworth (1968) used the detector to determine the attachment rate coefficient k_a for SF$_6$ and C$_7$F$_{14}$. The values they obtained for thermal electrons, 2.41×10^{-7}

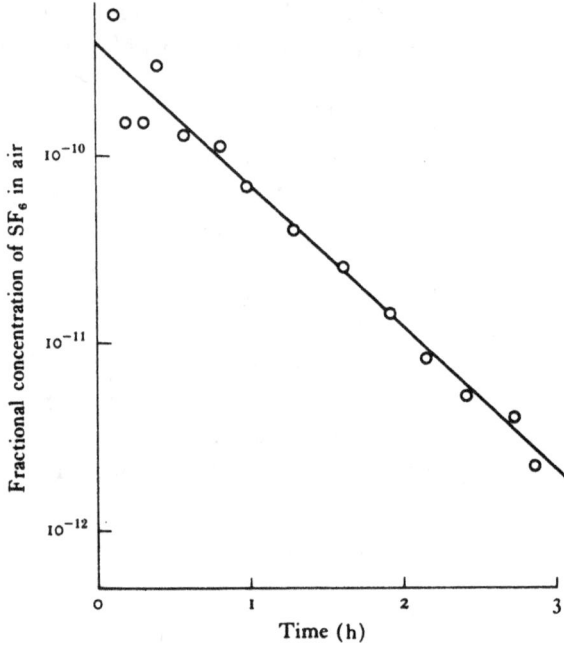

Fig. 15.24. Decay of SF$_6$ concentration in a room used as a large scale dilution vessel. —— calculated, ○ observed with an electron capture detector.

and 7.98×10^{-8} cm^3 s^{-1} respectively, agree quite well with those obtained by other methods (see Chapter 9, p. 379).

Even in its early form, with d.c. instead of pulse operation, the electron capture detector proved to be very valuable as a selective detector for chlorinated pesticides in crops. Thus Fig. 15.25 shows chromatograms for toxicant-treated and control crops taken by Goodwin, Goulden and Reynolds (1961). Concentrations of the toxicant as low as 1 part in 10^6 are readily detected. A little later Boettner and Dallos (1965) made a comparative study of the sensitivity of the detector to chlorinated and lead-substituted compounds. They found that, while for the latter it is no more sensitive than ionization detectors, it has the advantage of high selectivity – thus it was readily possible to use the electron capture detector

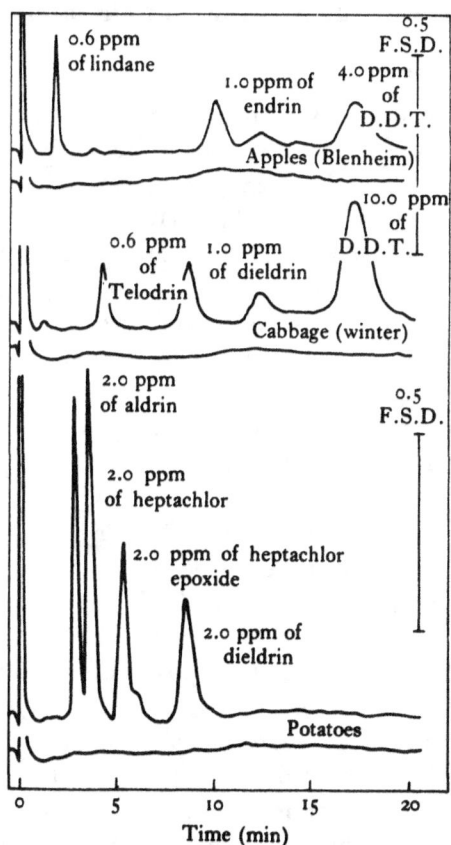

Fig. 15.25. Chromatograms of toxicant-treated and control crops obtained using an electron capture detector operated in the d.c. mode. The toxicants are the halogenated pesticides indicated. From Goodwin *et al.* (1961). F.S.D. denotes 'full scale deflection'.

to observe the lead compounds in the presence of much greater concentrations of unsubstituted hydrocarbons. This selectivity is also very valuable in dealing with halogenated compounds but, in addition, for compounds containing three or more halogen atoms the electron capture detector is also very much more sensitive than for ionization detectors. It is possible to take advantage of this sensitivity by 'tagging' otherwise non-reactive compounds with added or substituted halogen atoms in a quantitative way.

15.6 The use of negative ions in particle accelerators

Tandem accelerators

It was suggested by Bennett in 1937 that negative ions may be used to extend the energy to which ions may be accelerated in a van der Graaf machine. The principle is quite simple. Consider a machine which generates a voltage V. This may be used to accelerate, say, H^- ions from a source at ground potential to an energy V eV. If at this stage the accelerated beam is passed through a gas cell or thin foil, two electrons may be stripped from the ions to convert them to H^+ (see Chapter 13, p. 614) without reducing the energy of the beam or deviating it to a significant extent. The H^+ ions will then be further accelerated to reach a final energy $2V$ eV in passage from the stripping region to a target at ground potential. With more complex negative ions such as O^-, more than two electrons may be stripped off, leaving O^{n+} which will gain an additional energy nV eV by the time they have reached the grounded target, so their total kinetic energy will be $(n+1)V$ eV.

After the Second World War a considerable effort was devoted to the practical realization of these possibilities. This required the development of powerful negative-ion sources, an early contribution to which was the source developed by Weinman and Cameron (1956) which provided 25 μA of H^- ions. Many of the sources have received application in other areas of research. For optimum design it was also necessary to have available information on stripping cross-sections so that experiments on this subject were encouraged. The possibility of using metastable He^- ions in tandem accelerators was a major factor in the initiation of experimental studies of the lifetimes of these ions (see Chapter 5, p. 126). Work has proceeded on the development of strong sources of heavy negative ions. Although many of these developments were based on the generation of negative ions in gas discharges or through charge transfer from positive ions (see for example Dawton, 1969) a more or less universal source is now available based on production of negative ions on solid surfaces. As a result of this work, tandem accelerators are now in operation which provide H^+ beams of energy as high as 50 MeV.

Thus in the tandem generator designed for operation at the Daresbury Laboratory of the British Science Research Council, the high-voltage terminal will be maintained at 30 MV above ground. For insulation the terminal will be enclosed in a steel vessel filled with SF_6 (see p. 660). The accelerator is designed to produce multiply-charged ions and as many as 10 electrons may be removed from the negative ions on passage through the stripper.

For many experiments it is necessary to have available polarized H^+ beams. There is no difficulty in using the tandem principle for this purpose as a polarized beam of H^- may be produced by double capture of electrons from polarized H^+ (Grübler *et al.*, 1964, 1965, 1970; Haeberli, Grübler, Extermann and Schwandt, 1965), then stripped down again to polarized H^+ before the second stage of acceleration.

H^- ions in cyclotrons and synchrocyclotrons

A second quite different way in which negative ions may be used with advantage is in circular accelerators. By passing an accelerated beam of H^- ions through a stripping foil it will emerge as a well-defined proton beam which may be directly extracted from the machine because of the reversal of the curvature of the trajectory in the magnetic field.

This possibility was discussed in some detail by Wright (1957) but at first encountered difficulties because of stripping by residual gas. However, with improvement in vacuum technique, Risberg and Smythe (1962) were able to accelerate H^- ions in the 52 inch Colorado cyclotron. Since then there has been a steady increase in the number of circular accelerators operated in this way, sometimes using H^- beams polarized as described earlier. One of the more recent examples is the so-called meson factory constructed at the University of British Columbia which accelerates H^- ions up to an energy of 500 MeV. An important question which must be considered in the design of these circular machines is the loss of negative ions through detachment by the effective electric field which an ion experiences in moving with a finite velocity in the magnetic field – a field which could well be of the order of MV cm^{-1}. This has led to a considerable amount of experimental and theoretical research on this phenomenon, described in Chapter 11, p. 497, in order to specify design parameters for machines so that its effects are unimportant.

References

Chapter 1
Banyard, K. E. (1968), *J. Chem. Phys.* **48**, 2121.
Green, L. C., Lewis, M. N., Mulder, M. M., Wyeth, C. C. and Woll, J. W. (1954), *Phys. Rev.* **93**, 273.
Hart, A. and Herzberg, G. (1957), *Phys. Rev.* **106**, 79.
Hylleraas, E. A. (1930), *Z. Phys.* **60**, 624.
Pekeris, C. L. (1958), *Phys. Rev.* **112**, 1649.
Pekeris, C. L. (1962), *Phys. Rev.* **126**, 1470.
Shull, H. and Löwdin, P.-O. (1956), *J. Chem. Phys.* **25**, 1035.

Chapter 2
Bethe, H. A. and Goldstone, J. (1957), *Proc. Roy. Soc.* A **238**, 551.
Brueckner, K. A. (1954), *Phys. Rev.* **96**, 508.
Chisholm, C. D. H. and Öpik, U. (1964), *Proc. Phys. Soc.* **83**, 541.
Clementi, E. (1964), *Phys. Rev.* **135**, A 981.
Clementi, E. and McLean, A. D. (1964), *Phys. Rev.* **133**, A 419.
Clementi, E., McLean, A. D., Raimondi, D. L. and Yoshimine, M. (1964), *Phys. Rev.* **133**, A 1274.
Fock, V. (1930), *Z. Phys.* **61**, 126.
Hartree, D. R. (1928), *Proc. Camb. Phil. Soc.* **24**, 189.
Hartree, D. R. (1957), *The Calculation of Atomic Structures*, chapter 6, John Wiley and Sons, New York.
Hartree, D. R. and Hartree, W. (1936), *Proc. Roy. Soc.* A **156**, 45.
Hartree, D. R. and Hartree, W. (1948), *Proc. Roy. Soc.* A **193**, 299.
Hartree, D. R., Hartree, W. and Swirles, B. (1939), *Phil. Trans.* A **238**, 229.
Moser, C. M. and Nesbet, R. K. (1971), *Phys. Rev.* A **4**, 1336.
Nesbet, R. K. (1968), *Phys. Rev.* **175**, 2.
Nesbet, R. K. (1969), *Advan. Chem. Phys.* **14**, 1.
Norcross, D. W. (1974), *Phys. Rev. Lett.* **32**, 192.
Roothan, C. C. J. (1951), *Rev. Mod. Phys.* **23**, 69.
Schaefer, H. F. and Harris, F. E. (1968), *Phys. Rev.* **170**, 108.
Slater, J. C. (1929), *Phys. Rev.* **34**, 1293.
Weiss, A. W. (1968), *Phys. Rev.* **166**, 70.
Weiss, A. W. (1971), *Phys. Rev.* A **3**, 126.

Chapter 3

Bailey, T. L. (1955), *J. Chem. Phys.* **28**, 792.
Bakulina, I. N. and Ionov, N. I. (1957), *Dokl. AN. SSSR*, **116**, 41.
Bakulina, I. N. and Ionov, N. I. (1959), *Zh. Fiz. Chim.* **33**, 2069.
Bakulina, I. N. and Ionov, N. I. (1964), *Dokl. AN. SSSR*, **155**, 309.
Bates, D. R. (1947), *Proc. Roy. Irish Acad.* A **51**, 151.
Bates, D. R. and Moiseiwitsch, B. L. (1955), *Proc. Phys. Soc.* A **68**, 540.
Bernstein, R. B. and Metlay, M. (1951), *J. Chem. Phys.* **19**, 1612.
Branscomb, L. M., Burch, D. S., Smith, S. J. and Geltman, S. (1958), *Phys. Rev.* **111**, 504.
Catalán, M. H., Rohrlich, F. and Shenstone, A. G. (1954), *Proc. Roy. Soc.* A **221**, 421.
Crossley, R. J. S. (1964), *Proc. Phys. Soc.* **83**, 375.
Cubicciotti, D. (1959), *J. Chem. Phys.* **31**, 1646.
Cubicciotti, D. (1960), *J. Chem. Phys.* **33**, 1579.
Cubicciotti, D. (1961), *J. Chem. Phys.* **34**, 2189.
Doty, P. M. and Mayer, J. E. (1944), *J. Chem. Phys.* **12**, 323.
Edie, J. W. and Rohrlich, F. (1962), *J. Chem. Phys.* **36**, 623.
Edlèn, B. (1960), *J. Chem. Phys.* **33**, 98.
Feldmann, D., Rackwitz, R., Heinicke, E. and Kaiser, H. J. (1973), *Phys. Lett.* **45A**, 404.
Fowler, R. H. (1929), *Statistical Mechanics*, Cambridge University Press.
Froese, C. (1966), *J. Chem. Phys.* **45**, 1417.
Ginsberg, A. P. and Miller, J. M. (1958), *J. Inorg. Nucl. Chem.* **7**, 351.
Glockler, G. (1934), *Phys. Rev.* **46**, 111.
Glockler, G. and Calvin, M. (1935), *J. Chem. Phys.* **3**, 771.
Hotop, H., Bennett, R. A. and Lineberger, W. C. (1973), *J. Chem. Phys.* **58**, 2373.
Johnson, H. R. and Rohrlich, F. (1959), *J. Chem. Phys.* **30**, 1608.
Khovenstenko, V. I. and Dukel'skii, V. M. (1960), *Sov. Phys. JETP*, **10**, 465.
Langmuir, I. and Kingdon, K. H. (1925), *Proc. Roy. Soc.* A **107**, 61.
McCullum, K. J. and Mayer, J. E. (1943), *J. Chem. Phys.* **11**, 56.
Mayer, J. E. (1930), *Z. Phys.* **61**, 798.
Mayer, J. E. and Helmholz, L. (1932), *Z. Phys.* **75**, 19.
Mayer, J. E. and Maltbie, M. McL. (1932), *Z. Phys.* **75**, 748.
Metlay, M. and Kimball, G. E. (1948), *J. Chem. Phys.* **16**, 744.
Page, F. M. (1960), *Trans. Faraday Soc.* **56**, 1742.
Page, F. M. (1961), *Trans. Faraday Soc.* **57**, 359.
Politzer, P. (1968), *Trans. Faraday Soc.* **64**, 2241.
Scheer, M. D. (1970), *J. Res. Nat. Bur. Stand.* **74A**, 37.
Scheer, M. D. and Fine, J. (1969), *J. Chem. Phys.* **50**, 4343.
Sutton, P. P. and Mayer, J. E. (1934), *J. Chem. Phys.* **2**, 145.
Sutton, P. P. and Mayer, J. E. (1935), *J. Chem. Phys.* **3**, 20.
Vier, D. T. and Mayer, J. E. (1944), *J. Chem. Phys.* **12**, 28.

Zandberg, E. Ya., Kamenev, A. G. and Paleev, V. I. (1971), *Sov. Phys.-Tech. Phys.* **16**, 832.
Zandberg, E. Ya., Kamenev, A. G. and Paleev, V. I. (1972), *Sov. Phys.-Tech. Phys.* **16**, 1567.
Zandberg, E. Ya. and Paleev, V. I. (1970), *Sov. Phys.-Dokl.* **15**, 52.
Zollweg, R. J. (1969), *J. Chem. Phys.* **50**, 4251.

Chapter 4

Bardsley, J. N., Herzenberg, A. and Mandl, F. (1966), *Proc. Phys. Soc.* **89**, 305, 321.
Bates, D. R. and Moiseiwitsch, B. L. (1955), *Proc. Phys. Soc.* A **68**, 540.
Burke, P. G. and Taylor, A. J. (1966), *Proc. Phys. Soc.* **88**, 549.
Ehrhardt, H. and Willmann, K. (1967), *Z. Phys.* **203**, 1.
Eliezer, I., Taylor, H. S. and Williams, J. K. (1967), *J. Chem. Phys.* **47**, 2165.
Fano, U. (1961), *Phys. Rev.* **124**, 1866.
Fano, U. and Cooper, J. W. (1964), *Phys. Rev.* **137A**, 1364.
Feshbach, H. (1958), *Ann. Phys.* **5**, 357.
Feshbach, H. (1962), *Ann. Phys.* **19**, 287.
Hazi, A. U. and Taylor, H. S. (1970), *Phys. Rev.* A **1**, 1100.
Heddle, D. W. O., Keesing, R. G. W. and Kurepa, J. M. (1973), *Proc. Roy. Soc.* A **334**, 135.
Herzenberg, A. (1973), *Proc. Coral Gables, Conf. Fundamental Interactions in Physics*, Plenum Press, p. 261.
Herzenberg, A. and Mandl, F. (1963), *Proc. Roy. Soc.* A **274**, 253.
Massey, H. S. W., Burhop, E. H. S. and Gilbody, H. B. (1969), *Electronic and Ionic Impact Phenomena*, 2nd edition, Clarendon Press, Oxford.
Mott, N. F. and Massey, H. S. W. (1965), *The Theory of Atomic Collisions*, 3rd edition, Clarendon Press, Oxford.
Siegert, A. F. (1939), *Phys. Rev.* **56**, 750.
Simpson, J. A. (1964), *Rev. Sci. Inst.* **35**, 1698.
Stamatovic, A. and Schulz, G. J. (1970), *Rev. Sci. Instr.* **41**, 423.
Taylor, H. S. (1970), *Adv. Chem. Phys.* **18**, 91.
Wu, Ta-You (1936), *Phil. Mag.* **22**, 837.
Zollweg, R. S. (1969), *J. Chem. Phys.* **50**, 4251.

Chapter 5

Andrick, D. and Ehrhardt, H. (1966), *Z. Phys.* **192**, 99.
Andrick, D., Eyb, M. and Hofmann, M. (1972), *J. Phys.* B (G.B.) **5**, L15.
Baumann, H., Heinicke, E., Kaiser, H. J. and Bethge, K. (1971), *Nucl. Instr. Methods*, **95**, 389.
Bhatia, A. K. (1970), *Phys. Rev*, A **2**, 1667.

Blau, L. M., Novick, R. and Weinflash, D. (1970), *Phys. Rev. Lett.* **24**, 1268.
Bolduc, E., Quéméner, J. J. and Marmet, P. (1972), *J. Chem. Phys.* **57**, 1957.
Burke, P. G. and Taylor, A. J. (1966), *Proc. Phys. Soc.* **88**, 549.
Callaway, J., La Bahn, R. W., Pu, R. T. and Duxler, W. M. (1968), *Phys. Rev.* **168**, 12.
Chamberlain, G. E. and Heideman, H. G. M. (1965), *Phys. Rev. Lett.* **15**, 337.
Cunningham, D. L. and Edwards, A. K. (1973), *Phys. Rev. A* **8**, 2960.
Cvejanovic, S., Comer, J. and Read, F. H. (1973), *Proc. 8th Int. Conf. Physics of Electronic and Atomic Collisions*, Belgrade, Abstracts, p. 441.
Drake, G. W. F. (1970), *Phys. Rev. Lett.* **24**, 126.
Drake, G. W. F. (1973), *Astrophys. J.* **184**, 145.
Edwards, A. K., Risley, J. S. and Geballe, R. (1971), *Phys. Rev. A* **3**, 583.
Ehrhardt, H. L., Langhans, L. and Linder, F. (1968), *Z. Phys.* **214**, 179.
Eliezer, L. and Pan, Y. K. (1970), *Theor. Chim. Acta*, **16**, 63.
Estberg, G. N. and La Bahn, R. W. (1970), *Phys. Rev. Lett.* **24**, 1265.
Fano, U. (1961), *Phys. Rev.* **124**, 1866.
Fano, U. (1973), Private communication to Schulz, R. J.
Fano, U. and Cooper, J. W. (1965), *Phys. Rev.* **137**A, 1364.
Fleming, R. J. and Higginson, G. S. (1963), *Proc. Phys. Soc.* **81**, 974.
Fogel, Y. M., Kozlov, V. F. and Kalmykov, A. A. (1959), *Zh. Eksp. Teor. Fiz.* **36**, 1354 (*Sov. Phys. JETP* (1959), **9**, 963).
Fung, A. C. and Matese, J. J. (1972), *Phys. Rev. A* **5**, 22.
Gailitis, M. and Damburg, R. (1963), *Proc. Phys. Soc.* **82**, 192.
Gibson, J. R. and Dolder, K. T. (1969), *J. Phys. B* **2**, 1180.
Golden, D. E. and Bandel, H. W. (1965), *Phys. Rev. Lett.* **14**, 1010.
Golden, D. E. and Zecca, A. (1970), *Phys. Rev. A* **1**, 241.
Golden, D. E. and Zecca, A. (1971), *Rev. Sci. Instr.* **42**, 210.
Grissom, J. T., Compton, R. N. and Garrett, W. R. (1969), *Phys. Lett.* **30A**, 117.
Hart, A. and Herzberg, G. (1957), *Phys. Rev.* **106**, 79.
Haselton, H. H. (1973), *Bull. Amer. Phys. Soc.* **18**, 710.
Heddle, D. W. O., Keesing, R. G. W. and Kurepa, J. M. (1974), *Proc. Roy. Soc. A* **334**, 135.
Heddle, D. W. O., Keesing, R. G. W. and Kurepa, J. M. (1974), *Proc. Roy. Soc. A* **337**, 435.
Hiby, J. W. (1939), *Ann. Phys.* **34**, 473.
Holøien, E. and Midtal, J. (1955), *Proc. Phys. Soc. A* **68**, 815.
Kleinpoppen, H. and Raible, V. (1965), *Phys. Lett.* **18**, 24.
Kuyatt, C. E., Simpson, J. A. and Mielczarek, S. R. (1965), *Phys. Rev.* **138**, A 385.

Kwok, K. L. and Mandl, F. (1965), *Proc. Phys. Soc.* **86**, 501.
McGowan, J. W. (1967), *Phys. Rev.* **156**, 165.
McGowan, J. W., Clarke, E. M. and Curley, E. K. (1965), *Phys. Rev. Lett.* **15**, 917.
Mader, D. L. and Novick, R. (1972), *Phys. Rev. Lett.* **29**, 199.
Maier-Leibnitz, H. (1935), *Z. Phys.* **95**, 499.
Manson, S. T. (1971), *Phys. Rev.* A **3**, 147.
Massey, H. S. W., Burhop, E. H. S. and Gilbody, H. B. (1969), *Electronic and Ionic Impact Phenomena*, 2nd edition, pp. 619–29, Clarendon Press, Oxford.
Matese, J. J., Rountree, S. P. and Henry, R. J. W. (1973a), *Phys. Rev.* A **7**, 846.
Matese, J. J., Rountree, S. P. and Henry, R. J. W. (1973b), *Phys. Rev.* A **8**, 2965.
Mazeau, J. F., Gresteau, G., Joyez, G., Reinhardt, J. and Hall, R. I. (1972), *J. Phys.* B **5**, 1890.
Moores, D. L. and Norcross, D. W. (1972), *J. Phys.* B (G.B.) **5**, 1482.
Mott, N. F. and Massey, H. S. W. (1965), *The Theory of Atomic Collisions*, Clarendon Press, Oxford.
Nicholas, D. J., Trowbridge, C. W. and Allen, W. D. (1968), *Phys. Rev.* **167**, 38.
Nicolaides, C. A. (1972), *Phys. Rev.* A **6**, 2078.
Norcross, D. W. (1974), *Phys. Rev. Lett.* **32**, 192.
Novick, R. and Weinflash, D. (1970), *Proc. Int. Conf. Precision Measurements and Fundamental Constants*, Gaithersberg, Md. ed. Langenberg, D. N. and Taylor, B. N. Nat. Bureau Standards Spec. Publ. No. 343, p. 403.
O'Malley, T. F. and Geltman, S. (1964), *Phys. Rev.* **137**, A 1344.
Ormonde, S., Smith, K., Torres, B. W. and Davies, A. R. (1973), *Phys. Rev.* **8**, 262.
Peart, B. and Dolder, K. (1973), *J. Phys.* B **6**, 1497.
Pichanick, F. M. J. and Simpson, J. A. (1968), *Phys. Rev.* **168**, 64.
Priestley, H. and Whiddington, R. (1934), *Proc. Roy. Soc.* A **145**, 462.
Quéméner, J. J., Paquet, C. and Marmet, P. (1971), *Phys. Rev.* A **4**, 494.
Riviere, H. G. and Sweetman, D. R. (1960), *Phys. Rev. Lett,* **5**, 560.
Sanche, L. and Burrow, P. D. (1972), *Phys. Rev. Lett.* **29**, 1639.
Sanche, L. and Schulz, G. J. (1972), *Phys. Rev.* A **5**, 1672.
Schaefer, H. F. and Harris, F. E. (1968), *Phys. Rev. Lett.* **21**, 1561.
Schulz, G. J. (1963), *Phys. Rev. Lett.* **10**, 104.
Seiler, G. J., Oberoi, R. S. and Callaway, J. (1970), *Phys. Lett.* **31A**, 547.
Silverman, S. M. and Lassettre, E. N. (1964), *J. Chem. Phys.* **40**, 1265.
Simpson, J. A. and Fano, U. (1964), *Phys. Rev. Lett.* **11**, 158.
Simpson, J. A., Menendez, M. G. and Mielczarek, S. R. (1966), *Phys. Rev.* **150**, 76.

Sinfailam, A. L. and Nesbet, R. K. (1972), *Phys. Rev.* A **6**, 2118.
Taylor, H. S. and Thomas, L. D. (1972), *Phys. Rev. Lett.* **28**, 1091.
Temkin, A., Bhatia, A. K. and Bardsley, J. N. (1972), *Phys. Rev.* A **5**, 1663.
Walton, D. S., Peart, B. and Dolder, K. (1970), *J. Phys.* B **3**, L148.
Weiss, A. W. (1968), *Phys. Rev.* **166**, 70.
Wigner, E. (1948), *Phys. Rev.* **73**, 1002.
Wu, Ta-You (1936), *Phil Mag.* **22**, 837.
Young, A. D. (1968), *J. Phys.* B **1**, 1073.

Chapter 6

Baede, A. P. M., Auerbach, D. J. and Los, J. (1973), *Physica*, **64**, 134.
Bakulina, I. N. and Ionov, N. I. (1959), *Sov. J. Phys. Chem.* **33**, 286.
Bardsley, J. N., Herzenberg, A. and Mandl, F. (1966), *Proc. Phys. Soc.* **89**, 305.
Barsuhn, J. (1974), *J. Phys.* B **7**, 155.
Bennett, R. A. (1972), Thesis, University of Colorado.
Berkowitz, J., Chupka, W. A. and Gutman, D. (1971), *J. Chem. Phys.* **55**, 2733.
Berkowitz, J., Chupka, W. A. and Walter, T. A. (1969), *J. Chem. Phys.* **50**, 1497.
Birtwistle, D. T. and Herzenberg, A. (1971), *J. Phys.* B **4**, 53.
Boness, M. J. W. and Hasted, J. B. (1966), *Phys. Lett.* **21**, 526.
Boness, M. J. W., Hasted, J. B. and Larkin, I. W. (1968), *Proc. Roy. Soc.* A **305**, 493.
Brewer, L., Hicks, W. T. and Krikorian, O. H. (1962), *J. Chem. Phys.* **36**, 182.
Burch, D. S., Smith, S. J. and Branscomb, L. M. (1958), *Phys. Rev.* **112**, 171.
Burch, D. S., Smith, S. J. and Branscomb, L. M. (1959), *Phys. Rev.* **114**, 1652.
Burke, P. G. and Chandra, N. (1972), *J. Phys.* B **5**, 1696.
Burke, P. G. and Sinfailam, A. L. (1970), *J. Phys.* B **3**, 641.
Burt, J. A. (1972a), *Ann. Geophys.* **3**, 28.
Burt, J. A. (1972b), *J. Chem. Phys.* **57**, 4649.
Byerly, R. and Beaty, E. C. (1971), *J. Geophys. Res.* **76**, 4596.
Cade, P. E. (1967a), *J. Chem. Phys.* **47**, 2390.
Cade, P. E. (1967b), *Proc. Phys. Soc.* **91**, 842.
Celotta, R. J., Bennett, R. A. and Hall, J. L. (1974), *J. Chem. Phys.* **60**, 1740.
Celotta, R. J., Bennett, A., Hall, J. L., Siegel, M. W. and Levine, J. (1972), *Phys. Rev.* A **6**, 631.
Chantry, P. J. (1971), *J. Chem. Phys.* **55**, 2746.

Chupka, W. A., Berkowitz, J. and Gutman, D. (1971), *J. Chem. Phys.* **55**, 2724.
Conway, D. C. and Nesbitt, L. E. (1968), *J. Chem. Phys.* **48**, 509.
Curran, R. K. (1961), *J. Chem. Phys.* **35**, 1849.
De Corpo, J. J. and Franklin, J. L. (1971), *J. Chem. Phys.* **54**, 1885.
D'Orazio, L. A. and Wood, R. H. (1965), *J. Phys. Chem.* **69**, 2550.
Dunkin, D. B., Fehsenfeld, F. C. and Ferguson, E. E. (1970), *J. Chem. Phys.* **53**, 987.
Ehrhardt, H., Langhans, L., Linder, F. and Taylor, H. S. (1968), *Phys. Rev.* **173**, 222.
Ehrhardt, H. and Willmann, K. (1967), *Z. Phys.* **204**, 462.
Eliezer, I., Taylor, H. S. and Williams, J. K. (1967), *J. Chem. Phys.* **47**, 2165.
Farragher, A. L., Page, F. M. and Wheeler, R. C. (1964), *Disc. Faraday Soc.* **37**, 205.
Fehsenfeld, F. C., Albritton, D. L., Burt, J. A. and Schiff, H. I. (1969), *Can. J. Phys.* **47**, 1793.
Fehsenfeld, F. C., Ferguson, E. E. and Bohme, D. K. (1969), *Planet. Space Sci.* **17**, 1759.
Feldmann, D. (1970), *Z. Naturforsch.* **25**a, 621.
Ferguson, E. E. (1969), *Can. J. Phys.* **47**, 1815.
Ferguson, E. E., Dunkin, D. B. and Fehsenfeld, F. C. (1972), *J. Chem. Phys.* **47**, 1459.
Fischer, R., Neuert, H., Peuckert-Kraus, K. and Vogt, D. (1966), *Z. Naturforsch*, **21** A, 501.
Frosch, R. P. (1971), *J. Chem. Phys.* **54**, 2660.
Gilbert, L. and Wahl, A. C. (1971), *J. Chem. Phys.* **55**, 5247.
Gilmore, F. R. (1965), *J. Quant. Spectry. Radiat. Transfer*, **5**, 369.
Goubeau, J. and Klemm, W. (1937), *Z. Phys. Chem.* B **36**, 322.
Haas, R. (1957), *Z. Phys.* **148**, 177.
Hara, S. (1967), *J. Phys. Soc. Japan*, **22**, 710.
Hara, S. (1969), *J. Phys. Soc. Japan*, **27**, 1009, 1262.
Harries, W. and Hertz, G. (1927), *Z. Phys.* **46**, 177.
Hasted, J. B. and Awan, A. M. (1969), *J. Phys.* B **2**, 367.
Helbing, R. K. B. and Rothe, E. W. (1969), *J. Chem. Phys.* **51**, 1607.
Herzenberg, A. (1968), *J. Phys.* B **1**, 548.
Herzberg, G. and Lagerquist, A. (1968), *Can. J. Phys.* **46**, 2363.
Hughes, B. M., Lifshitz, C. and Tiernan, T. O. (1973), *J. Chem. Phys.* **59**, 3162.
Inoue, M. (1966), *J. Chim. Phys.* **63**, 1061.
James, H. M., Coolidge, A. S. and Present, R. D. (1936), *J. Chem. Phys.* **4**, 187.
Johnsen, R., Brown, H. L. and Biondi, M. A. (1971), *Bull. Amer. Phys. Soc.* **16**, 213.
Kay, J. and Page, F. M. (1966), *Trans. Faraday Soc.* **62**, 3081.

Krauss, M. and Mies, F. H. (1970), *Phys. Rev.* A **1**, 1592.
Krauss, M. D., Neumann, D., Wahl, A. C., Das, G. and Zemke, W. (1973), *Phys. Rev.* A **7**, 69.
Lacmann, K. and Herschbach, D. R. (1970), *Chem. Phys. Lett.* **6**, 106.
Larkin, I. W. and Hasted, J. B. (1972), *J. Phys.* B, **5**, 95.
Linder, F. and Schmidt, H. (1971*a*), *Z. Naturforsch.* **26**a, 1603.
Linder, F. and Schmidt, H. (1971*b*), *Z. Naturforsch.* **26**a, 1617.
Lineberger, W. C. and Patterson, T. A. (1972), *Chem. Phys. Lett.* **13**, 40.
Lifshitz, C., Hughes, B. M. and Tiernan, T. O. (1970), *Chem. Phys. Lett*, **7**, 469.
Lifshitz, C., Tiernan, T. O. and Hughes, B. M. (1973), *J. Chem. Phys.* **59**, 3182.
Milligan, D. E. and Jacox, M. E. (1969), *J. Chem. Phys.* **51**, 1952.
Morse, P. M. (1929), *Phys. Rev.* **34**, 57.
Moruzzi, J. L. and Phelps, A. V. (1966), *J. Chem. Phys.* **45**, 4617.
Mulliken, R. S. (1932), *Rev. Mod. Phys.* **4**, 17.
Mulliken, R. S. (1934), *Phys. Rev.* **46**, 549.
Mulliken, R. S. (1942), *Rev. Mod. Phys.* **14**, 204.
Mulliken, R. S. (1950), *J. Am. Chem. Soc.* **72**, 600.
Mulliken, R. S. (1958), *Can. J. Chem.* **36**, 10.
Nalley, S. J. and Compton, R. N. (1971), *Chem. Phys. Lett.* **9**, 529.
Nalley, S. J., Compton, R. N., Schweinler, H. C. and Anderson, V. E. (1973), *J. Chem. Phys.* **59**, 4125.
Nesbet, R. K. (1964), *J. Chem. Phys.* **40**, 3619.
O'Hare, P. A. G. (1971), *J. Chem. Phys.* **54**, 4124.
O'Hare, P. A. G. and Wahl, A. C. (1970), *J. Chem. Phys.* **53**, 2469.
O'Hare, P. A. G. and Wahl, A. C. (1971), *J. Chem. Phys.* **55**, 666.
O'Malley, T. F. (1967), *Phys. Rev.* **155**, 59.
Pack, J. L. and Phelps, A. V. (1966), *J. Chem. Phys.* **44**, 1870.
Page, F. M. and Goode, G. C. (1969), *Negative Ions and the Magnetron*, Wiley–Interscience, London.
Paulson, J. F. (1966), *Advan. Chem. Sci.* **58**, 28.
Pfeiffer, G. V. and Allen, L. C. (1969), *J. Chem. Phys.* **51**, 190.
Phelps, A. V. and Pack, J. L. (1961), *Phys. Rev. Lett.* **6**, 111.
Robb, M. A. and Csizmadia, I. G. (1971), *Int. J. Quant. Chem.* **5**, 605.
Sanche, L. and Schulz, G. J. (1973), *J. Chem. Phys.* **58**, 479.
Schulz, G. J. (1959), *Phys. Rev.* **116**, 1141.
Schulz, G. J. (1962), *Phys. Rev.* **125**, 229.
Schulz, G. J. (1964), *Phys. Rev.* **135**, A 988.
Spence, D. and Schulz, G. J. (1971), *Phys. Rev.* A **3**, 1968.
Taylor, H. S. and Harris, F. E. (1963), *J. Chem. Phys.* **39**, 1012.
Walsh, A. D. (1953), *J. Chem. Soc.* **3**, 2260.
Warneck, P. (1969), *Chem. Phys. Lett.* **3**, 532.
Wood, R. H. and d'Orazio, L. A. (1965), *J. Phys. Chem.* **69**, 2562.

Yatsimirskii, K. B. (1947), *Izvest. Akad. Nauk. SSSR, Otd. Khim. Nauk*, **411**, 453; *J. Gen. Chem. USSR*, **17**, 2019.
Young, L. B., Lee-Ruff, E. and Bohme, D. K. (1971), *Can. J. Chem.* **49**, 979.
Zemke, W. T., Das, G. and Wahl, A. C. (1972), *Chem. Phys. Lett.* **14**, 310.

Chapter 7

Chutjian, A., Truhlar, D. G., Williams, W. and Trajmar, S. (1972), *Phys. Rev. Lett.* **29**, 1580.
Comer, J. and Read, F. H. (1971*a*), *J. Phys.* B **4**, 368.
Comer, J. and Read, F. H. (1971*b*), *J. Phys.* B **4**, 1055.
Eliezer, L., Taylor, H. S. and Williams, J. K. (1967), *J. Chem. Phys.* **47**, 2165.
Golden, D. E. (1971), *Phys. Rev. Lett.* **27**, 227.
Heideman, H. G. M., Kuyatt, C. E. and Chamberlain, G. E. (1966), *J. Chem. Phys.* **44**, 355.
Joyez, G., Comer, J. and Read, F. H. (1973), *Proc. 8th Int. Conf. Physics of Electronic and Atomic Collisions*, Belgrade, Abstracts, p. 449.
Kisker, E. (1972), *Z. Phys.* **257**, 51.
Kuyatt, C. E., Mielczarek, S. R. and Simpson, H. A. (1964), *Phys. Rev. Lett.* **12**, 293.
Kuyatt, C. E., Simpson, J. A. and Mielczarek, S. R. (1966), *J. Chem. Phys.* **44**, 437.
Mazeau, J., Gresteau, G., Hall, R. I., Joyez, G. and Reinhardt, J. (1973), *J. Phys.* B **6**, 862.
Mazeau, J., Gresteau, G., Joyez, G., Reinhardt, J. and Hall, R. I. (1972), *J. Phys.* B **5**, 1890.
Pavlovic, Z., Boness, M. J. W., Herzenberg, A. and Schulz, G. J. (1972), *Phys. Rev.* A **6**, 676.
Sanche, L. and Schulz, G. J. (1971), *Phys. Rev. Lett.* **26**, 943.
Sanche, L. and Schulz, G. J. (1972), *Phys. Rev.* A **6**, 69.
Truhlar, D. G., Trajmar, S. and Williams, W. (1972), *J. Chem. Phys.* **57**, 3250.
Weingartshofer, A. H., Ehrhardt, H., Herman, V. and Linder, F. (1970), *Phys. Rev.* A **2**, 294.

Chapter 8

Bates, D. R. and Massey, H. S. W. (1943), *Phil. Trans. Roy. Soc.* A **239**, 269.
Berkowitz, J., Chupka, W. A., Guyon, P. M., Holloway, J. H. and Spohr, R. (1971), *J. Chem. Phys.* **54**, 5165.

Berkowitz, J., Chupka, W. A. and Walter, T. A. (1969), *J. Chem. Phys.* **50**, 1497.
Berry, R. S. and David, C. W. (1964), *Atomic Collision Processes*, p. 543, North-Holland, Amsterdam.
Berry, R. S. and Reimann, C. W. (1963), *J. Chem. Phys.* **38**, 1540.
Boldt, G. (1959a), *Z. Phys.* **154**, 319.
Boldt, G. (1959b), *Z. Phys.* **154**, 330.
Chupka, W. A. and Berkowitz, J. (1971), *J. Chem. Phys.* **54**, 5126.
Dibeler, V. H., Walker, J. A. and McCulloh, K. F. (1969a), *J. Chem. Phys.* **50**, 4592.
Dibeler, V. H., Walker, J. A. and McCulloh, K. F. (1969b), *J. Chem. Phys.* **51**, 4230.
Fuchs, R. (1951), *Z. Phys.* **130**, 69.
Iczkowski, R. P. and Margrave, J. L. (1959), *J. Chem. Phys.* **30**, 403.
Johns, J. W. C. and Barrow, R. F. (1959), *Proc. Roy. Soc.* A **251**, 504.
Lochte-Holtgreven, W. (1951), *Naturwiss.* **38**, 258.
Lochte-Holtgreven, W. and Nissen, W. (1952), *Z. Phys.* **133**, 124.
Morrison, J. D., Hurzeler, H., Inghram, M. G. and Stanton, H. E. (1960), *J. Chem. Phys.* **33**, 821.
Mück, G. and Popp, H.-P. (1968), *Z. Naturforsch.* **23**a, 1213.
Mulliken, R. S. (1940), *Phys. Rev.* **37**, 500.
Myer, J. A. and Samson, J. A. R. (1970), *J. Chem. Phys.* **52**, 716.
Nissen, W. (1954), *Z. Phys.* **139**, 638.
Peters, T. (1953), *Z. Phys.* **135**, 573.
Popp, H.-P. (1967), *Z. Naturforsch.* **22**, 254.
Stampfer, J. G. and Barrow, R. F. (1958), *Trans. Faraday Soc.* **54**, 1592.
Stricker, W. and Krauss, L. (1968), *Z. Naturforsch.* **23**A, 486.
Terenin, A. and Popow, B. (1932), *Z. Phys.* **75**, 338.
Venkateswarlu, P. (1969), *Bull. Amer. Phys. Soc.* **14**, 622.
Verma, R. D. (1960), *J. Chem. Phys.* **32**, 738.
Watanabe, K. (1957), *J. Chem. Phys.* **26**, 542.
Weber, O. (1958), *Z. Phys.* **152**, 281.

Chapter 9

Asundi, R. K., Craggs, J. D. and Kurepa, M. V. (1963), *Proc. Phys. Soc.* **82**, 967.
Bailey, V. A. (1925), *Phil. Mag.* **50**, 825.
Bailey, V. A. and Duncanson, W. E. (1930), *Phil. Mag.* **2**, 145.
Bailey, V. A. and Healey, R. H. (1935), *Phil. Mag.* **9**, 725.
Bailey, V. A., Makinson, R. E. and Somerville, J. M. (1937), *Phil. Mag.* **24**, 177.

REFERENCES: CHAPTER 9   703

Bardsley, J. N., Herzenberg, A. and Mandl, F. (1964), *Atomic Collision Processes* (ed. McDowell, M. R. C.), p. 415, North-Holland, Amsterdam.
Becker, R. H., Groth, W. and Schurath, U. (1971), *Chem. Phys. Lett.* **8**, 259.
Begun, G. M. and Compton, R. N. (1969), *J. Chem. Phys.* **51**, 2367.
Biondi, M. A. (1958), *Phys. Rev.* **109**, 2005.
Blaunstein, R. P. and Christophorou, L. G. (1968), *J. Chem. Phys.* **49**, 1526.
Blewett, J. P. (1936), *Phys. Rev.* **49**, 900.
Bloch, F. and Bradbury, N. E. (1935), *Phys. Rev.* **48**, 689.
Bortner, T. E. and Hurst, G. S. (1958), *Health Phys.* **1**, 39.
Bradbury, N. E. (1933), *Phys. Rev.* **44**, 883.
Bradbury, N. E. (1934), *J. Chem. Phys.* **2**, 827.
Buchdahl, R. (1941), *J. Chem. Phys.* **9**, 146.
Buchel'nikova, I. S. (1958), *Zh. Eksp. Teor. Fiz.* **35**, 1119; *Sov. Phys. JETP*, **8** (1959), 783.
Burrow, P. D. (1973), *J. Chem. Phys.* **59**, 4922.
Chanin, L. M., Phelps, A. V. and Biondi, M. A. (1959), *Phys. Rev. Lett.* **2**, 344.
Chanin, L. M., Phelps, A. V. and Biondi, M. A. (1962), *Phys. Rev.* **128**, 219.
Chantry, P. J. (1968), *Phys. Rev.* **172**, 125.
Chantry, P. J. (1969), *J. Chem. Phys.* **51**, 3369.
Chantry, P. J. (1972a), *J. Chem. Phys.* **57**, 3180.
Chantry, P. J. (1972b), private communication to A. V. Phelps.
Chantry, P. J. and Schulz, G. J. (1964), *Phys. Rev. Lett.* **12**, 449.
Chantry, P. J. and Schulz, G. J. (1967), *Phys. Rev.* **156**, 134.
Christophorou, L. G. and Stockdale, J. A. (1968), *J. Chem. Phys.* **48**, 1956.
Clarke, I. D. and Wayne, R. P. (1969), *Chem. Phys. Lett.* **3**, 73.
Claydon, C. R., Segal, G. A. and Taylor, H. S. (1970), *J. Chem. Phys.* **54**, 3799.
Compton, R. N. and Christophorou, L. G. (1967), *Phys. Rev.* **154**, 110.
Compton, R. N., Christophorou, L. S., Hurst, G. S. and Reinhardt, P. W. (1966), *J. Chem. Phys.* **45**, 4634.
Craggs, J. D. and McDowell, C. A. (1955), *Rep. Progr. Phys.* **18**, 374.
Crompton, R. W., Rees, J. A. and Jory, R. L. (1965), *Aust. J. Phys.* **18**, 541.
Curran, R. K. (1961), *J. Chem. Phys.* **35**, 1849.
Curran, R. K. and Fox, R. E. (1961), *J. Chem. Phys.* **34**, 1590.
Davis, F. J. and Nelson, D. R. (1969), *Chem. Phys. Lett.* **6**, 277.
Davis, F. J. and Nelson, D. R. (1970), *Chem. Phys. Lett.* **7**, 461.
De Corpo, J. J. and Franklin, J. L. (1971), *J. Chem. Phys.* **54**, 1885.
Doehring, A. (1952), *Z. Naturforsch*, **7a**, 253.
Dorman, F. H. (1966), *J. Chem. Phys.* **44**, 3856.

Doumont, M., Henglein, A. and Jäger, K. (1969), *Z. Naturforsch*, **24**a, 683.
Dunn, G. H. (1962), *Phys. Rev. Lett.* **8**, 62.
Dunn, G. H. and Kieffer, L. J. (1963), *Phys. Rev.* **132**, 2109.
Edelson, D., Griffiths, J. E. and McAfee, K. B. (1962), *J. Chem. Phys.* **37**, 917.
Engelhardt, A. C., Phelps, A. V. and Risk, C. G. (1964), *Phys. Rev.* **135**A, 1566.
Fehsenfeld, F. C. (1970), *J. Chem. Phys.* **53**, 2000.
Fite, W. L. and Brackmann, R. T. (1963), *Proc. 6th Int. Conf. Ionization Phenomena in Gases*, Paris, p. 21.
Fox, R. E. (1958), *Phys. Rev.* **109**, 2008.
Fox, R. E. and Curran, R. K. (1961), *J. Chem. Phys.* **34**, 1590.
Franklin, J. L., Hierl, P. M. and Whan, D. A. (1967), *J. Chem. Phys.* **47**, 3148.
Grünberg, R. (1969), *Z. Naturforsch.* **24**a, 1039.
Gunton, R. C. and Shaw, T. M. (1965), *Phys. Rev.* **140**, 748.
Hagstrum, H. D. and Tate, J. T. (1941), *Phys. Rev.* **59**, 354.
Hake, R. D. and Phelps, A. V. (1966), Westinghouse Research Report 66-1E2-P1.
Hanson, E. E. (1937), *Phys. Rev.* **51**, 86.
Healey, R. H. (1938), *Phil. Mag.* **26**, 940.
Healey, R. H. and Kirkpatrick, C. B. (1941), as quoted in Healey, R. H. and Reed, J. W., The behaviour of slow electrons in gases p. 10 (Amalgamated Wireless, Sydney, 1941).
Henderson, W. R., Fite, W. L. and Brackmann, R. T. (1969), *Phys. Rev.* **183**, 157.
Henis, J. M. S. and Mabie, C. A. (1970), *J. Chem. Phys.* **53**, 2999.
Hickam, W. M. and Fox, R. E. (1956), *J. Chem. Phys.* **25**, 642.
Jäger, K. and Henglein, A. (1968), *Z. Naturforsch.* **23**, 1122.
Kaufman, F. (1967), *J. Chem. Phys.* **46**, 2449.
Lagergren, C. R. (1955), Dissertation, Univ. Minnesota.
Lee, T. G. (1963), *J. Phys. Chem.* **67**, 360.
Loeb, L. and Cravath, R. (1929), *Phys. Rev.* **33**, 605.
Lozier, W. W. (1930), *Phys. Rev.* **36**, 1417.
Lozier, W. W. (1934), *Phys. Rev.* **46**, 268.
MacNeil, K. A. G. and Thynne, J. C. J. (1968), *Trans. Faraday Soc.* **64**, 2112.
Mahan, B. H. and Walker, I. C. (1967), *J. Chem. Phys.* **47**, 3780.
Mahan, B. H. and Young, C. E. (1966), *J. Chem. Phys.* **44**, 2192.
Marriott, J., Thorburn, R. and Craggs, J. D. (1954), *Proc. Phys. Soc.* B **67**, 437.
Massey, H. S. W. (1969), *Electronic and Ionic Impact Phenomena*, 2nd edition, vol. 2, Clarendon Press, Oxford.
Massey, H. S. W. and Burhop, E. H. S. (1969), *Electronic and Ionic*

Impact Phenomena, 2nd edition, vol. 1, Chap. 2, Clarendon Press, Oxford.
Mendas, I. and Stamatovic, A. (1973), *Proc. 8th Int. Conf. Physics of Electronic and Atomic Collisions*, Belgrade, Abstracts, p. 465.
Metlay, M. and Kimball, G. E. (1948), *J. Chem. Phys.* **16**, 744.
Mott, N. F. and Massey, H. S. W. (1965), *The Theory of Atomic Collisions*, 3rd edition, Clarendon Press, Oxford.
Odom, R. W., Smith, D. L. and Futrell, J. H. (1974), *Chem. Phys. Lett.* **24**, 227.
O'Malley, T. F. (1967), *Phys. Rev.* **155**, 59.
O'Malley, T. F. and Taylor, H. S. (1968), *Phys. Rev.* **176**, 207.
Pack, J. L. and Phelps, A. V. (1966), *J. Chem. Phys.* **44**, 1870.
Pack, J. L., Voshall, R. E. and Phelps, A. V. (1962), *Phys. Rev.* **127**, 2084.
Page, F. M. (1961), *Trans. Faraday Soc.* **57**, 359.
Parkes, D. A. and Sugden, T. M. (1972), *J.C.S. Faraday Trans.* II, **68**, 600.
Parr, J. E. and Moruzzi, J. (1972), *J. Phys. D* **5**, 514.
Phelps, A. V. and Voshall, R. E. (1968), *J. Chem. Phys.* **49**, 3246.
Puckett, L. J., Kregel, M. D. and Teague, M. W. (1971), *Phys. Rev. A* **4**, 1659.
Rapp, D. and Briglia, D. D. (1965), *J. Chem. Phys.* **43**, 1480.
Rapp, D., Englander-Golden, P. and Briglia, D. D. (1965), *J. Chem. Phys.* **42**, 4081.
Rapp, D., Sharp, T. E. and Briglia, D. D. (1965), *Phys. Rev. Lett.* **14**, 533.
Reese, R., Dibeler, V. H. and Mohler, F. L. (1956), *J. Res. Nat. Bur. Stand.* **57**, 367.
Ryzko, H. (1966), *Arkiv. Fys.* **32**, 1.
Schulz, G. J. (1959), *Phys. Rev.* **113**, 816.
Schulz, G. J. (1960), *J. Chem. Phys.* **33**, 1661.
Schulz, G. J. (1961), *J. Chem. Phys.* **34**, 1778.
Schulz, G. J. and Asundi, R. K. (1965), *Phys. Rev. Lett.* **15**, 946.
Schulz, G. J. and Dowell, J. T. (1962), *Phys. Rev.* **128**, 174.
Schulz, G. J. and Spence, D. (1969), *Phys. Rev. Lett.* **22**, 47.
Sharp, T. E. and Dowell, J. T. (1969), *J. Chem. Phys.* **50**, 3024.
Smith, R. A. (1936), *Proc. Camb. Phil. Soc.* **32**, 482.
Smith, S. J. and Branscomb, L. M. (1955), *J. Res. Nat. Bur. Stand.* **55**, 165.
Spence, D. and Schulz, G. J. (1969), *Phys. Rev.* **188**, 280.
Spence, D. and Schulz, G. J. (1972), *Phys. Rev. A* **5**, 724.
Spence, D. and Schulz, G. J. (1973a), *J. Chem. Phys.* **58**, 1800.
Spence, D. and Schulz, G. J. (1973b), *Proc. 8th Int. Conf. Physics of Electronic and Atomic Collisions*, Belgrade, Abstracts, p. 467.
Stamatovic, A. and Schulz, G. J. (1970), *J. Chem. Phys.* **53**, 2663.
Stamatovic, A. and Schulz, G. J. (1973), *Phys. Rev. A* **7**, 589.

Stelman, D., Moruzzi, J. L. and Phelps, A. V. (1972), *J. Chem. Phys.* **56**, 4183.
Stockdale, J. A., Compton, R. N. and Schweinler, H. C. (1970), *J. Chem. Phys.* **53**, 1502.
Tate, J. T. and Lozier, W. W. (1932), *Phys. Rev.* **39**, 254.
Tozer, B. A. (1958), *J. Electron. Contr.* **4**, 149.
Truby, F. K. (1968), *Phys. Rev.* **172**, 24.
Truby, F. K. (1971), *Phys. Rev.* A **4**, 613.
Vought, R. M. (1947), *Phys. Rev.* **71**, 93.
Van Brunt, R. J. and Kieffer, L. J. (1970), *Phys. Rev.* A **2**, 1899.
Warman, J. M. and Fessenden, R. W. (1968), *J. Chem. Phys.* **49**, 4718.
Warman, J. M., Fessenden, R. W. and Bakale, G. (1972), *J. Chem. Phys.* **57**, 2702.
Weller, C. S. and Biondi, M. A. (1968), *Phys. Rev.* **172**, 198.

Chapter 10

Baede, A. P. M. (1972), *Physica*, **59**, 541.
Baede, A. P. M., Auerbach, D. J. and Los, J. (1973), *Physica*, **64**, 134.
Baede, A. P. M. and Los, J. (1971), *Physica*, **52**, 422.
Bates, D. R. and Boyd, T. J. M. (1956), *Proc. Phys. Soc.* **69A**, 910.
Collins, L. E. and Stroud, P. T. (1967), *Proc. Phys. Soc.* **90**, 641.
Compton, R. N. and Cooper, C. D. (1973), *J. Chem. Phys.* **59**, 4140.
Curran, R. K. and Donahue, T. M. (1960), *Phys. Rev.* **118**, 1233.
Fogel, Ya. M., Ankudinov, V. A. and Pilipenko, D. V. (1959), *Sov. Phys. JETP*, **8**, 601.
Fogel, Ya. M., Ankudinov, V. A. and Pilipenko, D. V. (1960), *Sov. Phys. JETP*, **11**, 18.
Fogel, Ya. M., Ankudinov, V. A., Pilipenko, D. V. and Topolia, N. V. (1958), *Sov. Phys. JETP*, **7**, 400.
Fogel, Ya. M., Mitin, R. V., Koslov, V. G. and Romashko, N. D. (1959); *Sov. Phys. JETP*, **8**, 390.
Gilbody, H. B., Browning, R., Dunn, K. F. and McIntosh, A. I. (1969), *J. Phys.* B **2**, 465.
Gilbody, H. B., Dunn, K. F. and Browning, R. (1970), *J. Phys.* B, **3**, L19.
Helbing, R. K. B. and Rothe, E. W. (1969), *J. Chem. Phys.* **51**, 1607.
Lacmann, K. and Herschbach, D. R. (1970), *Chem. Phys. Lett.* **6**, 106.
Leffert, C. B., Jackson, W. M. and Rothe, E. W. (1973), *J. Chem. Phys.* **58**, 5801.
McClure, G. W. (1964), *Phys. Rev.* A **134**, 1226.
Mott, N. F. and Massey, H. S. W. (1965), *The Theory of Atomic Collisions*, 3rd edition, Clarendon Press, Oxford.
Nalley, S. J., Compton, R. N., Schweinler, H. C. and Anderson, V. E. (1973), *J. Chem. Phys.* **59**, 4125.

Oparin, V. A., Il'in, R. N., Serenkov, I. T., Solov'yev, E. S. and Fedorenko, N. V. (1971), *Proc. 7th Int. Conf. Physics of Electronic and Atomic Collisions*, Amsterdam, Abstracts, p. 796.
Schryber, V. (1966), *Helv. Phys. Acta*, **39**, 562.
Stier, P. M. and Barnett, C. F. (1956), *Phys. Rev.* **103**, 896.
Toburen, L. H. and Nakai, M. Y. (1969), *Phys. Rev.* **177**, 191.
Williams, J. F. (1966), *Phys. Rev.* **150**, 7.
Williams, J. F. (1967), *Phys. Rev.* **153**, 116.
Williams, J. F. and Dunbar, D. N. F. (1966), *Phys. Rev.* **149**, 62.

Chapter 11

Badger, R. M. (1934), *J. Chem. Phys.* **2**, 129.
Bely, O. and Schwartz, S. B. (1969), *J. Phys. B* **2**, 159.
Bennett, R. A. (1972), Thesis, University of Colorado.
Berry, R. S. and Reimann, C. W. (1963), *J. Chem. Phys.* **38**, 1540.
Berry, R. S., Reimann, C. W. and Spokes, G. N. (1961), *J. Chem. Phys.* **35**, 2237.
Boldt, G. (1959), *Z. Phys.* **154**, 319.
Branscomb, L. M. (1966), *Phys. Rev.* **148**, 11.
Branscomb, L. M., Burch, D. S., Smith, S. J. and Geltman, S. (1958), *Phys. Rev.* **111**, 504.
Branscomb, L. M. and Fite, W. L. (1954), *Phys. Rev.* **93**, 651A.
Branscomb, L. M. and Smith, S. J. (1955), *Phys. Rev.* **98**, 1028, 1127.
Branscomb, L. M. and Smith, S. J. (1956), *J. Chem. Phys.* **25**, 598.
Branscomb, L. M., Smith, S. J. and Tisone, G. (1965), *Proc. 4th Int. Conf. Physics of Electronic and Atomic Collisions*, Quebec, Abstracts, p. 106.
Brauman, J. I. and Smyth, K. C. (1969), *J. Amer. Chem. Soc.* **91**, 7778.
Brehm, B., Gusinow, M. A. and Hall, J. L. (1967), *Phys. Rev. Lett.* **19**, 737.
Burch, D. S., Smith, S. J. and Branscomb, L. M. (1958), *Phys. Rev.* **112**, 171.
Burch, D. S., Smith, S. J. and Branscomb, L. M. (1959), *Phys. Rev.* **114**, 1652.
Burt, J. A. (1972a), *Ann. Geophys.* **28**, 607.
Burt, J. A. (1972b), *J. Chem. Phys.* **57**, 4649.
Cade, P. E. (1967), *Proc. Phys. Soc.* **91**, 842.
Cahill, T. A., Richardson, J. R. and Verba, J. W. (1966), *Nucl. Instrum. Methods*, **39**, 278.
Celotta, R. J., Bennett, R. A. and Hall, J. L. (1974), *J. Chem. Phys.* **60**, 1740.
Celotta, R. J., Bennett, R. A., Hall, J. L., Siegel, M. W. and Levine, J. (1972), *Phys. Rev. A* **6**, 631.
Chandrasekhar, S. (1945), *Astrophys. J.* **102**, 223.

Cooper, J. W. and Martin, J. B. (1962), *Phys. Rev.* **126**, 1482.
Cooper, J. W. and Zare, R. N. (1968), *J. Chem. Phys.* **48**, 942.
Dance, D. F., Harrison, M. F. A. and Rundel, R. D. (1967), *Proc. Roy. Soc.* A **299**, 525.
Darewych, G. and Neamtam, S. M. (1963), *Nucl. Instrum. Methods*, **21**, 247.
Doughty, N. A., Fraser, P. A. and McEacharn, R. P. (1966), *Mon. Notic. Roy. Astron. Soc.* **132**, 255.
Feldmann, D. (1970), *Z. Naturforsch.* **25**a, 621.
Feldmann, D., Rackwitz, R., Heinicke, E. and Kaiser, H. J. (1973), *Phys. Lett.* **45**A, 404.
Geltman, S. (1962), *Astrophys. J.* **136**, 935.
Hall, J. L., Robinson, E. J. and Branscomb, L. M. (1965), *Phys. Rev. Lett.* **14**, 1013.
Hall, J. L. and Siegel, M. W. (1968), *J. Chem. Phys.* **48**, 943.
Herman, F. and Skillman, S. (1963), *Atomic Structure Calculations*, Prentice-Hall, New Jersey.
Herzberg, G. and Lagerquist, A. (1968), *Can. J. Phys.* **46**, 2363.
Hiskes, J. R. (1962), Quoted in Judd, D. L., *Nucl. Instrum. Methods*, **18**, (1962), 70.
Hotop, H., Bennett, R. A. and Lineberger, W. C. (1973), *J. Chem. Phys.* **58**, 2373.
Hotop, H. and Lineberger, W. C. (1973), *J. Chem. Phys.* **58**, 2379.
Hotop, H., Patterson, T. A. and Lineberger, W. C. (1973), *Phys. Rev.* A **8**, 762.
Inokuti, M. and Kim, Y.-K. (1968), *Phys. Rev.* **173**, 154.
Inokuti, M., Kim, Y.-K. and Platzman, R. L. (1967), *Phys. Rev.* **164**, 55.
Kaplan, S. N., Paulikas, G. A. and Pyle, R. V. (1963), *Phys. Rev.* **131**, 2574.
Lineberger, W. C. and Patterson, T. A. (1972), *Chem. Phys. Lett.* **13**, 40.
Lineberger, W. C. and Woodward, B. W. (1970), *Phys. Rev. Lett.* **25**, 424.
Lipsky, L. (1967), *Proc. 5th Int. Conf. Physics of Electronic and Atomic Collisions*, Leningrad, Abstracts, p. 617.
Macek, J. (1967), *Proc. Phys. Soc.* **92**, 365.
Mandl, A. and Hyman, H. A. (1973), *Phys. Rev. Lett.* **31**, 417.
Massey, H. S. W. and Bates, D. R. (1940), *Astrophys. J.* **91**, 202.
Moores, D. L. and Norcross, D. W. (1974), *Phys. Rev.* A **10**, 1646.
Mott, N. F. and Massey, H. S. W. (1965), *The Theory of Atomic Collisions*, 3rd edition, Clarendon Press, Oxford.
Mullen, B. and Vogt, E. W. (1968), unpublished communication to Stinson *et al.*, 1969.
Norcross, D. W. (1974), *Phys. Rev. Lett.* **32**, 192.
O'Malley, T. F. (1965), *Phys. Rev.* **137**A, 1668.
Oparin, V. A., Ill'in, R. N., Serenkov, I. T., Solov'yev, E. S. and Fedorenko, N. V. (1970), *JETP Lett.* **12**, 162.

Oppenheimer, J. R. (1928), *Phys. Rev.* **31**, 66.
Patterson, T. A., Hotop, H., Kasdan, A., Norcross, D. W. and Lineberger, W. C. (1974), *Phys. Rev. Lett.* **32**, 189.
Peart, B. and Dolder, K. T. (1973), *J. Phys.* B **6**, 1497.
Peart, B., Walton, D. S. and Dolder, K. T. (1970), *J. Phys.* B **3**, 1346.
Peart, B., Walton, D. S. and Dolder, K. T. (1971), *J. Phys.* B **4**, 88.
Robinson, E. J. and Geltman, S. (1967), *Phys. Rev.* **153**, 4.
Rau, A. R. P. and Fano, U. (1971), *Phys. Rev.* A **4**, 1751.
Schwartz, C. (1961), *Phys. Rev.* **123**, 1700.
Seman, M. L. and Branscomb, L. M. (1962), *Phys. Rev.* **125**, 1602.
Siegel, M. W., Celotta, R. J., Hall, J. L., Levine, J. and Bennett, R. A. (1972), *Phys. Rev.* A **6**, 607.
Sinnott, G. and Beaty, E. C. (1971), *Proc. 7th Int. Conf. Physics of Electronic and Atomic Collisions*, Amsterdam, Abstracts, p. 176.
Smirnov, B. A. and Chibisov, M. I. (1965), *Sov. Phys. JETP*, **22**, 585.
Smith, S. J. and Branscomb, L. M. (1955), *Phys. Rev.* **99**, 1657A.
Smith, S. J. and Branscomb, L. M. (1960), *Rev. Sci. Instrum.* **31**, 733.
Smith, S. J. and Burch, D. S. (1959), *Phys. Rev.* **116**, 1125.
Smyth, K. C. and Brauman, J. I. (1972a), *J. Chem. Phys.* **56**, 1132.
Smyth, K. C. and Brauman, J. I. (1972b), *J. Chem. Phys.* **56**, 4620.
Smyth, K. C., McIver, R. T., Brauman, J. I. and Wallace, R. W. (1971), *J. Chem. Phys.* **54**, 2758.
Steiner, B. (1968), *J. Chem. Phys.* **49**, 5097.
Steiner, B., Seman, M. L. and Branscomb, L. M. (1962), *J. Chem. Phys.* **37**, 1200.
Stinson, G. M., Olsen, W. C., McDonald, W. J., Ford, P., Axen, D. and Blackmore, E. W. (1969), *Nucl. Instr. Methods*, **74**, 333.
Swales, F. W. (1968), Rutherford High Energy Laboratory, Report PA/M/17.
Tisone, G. and Branscomb, L. M. (1966), *Phys. Rev. Lett.* **17**, 236.
Walton, D. S., Peart, B. and Dolder, K. T. (1971), *J. Phys.* B **4**, 1343.
Warneck, P. (1969), *Chem. Phys. Lett.* **3**, 532.
Weiss, A. W. (1968), *Phys. Rev.* **166**, 70.
Wigner, E. (1948), *Phys. Rev.* **73**, 1002.
Woo, S. B., Branscomb, L. M. and Beaty, E. C. (1969), *J. Geophys. Res.* **74**, 2933.

Chapter 12

Adams, N. G., Bohme, D. K., Dunkin, D. B., Fehsenfeld, F. C. and Ferguson, E. E. (1970), *J. Chem. Phys.* **52**, 3133.
Berkowitz, J., Chupka, W. A. and Gutman, D. (1971), *J. Chem. Phys.* **55**, 2733.

Berry, R. S., Cernock, T. M., Coplan, M. and Ewing, J. J. (1968), *J. Chem. Phys.* **49**, 127.
Chanin, L. M., Phelps, A. V. and Biondi, M. A. (1962), *Phys. Rev.* **128**, 219.
Chantry, P. (1969), *J. Chem. Phys.* **51**, 3380.
Chupka, W. A., Berkowitz, J. and Gutman, D. (1971), *J. Chem. Phys.* **55**, 2724.
Conway, D. C. and Nesbitt, L. E. (1968), *J. Chem. Phys.* **48**, 509.
Cottrell, T. L. (1958), *The Strength of Chemical Bonds*, 2nd edition, Butterworths, London.
Dalgarno, A. and Browne, J. C. (1967), *Astrophys. J.* **149**, 231.
Dunkin, D. B., Fehsenfeld, F. C. and Ferguson, E. E. (1972), *Chem. Phys. Lett.* **15**, 257.
Fehsenfeld, F. C., Albritton, D. L., Burt, J. A. and Schiff, H. I. (1969), *Can. J. Chem.* **47**, 1793.
Fehsenfeld, F. C. and Ferguson, E. E. (1968), *Planet. Space Sci.* **16**, 701.
Fehsenfeld, F. C. and Ferguson, E. E. (1972), *Planet. Space Sci.* **20**, 295.
Fehsenfeld, F. C. and Ferguson, E. E. (1970), *J. Chem. Phys.* **53**, 2614.
Fehsenfeld, F. C., Ferguson, E. E. and Bohme, D. K. (1969), *Planet. Space Sci.* **17**, 1759.
Fehsenfeld, F. C., Ferguson, E. E. and Schmeltekopf, A. L. (1966), *J. Chem. Phys.* **45**, 1844.
Ferguson, E. E. (1967), *Rev. Geophys.* **5**, 305.
Ferguson, E. E. (1969), *Can. J. Chem.* **47**, 1815.
Ferguson, E. E., Dunkin, D. B. and Fehsenfeld, F. C. (1972), *J. Chem. Phys.* **57**, 1459.
Ferguson, E. E., Fehsenfeld, F. C., Dunkin, D. B., Schmeltekopf, A. L. and Schiff, H. I. (1964), *Planet. Space Sci.* **12**, 1169.
Freund, R. S. (1971), *J. Chem. Phys.* **54**, 3125.
Herzenberg, A. (1967), *Phys. Rev.* **160**, 80.
Hughes, B. M., Lifshitz, C. and Tiernan, T. O. (1973), *J. Chem. Phys.* **59**, 3162.
Huxley, L. G. H., Crompton, R. W. and Bagot, C. H. (1959), *Aust. J. Phys.* **12**, 303.
Langevin, P. (1905), *Ann. Chem. Phys.* **5**, 245.
Lifshitz, C., Hughes, B. M. and Tiernan, T. O. (1970), *Chem. Phys. Lett.* **7**, 469.
Lifshitz, C., Tiernan, T. O. and Hughes, B. M. (1973), *J. Chem. Phys.* **59**, 3182.
McDaniel, E. W., Martin, D. W. and Barnes, W. S. (1962), *Rev. Sci. Instrum.* **33**, 2.
McFarland, M., Dunkin, D. B., Fehsenfeld, F. C., Schmeltekopf, A. L. and Ferguson, E. E. (1972), *J. Chem. Phys.* **56**, 2358.
McKnight, L. G. (1970), *Phys. Rev.* A **2**, 762.

McKnight, L. G. and Sawina, J. M. (1971), *Phys. Rev.* A **4**, 1043.
Mandl, A., Evans, E. W. and Kivel, B. (1970), *Chem. Phys. Lett.* **5**, 307.
Marx, R., Mauclaire, G., Fehsenfeld, F. C., Dunkin, D. B. and Ferguson, E. E. (1973), *J. Chem. Phys.* **58**, 3267.
Mauer, J. L. and Schulz, G. J. (1973), *Phys. Rev.* A **7**, 593.
Moruzzi, J. L., Ekin, J. W. and Phelps, A. V. (1968), *J. Chem. Phys.* **48**, 3070.
Moruzzi, J. L. and Phelps, A. V. (1966), *J. Chem. Phys.* **45**, 4617.
Moseley, J. T., Snuggs, R. M., Martin, D. W. and McDaniel, E. W. (1968), *Phys. Rev. Lett.* **21**, 873.
Pack, J. L. and Phelps, A. V. (1966a), *J. Chem. Phys.* **44**, 1870.
Pack, J. L. and Phelps, A. V. (1966b), *J. Chem. Phys.* **45**, 4316.
Parkes, D. A. (1972a), *J. Chem. Soc. (Lond.)*, **68**, 2103.
Parkes, D. A. (1972b), *J.C.S. Faraday Soc.* I, **68**, 2121.
Parkes, D. A. and Sugden, T. M. (1972), *J.C.S. Faraday Trans.* II, **68**, 600.
Paulson, J. F. (1966), *Adv. Chem. Ser.* **58**, 28.
Paulson, J. F. (1970), *J. Chem. Phys.* **52**, 963.
Phelps, A. V. and Pack, J. L. (1961), *Phys. Rev. Lett.* **6**, 111.
Puckett, L. J. and Lineberger, W. C. (1970), *Phys. Rev.* A **1**, 1635.
Snuggs, R. M., Volz, D. J., Gatland, I. R., Schummers, J. H., Martin, D. W. and McDaniel, E. W. (1971), *Phys. Rev.* A **3**, 487.
Snuggs, R. M., Volz, D. J., Schummers, J. H., Martin, D. W. and McDaniel, E. W. (1971), *Phys. Rev.* A **3**, 477.
Stockdale, J. A. D., Compton, R. N. and Schweinler, H. C. (1970), *J. Chem. Phys.* **53**, 1502.
Vogt, D. (1969), *Intern. J. Mass Spectrum. Ion Phys.* **3**, 81.
Wannier, G. H. (1951), *Phys. Rev.* **83**, 281.

Chapter 13

Andreev, E. P., Ankudinov, V. A. and Bobashev, S. V. (1966), *Sov. Phys. JETP*, **23**, 375.
Andreev, E. P., Ankudinov, V. A., Dukel'skii, V. M. and Orbeli, A. L. (1969), *Proc. 6th Int. Conf. Physics of Electronic and Atomic Collisions*, Cambridge, Mass., Abstracts, p. 800.
Bailey, T. L. (1961), *Proc. 2nd Int. Conf. Physics of Electronic and Atomic Collisions*, Boulder, Abstracts, p. 54.
Bailey, T. L. and Mahadevan, P. (1970), *J. Chem. Phys.* **52**, 179.
Bardsley, J. N. (1967), *Proc. Phys. Soc.* **91**, 300.
Bates, D. R. and Walker, J. C. G. (1967), *Proc. Phys. Soc.* **90**, 333.
Bydin, Yu. F. (1966a), *Sov. Phys. JETP*, **22**, 762.
Bydin, Yu. F. (1966b), *Sov. Phys. JETP*, **23**, 23.
Bydin, Yu. F. and Dukel'skii, V. M. (1957), *Sov. Phys. JETP*, **4**, 474.

Cunningham, D. L. and Edwards, A. K. (1973), *Phys. Rev.* A **8**, 2960.
Dalgarno, A. and McDowell, M. R. C. (1956), *Proc. Phys. Soc.* A **69**, 615.
Davidovic, D. M. and Janev, R. K. (1969), *Phys. Rev.* **186**, 89.
Demkov, J. N. (1952), *Uch. Zap. Leningrad*, **146**, 74.
Dimov, G. I. and Rosljakov, G. V. (1971), *Proc. 7th Int. Conf. Physics of Electronic and Atomic Collisions*, Amsterdam, Abstracts, p. 800.
Dukel'skii, V. M. and Fedorenko, N. V. (1956), *Sov. Phys. JETP*, **2**, 307.
Edwards, A. K., Risley, J. S. and Geballe, R. (1971), *Phys. Rev.* A **3**, 583.
Firsov, O. B. (1951), *Zh. Eksp. Teor. Fiz.* **21**, 1001.
Fite, W. L., Brackmann, R. T. and Snow, W. R. (1958), *Phys. Rev.* **112**, 1161.
Fogel, A. M., Ankudinov, V. A. and Slabospitskii, R. E. (1957), *Sov. Phys. JETP*, **5**, 382.
Geballe, R. and Risley, J. S. (1973), *Proc. 8th Int. Conf. Physics of Electronic and Atomic Collisions*, Belgrade, Abstracts, p. 834.
Gilbody, H. B., Dunn, K. F., Browning, R. and Latimer, C. (1970), *J. Phys.* B **3**, 1105.
Hasted, J. B. (1952), *Proc. Roy. Soc.* A **212**, 235.
Hasted, J. B. (1954), *Proc. Roy. Soc.* A **222**, 74.
Hasted, J. B. and Smith, R. A. (1956), *Proc. Roy. Soc.* A **235**, 349.
Hummer, D. G., Stebbings, R. F., Fite, W. L. and Branscomb, L. M. (1960), *Phys. Rev.* **119**, 668.
Keever, W. C., Lockwood, G. J., Helbig, F. H. and Everhart, E. (1968), *Phys. Rev.* **166**, 68.
McDowell, M. R. C. and Peach, G. (1959), *Proc. Phys. Soc.* **74**, 463.
Pedersen, E. H. and Hvelplund, P. (1973), *J. Phys.* B **6**, 2600.
Pilipenko, D. V., Gusev, V. A. and Fogel, Ya. M. (1966), *Sov. Phys. JETP*, **22**, 965.
Rapp, D. and Francis, W. E. (1962), *J. Chem. Phys.* **37**, 2631.
Roche, A. E. and Goodyear, C. C. (1969), *J. Phys.* B **2**, 191.
Rutherford, J. A. and Turner, B. R. (1967), *J. Geophys. Res.* **72**, 3795.
Ryding, G., Wittkower, A. B. and Rose, P. H. (1968), *Phys. Rev.* **174**, 149.
Sida, D. W. (1955), *Proc. Phys. Soc.* A **68**, 240.
Simpson, F. R. and Gilbody, H. B. (1972), *J. Phys.* B **5**, 1959.
Snow, W. R., Rundel, R. D. and Geballe, R. (1969), *Phys. Rev.* **178**, 228.
Stedeford, J. B. H. and Hasted, J. B. (1955), *Proc. Roy. Soc.* A **227**, 466.
Stier, P. M. and Barnett, C. F. (1956), *Phys. Rev.* **103**, 896.
Tisone, G. C. and Branscomb, L. M. (1964), J.I.L.A. Rept. Boulder, Colorado, No. 2.
Whittier, A. C. (1954), *Can. J. Phys.* **32**, 275.
Williams, J. F. (1967), *Phys. Rev.* **154**, 9.
Wynn, M. J., Martin, J. D. and Bailey, T. L. (1970), *J. Chem. Phys.* **52**, 191.

Chapter 14

Aberth, W. H. and Peterson, J. K. (1970), *Phys. Rev.* **1**, 158.
Aberth, W., Peterson, J. K., Lorents, D. C. and Cook, C. J. (1968), *Phys. Rev. Lett.* **20**, 979.
Bates, D. R. and Boyd, T. J. M. (1956), *Proc. Phys. Soc.* **69A**, 910.
Bates, D. R. and Flannery, M. R. (1968), *Proc. Roy. Soc.* A **302**, 367.
Bates, D. R. and Lewis, J. T. (1955), *Proc. Phys. Soc.* **68A**, 173.
Bates, D. R. and Massey, H. S. W. (1947), *Proc. Roy. Soc.* A **192**, 1.
Bates, D. R. and Moffett, R. J. (1966), *Proc. Roy. Soc.* A **291**, 1.
Gaily, T. D. and Harrison, M. F. A. (1970a), *J. Phys.* B (GB) **3**, L25.
Gaily, T. D. and Harrison, M. F. A. (1970b), *J. Phys.* B (GB) **3**, 1098.
Gardner, M. E. (1938), *Phys. Rev.* **53**, 75.
Greaves, C. (1964), *J. Electron. Control*, **17**, 171.
Hirsch, M. N., Halpern, G. M. and Wolf, N. S. (1968), *Bull. Amer. Phys. Soc.* **13**, 199.
Langevin, P. (1903), *Ann. Chim. Phys.* **28**, 289, 433.
McGowan, S. (1965), *Phys. Med. Biol.* **10**, 25.
McGowan, S. (1967), *Can. J. Phys.* **45**, 449, 439.
Mächler, W. von (1936), *Phys. Z.* **37**, 211.
Magee, J. L. (1952), *Dis. Faraday Soc.* **12**, 33.
Mahan, B. H. and Person, J. C. (1964), *J. Chem. Phys.* **40**, 392.
Moseley, J., Aberth, W. and Peterson, J. R. (1970), *Phys. Rev. Lett.* **24**, 435.
Natanson, G. L. (1959), *Zh. Tekh. Fiz.* **29**, 1373; *Sov. Phys. Tech. Phys.* **4**, 1263.
Rundel, R. D., Aitken, K. L. and Harrison, M. F. A. (1969), *J. Phys.* B (GB) **2**, 954.
Rutherford, E. and Thomson, J. J. (1897), *Phil. Mag.* **44**, 422.
Sayers, J. (1938), *Proc. Roy. Soc.* A **169**, 83.
Thomson, J. J. (1924), *Phil. Mag.* **47**, 337.
Trujillo, S. M., Neynaber, R. H. and Rothe, E. W. (1966), *Rev. Sci. Instr.* **37**, 1655.
Yeung, T. H. Y. (1958), *Proc. Phys. Soc.* **71**, 341; *J. Electron Control*, **5**, 307.

Chapter 15

Arnold, F., Kessel, J., Krankowsky, D., Wieder, H. and Zähringer, J. (1971), *J. Atmos. Terr. Phys.* **33**, 1169.
Bates, D. R. and Massey, H. S. W. (1947), *Proc. Roy. Soc.* A **192**, 1.
Božin, S. E. and Goodyear, C. C. (1967), *Brit. J. Appl. Phys.* **18**, 49.
Božin, S. E. and Goodyear, C. C. (1968), *J. Phys.* D **1**, 327.

Bhalla, M. S. and Craggs, J. D. (1962), *Proc. Phys. Soc.* **80**, 151.
Boettner, E. A. and Dallos, F. C. (1965), *J. Gas Chromatogr.* **3**, 190.
Bowman, M. R., Thomas, L. and Geisler, J. E. (1970), *J. Atmos. Terr. Phys.* **32**, 1661.
Boyd, R. L. F. and Morris, D. (1955), *Proc. Phys. Soc.* A **68**, 1.
Boyd, R. L. F. and Thompson, J. B. (1959), *Proc. Roy. Soc.* A **252**, 102.
Chandrasekhar, S. and Breen, F. H. (1946), *Astrophys. J.* **103**, 41.
Chen, E., George, R. D. and Wentworth, W. E. (1968), *J. Chem. Phys.*, **49**, 1973.
CIRA (Cospar Int. Ref. Atmosphere) 1965, North-Holland, Amsterdam.
Colegrove, F. D., Johnson, F. S. and Hanson, W. B. (1966), *J. Geophys. Res.* **71**, 2227.
Dawton, R. H. V. M. (1969), *Nucl. Instr. Method*, **67**, 341.
Deutsch, A. J. (1948), *Rev. Mod. Phys.* **20**, 388.
Devins, J. C. and Wolff, R. J. (1965), *Ann. Rep. Conf. Electrical Insulation*, Nat. Acad. Sci-NRC, Washington, D.C., Publication 1238.
Doughty, N. A. and Fraser. P. A. (1966), *Mon. Not. Roy. Soc.* **132**, 267.
Eméleus, K. G. and Sayers, J. (1938), *Proc. Roy. Irish Acad.* **44**, 87.
Evans, W. F. J., Hunten, D. M., Llewellyn, E. J. and Jones, A. V. (1968), *J. Geophys. Res.* **73**, 2885.
Frommhold, L. (1964), *Fortschr. Phys.* **12**, 597.
Geballe, R. and Reeves, M. L. (1953), *Phys. Rev.* **92**, 867.
Goldberg, R. A. and Blumle, L. J. (1970), *J. Geophys. Res.* **75**, 133.
Goodwin, E. S., Goulden, R. and Reynolds, J. G. (1961), *Analyst*, **86**, 697.
Grübler, W., Haeberli, W. and Schwandt, P. (1964), *Phys. Rev. Lett.* **12**, 595.
Grübler, W., König, V. and Schmelzbach, P. A. (1970), *Nucl. Instr. Method.* **86**, 127.
Grübler, W., Schwandt, P., Yule, T. J. and Haeberli, W. (1965), *Nucl. Instr. Method.* **41**, 245.
Haeberli, W., Grübler, W., Extermann, P. and Schwandt, P. (1965), *Phys. Rev. Lett.* **15**, 267.
Harrison, M. A. and Geballe, P. (1953), *Phys. Rev.* **91**, 1.
Langmuir, I. (1929), *Phys. Rev.* **33**, 954.
Le Levier, R. E. and Branscomb, L. M. (1968), *J. Geophys. Res.* **73**, 27.
Leu, M. T., Biondi, M. A. and Johnsen, R. (1973), *Phys. Rev.* A **7**, 292.
Lovelock, J. E. (1963), *Anal. Chem.* **35**, 474.
Lovelock, J. E. and Lipsky, S. R. (1960), *J. Amer. Chem. Soc.* **82**, 431.
Lovelock, J. E., Maggs, R. J. and Adlard, E. R. (1971), *Anal. Chem.* **43**, 1962.
Maggs, R. J., Joynes, P. L., Davies, A. J. and Lovelock, J. E. (1971), *Anal. Chem.* **43**, 1966.
Martyn, D. F. and Pulley, O. O. (1936), *Proc. Roy. Soc.* A **154**, 455.

Massey, H. S. W. (1969), *Electronic and Ionic Impact Phenomena*, 2nd edition, vol. 2, p. 997, Clarendon Press, Oxford.
Massey, H. S. W. (1971), *Electronic and Ionic Impact Phenomena*, 2nd edition, vol. 3, p. 2044, Clarendon Press, Oxford.
Massey, H. S. W. and Bates, D. R. (1940), *Astrophys. J.* **91**, 202.
Massey, H. S. W. and Bates, D. R. (1942), *Rept. Prog. Phys.* **9**, 62.
Massey, H. S. W. and Burhop, E. H. S. (1969), *Electronic and Ionic Impact Phenomena*, 2nd edition, vol. 1, p. 123, Clarendon Press, Oxford.
Meira, L. G. (1971), *J. Geophys. Res.* **76**, 202.
Mehr, F. J. and Biondi, M. A. (1969), *Phys. Rev.* **181**, 264.
Narcisi, R. S. and Bailey, A. D. (1965), *J. Geophys. Res.* **70**, 3687.
Narcisi, R. S., Bailey, A. D., Della Lucca, L., Sherman, C. and Thomas, D. M. (1971), *J. Atmos. Terr. Phys.* **33**, 1147.
Narcisi, R. S., Bailey, A. D., Wlodyka, L. E. and Philbrick, C. R. (1972), *J. Atmos. Terr. Phys.* **34**, 647.
Penning, F. M. (1938), *Ned. T. Naterurkde*, **5**, 33.
Prasad, A. N. and Craggs, J. D. (1961), *Proc. Phys. Soc.* **77**, 385.
Price, D. A., Lucas, J. and Moruzzi, J. L. (1972), *J. Phys. D* **5**, 1249.
Price, D. A. and Moruzzi, J. L. (1973), *Proc. 11th Int. Conf. Ionization Phenomena in Gases*, p. 49.
Raether, H. (1964), *Electron Avalanches and Breakdown in Gases*, Butterworth, London
Razzak, S. A. A. and Goodyear, C. C. (1968), *J. Phys. D* **1**, 1215.
Razzak, S. A. A. and Goodyear, C. C. (1969), *J. Phys. D* **2**, 1577.
Reid, G. C. (1970), *J. Geophys. Res.* **75**, 2551.
Risberg, M. E. and Smythe, R. (1962), *Nucl. Instr. Method*, **18–19**, 66.
Sechrist, C. F. (1972), *J. Atmos. Terr. Phys.* **34**, 1565.
Shimazaki, T. and Laird, A. R. (1970), *J. Geophys. Res.* **75**, 3221.
Spencer-Smith, J. L. (1935), *Phil. Mag.* **19**, 866.
Strittmatter, P. A. and Wickramasinghe, D. T. (1971), *Mon. Not. Roy. Astr. Soc.* **152**, 47.
Stromgren, B. (1940), *Festschrift für Elis Stromgren*, p. 218, Munksgaard, Copenhagen.
Thomas, L., Gondhalekar, P. M. and Bowman, M. R. (1973), *J. Atmos. Terr. Phys.* **35**, 397.
Thompson, J. B. (1959), *Proc. Phys. Soc.* **73**, 818.
Thompson, J. B. (1961a), *Proc. Roy. Soc. A* **262**, 503.
Thompson, J. B. (1961b), *Proc. Roy. Soc. A* **262**, 519.
Townsend, J. S. (1910), *The Theory of Ionization of Gases by Collision*, Constable and Co., London.
Wagner, K. H. (1971), *Z. Phys.* **241**, 258.
Warner, B. (1967), *Mon. Not. Roy. Astr. Soc.* **137**, 119.
Weinman, J. A. and Cameron, J. R. (1956), *Rev. Sci. Instr.* **27**, 288.

Wildt, R. (1939), *Astrophys. J.* **89**, 295.
Wildt, R. (1941), *Astrophys. J.* **93**, 47.
Wentworth, W. E., Chen, E. and Lovelock, J. E. (1966), *J. Phys. Chem.* **70**, 445.
Wheeler, J. A. and Wildt, R. (1942), *Astrophys. J.* **95**, 281.
Wright, B. T. (1957), *Archiv. for Math. Naturvidensk. B. Liv.* Nr. 2.

AUTHOR INDEX

Aberth, W. H.   633, 635, 638
Adams, N. G.   549, 578
Adlard, E. R.   687
Aitken, K. L.   713
Albritton, D. L.   184, 547
Allen, W. D.   126, 216, 217
Anderson, V. E.   221, 398
Andreev, E. P.   606, 608
Andrick, D.   124, 126, 148
Ankudinov, V. A.   402, 405, 606, 608, 615
Arnold, F.   671
Asundi, R. K.   280, 311, 312, 313
Auerbach, D. J.   211, 391
Awan, A. M.   187
Axen, D.   497

Badger, R. M.   481
Baede, A. P. M.   211, 391, 394, 395, 396, 398
Bagot, C. H.   554
Bailey, A. D.   669, 671
Bailey, T. L.   36, 40, 593, 594, 603, 604, 605, 614
Bailey, V. A.   290, 345, 351
Bakale, G.   364
Bakulina, I. N.   40, 43, 210
Bandel, H. W.   122, 125
Banyard, K. E.   9, 11
Bardsley, J. N.   107, 125, 176, 178, 179, 268, 584, 599, 600
Barnes, W. S.   522
Barnett, C. F.   402, 403, 404, 605, 606, 607
Barrow, R. F.   257, 259, 260
Barsuhn, J.   210
Bates, D. R.   49, 54, 60, 67, 254, 388, 421, 588, 599, 600, 605, 606, 607, 619, 621, 624, 625, 637, 664, 679

Baumann, H.   152, 153, 154
Beaty, E. C.   215, 431, 448, 489
Becker, R. H.   325
Begun, G. M.   376, 377
Bely, O.   503, 510
Bennett, R. A.   42, 183, 184, 190, 462, 476, 478, 482, 484, 485, 486, 690
Berkowitz, J.   184, 191, 210, 211, 215, 257, 258, 259, 260, 261, 262, 263, 537, 567, 568, 570, 577
Bernstein, R. B.   40
Berry, R. S.   251, 253, 260, 431, 443, 461, 566
Bethe, H. A.   26
Bethge, K.   152
Bhalla, M. S.   660, 661, 662
Bhatia, A. K.   121, 125
Biondi, M. A.   184, 295, 340, 342, 344, 530, 554, 669
Birtwistle, D. T.   201, 202, 203, 204
Blackmore, E. W.   497
Blau, L. M.   130, 132
Blaunstein, R. P.   377, 381
Blewett, J. P.   345
Bloch, F.   327, 329
Blumle, L. J.   669
Bobashev, S. V.   608
Boettner, E. A.   688
Bohme, D. K.   219, 549
Boldt, G.   253, 254, 453
Bolduc, E.   143
Boness, M. J. W.   187, 197, 198, 205, 217, 235
Bortner, T. E.   297
Bowman, M. R.   667, 672
Boyd, R. L. F.   644
Boyd, T. J. M.   388, 619
Bozin, S. E.   662

Brackmann, R. T.   286, 321, 322, 323, 324, 598
Bradbury, N. E.   292, 296, 327, 329, 345, 352
Branscomb, L. M.   41, 184, 277, 417, 431, 432, 433, 448, 449, 451, 453, 454, 457, 458, 459, 474, 475, 476, 478, 489, 504, 506, 510, 597, 615, 672
Braumann, J. I.   447, 486
Breen, F. H.   681
Brehm, B.   438, 450
Brewer, L.   207
Briglia, D. D.   278, 279, 280, 311, 316, 333, 334, 335, 336, 339, 352, 353, 358, 359, 360, 372
Brown, H. L.   184
Browne, J. C.   518, 519
Browning, R.   407, 409, 612
Brueckner, K. A.   26
Buchdahl, R.   344
Buchel'nikova, I. S.   347, 372, 376
Burch, D. S.   41, 184, 422, 449, 451, 478
Burhop, E. H. S.   76, 107, 139, 292, 649
Burke, P. G.   94, 117, 118, 120, 125, 202, 203, 204
Burrow, P. D.   117, 118, 120, 325, 326
Burt, J. A.   184, 449, 489, 547
Bydin, Yu. F.   605, 606, 612, 614
Byerly, R.   215

Cade, P. E.   180, 181, 182, 183, 213, 477, 478
Cahill, T. A.   497, 499, 500
Callaway, J.   117, 118, 128
Calvin, M.   35, 39
Cameron, J. R.   690
Catalán, M. H.   63
Celotta, R. J.   183, 184, 218, 476, 478, 479, 482, 484, 486, 487
Cernock, T. M.   566
Chamberlain, G. E.   135, 232
Chandra, N.   204
Chandrasekhar, S.   420, 681

Chanin, L. M.   295, 296, 327, 332, 530, 554
Chantry, P. J.   184, 218, 276, 278, 282, 285, 286, 317, 318, 319, 333, 334, 336, 339, 352, 353, 354, 356, 360, 361, 362, 363, 368, 369, 558, 559, 578
Chen, E.   685, 688
Chibisov, M. I.   500
Chisholm, C. D. H.   30
Christophorou, L. G.   347, 348, 349, 373, 377, 381
Chupka, W. A.   184, 210, 211, 257, 260, 261, 537, 538, 539, 567, 569, 570, 571
Chutjian, A.   237
Clarke, E. M.   117
Clarke, I. D.   325
Claydon, C. R.   351, 357
Clementi, E.   25, 58, 60, 62, 64, 65
Colegrove, F. D.   667
Collins, L. E.   407
Comer, J.   125, 223, 225, 226, 227, 232
Compton, R. N.   140, 184, 221, 347, 348, 349, 373, 376, 377, 398, 399
Conway, D. C.   220, 540, 541
Cook, C. J.   633
Coolidge, A. S.   166
Cooper, J. W.   82, 139, 399, 421, 424, 453, 454, 458, 459
Coplan, M.   566
Cottrell, T. L.   573
Craggs, J. D.   280, 376, 377, 657, 660, 661, 662
Cravath, R.   290, 292
Crompton, R. W.   350, 351, 554
Crossley, R. J. S.   58, 59, 60
Csizmadia, I. G.   219
Cubiccioti, D.   47
Cunningham, D. L.   146, 147
Curley, E. K.   117
Curran, R. K.   184, 353, 377, 400, 403
Cvejanovic, S.   125

Dalgarno, A.   518, 519, 582, 601

AUTHOR INDEX

Dallos, F. C. 688
Damburg, R. 116
Dance, D. F. 504, 506, 509, 510, 511
Darewych, G. 497
Das, G. 189
David, C. W. 251, 253
Davidovic, D. M. 582
Davies, A. R. 144
Davis, F. J. 373, 686
Dawton, R. H. V. M. 690
de Corpo, J. J. 211, 288, 346, 381
Della Lucca, L. 671
Demkov, J. N. 383
Deutsch, A. L. 678
Devins, J. C. 662
Dibeler, V. H. 257, 258, 259, 376
Dimov, G. I. 603, 612
Doehring, A. 295
Dolder, K. T. 125, 151, 504, 508, 511, 512
Donahue, T. M. 400, 403
d'Orazio, L. A. 184, 215
Dorman, F. H. 349, 350
Doty, P. M. 40
Doughty, N. A. 422, 681
Doumont, M. 349
Dowell, J. T. 326, 351
Drake, G. W. F. 121
Dukel'skii, V. M. 43, 606, 612, 614
Duncanson, W. E. 351
Dunkin, D. B. 184, 219, 402, 403, 528, 549, 559, 564, 572, 574, 577
Dunn, G. H. 271, 288, 406, 409
Dunn, K. F. 612
Duxler, W. M. 128

Edelson, D. 373
Edie, J. W. 57
Edlén, B. 58, 60
Edwards, A. K. 118, 119, 143, 144, 146, 147, 608
Ehrhardt, H. L. 112, 113, 124, 126, 135, 136, 192, 194, 197, 200, 201, 205, 206, 207, 223

Ekin, J. W. 553
Eliezer, L. 95, 125, 140, 175, 176, 178, 226, 227, 228
Eméleus, K. G. 640
Engelhardt, A. C. 377
Englander-Golden, P. 278
Estberg, G. N. 128
Evans, E. W. 566
Evans, W. F. J. 667
Everhart, E. 601
Ewing, J. J. 566
Extermann, P. 691
Eyb, M. 148

Fano, U. 82, 122, 139, 428, 430
Farragher, A. L. 191
Fedorenko, N. V. 415, 500, 615
Fehsenfeld, F. C. 184, 219, 309, 372, 373, 528, 547, 548, 549, 554, 559, 560, 561, 564, 574, 577
Feldmann, D. 42, 183, 190, 208, 210, 218, 472, 477, 484, 485, 486, 487, 488
Ferguson, E. E. 215, 219, 528, 547, 548, 549, 558, 559, 560, 561, 564, 567, 575, 576, 577, 578
Feshbach, H. 83
Fessenden, R. W. 364
Fine, J. 46
Firsov, O. B. 583, 602
Fischer, R. 184
Fite, W. L. 286, 321, 322, 323, 324, 417, 432, 449, 597, 598
Flannery, M. R. 621, 625
Fleming, R. J. 122
Fock, V. 19
Fogel, Y. M. 144, 402, 403, 404, 405, 411, 412, 414, 605, 615
Ford, P. 497
Fowler, R. H. 34
Fox, R. E. 342, 343, 344, 358, 370, 371, 376
Francis, W. E. 583, 586, 602
Franklin, J. L. 211, 288, 346, 381, 382
Fraser, P. A. 422, 681
Freund, R. S. 572

Froese, C. 61
Frommhold, L. 659
Frosch, R. P. 209
Fuchs, R. 249
Fung, A. C. 150
Futrell, J. H. 373

Gailitis, M. 116
Gaily, T. D. 635, 636, 639
Gardner, M. E. 628
Garrett, W. R. 141
Gatland, I. R. 548
Geballe, R. 118, 599, 604, 605, 608, 609, 610, 655, 656, 657, 660, 661
Geisler, J. E. 667
Geltman, S. 41, 117, 118, 422, 424, 425, 451, 453, 458, 459, 493
George, R. D. 688
Gibson, J. R. 125
Gilbert, L. 211, 212, 213
Gilbody, H. B. 76, 107, 139, 406, 407, 408, 409, 410, 411, 611, 612
Gilmore, F. R. 191
Ginsberg, A. P. 58, 59, 60, 61, 62
Glockler, G. 35, 39, 49, 59
Goldberg, R. A. 669
Golden, D. E. 122, 125, 139, 226
Goldstone, J. 26
Gondhalekar, P. M. 672
Goode, G. C. 219
Goodwin, E. S. 688, 689
Goodyear, C. C. 595, 596, 613, 614, 662
Goubeau, J. 181
Goulden, R. 688
Greaves, C. 629, 630, 631
Green, L. C. 10
Greenstein, J. L. 682
Gresteau, G. 125, 231
Griffiths, J. E. 373
Grisson, J. T. 140, 143
Groth, W. 325
Grübler, W. 691
Grünberg, R. 296, 297, 328

Gunton, R. C. 340
Gusev, V. A. 605
Gusinow, M. A. 438
Gutman, D. 184, 211, 537, 567
Guyon, P. M. 257

Haas, R. 197
Haeberli, W. 691
Hagstrum, H. D. 316, 333, 339
Hake, R. D. 332, 358
Hall, J. L. 183, 184, 438, 449, 453, 459, 476, 478, 482, 484, 489, 490, 491, 492, 493
Hall, R. I. 125, 231
Halpern, G. M. 632
Hanson, E. E. 339
Hanson, W. B. 667
Hara, S. 177
Harries, W. 197
Harris, F. E. 28, 145, 175
Harrison, M. F. A. 504, 635, 636, 639, 655, 656, 657
Hart, A. 7, 10, 121
Hartree, D. R. 18, 19, 22
Hartree, W. 23
Haselton, H. H. 143
Hasted, J. B. 187, 197, 198, 205, 217, 605, 606, 612, 613
Hazi, A. U. 102, 103
Healey, R. H. 327, 344, 345
Heddle, D. W. O. 114, 137, 138, 139
Heideman, H. G. M. 135, 232
Heinicke, E. 42, 152, 472
Helbig, F. H. 601
Helbing, R. K. B. 211, 393, 396, 397
Helmholz, L. 47
Henderson, W. R. 286, 287, 321, 322, 323
Henglein, A. 349, 377
Henis, J. M. S. 373
Henry, R. J. W. 144, 146
Herman, V. 223, 425
Herschbach, D. R. 184, 191, 397, 398
Hertz, G. 197
Herzberg, G. 7, 10, 121, 208, 495

Herzenberg, A. 97, 106, 107, 176, 201, 202, 203, 204, 235, 268, 518
Hiby, J. W. 126
Hickam, W. M. 342, 370, 371
Hicks, W. T. 207
Hierl, P. M. 382
Higginson, G. S. 122
Hirsch, M. N. 632
Hiskes, J. R. 497, 499, 500
Hofmann, M. 148
Holloway, J. H. 257
Holøien, E. 126
Hotop, H. 42, 437, 438, 455, 456, 457, 461, 462, 463, 465, 467, 474, 476
Hughes, B. M. 211, 217, 221, 567, 568, 569, 570, 574
Hummer, D. G. 596, 598, 599, 600, 601
Hunten, D. M. 667
Hurst, G. S. 297, 373
Hurzeler, H. 255
Huxley, L. G. H. 554
Hvelplund, P. 612
Hyman, H. A. 461
Hylleraas, E. A. 6, 7

Iczkowski, R. P. 257
Il'in, R. N. 415, 500
Inghram, M. G. 255
Inokuti, M. 501, 502, 503, 509, 510
Inoue, M. 210
Ionov, N. I. 40, 43, 210

Jackson, W. M. 398
Jacox, M. E. 209
Jäger, K. 349, 377
James, H. M. 166
Janev, R. K. 582
Johns, J. W. C. 257, 260
Johnsen, R. 184, 669
Johnson, F. S. 667
Johnson, H. R. 49
Jones, A. V. 667
Jory, R. L. 351
Joyez, G. 125, 227, 231
Joynes, P. L. 686

Kaiser, H. J. 42, 152, 472
Kalmykov, A. A. 144
Kamenev, A. G. 42
Kaplan, S. N. 497
Kasdan, A. 467
Kaufman, F. 359
Kay, J. 181
Keesing, R. G. B. 114, 137
Keever, W. C. 601
Kessel, J. 671
Khovenstenko, V. I. 43
Kieffer, L. J. 288, 289, 320
Kim, Y.-K. 501, 502, 503, 509, 510
Kimball, G. E. 41, 277
Kingdon, K. H. 38
Kirkpatrick, C. B. 327
Kisker, E. 233
Kivel, B. 566
Kleinpoppen, H. 117, 118
Klemm, W. 181
König, W. 691
Kozlov, V. F. 144
Krankowsky, D. 671
Krauss, L. 257
Krauss, M. 190, 198, 199, 201, 202
Kregel, M. D. 306, 340
Krikorian, O. H. 207
Kurepa, J. M. 114, 137
Kurepa, M. V. 280
Kuyatt, C. E. 122, 125, 137, 139, 142, 143, 223, 226, 232
Kwok, K. L. 125

La Bahn, R. W. 128
Lacmann, K. 184, 191, 397, 398
Lagergren, C. R. 337, 338
Lagerquist, A. 208, 495
Laird, A. R. 666
Langevin, P. 621, 623
Langhans, L. 135, 205
Langmuir, I. 38
Larkin, I. W. 217
Lassettre, E. N. 139
Latimer, C. 612
Lee, T. G. 377
Lee-Ruff, E. 219
Leffert, C. B. 398

AUTHOR INDEX

LeLevier, R. E. 672
Leu, M. T. 669
Levine, J. 184, 478, 482
Lewis, J. T. 619, 637
Lewis, M. N. 10
Lifshitz, C. 211, 221, 567, 572, 573, 574
Linder, F. 135, 186, 187, 188, 205, 223
Lineberger, W. C. 42, 209, 430, 435, 437, 438, 455, 457, 461, 462, 463, 465, 467, 474, 476, 494, 529, 560
Lipsky, L. 421
Lipsky, S. R. 683
Llewellyn, E. J. 667
Lochte-Holtgreven, W. 249
Lockwood, G. J. 601
Loeb, L. 290, 292
Lorents, D. C. 633
Los, J. 211, 391, 396
Lovelock, J. E. 683, 685, 686, 687
Löwdin, P-O. 9
Lozier, W. W. 274, 275, 310, 316, 333
Lucas, J. 657

Mabie, C. A. 373
McAfee, K. B. 373
McCallum, K. J. 40
McClure, G. W. 402, 403
McCulloh, K. F. 257
McDaniel, E. W. 522, 523, 525, 541
McDonald, W. J. 497
McDowell, C. A. 377
McDowell, M. R. C. 582, 587, 599, 600, 601, 607
McEacharn, R. P. 422
Macek, J. 423, 424
McFarland, M. 564, 565
McGowan, J. W. 117, 118, 125
McGowan, S. 627, 628
Mächler, W. von 627, 628
McIntosh, A. I. 406
McIver, R. T. 447
McKnight, L. G. 542, 548
McLean, A. D. 25

MacNeil, K. A. G. 376
Mader, D. L. 134
Magee, J. L. 621
Maggs, R. J. 686, 687
Mahadevan, P. 603
Mahan, B. H. 365, 366, 372, 377, 621, 629
Maier-Leibnitz, H. 122
Makinson, R. E. 345
Maltbie, M. McL. 48
Mandl, A. 461, 566
Mandl, F. 106, 107, 125, 176, 268, 518
Manson, S. T. 127, 128, 134
Margrave, J. L. 257
Marmet, P. 141, 143
Marriott, J. 377
Martin, D. W. 522, 523, 541
Martin, J. B. 424, 453, 458
Martin, J. D. 594
Martyn, D. F. 663
Marx, R. 559, 560
Massey, H. S. W. 76, 84, 94, 107, 108, 109, 139, 148, 254, 265, 274, 292, 333, 387, 421, 426, 501, 621, 649, 664, 679
Matese, J. J. 144, 145, 146, 150
Mauclaire, G. 559
Mauer, J. L. 534, 535, 546, 547, 552, 555, 556, 557
Mayer, J. E. 35, 39, 40, 46, 47, 48
Mazeau, J. F. 125, 231, 232
Mehr, F. J. 669
Meira, L. G. 667
Mendas, I. 356
Menendez, M. G. 140
Metlay, M. 40, 41, 277
Midtdal, J. 126
Mielczarek, S. R. 122, 140, 223, 226
Mies, F. H. 198, 199, 201, 202
Miller, J. M. 58, 59, 60, 61, 62
Milligan, D. E. 209
Moffett, R. J. 621, 624, 625
Mohler, F. L. 376
Moiseiwitsch, B. L. 54, 67
Moores, D. 148, 425, 427, 428, 429, 466, 467, 468, 469, 470, 471

Morrison, J. D.   255, 256
Morse, P. M.   163
Moruzzi, J. L.   220, 350, 351, 366, 525, 526, 540, 549, 553, 554, 657, 659
Moseley, J. T.   523, 635, 637
Moser, C. M.   27, 28
Mott, N. F.   76, 84, 94, 148, 265, 274, 387, 426, 501
Mück, G.   250, 251, 252, 253
Mulder, M. M.   10
Mullen, B.   497, 499, 500
Mulliken, R.   162, 166, 184, 194, 211, 213, 214, 255
Myer, J. A.   256

Nakai, M. Y.   412, 414
Nalley, S. J.   184, 221, 398
Narcisi, R. S.   669, 671
Natanson, G. L.   621
Neamtam, S. M.   497
Nelson, D. R.   373
Nesbet, R. K.   26, 27, 28, 125, 168, 204
Nesbitt, L. E.   220, 540, 541
Neuert, H.   184
Neumann, D.   189
Neynaber, R. H.   633
Nicholas, D. J.   126
Nicolaides, C. A.   140
Nissen, W.   249
Norcross, D. W.   29, 148, 150, 425, 426, 427, 428, 429, 466, 467, 468, 469, 470, 471
Novick, R.   130, 132, 133, 134

Oberoi, R. S.   117, 118
Odom, R. W.   373, 375
O'Hare, P. A. G.   213
Olsen, W. C.   497
O'Malley, T. F.   117, 118, 189, 190, 272, 322, 323, 324, 326, 419
Oparin, V. A.   415, 500
Öpik, U.   30
Oppenheimer, J. R.   496
Orbeli, A. L.   606
Ormonde, S.   118, 144

Pack, J. L.   184, 185, 220, 330, 351, 532, 533, 534, 543, 544, 545, 549, 551, 554, 564
Page, F. M.   38, 181, 191, 219, 278
Paleev, V. I.   39, 42
Pan, Y. K.   125, 140
Paquet, C.   141
Parkes, D. A.   340, 341, 559, 561, 562, 563, 564, 565, 567
Parr, J. E.   350, 351
Patterson, T. A.   209, 457, 467, 468, 470, 471, 494
Paulikas, G. A.   497
Paulson, J.   218, 559
Pavlovic, Z.   235
Peach, G.   587, 599, 600, 607
Peart, B.   151, 504, 505, 506, 508, 509, 510, 511, 512
Pedersen, E. H.   612
Pekeris, C. L.   7, 9, 10, 423
Penning, F. M.   654
Person, J. C.   628, 629
Peters, T.   249
Peterson, J. K.   633, 635, 638
Peuckert-Kraus, K.   184
Pfieffer, G. V.   216, 217
Phelps, A. V.   184, 185, 220, 295, 330, 332, 351, 358, 364, 366, 377, 525, 526, 530, 532, 533, 534, 540, 543, 544, 545, 549, 551, 553, 554, 564
Philbrick, C. R.   671
Pichanick, F. M. J.   137, 143
Pilipenko, D. V.   402, 405, 605, 607
Platzman, R. L.   501
Politzer, P.   61
Popow, B.   255
Popp, H. P.   250, 251, 252, 253, 260
Prasad, A. N.   657
Present, R. D.   166
Price, D. A.   657, 659
Priestley, H.   138
Pu, R. T.   128
Puckett, L. J.   306, 307, 308, 340, 366, 529, 560
Pulley, O. O.   663

Pyle, R. V. 497

Quéméner, J. J. 141, 142, 143

Rackwitz, R. 42, 472
Raether, H. 659
Raible, V. 117, 118
Raimondi, D. L. 25
Rapp, D. 278, 279, 280, 311, 316, 333, 334, 335, 336, 339, 352, 353, 358, 359, 360, 372, 583, 586, 602
Rau, A. R. P. 428, 430
Razzak, S. A. A. 662
Read, F. H. 125, 223, 225, 226, 227, 232
Rees, J. A. 351
Reese, D. 376
Reeves, M. L. 655, 660, 661
Reid, G. C. 666, 669, 670, 671, 672
Reimann, C. W. 260, 431, 461
Reinhardt, J. 125, 231
Reinhardt, P. W. 373
Reynolds, J. G. 688
Richardson, J. R. 497
Risberg, M. E. 691
Risk, C. G. 377
Risley, J. S. 118, 608, 609, 610
Riviere, H. G. 126
Robb, M. A. 219
Robinson, E. J. 424, 425, 453, 458, 459, 489, 493
Roche, A. E. 595, 596, 613, 614
Rohrlich, F. 49, 56, 63
Roothan, C. C. J. 23
Rose, P. H. 611
Roslijakov, G. V. 603, 612
Rothe, E. W. 211, 393, 396, 397, 398, 633
Rountree, S. P. 144, 146
Rundel, R. D. 504, 599, 604, 605, 635, 636
Rutherford, E. 625
Rutherford, J. A. 603, 604, 605
Ryding, G. 611, 616
Ryzko, H. 350, 351

Samson, J. A. R. 256

Sanche, L. 117, 118, 120, 122, 124, 125, 137, 139, 141, 142, 143, 217, 223, 224, 229, 232, 234, 235, 236, 237, 238, 239, 241
Sawina, J. M. 542
Sayers, J. 626, 627, 628, 640
Schaefer, H. F. 28, 145
Scheer, M. D. 39, 46, 65
Schiff, H. I. 184, 528, 547
Schmeltekopf, A. L. 528, 547, 564
Schmelzback, P. A. 691
Schmidt, H. 186, 187, 188
Schryber, V. 412
Schulz, G. J. 111, 117, 118, 122, 124, 125, 137, 139, 141, 142, 143, 192, 193, 194, 197, 217, 223, 224, 229, 232, 234, 235, 236, 237, 238, 239, 241, 276, 278, 281, 282, 284, 285, 310, 311, 313, 317, 318, 319, 323, 326, 330, 333, 335, 337, 348, 349, 353, 354, 355, 356, 358, 359, 360, 372, 379, 380, 483, 534, 535, 546, 547, 552, 555, 556, 557
Schummers, J. H. 541
Schurath, U. 325
Schwandt, P. 691
Schwartz, C. 423
Schwartz, S. B. 503, 510
Schweinler, H. C. 221, 376, 398
Sechrist, C. F. 667, 668
Segal, G. A. 351
Seiler, G. J. 117, 118
Seman, M. L. 458, 459
Serenkov, I. T. 415, 500
Sharp, T. E. 311, 351
Shaw, T. M. 340
Shenstone, A. G. 63
Sherman, C. 671
Shimazaki, T. 666
Shull, H. 9
Sida, D. W. 587
Siegel, M. W. 184, 439, 449, 459, 478, 482, 483
Siegert, A. F. 104, 168
Silverman, S. M. 139

Simpson, F. R. 611
Simpson, J. A. 110, 122, 137, 140, 143, 223, 226
Sinfailam, A. L. 125, 202, 203
Sinnott, G. 489
Skillman, S. 425
Slabospitskii, R. E. 615
Slater, J. C. 18, 22
Smirnov, B. A. 500
Smith, D. L. 373
Smith, K. 144
Smith, R. A. 266, 605, 612, 613
Smith, S. J. 41, 184, 277, 422, 432, 433, 449, 451, 454, 457, 474, 475, 478
Smyth, K. C. 447, 486
Smythe, R. 691
Snow, W. R. 598, 599, 602, 604, 605
Snuggs, R. M. 523, 541, 542, 548
Solov'yev, E. S. 415, 500
Somerville, J. M. 345
Spence, D. 192, 193, 194, 281, 323, 330, 353, 354, 355, 356, 372, 379, 380, 483
Spencer-Smith, J. L. 641, 651
Spohr, R. 257
Spokes, G. N. 431
Stamatovic, A. 111, 284, 285, 333, 335, 337, 355, 356
Stampfer, J. G. 257, 259
Stanton, H. E. 255
Stebbings, R. F. 597
Stedeford, J. B. H. 605, 606
Steiner, B. 460, 477
Stelman, D. 366, 367, 369
Stier, P. M. 402, 403, 404, 605, 606, 607
Stinson, G. M. 497, 499, 500
Stockdale, J. A. 376, 377, 559, 572
Stricker, W. 257
Strittmatter, P. A. 682
Stromgren, B. 678
Stroud, P. T. 406
Sugden, T. M. 340, 341, 561, 562, 563, 564
Sutton, P. P. 35, 39, 40

Swales, F. W. 499
Sweetman, D. R. 126
Swirles, B. 23

Tate, J. T. 274, 275, 316, 333, 339
Taylor, A. J. 94, 117, 118, 120, 125
Taylor, H. S. 95, 102, 103, 151, 175, 176, 205, 227, 272, 351
Teague, M. W. 306, 340
Temkin, A. 125
Terenin, A. 255
Thomas, D. M. 671
Thomas, L. 151, 667, 672
Thompson, J. B. 642, 644, 646, 647, 648, 649, 650
Thomson, J. J. 621, 623, 626
Thorburn, R. 377
Thynne, J. C. J. 376
Tiernan, T. O. 211, 221, 567
Tisone, G. 504, 506, 510, 615
Toburen, L. H. 412, 414
Topolia, N. V. 402
Torres, B. W. 144
Townsend, J. S. 653
Tozer, B. A. 276
Trajmar, S. 237
Trowbridge, C. W. 126
Truby, F. K. 302, 303, 304, 341, 344, 345
Truhlar, D. G. 237
Trujillo, S. M. 633
Turner, B. R. 603, 604, 605

van Brunt, R. J. 288, 289, 320
Venkateswarlu, P. 255
Verba, J. W. 497
Verma, R. D. 255
Vier, D. T. 41
Vogt, D. 184, 572
Vogt, E. 497, 499, 500
Volz, D. J. 541
Voshall, R. E. 351, 364
Vought, R. M. 377

Wagner, K. H. 659
Wahl, A. C. 189, 211, 212, 213
Walker, I. C. 365, 366

Walker, J. A. 257
Walker, J. C. G. 588, 599, 600, 605, 606, 607
Wallace, R. W. 447
Walsh, A. D. 214
Walter, T. A. 210, 261
Walton, D. S. 151, 504, 508, 511
Wannier, G. H. 521
Warman, J. M. 364
Warneck, P. 217, 488, 489
Warner, B. 682
Watanabe, K. 255
Wayne, R. P. 325
Weber, O. 251, 252
Weinflash, D. 130, 132, 133
Weingartshofer, A. H. 223, 224, 226
Weinman, J. A. 690
Weiss, A. W. 27, 28, 127, 128, 425
Weller, C. S. 340
Wentworth, W. E. 685, 688
Whan, D. A. 382
Wheeler, J. A. 680
Wheeler, R. C. 191
Whiddington, R. 138
Whittier, A. C. 605, 607
Wickramasinghe, D. T. 682
Wieder, H. 671
Wigner, E. 148, 426
Wildt, R. 678, 680
Williams, J. F. 402, 403, 404, 412, 414, 605, 615, 616

Williams, J. K. 95, 176, 227
Williams, W. 237
Willmann, K. 112, 192, 194, 197, 200, 201
Wittkower, A. B. 611
Wlodyka, L. E. 671
Wolf, N. S. 632
Wolff, R. J. 662
Woll, J. W. 10
Woo, S. B. 431, 448, 482
Wood, R. H. 184, 215
Woodward, B. W. 430, 435, 457
Wright, B. T. 691
Wu, Ta-You 77, 126
Wyeth, C. C. 10
Wynn, M. J. 594, 595, 613

Yatsimirskii, K. B. 217
Yeung, T. H. Y. 630
Yoshimine, M. 25
Young, A. D. 125
Young, C. E. 372, 377
Young, L. B. 219,
Yule, T. J. 691

Zähringer, J. 671
Zandberg, E. Ya. 39, 42
Zare, R. N. 421, 454, 459
Zecca, A. 125, 139
Zemke, W. T. 189
Zollweg, R. J. 50, 53, 57, 58, 59, 61, 62, 63, 64, 65, 67

SUBJECT INDEX

Absorption of light by negative ions
  in drift tubes 448
  in free–free transitions 681
  in ion cyclotron resonance spectrometer 447
  in shock-heated alkali halides 442
Affinity, electron, see Electron affinity
  spectrum 242; methods for observation of 248
Afterglow
  flowing: application to measurement of 309, 527, 547, 558, 560
  static: measurement of attachment coefficients in 299, 306
Ag, electron affinity of 43, 463
Ag^-
  energy of terms of ground configuration of 56
  photodetachment from 463, 464
Al, electron affinity of 58
Al^-
  energy of terms of ground configuration of 55
  HFR field for 25
Angular distribution
  of ionic momenta in dissociative attachment 271–2, 288, 320
  of photoelectrons 421, 438
  of scattered electrons 87, 124, 136, 149, 170, 187, 204, 206
  measurement of 112
Ar^-, autodetaching states of 142
As, electron affinity of 62, 472
As^-, energy of terms of ground configuration of 56

AsH_2, electron affinity of 218, 486
AsH_2^-, photodetachment cross-section 486
Asymmetry parameter in angular distribution of photoelectrons 421
At, electron affinity of 54
Atmosphere
  solar 678, 679, 680, 681
  stellar, effect of negative ions in 682
  terrestrial, concentration in 665, 666
Attachment, see Dielectronic attachment, Dissociative attachment, Radiative capture
Attachment coefficient
  definition 291
Attachment frequency 291
Attachment rate coefficient 292
Autodetaching states
  calculation of energies of 92
  lifetime of, theory of 79
  metastable 80
  observation of 81, 88, 107
  of diatomic negative ions 168, 222
  relation to: elastic scattering of electrons, 82, 88; inelastic scattering of electrons, 90; photodetachment, 92; Siegert states, 104
  type I and type II 107
Autodetachment
  definition of 77
  effect of, on: associative detachment, 517; dissociative attachment, 266; electron capture in 3-body collisions, 269

Au, electron affinity of   43, 462
Au⁻
  energy of terms of ground
    configuration of   56
  photodetachment cross-section
    of   462

B
  electron affinity of   28, 58
  single electron capture by   414
B⁻
  energy of terms of ground
    configuration of   55
  formation of, from B   414
Ba, electron affinity of   64
Be, electron affinity of   58
Be⁻
  configuration of ground state of
    60, 67
  energy of terms of $(2s^2)2p$
    configuration of   55
Bi, electron affinity of   42, 62, 472
Bi⁻
  double detachment from,
    observed   615
  energy of terms of ground
    configuration of   56
Bi^{2-}, observation of   155
Br
  electron affinity of   40, 44, 253,
    255, 461
  fine structure of   461
Br⁻
  absorption of light by   461
  charge transfer collisions of
    569, 570, 573, 603
  detachment, single, from   566,
    612
  detachment, double, from   615
Br^{2-}, observation of   155
Br_2^+, yield of, from photoionization
  of Br_2 256
Br_2
  current build-up and electrical
    breakdown in   662
  electron affinity of   381, 396,
    569
  photoionization of   256
  polar photodissociation of   255

Br_2^+ yield of, from photoionization
  of Br_2 256

C
  electron affinity of   28, 58
  single electron capture by   414
C⁻
  associative detachment reactions
    of   561
  bound excited states of   66,
    415, 459
  charge transfer reactions of
    602
  energy of terms of ground
    configuration of   55
  formation of   337, 356, 414
  HFR field of   25
  importance of, in carbon stars
    682
  photodetachment from   458,
    459
C⁺, double electron capture by
  414
C_2, electron affinity of, 208: from
  photodetachment, 482
C_2^-
  band spectrum of   208, 494
  bound excited states of   209,
    210
  ground state of   207
  potential-energy curves for
    209, 210
  photodetachment cross-section
    of   494
  two-photon detachment from,
    494; variation of with laser
    flux, 495
Ca, electron affinity of   64
Capture of electrons
  by fast neutral atoms in gases
    399
  by molecules   266, 269
  double, by fast positive ions in
    gases   414
  in three-body collisions   265
  radiative   242
Cd, electron affinity of   62
Cd⁻, energy of $(5s)^2 5p$ terms of
  56

SUBJECT INDEX

CF, electron affinity of 213
CF^-, HFR field for 213
CH, electron affinity of 183, 478
CH^-
   HFR field for 183
   photodetachment from 478
Charge transfer
   endothermic 567
   symmetrical 542, 579, 583, 584, 602
   unsymmetrical, 586, 602
Charge transfer source of low-energy Cs atoms 393
Cl
   affinity spectrum of 250, 251, 253
   electron affinity of 40, 44, 58, 253, 461
   fine structure of 253, 461
Cl^-
   absence of bound excited state of 67
   absorption of light by 444, 461
   associative detachment reactions of 561
   autodetaching states of 145
   charge transfer reactions of 603
   double detachment from 615
   HFR field of 25
Cl^{2-}, observation of 155
Cl_2
   current build-up and electrical breakdown in 662
   electron affinity of 381, 396, 569
   HFR field of 213
Cl_2^-
   HFR field of 213
   production of 539
ClO_2, geometrical configuration of ground state of 214
Close-coupling expansion 94, 148
Cluster ions 214
   importance in D region of ionosphere 214
   exchange reactions involving 578

CN, electron affinity of 210, 263
CN^-
   dissociation energy of 210
   yield of, from polar photo-dissociation of HCN 261
Co, electron affinity of 64
Co^-, HFR field of 25
CO
   dissociative attachment in 279, 333, 334, 335, 336, 337
   vibrationally excited, production of, from dissociative attachment 356
CO^-
   autodetaching states of 205, 235, 236, 237, 238
   ground state of 204
CO_2, dissociative attachment in 331, 352–8, 659
CO_2^-
   geometrical configuration of ground state of 214, 218, 257, 557
   observation of, in ionic reactions 218
CO_3, electron affinity of 220
CO_3^-
   formation of 550
   in lower ionosphere 220, 670, 671
   photodetachment from 489
CO_4^-
   cluster exchange reactions of 578
   dissociation energy of 221, 552
   equilibrium constant for production from O_2^- and CO_2 220, 551
   formation of, in cluster exchange reactions 578
   in lower ionosphere 220, 671
Collision stabilization of negative-ion complexes 269
Configuration of electrons
   in atoms 16, 52, 53
   in diatomic molecules 160
Core polarization and structure of negative ions of alkali metal atoms 29

Correlation between electrons in atoms
  importance of, for electron affinity calculations 9, 25
  calculation of: in general, 26: for F^- 27; for B^-, C^-, Li^-, N^-, Na^-, O^-, K^-, 28; for $He^-(^4P)$, 127; for $N^-(^1D$ and $^4S)$, 145
  empirical estimation of: for atoms, 25; for diatomic hydrides, 181
Cr, electron affinity of 64, 472
Cr^-, HFR field for 25
Cross-sections
  for capture of electrons 243, 247, 264, 403, 404
  for dissociative attachment 268, 278
  for elastic scattering 69, 71, 87, 88
  for inelastic scattering 91
  for ionic reactions 515
  for ion pair production in neutral–neutral collisions 387
  for photodetachment 417, 420
  for production of He^- from He (2^3S) 410
  for symmetrical charge transfer 582, 583, 584
  per unit light intensity, for two-photon detachment 492
Cs, electron affinity of 30, 64, 469
Cs^-
  autodetaching states of 150, 470
  photodetachment cross-section of 470
  $p^2\,^3P_e$ state of 150
CS_2^-, charge transfer reaction of: with NO_2, 571; CS_2, 574
Cu, electron affinity of 43, 463
Cu^-
  angular distribution of photoelectrons from 463
  energy of terms of ground configuration of 55
  HFR field of 25
  photoelectron spectrum of 463
Current build-up in gases
  effect of attachment on 655
  in halogen-containing substances 660
  in O_2 and $O_2 + H_2$ 657
  in SF_6 660
  Townsend theory of 653
Cyclotrons, accelerating H^-
  500 MeV at British Columbia 500, 691
  52 inch at Colorado 691
Cyclotron-resonance spectrometer, for measurement of:
  lifetimes 373
  ionic reaction rates 559, 560
  photodetachment cross-section 443

D_2, dissociative attachment in 310, 313
Detachment of electrons from negative ions
  associative 517, 530
  by electron impact 501, 503
  double, measurement of 614
  effect of, on charge transfer cross-sections 584
  in collisions with neutral atoms and molecules 587
  semi-classical theory of 588
  comparison of theoretical and observed results for 606–7
  in electrostatic fields 495, 497, 500
  in ionic reactions 516, 530
  in photon collisions 417, 489
  summary of possibilities 416
Dielectronic attachment 254
Diffusion coefficient, ambipolar 299
  in iodine afterglow 305
  in presence of negative ions 645, 647
  of electrons and ions in afterglows 300
  of ions in drift tube 521

SUBJECT INDEX 731

Dissociative attachment
  energy relations in 270, 276
  experimental methods using homogeneous electron beams 274, 276, 278, 282, 284, 288 325
  general theoretical considerations 266
  identification of upper state in 271
  to polyatomic molecules 345, 346
Drift velocity: of electrons, 291; of ions, 293

Electrical breakdown in gases
  effect of attachment on 654
  in air and O_2 657
  in SF_6 661
  Townsend theory of 653
Electron affinity
  of atoms, calculation of, 6, 7, 28; empirical determination of, 49, 58, 59, 61; measurement of, 32, 33, 34, 38, 46
  of the elements, well-determined values for 44–5
  of diatomic molecules 172, 181
  of molecules in general 166
  of polyatomic molecules 215
Electron-capture detector 683
  for assaying strongly attaching substances 687
  for detecting halogen-containing pesticides 688
Electron filter, application of to attachment measurements 292
Emission spectrum of the sun
  observed 678
  theory of 673
Energy distribution
  of electrons 80, 113, 119, 143, 146, 438, 610, 613
  of ions 260, 276, 282, 317, 336, 339, 353, 360
  of negative particles in a discharge in O_2 643–4

F
  affinity spectrum of 253
  electron affinity of 27, 40, 44, 58, 253, 461
  single electron capture by 414
F^+, from photoionization of F_2 259, 261
F^-
  absence of bound excited state of 67
  absorption of light by 461
  direct detachment from 566
  fine structure of 461
  HFR field for, 25; including correlation, 27
  yield of, from photodissociation of F_2 258
F^{2-}, observation of 155
F_2
  electron affinity of 381, 396, 569
  HFR field of 213
  polar photodissociation of 257
F_2^-
  ground state of 210
  HFR calculation of potential-energy curves for 210
  vertical detachment energy of 213
Fe, electron affinity of 64
Fe^-, HFR field for 25
Franck–Condon principle 165
  importance of, in: dissociative attachment, 268; photodetachment, 437; photoelectron spectrum, 482; polar dissociation, 273

Ga
  electron affinity of 62
  energy of terms of ground configuration of 56
Gas chromatography 683
Ge, electron affinity of 42, 62, 472
Ge^-
  energy of terms of ground configuration of 56
  metastable state of 66

SUBJECT INDEX

Glow discharges, effects of negative ions on 641, 648

H
affinity spectrum of, observation of 249, 251, 252
charge distribution in 11
cross-section of, for radiative capture of electrons 247
detachment collisions of, with H^-, theory of 516
electron affinity of 6, 9, 43
free–free collisions of electrons in 680
mean static field of 2

H^-
charge distribution in 11
charge transfer reactions of with: H, 579, 599, 600, 601; NO_2, 588
concentration of, in solar atmosphere 679
detachment from, on collision with atoms and molecules 516, 587–8, 606–10
detachment from, by electron impact 503, 504, 511
detachment from, in electrostatic fields 497, 498, 691
diamagnetic susceptibility of 15
double detachment from 508, 512, 515
doubly excited states of 90, 115, 116, 118, 120, 121
formation of 348, 403–4, 414
form factor of 15
mean values of r^n for 13
mutual neutralization of: with H^+, 619, 635; He, 639
photodetachment from 421, 423, 424, 449, 681
polarized beams of 690
probability, of H^+ production in H^-–H collisions 601
two-electron radial density function of 12
use of in: circular accelerators, 498, 690; tandem accelerators, 690

variational wave functions for, 7
H^{2-}, autodetaching states of 151, 511

H^+
double electron capture by, in He and Ar 414
hydrated, in lower ionosphere 669
yield of, from photodissociation of HCN 261

H_2
dissociative attachment in 310, 311, 312
effect of, in current build-up in O_2 657
polar dissociation of 316

H_2^-
states of, dissociating into normal H and H^- 174, 176, 312, 315, 416
excited autodetaching states of 223, 226, 227

Halogen-containing molecules attachment of electrons to 368, 376
detection of trace amounts of 688

Halogen-containing pesticides, detection of 688

Hartree–Fock (HF) method, for calculating atomic structures 19

Hartree–Fock–Roothan (HFR) procedure 23
application to: various atomic negative ions, 25; molecules, 166; OH and OH^-, 180

HCN, polar photodissociation of 261

HD, dissociative attachment in 310, 313

He
effectiveness of, as third body in attachment to O_2 331
free–free transitions in, importance of in white dwarf stars 682

2^3S state of, capture of electrons to 406
He⁻
 non-metastable autodetaching states of, $1s2s^{2\,2}S$ state, 122, 124; $2\,^2P$ and $2\,^2D$ states, 135; 3-quantum and higher states, 137; triply-excited states, 138
 metastable doubly-excited states of, $1s2p^{2\,2}P_e$ state, 115; $1s2s2p\,^4P$ state, 80, 127, 134, 405, 415, 450, 611, 612, 616
Hf, electron affinity of 62
HF
 dissociation energy of 257
 polar photodissociation of 257
Hg, electron affinity of 62
Hg⁻, energy of terms of ground configuration of 56
H_2O
 effectiveness of as third body in attachment to O_2 331
 dissociative attachment in 347, 349, 350
 swarm experiments in 351

I, electron affinity of 40, 44, 255, 461, 480
I⁻
 associative detachment from, in collisions with I 652
 charge transfer reactions of 538, 567, 568, 569, 570, 603
 detachment from, in fast collisions 612
 double detachment from, in fast collisions 615
 energy of terms of ground configuration of 56
 photodetachment cross-section of 460, 461
 production of, in shock-heated alkali iodide vapour 442
 two-photon detachment from 489, 491, 493
I^{2-}, observation of 153
I⁺, yield of, from photoionization of I_2 256

I_2
 attachment rate coefficient of electrons in 302, 341
 electron affinity of 213, 381, 396, 569
 mutual neutralization of ions in 630
 polar photodissociation of 255
IBr, electron affinity of 396, 570
In, electron affinity of 62
In⁻, energy of terms of ground configuration of 56
Ionic reactions
 classification of 513
 effect of, on current build-up in air and O_2 658
 of oxygen ions in oxygen 540, 543
 of oxygen ions with other neutral species 549, 550
Ionization potentials
 isoelectronic extrapolation of 49
 regularities in 50
Ionosphere, terrestrial 663
 E and F regions of 665
 lower 666, 669, 670, 671, 672
Ion pair production
 in alkali metal atom collisions with: halogen molecules, 394; NO and NO_2, 398; O_2, 397; SF_6 and TeF_6, 399
 in neutral–neutral collisions 383, 384, 387
Isoelectronic extrapolation
 of centres of configurations 54–6
 of ionization potentials and electron affinities 49
Isotope effect
 in charge transfer collisions 574
 in dissociative attachment 311, 312, 350
Ir, electron affinity of 64

K, electron affinity of 29, 64, 469

K⁻
  autodetaching states of 148
  p² ³P_e state of 150
  HFR field for, 25; including core polarization, 29
  photodetachment from 426, 429, 467, 468
Kr⁻, autodetaching states of 142
La, electron affinity of 64
Lasers
  argon ion, use of in photoelectron spectroscopy 438
  tunable dye, use of in measurement of photodetachment cross-sections 435
Level widths and lifetimes of autodetaching states 86
Li
  electron affinity of 28, 58, 469
  scattering of electrons by and autodetaching states of Li⁻ 148
  single electron capture by, in collisions with H_2 414
Li⁻
  autodetaching states of 150
  formation of 414
  HFR field for 25
  photodetachment from 426, 429
Li⁺, double electron capture by 414
Methods, empirical for determining electron affinities 48
Methods experimental
  for analysing a glow discharge in O_2 642
  for detection of trace amounts of strongly-attaching substances 683, 686, 687
  for measurement of: cross-sections and threshold energies, for ion pair production in neutral–neutral collisions, 389; cross-sections and threshold energies for polar photodissociation, 255, 257; cross-sections for charge transfer and detachment in fast collisions, 591, 593, 595, 597; cross-sections for double detachment by electron impact, 508; cross-sections for double electron capture by positive ions in gases, 414; cross-sections for electron capture by fast H atoms in gases, 399, 401; cross-section for photodetachment, 432, 442, 443, 448; cross-sections for production of He⁻(⁴P), 406; cross-sections for single detachment by electron impact, 504, 509; cross-sections for three-body attachment to O_2, 281; electron affinities of atoms, 31–48; electron affinities of molecules, diatomic, 172; polyatomic, 215; energy and angular distribution of photoelectrons, 438; fine structure separation in He⁻(⁴P), 134; lifetimes of He⁻(⁴P), 130, SF_6^-, 373; mobilities and reaction rates of negative ions, 522, 526, 529, 530; rate of detachment in electrostatic fields, 497; recombination and mutual neutralization of negative ions, 625, 628, 629, 630, 632, 634; threshold energies for charge transfer, 537
  for observation of: affinity spectrum, 249, 251, 252; doubly-charged negative ions, 151; non-metastable autodetaching states, 109, 112, 114, 115
  for studying: attachment in electron swarms, 292, 297, 299, 306, 309; dissociative attachment, 274, 276, 278, 282, 284, 288, 325; ionosphere, by rocket equipment, 664

Methods theoretical
　for calculating energies of states of negative atomic ions　6, 19, 93–4
　for calculating properties of ground states of diatomic molecules　166
Methods, variational, *see* Variational methods
Mg, electron affinity of　58
Mg⁻
　configuration of ground state of　60, 67
　energy of terms of $(3s)^2 3p$ configuration of　55
Microwaves, use of, for probing afterglows　302
Mn, electron affinity of　64
Mn⁻, HFR field for　25
Mo, electron affinity of　45, 64
Mobilities of ions in gases　299, 519
　and rate of three-body recombination between ions　624
　reduced　520
　relation to charge transfer cross-section　584
　relation to diffusion coefficient　519
　theory of　520
Molecules, diatomic
　calculation of properties of ground states of　166
　determination of electron affinities of　172
　enumeration and properties of electronic states of　160
　potential-energy curves for　156
　vibrational and rotational states of　162
Morse potential　163
Mutual neutralization of positive and negative ions　618
　measurement of　628, 632
　in lower ionosphere　669

N
　affinity spectrum of　249, 254
　electron affinity of　28, 58
　production of, for flowing afterglow experiments　527
　titration of, with NO　527
N⁻
　energy of terms of ground configuration of　55
　HFR field for　25
　$2p^4\,^1D$ state of　81, 144, 145, 254
N₂, effectiveness of, in three-body attachment in O₂　331
N₂⁻
　autodetaching states of　228, 231, 232, 233, 235, 240
　ground state of　194, 198, 202
Na
　electron affinity of　28, 58, 469
　scattering of electrons by, and autodetaching states of Na⁻　148
Na⁻
　autodetaching states of　148, 150
　detachment from, in fast collisions: single, 614; double, 615
　HFR field for, 25; including core polarization, 29
　photodetachment from　427, 429, 469
Nb, electron affinity of　64
Ne, effect of chlorine on a glow discharge in　641
Ne⁻, autodetaching states of　142
Negative ions
　doubly-excited states of, and autodetachment　79
　doubly charged　150
　destruction of, *see* Detachment
　effect of in: current build-up and electrical breakdown in gases, 653; ionosphere, 665, 668; solar atmosphere, 678; stellar atmospheres, 682
　formation of, by: capture of electrons in three-body collisions, 265, 269; dissociative attachment, 266; double capture of electrons

Negative ions—*contd.*
  formation of—*contd.*
    by positive ions, 414; excited atoms, 77; polar dissociation, 272; polar photodissociation, 255; radiative capture of electrons, 242
  molecular, diatomic 158, 174
  singly-excited states of 66, 67
NH, electron affinity of 183, 477, 478
NH$^-$
  HFR field for 183
  photodetachment from 478
NH$_2$, electron affinity of 218, 486
NH$_2^-$
  charge transfer reaction of, with NO$_2$ 567
  photodetachment cross-section of 486
  photoelectron spectrum of 486
NH$_3$, dissociative attachment in 351
Ni, electron affinity of 64
Ni$^-$
  bound excited state of 67
  HFR field for 25
NO
  attachment in afterglow in 306, 309, 341, 562
  dissociative attachment in 339
  electron affinity of 398, 484, 564, 565, 568
  swarm experiments in 340
  use of, for N titration, O production 527
NO$^-$
  autodetaching states of 191, 237, 240, 341
  charge transfer reaction of, with N$_2$O 578
  direct detachment from 460, 562, 565
  formation of, in dissociative attachment experiments in N$_2$O 558
  ground state of, properties of p.e. curves of 193, 483

  photoelectron spectrum of 483
  production of, in flowing afterglow experiments 558, 561
  rearrangement collisions with HCl 575
N$_2$O
  dissociative attachment in 353, 359, 362
  electron affinity of 218
  formation of NO$^-$ in dissociative attachment experiments in 558, 559
  ground state of, geometrical configuration of 214
  swarm experiments in 364
N$_2$O$^-$
  geometrical configuration of ground state of 214, 218
  observation of: in charge transfer reactions between NO$^-$ and N$_2$O, 578; in mass spectra, 218
NO$_2$
  attachment to, rate constant for 309, 365
  electron affinity of 217, 393, 570
  charge transfer reactions of, with negative ions 571
  ground state of, geometrical configuration of 214
NO$_2^-$
  autodetaching states of 217
  charge transfer reactions of with Cl$_2$, 572; NO$_2$, 574
  ground state of 214, 216, 362, 365
  photodetachment cross-section of 488
  production of, in NO afterglow 341
  rate of hydration of 529, 560
NO$_3$, electron affinity of 219, 576, 578
NO$_3^-$
  formation of 575, 578
  hydrated, in lower ionosphere 671
  importance of, in lower ionosphere 219

SUBJECT INDEX 737

rearrangement collisions of 575–6
NS, electron affinity of 213
NS⁻, HFR field for 213

O
  affinity spectrum of 249, 253
  dielectronic attachment to 254
  electron affinity of 28, 41, 58, 277, 318, 451
  radiative capture of electrons by 247
  single electron capture by, in fast collisions 414

O⁻
  absence of bound excited states of 67
  angular distribution of, in dissociative attachment to O_2 272
  associative detachment reactions of 532, 547, 550, 552
  autodetaching states of 143
  charge transfer reactions:
    dissociative, with SF_6, 573;
    in fast collisions, 602, 603; in slow collisions, 548, 550, 558, 572
  detachment from, in slow collisions 546, 555
  energy distribution of, from dissociative attachment 336, 339, 349, 353, 360
  energy of terms of ground configuration of 55
  formation of 414
  HFR field for 25
  importance of, in: glow discharges in O_2, 650; ionosphere, 664
  mobility of, in O_2 541
  mutual neutralization collisions of with: N⁺ (exp.), 638; O⁺ (theory), 621
  photodetachment from 451, 453, 454
  production of, for flowing afterglow experiments 528
  reactions of, in lower ionosphere 668
  reaction rates of with O, O_2 and O_3 543
  reactions of, in lower ionosphere 668
  rearrangement collisions of with N_2O 550, 558, 559
  three-body associative reactions of 542, 550
  use of, in tandem accelerators 690

O^{2-}
  observation of 155
  total energy of, from energy relations in cyclic processes 48

O⁺
  double electron capture by, in fast collisions 414
  rate of production of, in O_2 discharge 649

O_2
  $a^1\Delta_g$ state of, dissociative attachment to 325, 547
  attachment coefficient in 327, 328, 329, 330, 331, 665, 668
  current build-up in 658, 659
  dissociative attachment of electrons to 272, 316, 317, 318, 319, 320, 321, 323, 649
  electron affinity of 184, 397, 481, 546, 568, 645, 649

O_2^-
  autodetaching states of 186, 237
  charge transfer reactions of 548, 550, 572, 603
  detachment from, in collisions with O_2 614
  direct detachment from 189, 190, 320, 323, 543, 546, 547
  formation of O_4^- from 541
  ground state of 184, 189, 329, 481, 482
  hydrated 578
  in ionosphere 668, 671
  mobility of, in O_2 541

738   SUBJECT INDEX

O_2^-—contd.
  mutual neutralization collisions of 638
  photodetachment from 478, 479, 482
  production of 356, 528
  reaction rates of, with O, O_2 and O_3 543
  rearrangement collisions of, with N 550
  three-body association reactions of 549, 552
O_3
  attachment coefficient in 366, 367, 368
  electron affinity of 215, 570
  ground state of, geometrical configuration of 214
O_3^-
  mobility of, in O_2 542
  photodetachment cross-section for 489
  reactions of, in lower ionosphere 668
  rearrangement collisions of 550
O_4^-
  dissociation energy of 220, 541
  mobility of, in O_2 542
  observation of 220
  photodetachment from 489
  reaction rates of, with O 643
  rearrangement collisions of 548, 550
  reactions of, in lower ionosphere 618
OD, electron affinity of 475
OD$^-$, photodetachment from 474, 476
OF, electron affinity of 213
OF$^-$, HFR field for 213
OH
  dipole and quadrupole moments of 182
  electron affinity of 180, 181, 476
OH$^-$
  associative detachment reactions of 561
  charge transfer reactions with NO_2 567, 603
  ground state of 180, 182
  photodetachment 474, 476
  production of, from dissociative attachment to H_2O 349
Optimized valence orbital (OVO) method 167
Orbitals 16
  basis trial functions for: for atoms, 23; F$^-$ and Cl$^-$, 24; molecules, 166
  bonding and antibonding in molecules 161
Orbiting in collisions 514
  importance of, for ionic reaction rates 515
Os, electron affinity of 64

P, electron affinity of 58, 472
P$^-$
  energy of terms of ground configuration of 55
  HFR field for 25
Pauli principle
  and electron–atom scattering 76
  and horizontal variation of ionization energies and electron affinities 51
  and spin–orbit coupling 17
  and stability of negative ions 4
Pb, electron affinity of 42, 62
Pb$^-$, energy of terms of ground configuration of 57
Pd, electron affinity of 64
Pd$^-$, bound excited state of 67
PH
  electron affinity of 183
  HFR field for 183
PH$^-$, HFR field for 183
PH$_2$, electron affinity of 486
PH$_2^-$, photodetachment from 447, 486
Phase shift in scattering theory
  for elastic scattering 70, 71, 72, 86, 101, 177, 202
  for inelastic scattering 75
  for symmetrical charge transfer 582, 585

SUBJECT INDEX 739

Photodetachment
  effect of autodetaching states on 92
  from ions aged in a drift tube 448
  measurement of cross-sections for 432, 435, 442, 445
  theory of 417, 418, 420
Photoelectron spectra, measurement of 438
Po, electron affinity of 62
Po⁻, energy of terms of ground configuration of 57
Polar dissociation of molecules 272
  energy relations in 273
Polar photodissociation
  definition of 242
  of Br_2, 256; F_2 and HF, 256; HCN, 262; I_2, 255
Polyatomic molecules
  attachment to 345, 346, 376, 379, 381
  determination of electron affinities of 215
Polyatomic negative ions 214
  geometrical configuration of ground states of 214
  vertical detachment energy of 214
Positive column in glow discharge, effect of negative ions on 641
Projection method for calculating energies of autodetaching states 93
Pseudo-crossing of potential-energy curves 385
  and theory of: ion pair formation in neutral–neutral collisions, 387; mutual neutralization, 618
  probability of crossing 386
Pseudo-natural orbital (PNO) method 167
Pt, electron affinity of 64, 466
Pt⁻
  bound excited states of 67, 466
  photodetachment from 464, 465

Pulse methods
  for measuring attachment coefficients 293
  for measuring detachment rates 530

Radiative capture
  of electrons by atoms 254
  theory of 243
  threshold laws for 244
Rare earth atoms, electron affinity of 65
Rb, electron affinity of 30, 469
Rb⁻
  autodetaching states of 150
  $p^{2\,3}P_e$ state of 150
  photodetachment from 472
  structure of, including core polarization 30
Re, electron affinity of 46, 64
Recombination
  of electrons, in ionosphere 663
  of ions: in afterglow, 301; methods for measurement of rates of, 625, 628; radiative, 617; three-body, 617, 623, 624
Rh, electron affinity of 64
Ru, electron affinity of 64
Russell–Saunders coupling 17

S, electron affinity of 58, 454, 457
S⁻
  charge transfer reactions of, with NO_2, 571; SF_6 572, 573; SO_2, 569
  energy of terms of ground configuration of 55
  photodetachment from 454, 457
S^{2-}, total energy of, from energy relations in cyclic processes 48
S_2, electron affinity of 190, 484
S_2^-, photoelectron spectrum of 484
Saturation in laser photodetachment, for Ag⁻ 437

Sb, electron affinity of 42, 62, 472
Sb⁻
  double detachment from 615
  energy of terms of ground configuration of 55
Sc, electron affinity of 64
Sc⁻, HFR field for 25
Scattering
  elastic 68, 73, 82, 97
  inelastic 90
  involving vibrational excitation 170
SD⁻, photodetachment cross-section of 477
Se, electron affinity of 62, 457
Se⁻
  energy of terms of ground configuration of 55
  photodetachment from 454, 456, 458
SF_5 electron affinity of 399
SF_5^-, charge transfer reactions of, with NO_2 573
  formation of 370, 573
SF_6
  attachment to, measurement of 309, 370, 688
  application of 342, 369, 370
  assay of, using electron capture detector 687
  charge transfer reactions of 572, 574
  current build-up in 660
  electrical breakdown in 661
  electron affinity of 399, 572
SF_6^-
  charge transfer reactions of 572
  lifetime of 373
  production of by attachment to SF_6 370
SH
  electron affinity of 132, 477
  HFR field for 183
SH⁻
  charge transfer reactions of 571
  ground state of 183
  photodetachment from 477

Shape resonances 97
  and Siegert states 104
  stabilization method for calculating energies of 102
Si, electron affinity of 42, 58
Si⁻
  bound excited state of 66
  detachment from, in electrostatic fields 415
  energies of terms of ground configuration of 55
  formation of by double electron capture 414
  HFR field for 25
Siegert states 104
SiF, electron affinity of 213
SiF⁻, HFR field for 213
SiH, electron affinity of 132
SiH⁻, HFR field for 183
Sn, electron affinity of 42, 62, 472
Sn⁻
  bound excited state of 66
  energy of terms of ground configuration of 55
  photodetachment from 472
SO, electron affinity of 190, 486
SO⁻, photoelectron spectrum of 486
SO_2
  electron affinity of 488, 569
  ground state of, geometrical configuration of 214
SO_2^-
  charge transfer reactions of 571, 574
  ground state of 215, 488
  photodetachment cross-section of 488
  photoelectron spectrum of 487
Space potential, radial variation of, in glow discharges in N_2 and O_2 648
Sputter sources of neutral atoms 389
Sr, electron affinity of 64
Stabilization method for calculating energies of:
  autodetaching states, 93;
  shape resonances, 102

Stabilization of negative-ion
    complexes by collision, 269;
    in O_2, 329; of SF_6^-, 370, 373
Static afterglows
    attachment experiments in
        299, 306
    for measuring ionic reaction
        rates 529
Sum rules for photodetachment
    cross-sections, 420; for H^-,
    423; O^-, 454

Ta, electron affinity of 46, 64
Tandem accelerators 690
Tc, electron affinity of 64
Te, electron affinity of 62, 472
Te^-, energy of terms of ground
    configuration of 56
Te^{2-}, observation of 155
TeF_6, electron affinity of 221,
    399
Terms
    in complex spectra 18
    of ground configurations of
        negative ions 55–7
Threshold behaviour of cross-
    sections
    for photodetachment 418, 424,
        426–8
    for radiative capture of
        electrons by atoms 244
Threshold energy, determination
    of
    for dissociative attachment
        276, 346, 381
    for ion pair production in
        neutral–neutral collisions 390
    for endothermic charge transfer
        reactions 537
    for polar photodissociation
        258
Ti, electron affinity of 64
Ti^-, HFR field for 25
Tl, electron affinity of 64
Tl^-, energy of terms of ground
    configuration of 57

V, electron affinity of 64
V^-, HFR field for 25
Variational method for calculating
    ground state energies 6, 23
    energies of autodetaching
        states 93, 106
Vertical detachment energy 166
Vibrational excitation of
    molecules by electrons via
        autodetaching
    states for: CO, 206; H_2, 178,
        225; N_2, 197; NO, 192; NO_2,
        217; O_2, 186

W, electron affinity of 46, 64
Window resonances in photo-
    detachment, observed for:
    Cs^-, 471; Rb^-, 472

Xe^-, autodetaching states of 142

Zn^-, energy of terms of $(4s)^2 4p$
    configuration of 55
Zr, electron affinity of 64

For EU product safety concerns, contact us at Calle de José Abascal, 56–1°,
28003 Madrid, Spain or eugpsr@cambridge.org.

www.ingramcontent.com/pod-product-compliance
Lightning Source LLC
LaVergne TN
LVHW021649060526
838200LV00050B/2280